W9-BKJ-253

Statistics

THIRD EDITION

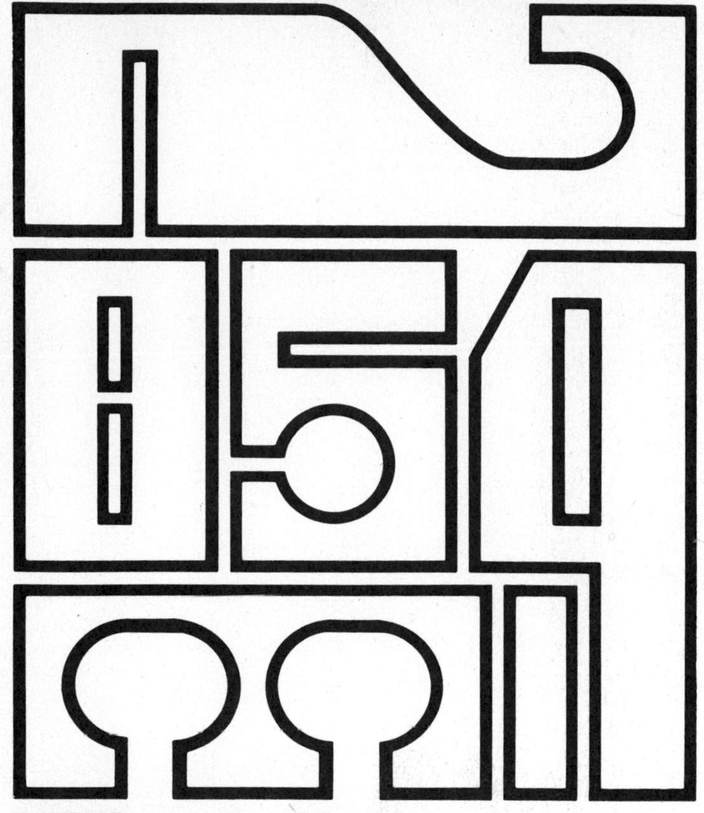

William L. Hays
The University of Texas at Austin

Holt, Rinehart and Winston

New York Chicago San Francisco Philadelphia
Montreal Toronto London Sydney Tokyo
Mexico City Rio de Janeiro Madrid

Library of Congress Cataloging in Publication Data

Hays, William Lee, 1926–
Statistics.

Second ed. published in 1973 under title: Statistics for the social sciences.
Bibliography: p. 669
Includes index.
1. Social sciences—Statistical methods. 2. Statistics. 3. Psychometrics. I. Title.
HA29.H352 1981 519.5′0243 80-27220

ISBN 0-03-056706-8

Address correspondence to: 383 Madison Avenue, New York, N.Y. 10017

Printed in the United States of America
Published simultaneously in Canada
2 3 4 144 9 8 7 6 5 4 3

CBS COLLEGE PUBLISHING
Holt, Rinehart and Winston
The Dryden Press
Saunders College Publishing

Preface

In 1963, when this book was first published under the title, *Statistics for Psychologists,* its aims were set forth as follows:

> This book represents an attempt to give the elements of modern statistics in a relatively nonmathematical form, but in somewhat more detail than is customary in texts designed for psychologists, and with considerably more emphasis on the theoretical rather than the applied aspects of the subject. It is designed as a text of at least an intermediate level of difficulty. I have felt for some time that the serious student in experimental psychology both needs and wishes to know somewhat more of the language and concepts of theoretical statistics than is provided in the usual "cookbook" of statistical methods. Granted that a real understanding and appreciation of mathematical statistics requires a considerable degree of mathematical training and sophistication, a great deal of statistical theory can be got across to the serious student familiar only with elementary algebra, provided that a relatively detailed exposition of the concepts accompanies the mathematical treatment, and provided that the student is genuinely interested in acquiring a grasp of this subject.

The use that the book has received in the social sciences generally over the years apparently justifies my hope that there are students and teachers in many fields who find this approach worthwhile. Essentially, the aims of this third edition remain the same.

Nevertheless, when I set about preparing this new edition, my chief concern was to make it somewhat more concise. Among other things, having returned to regular teaching after a 15-year hiatus spent in university administration, I have recently grown much more sensitive to the need to make this text portable! To this end, I have shortened a great many sections, and eliminated a fair number, while trying to maintain the essential coverage of the earlier editions. Three entire chapters have been omitted: the chapter on set and function theory covered material that is now such a commonplace of school mathematics that it could hardly serve its original purpose; the chapter on joint variables and independence could be split up among other chapters; and the chapter on Bayesian methods, though still representing an approach I believe to be potentially important for the social and behavioral sciences, could not treat the subject fully in the limited space available. Furthermore, I am neither so sanguine nor quite so choleric about certain issues as I was a few years ago, and thus feel no need to waste further words on them. In all, then, a good many things have been shortened or omitted, with, I trust, little damage to the coverage of topics that this book attempts.

In addition, a good bit of reworking of the chapters on analysis of variance and on regression has been done. The ready availability of computer programs for multiple regression and for multivariate analysis generally is giving such methods a far more ubiquitous role in research than they formerly enjoyed. In order to understand and to take advantage of the many options these methods present, the student needs some early groundwork in the general linear model, and especially in the essential connections between multiple regression and analysis of variance. Again, because of the limitations of space and the level of preparation expected of students using this book, the coverage of these topics has to be rather cursory. Any adequate treatment of linear models and regression theory almost demands the use of matrix and vector theory, and I was tempted to go in this direction. Although I have included a small section on vector operations in Appendix C, in order to introduce the notions of orthogonalization, I resisted the impulse to go further on the grounds that a treatment by matrix theory belongs in another book, where the concepts can be introduced early and thoroughly.

I have also tried to include more, and simpler, problems in this edition than in the last. The 570 or so problems included represent an abosolute increase of 25 percent over the second edition, and, for the chapters included in both editions, nearly a 50 percent increase in number. Solutions to the odd-numbered exercises are printed at the end of the book. A separate answer key with solutions to the even-numbered exercises is available to instructors who adopt the third edition.

The almost universal use of hand-held calculators also had some bearing on my decisions about what to include. Thus, a lengthy table of squares and square roots now has about as much use in a statistics text as the multiplication tables. Although I have retained some other tables, such as factorials and powers of e, these are now such common operations on calculators that they probably are superfluous as well. A technique such as the log-linear analysis in Chapter 15 looks computationally formidable, but these computations are now relatively easy to do on a calculator with a natural log function, which a great many have. Details of the computation of means, variances, and even correlation coefficients become relatively less important in an age when these operations are preprogrammed on many inexpensive calculators.

As before, this book is aimed primarily at the first-year graduate student in one of the social or behavioral sciences. I have therefore assumed that the student probably will have had at least one undergraduate statistics course, and that the present course will be followed by more specialized advanced courses, such as experimental design. However, I believe that the level is generally elementary enough to be followed by an apt student without specific preparation in statistics, and yet advanced enough to give the student a start in relatively simple research and data analysis. Ideally, the text will be used in a two-semester course. However, many of the sections, especially in the early part of the book, are sufficiently self-contained that they can be omitted without serious loss of continuity. Thus, it is entirely possible for the teacher to cut and tailor the topics covered to fit the requirements of a one-semester course, especially if the students already have some background in this area.

I wish to indicate my indebtedness and to offer my thanks to the very many students and teachers who have contacted me through the years with comments and suggestions for improvement of this text. I am especially grateful to all of those who have identified and helped me to correct errors of various kinds. I hope that a new generation of students and teachers will be willing to give me the same assistance.

Once again I extend my sincere thanks to the late Professor E. S. Pearson and the trustees of *Biometrika* for their kind permission to use tables from the *Biometrika Tables for Statisticians,* (Vol. 1, 3rd ed.) and to Professor R. S. Burington and the McGraw-Hill Company for graciously allowing me to reprint the table of binomial probabilities from R. S. Burington and D. C. May, *Handbook of Probability and Statistics with Tables* (2nd ed.). I would also like to thank the reviewers of this text: Robert G. Malgady and Stanley A. Mulaik. Finally, to my wonderful, and wonderfully understanding, Palma, Lecann, and Scott, goes more appreciation than I can ever adequately express.

W. L. H.
Austin, Texas
December 1980

Contents

INTRODUCTION

ON THE NATURE AND THE ROLE OF INFERENTIAL STATISTICS

The word, ''statistics,'' came into English from Latin and German, and ultimately derives from the same Indo-European root which gave us ''standing,'' ''status,'' ''state,'' and even ''understand.'' In the minds of most people, ''statistics'' has a lot in common with these related words, meaning roughly, a description of ''how things are.'' It is, of course, true that a part of the theory of statistics concerns effective ways of summarizing and communicating masses of information which describe some situation. This part of the overall theory and set of methods is usually known as ''descriptive statistics.''

Although descriptive statistics form an important basis for dealing with data, a major part of the theory of statistics is concerned with another question: How does one go beyond a given set of data, and make general statements about the large body of potential observations, of which the data collected represent but a sample? This is the theory of *inferential statistics,* with which this book is mainly concerned.

Applications of inferential statistics occur in virtually all fields of research endeavor—the physical sciences, the biological sciences, the social sciences, engineering, market and consumer research, quality control in industry, and so on, almost without end. Although the actual methods differ somewhat in the different fields, the applications all rest on the same general theory of statistics. By examining what the fields have in common in their applications of statistics we can gain a picture of the basic problem studied in mathematical statistics. *The major applications of statistics in any field all rest on the possibility of repeated observations or experiments made under essentially the same conditions.* That is, either the researcher actually can observe the same process repeated many times, as in industrial quality control, or there is the *conceptual* possibility of repeated observation, as in a scientific experiment that might, in principle, be repeated under identical conditions. However, in any circumstance where repeated observations are made, even though every precaution is taken to make conditions exactly the same the results of observations will vary, or tend to be different, from trial to trial. The researcher has control over some, but not all, of the factors that make outcomes of observations tend to differ from each other.

When observations are made under the same conditions in one or more respects, but they give outcomes differing in other ways, then there is some *uncertainty* connected with observation of any given object or phenomenon. Even though some things are known to be true about that object in advance of the observation, the experimenter cannot predict with complete certainty what its other characteristics will be. Given enough repeated observations of the same object or kind of object a good bet may be formulated about what the other characteristics are likely to be, but one cannot be completely sure of the status of any given object.

This fact leads us to the central problem of inferential statistics: *in one sense, inferential statistics is a theory about uncertainty, the tendency of outcomes to vary when repeated observations are made under identical conditions*. Granted that certain conditions are fulfilled, theoretical statistics permits deductions about the *likelihood* of the various possible outcomes of observation. The essential concepts in statistics derive from the theory of probability, and the deductions made within the theory of statistics are, by and large, statements about the probability of particular kinds of outcomes, given that initial, mathematical, conditions are met.

Mathematical statistics is a formal mathematical system. Any mathematical system consists of these basic parts:

1. A collection of undefined "**things**" or "**elements,**" considered only as abstract entities;
2. A set of undefined **operations,** or possible relations among the abstract elements;
3. A set of **postulates** and **definitions,** each asserting that some specific relation holds among the various elements, the various operations, or both.

In any mathematical system the application of logic to combinations of the postulates and definitions leads to *new* statements, or theorems, about the undefined elements of the system. *Given* that the original postulates and definitions are true, then the new statements *must* be true. Mathematical systems are purely abstract, and essentially undefined, **deductive** structures. In the first chapter we will see that the abstract system known as the theory of probability has this character.

Mathematical systems are not really "about" anything in particular. They are systems of statements about "things" having the formal properties given by the postulates. No one may know what the original mathematician really had in mind to call these abstract elements. Indeed, they may represent absolutely nothing that exists in the real world of experience, and the sole concern may be in what one can derive about the other necessary relations among abstract elements given particular sets of postulates. It is perfectly true, of course, that many mathematical systems originated from attempts to describe real objects or phenomena and their interrelationships: historically, the abstract systems of geometry, school algebra, and the calculus grew out of problems where something very practical and concrete was in the back of the mathematician's mind. However, as *mathematics* these systems deal with completely abstract entities.

When a mathematical system is interpreted in terms of real objects or events, then the system is said to be a **mathematical model** for those objects or events. Somewhat more precisely, the undefined terms in the mathematical system are identified with particular, relevant, **properties** of objects or events; thus, in applications of arithmetic, the number symbols are identified with magnitudes or amounts of some particular property that objects possess, such as weight, or extent, or numerosity. The system of arithmetic need not apply to other characteristics of the same objects, as, for example, their colors. Once this identification can be made between the mathematical system and the relevant properties of objects, then anything that is a logical consequence in the system is a true statement about objects in the model, *provided,* of course, *that the formal characteristics of the system actually parallel the real characteristics of objects in terms of the particular properties considered.* In short, in order to be useful as a mathematical model, a mathematical system must have a formal structure that "fits" at least one aspect of a real situation.

Probability theory and statistics are each both mathematical systems *and* mathematical models. Probability theory deals with elements called "events," which are completely abstract. Furthermore, these abstract things are paired with numbers called "probabilities." The theory itself is the system of logical relations among these essentially undefined things. The experimenter uses this abstract system as a mathematical model: the experiment produces a real outcome, which is called an event, and the model of probability theory provides a value which is interpreted as the relative frequency of occurrence for that outcome. If the requirements of the model are met, this is a true, and perhaps useful result. If the experiment really does not fit the requirements of probability theory as a system, then the statement made about the actual result need not be true. (This point must not be overstressed, however. We will find that often a statistical method can yield practically useful results even when its requirements are not fully satisfied. Much of the art in applying statistical methods lies in understanding when and how this is true.)

Mathematical systems such as probability theory and the theory of statistics are, by their very nature, deductive. That is, formal assertions are postulated as true, and then by logical argument true conclusions are reached. All well-developed theories have this formal, logico-deductive character.

On the other hand, the problem of the empirical scientist is essentially different from that of the logician or mathematician. Scientists search for general relations among events; these general relations are those which can be expected to hold whenever the appropriate set of circumstances exists. The very name "empirical science" asserts that these laws shall be discovered and verified by the actual observation of what happens in the real world of experience. However, no mortal scientist ever observes all the phenomena about which a generalization must be made. Scientific conclusions about what would happen for *all* of a certain class of phenomena always come from observations of only a very few particular cases of that phenomenon.

The student acquainted with logic will recognize that this is a problem of **induction.** The rules of logical *deduction* are rules for arriving at true consequences from true premises. Scientific theories are, for the most part, systems of deductions from basic principles held to be true. If the basic principles are true, then the deductions must be true. However, how does one go about arriving at and checking the truth of the initial propositions? The answer is, for an empirical science, observation and inductive generalization—going from what is true of some observations to a statement that this is true for *all possible* observations made under the same conditions. Any empirical science begins with observation and generalization.

Furthermore, even after deductive theories exist in a science, experimentation is used to check on the truth of these theories. Observations that contradict deductions made within the theory are prima facie evidence against the truth of the theory itself. Yet, how does the scientist know that the results are not an accident, the product of some chance variation in procedure or conditions over which there is no control? Would the result be the same in the long run if the experiment could be repeated many times?

It takes only a little imagination to see that this process of going from the specific to the general is a very risky one. Each observation the scientist makes is different in some way from the next. Innumerable influences are at work altering—sometimes minutely, sometimes radically—the similarities and differences the scientist observes among events. Controlled experimentation in any science is an attempt to minimize at least part of the accidental variation or "error" in observation. Precise techniques of measurement are aids to scientists in sharpening their own rather dull powers of observation and comparison among events. So-called "exact sciences," such as physics and chemistry, have thus been able to remove a substantial amount of the unwanted variation among observations from time to time, place to place, observer to observer, and hence are often able to make general statements about physical phenomena with great assurance from the observation of quite limited numbers of events. Observations in these sciences can often be made in such a way that the generality of conclusions is not a major point at issue. Here, there is relatively little reliance on probability and statistics. (However, as even these scientists delve into the molecular, atomic, and subatomic domain, negligible differences turn into enormous unpredictabilities and statistical theories become an important adjunct to their work.)

In the biological, behavioral, and social sciences, however, the situation is radically different. In these sciences the variations between observations are not subject to the precise experimental controls that are possible in the physical sciences. Refined measurement techniques have not reached the stage of development that they have attained in physics and chemistry. Consequently, the drawing of general conclusions is a much more dangerous business in these fields, where the sources of variability among living things are extremely difficulty to identify, measure, and control. And yet the aim of the social or biological scientist is precisely the same as that of the physical scientist—arriving at general statements about the phenomena under study.

Faced with only a limited number of observations or with an experiment that can be conducted only once, the scientist can reach general conclusions only in the form of a "bet" about what the true, long run, situation actually is like. Given only sample evidence, the scientist is always unsure of the "goodness" of any assertion made about the true state of affairs. The theory of statistics provides ways to assess this uncertainty and to calculate the probability of being wrong in deciding in a particular way. *Provided that the experimenter can make some assumptions about what is true, then the deductive theory of statistics tells us how likely particular results should be*. Armed with this information, the experimenter is in a better position to decide what to say about the true situation. Regardless of what one decides from evidence, it *could* be wrong; but deductive statistical theory can at least determine the probabilities of error in a particular decision.

In recent years, a branch of mathematics has been developed around this problem of decision making under uncertain conditions. This is sometimes called "statistical decision theory." One of the main problems treated in decision theory is the choice of a decision rule, or "deciding how to decide" from evidence. Decision theory evaluates rules for deciding from evidence in the light of what the decision maker wants to accomplish. As we shall see in later chapters, mathematics can tell us wise ways to decide how to decide under some circumstances, but it can never tell the experimenter how a decision must be reached in any particular situation. The theory of statistics supplies one very important piece of information to the experimenter: the probability of sample results *given* certain conditions. Decision theory can supply another: optimal ways of using this and other information to accomplish certain ends. Nevertheless, neither theory tells the experimenter *exactly* how to decide—how to make the inductive leap from observation to what is true in general. This is the experimenter's problem, and the answer must be sought outside of deductive mathematics, and in the light of what the experimenter is trying to do.

ABOUT THIS BOOK

This book is addressed to upper division or graduate students in the social and behavioral sciences. Such students are the people who will produce the significant social and behavioral science research in years to come, and who will make up the audience for much of this research. As a part of their professional equipment, these students need to know statistics, at a level beyond an undergraduate course, and just short of the specialized research design and methodology courses needed to round out their graduate programs. Such students are the "you" in this book.

You will soon discover that the main concern in this book is with the theory underlying inferential methods, rather than with a detailed exposition of all the different methods social scientists and others find useful. The author had no intention of writing a "cookbook" that would equip students to meet every possible situation they might encounter. Many methods will be introduced, it is true, and we will, in fact, discuss most of the elementary techniques for statistical inference currently in use. However, in the past few years the concerns of the social scientist have begun to grow increasingly complicated. Theory is growing, and social scientists are turning their attention to new problems and techniques for data analysis that are becoming much more sophisticated than in the past. The statistical analyses required in many such studies are simply not in the "cookbooks." From all indications, this trend will continue, and by the time that you, the student, are in the midst of your professional career it may well be the case that entirely new statistical methods will be required, replacing many of the methods currently found useful.

Furthermore, a true revolution has occurred in the past two decades, deeply affecting the application and the teaching of statistical methodology. This has been brought about by the new generations of computers, which are faster, more flexible, and cheaper to use than anyone would have dreamed only a few years back. Large-scale statistical analysis is now done by computer in almost all research settings.

Even the beginner in research turns to the computer for all but the simplest computational routines. Techniques that were once viewed as impracticable because of the amount of computation involved as are now quite routine parts of the social scientist's methodology. Prepackaged programs for all of the common statistical techniques are now available in virtually all computing centers. Among the best known of these sets of prepackaged programs is the *Statistical Package for the Social Sciences* (1975) *SPSS*. There also are a number of others available, giving a wide range of possibilities to the social and behavioral scientist.

Prior to this development, when most computations were done by hand with the assistance of a desk calculator, it was necessary to devote a large part of statistical instruction to computational methods, with emphasis on shortcuts to lessen the computational burden. Now all has changed, and it is no longer necessary to dwell so long on the how-to-do-it aspects. Even inexpensive pocket calculators are having impact on how statistics can be taught and learned. These pocket or hand calculators provide an amazing range of functions formerly available only in tables, in addition to basic arithmetic. Pocket calculators have taken us light-years away from even the most sophisticated desk calculators, and cost a fraction of what one paid for the old machines. One of the best investments a student of statistics can make is in a really good pocket calculator, which includes such functions as square roots, reciprocals, logarithms, powers, etc. Many calculators have common statistical analyses built in, which can be a big labor-saving device, while others can be programmed for unusual operations required for repeated, short-term, use.

For all of these reasons, the problems of learning (and of teaching) statistics are changing rapidly. With many new techniques at the disposal of the social scientist, and with the computer ready to give almost any analysis we request, the problem is shifting from "how-to-do-it" to "what can be done."

As social and behavioral research becomes more sophisticated and more and more methods are made available for use in particular situations, a point is rapidly being reached where the research worker simply cannot be familiar with all the statistical methods that might pertain to a given problem. It seems unfair to demand that each competent researcher must also be a competent statistician as well, although a few gifted individuals (*not* including the author) have somehow found time and brainpower to be both. In short, the days when each researcher was his own "do it yourself" statistician, relying on a handy "cookbook" of set methods, are about over.

Instead, it is becoming increasingly important for the research worker to understand some of the principles that unite all methods for statistical inference, and some of the general models that are capable of almost limitless adaptation and modification to fit particular research requirements. Secondly, it is important for the serious worker in research to form consultative relationships with statisticians, who can give guidance when problems begin to require solutions that extend beyond familiar methods. A large part of the work of most applied statisticians consists of consultation on problems of design and analysis of experiments, and many are available for such consultation on a professional basis. The statistician can usually provide answers to the research worker's questions, *provided that the statistician is asked about the problem before the data are collected, and can participate in the efficient and logical planning of the experiment*. It is most unreasonable to expect the statistician to reach in a hat and pull out a method that will extract meaning from a poorly designed or executed study.

The statistician does not expect the scientist to know all about theoretical statistics, nor does the scientist expect the statistician to know all about a particular problem. But to work together effectively, each must have some idea of the basic concepts the other uses. This is one reason for theoretical emphasis in this book. At the very outset, the student needs to know something about the points of view and concepts of theoretical statistics in order to appreciate its resources and not become lost in the complexities of using the statistical language.

This book is not, nor does it pretend to be, a first course in mathematical statistics. Ideally, the serious student in the social or behavioral sciences should take at least one such course, although there are two practical difficulties: The content and the organization of courses in mathematical statistics are framed for the training of statisticians, not behavioral scientists, and the peculiar problems of these research areas are not emphasized in such courses: This is as it should be. In the second place, to become a really good researcher is a full-time job, and the student may not have the time to devote to the mathematical statistics courses and their prerequisites in order to gain the essential background needed.

Thus, this book contains some of the concepts, results, and theoretical arguments that come from mathematical statistics, but these results and arguments are given at a far more intuitive and informal level than would be the case for a student in mathematical statistics. Only very seldom will the level of mathematics used rise above the high school level, although the mathematical concepts used will occasionally be unfamiliar to some students. Occasionally we will use some results coming from the application of the calculus, especially results having to do with the idea of a "limit"; these ideas really cannot be treated adequately at an elementary level. Furthermore, in one instance (Appendix C) we will make a brief foray into the theory of vector spaces in order to demonstrate a concept of great importance for some of the methods to be described. From a mathematician's technical point of view, many of our statements are incomplete, poorly framed, or imprecise. On the other hand, many of these ideas can be grasped intuitively by the serious student, and the author feels that this intuitive understanding is better than no understanding at all, *provided* that the student understands the limitations of a presentation such as this.

THE ORGANIZATION OF THE TEXT

A glance at the table of contents reveals the topics covered, and there is little point in a detailed listing here. However, it should be pointed out that the chapters in this book fall roughly into two sections: Chapters 1 through 6 deal largely with the essential ideas of probability and of distributions, the two central notions of theoretical statistics. The first chapter lays a foundation for these topics by introducing very fundamental concepts of probability. A clear idea of these concepts can do a great deal to clarify the remainder of the book. Chapters 1 through 6 are very closely related in the topics covered, and each succeeding chapter builds on the concepts introduced in the preceding ones.

Chapter 7 develops some of the issues connected with the actual use of results from statistics, particularly the problem of making up one's mind from data. Chapters 10 through 14 discuss methods for various kinds of inferences to be made in simple experimental situations. The methods are closely linked, as each is a special instance of a general linear model relating an observed value to factors that may influence it. Chapter 15 deals with qualitative data, and Chapter 16 gives some of the basic ideas of order statistics, an alternate approach to many problems.

A theme that runs throughout this book is the search for relationships. A statistical relation will be said to exist when knowledge of one property of an object or event *reduces our uncertainty* about another property that object or event will show. A statistical relation occurs when things tend to "go together" in a systematic way. This theme will recur *ad nauseam* in the chapters to follow, but it is an important one.

Finally, Appendixes A and B, rules for the manipulation of summations and of expectations, are very important, since we will use these rules to considerable extent in our simple derivations of results. Appendix C concerns the principles underlying bivariate distributions and linear combinations of variables. These are more advanced matters that will be useful to know about in the later chapters of this book. Appendix D contains various useful tables.

Many mathematical expressions occur throughout this text. These are of three kinds: one, algebraic equivalences serving as steps in some derivation; two, actual definitions or principles stated mathematically; and three, computational formulas useful in some method. Some of the mathematical expressions are numbered; ordinarily this occurs when some reference will be made to that expression at a later point. If the number for any expression is followed by an asterisk (*), then this is an important definition or relationship that is worthy of your special attention.

Every chapter is followed by a set of problems which cover most, though not necessarily all, of the concepts and the methods introduced in that chapter. An effort has been made to make these problems relatively interesting in content, and to keep them at a level which will not be beyond the ability of a student who has read the text carefully. Solutions to the odd-numbered problems may be found in the back portion of the book.

Now we shall turn to an exploration of some of the theory and some of the methods of statistics. We will start with the key concept underlying statistical theory, the notion of probability.

Chapter 1
ELEMENTARY PROBABILITY THEORY

Statistical inference involves statements about probability. Everyone's vocabulary includes the words ''probable'' and ''likely,'' and most of us have some notion of the meaning of statements such as ''The probability is one-half that the coin will come up heads.'' However, in order to understand the methods of statistical inference and use them correctly, one must have some grasp of probability theory. This is a mathematical system, closely related to the algebra of sets now taught in most elementary schools. Like many mathematical systems, probability theory becomes a useful model when its elements are identified with particular things, in this case the *outcomes of real or conceptual experiments*. Then the theory lets us deduce propositions about the likelihood of various outcomes, if certain conditions are true.

Originally, of course, the theory of probability was developed to serve as a model of games of chance—in this case the ''experiment'' in question was rolling a die, or spinning a roulette wheel, or dealing a hand from a deck of playing cards. As this theory developed, it became apparent that it could also serve as a model for many other kinds of things having little obvious connection with games, such as results in the sciences. One feature is common to most applications of this theory, however: the observer is uncertain about what the outcome of some observation will be, and must eventually infer or guess what will happen. In particular, the observer needs to know what will happen *in the long run* if observations could be made indefinitely.

In this respect the scientist is like the gambler keeping track of the numbers coming up on a roulette wheel. At any given opportunity only the tiniest part of the total set of things the scientist would like to know about can be observed. Given our human frailty as observers we must fall back on logic; *given* that certain things are true, we can make deductions about what *should* be true in the long run. The logical machinery of "in the long run" is formalized in the theory of probability. The scientist's statements about *all* observations of such and such phenomena are on a par with the gambler's; both are deductions about what should be true, if the initial conditions are met. Furthermore, the gambler, the businessman, the engineer, the scientist, and, indeed, every person must make decisions based on incomplete evidence. Each does so in the face of risk about how good those decisions will turn out to be. Probability theory alone does not tell any of these people how they should decide, but it does give ways of evaluating the degree of risk one takes for some decisions.

The theory of probability is erected on a few very simple concepts. Indeed, the main terms in the theory, the "events," are simply sets of possibilities. However, before we develop this idea further, we need to talk about the ways that events are made to happen: simple experiments.

1.1 / SIMPLE EXPERIMENTS

Modern probability theory starts with the notion of a *simple experiment*. We shall mean nothing very fancy by the term "simple experiment," and there is no implication that a simple experiment need be anything even remotely resembling a laboratory experiment. **A simple experiment is some well-defined act or process that leads to a single well-defined outcome.** Some simple experiments are tossing a coin to see whether heads or tails comes up, cutting a deck of cards and noting the particular card that appears on the bottom of the cut, opening a book at random and noting the first word on the right-hand page, running a rat through a **T** maze and noting whether it turns to the right or the left, lifting a telephone receiver and recording the time until the dial tone is heard, asking an individual's political preference, giving a person an intelligence test and computing the score, counting the number of colonies of bacteria seen through a microscope, and so on, literally without end. The simple experiment may be real (actually carried out) or conceptual (completely idealized), but it must be capable of being described and be repeatable (at least in principle). We also require that each performance of the simple experiment have one and only one outcome, that we can know when it occurs, and that the set of all possible outcomes can be specified. Any single performance, or trial, of the experiment must eventuate in exactly one of these possibilities.

Obviously, this concept of a simple experiment is a very broad one, and almost any describable act or process can be called a simple experiment. There is no implication that the act or process even be initiated by the experimenter, who need function only in the role of an observer. On the other hand, it is essential that the outcome, whatever it is, be unambiguous and capable of being categorized among all possibilities.

1.2 / EVENTS

The basic elements of probability theory are the possible distinct outcomes of some idealized simple experiment. The set of all possible distinct outcomes for a simple experiment will be called the **sample space** for that experiment. Any member of the sample space is called a **sample point,** or an **elementary event.** Every separate and "thinkable" result of an experiment is represented by *one and only one* sample point or elementary event. Any elementary event is *one possible result* of a single trial of the experiment.

For example, we have a standard deck of playing cards. The simple experiment is drawing out one card, haphazardly, from the shuffled deck. The sample space consists of the fifty-two separate and distinct cards that we might draw. If the experiment is stopping a person on the street, then the sample space consists of all the different individuals we might possibly stop. If the experiment is reading a thermometer under particular conditions, then the sample space is all the different numerical readings that the thermometer might show.

Seldom, however, is the *particular* elementary event that occurs on a trial of any special interest. Rather the actual outcome takes on importance only in relation to all possible outcomes. Ordinarily we are interested in the *kind* or *class* of outcome an observation represents. The outcome of an experiment is measured at least by allotting it to some qualitative class. For this reason, the main concern of probability theory is with *sets* of elementary events. **Any set of elementary events is known simply as an event, or an event-class.**

Imagine, once again, that the experiment is carried out by drawing playing cards from a standard pack. The fifty-two sample points (the distinct cards) can be grouped into sets in a variety of ways. The suits of the cards make up four sets of thirteen cards each. The event ''spades'' is the set of all card possibilities showing this suit, the event ''hearts'' is another set, and so on. The event ''ace'' consists of four different elementary events, as does the event ''king,'' and so on.

If the experiment involves noting the eye color of some person stopped on the street, the experimenter may designate seven or eight different eye-color classes. Each such set is an event. The event ''blue eyes'' is said to ''occur'' when we encounter a person who is a member of the class ''blue eyes.'' If, on the other hand, we find the weight of each person we stop, then there may be a vast number of ''weight events,'' different weight numbers standing for classes into which people may fall. If a person is stopped who weighs exactly 168 pounds, then the event ''168 pounds'' is said to occur.

In short, **events are sets, or classes, having the elementary events as members.** The elementary events are the raw materials that make up event-classes. The occurrence of any member of event-class A makes us say that event A has occurred. Since *any* subset of the sample space is an event, then some event *must* occur on each and every trial of the experiment.

1.3 / EVENTS AS SETS OF POSSIBILITIES

The elementary events making up any sample space are, as we have seen, those separate, distinct, and identifiable outcomes to a given simple experiment. However, as we have also seen, interest is usually focused on groupings or sets of elementary events, according to one or more characteristics that the elementary events show. Thus, it is convenient to use some of the language of sets in discussing events.

In the following the symbol $\mathcal{S}$ will be used to stand for the sample space, which is the set of all possible elementary events for the simple experiment. Then capital letters such as A, B, and so on will represent events, each of which is a subset, or subgrouping, of elementary events in $\mathcal{S}$. For a given simple experiment, every event discussed must be a subset of $\mathcal{S}$.

The set $\mathcal{S}$ may contain either a finite or an infinite number of elementary events. In this initial discussion the number of elements in $\mathcal{S}$ may be assumed to be finite. Nevertheless, in many of the most important parts of the theory of probability the sample space is assumed to be infinitely large.

What are the possible events making up a sample space $\mathcal{S}$? First of all, the set $\mathcal{S}$ is an event: $\mathcal{S}$ is the "sure" event, since it is bound to occur (that is, some elementary event must occur on every trial of the simple experiment).

The event $\varnothing$ is called "the impossible event," since it cannot occur. However, the "null set" $\varnothing$ is defined to be a subset of every other set. Hence, it is quite possible to think even of $\varnothing$ as an event, or subset of $\mathcal{S}$.

Now suppose that there are two events, A and B, each composed of elementary events in $\mathcal{S}$, and thus each subsets of $\mathcal{S}$. Then we can imagine an outcome to the simple experiment which is a member *both* of A and of B. In this instance we say that the event "A and B" has occurred. The outcome to the experiment qualifies both as an event A and as an event B.

Similarly, if A is an event and B is an event, we can imagine an outcome which is *either* an event A *or* an event B. Such an outcome represents an occurrence of the event "A or B." This event occurs when the elementary event qualifies as an A, or it qualifies as a B, or when the elementary event qualifies *both* as an A and a B. Notice that the occurrence of "A and B" is thus automatically an occurrence of the event "A or B." However, the reverse is not true. (In discussions of such *joint events,* a little of the notation of set theory becomes useful. We shall use the symbol $\cap$, read as "intersection," to stand for the word "and" in discussing an event such as "A and B." Thus,

$$(A \text{ and } B) = (A \cap B).$$

The symbol $\cup$, read "union," will stand for the word "or" when we discuss events such as "A or B." Thus,

$$(A \text{ or } B) = (A \cup B).)$$

Suppose that there is some simple experiment with sample space $\mathcal{S}$; also suppose that A is some event in that sample space. Now an elementary event occurs which does *not* qualify for the event A. Then we say that the event "not A" has occurred. The event "not A" is often symbolized by $\bar{A}$, the letter standing for the event, but with a horizontal bar above it. Given any event such as A, then the event $\bar{A}$ is called the *complement* of the event A. Every event in any sample space $\mathcal{S}$ will have a complement.

A very useful way to discuss a sample space $\mathcal{S}$ and the events in that space is by use of a group of figures known as a Venn Diagram, after the logician J. Venn (1834–1923). Figure 1.3.1 pictures the sample space $\mathcal{S}$ as a rectangle. Let us imagine that each and every point within the rectangle is a possible elementary event in $\mathcal{S}$. Now within the rectangle $\mathcal{S}$ there is a circle marked A. This circle symbolizes the event A, and every point within A represents an elementary event which is an occurrence of the event A as well. Now look at Figure 1.3.2. Here there is once again the sample space $\mathcal{S}$, shown as a rectangle, and also two events, A and B. Every point in circle A represents an elementary event qualifying as an "A" event, and every point within B is a possible elementary event qualifying as a "B." Notice

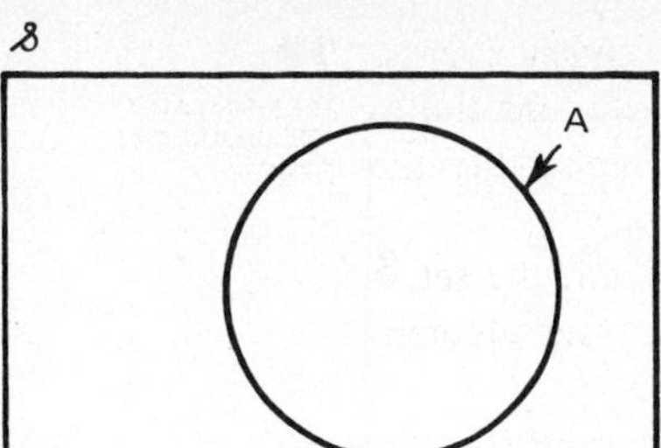

FIGURE 1.3.1.

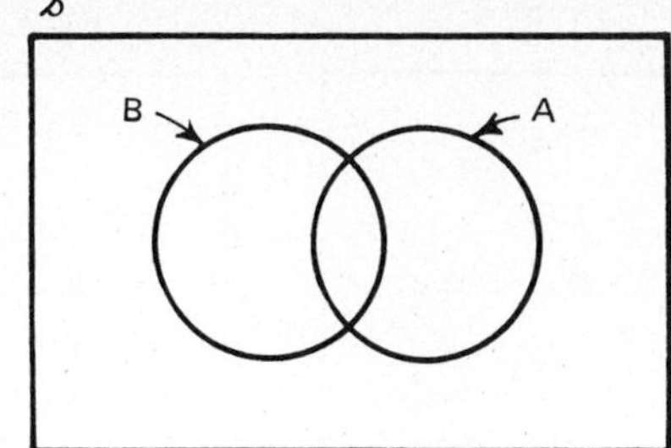

FIGURE 1.3.2.

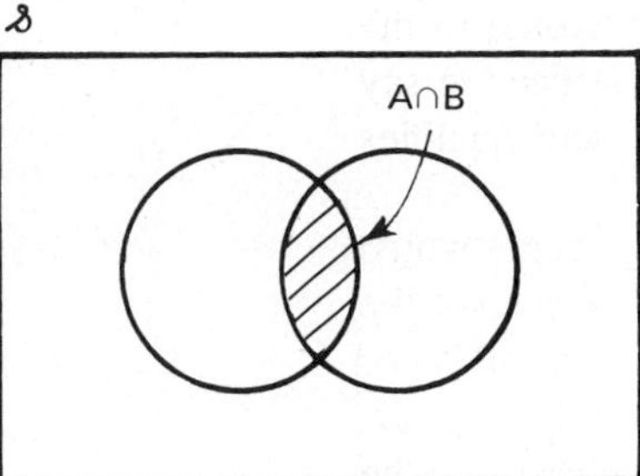

FIGURE 1.3.3.

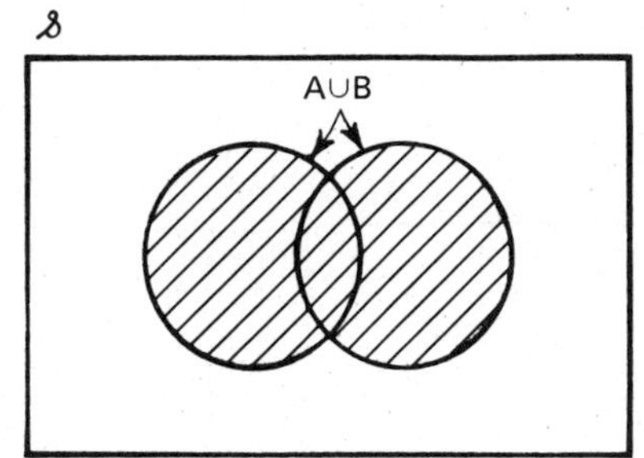

FIGURE 1.3.4.

that some events qualify *both* for event A and for event B. This is the set making up the event $A \cap B$. This event $A \cap B$ is shown as the shaded area in Figure 1.3.3.

Recall that the event $A \cup B$ includes all elementary events that qualify for A, or for B, or for both. The Venn Diagram of Figure 1.3.4 shows the event $A \cup B$ as a shaded area. Notice that when an elementary event qualifies for $A \cap B$ it automatically qualifies for $A \cup B$ but that the reverse is not true, since it is entirely possible for an elementary event to qualify for $A \cup B$ even though it does not qualify for $A \cap B$.

The complementary event $\overline{A}$ for event A is shown as the shaded area in Figure 1.3.5. An elementary event qualifying as a member of $\overline{A}$ is a member of $\mathcal{S}$, of course, but not a member of A. Figure 1.3.6 shows the event $A \cap \overline{B}$, or "A and not B." This event includes all elementary events that qualify for A but do not qualify for B.

An important situation exists when it is impossible for an elementary event to qualify both for event A and for event B. In this case, $A \cap B = \varnothing$, as shown in Figure 1.3.7. Two sets such as A and B in this figure, for which $A \cap B = \varnothing$, are said to be *mutually exclusive:* if one of these events occurs, the other event cannot occur.

It is possible to have three or more different events in the same sample space, $\mathcal{S}$. For example, the three circles in Figure 1.3.8 stand for the three events A, B, and C in the same sample space $\mathcal{S}$. The shaded portion of this figure represents the event $A \cup B \cup C$, which means "A or B or C." Notice that in this example an elementary

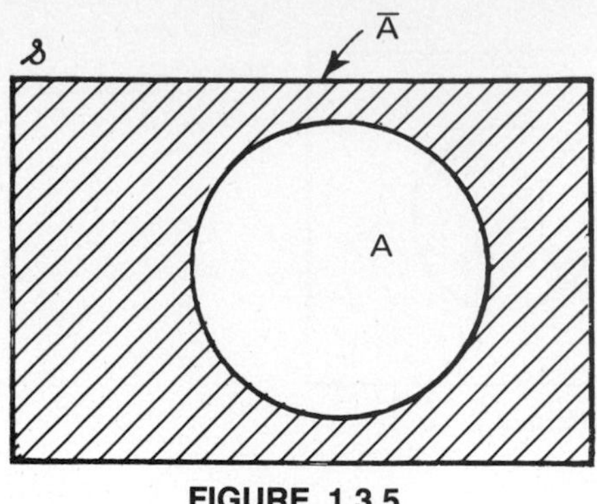

FIGURE 1.3.5.

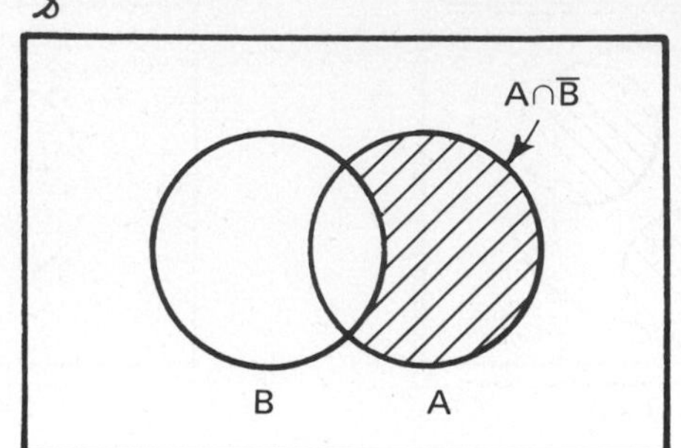

FIGURE 1.3.6.

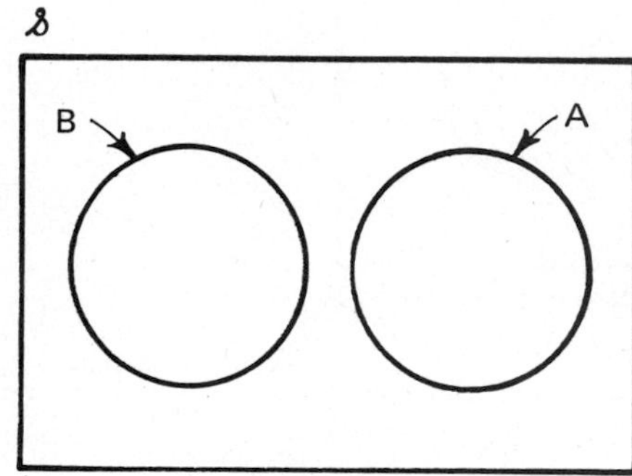

FIGURE 1.3.7.

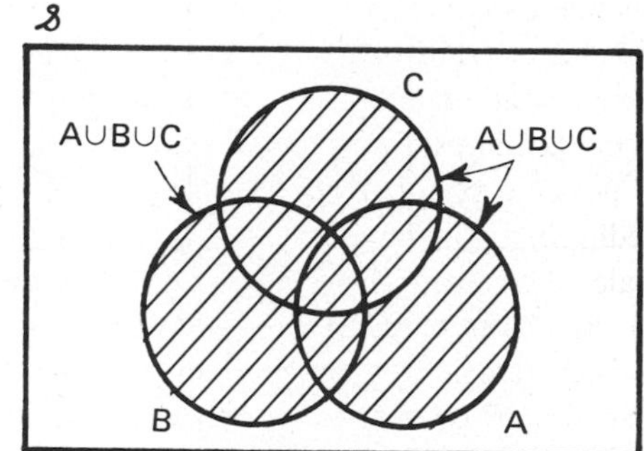

FIGURE 1.3.8.

event might qualify for events such as $A \cap B$, $A \cap C$, $B \cap C$, $A \cap B \cap C$, and so on for all of the events A, B, C, as well as their complementary events.

The use of Venn Diagrams such as those shown above can sometimes be helpful in thinking about probabilities, and even in simple probability calculations. It is good practice to learn to visualize events within a sample space in this way, especially when one is dealing with a number of different possible events.

A Venn Diagram can also help to explain another pair of concepts that occur frequently in probability theory. We have just seen that two events are said to be mutually exclusive if no elementary event can qualify for their intersection, so that $A \cap B = \varnothing$. More than two events such as A, B, and C may also be mutually exclusive. This means simply that any *pair* of these events, such as A and B, or B and C, are themselves mutually exclusive. This situation is illustrated by the shaded circles in Figure 1.3.9. In addition, two or more events may be described as *exhaustive,* which means that each elementary event in $\mathcal{S}$ must qualify for one (or more) of these events. Three *mutually exclusive and exhaustive* events $\{A, B, C\}$ are shown in Figure 1.3.10. Such a set of mutually exclusive and exhaustive events is called a *partition* of the sample space $\mathcal{S}$. A glance at Figure 1.3.10 should show

FIGURE 1.3.9.

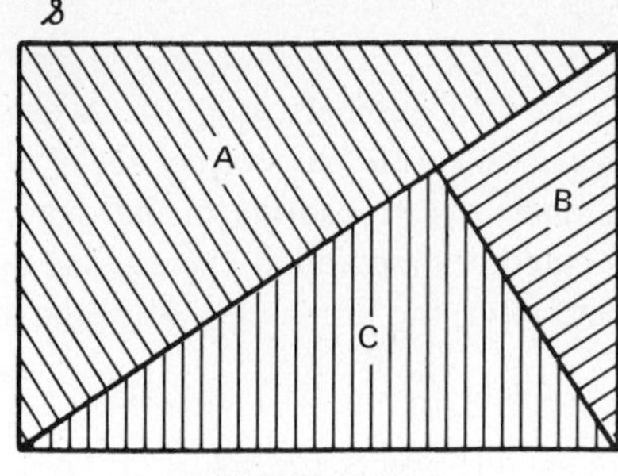

FIGURE 1.3.10.

you why this name is appropriate.

Let us take a concrete example of some events. Suppose that we had a list containing the name of every living person in the United States. We close our eyes, point a finger at some spot on the list, and choose one person to observe. The elementary event is the actual person we see as the result of that choice, and the set $\mathcal{S}$ is the total set of possible persons. Suppose that the event A is the set "female," and B is the set "redheaded," among this total set of persons. If our chosen person turns out to be female, then event A occurs; if not, event $\overline{A}$. If the individual turns out to be redheaded, this is an occurrence of event B; if not, event $\overline{B}$. If the observation shows up as both female and redheaded, $A \cap B$ occurs; if either female or redheaded, then event $A \cup B$ occurs. If the person is female but not redheaded, this is an occurrence of the event $A \cap \overline{B}$. One thing is sure; we will observe a person living in the United States, and the event $\mathcal{S}$ must occur.

As another example, imagine that we choose one student from a junior high school in some city. All of the different individual students we might choose make up the sample space $\mathcal{S}$. This school contains three ethnic groups: Black, Mexican-American, and Anglo. The event "Black" occurs when we draw a person belonging to that ethnic group, the event "Mexican-American" occurs when a person is drawn belonging to that classification, and so on. Let us suppose that the three ethnic groups are mutually exclusive and exhaustive, so that for instance, a person cannot be both Black and Mexican-American, and each student must belong to one of the three groups. Then the three events form a partition of the sample space $\mathcal{S}$. (Try to visualize both of these examples in terms of a Venn Diagram such as those in the preceding figures.)

In short, each and every subset that can be formed from the elementary events in a sample space $\mathcal{S}$ is an event. How many events then are there for a given sample space? It can be shown that if a sample space contains exactly N elementary events (meaning that there are exactly N discernibly different outcomes to the simple experiment) then there are exactly 2^N different events, or subsets, in $\mathcal{S}$, including both $\varnothing$ and $\mathcal{S}$ itself. This group of all possible subsets of the sample space $\mathcal{S}$ is known as the *family of events* in $\mathcal{S}$. This concept of the family of events in some sample space $\mathcal{S}$ is useful when we wish to talk about some event that is possible in this $\mathcal{S}$, but without having to specify exactly which one. We will use this idea in the next section.

1.4 / PROBABILITIES

Formally, probabilities are just numbers assigned to each and every event in some sample space. That is, each possible outcome or set of outcomes for a simple experiment is assigned a number, the probability of that particular outcome or event. As we have seen, sets of outcomes are called events, and so each possible event, such as A, is given a number, $p(A)$, the probability of event A. Thus,

$p(A)$ = probability of event A.

Recall that the entire sample space, the set of all possible distinct outcomes of the simple experiment, is denoted by the symbol $\mathcal{S}$. Even the set $\mathcal{S}$ itself can be thought of as an event ("some outcome occurred"), and the event $\mathcal{S}$ is associated with probability $p(\mathcal{S})$.

On the other hand, some outcomes are simply impossible in any given simple experiment. These impossible outcomes are symbolized by the event $\varnothing$, with probability $p(\varnothing)$.

In exactly this same way any other event which is a subset of $\mathcal{S}$ is associated with a probability. Thus, if there are two events A and B, both subsets of $\mathcal{S}$, then we can speak of probabilities such as $p(A)$, $p(B)$, $p(A \cap B)$, $p(A \cup B)$, and so on. Exactly this same idea applies to any event which is a part of the family of events that can be constructed from the elementary events of $\mathcal{S}$. Each and every such event has a number, its probability.

We are now ready for a somewhat more formal definition of these numbers known as probabilities. Modern probability theory is constructed from a small group of basic statements, or axioms, giving the properties that these probability numbers must exhibit. Although a truly sophisticated treatment of this topic is beyond our scope, a somewhat simplified version of these axioms can be stated as follows:

DEFINITION: Given the sample space $\mathcal{S}$, and the family of events in $\mathcal{S}$, a probability function associates with each event A a real number $p(A)$, the probability of event A, such that the following axioms are true:

1. $p(A) \geq 0$ for every event A.
2. $p(\mathcal{S}) = 1$.
3. If there exists some countable set of events, $\{A_1, A_2, \cdots, A_N\}$, and if these events are all mutually exclusive, then $p(A_1 \cup A_2 \cup \cdots \cup A_N) = p(A_1) + p(A_2) + \cdots + p(A_N)$ (the probability of the union of mutually exclusive events is the sum of their separate probabilities).

(In this connection, the word "countable" means simply that it is possible to associate each and every distinct event in the set with one and only one of the "natural" or counting numbers. Later, when we deal with uncountably infinite sets of events, the definition will be extended accordingly.)

In essence, this definition states that paired with each event A is a nonnegative number, probability $p(A)$, and that the probability of the sure event $\mathcal{S}$, or $p(\mathcal{S})$, is always 1.00. Furthermore, if A and B are any two *mutually exclusive* events in the sample space, the probability of their union, $p(A \cup B)$, simply the sum of their two probabilities, $p(A) + p(B)$. In the same way, if A, B, and C are all mutually exclusive, then $p(A \cup B \cup C)$ is equal to $p(A) + p(B) + p(C)$, and so on for any number of mutually exclusive events.

It is important to remember at this stage that this is a purely formal definition of probability, in terms of numbers associated with sets of possibilities making up a sample space $\mathcal{S}$. *Events* have probabilities, and in order to discuss probabilities we must always specify the events to which these probabilities belong. When we speak of the probability that a citizen of the United States has red hair, we are referring to the probability number associated with the event "red hair" in the sample space, "all citizens of the United States at a given moment in time." Similarly when we say that the probability that a coin will come up heads is .50, we are stating that the number .50 is paired with the event "heads" in the sample space "all possible results of tossing a particular coin."

At first this may seem an extremely unmotivated and arbitrary way to discuss probability. Everyone knows that the word "probability" means more than a mere number assigned to a set, and we shall certainly give these probability numbers additional meaning in sections to come. For the present, however, let us accept this formal definition at face value.

1.5 / SOME SIMPLE RULES OF PROBABILITY

The concepts of simple experiment, sample space, and events, taken together with the three axioms given above, can be used to deduce any number of new consequences. These consequences, supplemented by new definitions and further consequences or theorems, are what is meant by the theory of probability. Each of these mathematical results gives some relationship that probability numbers must exhibit. Along with the three basic axioms, these derived relationships can be thought of as the rules of probability. A few of these rules, which follow directly from the axioms, will now be discussed.

Recall that for any event A there will also be a complementary event "not A," or $\bar{A}$. Then for any event A in $\mathcal{S}$,

---

Probability rule 1: $p(\bar{A}) = 1 - p(A)$

(This is sometimes called *the rule of complementary probability*.) This simply says that the probability of the event $\bar{A}$ (not A) is one minus the probability of the event A. Why should this be true? First notice that the two events A and $\bar{A}$ are exhaustive, since logically any elementary must be a member of one of these classifications. Consequently, $A \cup \bar{A} = \mathcal{S}$. Note further that A and $\bar{A}$ are mutually exclusive, since an elementary event cannot be a member of both A and $\bar{A}$. On putting these ideas together with Axioms 3 and 2 we have

$$p(A \cup \bar{A}) = p(A) + p(\bar{A})$$

and $$p(A \cup \bar{A}) = p(\mathcal{S}),$$

so that $$p(A) + p(\bar{A}) = 1.00$$

and thus $$p(\bar{A}) = 1 - p(A).$$

Within what range of values must any probability lie? The answer is given by

Probability rule 2: $0 \leq p(A) \leq 1.00$ for any event A.

(this may be thought of as *the probability range*.) In other words, only positive numbers lying between zero and one, inclusive, may be used to signify probabilities.

This too is easy to deduce by use of the axioms and Rule 1. We have just seen that the probability of an event A plus the probability of its complement $\overline{A}$ must sum to 1.00. Now, just suppose that rule 2 were not true, and that some event A *could* have a probability greater than 1.00. However, then event $\overline{A}$ would have to have a probability less than 0 in order for Rule 1 to be true. This would contradict Axiom 1, which says that no event may have a probability less than 0. Since axioms may not be contradicted, then Rule 2 must be true.

We already know from Axiom 2 that $p(\mathcal{S})$ must have the value 1.00. However, what event has probability of zero? This is given by

Probability rule 3: $p(\varnothing) = 0$, for any $\mathcal{S}$.

(You may wish to think of this as the *rule of the impossible event*.) That is, the impossible event $\varnothing$ always receives probability 0.

This is also easy to demonstrate. For any simple experiment, the sample space $\mathcal{S}$ contains all of the possible outcomes or elementary events. For an outcome to occur which is not in $\mathcal{S}$ is impossible. Therefore, the event "not $\mathcal{S}$", or $\overline{\mathcal{S}}$, is the same as $\varnothing$. From Rule 1 it is known that

$$p(\mathcal{S}) + p(\overline{\mathcal{S}}) = 1.00.$$

It is also known from Axiom 2 that $p(\mathcal{S}) = 1.00$. Hence $p(\varnothing) = 0$.

A very important rule applies for any two event A and B, whether or not they are mutually exclusive. In other words, this rule applies even if $p(A \cap B)$ is not equal to zero. Thus,

Probability rule 4: For any two event A and B in $\mathcal{S}$,

$$p(A \cup B) = p(A) + p(B) - p(A \cap B).$$

(This is often called *the "or" rule of probability*.) That is, the probability of the even "A or B" is always equal to the probability of event A plus the probability of event B, minus the probability of the event "A and B." Look at Figure 1.3.4 and see if you can tell why this must be true.

Notice that Axiom 3 provides a special case of the "or" rule: When A and B are mutually exclusive, then $p(A \cap B) = 0$, and $p(A \cup B) = p(A) + p(B)$.

(It is also possible to generalize Rule 4 to three or more events and we shall do so in Section 1.11.)

Finally, a useful rule deals with a *partition* of the sample space $\mathcal{S}$. Recall that a partition of the sample space $\mathcal{S}$ consists of two or more events, which are all mutually exclusive, and which exhaust the possibilities in $\mathcal{S}$. Suppose there were a set of two or more events in $\mathcal{S}$, represented by A for the first event, and by L for the last event. Then

Probability rule 5: If the set of events $A, \cdots, L$ constitute a partition of $\mathcal{S}$, then $p(A \cup \cdots \cup L) = p(A) + \cdots + p(L) = 1.00$.

(If you like, this may be described as the *partition rule*.) In other words, when any set of events are mutually exclusive and exhaustive, and thus form a partition of $\mathcal{S}$, the sum of their probabilities must be equal to 1.00.

It is a simple matter to illustrate these elementary rules of probability. Take once again the example of drawing playing cards one at a time from a standard deck, which is well shuffled before each draw. The sample space $\mathcal{S}$ here consists of all of the fifty-two different cards that we might draw. Some of the possible events are "the card has the number 10," "the card is a diamond," "the card is a queen of spades," "the suit of the card is red," "the card is an odd number in a black suit, but not a picture card," and so on. Some outcomes cannot occur: the event "card is a gorilla" is (or surely ought to be) impossible, and thus is the same as $\varnothing$.

Suppose that the event A in this example is "card is a club," and that we happen to know in this example that $p(A) = 1/4$. What is the event $\overline{A}$? This would be the event "card is not a club," and we could find its probability by use of Rule 1 by taking $p(\overline{A}) = 1 - (1/4) = 3/4$.

If the event A is taken to be "card is a king," and the event B is "card is a jack," then we could also consider the event $A \cup B$, "the card is a king or a jack." For this playing card experiment it happens that $p(A) = 1/13$ and that $p(B) = 1/13$. However, a card cannot be both a king and a jack, so that $p(A \cap B) = 0$. Then by Axiom 3,

$$p(\text{king or jack}) = \left(\frac{1}{13}\right) + \left(\frac{1}{13}\right) = \frac{2}{13}.$$

On the other hand, suppose that event A is "card is a king," and that event C is "card is a heart." Here it happens that $p(A) = 1/13$, and $p(C) = 1/4$. Furthermore, $p(A \cap C)$ is not zero here, but rather is 1/52. Then, in order to find $p(A \cup C)$, we use Rule 4 to obtain

$$p(\text{king or heart}) = \left(\frac{1}{13}\right) + \left(\frac{1}{4}\right) - \left(\frac{1}{52}\right) = \left(\frac{16}{52}\right).$$

What is the probability of observing a "seven," an "eight," or a "nine" for the card drawn? Among playing cards each of these events has probability of 1/13, and, since these are mutually exclusive events, we know from Axiom 3 that

$$p(\text{seven or eight or nine}) = \left(\frac{1}{13}\right) + \left(\frac{1}{13}\right) + \left(\frac{1}{13}\right) = \frac{3}{13}.$$

For the simple experiment of drawing playing cards, the four suits, (spades, hearts, diamonds, and clubs) form a partition of the sample space. That is, these suit events are mutually exclusive, since if a card belongs to one suit it cannot belong to any other. Furthermore, the suits are exhaustive: each card drawn from a standard deck must belong to one and only one of these four suits. Then, by Rule 5,

$$p(\text{spades}) + p(\text{hearts}) + p(\text{clubs}) + p(\text{diamonds}) = 1.00.$$

Many more examples of these simple manipulations are given in the exercises at the end of this chapter.

1.6 / EQUALLY PROBABLE ELEMENTARY EVENTS

So far probabilities of events have been discussed only in a rather formal sense, and we have not really gone into how these probabilities are to be calculated, nor what these probability numbers mean. Now we turn to the problem of calculating probabilities for a given simple experiment.

Bear in mind that each distinct outcome making up the sample space $\mathcal{S}$ can itself be thought of as an event, and that each of these elementary events is associated with a probability. Any pair of elementary events must be mutually exclusive since each is a distinct outcome. Now suppose that for some simple experiment we were actually given the probability for each elementary event. Then we would be able to calculate the probability for any event A. *The probability of any event A in the sample space $\mathcal{S}$ is simply the sum of the probabilities of all of the elementary events in $\mathcal{S}$ qualifying for the event A.* In other words, the probability of event A is nothing more than the sum of the probabilities of the elementary events that A contains. This follows directly from Axiom 3.

This principle for finding the probability of an event is certainly simple enough, but there is one obvious hitch: How does one know the probabilities of the elementary events in the first place? Fortunately, a great many of the simple experiments for which we need to calculate probabilities, and particularly games of chance, have a feature that does away with this problem. *A great many simple experiments are conducted in such a way that it is reasonable to assume that each and every distinct elementary event has the same probability.* When tickets are drawn in a lottery, for example, great pains are taken to have the tickets well shaken up in a tumbler before each one is drawn; this mixing operation makes it reasonable to believe that any particular ticket has the same chance of being drawn as any other. Cards are thoroughly shuffled and cut, perfectly balanced dice are thrown in a dice-cage, and, in fact, almost all gambling situations have some feature which makes this equal-chances assumption reasonable. As we shall see presently, even experiments that are not games of change also are carried in such a way that each and every elementary event should have the same probability.

When there is some finite number of elementary events in a sample space, where each elementary event has exactly the same probability, then the probability of any event is particularly easy to compute. Imagine a sample space containing N distinct elementary events. For any event A, let the number of elementary events that are members of A be denoted by $n(A)$. The following principle applies:

If all elementary events in the sample space $\mathcal{S}$ have exactly the same probability, then the probability of any event A is given by

$$p(A) = \frac{\text{number of elementary events in } A}{\text{total number of elementary events in } \mathcal{S}}$$

$$= \frac{n(A)}{N}.$$

For equally likely elementary events, the probability of an event A is its relative frequency in the sample space.

As an illustration, suppose that a box contains ten marbles. Five of these marbles are white, three are red, and two are black. We perform the simple experiment of drawing a marble out of the box (without looking) in such a way that the marbles are well mixed up, and that there is no reason for any given marble to be favored in our drawing. Now we can identify our experiment with the model of probability we have been discussing. An outcome is the result of our drawing a marble and looking at it. An elementary event is a particular marble, in this instance, and there are exactly ten distinct marbles; hence there are ten elementary events making up the sample space $\mathcal{S}$. The marble-events are mutually exclusive, and each has probability 1/10.

We are concerned with the three events: ''white,'' ''red,'' and ''black.'' Notice that these three events are subsets of $\mathcal{S}$, containing five, three, and two elementary events (marbles), respectively. What is the probability of our drawing a red marble? The answer is given by

$$p(\text{red}) = \frac{\text{number of red marbles}}{\text{total number of marbles}}$$

$$= \frac{3}{10}$$

so that we may say correctly that the probability of the event ''red'' in this experiment is .30. In the same way, one can find

$p(\text{white}) = .50$, and $p(\text{black}) = .20$.

Furthermore, it is easy to see, by applying the rules of Section 1.5 that

$p(\text{no color}) = .00$
$p(\text{red or white}) = .30 + .50 = .80$
$p(\text{red and white}) = .00$
$p(\text{red or black}) = .30 + .20 = .50$
$p(\text{red or white or black}) = .30 + .20 + .50 = 1.00$

and so on, for any other event.

Consider another example. A teacher has thirty children in a classroom. In some completely ''random'' and unsystematic way, the teacher chooses a child and notes the father's occupation. The children are each equally likely to be chosen by the teacher. In this example the elementary events are the different children who might be chosen: there are thirty such elementary events possible, making up the sample space $\mathcal{S}$. Now suppose that there are only four classes of occupation represented in the room: professional, white-collar, skilled labor, and unskilled labor. These four classes will be labeled events A, B, C, and D. If three children represent group A, fifteen, group B, ten, group C, and two, group D, what is the probability that a child chosen will fall into a given group? In other words, what are the probabilities of the four different events? Once again, these probabilities are given by the relative frequencies:

$$p(A) = \frac{\text{number of elementary events in } A}{\text{total number of elementary events}}$$

$$= \frac{3}{30} = .10.$$

In the same way we find:

$$p(B) = \frac{15}{30} = .50$$

$$p(C) = \frac{10}{30} = .33$$

$$p(D) = \frac{2}{30} = .07.$$

Finally, suppose that the simple experiment consisted of giving some poor unprepared graduate student a choice of six essay topics on an examination. These topics are numbered in order from "1" through "6." Suppose further that the student knows nothing about any of these topics, so that they are all equally likely to be picked. Then the probability that item 1 is picked should be 1/6, item 2 has probability of 1/6, and so on. What then is the probability that the student picks an odd-numbered topic? This event occurs when the choice falls on topic 1, topic 3, or topic 5. Consequently, the desired probability is

$$p(1 \text{ or } 3 \text{ or } 5) = p(\text{odd number}).$$

Since these events are mutually exclusive, the probability is found from Axiom 3 to be

$$p(1 \text{ or } 3 \text{ or } 5) = \frac{1}{6} + \frac{1}{6} + \frac{1}{6} = .50.$$

What is the probability that the student chooses the first or the last topic? This event involves the choice of topic 1 with probability of 1/6, or topic 6, also with a probability of 1/6. Thus

$$p(\text{first or last topic}) = \frac{1}{3}.$$

In summary, the simplest approach to the calculation of probabilities rests on the assumption that all of the elementary events in a given sample space $\mathcal{S}$ have the *same* probability. However, there is nothing in the formal theory of probability that makes this assumption necessary, and one may talk perfectly well about probabilities even in situations where this assumption may not be true. Even so, the calculation of probabilities of events is made very much simpler when this assumption is made, and in virtually all that follows we will confine our discussion to this case.

1.7 / "IN THE LONG RUN"

So far we have seen how the outcomes of an experiment are identified with the elementary events in a sample space, and how probability values may be assigned to each event. In the situation where the elementary events are themselves equally probable, then the probability of any event is just its relative frequency in the space. However, we still have not reached an answer to the question of what these probabilities mean. How are probabilities related to the things one observes?

In the first example of the preceding section, if one were to keep drawing marbles out of the box, and after each draw were to replace the marble in the box before drawing again, then in the *long run,* after very many observations, *what proportion of the marbles drawn should be red?* It should take only a very little thought to arrive at the conclusion that one should *expect* about 3 in 10 such observations to be red. In the same way, in the second example, suppose that the teacher keeps choosing children in this unsystematic way, and any given child can be chosen over and over again. If each child is equally likely to be chosen at any time, then it seems reasonable that in the long run, over a large number of such observations, about .10 or one in ten of the children chosen should have fathers in occupation group A.

In the third example, suppose that many students were asked to choose one essay from the six topics, and that the topics were equally likely to be chosen by any student. Then the first topic should be chosen by about 1/6 of the students, the second by 1/6, and so on. Similarly, we should expect about half of the students to choose an odd-numbered topic, and about 1/3 to choose either the first or last topic.

Despite the different contexts of these three examples, they all illustrate the same principle:

The probability of an event denotes the relative frequency of occurrence of that event to be expected in the long run.

Simple examples of this sort all involve equally likely elementary events, and in many applications of probability theory, especially to games of chance, this assumption is made. On the other hand, the same idea applies even when elementary events are *not* equally likely: *in the long run, the relative frequency of occurrence approaches the probability for any event.*

Thus, the idea of relative frequency is connected with probability in two ways:

1. If elementary events are equally probable, the probability of an event is its relative frequency in the sample space.
2. The *long-run* relative frequency of occurrences of event A over trials of the experiment *should approach* $p(A)$. This is true regardless of whether elementary events are equally probable or not, provided that observations are made "independently" and "at random."

(Shortly we will give more formal meanings to the terms "independently" and "at random." However, for the moment think of "independent" as meaning "each has nothing to do with any other," and "at random" as meaning, "by chance alone.")

The connection between probability and long-run relative frequency is both a simple and appealing one, and it does form a tie between the purely formal notion, "the probability of an event," and something we can actually observe, a relative frequency of occurrence. A statement of probability tells us what to *expect* about the relative frequency of occurrence of an event given that enough sample observations are made at random: In the long run, the relative frequency of occurrence of event X should approach the probability of this event, if independent trials are made at random over an indefinitely long sequence. As it stands, this idea is a familiar and reasonable one, but it will be useful to have a more exact statement of this principle for use in future work. The principle was first formulated and proved by James Bernoulli in the early eighteenth century, and it often goes by the name "Bernoulli's theorem." A more or less precise statement of Bernoulli's theorem goes as follows:

If the probability of occurrence of the event X is $p(X)$, and if N trials are made, independently and under exactly the same conditions, then the probability that the relative frequency of occurrence of X differs from $p(X)$ by any amount, however small, approaches zero as the number of trials grows indefinitely large.

In effect, the theorem says this: Imagine some N trials made at random and in such a way that the outcome on any one trial cannot possibly influence the probability of outcomes on any other trial; that is, N independent trials are made. This act of taking N independent trials of the same simple experiment may itself be regarded as another experiment in which the possible outcomes are the numbers of times event X occurs out of N trials. Each possible such outcome has a probability. Then as N becomes very large, the probability that the relative frequency of X differs from $p(X)$ in any way becomes very small. This does not mean, however, that the proportion of X occurrences among any N trials *must* be $p(X)$; the proportion actually observed might be any number between 0 and 1. Nevertheless, given more and more trials, the relative frequency of X occurrences may be expected to come closer and closer to $p(X)$.

In practical terms, Bernoulli's theorem says that even if we only have a limited number of trials, we should expect the probability of any event to be reflected in the relative frequency we actually observe for that event. In the long run, such an observed relative frequency should approach the true probability. Although Bernoulli's theorem does work for us in this way, it does so not because of any necessary compensation for early "misses" by "hits" later on. Rather, the theorem holds because departures from what one expects to occur are simply *swamped out* as the total number of trials or observations becomes very large.

Please do not fall into the error of thinking that an event is ever due to occur on any given trial. If you toss a coin 100 times, you expect about 50 percent of the tosses to show the event "heads," since the theoretical probability of that event for a fair coin is .50. This does not mean, however, that for any given one-hundred trials (or one-thousand, or million, or billion trials) the coin must show 50 percent heads. This need not be true at all. Every one of your 100 tosses *could* result in the event heads, or none of them might result in this event—the coin does not ever have to come up heads in any finite number of tosses. Only in an infinite number of tosses must the relative frequency equal .50. The only thing which we can say with assurance is that .50 is the relative frequency of heads we should *expect* to observe in any given number N of tosses, and that it is increasingly probable that we observe close to 50 percent heads as the N grows larger. But on any finite number of tosses of the coin, the relative frequency of heads observed can be anything. This same is true for the occurrence of any event in any simple experiment in which observations are made at random and in which the probability $p(X)$ of the event X is other than 1 or 0.

Although it is true that the relative frequency of occurrence of any event must exactly equal the probability only for an infinite number of independent trials, this point must not be overstressed. Even with relatively small numbers of trials we have very good reason to expect the observed relative frequency to be quite close to the probability. The rate of convergence of the relative frequency to the probability is very rapid, even at the lower levels of sample size, although the probability of a small discrepancy between relative frequency and probability is much smaller for extremely large than for extremely small samples. A probability is not a curiosity requiring unattainable conditions, but rather a value that can be estimated with considerable accuracy from a sample. Our best bet about the probability of an event is the actual relative frequency we have observed from some N trials, and the larger N is, the better the bet.

1.8 / AN EXAMPLE OF SIMPLE STATISTICAL INFERENCE

Imagine that you have a friend who claims to possess the power of extrasensory perception (ESP). You have some doubts about this, but wish to check out this claim. As it happens, there is a standard deck of cards often used in such experiments. This deck is composed of cards printed with circles, squares, and triangles. There are, lets say, sixty cards in all, with twenty of each kind.

The experiment goes as follows: You shuffle the cards thoroughly, and draw one card unsystematically from the deck, completely out of sight of the subject. You then concentrate intently on the figure for the card drawn, and your friend uses the alleged ESP to guess the figure represented. Any card drawn is then replaced, and the experiment repeated for as many trials as desired.

The sample space of elementary events can be represented by all combinations of card drawn and figure guessed, as shown below:

		figure guessed		
		circle	*square*	*triangle*
card	circle	1/9	1/9	1/9
	square	1/9	1/9	1/9
	triangle	1/9	1/9	1/9

If nothing but chance is going on in this experiment, it is reasonable to suppose that each of these card-guess events has the same probability, or 1/9. Furthermore, the event "correct," since it involves three of these possibilities, has probability $p(\text{correct}) = 1/3$.

On the other hand, suppose that ESP actually is operating here. This should lead to more correct guesses than otherwise, or more than 1/3 correct. In other words, there are really two theories being compared: "no ESP," represented by the situation with $p(\text{correct}) = 1/3$, and "some ESP," represented by $p(\text{correct})$ greater than 1/3.

Now let us say that you make a series of trials of the experiment, where a card is drawn and a guess made on each trial. Suppose that your friend is correct on the first trial. Does this say anything about ESP? Not very much, since we expect one out of three trials to be correct anyway, even without ESP. However, suppose that you keep making trials, so that after 12 cards have been drawn, there are eight correct guesses. Here, one expects 1/3 of 12, or only four correct guesses, so that this result of eight correct is somewhat above the expected number. Such a result is not very likely to have occurred when $p(\text{correct}) = 1/3$, but is more likely when $p(\text{correct}) > 1/3$. The data so far accord better with the theory "some ESP" than with "no ESP." Still, you keep going until, perhaps, 900 trials have been completed. Here imagine that 600 correct guesses out of 900 trials were recorded. This is *very* far from the (900)1/3 or 300 correct expected by chance alone, and is very unlikely to have occurred if the true probability of a correct guess is 1/3. Your friend's claim begins to look pretty good at this point. Your best guess about $p(\text{correct})$ is 2/3, not 1/3.

Nevertheless, 600 correct out of 900 trials is not impossible even when the probability of a correct guess is only 1/3. Perhaps your friend simply got lucky. Regardless of how many trials that are made, short of an infinite number, the number of correct guesses could be anything, from none correct to all correct. Even so, large departures from expectation are themselves unlikely, and make us suspect that we were not expecting the right thing in the first place.

By the same token, even though your friend actually has ESP which is slightly less than perfect, most or even all of the guesses *could* have been wrong, in any limited series of trials. Nevertheless, this is not what one expects to happen if the ESP theory is true. The mere fact of many failures would make you very skeptical of the ESP claim, and the more these failures piled up over many trials, the more skeptical you would tend to become.

Please understand that this little experiment was not described in order to show how to do experiments on ESP. Rather, it merely illustrates in fairly simple form the process that is often applied in statistical inference. In such situations, empirical evidence is to be used to check the validity of some theoretical state of affairs. The theory can itself be used to determine the probability with which a certain kind of event should be observed. Thereupon an experiment is conducted in which the event or events in question are possible outcomes, and some N observations are made at random and independently. The theoretical probabilities tell what to expect with regard to the events of interest, and the relative frequencies are compared to those that are expected. *To the extent that the obtained relative frequencies depart widely from the theoretical probabilities, then there is evidence that the theory is not true.* Moreover, the larger the number of observations, N, the more weight is given to any extent of departure from expectation as evidence against the theory. On the other hand, one can never be completely sure that the theory is false unless an infinite number of observations have been made.

By the same token, even though the evidence appears to agree well with a theory, this in no way implies that the theory must be true. Only after an infinite number of trials, when the relative frequency of occurrence of events matches the theoretical probability exactly, can one assert with complete confidence that the theory is true. Increased numbers of trials lend confidence to our judgments about the true state of affairs, but we can always be wrong, short of an infinite number of observations.

The principle embodied in Bernoulli's theorem can be used not only to compare empirical results with theoretical probabilities, but also to *estimate* the true probabilities of events by observing their relative frequencies over some limited numbers of trials. For example, if 67 percent of your friend's guesses are correct, the best estimate of the actual probability is 2/3, not 1/3.

As another example, visualize a box containing marbles of different solid colors. The number of marbles in the box, and how many different colors they have are both unknown, but the task is to find out the probability of each color when the marbles are drawn individually from the box. The simple experiment consists of mixing the marbles well, drawing one out, observing its color, and then putting it back in the box (random sampling with replacement). Now suppose that the first marble is white. Then the observed relative frequencies of colors are white 1.00, other .00. We wish to decide about the probabilities in such a way that our estimate would, if correct, have made the occurrence of this particular sample result have the greatest prior probability. What probability of the event "white" would make drawing a white marble most likely? The answer is $p(\text{white}) = 1.00$, and $p(\text{other}) = .00$, so that this is our first estimate of the probabilities of the colors of the marbles.

However, we now draw a second marble at random with replacement. This marble turns out to be red, and our best guess about the probabilities is now white .50, red .50, other .00, as these probabilities would make the occurrence of the obtained sample of two marbles most likely. If we kept on drawing, observing, and replacing marbles for 100 trials, shaking the box so that the marbles (elementary events) are equally likely to be chosen on a trial, we might begin to get a reasonably clear picture of the probabilities of the colors:

Color	*Relative frequency*
white	.24
red	.50
blue	.26

By the time we had made 10,000 observations, we could be even more confident that the probabilities are close to:

Color	*Relative frequency*
white	.24
red	.50
blue	.24
green	.02

and so on. Given an infinite number of trials, we could specify the relative frequencies (that is, the probabilities) of the marbles in the box precisely. The larger the number of observations, the less do we expect to "miss" in our estimates of what the box contains. However, for fewer than an infinite number of observations, the observed relative frequencies *need not* reflect the probabilities of the colors exactly, though they may well do so.

It is important to note that since we are drawing samples *with replacement* in this situation, the fact that there may be only some finite numbers of distinct elementary events (marbles in this case) in no way prevents us from taking a very large or even an infinite number of sample observations with replacement after each trial. Thus Bernoulli's theorem applies perfectly well to situations in which there may be a very small number of distinct elementary events possible, so long as one can draw an unlimited number of samples or make an unlimited number of trials of the experiment (as in coin tossing or card drawing).

The scientist making observations is doing something quite similar to drawing marbles from a box. Because of limited powers of observation the scientist cannot "see into the box" and generalize from observing all such phenomena. Nevertheless, sample observations can be made and generalizations can be drawn from what is actually observed. Such a generalization is much like a bet, and how good the bet is largely depends on how many observations the scientist is able to make. The scientist can never be sure that the generalizations drawn are the correct ones. Yet, the risk of being wrong can be made quite small if a sufficiently large number of observations are made.

1.9 / PROBABILITIES AND BETTING ODDS

One common way of expressing the probabilities of two mutually exclusive events is in terms of betting odds. The formal connection between betting odds and probabilities is as follows:

If the probability of an event is p, then the odds in favor of the event are p to $(1 - p)$.

Thus, if some event A has a probability $p(A) = 3/4$, then the odds in favor of A are 3/4 to 1/4, or 3 to 1. If $p(A) = 1/8$, then the odds in favor of A are only 1/8 to 7/8, or 1 to 7. When two fair dice are tossed, the probability of the event "7" is 1/6, so that the odds in favor of a "7" are 1 to 5. The odds against the occurrence of a "7" are 5 to 1. As another example, suppose that a class contains 15 girls and 11 boys. A student is selected at random from the class. What are the odds in favor of the selection of a girl? Since the probability of the event "girl" is 15/26, the odds in favor of the selection of a girl are 15 to 11.

It is equally simple to convert statements of odds into statements of probability:

If the odds in favor of some event A are x to y, then the probability of that event is given by $p(A) = x/(x + y)$.

If the odds for some event are 9 to 2, then the probability of that event is $9/(9 + 2)$ or 9/11. If the odds for some event are 1 to 1, or "even," then the probability of that event must be 1/2.

In betting situations individuals often give or accept odds. This is true not only in games of chance but also in situations where the "objective" probability of the event in question would be difficult to determine. In such situations, a statement of acceptable odds can sometimes be taken as a reflection of an individual's judgment of probability. When someone says that he believes the odds are 2 to 1 that the Pittsburgh Steelers will beat the Los Angeles Rams in their impending game, he is saying that the probability is, for him, $p(\text{Steelers beat Rams}) = 2/3$. When the weather forecaster says that the odds are 5 to 2 against rain tomorrow, she is saying that her judged probability $p(\text{no rain tomorrow}) = 5/7$, or $p(\text{rain tomorrow}) = 2/7$.

Odds are often stated in monetary terms as well. Suppose that bettor I and bettor II agree that the odds are 2 to 1 in favor of the Pittsburgh Steelers against the Los Angeles Rams (there will be a sudden-death playoff, so that no tie is possible in this game). Bettor I picks the Steelers and II picks the Rams. How much should each put up in order to make this a fair bet? The first bettor has chosen the event he believes to be the more likely, and thus he should stand to gain less if he wins than bettor II, who has the less likely event. The bet becomes fair when bettor I agrees to put up \$2 to \$1 for bettor II. In this way bettor I gets only \$1 if he wins, whereas bettor II gets \$2. In general, a bet is fair when the ratio of the moneys put up is the same as the odds. In this case, the bettor on the more likely event puts up \$2 for each \$1 on the less likely event, since the odds are 2 to 1. Looked at in another way, *a bet is fair if the following relationship holds:*

$$(\text{amount won if } A \text{ occurs})[p(A)] = (\text{amount lost if } \bar{A} \text{ occurs})[1 - p(A)].$$

When a bet is fair, it is also true that

$$\frac{(\text{amount bet on event } A)}{(\text{amount bet against event } A)} = \text{odds in favor of } A.$$

The amounts bet in a fair bet thus give the odds, and consequently the probabilities involved. When someone says "Five dollars will get you ten dollars that such and such will occur," he is actually saying that he wants to bet on an event with odds of 2 to 1, so that he is willing to put up \$2 for every \$1 you put up. This is another way of saying that the probability is, in his judgment, 2/3.

Strictly speaking, in a bet such as we have been describing, the two bettors agree only that true odds are *at least* as favorable as those accepted by the person who bets on the event, and, *at most* as favorable as those accepted by the person who bets against it. If bettor I accepts odds of 2 to 1 in the example given above, but really believes that the odds are, say, 4 to 1, he should also believe that the bet is biased in his favor. On the other hand, if bettor II believes that the odds are less in favor of the event, say 1 to 1 rather than 2 to 1, and he nevertheless accepts the 2-to-1 bet, he should also believe the bet to be biased in his favor. In short, when two people agree on a bet with odds of x to y, we can say only that the judged probability of the person who bets on event A is $p(A) \geq x/(x + y)$, and that the judged probability of the person who bets against event A is $p(A) \leq x/(x + y)$. Even so, the fact that two individuals can agree on a bet gives information on the degree of belief each holds about the occurrence of an event.

The expression of probabilities in terms of odds is a quite common feature of daily life. The fact that people can make these judgments about events, and act on the basis of such judgments, is one of the bases for an alternative interpretation, which we will now explore.

1.10 / OTHER INTERPRETATIONS OF PROBABILITY

Long-run relative frequency is but one interpretation that can be given to the formal notion of probability. It is important to remember that this is an *interpretation* of the abstract model. The model per se is a system of relations among and rules for calculating with numbers that happen to be called probabilities. There is no "true meaning" of probability, any more than there is a true meaning of the symbol "x" as used in school algebra. Probability is an abstract mathematical concept that takes on meaning in the ordinary sense only when it is identified with something in our experience, such as relative frequency of real events.

The probability concept acquired its interpretation as relative frequency because it was originally developed to describe certain games of chance where plays (such as spinning a roulette wheel or tossing dice or dealing cards) are indeed repeated for very many trials. Similarly, there are situations in which the scientist makes many observations under the same conditions, and so the mathematical theory of probability is given a relative frequency interpretation here as well.

On the other hand, there are some students of the matter who object to this as the exclusive interpretation of probability, and who have shown that quite different interpretations can be advanced that do not identify formal probability exclusively with relative frequency. Indeed, an everyday use of the probability concept does not have this relative frequency connotation at all. We say "It will probably rain tomorrow," or "The Yankees will probably win the pennant," or "I am unlikely to pass this test"; our hearers have no difficulty in understanding what we mean in each instance, but it is very difficult to see how these statements describe long-run relative frequencies of outcomes of simple experiments repeated over and over again. Each such statement describes the speaker's *certainty* or *degree of belief* about an event that will occur once and once only. Our inclination to use naive notions of probability in this way is one of the reasons theorists have sought other interpretations.

For many years some theorists of probability and statistics have studied the implications of a subjective, or personalistic, approach to the interpretation of probability. Such an interpretation is not so much an alternative to, as an extension of, the usual relative frequency interpretation of probability. Under this subjective approach, a probability is seen as a measure of individual degree of belief, the quantified judgment of a particular individual with respect to the occurrence of a particular event. An individual is assumed to hold, with some measurable degree of confidence, the belief that a particular event will or will not occur. Then the measure of that degree of belief about the event is represented as a probability. Under this conception, it is perfectly reasonable to assign a probability value to an event that can occur only once. The event "rain tomorrow" for a particular day is such a nonrepetitive event. That particular "tomorrow" will occur once and only once, and it will or will not rain at the location in question. When an individual says that "the probability is 1/5 that it will rain at this place tomorrow," this is a value reflecting the strength of this person's belief that the event in question will occur. It is very difficult to interpret such a statement in terms of relative frequency, but it does make intuitive sense when viewed as an individual's subjective assessment of a situation. We make similar statements all the time without a moment's second thought.

Degree of belief, and thus subjective probability, can be inferred from the choice behavior of an individual. It is not from what one says, but rather from the choices that one makes, that a measure of subjective probability can be gained. It is true that, in order to find these measures, we must make certain assumptions about behavior. Certain reasonable "axioms of choice" or "axioms of coherence" must be satisfied if the values inferred from individual behavior are to be treated as probabilities satisfying the mathematical properties given above. In this rapid overview we cannot take the time to go into these axioms. Suffice it to say that if an individual's behavior is consistent in the ways specified by these axioms, then probabilities inferred from individual behavior *can* be treated by the full machinery of probability theory.

It is also quite possible to apply thc subjective interpretation of probability to events that are repetitive, and which thus lend themselves to a relative frequency interpretation as well. Again consider a game of dice. A player might very well have a definite degree of belief that the dice will turn up "7" on the next play. Given certain assumptions, such that each die is "fair," that the method of tossing does not influence the outcome, and so on, an intelligent player might well behave as though the chances were 1/6 that the dice would total 7 on the next throw. This is the same value that simple probability calculations dictate under these assumptions. After all, remember that the probability computations are absolutely neutral with respect to the interpretation to be placed on the probabilities. The difference between the subjective and the relative frequency approaches lies only in the way the probability values are interpreted, not in the way they are treated mathematically. In one view, probability calculations can be thought of as ways of settling on an appropriate degree of belief, given certain assumptions and information about the circumstances of the event in question.

Critics of this approach are quick to point out that because they stand for degrees of belief, subjective probabilities may vary from individual to individual for the same event. Since individuals vary in their backgrounds, knowledge, and available information, it is quite reasonable to suppose that they will vary in the degree of belief they hold for the occurrence of a given event. If this were not the case, professional gamblers, as well as thousands of perfectly respectable businesspeople, would soon be out of work. However, such individual differences in degrees of belief do not unduly trouble advocates of the subjective approach, provided that differing probability values accurately reflect differences in judgment among individuals. Students of the subjective approach to probability are especially interested in the choice or preference behavior of an individual among bets or lotteries. Furthermore, they are concerned with changes in degree of belief by individuals, as new information is gained and as information is shared. Subjective probabilities can be assessed from such individual choice behavior, and in this sense they are just as "objective" as probabilities interpreted in other ways.

In an elementary discussion such as this, probability can usually be interpreted as relative frequency. Most applications of statistics deal with situations where sampling can be repeated many times, at least in theory, and in these situations the relative-frequency interpretation does make sense. However, when we deal with decision-making based upon statistical information, we will have occasion to discuss subjective probabilities once again.

Generally, in the material to follow, we will work with probabilities without specifying whether these are to be interpreted in one way or the other. We can do this because, essentially, probability is an abstract mathematical concept, for which certain consequences must follow from certain premises, regardless of what the "real" interpretation of the concept may be.

1.11 / MORE ABOUT JOINT EVENTS

As we have seen, the elementary events making up a sample space may be grouped into events of various kinds. Given some event A, then the occurrence of some elementary event either does or does not qualify as an occurrence of A. Furthermore, given that an elementary is an A, it may also qualify for some event B as well. In that instance, when the elementary event occurs we say that the *joint event* $A \cap B$ occurs.

Provided that the elementary events in $\mathcal{S}$ are equally likely to occur, then the probability of $A \cap B$ is found just as for any other event:

$$p(A \cap B) = \frac{\text{number of elementary events in } A \cap B}{\text{total number of elementary events}}. \qquad [1.11.1^*]$$

Furthermore, Bernoulli's theorem holds for such joint events just as it does for any events: the long-run relative frequency of any joint event should approach the probability of that event. The best estimate of the probability of a joint event is the relative frequency we actually observe in some N random and independent trials of the simple experiment.

Given the joint event "A and B" for some sample space $\mathcal{S}$, it is also possible to define three other joint events in terms of "not A" and "not B." The relations among these four events and their probabilities are given by the "or" rule for probabilities (Rule 4) of Section 1.5. Recall that for any event A there is also an event $\overline{A}$, consisting of all of the elementary events not in the set A, so that

$$p(A) + p(\overline{A}) = 1$$

Then along with $A \cap B$ we can discuss the other joint events:

$A \cap \overline{B}$, consisting of all elementary events in A and not in B;
$\overline{A} \cap B$, consisting of all elementary events not in A but in B;
$\overline{A} \cap \overline{B}$, consisting of all elementary events neither in A nor in B.

These four events are all mutually exclusive, so that

$$\begin{aligned} p(A \cap B) + p(A \cap \overline{B}) &= p(A) \\ p(A \cap B) + p(\overline{A} \cap B) &= p(B) \\ p(\overline{A} \cap B) + p(\overline{A} \cap \overline{B}) &= p(\overline{A}) \\ p(A \cap \overline{B}) + p(\overline{A} \cap \overline{B}) &= p(\overline{B}). \end{aligned} \qquad [1.11.2]$$

Rule 4 of Section 1.5 also covers the situation where two events A and B are not mutually exclusive:

$$p(A \cup B) = p(A) + p(B) - p(A \cap B).$$

By use of the complementary events $\overline{A}$ and $\overline{B}$ and the relationships given above, we can also write the "or" rule in still another way. Since, from 1.11.2 above,

$$p(A) + p(B) = p(A \cap \overline{B}) + p(\overline{A} \cap B) + 2p(A \cap B),$$

then

$$p(A \cup B) = p(A \cap \overline{B}) + p(\overline{A} \cap B) + p(A \cap B). \qquad [1.11.3]$$

This says that the probability of A or B or both is the sum of the probabilities of the events "A and not B," "not A and B," and "A and B."

As an illustration of the use of the *or* rule for combining probabilities, think of a city school system, where we are going to sample one pupil at random. In this school system, we know that 35 percent of the pupils are left-handed, so that we know also that the probability is .35 for the event "left-handed." In the same way we know that the probability is .51 of our observing a girl, and that the probability is .10 of observing a girl who is left-handed. What is the probability of observing *either* a left-handed student *or* a girl (or both)?

Let the event A be "girl," and let the event B be "left-handed." Then we know that

$$p(A) = .51,\ p(B) = .35,\ p(A \cap B) = .10.$$

Thus,

$$p(A \cup B) = p(A) + p(B) - p(A \cap B) = .51 + .35 - .10 = .76.$$

What is the probability of a girl who is right-handed? We know from 1.11.2 above that

$$p(A \cap \overline{B}) + p(A \cap B) = p(A)$$

so that

$$p(A \cap \overline{B}) = p(A) - p(A \cap B) = .51 - .10 = .41.$$

In a similar way we can find the probability of a boy who is left-handed:

$$p(\overline{A} \cap B) + p(A \cap B) = p(B)$$

so that

$$p(\overline{A} \cap B) = p(B) - p(A \cap B) = .35 - .10 = .25.$$

What is the probability of a boy who is right-handed? Since

$$p(\overline{A} \cap \overline{B}) + p(\overline{A} \cap B) = p(\overline{A}) = 1 - p(A),$$

then

$$p(\overline{A} \cap \overline{B}) = 1 - p(A) - p(\overline{A} \cap B) = 1 - .51 - .25 = .24.$$

Finally, what is the probability of a boy *or* right-handed?

$$\begin{aligned} p(\overline{A} \cup \overline{B}) &= p(\overline{A}) + p(\overline{B}) - p(\overline{A} \cap \overline{B}) \\ &= 1 - p(A) + 1 - p(B) - p(\overline{A} \cap \overline{B}) \\ &= .49 + .65 - .24 = .90. \end{aligned}$$

This example illustrates that given the probabilities of some of the joint events, it may be possible to find the probabilities that other joint events must have, because of the relationships shown in 1.11.2, and because of the "or" rule of probability. This can be very important in problems in which the probabilities of only some of the joint events can be assessed, and one wishes to deduce the probabilities that other joint events must show. When one is faced with this sort of problem, drawing a Venn Diagram is often extremely helpful.

These notions about joint events extend quite readily to more than two events. Perhaps there are three events A, B, and C. Then we might be interested in the joint event "A and B and C," symbolized by $A \cap B \cap C$. Thus, suppose that students on a large university campus were sampled at random. For each person observed, a record is made of age, sex, and year in college. Furthermore, suppose that the event A is "21 years or older," the event B is female," and the event C is "senior." Then the event "21 years or older and female and senior" is the joint event $A \cap B \cap C$. The probability that such an event will occur when we sample a student at random is

$$p(A \cap B \cap C) = \frac{\text{number of students in } A \cap B \cap C}{\text{total number of students}}.$$

There are, of course, only four possibilities for the joint occurrence of two sets and their complements: $A \cap B$, $A \cap \bar{B}$, $\bar{A} \cap B$, and $\bar{A} \cap \bar{B}$. However, when one is dealing with three events, A, B, and C, there are eight possible joint events involving all three. These are diagrammed in Figure 1.11.1.

Notice the relationships among these triple events and the pairwise joint events. Since the eight joint events shown in Figure 1.11.1 are all mutually exclusive, it must be true, for example, that

$$p(A) = p(A \cap B \cap C) + p(A \cap B \cap \bar{C}) + p(A \cap \bar{B} \cap C) + p(A \cap \bar{B} \cap \bar{C})$$

and that

$$p(A \cap B) = p(A \cap B \cap C) + p(A \cap B \cap \bar{C}),$$

and so on for events B and C.

Such relationships may also be used to deduce the probabilities of joint events other than the ones we already know.

Again using the college example begun above, suppose that we know that

$$p(A) = .60, \; p(B) = .51, \; p(C) = .44.$$

In addition we know that

$$p(A \cap B) = .30, \; p(A \cap B \cap C) = .05, \; p(\bar{A} \cap \bar{B} \cap C) = .09,$$

$$p(A \cap \bar{B} \cap \bar{C}) = .10.$$

What then is the probability of a female nonsenior 21 or over?

Since

$$p(A \cap B \cap C) + p(A \cap B \cap \bar{C}) = p(A \cap B)$$

then

$$p(A \cap B \cap \bar{C}) = p(A \cap B) - p(A \cap B \cap C) = .30 - .05 = .25.$$

What is the probability of a male senior over 21? We note from the figure that it must be true that

$$\begin{aligned} p(A \cap \bar{B} \cap C) &= p(A) - p(A \cap B \cap C) - p(A \cap B \cap \bar{C}) - p(A \cap \bar{B} \cap \bar{C}) \\ &= .60 - .05 - .25 - .10 = .20. \end{aligned}$$

Figure 1.11.1

Eight joint events in a sample space

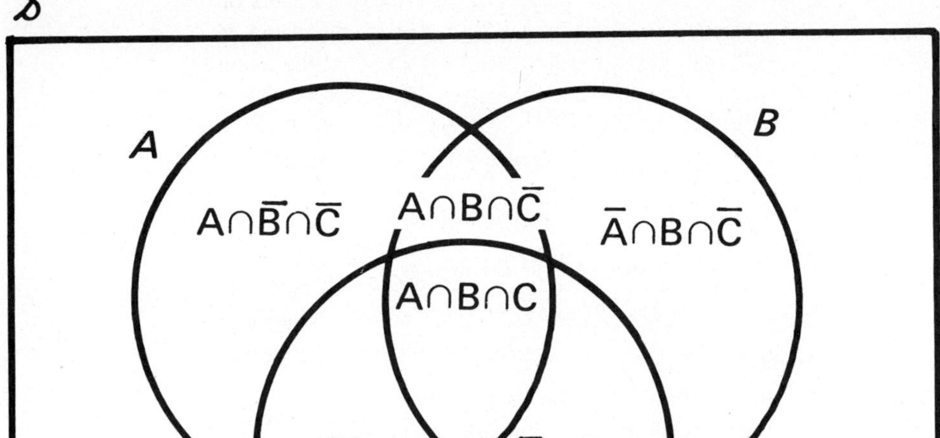

We can, of course, proceed in this same way for any other combination for which the relevant information is given.

In the diagram of Figure 1.11.1, an elementary event qualifies for the event $A \cup B \cup C$ if it is represented by a point within any of the three circles. The events A, B, and C thus include among them seven possible joint events involving all three. Hence

$$\begin{aligned} p(A \cup B \cup C) = {} & p(A \cap B \cap C) + p(A \cap B \cap \overline{C}) + p(A \cap \overline{B} \cap C) \\ & + p(\overline{A} \cap B \cap C) + (A \cap \overline{B} \cap \overline{C}) + p(\overline{A} \cap B \cap \overline{C}) \\ & + p(\overline{A} \cap \overline{B} \cap C). \end{aligned} \qquad [1.11.4]$$

This is equivalent to

$$p(A \cup B \cup C) = p(A) + p(B) + P(C) - p(A \cap B) - p(A \cap C) - p(B \cap C) + p(A \cap B \cap C). \quad [1.11.5]$$

Expression 1.11.5 is the generalization of the "or" rule of Section 1.5 for three events.

By exactly this same method we could show the probability of $A \cup B \cup C \cup D$ to be

$$\begin{aligned} p(A \cup B \cup C \cup D) = {} & p(A) + p(B) + p(C) + p(D) - p(A \cap B) \\ & - p(A \cap C) - p(A \cap D) - p(B \cap C) \\ & - p(B \cap D) - p(C \cap D) + p(A \cap B \cap C) \qquad [1.11.6] \\ & + p(A \cap B \cap D) + p(A \cap C \cap D) + p(B \cap C \cap D) \\ & - p(A \cap B \cap C \cap D), \end{aligned}$$

and so on for any number of events.

Once again, given part of the information about the probabilities of joint events, it is possible to deduce other probabilities by the relationships given above.

So far, we have spoken of joint events where the simple experiment produces elementary events that qualify simultaneously for two or more events in $\mathcal{S}$. However, there is another situation that also produces joint events. This is when two (or more) simple experiments are done at the same time, with each one producing an outcome. Then we may be interested in a joint outcome, "what happens on experiment I *and* what happens on experiment II."

For example a simple experiment (experiment I) here consists of rolling a die of six sides. At the same time, a simple experiment (experiment II) consists of tossing a penny. Now we are interested in the joint outcome of rolling the die *and* tossing the penny. For this joint experiment each elementary event consists of the number that comes up on the die and the side that comes up on the penny. The sample space is shown below:

(II) Side of penny							
	H	•	•	•	•	•	•
	T	•	•	•	•	•	•
		1	2	3	4	5	6

(I) Number on die

Each point in the diagram is one elementary event, a possible outcome of this experiment, and there are exactly (2)(6) = 12 such outcomes.

Consider the joint event "coin comes up heads, and die comes up with an odd number." There are exactly three elementary events that qualify: (heads, 1) (heads, 3), and (heads, 5). Since there are twelve elementary events in all,

$$p(\text{H and odd}) = \frac{3}{12} = \frac{1}{4}.$$

This simple example illustrates that sometimes an elementary event can also be a joint event: the event (head, 1) is the intersection of the event "heads" and the event "1," and so on.

Suppose that in sampling children in a school system, we restrict our observations only to girls. Here, there is a new sample space including only part of the elementary events in the original set. In this new sample space, consisting only of girls, what is the probability of observing a left-handed person? If all girls in the school system are equally likely to be observed,

$$p(\text{left-handed given girl}) = \frac{\text{number left-handed girls}}{\text{total number of girls}}.$$

This probability can be found from the *original* probabilities for the total sample space: first of all we know from the example of 1.11 that

$$p(\text{left-handed and girl}) = \frac{\text{number left-handed girls}}{\text{total number of pupils}} = .10$$

and that

$$p(\text{girl}) = \frac{\text{total number of girls}}{\text{total number of pupils}} = .51.$$

It follows that

$$p(\text{left-handed given girl}) = \frac{p(\text{left-handed and girl})}{p(\text{girl})} = \frac{.10}{.51} = .196$$

since the two probabilities put into ratio each have the same denominator

The probability just found, based on only a part of the total sample space, is called a conditional probability. A more formal definition follows:

Let A and B be events in a sample space made up of a finite number of elementary events. Then the conditional probability of B given A, denoted by $p(B|A)$, is

$$p(B|A) = \frac{p(A \cap B)}{p(A)}, \qquad [1.12.1^*]$$

provided that $p(A)$ is not zero.

Notice that the conditional probability symbol, $p(B|A)$, is read as "the probability of *B given A*." In the example it was given that the observation would be a girl; the value desired was the probability of "left-handed," given that a girl were observed.

For any two events A and B, there are two conditional probabilities that may be calculated:

$$p(B|A) = \frac{p(A \cap B)}{p(A)} \qquad [1.12.2]$$

and

$$p(A|B) = \frac{p(A \cap B)}{p(B)}. \qquad [1.12.3]$$

We might find the probability of left-handed, given girl (.10/.51 = .20) or the conditional probability of girl, given left-handed (.10/.35 = .29). In general, these two conditional probabilities will not be equal, since they represent quite different sets of elementary events.

As another example, take the probabilities found in the example for a coin and a die in Section 1.11. Here, the probability of a tail coming up on the coin is .50, and the probability of an odd number on the die *and* a tail on the coin is .25. The probability that the die comes up with an odd number, *given* that the coin comes up tails is

$$p(\text{odd}|\text{tails}) = \frac{.25}{.50} = .50.$$

The probability that the coin comes up tails and the die comes up with a 1 or a 2 can be found from the example to be 1/6 or .167. In order to know the probability that the die comes up with a 1 or a 2 given that the coin comes up tails, we take

$$p(1 \text{ or } 2|\text{tails}) = \frac{.167}{.50} = .33.$$

What are some of the properties that conditional probabilities must exhibit? In order to answer this question, think once again of two events A and B for which none of the probabilities $p(A)$, $p(B)$, and $p(A \cap B)$ is equal to zero. Then the conditional probability $p(B|A)$ can either be larger or smaller than $p(B)$; there is no necessary relation between the size of $p(B)$ and the size of $p(B|A)$, other than that they both must lie between zero and one.

Now consider the complementary event to B, or $\overline{B}$. What can we say about its conditional probability, given A. For any event B and its complement $\overline{B}$,

$$p(B|A) + p(\overline{B}|A) = 1. \qquad [1.12.4]$$

That is, the sum of the probability of B given A and the probability of not-B given A must be equal to 1.00. This is true because

$$p(A \cap B) + p(A \cap \overline{B}) = p(A)$$

so that

$$\frac{p(A \cap B)}{p(A)} + \frac{p(A \cap \overline{B})}{p(A)} = 1. \qquad [1.12.5]$$

If we know the conditional probability $p(B|A)$ and also $p(A)$, then we can also find the joint probability $p(A \cap B)$:

$$p(A)p(B|A) = p(A \cap B) \qquad [1.12.6]$$

and

$$p(B)p(A|B) = p(A \cap B),$$

since, by definition,

$$p(B|A) = \frac{p(A \cap B)}{p(A)}$$

and

$$p(A|B) = \frac{p(A \cap B)}{p(B)}.$$

It follows that

$$p(A)p(B|A) = p(B)p(A|B)$$

and that

$$\frac{p(B|A)}{p(A|B)} = \frac{p(B)}{p(A)}.$$

Why should we be concerned with conditional probabilities? In the first place, in a sense all probabilities are conditional probabilities. Whenever we specify the circumstances surrounding, and the assumptions underlying, a given simple experiment, we are laying down certain conditions. These conditions could, if we wished, be represented through the use of conditional probabilities. For example, many times in preceding sections the experiment of drawing one card from a well-shuffled standard deck of playing cards has been used. In view of the stated condition "well-shuffled deck of standard playing cards," an event such as "king of spades" actually has a conditional probability given by

$$p(\text{king of spades}|\text{well-shuffled standard deck of playing cards}).$$

The sample space is limited by the condition "well-shuffled deck." Usually, however, we simply assume that a well-shuffled standard deck has been chosen, and then proceed to ignore this condition in the remainder of the discussion.

Secondly, most of the information we deal with in everyday life, as the basis for the choices we must make, has a conditional character. We must constantly make choices and decisions, from the most trivial all the way to some that are of tremendous importance. These decisions are all predicated upon things that may or may not be true, on things that may or may not happen. The very best information we have to go on is usually no more than a probability. These probabilities are conditional, since virtually all of our information is of an "if-then" character. "If so and so is true, then the probability of this event must be such and such." It is not surprising that a concern with conditional probability, and with decisions based upon such probabilities, has been an important part of probability theory since its very beginning.

In fact, the theorem about conditional probabilities to be discussed in the next section has had great influence on the way people think about conditional probabilities, and especially how probabilities can change, given additional information.

1.13 / BAYES' THEOREM

The relations among various conditional probabilities are embodied in the theorem named for Thomas Bayes, an English clergyman who did early work in probability and decision theory. The simplest version of this theorem follows. *For two events A and B, where none of the probabilities $p(A)$, $p(B)$, and $p(A \cap B)$ is either 1.00 or 0, then the relation must hold:*

$$p(A|B) = \frac{p(B|A)p(A)}{p(B|A)p(A) + p(B|\bar{A})p(\bar{A})}. \qquad [1.13.1]$$

Bayes' theorem gives a way to find the conditional probability of event A given event B, provided that you know the probability of A, the conditional probability of B given A, and the conditional probability of B given $\overline{A}$ [once you know the probability of A you already know the probability of $\overline{A}$, or $1 - p(A)$]. This theorem, which must be true for any pair of events with probabilities as specified above, summarizes the connection between the conditional and unconditional probabilities. In addition, Bayes' theorem has a number of important practical uses.

Consider this example: at a university it is decided to try out a new placement test for admitting students to a special mathematics class. Experience has shown that in general only 60 percent of students applying for admission actually can pass this course. Heretofore, each student applying for the course was admitted. Of the students who passed the course, some 80 percent passed the placement test beforehand, whereas only 40 percent of those who failed the course could pass the placement test initially. If this test is to be used for placement, and only students who pass the test are to be admitted to the course, what is the probability that such a student will pass the course?

First of all, event A is defined to be "passes the course," and it is assumed that $p(A) = .60$. The event B is "passes the test." The conditional probabilities are taken to be

$$p(B|A) = .80 \qquad p(\overline{B}|A) = .20$$

$$p(B|\overline{A}) = .40 \qquad p(\overline{B}|\overline{A}) = .60.$$

However, we want to know the conditional probability, $p(A|B)$, that a student will pass the course, *given* that he has passed the test. Bayes' theorem shows

$$p(A|B) = \frac{(.80)\,(.60)}{(.80)\,(.60) + (.40)\,(.40)}$$

$$= \frac{.48}{.48 + .16} = .75.$$

Thus, the probability is .75 for a student's passing the course *given* that the test was passed.

If we wanted to find other conditional probabilities, we could also apply the theorem. For example,

$$p(\overline{A}|\overline{B}) = \frac{(.60)\,(.40)}{(.60)\,(.40) + (.20)\,(.60)}$$

$$= \frac{.24}{.24 + .12} = .67.$$

Notice that the probability of a student's passing the course given a passing score on the test is greater than the probability in general of passing the course. Any student is a better bet to pass the course given that the placement test was passed. What is the probability of being *right* if a student is admitted or refused the course strictly on the basis of this test? The probability of a correct decision can be found by

$$p(A \cap B) + p(\overline{A} \cap \overline{B}) = p(\text{correct}).$$

By 1.12.6, this is

$$p(\text{correct}) = p(B|A)p(A) + p(\overline{B}|\overline{A})p(\overline{A})$$
$$= (.80)(.60) + (.60)(.40) = .72.$$

In other words, the administrator will have a probability of .72 of being right about a student's proper placement by use of the test. If the administrator does not use the test and simply admits all students to the course, the probability of being right (of the student passing the course) is only .60. Thus, the test does something for the administrator; its use allows an increase of the probability of being right about a given student selected at random.

Bayes' theorem and other calculations with conditional probability are often used in this way, especially in questions of selection or diagnosis of subjects where initial probabilities are known. Good selection or diagnostic procedures are those permitting an increase in the probability of being right about an individual given some prior information, and such *conditional* probabilities can often be calculated by Bayes' theorem.

As a mathematical result, Bayes' theorem is necessarily true for conditional probabilities satisfying the basic axioms of probability theory. In and of itself Bayes' theorem is in no sense controversial. However, the question of its appropriate use has, in years past, been a focal point in the controversy between those who favor a strict relative-frequency interpretation of probability and those who would admit a subjective interpretation as well. The issue emerges quite clearly when some of the probabilities figuring in Bayes' theorem are associated with "states of nature" or with nonrepetitive events. As we have seen, it is usually quite difficult to give meaningful relative-frequency interpretations to probabilities for such states or events. The difficulty was compounded by the fact that in some past applications, Bayes' theorem yielded rather ridiculous results. This had the effect of casting many applications of the theorem into dispute. The history of the development of statistics in the late nineteenth and early twentieth centuries demonstrates some rather elaborate attempts to ignore or to "finesse" the problems of prior information and subjective probabilities.

In recent years, some theorists have shown that subjective probabilities, including those for nonrepetitive events, can be given an axiomatic status on a par with probabilities subject only to relative-frequency interpretations. Furthermore, a basis exists for the experimental determination of subjective probabilities through preferences among betting odds. Consequently, there has been a renewed interest in the application of Bayes' theorem. In particular, Bayes' theorem occupies a central place in the theory of how an individual's subjective probabilities change in the face of accumulating information.

Not all statisticians agree that some of these applications are proper, just as not all agree that a subjective interpretation of probability is meaningful. Nevertheless, the use of Bayes' theorem is becoming a feature of a considerable portion of the modern theory of statistics and of decision-making, as these theories begin to allow once again for the subjective interpretation of probabilities. We will examine this approach more closely in Chapter 7.

1.14 / INDEPENDENCE OF EVENTS

Now we are ready to take up the topic of independence of events, and give it a more precise formulation. The general idea used heretofore is that independent events are those having nothing to do with each other: the occurrence of one event in no wise affects the probability of the other event. But how does one know if two events A and B are independent? If the occurrence of the event A has nothing whatever to do with the occurrence of event B, then we should expect the conditional probability of B given A to be exactly the same as the probability of B, $p(B|A) = p(B)$. Likewise, the conditional probability of A given B should be equal to the probability of A, or $p(A|B) = p(A)$. The information that one event has occurred does not affect the probability of the other event, when the events are independent.

The condition of independence of two events may also be stated in another form: if $p(B|A) = p(B)$, then

$$\frac{p(A \cap B)}{p(A)} = p(B)$$

so that

$$p(A \cap B) = p(A)p(B).$$

In the same way,

$$p(A|B) = p(A)$$

leads to the statement that

$$p(A \cap B) = p(A)p(B).$$

This fact leads to the usual definition of independence:

If events A and B are independent, then the joint probability $p(A \cap B)$ is equal to the probability of A times the probability of B,

$$p(A \cap B) = p(A)p(B). \qquad [1.14.1*]$$

For example, suppose that you go into a library and at random select one book (each book having an equal likelihood of being drawn). The sample space consists of all distinct books that a person might select. Suppose that the proportion of books then on the shelves and classified as "fiction" is exactly .15, so that the probability of selecting such a book is also .15. Furthermore, suppose that the proportion of books having red covers is exactly .30. If the event "fiction" is independent of the event "red cover," the probability of the joint event is found very easily:

$$p(\text{fiction and red cover}) = p(\text{fiction})p(\text{red cover}) = (.15)\,(.30) = .045.$$

You should come up with a red-covered piece of fiction in about forty-five out of one thousand random selections.

It is certainly not true that all events must be independent, and it is very easy to give examples where the joint probability of two events is not equal to the product of their separate probabilities. For example, suppose that children were selected at random. A child is observed and hair color noted, and also whether or not this child is freckled. For our purposes, the event A consists of the set of all children having red hair, and the event B is the set of all children who are freckled. Is it reasonable that the event A will be independent of the event B? The answer is no; everyone knows that among red-heads freckles are much more common than among children in general. One would expect in this case $p(B|A)$ to be *greater* than $p(B)$, so that it should *not* be true that $p(A \cap B) = p(A)p(B)$. The events "red hair" and "freckled" *do* tend to occur together and are not ordinarily regarded as independent.

This example suggests one of the uses of the concept of independent events. The definition of independence permits us to decide whether or not events *are associated* or *dependent* in some way:

If, for two events A and B, $p(A \cap B)$ is not equal to $p(A)p(B)$, then A and B are said to be associated or dependent.

For example, consider a sample space with elementary events consisting of all the employed adult persons in the United States. We wish to answer the question, "Is making over thirty-thousand dollars a year independent of having a college education?" Let us call the event A "makes over thirty-thousand dollars a year," and the event B "has a college education." Now suppose that the proportion of employed adults in the United States who make more than thirty-thousand a year is .36, so that if each such person is equally likely to be selected, the probability of event A is .36. Suppose also that the proportion of employed adults with a college education is .41, so that $p(B) = .41$. Finally, the probability of the event $A \cap B$, our observing an employed adult making over thirty thousand a year who has a college education is .22.

Now if the events A and B were independent, this probability $p(A \cap B)$ should be $(.36)(.41) = .15$. The actual probability is, say, .22. This leads immediately to the conclusion that the two events are *not* independent, or *are* associated. We can also see that the probability of occurrence of event A given event B is much larger than it should be if A and B were independent: That is,

$$p(A|B) = p(A \cap B)/p(B) = (.22)/(.41) = .54.$$

If A and B were independent, it should have been true that

$$p(A|B) = p(A) = .36.$$

Incidentally, some students tend to confuse "independent" with "mutually exclusive" events. These are by no means the same! In fact, **two events *A* and *B* that are mutually exclusive cannot be independent unless one or both events have zero probability.** This is easy to show: Consider two mutually exclusive events, A and B. Since they are mutually exclusive,

$$A \cap B = \varnothing,$$

so that $p(A \cap B) = 0$. Now if A and B were independent, it would be true that $p(A \cap B) = p(A)p(B)$, but since $p(A \cap B) = 0$, this cannot be unless $p(A)$ or $p(B)$ or both are zero. Even at a commonsense level the difference between mutually exclusive and independent events is easy to show. Suppose that all men are either members of the class "balding" or the class "full head of hair." These two classes are mutually exclusive. Then if the events "balding" and "full head of hair" were independent, among those who are balding there would be the same proportion of men with a full head of hair, as the proportion of men with full heads of hair generally [i.e., $p(B|A) = p(B)$]. You must admit that this is a little hard to visualize.

The idea of independence is very important to many of the statistical techniques to be discussed in later chapters. The question of association or dependence of events is central to the scientific question of relationships among the phenomena we observe (literally, the question of "what goes with what"). Many techniques for studying the presence and degree of relatedness among observations depend directly upon this idea of comparing probabilites for joint events with the probabilities when events are independent.

Even at this early stage of the game, a little warning is in order: the concept of association of events must not be confused with the idea of causation. Association, as used in statistics, means simply that the events are not independent, and that a certain correspondence between their joint and separate probabilities does not hold. When events A and B are associated, it need not mean that A causes B or that B causes A, but only that the events occur together with probability different from the product of their separate probabilities. This warning carries force whenever we talk of the association of events; association may be a consequence of causation, but this need not be true.

1.15 / REPRESENTING JOINT EVENTS IN TABLES

Sometimes it is convenient to list joint events by means of a table, with each cell representing one joint possibility. For example, consider a sample space consisting of all individual students at a particular college campus. Each student is either male or female, of course, and each responds either "yes" or "no" to an opinion question. The possible joint events are as follows:

Male	(male and yes)	(male and no)
Female	(female and yes)	(female and no)
	Yes	No

It might be that on this particular campus, the probability of our observing a male student is .55 and of observing a female .45. Furthermore, suppose that the probability of obtaining a "yes" answer is .40, and .60 for obtaining a "no." Then the table of probabilities follows this general pattern:

p(Male) = .55	p(Male and Yes)	p(Male and No)
p(Female) = .45	p(Female and Yes)	p(Female and No)
	p(Yes) = .40	p(No) = .60

The values of p(Male), p(Female), p(Yes,) and p(No) are called the **marginal prob-**

abilities (they get the name from the obvious circumstance of appearing in the *margin* of the table). Each marginal probability is a sum of all the joint probabilities in some particular row or column of the table:

$$\begin{aligned} p(\text{Male}) &= p(\text{Male and Yes}) + p(\text{Male and No}); \\ p(\text{Female}) &= p(\text{Female and Yes}) + p(\text{Female and No}); \\ p(\text{Yes}) &= p(\text{Male and Yes}) + p(\text{Female and Yes}); \\ p(\text{No}) &= p(\text{Male and No}) + p(\text{Female and No}). \end{aligned}$$

Ordinarily the event classes that appear together along any margin of the table are *mutually exclusive and exhaustive*. That is, the set of event classes forms a partition of the total sample space. Thus, the events "Yes" and "No" are mutually exclusive and exhaustive and so are the events "Male" and "Female." Any set of mutually exclusive and exhaustive event classes that make up one margin of the table can be called an **attribute** or a **dimension.** This table embodies two attributes, "the sex of the student" and "the response of the student," respectively.

Suppose that the two attributes of the table were independent: this means that any event along one margin must be independent of every event along the other margin. Here, this is tantamount to saying that the sex of the student has absolutely nothing to do with how the question was answered. **If independence exists, the probability of each joint event** (as given in a cell of the table) **must be equal to the product of the probabilities of the corresponding marginal events.**

For example, if the two attributes are independent, then the probability $p(\text{Male and Yes})$ must be equal to the product of the two marginal probabilities,

$$p(\text{Male and Yes}) = p(\text{Male})p(\text{Yes}) = (.55)\,(.40) = .22.$$

In the same way the joint probabilities for each of the other cells may be found from products of marginal probabilities, and the following table should be correct:

$p(\text{Male}) = .55$	$p(\text{Male})p(\text{Yes}) = (.55)(.40) = .22$	$p(\text{Male})p(\text{No}) = (.55)(.60) = .33$
$p(\text{Female}) = .45$	$p(\text{Female})p(\text{Yes}) = (.45)(.40) = .18$	$p(\text{Female})p(\text{No}) = (.45)(.60) = .27$
	$p(\text{Yes}) = .40$	$p(\text{No}) = .60$

It is important to remember, however, that these will be the correct joint probabilities only if the attributes *are* independent. It might be that the following table is the true one:

$p(\text{Male}) = .55$	$p(\text{Male and Yes}) = .10$	$p(\text{Male and No}) = .45$
$p(\text{Female}) = .45$	$p(\text{Female and Yes}) = .30$	$p(\text{Female and No}) = .15$
	$p(\text{Yes}) = .40$	$p(\text{No}) = .60$

When this is true it is safe that for this sample space the sex of the student *is* associated with the answer to the question. Look at the conditional probabilities: here, the probability of a male students' answering "yes" is

$$p(\text{Yes}|\text{Male}) = \frac{.10}{.55} = .18$$

and the probability of a "yes" answer if the student is a female is

$$p(\text{Yes}|\text{Female}) = \frac{.30}{.45} = .67.$$

Thus, a female is much more likely to answer "yes" to this question than is a male. On the other hand, had the two attributes been independent, these two conditional probabilities *should* have been the same:

$$p(\text{Yes}|\text{Male}) = \frac{.22}{.55} = .40$$

$$p(\text{Yes}|\text{Female}) = \frac{.18}{.45} = .40,$$

indicating that sex gives no information about how a person tends to respond to the question.

Given such a table of joint events and their relative frequencies, we can always calculate the relative frequencies (or probabilities) that *should* have appeared in the various cells *if* the two attributes actually had been independent. We can also examine the relative frequencies or probabilities that actually did appear in each cell. Then, if the relative frequencies *given* independence are different from the relative frequencies obtained for each cell, we have a basis for saying that the two attributes are associated, or are dependent, to some extent. Many methods for assessing the relationship between attributes, and particularly those to be discussed in Chapter 15, are based directly on this idea.

1.16 / RANDOM SAMPLES AND RANDOM SAMPLING

Given some simple experiment and the sample space of elementary events, the set of outcomes of N separate trials is a **sample.** When the sample is drawn *with replacement,* the same elementary event can occur more than once. When the sample is drawn *without replacement,* the same elementary event can occur no more than once in a given sample. Ordinarily, probability and statistics deal with **random samples.** The word "random" has been used rather loosely in the foregoing discussion, and now the time has come to give it a more restricted meaning in connection with random samples.

It has already been suggested that probability calculations can be made quite simple when the elementary events in a sample space have equal probabilities. The theory of statistics deals with samples of size N from a specified sample space, and here, too, great simplification is introduced if each distinct sample of a particular size can be assumed to have equal probability of selection. For this reason, the elementary theory of statistics is based on the idea of simple random sampling:

A method of drawing samples such that each and every distinct sample of the same size N has exactly the same probability of being selected is called simple random sampling.

In most social science applications of sampling, the sample space consists simply of some large group of individuals that we might select and observe. Random sampling of one individual at a time means that every possible individual in the large group has an equal probability of being drawn. If we take N such individuals, this is simple random sampling only if each possible set of set of N has the same probability of being selected on such an occasion. Most of what follows will deal with simple random samples. In other words, our discussion will be confined to the situation where all possible samples of the same size have exactly the same probability. For us, sampling ''at random'' will always mean simple random sampling, as defined above. This does not mean, however, that the theory of statistics does not apply to situations where samples have unequal probabilities of occurrence; in more advanced work the theory and methods can be extended to any sampling scheme where the probabilities of the various samples are *known,* even though they are unequal.

An alternative definition of simple random sampling in terms of elementary events can also be given: ''Simple random sampling is a process of selecting elementary events for observation in such a way that each and every elementary event has precisely the same probability of being included in any sample of N observations.'' In *random sampling with replacement,* each elementary event has exactly the same probability of occurring on each trial. In random sampling *without replacement,* the composition of sample space $\mathcal{S}$ changes with each trial since an elementary event can occur only once in N trials. However, we shall assume that among the elementary events available for selection on a given trial, the probabilities are equal.

We shall also have many occasions to require **independent random sampling.** You may recall that independence was specified in Bernoulli's theorem. Elementary events are sampled independently when the occurrence of one elementary event has *absolutely no connection* with the probability of occurrence of the same or another elementary event on a subsequent trial. A series of tosses of a coin can be thought of as a random and independent sampling of events: what happens on one toss has no conceivable connection with what happens on any other. Each trial of a simple experiment is a sampling of elementary events, and the trials are independent when no connection exists between the particular outcomes of different trials. In independent random sampling of individuals for observation from some large ''population'' of such individuals, the inclusion of one individual in the sample has absolutely nothing to do with the possible inclusion or noninclusion of anyone else.

Take care to notice that our assuming samples of size N thus to be equally probable does not imply that *events* must be equally probable. Thus, in the marble example above, samples of N marbles were assumed equally probable, but the probabilities of the colors of the marbles were not equal. The assumption of equal probabilities is simply a way of saying that any elementary event has just as good an opportunity as any other to serve as a sample on any given trial.

1.17 / RANDOM NUMBERS

In practical situations there are a number of schemes for making sure that each unit drawn for observation, or each elementary event in the sample space, has equal probability. Until recently, the most common was use of a table of random numbers. Random number tables consist of many pages filled with digits, from 0 through 9. These tables have been composed in such a way that each digit is approximately equally likely to occur in any spot in the table, and there is no systematic connection between the occurrence of any digit in the sequence and any other. Many books on the design of experiments contain pages of random numbers, and very extensive tables of random numbers may be found in a book prepared by the RAND Corporation (1955).

At present, computers are most commonly used to generate random numbers. Most computing centers have programs available for the generation of as many random numbers as may be needed on a given occasion. Quite often a file of the potential units for observation is placed in the computer memory storage, and the computer itself selects the sample in a random way. These methods have pretty well replaced the use of random number tables in other than the smallest research settings.

These methods share a common drawback: A listing of the members of the sample space (all potential units for observation) must be possible. Except in some very restricted situations this is very difficult, or even impossible, to do. What usually happens in the social and behavioral sciences is that the investigator utilizes a sample space much more restricted than the one actually of interest. For example, an experimenter might like to make statements about behaviors of all people of a certain age, but the only group in reach for sampling consists of college sophomores in a particular locale. Then one of two things must occur, either it is assumed that the sample space employed is itself a random sample from some larger space, or inferences are confined to the group that can be sampled at random.

Random numbers are used in much the same way to achieve randomization in experiments. For example, an experimenter is going to administer three different treatments to a sample of rats, with one-third of the rats getting each treatment. The members of the total sample are listed and then assigned to the different groups by random numbers. This randomization is an extremely important part of experimental procedure, as we shall see, even though the sample itself is a random selection from some sample space.

Quite apart from their utility in drawing random samples, and in randomization, random numbers are interesting as another instance of the notion of ''randomness'' that is idealized in the theory of probability. A random process is one in which only chance factors determine the exact outcome of any particular trial of an experiment, or result of an observation. Although the possible outcomes may be known in advance, the particular outcome of a given trial is not. Nevertheless, built into the process is some regularity, so that each class of outcome (that is, each event) can reasonably be assigned a probability, representing its long-run relative frequency. Perhaps the simplest example of a random phenomenon is the result of tossing a coin. Only chance factors determine whether heads or tails will come up on a particular toss, but a fair coin is so constructed that there is just as much physical reason for heads to appear as for tails, and so we have the justification we need in order to say that, in the long run, heads will appear just as often as tails. This property of the coin is idealized when we say that the probability of heads is .50. In practice, of course, random numbers are generated electronically, but the general idea is the same: the physical process is such that no single number is favored for any particular outcome, and chance alone actually dictates the number that does occur in any place in a sequence. When we use random numbers to select samples, we are using this property of randomness, which was inherent in the process by which the numbers themselves were generated. The point is that the ideas of randomness and of probability are not just misty abstractions; we know how to create processes and devise operations that are approximately "random" and we *do* know how to manufacture events with particular probabilities, at least with probabilities that are approximately known.

EXERCISES

1. Describe an elementary event for each of the following simple experiments.

(a) Choosing a telephone number from the Moline, Illinois, directory for the current year.
(b) Selecting a student currently enrolled in a college or university in the United States.
(c) Opening a box of cookies taken down from a shelf of a certain grocery store and noting the number of cookies it contains.
(d) Selecting a current United States Senator to call for an interview.
(e) Counting the number of fast-food resturants in a given 50-mile stretch of a certain highway.
(f) Picking a particular day of a given year, and noting by how many points the New York stock market changed on that day, according to the Dow-Jones index.
(g) Choosing a particular area of India and noting the percentage of persons infected with leprosy.

2. For which of the simple experiments of exercise 1 might the following events reasonably occur?

(a) The individual selected comes from the state of Wisconsin.
(b) The number is 28.5.
(c) The number is -10.2
(d) The person chosen is twenty years old.
(e) The number represents a hospital.
(f) The number is zero.
(g) The number begins with the digit "3."

3. Write out in symbols and words the events represented by the following Venn diagrams:

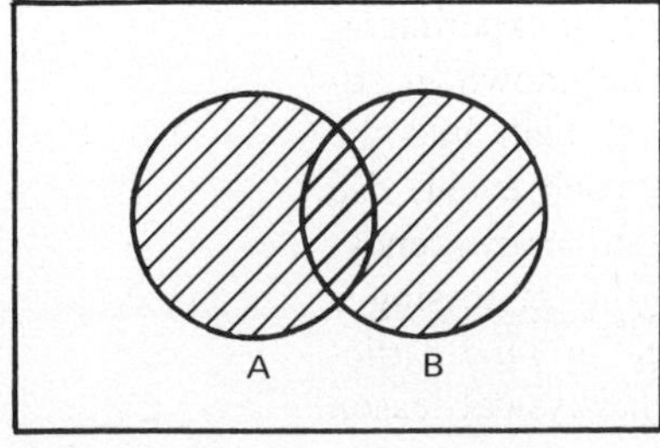

a

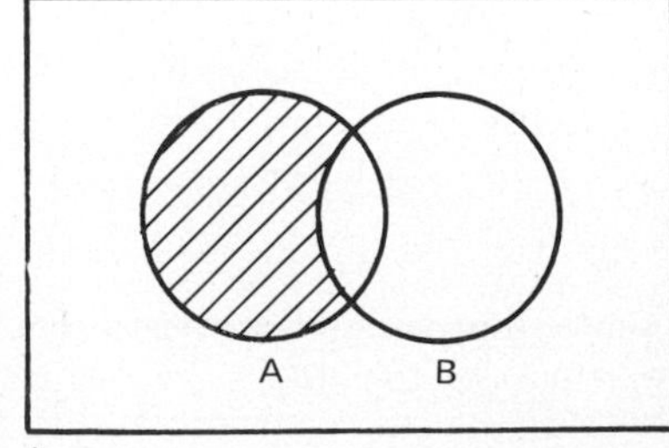

b

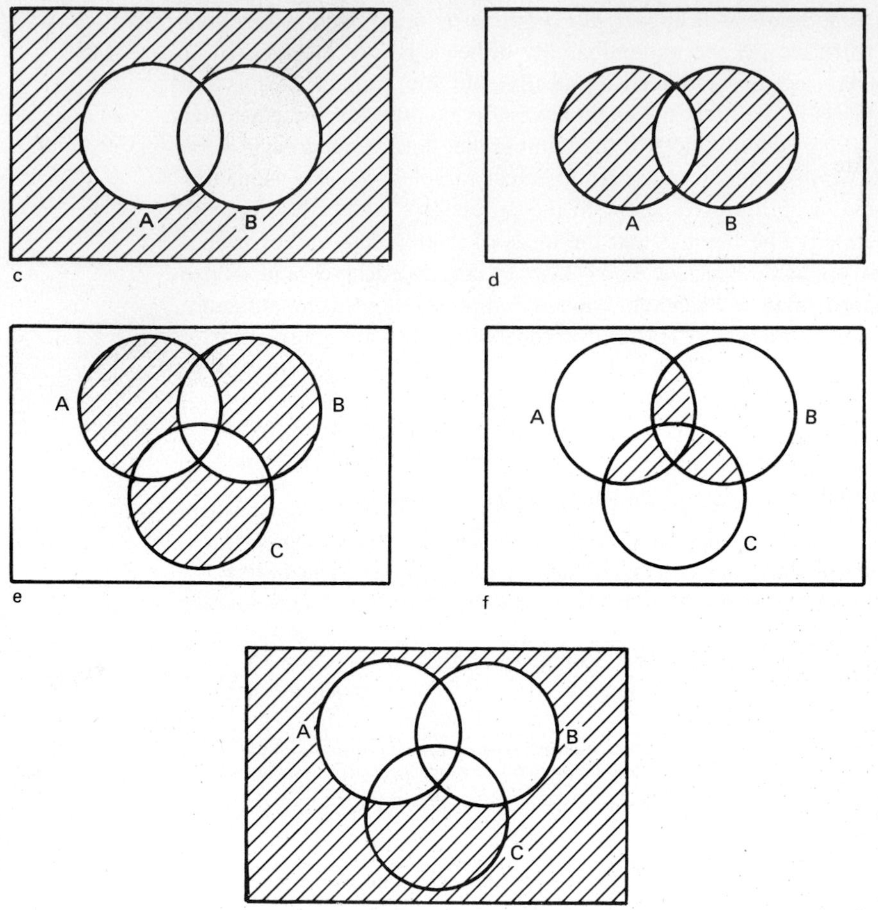

4. A bag contains pennies, nickels, and dimes. You draw one coin from the bag in such a way that any coin is equally likely to be selected. Let A be the event "penny," let B be the event "nickel," and let C be the event "dime." These events are mutually exclusive, of course. Give a verbal description of the the following events:

(a) $\mathcal{S}$
(b) $A \cup B$
(c) $A \cap B$
(d) $A \cup C$
(e) not A
(f) $B \cup$ not C
(g) not $B \cap$ not C
(h) $A \cup B \cup C$

5. Suppose that for the simple experiment of drawing a coin from a bag, as in exercise 4 above, the probabilities were $p(A) = 1/4$, $p(B) = 3/5$, and $p(C) = 3/20$. Find the probabilities of the events:

(a). $\mathcal{S}$

(b) $\varnothing$
(c) $A \cup B$
(d) $B \cup C$
(e) $A \cap B$
(f) not B
(g) $A \cap$ not B (**Hint:** Are "A" and "not B" mutually exclusive?)
(h) $C \cap$ not B
(i) $C \cup$ not B

6. A simple experiment consists of drawing exactly one playing card from a well-shuffled standard deck. The suit and value of the card is then noted. How many elements qualify for the following events?

(a) Ace
(b) Red suit
(c) A face card (i.e. king, queen, jack)
(d) An even-valued card, not a face card
(e) A spade or a diamond
(f) A 10 in a red suit

7. If the cards in the deck of exercise 6 are each equally likely to be drawn, so that each has probability of 1/52, find the probabilities of the events listed in exercise 6.

8. A person is selected at random from the list of registered voters in some community. Let the event A be "votes Democratic," let the event B be "is 60 years of age or older," and let the event C be "female." Give a verbal description of the following events:

(a) $A \cup B$
(b) $A \cap C$
(c) $A \cap$ not B
(d) A $\cup$ not $B \cap$ not C
(e) $A \cap B \cap C$

9. A part-time secretarial service has nine persons on its roster, as follows:

Name	*Sex*	*Hair color*	*Height in inches*
Mary	F	blonde	63
Susan	F	brunette	66
Jane	F	blonde	65
Alice	F	red	68
Eleanor	F	brunette	67
Bill	M	brunette	71
Lorna	F	blonde	63
John	M	red	70
Henry	M	blonde	73

Each individual is equally likely to be sent out on assignment.
What is the probability that an individual is sent who is

(a) Female
(b) Brunette
(c) Red-haired
(d) A red-haired male
(e) 68 or more inches tall
(f) A blonde female less than 65 inches tall

10. Suppose that over a period of time the secretarial service of exercise 9 filled 150 different assignments, and that each person was equally likely to be sent on any assignment. How often should we expect each of these events to have occurred?

(a) A male was sent
(b) A person over 65 inches tall was sent
(c) A brunette female was sent
(d) A brunette female or a blonde male was sent
(e) Either Mary or Alice was sent, but not both
(f) A redhead over 70 inches tall was sent

11. Two dice are rolled simultaneously. One die is white with black spots, and the other die is black with white spots. Let an elementary event be the occurrence of a pair of numbers (x,y), where x is the number of spots coming up on the white die, and y is the number of spots coming up on the black die. List all of the possible elementary events in the sample space $\mathcal{S}$. How many elementary events are there? Find the number of elements or (x,y) pairs that make up the following events:

(a) $x + y = 7$
(b) $x + y > 7$
(c) $x = y$
(d) $x \neq y$
(e) x is odd and y is even
(f) x is even *or* y is even
(g) x times $y \geqslant 8$

12. Given that each of the elementary events found in exercise 11 is equally likely, find the probabilities of the events listed.

13. Suppose that one letter is chosen from among the twenty-six letters making up the English alphabet. This is done is such a way that each letter is equally likely to be chosen. Find the probabilities for the following events:

(a) The letter is a vowel (include *y* as a vowel).
(b) The letter is a consonant other than *s* or *t*.
(c) The letter is a consonant from the first half of the alphabet.
(d) The letter chosen appears in the word EXODUS.
(e) The letter chosen appears either in the word BORN or in the word FILM.
(f) The letter appears either in the word LEAD or the word LOAD.

14. Suppose that a letter is drawn from the word SENSATION in such a way that any of the letters are equally probable. Then that letter is replaced and a second letter is drawn. Find the probabilities of the events listed below. (**Hints:** Treat the the two occurrences of the letter *S* in SENSATION as though they were two separate elementary events, and the same for *N*.) A simple graphic plot of all of the possible joint events (first letter, second letter) will be helpful.

(a) The two letters drawn are the same.
(b) The two letters drawn are both consonants.
(c) The letters drawn are both vowels.
(d) The letters drawn are two different consonants.
(e) The letters drawn in order form any of the words SO, AS, or IS.
(f) The letters drawn in order form any of the words, IN, ON, AT, or AN.
(g) If this simple experiment were repeated 162 times, how many times would you expect to get a consonant followed by a vowel?

15. Thirty-six poker chips are placed in a box. One-third of the chips are red, one-third blue, and one-third white. The chips of each color are numbered from 1 through 12. If one chip is drawn from the box, and if any chip is equally likely to be drawn, find the probabilities of the following events:

(a) The number drawn is a 3.
(b) The chip is red with an even number.
(c) The chip is blue with a number of 5, or more.
(d) The chip is red or the number is 8.
(e) The chip is white and the number is odd.
(f) The chip is white or the number is odd.

16. Suppose that the simple experiment described in exercise 15 is carried out 360 times, with the poker chip being replaced in the box after each drawing. How many times should you expect the following events to occur?

(a) The chip is not blue.
(b) The number is 3 or 5.
(c) The chip is red and the number is 12.
(d) The number drawn is less than 5.
(e) The number drawn is less than 2 or greater than 11.
(f) The chip is white with the number 5, 6, or 7.
(g) The chip is red with a number which is a perfect square.

17. The graduating class of a certain high school contains 52 percent boys and 48 percent girls. Among the boys, 37 percent are 19 years old or older, while among the girls only 12 percent are 19 years old or older. Suppose that a student is drawn at random from this class.

(a) What is the probability that he or she is less than 19 years of age?
(b) Are sex and age of student independent in this class? How do you know?
(c) Given that a student is 19 years old or older, what is the conditional probability that this is a boy?
(d) Given that a student has an age under 19 years, what is the conditional probability that this is a girl?

18. From the information provided in exercise 9 above, find the following conditional probabilities:

(a) p (male|blonde).
(b) p(red-headed|female).
(c) p(female|height under 70 inches).
(d) p(height under 70 inches|female).
(e) p(Alice|red hair).

19. Each subject in a behavioral experiment was given one of three experimental tasks. Let these three tasks be symbolized by A_1, A_2, and A_3. The number of errors each subject made in performing the task was recorded by used of the categories "0 errors" "1 error," and "2 or more errors." Thus, any subject was classified in two ways: "task," and "number of errors." The following table shows the probabilities for these joint events, for any subject drawn at random from the experimental group. Use these probabilities to find:

(a) The marginal probabilities of the task events, A_1, A_2, and A_3.
(b) The marginal probabilities for numbers of errors.
(c) The conditional probability for 0 errors given task A_1.
(d). The conditional probability for 2 or more errors given task A_2.
(e) The probability of task A_3, given no errors.

		⟨task⟩	
errors	A_1	A_2	A_3
0	.05	.02	.13
1	.08	.17	.10
2 or more	.20	.15	.10

20. A child's game has a metal spinner on a card. This card consists of a circular area divided into six colored sectors of equal area: black, white, red, yellow, green, and blue. The spinner is attached to the center of the card, and is free to come to rest any point along the circle. If the spinner is equally likely to come to rest in any sector, then what is the probability of the following:

(a) Black or white.
(b) Yellow.
(c) Neither white nor yellow.
(d) Either green or blue.
(e) Blue or not black.

21. Suppose that the card described in the exercise above were such that black, white, and yellow sectors were each twice the size of the individual sectors devoted to red, blue, and green. Calculate the probabilities of the events listed above under these circumstances, assuming that the pointer has equal probability of stopping at any point on the circle.

22. An archeologist is interested in three geographical areas containing ancient village sites. The area shown as the circle below came under the influence of culture A, the rectangular area under the influence of culture B, and the triangular area was influenced by culture C. The black dots show the possible sites. Suppose that the archeologist picks one of these sites at random to excavate. What is the probability that the site site selected came under the influence of:

(a) culture A?
(b) C?
(c) B?

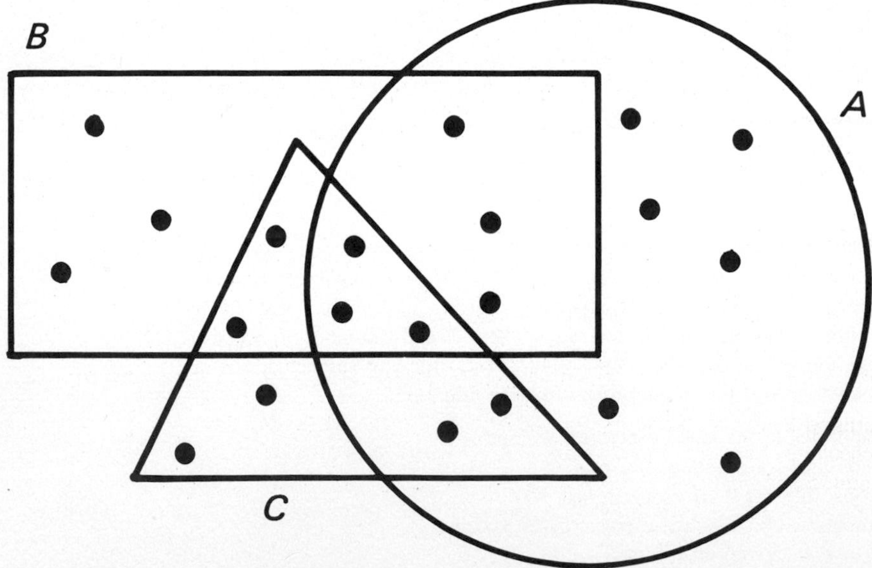

(d) A and B?
(e) A or B?
(f) B and C?
(g) B or C?
(h) A and B and C?
(i) B and C but not A?
(j) A and not B and not C?

23. In exercise 22 above, what is the probability that the site selected came under the influence of exactly one culture? Exactly two cultures? Three cultures? More than one culture? Less than three cultures?

24. The archeologist of exercise 22 knows that some of the sites that might be explored are productive in terms of artifacts, whereas others that might be explored will be nonproductive. The chart below shows the productive sites with an X and the nonproductive with a O. If a site is productive, what is the probability that it was influenced by:

(a) culture A?
(b) B?
(c) A and B?
(d) C but not A?
(e) two cultures only?
(f) A or B?
(g) B and C but not A?

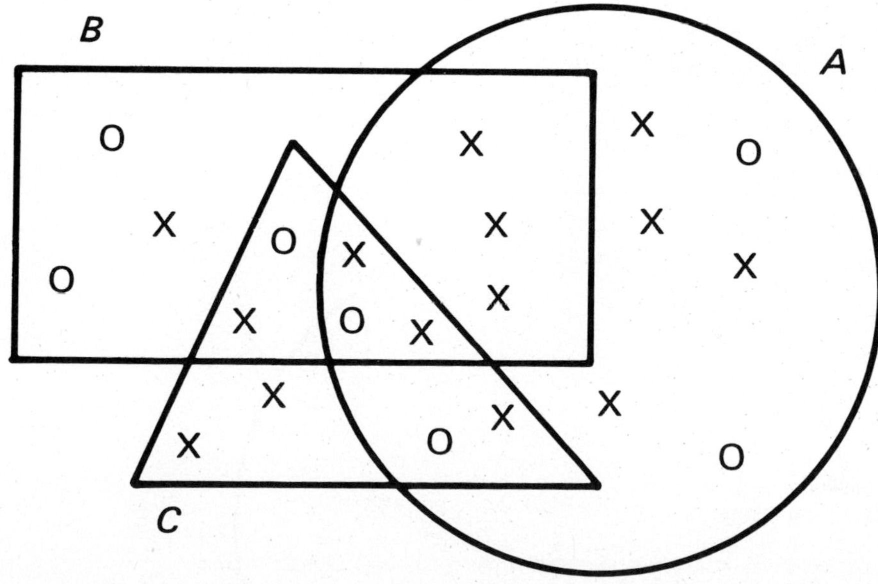

25. A native worker returns to the archeologist of exercise 24 with a report of finding artifacts. What would the archeologist be inclined to infer about the chances for a productive dig if the artifacts show the influence of

(a) Culture A?
(b) Culture C?
(c) Culture B but not A or C?

(d) Culture A but not B or C?
(e) Cultures A and B and C?

26. State the probabilities corresponding to the following odds and bets:

(a) 5 to 3 in favor of event a.
(b) 2 to 9 against event a.
(c) 6 to 4 in favor of event a.
(d) 17 to 15 in favor of a.
(e) \$5 to \$2 bet on a.
(f) \$9 to \$4 bet against a.
(g) \$1 to \$49 bet on a.

27. The following events can occur only once. If an individual regards the bets listed as fair bets, estimate the subjective probability held for these events:

(a) A bet of \$.50 to \$.10 that it will rain tomorrow.
(b) A bet of \$3 to \$7 that the Big Ten representative will win or tie in the Rose Bowl game next New Year's Day.
(c) A bet of \$1 to \$1 that this individual will fail a psychology course this term.

28. A sociologist identified 270 distinct neighborhoods in a large American city. In terms of ethnic balance, these neighborhoods were classified as being "mostly minority," "evenly mixed," and "mostly nonminority." Each neighborhood was also classified according to average income level, as "low income," "medium income," and "high income." The numbers of neighborhoods in each combination of classifications is shown in the following table.

			income	
		low	*medium*	*high*
ethnic balance	minority	79	9	3
	mixed	22	20	10
	nonminority	18	63	46

Suppose that a neighborhood is selected at random from this city. What is the probability that it will have these characteristics:

(a) Low income.
(b) Minority.
(c) Mixed.
(d) Low income or minority.
(e) Not minority.
(f) Not minority and high income.
(g) Low income and mixed.
(h) Not low income and not minority.

29. If a neighborhood is selected, as in exercise 28, what is the conditional probability that it will have the characteristic of:

(a) Low income, given that it is minority.
(b) High income, given that it is minority.
(c) Nonminority, given high income.
(d) Medium income, given mixed.
(e) Not middle income, given mixed.

30. Assume for the moment that the ethnic balance of a neighborhood in the city of exercise 28 is independent of the neighborhood income. What then should the probabilities for the nine possible joint events in the table have been? How many neighborhoods of each ethnic-balance and income combination should there have been? How do these compare with the actual numbers given? What does this suggest about the relation between ethnic balance and income in these neighborhoods?

31. The neighborhoods of exercise 28 were also classified according to whether they fell in a three mile radius of the city center, or whether they were beyond three miles. The table showing all three classifications is as follows:

	Within three miles			*Beyond three miles*		
	low	*medium*	*high*	*low*	*medium*	*high*
minority	69	5	0	10	4	3
mixed	14	4	1	8	16	9
nonminority	2	1	0	16	62	46

Once again, consider a neighborhood selected at random. What is the probability that it is:

(a) Within three miles of the city center.
(b) Nonminority and beyond three miles.
(c) Minority and within three miles.
(d) High income and beyond three miles.
(e) Low income and minority, given beyond three miles.
(f) Within three miles, given minority.
(g) Low income and within three miles, given minority.
(h) Low income, given minority and within three miles.

32. A certain test of vocational interests provides one of five "interest patterns" for any individual taking the test. Call these patterns A, B, C, D, and E. The test was given to 1000 persons, consisting of 294 engineers, 264 physicians, 187 writers, and 255 accountants. The table below gives the numbers in each occupational group giving each of the possible patterns.

	pattern A	B	C	D	E
engineer	117	58	70	33	16
physician	52	125	20	38	29
writer	13	24	12	37	101
accountant	79	33	94	35	14

If one individual is selected at random from this group, state the probability of observing:

(a) Pattern A.
(b) Pattern A or B.
(c) An engineer.
(d) A writer or an accountant.
(e) A physician with pattern D.
(f) A writer with pattern A.
(g) An accountant with pattern C.
(h) An engineer with pattern A or C.
(i) A physician with pattern B or an accountant with pattern C.

33. For the situation of exercise 32, what is the probability of observing:

(a) An engineer, given pattern A.
(b) A physician, given pattern B.
(c) A writer, given pattern E.

(d) Pattern D, given an accountant.
(e) Pattern A or pattern B, given a physician.
(f) An engineer, given pattern A or pattern B.
(g) A writer, given pattern A.
(h) An accountant, given any pattern except E.

34. Once again, an individual is drawn at random from the group in exercise 32. Under each of the following conditions, what is your best bet about the pattern that individual will exhibit?

(a) There is no information given about the person's occupation.
(b) The person is an accountant.
(c) The person is an engineer.
(d) The occupation is writer.
(e) The individual is a physician.
(f) Occupation is engineer or accountant.
(g) The person is not a writer.

35. An individual is selented at random from the group in exercise 32. Find out what the odds are that this person is

(a) A writer (as opposed to "not a writer").
(b) A physician.
(c) An engineer.
(d) An accountaint.
(e) An engineer or a physician.

36. Answer the questions of exercise 35 once again, but this time suppose that we know that the individual selected exhibits pattern B. What happens to the odds when one is provided with some information about the pattern, as opposed to no information? What would the addition of information about pattern do to the odds for occupations if pattern and occupation were actually independent? Is there evidence that pattern and occupation are independent for this group of people?

37. Now suppose that two individuals are drawn at random and independently (with replacement) from the group of exercise 32. What is the probability that

(a) The first is an engineer and the second a physician?
(b) Both are accountants?
(c) Both are writers?
(d) The first shows pattern A and the second pattern B?
(e) Both are physicians showing pattern B?
(f) Both are writers showing pattern E or both are accountants showing pattern C?
(g) Neither is a physician?

Chapter 2
FREQUENCY AND PROBABILITY DISTRIBUTIONS

Given the definitions of "event" and of "probability" in the last chapter, we are now ready to take up the first major set of concepts in statistics.

The idea of a frequency distribution for sets of observations will be introduced, together with some of the mechanics for constructing distributions of data. Then this idea of a frequency distribution will be paralleled by that of a probability distribution. The important special case of numerical events will be introduced under the heading "random variables."

Prior to any concerns with summarizing the data, or with comparing the obtained results with some theoretical state of affairs, the investigator must first measure the phenomena under study. It therefore seems appropriate at this point to mention the process of measurement, and to indicate the role that measurement considerations will play in the remainder of this text. A thorough discussion of this topic will not even be attempted, but we will try to examine its relevance to statistics.

2.1 / MEASUREMENT SCALES

Whenever the scientist makes observations of any kind, some classifying and recording scheme must be used. Any phenomenon or "thing" will have many distinguishable characteristics or attributes, but *the scientist must first single out those properties relevant to the question being studied*. For example, a scientist tries out a new serum on laboratory rats infected with a disease. Any single rat differs from any other rat in innumerable ways: coat markings, heart rate, length of tail, exact age, and so on. The scientist is interested, however, in only one thing: did the rat recover or did it die of the disease? Or, perhaps, the intelligence level of a group of boys is being compared with that of a group of girls. Individual boys and girls differ and same-sex groups differ among themselves in ways as diverse as body temperature, size of head, weight, color of hair, father's income, name, and so on ad infinitum. All these properties are ignored by the experimenter as immaterial to the immediate purpose, which is giving each child a number representing intelligence.

The classifying and recording scheme the scientist uses is based upon differences in some *particular* property or attribute that objects of observation exhibit. The scientist simply cannot pay attention to all the ways that things differ from each other. Many of these differences obviously are irrelevant to the purpose, and other potential differences are controlled by the scientist in making the observations: only rats (not all kinds of animals) are given the serum, and each child is given the same intelligence test in the same way. Still other differences that are germane to the conclusions, but not specifically controlled, are treated as statistical ''error.''

Once the scientist has singled out the property or properties to be studied, and established controls for the others, the classifying scheme is applied to each observation. Such a scheme is essential in order to record, categorize, and communicate the things observed. At its very simplest, *this scheme is a rule for arranging observations into equivalence classes, so that observations falling into the same set are thought of as qualitatively the same and those in different classes as qualitatively different in some respect.* In general, *each observation is placed in one and only one class, making the classes mutually exclusive and exhaustive.*

The process of grouping individual observations into qualitative classes is measurement at its most primitive level. Sometimes this is called **categorical or nominal scaling.** The set of equivalence classes itself is called a **nominal scale.** There are many areas in science where the best one can do is to group observations into classes, each given a distinguishing symbol or name. For example, taxonomy in biology consists of grouping living things into phyla, genera, species, and so on, which are simply sets or classes. Psychiatric nosology contains classes such as schizophrenic, manic-depressive, paretic, and so on; individual patients are classified on the basis of symptoms and the history of their disease.

At this level of measurement, what the classes happen to be called is quite irrelevant. Any set of equivalence classes can be transformed into another set of equivalence classes (that is, renamed), provided that common class membership, or equivalence, is strictly preserved. Thus, consider a set of seniors at a given university in a given year. They might be classified as ''liberal arts major,'' ''engineering major,'' ''education major,'' ''business major,'' and ''other major.'' However, they could just as well be classified as ''group alpha,'' ''group beta,'' ''group gamma,'' ''group delta,'' and ''group omega'' so far as the measurement process is concerned, provided that all members of a group in the first instance wind up as members of the same group in the second instance. Only the names of the classifications are changed. Any such equivalence-preserving transformation is permissible in nominal measurement.

The word ''measurement'' is usually reserved, however, for the situation where each individual is assigned a *number;* this number reflects a magnitude of some *quantitative* property. A definite rule must exist associating one and only one number with each individual, and the measurement procedure fills the role of this rule.

There are at least three kinds of numerical measurement that can be distinguished: these are often called *ordinal scaling, interval scaling,* and *ratio scaling*. A really accurate description of each kind of measurement is beyond the scope of this book, but we can gain some idea of these basic distinctions.

Imagine a set $\mathcal{O}$ consisting of N distinct objects. These objects may be given the labels o_1, o_2, and so on up to o_N. Now imagine any distinct pair of these objects, which we will call o_i and o_j.

Suppose that there is some property that pertains to each object in the set $\mathcal{O}$, such as temperature, weight, length, age, intelligence, or motivation. Each object has a certain amount or degree of that property. In principle, to any object o_i we could assign a number $t(o_i)$, standing for the amount that o_i actually "has" of that characteristic. In the same way, any other object o_j in the set could also be given its value $t(o_j)$.

Ideally, in measuring an object o_i we should be able to determine its $t(o_i)$ value directly. Unfortunately, this is not always (or even usually) possible. Rather, what we must do is to devise a procedure for pairing each o_i with another number, say $m(o_i)$, that we call its numerical measurement. The actual procedure we use to assign the $m(o_i)$ value constitutes a measurement rule. However, just any old procedure, regardless of how precise and "scientific" it appears, will not do: we want the various values of $m(o_i)$ assigned to the various possible objects o_i at least to reflect the $t(o_i)$ values showing the different degrees of the property. A measurement rule would surely be nonsense if it gave numbers having no connection at all with the true amounts of some property that different objects possess. Even though we may never be able to determine the $t(o_i)$ value for any object o_i exactly, we at least hope to find numbers $m(o_i)$ that will be related to these true values in a systematic way. The measurement numbers we obtain must be good reflections of the true quantities, so that information about magnitudes or amounts of the property can be at least inferred from the values observed.

Measurement operations or procedures differ in the information that the numerical measurements themselves provide about the true magnitudes. Some ways of measuring permit us to make very strong statements about what the differences or ratios among the true magnitudes must be, and thus about the actual differences in, or proportional amounts of, some property that different objects possess. On the other hand, some measurement operations permit only the roughest inferences to be made about true magnitudes from the measurement numbers themselves.

Now suppose that we have a measurement procedure that gives a number $m(o_i)$ to any object o_i, and also gives a number $m(o_j)$ to a different object o_j. Then we say that this is measurement at the *ordinal level* if the following statements are true:

---

1. $m(o) \neq m(o_j)$ implies that $t(o_i) \neq t(o_j)$.
2. $m(o_i) > m(o_j)$ implies that $t(o_i) > t(o_j)$.

In other words, when we measure the ordinal level, we can say at least that if two measurements are unequal the true magnitudes are unequal, and if one measurement is larger than another one magnitude exceeds another. However, we really cannot say *how much* the objects truly differ on the property in question.

For example, suppose that the objects in question were minerals of various kinds. Each mineral has a certain degree of *hardness,* represented by the quantity $t(o)$. We have no way to know these quantities directly, and so we devise the following measurement rule: take each pair of minerals and find if one *scratches* the other. Presumably, the harder mineral will scratch the softer in each case. When this has been done for each pair of minerals, give the mineral scratched by everything some number, the mineral scratched by all but the first a higher number, and so on, until the mineral scratching all others but scratched by nothing else gets the highest number of all. In each pair the ''scratcher'' gets a higher number than the ''scratchee.'' (Here we are assuming tacitly that ''*a* scratches *b*'' and ''*b* scratches *c*'' implies that ''*a* scratches *c*,'' and that we will get a *simple ordering* of ''what scratches what.'')

This measurement procedure gives an example of *ordinal* scaling. The possible numerical measurements themselves might be some set of numbers, such as (1, 2, 3, 4, . . .) or (10, 17, 24, . . .) or even other arbitrary symbols having some conventional order, as (*A*, *B*, *C*, . . .). In any case, the assignment of numbers or symbols to objects is only a form of ranking, showing which is ''more'' something. In the example, if one mineral gets a higher number than another then we can say that the first mineral is harder than the second. Notice, however, that this is really all we can say about the degree or amount of hardness each possesses. *Although the numbers standing for ordinal measurements may be manipulated by arithmetic, the answer cannot necessarily be interpreted as a statement about the true magnitudes of objects, nor about the true amounts of some property.*

Other measurement procedures give functions pairing objects with numbers where much stronger statements can be made about the true magnitudes from the numerical measurements. Suppose that the following statement, in addition to statements 1 and 2, were true:

---

3. For any object o_i, $t(o_i) = x$ if and only if $m(o_i) = ax + b$, where $a \neq 0$.

That is, the measurement number $m(o)$ is some *linear function* of the true magnitude x (the rule for a linear function is x multipled by some constant a and added to some constant b). When the statements 1, 2, and 3 are all true, the measurement operation is called **interval scaling,** or measurement at the **interval-scale level.**

Much stronger inferences about magnitudes can be made from interval-scale measurements than from ordinal measurements. In particular, we can say something precise about *differences* in objects in terms of magnitude. Consider two objects, o_i and o_j, once again. Then if we find

$$m(o_i) - m(o_j) = 4,$$

we can conclude that

$$t(o_i) - t(o_j) = 4/a,$$

the difference between the magnitudes of the two objects is 4 units, where a is simply some constant changing measurement units into ''real'' units, whatever they may be.

For example, finding temperature in Fahrenheit units is measurement on an interval scale. If object o_i has a reading of 180° and o_j has 140°, the difference (180 − 160) times a constant actually *is* the difference in ''temperature-magnitude'' between the objects. It is perfectly meaningful to say that the first object has twenty units more temperature than the second.

Given some measurement operation yielding an interval scale of some property, then *any other measurement operation for the same property also gives an interval scale, provided that the second way of measuring yields numbers that are a linear function of the first.*

A familiar example of this principle is temperature measurement in Fahrenheit and in Celsius degrees; each is an interval-scale measurement procedure, and the reading on one scale is a linear function of the reading on the other:

$$C° = (5/9)\,(F° - 32).$$

When measurement is at the interval-scale level, any of the ordinary operations of arithmetic may be applied to the differences between numerical measurements, and the result interpreted as a statement about magnitudes of the underlying property. The important part is this interpretation of a numerical result as a quantitative statement about the property shown by the objects. This is not generally possible for ordinal-scale measurement numbers, but it *can* be done for differences between interval-scale numbers. In very simple language: you can do arithmetic to your heart's content on any set of numbers, but your results are not necessarily true statements about amounts of some property objects possess unless interval-scale requirements are met by the procedure for obtaining those numbers.

Interval scaling is about the best one can do in most scientific work, and even this level of measurement is all too rare in social and behavioral sciences. However, especially in physical science, it is sometimes possible to find measurement operations making the following statement true:

---

4. For any object o_i, $t(o_i) = x$ if and only if $m(o_i) = ax$, where $a > 0$.

When the measurement operation defines a function such that statements 1 through 4 are all true, then measurement is said to be at the **ratio-scale level.** For such scales, *ratios* of numerical measurements can be interpreted directly as ratios of magnitudes of objects:

$$\frac{m(o_i)}{m(o_j)} = \frac{t(o_i)}{t(o_j)}.$$

For example, the usual procedure for finding the length of objects provides a ratio scale. If one object has a measurement value of 10 feet, and another a value of 20 feet, then it is quite legitimate to say that the second object has twice as much length as the first. Notice that this is not a statement one ordinarily makes about the *temperatures* of objects (on an interval scale): if the first object has a temperature reading of 10° and the second 20°, we do not ordinarily say that the second has twice the temperature of the first. Only when scaling is at the ratio level can the full force of ordinary arithmetic be applied directly to the measurements themselves, and the results reinterpreted as statements about magnitudes of objects. An important connection exists between interval scaling and ratio scaling. **When objects are measured on an interval scale, then differences between objects are measured on a ratio scale.** The concept of zero difference, or zero "distance," does have a fixed and nonarbitrary definition, and differences between objects can be treated by any of the methods available for ratio-scale values, provided that the original objects were measured on an interval scale. This accounts, in part, for the considerable preoccupation with differences among measurements that one encounters not only in the physical sciences but also in statistics.

Only one sort of transformation is permissible for values measured on a ratio scale. This is multiplication of each value by the same positive constant. In other words, the only permissible change in a ratio scale is in the unit of measurement employed.

As a related point, observe that an interval scale has an arbitrary origin or "zero point," which is determined by the constant b in statement 3. However, any ratio scale has a fixed zero point, which cannot be modified. One can alter the point on a thermometer that represents zero degrees of temperature, but one cannot change the point on a ruler that stands for zero units of length.

There are any number of examples that could be adduced to illustrate the differences among these levels of measurement, and other possible intermediate levels as well. The student who is interested in this topic enough to explore it further is urged to look into the initial chapters of the books by Stevens (1951), by Thrall, Coombs, and Davis (1954), and by Torgerson (1958); each gives a slightly different perspective on psychological measurement. Our immediate concern, however, is to see the implications of different levels of measurement for probability and statistics.

The problem of measurement, and especially of attaining interval scales, is an extremely serious one for the social and behavioral sciences. It is unfortunate that in their search for quantitative methods researchers sometimes overlook the question of level of measurement and tend to read quite unjustified meanings into their results. This has brought about a reaction in some quarters, where some people insist that such investigators are not justified in using most of the machinery of mathematics on their generally low-level measurements.

However, the core problem of level of measurement really lies outside the province of mathematics and statistics. Statistics deals with numbers, and statistical methods yield conclusions based on numbers, but there is absolutely nothing in the mathematical machinery that whistles and waves a flag to show that the numbers supplied are not really interval-scale measurements of some property of objects or human beings. The machinery works the same way and gives the same result regardless of whether the numbers are made up from the whole cloth or are the product of the most refined measurement procedures imaginable. Only the users of the statistical result, the investigators and their readers, can judge the reinterpretability of the numerical result into a valid statement about properties of things. Sometimes the investigator regards the numerical conclusion as a statement about *scores,* and does not go beyond this statement to a conclusion about the real magnitudes of some property. In other instances the whole investigation makes no sense at all unless the numerical conclusion can refer directly to magnitudes of some property. Surely the individual researcher knows his problem and his measurement procedures better than a stranger dogmatizing about hypothetical situations in a textbook! The scientist must face the problem of the interpretation of statistical results *within* the context of the scientific discipline and on *extramathematical* grounds.

Thus, it seems to this author that statistics qua statistics is quite neutral on this issue. In developing procedures, mathematical statisticians have assumed that techniques involving numerical scores, orderings, or categorization are to be applied where these numbers or classes are appropriate and meaningful within the experimenter's problem. If the statistical method involves the procedures of arithmetic used on numerical scores, then the numerical answer is formally correct. Even if the numbers are the purest nonsense, having no relation to real magnitudes or the properties of real things, the answers are still right *as numbers*. The difficulty comes with the interpretation of these numbers back into statements about the real world. If nonsense is put into the mathematical system, nonsense is sure to come out.

For these reasons, very little more will be said about scales of measurement in this book. Everything depends on what is being measured in whatever way, what is to be found out, and, most of all, what the investigator wishes to say about the "real properties" underlying the numerical measurements. From time to time, statistical techniques will be examined which seem especially appropriate to one or another of the scales of measurement, and this will, of course, be pointed out. Thus, in Chapter 15 techniques will be examined which lend themselves quite naturally to nominally scaled data, and in Chapter 16 the emphasis will be on data at the ordinal level of measurement. For many other methods, interval scale measurement may be tacitly assumed, even though it is completely obvious that this level of measurement cannot be attained for some things. In these cases the user of the statistical technique should understand the limitations imposed on the conclusions by the measuring device used (*not* by statistics).

2.2 / FREQUENCY DISTRIBUTIONS

As we have just seen, even in the simplest instances of measurement the experimenter makes some N observations and classifies them into a set of qualitative measurement classes. These qualitative classes are mutually exclusive and exhaustive, so that each observation falls into one and only one class.

As a summarization of these observations, the experimenter often reports the various possible classes, together with the number of observations falling into each. This may be done by a simple listing of classes, each paired with its frequency number, the number of cases observed in that category. The same information may be displayed as a graph, perhaps with the different classes represented by points or segments of a horizontal axis, the frequency shown by a point or a vertical bar above each class. Regardless of how this information is displayed, such a listing of classes and their frequencies is called a *frequency distribution. Any representation of the relation between a set of mutually exclusive and exhaustive measurement classes and the frequency of each is a frequency distribution.*

A frequency distribution is simply a function in which each of a set of classes is paired with a number, its frequency. Thus, in principle, a frequency distribution of real or theoretical data can be shown by any of the three ways one specifies any function: an explicit listing of class and frequency pairs, graphically, or by the statement of a rule for pairing a class of observations with its frequency. In describing actual data, the first two methods are almost always used alone, but there are circumstances where the experimenter wants to describe some theoretical frequency distribution, and in this case the mathematical rule for the function sometimes is stated.

The set of measurement classes may correspond to a nominal, an ordinal, an interval, or a ratio scale. Although the various possible classes will be qualitative in some instances and number classes in others, a frequency distribution can always be constructed, provided that each and every observation goes into one and only one class. Thus, for example, suppose that some N native United States citizens are observed, and the state in which each was born noted. This is like nominal measurement, where the measurement classes consist of the fifty states (and the District of Columbia) into which subjects could be categorized. On the other hand, the subjects might be students in a course and graded according to A, B, C, and so on. Here the frequency distribution would show how many got A, B, and so on. Notice that in this case the measurement actually is ordinal: A is better than B, B better than C. Nevertheless, the frequency distribution is constructed in the same way, except that the various classes are displayed in their proper order.

Even when measurement is at the interval- or ratio-scale level, one reports frequency distributions in this general way. Suppose that N college students were each weighed, and the weights noted to the nearest pound. The set of weight classes into which students are placed would perhaps consist of fifty or sixty different numbers; a weight class is the set of students getting the same weight number. There may be only one, or even no students, in a particular weight class. Nevertheless, the frequency distribution would show the pairing of each possible class with some number from zero to N, its frequency.

As we shall see, when the number of possible classes is very large, or even potentially infinite, the task of constructing a frequency distribution is made easier by a process of combining classes. The principle is, however, always the same: a display of the relation between classes and their frequencies.

2.3 / FREQUENCY DISTRIBUTIONS WITH A SMALL NUMBER OF MEASUREMENT CLASSES

The reasons for dealing with frequency distributions rather than raw data are not hard to see. Raw data almost always take the form of some sort of listing of pairs, each consisting of an object and its measurement class or number. Thus, if we were noting the hair color of eleven persons, our raw data might be something like this:

John Jones	black
Mary Smith	blonde
Jim Hardy	brown
Horace Goodman	brown
Alice Adams	blonde
Ann Wilk	red
William Thomas	brown
Bert Fox	red
Homer Giddens	black
Shirley Snider	brown
Richard Rowan	blonde

This set of pairs does contain all the relevant information, but if there are a great many objects being measured such a listing is not only laborious but also very confusing to anyone who is trying to get a picture of the set of observations as a whole. If we are interested only in the "pattern" of hair color in the group the names of the persons are irrelevant, and all that is necessary is the number of individuals having each hair color.

There are only four measurement classes in this example that show one or more cases, and so the frequency distribution can be displayed in this way:

Hair color	*f**
black	2
red	2
blonde	3
brown	4
	11 = *N*

*The letter *"f"* symbolizes *"frequency"* throughout.

It is possible to illustrate this idea of a frequency distribution in any number of ways: consider a study done on a group of twenty-five males, in order to determine their blood types. The subjects were classified variously as having blood types "A," "B," "AB," or "O." Table 2.3.1 lists the measurements of these twenty-five men. These data condense into the frequency distribution that follows the table.

Table 2.3.1

Blood types of twenty-five men.

Man number	*Blood type*
1	A
2	B
3	A
4	A
5	A
6	AB
7	O
8	A
9	A
10	A
11	O
12	B
13	O
14	B
15	A
16	B
17	O
18	B
19	O
20	A
21	B
22	B
23	A
24	A
25	O

Class	*f*
A	11
B	7
AB	1
O	6
	25 = *N*

In another study, 2000 families in some city were interviewed about their preferences in art, music, literature, recreation, and so forth. After each interview, the family was characterized as having "highbrow," "middlebrow," or "lowbrow" tastes. This we can consider as a case of ordinal-scale measurement, since these three "taste" categories do seem to be ordered: "highbrow" and "lowbrow" should be more different from each other than "middlebrow" is from either. A listing of families in terms of these classes would be confusing, if not grounds for libel suits. However, just as before, we can construct a frequency distribution summarizing these data:

Class	f
highbrow	50
middlebrow	990
lowbrow	960
	2000 = N

Notice how much more clearly the characteristics of the group as a whole emerge from a frequency distribution such as this than they possibly could from a listing of 2000 names each paired with a rating.

It is appropriate to point out that a frequency distribution provides clarity at the expense of some information in the data. It is not possible to know from the frequency distribution alone whether the family of John Jones at 2193 Spring Street is high-, middle-, or low-browed. Such information about *particular* objects is sacrificed in a frequency distribution to gain a picture of the group of measurements *as a whole*. This is true of all descriptive statistics; we want clear pictures of large numbers of measurements, and we can do this only by losing detail about particular objects. The process of weeding out particular qualities of the object that happen to be irrelevant to our purpose, begun whenever we measure, is continued when we summarize a set of measurements.

2.4 / GROUPED DISTRIBUTIONS

In the distributions shown above, there were a few "natural" categories into which all the data fit. It often happens, however, that data are measured in some way giving a great many categories into which a given observation might fall. Indeed, the number of potential categories of description often vastly exceeds the number of cases observed, so that little or no economy of description would be gained by constructing a distribution showing each *possible* class. The most usual such situation occurs when the data are measured in numerical terms. For instance, in the measurement of height or weight, there are, in principle, an infinite number of numerical classes into which an observation might fall: all the positive real numbers. Even when the measurements result in fewer than an infinite number of possibilities, it is still quite common for the number of available categories to be very large.

For this reason, it is necessary to form **grouped** frequency distributions. Here, frequencies are assigned not to each possible measurement category, but rather to intervals or groupings of categories.

For example, consider a number of people who have been given an intelligence test. The data of interest are the numerical IQ scores. Immediately, we run into a problem of procedure. It could very well turn out that for, say 150 individuals, the IQ score would be different for each case, and thus a frequency distribution using each different IQ number as a measurement class would not condense that data any more than a simple listing.

The solution to this problem lies in grouping the possible measurement classes into new classes, called **class intervals,** each including several score possibilities. Proceeding in this way, if the IQs 105, 104, 100, 101, and 102 should turn up in the data, instead of listing each in a different class, we might put them all into a single class, 100–105, along with any other IQ measurements that fall between these limits. Similarly, we group other sets of numbers into class intervals. On our doing this, one way that the frequency distribution might look for a group of 150 persons is shown in Table 2.4.1.

Table 2.4.1

Intelligence Quotients for a group of 150 persons, arranged in class intervals.

Class interval	*Midpoint*	*f*
124–129	126.5	8
118–123	120.5	0
112–117	114.5	10
106–111	108.5	20
100–105	102.5	65
94– 99	96.5	22
88– 93	90.5	23
82– 87	84.5	2
		150 = N

Forming class intervals has enabled us to condense the data so that a simple statement of the frequency distribution can be made in terms of only a few classes. In this distribution, the various class intervals are shown in order in the first column on the left. The extreme right column lists the frequencies, each paired with the class in that row of the table. The sum of the frequencies must be N, the total number of observations. The middle column contains the midpoint of each class interval; these will be discussed further in Section 2.7.

2.5 / CLASS INTERVAL SIZE AND CLASS LIMITS

The first problem in constructing a grouped frequency distribution is deciding how big the intervals shall be. What are the largest and the smallest numbers that may go into an interval? One calls the difference between the largest and the smallest number that may go into any class interval the *interval size*. We will let the symbol i denote this interval size.

In the example above, as in most examples to follow, the size i is the same for each class interval. This is the accepted convention for most work in social sciences. (Some exceptions will be mentioned in Section 2.9, however.) For the example, $i = 6$, which means that the largest and the smallest score going into a class interval will differ by six units. Take the class interval labeled 100–105. It may seem that the smallest number going into this interval is 100, and the largest 105; this is not, however, true. Actually, the numbers 99.5, 99.6, 99.7 would also go in this interval, should they occur in the data. Also, the numbers 105.1, 105.2 are included. On the other hand, 99.2 or 105.8 would be excluded. The interval actually includes any number *greater than or equal to* 99.5 *and less than* 105.5. These are called the **real limits** of the interval, in contrast to those actually listed, which are the **apparent limits.** Thus, the real limits of the interval 100–105 are

99.5 to 105.5

and

$$i = 105.5 - 99.5 = 6.$$

In general,

real lower limit = apparent lower limit minus .5 (unit difference)
real upper limit = apparent upper limit plus .5 (unit difference).

The term "unit difference" demands some explanation. In measuring something we usually find some limitation on the accuracy of the measurement, and seldom can one measure with *any* desired degree of accuracy. For this reason, measurement in numerical terms is always rounded, either during the measurement operation itself, or after the measurement has taken place. In measuring weight, for example, we obtain accuracy only within the nearest pound, or the nearest one tenth of a pound, or one hundredth of a pound, and so on. If one were constructing a frequency distribution where weight had been rounded to the nearest pound, then a unit difference is one pound, and the real limits of an interval such as 150–190 would be 149.5–190.5. The i here would be 41 pounds.

On the other hand, suppose that weights were accurate to the nearest one tenth of a pound; then the unit difference would be .1 pound, and the real class interval limits would be

149.95–190.05.

The way that rounding has been carried out will have a real bearing on how the successive class intervals will be constructed. Suppose that we wanted to construct the distribution of annual income for a group of American men, where the income figures have been rounded to the nearest thousand dollars. We have decided that the apparent upper limit of the top interval shall be 25,000 dollars, and that i should equal 5,000. Then the classes would have the following limits:

Real	*Apparent*
20,500–25,500	21,000–25,000
15,500–20,500	16,000–20,000
10,500–15,500	11,000–15,000
5,500–10,500	6,000–10,000
500– 5,500	1,000– 5,000

Here, one half of a unit difference is 500 dollars.

Now suppose that each man's income has been rounded to the nearest one hundred dollars. Once again, with i = $5,000 and with the top apparent limit $25,000, we form class intervals, but this time the class limits are:

Real	*Apparent*
20,050–25,050	20,100–25,000
15,050–20,050	15,100–20,000
10,050–15,050	10,100–15,000
5,050–10,050	5,100–10,000
50– 5,050	100– 5,000

This difference in unit for the two distributions could make a difference in the "picture" we get of the data. For example, a man making exactly \$20,100 would fall into the top interval in the second distribution, but into the second from top interval in the first. It is important to decide upon the accuracy represented in the data before the real limits for the class intervals are chosen. Is every digit recorded in the raw data regarded as significant, or will the data be rounded to the "nearest" unit?

2.6 / INTERVAL SIZE AND THE NUMBER OF CLASS INTERVALS

One can use any number for i in setting up a distribution. However, it should be obvious that there is very little point in choosing i smaller than the unit difference (for example, letting i = .1 pound when the data are in nearest whole pounds), or in choosing i so large that all observations fall into the same class interval. Furthermore, i is usually chosen to be a *whole number* of units, whatever the unit may be. Even within these restrictions, there is considerable flexibility of choice. For example, the data of Section 2.4 could have been put into a distribution with i = 3, as shown in Table 2.6.1. Notice that this distribution with i =3 is somewhat different

Table 2.6.1

Intelligence Quotient scores of 150 persons, arranged into intervals of three units.

Class interval	*Midpoint*	*f*
126–128	127	5
123–125	124	3
120–122	121	0
117–119	118	0
114–116	115	7
111–113	112	3
108–110	109	8
105–107	106	13
102–104	103	44
99–101	100	20
96– 98	97	10
93– 95	94	12
90– 92	91	14
87– 89	88	9
84– 86	85	0
81– 83	82	2
		150 = N

from the distribution with i = 6, even though they are based on the same set of data. For one thing, here there are sixteen class intervals, whereas there were eight before. This second distribution also gives somewhat more detail than the first about the original set of data. We now know, for example, that there was no one in the group of persons who got an IQ score between the real limits of 83.5 and 86.5, whereas we could not have told this from the first distribution. We can also tell that more cases showed IQs between 101.5 and 104.5 than between 98.5 and 101.5; conceivably this could be a fact of some importance. On the other hand, the first distribution gives a simpler and, in a sense, a neater picture of the group than does the second.

The decision facing the maker of frequency distributions is, "Shall I use a small class-interval size and thus get more detail, or shall I use a large class-interval size and get more condensation of the data?" There is no fixed answer to this question; it depends on the kind of data and the uses to which they will be put. However, a convention does exist which says that for most purposes *ten to twenty class intervals* give a good balance between condensation and necessary detail, and in practice one usually chooses class-interval size to make about that number of intervals.

A handy rule of thumb for deciding about the size of class intervals is given by

$$i = \frac{\text{highest score in data} - \text{lowest score}}{\text{number of class intervals}}.$$

After deciding on some convenient number of class intervals, you divide this number into the difference between the highest and lowest scores, or the **range** of scores, and find the size that i will have to be. In practice, this may give an i that is not a whole number, in which case you simply use the nearest whole number for i. In the example just given, the data showed 128 as the highest and 82 as the lowest scores, with a range of 46. To find i giving about 16 class intervals we would divide the range by 16:

$$\frac{128 - 82}{16} = 2.87.$$

Since this is a decimal number, we choose the nearest whole number, or 3.

Note that in the last distribution some intervals show a frequency of 0; it is not absolutely necessary that such intervals be listed. However, intervals with zero frequency are usually listed when they fall between other intervals that do not have a zero frequency. We could have had an interval 129–131 having a zero frequency in the last distribution; however, we did not list this interval because, unlike the interval 120–122, it did not fall between other nonzero intervals. In addition, it is important to take the total N into account when deciding upon the number and size of class intervals. It is really not very interesting to look at a distribution with fewer than five or so observations per class interval on the average, although in some exceptional situations this might be allowable. Usually, then, when fairly small numbers of observations are involved, it will pay to examine the average number of cases per class interval. This can be found from N/C, where C is the number of class intervals, or from iN/range. If this number comes out to be as small as 5 or less, consideration should be given to a larger interval size, or perhaps even to an unequal interval size over some intervals of values.

Two other conventions are useful in making distributions from data. The first is that, in the social sciences, the class intervals are usually listed starting with low numbers at the bottom of the list and with high numbers at the top, as shown in the example. Another convention followed here is to start figuring the class intervals by listing the *highest score* in the data as the *apparent upper limit of the top interval;* then it is a simple matter to find the other apparent upper limits by subtracting the class-interval size successively from this number. The apparent *lower* limit of each interval is then *one unit more* than the *upper* apparent limit of the interval below it. For example, suppose 128 were the top score in the data, and that i were 3. Taking 128 as the highest apparent limit, the upper limit of the next interval would be 128 − 3, or 125, next would be 125 − 3 or 122, and so on, until all upper limits were found. Then, given an interval with apparent upper limit of 122, the lower limit of the next interval up would be 123, and given the interval with upper limit 125, the lower limit of the next interval up would be 126, and so on.

2.7 / MIDPOINTS OF CLASS INTERVALS

There is one more feature of frequency distributions that has not yet been discussed. On putting a set of measurements into a frequency distribution, one loses the power to say exactly what the original numbers were. For the example of Section 2.6 a person in the group of 150 cases who has an IQ of 102 is simply counted as one of the 44 individuals who make up the frequency for the interval 102–104. You cannot tell from the distribution exactly what the IQs of those 44 individuals were, but only that they fell within the limits 102–104. What do we call the IQs of these 44 individuals? We call them all 103, the midpoint of the interval. **The midpoint of any class interval is that number which would fall exactly halfway among the possible numbers included in the interval.** A moment's thought will convince you that 103 falls halfway between the real limits of 101.5 and 104.5. In a like fashion, all 5 cases falling into the class interval 126–128 will be called by the midpoint 127; all 3 cases in the interval 123–125 will be called 124, and so on for each of the other class intervals.

The real limits can also be defined in terms of the midpoint:

$$\text{real limits} = \text{midpoint} \pm .5i.$$

Thus, given only the midpoints of the distribution, one can find the class limits.

In computations using grouped distributions the midpoint is used to substitute for each raw score in the interval. For this reason, it is convenient to choose an odd number for i whenever possible; this makes the midpoint a whole number of units and simplifies computations. (Actually, this is a far less important consideration than in former times. Efficient calculators and computers have almost eliminated the need for calculations from such grouped data, even for a large N value.)

2.8 / ANOTHER EXAMPLE OF A GROUPED FREQUENCY DISTRIBUTION

Suppose that seventy-five students in high school had been used as subjects in an experiment on verbal memory. Each student was given a list containing forty-eight pairs of words and allowed to study the list as a whole for ten minutes. The first member of each pair was then shown in order to the individual student, and his task was to recall the second word. The raw data are shown in Table 2.8.1. The largest number of words recalled by any student was thirty-five, and the smallest number was ten. The range was thus 35 − 10 or 25.

Table 2.8.1 Scores of seventy-five students in one trial of paired-associate learning.

Student	*Score*	*Student*	*Score*	*Student*	*Score*
1	19	26	18	51	15
2	16	27	14	52	15
3	11	28	26	53	14
4	13	29	16	54	11
5	12	30	19	55	23
6	20	31	21	56	13
7	11	32	17	57	19
8	24	33	20	58	17
9	19	34	12	59	21
10	16	35	22	60	15
11	12	36	11	61	20
12	24	37	13	62	16
13	19	38	16	63	12
14	17	39	10	64	10
15	18	40	17	65	10
16	25	41	17	66	15
17	16	42	19	67	18
18	20	43	11	68	10
19	17	44	10	69	19
20	19	45	15	70	14
21	15	46	10	71	12
22	35	47	13	72	18
23	16	48	13	73	14
24	11	49	14	74	15
25	26	50	10	75	20

Now we want to make a frequency distribution having about ten class intervals. According to our rule of thumb, given above:

$$i = \frac{\text{range}}{\text{number of intervals}} = \frac{25}{10} = 2.5.$$

Thus $i = 3$ should provide us with a convenient class-interval size.

We start with the largest number, 35, and make it the upper apparent limit of the highest class interval. That interval must then have real limits of 32.5 and 35.5 with apparent limits of 33–35. The real limits of the second from highest interval must be 29.5–32.5, and so on. Each time, the difference between the real limits to an interval must be i, or 3, and the differences between the successive lower apparent limits (and also successive upper limits) must also be 3. Proceeding in this way, we find the class intervals shown in Table 2.8.2.

The table is completed by inspecting the raw data to find the number of cases that fall into each class interval. Only one case falls between the real limits 32.5–35.5, and so the frequency for the highest interval is 1. The frequency for the lowest interval is 13, since exactly thirteen individuals showed scores between the real limits 8.5 and 11.5. Just as in the preceding examples, the midpoint for each interval stands exactly midway between the two real limits.

What can the experimenter tell from looking at this distribution that would not have been obvious from the raw data? First of all the "typical" range of performance is really a fairly short interval of scores; the large majority of students scored in the range of 11 points from 9 through 20. The most ''popular'' score interval is 15–17. Note that this concentration of cases lies at the low end of the *conceivable range* of scores 0 through 48. If the experimenter thinks of a score as reflecting the student's ability to memorize paired associates, then the task is, by and large, a hard one for the students. The single individual scoring in the interval 33–35 is really most atypical. This student's score falls very far from the bulk of the cases in the distribution.

Table 2.8.2

Number of paired associates recalled on the first trial by a sample of seventy-five high school students.

Class interval	*Midpoint*	*f*
33–35	34	1
30–32	31	0
27–29	28	0
24–26	25	5
21–23	22	4
18–20	19	17
15–17	16	20
12–14	13	15
9–11	10	13
		75 = *N*

Naturally, there are other, more precise statements that the experimenter can make about what these data show. We will discuss these summary indices in Chapter 4. For the moment, however, it should be clear that a distribution does communicate information about a set of observations *as a whole* even without further analysis of the data.

2.9 / FREQUENCY DISTRIBUTIONS WITH OPEN OR UNEQUAL CLASS INTERVALS

Quite often the data are such that it is not possible to make a frequency distribution with intervals of a constant size. This most commonly occurs when exact scores are not known for some of the cases. For example, in a study of the trials that it takes an animal to learn a discrimination problem the experimenter found that out of 100 animals, 5 could not learn the problem in sixty trials. It was felt that some of the animals would never learn the problem at all, and running these animals ceased at the sixtieth trial. This means, however, that the 5 animals could not be given an exact score, and could only be put in the top class of the score distribution, in an interval called ''sixty or more.'' This interval is **open,** since there is no way to determine its upper real limit. Furthermore, there is no way to give such an open interval a midpoint.

In other instances, it may be that there are extreme scores in a distribution that are widely separated from the bulk of the cases. When this is true, a class-interval size that is small enough to show ''detail'' in the more concentrated part of the distribution will eventuate in very many classes with zero frequency before the extreme scores are included. Enlarging the class-interval size will reduce the number of unnecessary intervals, but will also sacrifice detail. When this situation arises, it is often wise to have a varying class-interval size, so that the class intervals are narrow where the detail is desired, but rather broad toward the extreme or extremes of the distribution. In this case, class-interval limits and midpoints can be found in the usual way, provided that one takes care to notice where the interval size changes.

2.10 / GRAPHS OF DISTRIBUTIONS: HISTOGRAMS

Now we direct our attention for a time to the problem of putting a frequency distribution into graphic form. Often a graph shows that the old bromide, "a picture is worth a thousand words," is really true. If the purpose is to provide an easily grasped summary of data, nothing is so effective as a graph of a distribution.

There are undoubtedly all sorts of ways any frequency distribution might be graphed. A glance at any national news magazine will show many examples of graphs which have been made striking by some ingenious artist. However, we shall deal with only three "garden varieties": the histogram, the frequency polygon, and the cumulative frequency polygon.

The histogram is really a version of the familiar "bar graph." In a histogram, each class or class interval is represented by a portion of the horizontal axis. Then over each class interval is erected a bar or rectangle equal in height to the frequency for that particular class or class interval. The histograms representing the frequency distributions in Sections 2.3 and 2.6 are shown in Figures 2.10.1 to 2.10.3. It is customary to label both the horizontal and the vertical axes as shown in these figures, and to give a label to the graph as a whole. In the case where class intervals are employed to group numerical measurements, a saving of time and space may be

Figures 2.10.1/2.10.2

(left). Blood types of twenty-five American males. *(right).* Taste ratings for two thousand, American families.

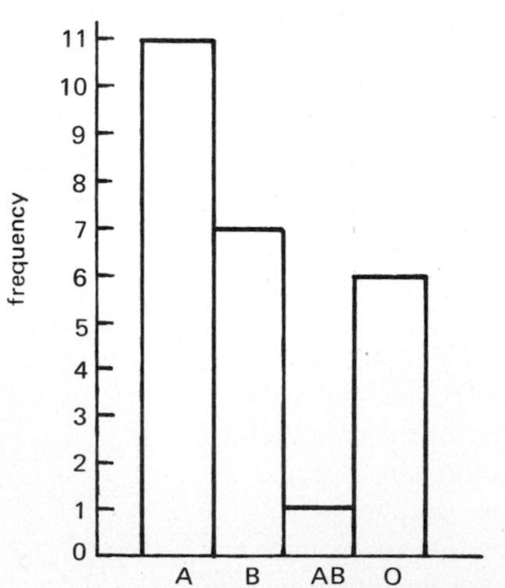

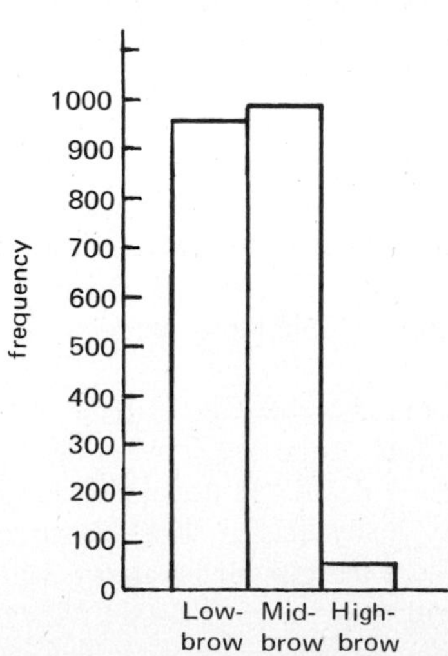

Figure 2.10.3

Intelligence quotients of a sample of one hundred and fifty subjects, with $i = 3$.

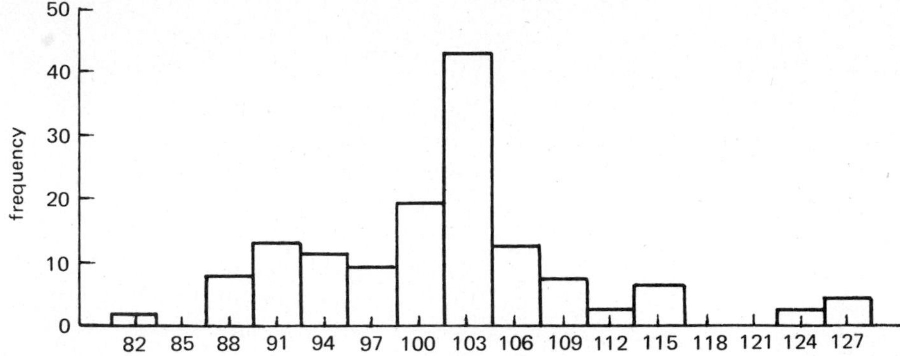

accomplished by labeling the class intervals by their midpoints (as in Figure 2.10.3) rather than by their apparent or real limits. With categorical measurements, this problem does not arise, of course, since each class can only be given its name or symbol.

If you will look at the two histograms shown in Figures 2.10.4 and 2.10.5 you will get very different first impressions, even though they represent the same frequency distribution. These two figures illustrate the effect that proportion can have on the viewer's first impressions of graphed distributions. Various conventions about graphs of distributions exist in different fields, of course. In fact, those who make a business of using statistics for persuasion often choose a proportion designed to give a certain impression. However, it is most usual to find graphs arranged so that the vertical axis is about three fourths as long as the horizontal. This usually will give a clear and esthetically pleasing picture of the distribution. (An amusing and instructive recital of ways to ''adjust'' graphs and other statistical presentations so as to get a desired effect is given in the little book by Huff, *How To Lie with Statistics*.)

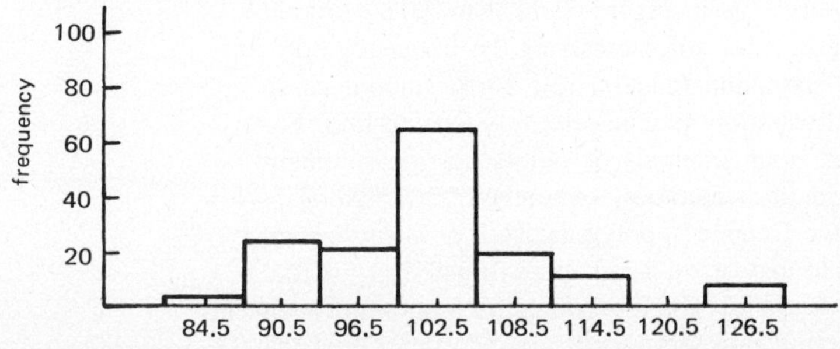

Figure 2.10.4

Intelligence quotients of a sample of one hundred and fifty subjects, with $i = 6$.

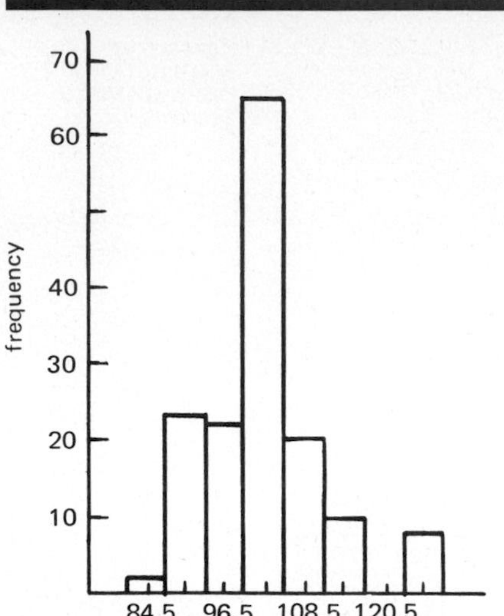

Figure 2.10.5

Another version of Figure 2.10.4

2.11 / FREQUENCY POLYGONS

The histogram is a useful way to picture any sort of frequency distribution, regardless of the scale of measurement of the data. However, a second kind of graph is often used to show frequency distributions, particularly those that are based upon numerical data: this is the frequency polygon. In order to construct the frequency polygon from a frequency distribution one proceeds exactly as though one were making a histogram; that is, the horizontal axis is marked off into class intervals and the vertical into numbers representing frequencies. However, instead of using a bar to show the frequency for each class, a point on the graph corresponding to the midpoint of the interval and the frequency of the interval is found. These points are then joined by straight lines, each being connected to the point immediately succeeding and the point immediately following, as in Figure 2.11.1.

The frequency polygon is especially useful when there are a great many potential class intervals, and it thus finds its chief use with numerical data that could, in principle, be shown as a smooth curve, given enough observations. This would be true of a distribution based on a very large number of numerical measurements on a potentially continuous scale. If we maintained the same proportions in our graph, but employed a much greater number of class intervals, the frequency polygon would provide a picture more like a smooth curve, as in Figures 2.11.2 to 2.11.4, based on several thousand cases. Although the function rule describing the frequency polygon may be extremely complicated to state, the function rule for a smooth curve approximately the same as that of the distribution may be relatively easy to find. For this reason, a frequency polygon based upon interval- or ratio-scale measurement with a relatively large number of classes is sometimes "smoothed" by creating a curve that approximates the shape of the frequency polygon. This new curve may have a rule that is simple to state and that may serve as an approximation of the rule describing the frequency distribution itself. While the procedures for smoothing such a frequency polygon and finding a function rule that comes close to describing the distribution are outside the scope of this book, they may be found in advanced texts

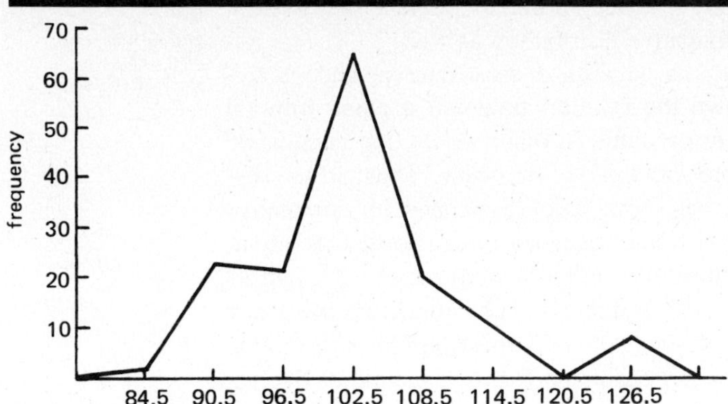

Figure 2.11.1 Intelligence quotients of a sample of one hundred and fifty subjects.

on the analysis of experimental data. Suffice it to say, for the moment, that when we come to a discussion of the "shapes" of distributions, we shall often refer to smooth curves as examples of types of distributions, when actually the frequency distributions themselves could only be represented by "jagged" polygons composed of straight lines. Nevertheless we may use the smooth curves as good approximations to the picture that the frequency polygon presents of the distribution.

Figures 2.11.2–2.11.4

Frequency polygons for a large number of cases arranged into successively smaller class intervals.

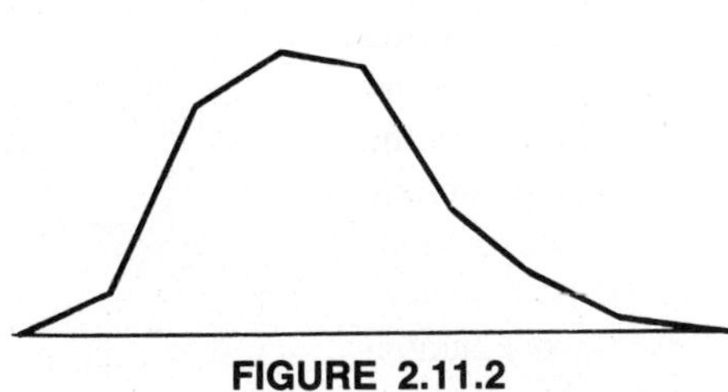

FIGURE 2.11.2

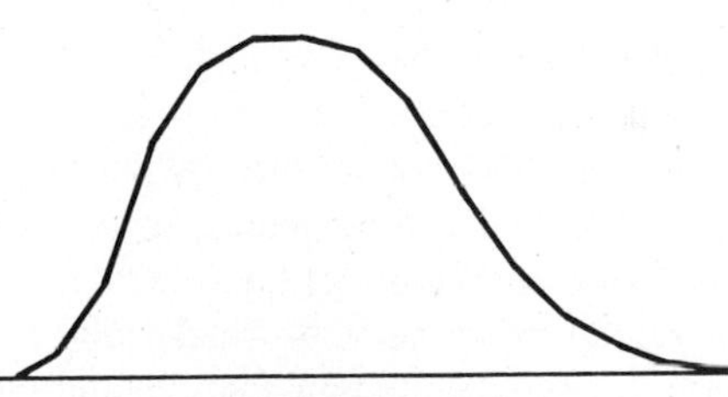

FIGURE 2.11.3

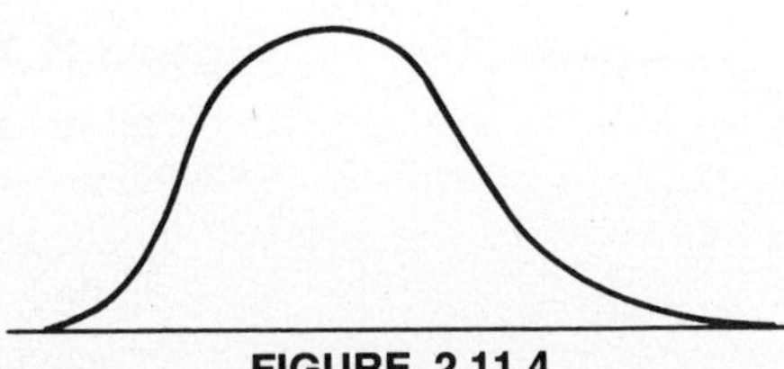

FIGURE 2.11.4

2.12 / CUMULATIVE FREQUENCY DISTRIBUTIONS

For some purposes it is convenient to make a different arrangement of data into a distribution, called a cumulative frequency distribution. For example, when it is desired to find where certain scores rank relative to all of the others ("percentile ranks," as discussed in 4.19), a cumulative distribution may be used. Often a learning curve may be shown as a form of cumulative distribution as well.

Instead of showing the relation that exists between a class interval and its frequency, **a cumulative distribution shows the relation between a class interval and the frequency at or below its real upper limit.** In other words, the cumulative frequency shows how many cases fall into *this interval or below*. Thus, in the distribution in Section 2.4 the lowest interval shows two cases, and its cumulative frequency is 2. Now the next class interval has a frequency of 23, so that its cumulative frequency is 23 + 2, or 25. The third interval has a frequency of 22, and its cumulative frequency is 22 + 23 + 2, or 47, and so on. The cumulative frequency of the very top interval must always be N, since all cases must lie either in the top interval or below. The cumulative frequency distribution for the example of Section 2.4 is given by Table 2.12.1.

Table 2.12.1

Cumulative frequency distribution for intelligence score data of Section 2.4.

Class interval	*Cumulative frequency*
124–129	150
118–123	142
112–117	142
106–111	132
100–105	112
94– 99	47
88– 93	25
82– 87	2

Cumulative frequency distributions are often graphed as polygons. The axes of the graph are laid out just as for a frequency polygon, except that the class intervals are labeled by their *real upper limits* rather than by their midpoints. The numbers on the vertical axis are now cumulative frequencies, of course. The graph of the example from Section 2.4 is shown in Figure 2.12.1.

Estimates of percentile ranks may be read from a cumulative frequency polygon, such as Figure 2.12.1, by finding the cumulative frequency for any number

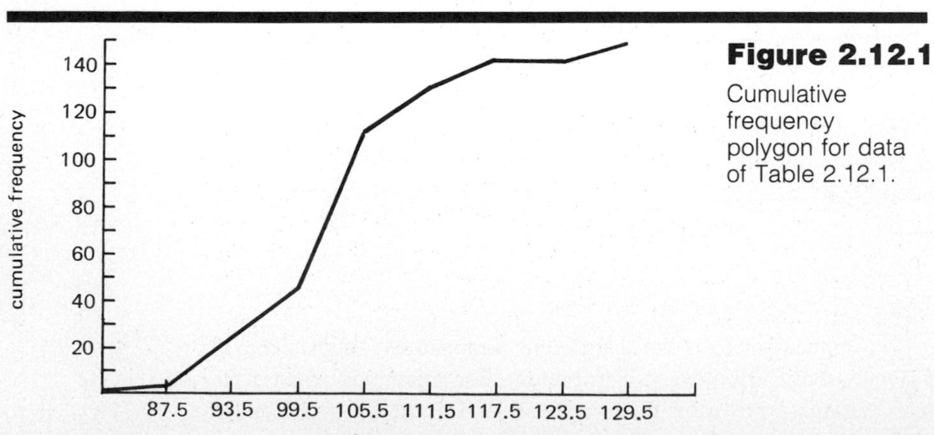

Figure 2.12.1

Cumulative frequency polygon for data of Table 2.12.1.

and then multiplying this cumulative frequency by $100/N$. Thus, for example, the percentile rank corresponding to 99.5 is 47(100)/150 or 31.33. The cumulative frequency for a score of 92 is about 19, so that its percentile rank is about 19(100)/150 = 12.7, and so on. Methods of 4.19 may be used for exact values.

Cumulative frequency distributions often have more or less this characteristic **S** shape, although the slope and the size of the "tails" on the **S** will vary greatly from distribution to distribution. Like a frequency polygon, a cumulative frequency polygon is sometimes smoothed into a curve as similar as possible to the polygon. You sometimes encounter a reference to such a smoothed cumulative distribution curve under the term **ogive.**

2.13 / PROBABILITY DISTRIBUTIONS

Suppose that there were some finite set of objects, and that we sampled these objects at random with replacement. Each object sampled is measured in some way. The measurement may be at the nominal-, ordinal-, interval-, or ratio-scale level, so long as each object is measured by the same procedure. *Now notice that the occurrence of any object in a sample is an elementary event, and that the actual measurement class or number assigned any object is an event.* Since many different elementary events can be allotted the same number or classification, each measurement class is an event class. These measurement events are mutually exclusive and exhaustive, each object being given one and only one number or category.

Furthermore, since the different possible measurement classes define events, there is a probability that can be assigned to each; there is, for example, some probabilty that the score of an object will be 10, or that the object will be assigned to the class "high need for achievement." In short, given the sample space $\mathcal{S}$ and a set of measurement events, we can think of a probability function pairing some probability with each possible measurement class.

Any statement of a function associating each of a set of mutually exclusive and exhaustive events with its probability is a probability distribution.

Notice how the idea of probability distribution exactly parallels that of frequency distribution: each type is based on a set of mutually exclusive and exhaustive measurement classes. In a frequency distribution, each measurement class is paired with a frequency, and in a probability distribution each is paired with its probability. Just as the sum of the frequencies must be N in a distribution, so must the sum of the probabilities be 1.00.

As a simple example of probability distribution, imagine a sample space of all girls of high-school age. Such a girl is selected at random and classified as "right-handed," "left-handed," or "ambidextrous." The probability distribution might be:

Class	*p*
right	.60
left	.30
ambidextrous	.10
	1.00

Or, perhaps, the height of a boy drawn at random from all U.S. boys is measured. Some seven class intervals are used to record the height of any boy observed, and the probability distribution might be:

Height in inches	*p*
78–82	.002
73–77	.021
68–72	.136
63–67	.682
58–62	.136
53–57	.021
48–52	.002
	1.000

Here, each class interval is an event.

Corresponding to any frequency distribution of N cases there is a probability distribution, if single cases are selected from the total group of N cases at random with replacement. Thus, imagine a set of 50 persons, whose scores on some test formed the following distribution:

Scores	*f*
90–99	3
80–89	8
70–79	15
60–69	14
50–59	10
	50

One person is drawn at random from among this set. What are the probabilities associated with the various intervals he might fall into? If the individuals are all assumed equally likely to be observed, then for any class interval *A*,

$$p(A) = \frac{\text{frequency of } A}{\text{total frequency}}.$$

Thus, the class interval 90–99 has a probability of 3/50, or .06, and the other probabilities can be found in the same way, giving

Scores	*p*
90–99	.06
80–89	.16
70–79	.30
60–69	.28
50–59	.20
	1.00

In short, any group of cases summarized in a frequency distribution can also be regarded as a sample space of *potential* observations. If a single case is sampled at random, then the probabilities of the various classes are the same as the relative frequencies of those classes.

Furthermore, given some theoretical probability distribution, and N observations made independently and at random with replacement, then there is a *theoretical frequency distribution* as well, where for any event class A

$$\text{theoretical frequency} = Np(A).$$

This is the frequency distribution we *expect* to observe, given the theoretical probability distribution. For example, suppose that the probability distribution of heights of boys given above were true for some specific population of boys. What is the theoretical frequency distribution of heights for 1000 boys sampled independently and at random? The answer is

Height in inches	$f = Np(A)$
78–82	2
73–77	21
68–72	136
63–67	682
58–62	136
53–57	21
48–52	2
	1000

Given that the probability distribution is correct, then for a sample of 1000 boys we should expect our obtained frequency distribution to show 682 boys in the interval 63–67 inches, only 2 boys in the interval 78–82, and so on. This distribution might never actually occur in practice, but it is our best guess about what will occur if the probability distribution is a correct statement. Whereas the frequencies in real distributions must always be whole numbers, in theoretical distributions it is possible to have fractional frequencies; if only 100 boys were observed in this example, then the frequencies for the top and bottom intervals would be, in theory, .2 each.

Like frequency distributions, a probability distribution may be specified either by a listing such as we have shown here, or in graphic form. However, unlike actual obtained distributions of data, the most important probability distributions are theoretical ones that can be given a mathematical rule. In order to discuss such distributions we need the concept of a ''random variable.''

2.14 / RANDOM VARIABLES

A special terminology is useful for discussing probability distributions of numerical scores. Imagine that each and every possible elementary event in some $\mathcal{S}$ is assigned a number. That is, various elementary events are paired with various values of a variable. Thus, an elementary event may be a person, with some height in inches; or the elementary event may be the result of tossing a pair of dice, with the assigned number being the total of the spots that come up; or the elementary event may be a rat, with the number standing for the trials taken to learn a maze. Each and every elementary event is thought of as getting one and only one such number.

Let X represent a function that associates a real number with each and every elementary event in some sample space $\mathcal{S}$. Then X is called a random variable on the sample space $\mathcal{S}$.

Bear in mind the distinction between the elementary events themselves, which may or may not be numbers, and the values associated with the various elementary events, which are the values of the random variable. Thus, for example, consider the set of American males 21 years or older. An individual is drawn at random from this set. The sample space here consists of the entire set of American males of the specified age, and each such male is an elementary event in this sample space. Now we can associate with each elementary event a real value, the income of the man during the current calendar year. The values that the random variable X can thus assume are the various income values associated with the men. The particular value x occurs when a man is chosen who has income x.

On the other hand, suppose that a box were filled with slips of paper, each inscribed with a real number. Here the elementary events are the slips bearing the numbers. The value x is associated with a particular slip, so that x occurs when the particular slip is drawn at random. However, there are many different assignments of numbers to slips that might be chosen for X. Thus, we might define X as the square of the number on a slip, so that $X = 4$ occurs when a slip is drawn bearing the number 2. The point is that a random variable represents values that are associated in some way with elementary events, so that particular values of X occur when the appropriate elementary events occur. The way in which the numerical values get associated with the elementary events is open to the widest latitude of definition, however.

Although the term "random variable" is rather awkward, it will be used here because of its popularity in statistical writing. The terms "chance variable" or "stochastic variable" are sometimes used. These all mean precisely the same thing, however; a symbol for number events each having a probability.

Given the random variable X, other events may be defined. Thus, given two numbers a and b,

$$(a \leq X \leq b)$$

is the event of some value of X lying between the numbers a and b (inclusive). This event has some probability $p(a \leq X \leq b)$. Furthermore, there is the event

$$(a \leq X),$$

a value greater than or equal to a; this has probability $p(a \leq X)$.

As another example of a random variable, suppose that X symbolizes the height of an American man, measured to the nearest inch. Here, there is some probability that $X \leq 60$, or $p(X \leq 60)$, since there is presumably some set of American men with heights less than or equal to 60 inches. Furthermore,there is some probability $p(70 \leq X \leq 72)$, since there is a set of American men having heights between five feet ten inches and six feet, inclusive. Here, X symbolizes the numerical value (height in inches) assigned to an American man (an elementary event). This symbol X represents any one of many different such values, and for any arbitrary number a there is some probability that the particular value x is less than or equal to the value of a.

Although the notation employed for random variables varies considerably among various authors, we will find it convenient to use capital letters, such as X, Y, Z, to denote random variables, and lowercase letters, such as x, y, z and a, b, c, to denote *particular values* that the random variable in question can take on. (Occasionally, this convention becomes awkward, and it will be violated from time to time in future sections. However, the context will usually make the random variable notation clear in these instances.) The expression $p(X = x)$ symbolizes the probability that the random variable X takes on the particular value x. Often, for convenience, this will be written simply as $p(x)$. In the same way, $p(a \leqslant X)$ stands for the probability that X takes on some particular value x greater than or equal to the value symbolized by a, $p(a \leqslant X \leqslant b)$ symbolizes the event in which X takes on some value lying between the two values symbolized by a and b, and so on.

2.15 / DISCRETE RANDOM VARIABLES

In a great many situations, only a limited set of numbers can occur as values of a random variable. That is, some values have zero probability, even though there may be values to either side that have probability greater than zero. Quite often, the set of numbers that can occur as values of the random variable is relatively small, or at least finite in extent. As an example of a limited set of values assumed by a random variable, suppose that a volume of the *Encyclopedia Britannica* is opened at random, and the page number noted. In this instance, the values of the random variable are all of the different page numbers that might occur. These would be all of the whole numbers between 1 and, say, 1009 (perhaps with a few Roman numbers thrown in). The number of possibilities is thus finite, and, indeed, quite limited with respect to all possible numbers. Real numbers such as -34.6, 1.1103, and the square root of 2 would not occur, of course.

Some random variables can assume what is called a "countably infinite" set of values. One example of a countably infinite set would be the ordinary counting numbers themselves, where the count goes on without end. A simple experiment in which one counts the number of trials until a particular event occurs would give a random variable taking on these counting values.

In either of these situations, the random variable is said to be discrete. **If a random variable *X* can assume only a particular finite or countably infinite set of values, it is said to be a discrete random variable.** As we shall see, not all random variables are discrete, but a large number of random variables of practical and theoretical interest to us will have this property.

If cases are sampled from any frequency distribution that has been grouped into class intervals, then the random variable X is usually regarded as symbolizing the midpoints of the intervals. For a finite number of class intervals, such an X is a discrete random variable. Also, when measurements have been rounded to the *nearest* unit, as in height to the nearest inch, this is like forming class intervals, and the random variable standing for such a rounded measurement is often thought of as discrete. In fact, almost any practical situation involving numerical data can be thought of in terms of a discrete random variable.

Probability calculations are often very simple when one is dealing with a discrete random variable where only very few values can occur. For example, take the simple experiment of rolling a pair of fair dice. Each die has six sides, and each side contains one to six spots. Then the sample space $\mathcal{S}$ consists of these possibilities:

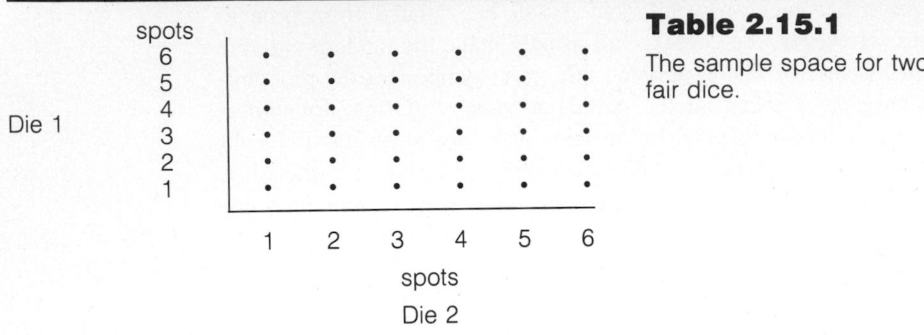

Table 2.15.1

The sample space for two fair dice.

Table 2.15.1 shows the sample space $\mathcal{S}$ for the simple experiment of rolling two fair dice. Each dot is an elementary event. Thus the dot corresponding to 5 for die 1 and 4 for die 2, is the elementary event "die 1 comes up 5 and die 2 comes up 4." You will notice that there are exactly 36 possibilities, or 36 elementary events, in this sample space.

Now let us define the random variable X to be the sum of the spots on the two dice. That is,

$$X = (\text{spots on die 1}) + (\text{spots on die 2}).$$

Table 2.15.2 then gives the value of X associated with each elementary event in $\mathcal{S}$.

Die 1						
6	7	8	9	10	11	12
5	6	7	8	9	10	11
4	5	6	7	8	9	10
3	4	5	6	7	8	9
2	3	4	5	6	7	8
1	2	3	4	5	6	7
Die 2	1	2	3	4	5	6

Table 2.15.2

A random variable X defined on the sample space of Table 2.15.1.

Here, we are assuming that the dice are both fair, so that the probability of each elementary event is the same, or 1/36. Then what are the probabilities of the various values of the random variable associated with these elementary events? Notice that the value "2" is associated with exactly one elementary event, so that its relative frequency, or probability, is 1/36. The value "3" is associated with two elementary events, so that the probability of a "3" is 2/36. The value of "7" has the highest relative frequency, so that the probability of a "7" is 6/36. Proceeding in this way we can work out the following probability distribution of the particular values x of the random variable X, shown in Table 2.15.3.

Table 2.15.3

Probability distribution for the number of spots appearing on two fair dice.

x	p
12	1/36
11	2/36
10	3/36
9	4/36
8	5/36
7	6/36
6	5/36
5	4/36
4	3/36
3	2/36
2	1/36
	36/36

On the basis of this distribution of the random variable X, any number of other questions may be answered. For example, often one is interested in finding the probability that the obtained value of some random variable X will fall *between* two particular values, or in some interval. In this instance, what is the probability that X, the number of spots, is between 3 and 5 inclusive?

In order to find such probabilities, we rely once again on rule 4 of Section 1.5. The various possible values of X are mutually exclusive events, and so

$$p(3 \leqslant X \leqslant 5) = p(3 \cup 4 \cup 5) = p(3) + p(4) + p(5).$$

Using these probabilities, we find

$$p(3 \leqslant X \leqslant 5) = p(X = 3) + p(X = 4) + p(X = 5)$$
$$= \frac{2 + 3 + 4}{36} = \frac{1}{4}.$$

Similarly,

$$p(9 \leqslant X \leqslant 10) = p(X = 9) + p(X = 10) = \frac{7}{36}.$$

In general, the probability that X falls in the interval between any two numbers a and b, inclusive, is found by the sum of probabilities for X over *all possible* values between a and b, inclusive:

$$p(a \leqslant X \leqslant b) = \text{sum of } p(X = c) \text{ for all } c \text{ such that } a \leqslant c \leqslant b.$$

By the same argument,

$$p(a \leqslant X) \text{ sum of } p(X = c) \text{ for all } c \text{ such that } a \leqslant c.$$

Furthermore,

$$p(X < a) = 1 - p(a \leqslant X).$$

For this example,

$$p(5 \leqslant X) = p(X = 5) + \cdots + p(X = 12) = \frac{4}{36} + \cdots + \frac{1}{36} = \frac{30}{36} = \frac{5}{6}$$

On the other hand;

$$p(X < 5) = p(X = 4) + p(X = 3) + p(X = 2) = \frac{3}{36} + \frac{2}{36} + \frac{1}{36} = \frac{1}{6}$$

This example, simple as it is, illustrates the main concepts underlying any discrete random variable. A simple experiment generates a sample space of possible outcomes or elementary events, such as that illustrated in Table 2.15.1. Then, each elementary event is thought of as associated with a number, such as the sum of the number of spots on the dice as in the example. Such numerical values are themselves events, and have probabilities. The value-events, or values of the random variable, can then be displayed along with their probabilities in the form of a probability distribution. The probabilities for other numerical events, such as intervals of values, can be calculated from the distribution.

Notice that this example is purely theoretical. It rests on the theory that the two dice in question are fair, so that each side of each die is equally likely to come up. This need not be true, of course. However, we could use this theoretical distribution of a discrete random variable in order to check on the fairness of an actual pair of dice. Thus, if we toss a real pair of dice 100 times, we expect that (100) (6/36) or about 17 percent of the tosses should show the event "7," and in the same way we could work out the other frequencies we expect, if the dice are fair. A comparison of the frequencies we obtain with the frequencies we expect then gives us evidence about the fairness or unfairness of the dice.

Once again, this is an illustration of very simple statistical inference. We have a question, "Are these particular dice fair, or are they loaded?" That question is to be answered by actually examining the behavior of the pair of dice. The theory of probability tells us the probabilities of the various outcomes for tossing a pair of fair dice, and the idea of a random variable lets us construct the theoretical distribution of values that a pair of fair dice should show, along with their probabilities. These probabilities tell the long-run relative frequencies that we should expect any pair of fair dice to show. For any N actual trials, we also know the expected frequency of any value by taking N times the probability of that value. Thus, we can construct a distribution of expected frequencies from the probability distribution. Now we actually toss the dice N times. For any N we can construct a frequency distribution of the results. A comparison of the frequency distribution we *obtain* with the frequency distribution we *expect* gives evidence in answer to our original question. If the obtained frequencies are very different from those that we expect, then we have evidence that the dice are not fair, and we reject our theory that they are. On the other hand, if the expected and the obtained frequencies are very close, we have no good basis for saying that the dice are anything but fair.

Most of the things to be discussed in succeeding sections of this book are simply ways of formalizing this basic set of procedures, and extending them to more complex situations. Moreover, this general idea will underlie almost all of the methods of statistical inference still to be introduced.

2.16 / GRAPHS OF PROBABILITY DISTRIBUTIONS

The same general procedure is used for graphing either a frequency distribution or the probability distribution of a discrete random variable. The two most common forms are histograms and probability mass functions.

Figure 2.16.1

Relative frequencies for the sum of the spots on two fair dice.

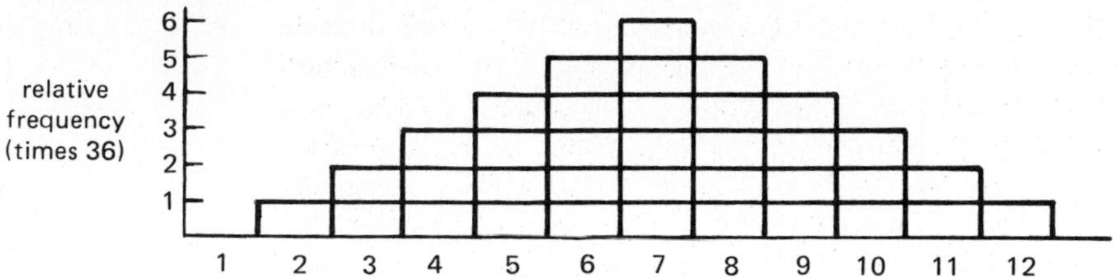

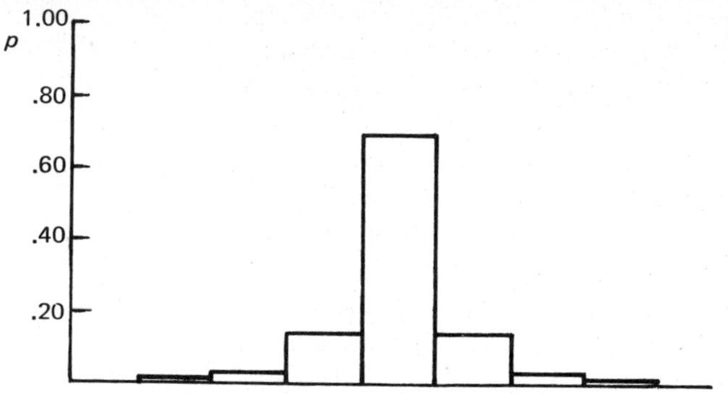

Figure 2.16.2

A hypothetical probability distribution graphed as a histogram.

First, let us deal with the histogram. When the random variable is discrete and can take on only a few values, it is customary to show each value as an interval on the X axis, with the height of the bar above that interval indicating the probability. For example, the distribution in Table 2.15.3 gives the histogram of Figure 2.16.1. Notice that each of the blocks making up the column above a value of X has exactly the same area, and the number of blocks divided by the total number gives the probability. In other words, if each block had area of 1/36, the total area would be 1.00, and the area of any column would be the probability of that value of X.

This idea extends to the histogram of any probability distribution: the total area covered by all columns is regarded as 1.00, and the area of any single column representing an interval or class is thought of as a probability for that interval or class. Another example is given by Fig. 2.16.2.

The use of a histogram for the distribution of a discrete random variable emphasizes the analogy between probabilities and relative areas, with total probability corresponding to total area in the graph. The values on the X axis to either side of any midpoint of an interval are treated as though they could occur with the same probability as the midpoint itself. This is, of course, not generally the case with a discrete random variable, although this convenient fiction is adopted in order that the connection between areas and probabilities can be seen.

An alternate method of graphing the probability distribution for a discrete random variable emphasizes the discreteness of the random variable at the expense of the

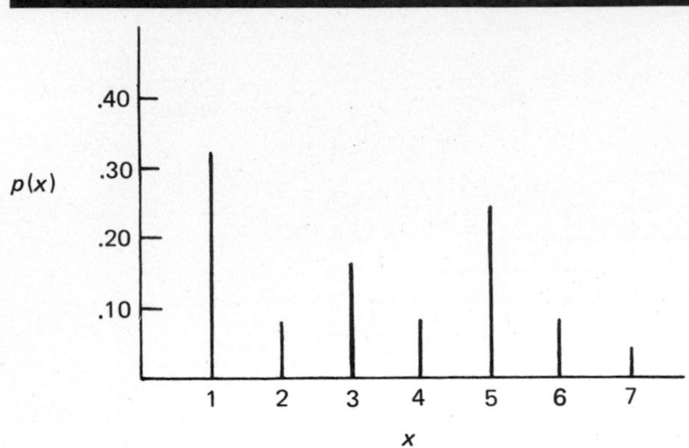

Figure 2.16.3

Probability mass function for a discrete random variable.

analogy between probability and area. This alternate form of graph is called a **probability mass function.** An example of a probability mass function is shown in Figure 2.16.3.

Note that in the probability mass function graphed in Figure 2.16.3 the same basic features exist as in a histogram. The X axis of the graph is marked off with the values that the random variable can actually assume. The Y axis, which may or may not be explicitly shown, is a measure of units of probability. Then above each possible value x of X a vertical line is raised to the height corresponding to $p(X = x)$. The discreteness of the random variable is emphasized by the gaps between the vertical lines standing for the various possible values of X. A value corresponding to a point in one of these gaps has zero probability.

The probabilities, or masses, in any such probability mass function must be nonnegative, of course, and the sum of the probabilities over all possible values of X must be equal to 1.00.

2.17 / FUNCTION RULES FOR DISCRETE RANDOM VARIABLES

Recall that a probability function is simply a way of pairing each of a set of events with its probability. When the events are themselves numerical values, we call the function a random variable. Random variables can be specified in at least three ways: by listing each possible numerical event along with its probability, by graphing this relationship, and by stating a rule that tells how to find the probability for each possible value.

In many instances, particularly in theoretical statistics, it will be much more convenient to specify the distribution of a discrete random variable by its rule, rather than by a simple listing or by a graph. Some very simple examples will show the form that such function rules often take. Later we will deal with much more complicated function rules, but the general ideas presented here will still obtain.

Suppose that we are interested in a discrete random variable X, which can take on only one of 6 different values on any occurrence, the values $1, 2, \cdots, 6$. Furthermore, the probability for the occurrence of any particular value from among these 6 possible values is exactly the same as for any other. How would one symbolize the rule for this random variable? The answer is

$$p(x) = \begin{cases} 1/6 & \text{if } x = 1, 2, 3, \cdots, 6, \\ 0 & \text{otherwise.} \end{cases}$$

You can check for yourself to see that the requirements of a probability function are met for this random variable: the probability of any value is nonnegative, and the sum of probabilities over all possible values is 1.00. Figure 2.17.1 shows the probability mass function for this discrete random variable.

As another example, consider the discrete random variable X, which, once again, can assume only the values 1, 2, 3, 4, 5, and 6. This time, however, the probability function for the random variable obeys the following rule:

$$p(x) = \begin{cases} x/21 & \text{if } x = 1, 2, 3, 4, 5, 6, \\ 0 & \text{otherwise.} \end{cases}$$

Thus $p(X = 1) = 1/21$, $p(X = 2) = 2/21$, and so on. The probability mass function is graphed as Figure 2.17.2.

Notice that the sum of the probabilities over all of the different values of the random variable is

$$\frac{1 + 2 + 3 + 4 + 5 + 6}{21} = \frac{21}{21} = 1.00$$

just as it should be for any discrete random variable.

Finally, once again consider a random variable X that can take on the values 1, 2, 3, 4, 5, and 6. This time, however, let the function rule be as follows:

$$p(x) = \begin{cases} x/12 & \text{if } x = 1, 2, 3, \\ (7 - x)/12 & \text{if } x = 4, 5, 6, \\ 0 & \text{otherwise.} \end{cases}$$

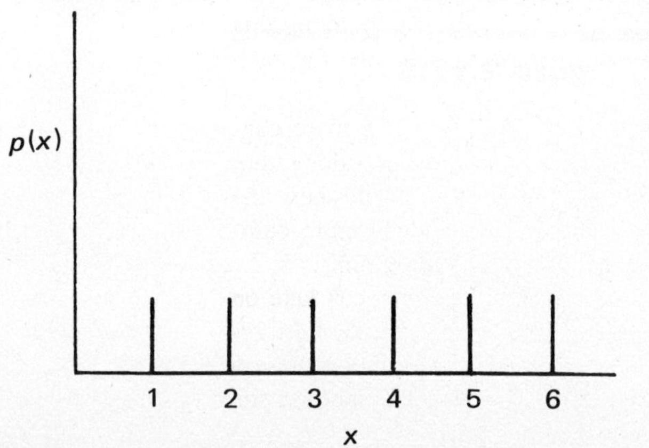

Figure 2.17.1

Probability mass function for a discrete random variable, where $p(x) = 1/6$, for $x = 1, 2, \cdots 6$.

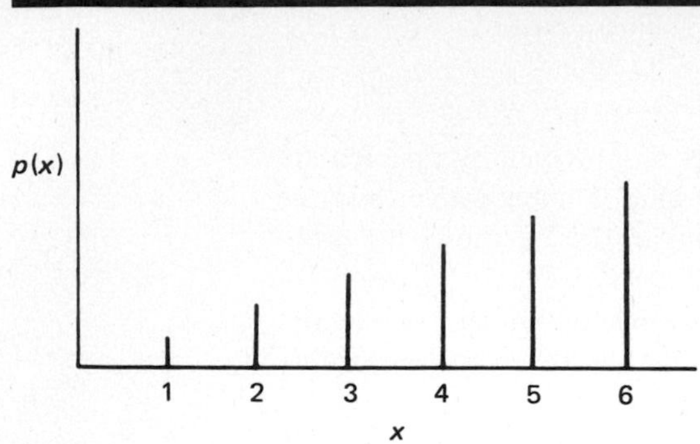

Figure 2.17.2

Probability mass function where $p(x) = x/21$ for $x = 1, 2, 3, 4, 5, 6$

This rule gives the following probabilities to values of the random variable:

$p(X = 1) = 1/12$
$p(X = 2) = 2/12$
$p(X = 3) = 3/12$
$p(X = 4) = (7 - 4)/12 = 3/12$
$p(X = 5) = (7 - 5)/12 = 2/12$
$p(X = 6) = (7 - 6)/12 = 1/12$

Once again, as they should, these probabilities sum to 1.00. This probability mass function is shown as Figure 2.17.3. Note how this rule produces a very different picture from those given by the first two rules.

These are among the simplest of all function rules for discrete random variables, of course. Those of ultimate interest for us will tend to be more elaborate. However, even these simple examples show how hypothetical random variables can be described in terms of their rules, even when the rules are really quite arbitrary, and completely hypothetical, as these examples happen to be. Even though the possible function rules for discrete random variables are limitless in their variety, they all show the same basic features. These are rules for pairing with each possible value of the random variable, $X = x$, a probability $p(X = x)$ that the random variable

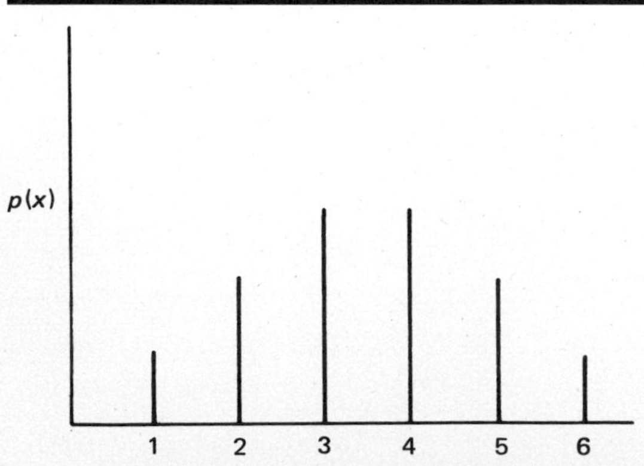

Figure 2.17.3

Probability mass function where $p(x) = x/12$ for $x = 1, 2, 3$ and $p(x) = (7 - x)/12$ for $x = 4, 5, 6$.

takes on that value. Each probability provided by the rule is nonnegative, and the sum of the probabilities over all possible values of X must be equal to 1.00.

2.18 / CONTINUOUS RANDOM VARIABLES

As we have just seen, the graph for a discrete random variable shows sudden jumps or breaks in the plot of the values and their probabilities. However, the idea of a continuous random variable should suggest a smooth curve. Moreover, the distribution of a continuous random variable has a slightly different interpretation from that of a discrete variable, and thus demands a somewhat different terminology.

We can set the stage for the discussion of continuous random variables by considering a variable X whose values are grouped into intervals. Consider once again the examples of heights from Section 2.13. In Figure 2.18.1 each interval of values is shown as the corresponding interval on the X axis. If the total area of the histogram is equal to 1.00, as it should be, then the area of any column is the probability of the corresponding interval.

Now suppose that we were able to measure height to any degree of precision, regardless of how many decimal places this might involve. In other words, suppose that our measurements, and the population being measured, were such that we could choose any class interval size i for the random variable X, and still have a nonzero probability that values in the interval would occur, so long as the interval lay between two reasonable extreme values, say u and v. In this instance we could arrange measurements of height according to a scheme of class intervals of any size, however small, and still have a chance of observing a case in any interval.

For an interval size smaller than the $i = 5$ used above, say $i = 1$, our histogram might look like that shown in Figure 2.18.2. Here the number of intervals is larger and the histogram area for each is smaller. Let us use the symbol ΔX to stand for a

Figure 2.18.1

Probability distribution of height in inches for a specific population of boys.

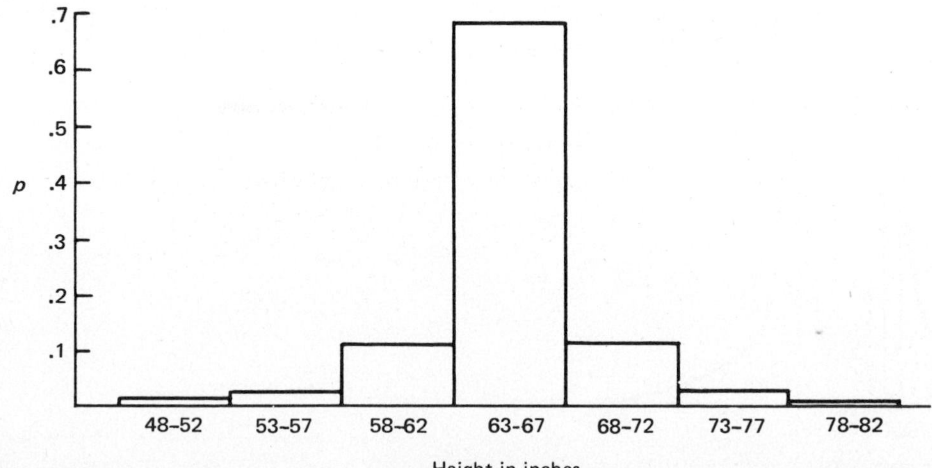

very small positive number, representing a very small difference in the X values. Then suppose that the class interval size were made *extremely* small, $i = \Delta X$, still under the assumption that any class interval within the limits u and v will have a probability greater than zero. In this circumstance each class interval will have an extremely small corresponding area, and the area for the histogram as a whole would be very nearly the same as enclosed in a smooth curve (Fig. 2.18.3). The width of the bar shown on the graph is a representative class interval. Notice that the area cut off by the interval with width ΔX under the curve is almost the same as the area of the histogram bar itself; that is, the area under the curve for that interval is almost the same as the probability.

Finally, imagine that the class-interval size is made *very nearly zero*. As the class-interval size ΔX approaches zero, the probability associated with any class interval and the area cut off by the interval under the curve should agree increasingly well. A random variable that can be represented in terms of arbitrarily small class intervals of size ΔX, with the probability of any interval corresponding *exactly* to the area cut off by the interval under a smooth curve, is called **continuous.** Notice, however, that as the class-interval size ΔX approaches zero, the probability associated with any class interval must also approach zero, since the corresponding area under the

Figures 2.18.2/2.18.3

The approximation of a histogram by a smooth curve, when the number of cases is very large, and the interval size is made very small.

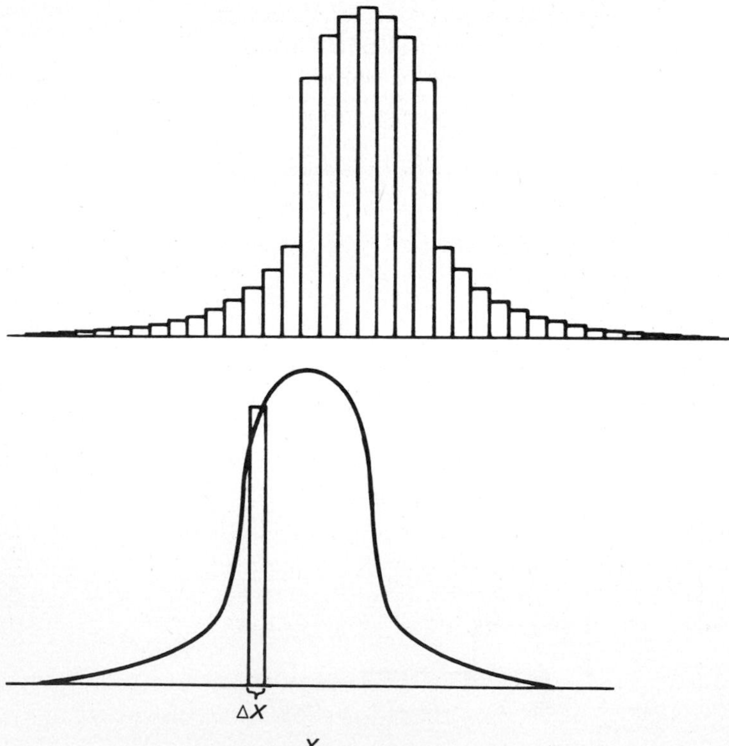

curve is steadily reduced. Next we will discuss some of the ramifications of this property of a continuous random variable.

In Section 2.15 it was possible for us to discuss discrete random variables in terms of probabilities $p(x)$, the probability that the random variable takes on exactly some value. However, the fact we have just illustrated, that the area in any interval under a smooth curve approaches zero when interval size goes toward zero, makes it necessary to discuss continuous random variables in a special way.

Consider the example of an infinite set of men weighed with any degree of accuracy. What is the probability that a man will be observed at random weighing *exactly* 160 pounds? This event is an interval of values smaller than 159.5–160.5, or 159.95–160.05, or any other interval $160 \pm .5i$, where i is not exactly zero. Since the smaller the interval, the smaller the probability, the probability of exactly 160 pounds is, in effect, zero. **In general, for continuous random variables, the occurrence of any exact value of X may be regarded as having zero probability.**

For this reason, one does not usually discuss the probability per se for a value of a continuous random variable; probabilities are discussed only for *intervals* of X values in a continuous distribution. Instead of the probability that X takes on some value a, we deal with so-called **probability density** of X at a, symbolized by

$$f(a) = \text{probability density of } X \text{ at } a.$$

(Notice that this use of the symbol $f(a)$ to denote a probability density is different from the use of the letter "f" alone, which stands for "frequency.")

Loosely speaking, one can say that the probability density at a is the *rate of change* in the probability of an interval with lower limit a, for minute changes in the size of the interval. This rate of change will depend on two things: the function rule assigning probabilities to intervals such as $(X \leq a)$ and the particular "region" of X values we happen to be talking about.

Rather than talk about probabilities of X values per se, for continuous random variables it is mathematically far more convenient to discuss the probability density and reserve the idea of probability only for the discussion of *intervals* of X values. This need not trouble us especially, as usually we will be interested only in intervals of values in the first place. We will even continue to speak loosely of the probabilities for different X values. However, distribution functions plotted as smooth curves are really plots of probability *densities,* and the only probabilities represented are the areas cut off by nonzero intervals under these curves. Furthermore, when we come to look at a function rule for a continuous random variable, we will actually be looking at the rule for a density function.

If the random variable is discrete, then we can say that

$$p(X = a) = f(a) = \text{the probability density of } X \text{ at value } a.$$

The terms "probability at value x" and "density at value x" can be used as though they were synonomous in the case of discrete random variables. On the other hand, these terms are not interchangeable in the case of continuous random variables. For a continuous random variable probabilities are defined only for intervals of values: e.g., $p(X \leq a)$.

Although probability densities such as $f(x)$ are not probabilities, **intervals of values can always be assigned probabilities,** regardless of whether the variable is discrete or continuous.

We have already seen that for discrete variables, the probability of an interval, say $a \leqslant X \leqslant b$, is simply the sum of the probabilities for all values of X such that $a \leqslant X$ and $X \leqslant b$. For continuous random variables, the probability of any interval depends on the probability density associated with each value in the interval. The probability of any continuous interval is given by

$$p(a \leqslant X \leqslant b) = \int_a^b f(x)\, dx.$$

The mathematically unsophisticated reader need not worry over the symbolism used here; **it suffices to say that the probability of an interval is the same as the area cut off by that interval under the curve for the probability densities, when the random variable is continuous, and the total area is equal to 1.00.** The expression $f(x)$ may be thought of as the height, and dx can be thought of as the width, of the area created by a minute interval with midpoint $X = x$, somewhere between a and b. When the number of such intervals approaches an infinite number and their size approaches zero, the sum of all these areas is the *entire* area cut off by the limits a and b. Since there is an infinite number of such intervals, this sum is expressed by the definite integral sign $\int_a^b$. *This agrees in general form with the definition of the probability of an interval for a discrete variable, which also is a sum, though of probabilities rather than probability densities. Furthermore, in a histogram for a discrete variable, the area in an interval is a sum of areas, and this too is like the summing of areas under a smooth curve, yielding the probability of an interval for a continuous variable.*

A continuous distribution of a random variable is a theoretical state of affairs. Continuous distributions can never be observed, but are only idealizations. In the first place, no set of potential observations is actually infinite in number; the mortal scientist can deal only with finite numbers of real things, and so obtained data distributions are always discrete. Regardless of the size of the sample, it will never be large enough to permit each of the possible real numbers in any interval to be observed as values of the random variable. In the second place, measurements are imprecise. Not even in the most precise work known can accuracy be obtained to *any* number of decimals. This puts a limit to the actual possibility of encountering a continuous distribution in practice.

Why, then, do statisticians so often deal with these idealizations? The answer is that, mathematically, continuous distributions are far more tractable than discrete distributions. The function rules for continuous distributions are relatively easy to state and to study using the full power of mathematical analysis. This is not usually true for discrete distributions. On the other hand, continuous distributions are very good approximations to many truly discrete distributions. This fact makes it possible to organize statistical theory about a few such idealized distributions and find methods that are good approximations to results for the more complicated discrete situations. Nevertheless, the student should realize that these continuous distributions are mathematical abstractions that happen to be quite useful; they do not necessarily describe "real" situations.

2.19 / CUMULATIVE DISTRIBUTION FUNCTIONS

As we have seen, the distribution of a discrete random variable may be represented by a histogram or by a probability mass function, and that of a continuous random variable by the smooth curve of its probability density function. There is also another important way to describe the distribution of a random variable. This is through use of the so-called "cumulative distribution function." Any frequency distribution can be converted into a cumulative frequency distribution, which shows the relation between a class interval and the frequency of cases falling at or below the interval's upper limit. In much the same way, a probability or density function can be converted into a cumulative probability distribution. A cumulative probability distribution shows the relation between the possible values a of a random variable X, and the probability that the value of X is less than or equal to a. That is, the cumulative probability distribution is a function relating all possible values a to the probabilities $p(X \leqslant a)$.

The probability that a random variable X takes on a value less than or equal to some particular value a is often written as $F(a)$:

$$F(a) = p(X \leqslant a).$$

The symbol $F(a)$ denotes the particular probability for the interval $X \leqslant a$; the general symbol $F(X)$ is sometimes used to represent the existence of the function relating the various values of X to the corresponding *cumulative* probabilities.

The probability that a continuous random variable takes on any value between limits a and b can be found from

$$p(a \leqslant X \leqslant b) = F(b) - F(a).$$

This is seen easily if it is recalled that $F(b)$ is the probability that X takes on value b or below, $F(a)$ is the probability that X takes on value a or below; their difference must be the probability of a value between a and b.

In the inequalities such as $a \leqslant X \leqslant b$, the "less than or equal to" sign is used. These could just as well be written with "less than" signs alone, however, and the statements would still be true for *a continuous* distribution. The reason is that for a continuous random variable the probability that X equals any exact number is regarded as zero, and thus the probabilities remain the same whether or not a or b or both are considered inside or outside of the interval. However, for discrete variables $<$ and $\leqslant$ signs may lead to different probabilities.

The symbol $F(a)$ can be used to represent the cumulative probability that X is less than or equal to a either for a continuous or a discrete random variable. All random variables have cumulative distribution functions. Occasionally in mathematical statistics the terms "distribution function" or "probability function" are used to refer only to the *cumulative* distribution of a random variable, and "density function" is used where we have used "probability distribution." However, the author believes the terms used here are simpler for the beginning student to learn and use.

Given the probabilities associated with the possible values of a discrete random variable, it is merely a matter of addition to find the $F(a)$ values.

As an example of a cumulative distribution function consider a discrete random variable with the following rule:

$$p(x) = \begin{cases} 1/4 & \text{for } x = 0, \\ 1/2 & x = 1, \\ 1/4 & x = 2, \\ 0 & \text{otherwise.} \end{cases}$$

Now to determine the cumulative distribution function, we need only the values $F(0)$, $F(1)$, and $F(2)$, since these are the only values of X with nonzero probabilities. For $X < 0$, the probability is zero, of course, and for $X = 0$, the probability is 1/4. Hence

$$F(0) = 0 + 1/4 = 1/4.$$

For $F(1)$ we have

$$F(1) = F(0) + p(1) = 1/4 + 1/2 = 3/4.$$

For $F(2)$, we find

$$F(2) = F(1) + p(2) = 3/4 + 1/4 = 1.00.$$

For any value of X above 2, the value of $F(x) = 1.00$.

The cumulative distribution function for this random variable is now fully determined. We can plot this function as in Figure 2.19.1. Note that in this plot the same $F(x)$ value occurs for all points in the interval $0 \leq X < 1$, the same value holds for all points in the interval $1 \leq X < 2$, and so on. This gives the graph the distinctive appearance of a set of plateaus, or "steps," and for this reason such a graph is called a "step function." Graphing the cumulative distribution function of any discrete variable will produce such a step function.

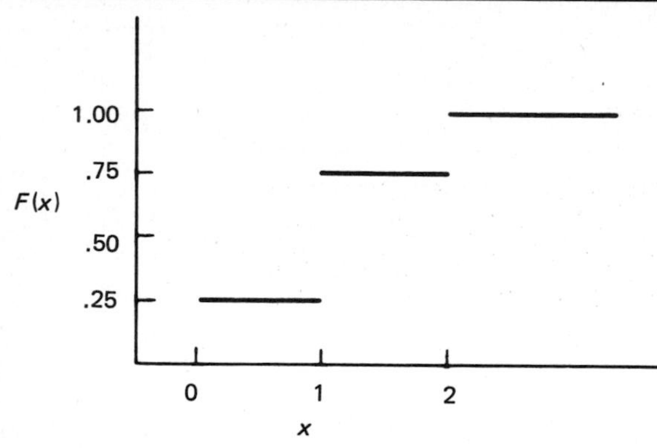

Figure 2.19.1
A step function, showing the cumulative probabilities for values of a discrete random variable.

2.20 / GRAPHIC REPRESENTATIONS OF CONTINUOUS DISTRIBUTIONS

A continuous probability distribution is always represented as a smooth curve erected over the horizontal axis, which itself represents the possible values of the random variable X. A curve for some hypothetical distribution is shown in Figure 2.20.1. The two points a and b marked off in the horizontal axis represent limits to some interval. The shaded portion between a and b shows

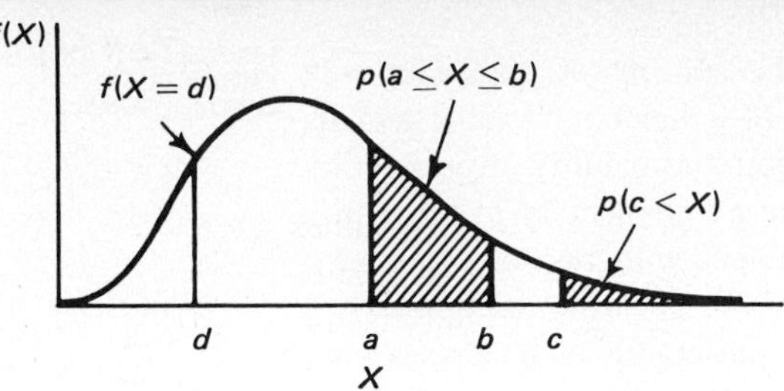

Figure 2.20.1

Probability densities, areas, and probabilities in a continuous distribution.

$$p(a \leqslant X \leqslant b),$$

the probability that X takes on a value between the limits a and b. Remember that this probability corresponds to an *area* under the curve: in any continuous distribution, the probability of an interval can be represented by an area under the distribution curve. The total area under the curve represents 1.00, the probability that X takes on *some* value.

In Figure 2.20.1 the shaded portion of the curve to the right of the point c is the probability $p(c < X)$. This is found by first taking

$$F(c) = p(X \leqslant c).$$

Since the total probability is 1.00, and since $(X \leqslant c)$ and $(c < X)$ are mutually exclusive events, then

$$p(c < X) = 1.00 - F(c).$$

Cumulative distribution curves often have more or less the characteristic **S** shape shown in Figure 2.20.2. Here, the horizontal axis shows possible values of X; for any point on the axis, a, the *height* of the curve, $F(a)$, is the probability that X is less than or equal to a.

Tables of theoretical probability distributions are most often given in terms of the cumulative distribution or $F(X)$, for a random variable. Given these cumulative probabilities for various values of X it is easy to find the probability for any interval by subtraction. When we come to use the table of the so-called "normal" distribution, we will find that cumulative probabilities are shown.

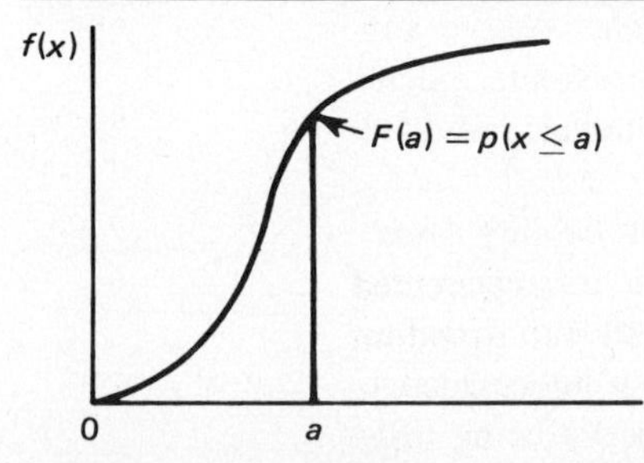

Figure 2.20.2

A continuous cumulative function, with $F(a)$ value.

2.21 / JOINT DISTRIBUTIONS OF RANDOM VARIABLES, AND APPENDIX C

Just as the basic idea of an event was extended to the concept of joint events in Chapter 1, it is perfectly possible to extend the idea of a random variable and its distribution to that of *joint random variables* having a joint probability or probability density function. This means that instead of identifying an elementary event with a single numerical value, we identify each elementary event with two or more such values, each representing a random variable.

The theory of joint random variables, both discrete and continuous, plays a very large role in the theory of statistics. On a number of occasions in future sections we will have to draw on principles from this theory. A few of the elementary ideas about joint random variables are given in the first section of Appendix C. Since we will be appealing to one or another of these principles from time to time in sections to come, the student may at this point find it worthwhile to start becoming familiar with the materials in the first section of this appendix. Most of it is a direct translation of the ideas of single-variable distributions to distributions involving joint variables.

2.22 / FREQUENCY AND PROBABILITY DISTRIBUTIONS IN USE

Both frequency and probability distributions exhibit the same basic features: each is a statement of a relation between the various possible measurement classes into which observations may fall and a number attached to each class. In the case of frequency distributions, this number is the frequency of cases actually observed in a particular class or class interval. For probability distributions, each class or interval is accompanied by its probability. In future chapters we will see that the same language is applied in the further summarization of either kind of distribution.

However, there is an important difference between frequency distributions of data and probability distributions. The latter almost invariably represent *hypothetical* or *theoretical* distributions. A probability distribution is some idealization of the way things might be if we only had all of the information. The frequency distribution represents what we have actually seen to be true from some limited number of observations.

The connection between frequency and probability distributions should be obvious by now: the probability distribution dictates what we should *expect* to observe in a frequency distribution, if the given state of affairs is true. Thus, if theoretically the random variable X has probability of .30 of taking on a value in the interval between 100 and 102, then given a frequency distribution of 500 sample observations we should expect the class interval 100–102 to contain (500) (.30) or 150 cases. This does not mean that this will be true for any given sample; we *might* observe any frequency between 0 and 500. Nevertheless, in the long run, if we sample indefinitely, 30 percent of all cases sampled should show scores in that interval, provided that our hypothetical probability distribution is right.

This is the reason for the parallel development of frequency and probability distributions carried on here. Each form of distribution plays a role in inference: the probability distribution specifies what to expect when the data are put into distribution form, and the obtained frequency distribution supplies us with our best evidence about the probability distribution (the *relative* frequencies of occurrence being estimates of the true probabilities). This parallel discussion will be continued in other chapters. However, the immediate goal is to show how theoretical probability distributions can be constructed. In order to do this we must pick up a few more rudiments of probability theory. These will be presented in the next chapter.

EXERCISES

1. Given data of the following types, state the scale of measurement which each type appears most clearly to represent.

- **(a)** The nationality of an individual's male parent.
- **(b)** Hand pressure as applied to a flexible bulb (i.e., on a dynamometer).
- **(c)** Memory ability, as measured by the number of words recalled from an initially memorized list.
- **(d)** The excellence of baseball teams, as determined by their won-lost records at the end of the season.
- **(e)** The time to clotting of samples of human blood.
- **(f)** Air distance between New York and other cities in the United States.
- **(g)** Time as measured before and after the discovery of the New World by Columbus in 1492.
- **(h)** Reading ability of fifth-grade children, as shown by their test performance relative to a national norm group.
- **(i)** U.S. Department of Agriculture classifications of fresh meats (''choice,'' etc.).
- **(j)** Social Security numbers.

2. For which sorts of measurement scales are the following manipulations most appropriate:

- **(a)** Taking the arithmetic difference between two values.
- **(b)** Changing the labels assigned to a set of objects to another arbitrarily chosen set of labels.
- **(c)** Stating that one one value represents a higher level of some property than does another value.
- **(d)** Taking the ratio of two values.
- **(e)** Taking the ratio of two differences between values.
- **(f)** Multiplying each value by a constant and then adding a constant to each.

3. A well-known magazine provides evaluative reports to consumers on the quality of various products. In a report on television sets of different brands, the ratings were as follows:

Brand	*Rating*
A	Acceptable
B	Good
C	Good
D	Acceptable
E	Poor
F	Acceptable
G	Acceptable
H	Poor
I	Very Good
J	Good
K	Acceptable
L	Acceptable
M	Poor
N	Very Good
O	Acceptable
P	Poor

Construct a frequency table displaying these data.

4. The mental health clinic of a university uses the following codes for the major types of problems that bring students in for assistance.

(a) General anxiety.
(b) General depression.
(c) Sexually related problems.
(d) Alcohol or drug-related problems.
(e) Problems of social adjustment.
(f) Family problems.
(g) Other problems.

During one day, fifty-four students visited the clinic, and the classifications used (one per student) were as follows:

A	B	B	E	B	D
B	C	E	B	B	A
C	F	D	G	G	D
G	A	G	A	F	G
B	B	B	G	G	G
G	G	B	D	B	B
F	B	G	C	C	F
E	G	B	G	G	B
B	B	D	G	A	B

Construct a frequency distribution showing the numbers receiving each code.

5. A questionnaire consisted of only three items. On each item the respondent was asked to circle one of the numbers from 1 through 5, where "1" indicated "strongly agree," "2" indicated "agree somewhat," "3" was "neither agree nor disagree," "4" stood for "disagree somewhat," and "5" represented "disagree strongly." The responses for thirty people are listed below for all three items. Form a frequency distribution of responses to item I, and then do the same thing for items II and III. On the basis of three distributions, what would one tend to conclude about the responses of these subjects?

Individual	Item I	Item II	Item III
1	1	2	2
2	2	4	4
3	2	1	1
4	1	3	4
5	3	5	3
6	3	3	3
7	4	2	5
8	1	5	4
9	4	3	3
10	2	4	4
11	4	2	2
12	1	3	4
13	2	4	4
14	5	3	3
15	3	1	1
16	2	3	4
17	2	4	5
18	2	2	2
19	1	4	1

20	3	2	2
21	5	3	5
22	2	3	3
23	5	4	4
24	1	2	5
25	4	3	5
26	3	3	3
27	3	3	5
28	2	4	4
29	4	2	2
30	3	3	3

6. Put each of the distributions found in exercise 5 above into the form of a histogram. Are the histograms of similar appearance? Why?

7. In the calendar year 1979, a group of elderly people had the following incomes. Form a frequency distribution of these data.

3881	3350	3509	3427	3169
3813	3563	3587	3275	3308
2808	3474	3710	3352	2955
3650	3456	3457	3137	2978
3547	3568	3913	3133	3093

(Suggestion: Employ a class-interval size of 121.)

8. For the distribution of exercise 7 above, draw a frequency polygon.

9. On a test of mechanical aptitude, the score is the number of seconds required in order to finish a certain task. A group of 52 students in the second grade received the following scores:

76.3	90.1	57.5	57.8
66.0	92.0	55.0	59.0
80.0	87.4	61.9	55.0
82.0	91.0	59.2	61.1
76.2	94.8	60.0	59.9
80.0	95.0	61.3	60.0
84.7	91.9	74.0	61.7
86.0	94.5	90.4	59.0
80.6	93.2	48.0	52.0
86.5	57.1	57.2	53.0
85.0	53.6	53.5	44.0
87.0	55.7	55.6	51.5
74.0	56.0	56.0	45.0

Construct a frequency distribution table for these data.

10. Suppose that a certain state contains 84 counties. For a given year the numbers of residents by county (shown in thousands) was as follows. Make a frequency distribution for these data, using $i = 10$ thousand. (**Hint:** Allow for one or more open or unequal intervals.)

3.2	415.8	23.5	91.6	19.0	4.1
29.1	22.1	27.2	5.1	40.9	23.8
31.1	57.5	14.1	32.5	24.4	15.6
13.2	42.3	52.2	14.1	32.1	28.0
41.2	104.6	27.7	43.3	1001.2	13.3
31.7	48.1	52.0	12.5	56.0	44.4
6.7	22.3	18.8	47.9	26.0	16.5

55.5	30.5	31.4	7.6	53.4	35.0
26.7	12.2	84.7	25.7	57.1	2.3
11.9	34.6	25.9	40.1	110.2	66.6
22.6	18.3	70.7	27.9	78.2	42.7
8.3	43.3	30.4	11.7	57.2	34.2
58.8	23.6	88.8	28.3	28.1	11.8
40.8	7.2	14.3	75.1	53.7	51.3

11. Plot the distribution of problem 10 as a frequency polygon.

12. Plot the cumulative frequency polygon for the data of exercise 9. Then see if you can estimate the frequencies corresponding to the intervals in **(a)–(c)**. (**Hint:** Treat the cases in any given class interval as though they were evenly spread across that interval.)

(a) $x \leqslant 70$
(b) $83 \leqslant x$
(c) $50 \leqslant x \leqslant 60$
(d) Find the x value that cuts off the bottom 50 percent of cases in the distribution.
(e) Between what two values do the middle 50 percent of cases seem to fall?

13. A student of the history of the English language noted the dates of the first recorded appearance in written English of 1000 words of foreign origin. The dates were as follows:

DATES	FREQUENCY	DATES	FREQUENCY
Before 1050	2	1451–1500	76
1051–1100	2	1501–1550	84
1101–1150	1	1551–1600	91
1151–1200	15	1601–1650	69
1201–1250	64	1651–1700	34
1251–1300	127	1701–1750	24
1301–1350	120	1751–1800	16
1351–1400	180	1801–1850	23
1401–1450	70	1851–1900	2

Plot the cumulative frequency distribution for these data.

14. Suppose that a word were drawn at random from among the 1000 words given in the distribution of exercise 13 above. Find the probability that the following events occur:

(a) The word drawn is one which appeared in English before 1301.
(b) The word drawn appeared in English after 1601.
(c) The word drawn appeared in English between 1501 and 1700 inclusive.
(d) The word drawn appeared in English either before 1201 or after 1801.

15. For the group of students in exercise 4 suppose that one of these these individuals were selected at random. Determine the probability that this individual would represent

(a) Code *A*.
(b) Code *B*.
(c) Code *E* or code *F*.
(d) Any code except *G*.

16. Suppose that ten students had been drawn from the group of exercise 4, with each student being replaced before the next student is drawn. Give the frequencies that one should expect over the seven categories.

17. The following data represent weights, to the nearest pound, of 78 boys in junior high school. Arrange these data into a frequency polygon with $i = 5$.

122	122	110	118	120	111
117	122	107	127	146	113
116	119	108	118	127	116
114	118	153	125	138	126
110	117	148	119	133	113
109	112	134	125	134	106
107	108	140	119	128	118
105	117	108	124	144	115
103	117	126	120	132	119
102	126	103	121	137	133
134	123	136	128	136	148
126	118	112	116	146	137
123	113	118	127	152	124

18 Imagine that the 78 boys in exercise 17 is but a sample from a much larger group of junior high school boys. If this sample had been drawn at random, what is our best guess about the probability of each of the intervals of values as used above?

19. Suppose that a discrete random variable can take on only the values 1, 2, 3, 4, and 5. The rule giving the probability for each of these values is:

$$p(x) = \begin{cases} \dfrac{x^2}{55} & \text{for } x = 1, 2, 3, 4, 5 \\ 0 & \text{otherwise} \end{cases}$$

(a) Plot the probability mass function for this random variable.
(b) Plot the cumulative distribution function (step function) for this variable.
(c) What is the probability that the variable takes on the values 2,3, or 4?
(d) What is the probability that the random variable takes on *some* value?

20. A theoretical random variable can take on only the whole number values between -3 and 3, inclusive. The probability for each value is given by the rule:

$$p(x) = \begin{cases} \dfrac{4 + x}{16} & \text{for } x = -3, -2, -1, 0 \\ \dfrac{4 - x}{16} & \text{for } x = 1, 2, 3 \\ 0 & \text{otherwise} \end{cases}$$

(a) Find the probability mass function for this variable.
(b) Plot the cumulative distribution function, or step function, for this variable.
(c) If 32 independent values of this random variable should be obtained, what should we expect the frequency distribution to be like?

21. A random variable has a probability function defined as follows:

$$f(x) = \begin{cases} \frac{1}{10} & -3 \leq x \leq 1 \\ \frac{2}{5} & -1 < x \leq 0 \\ \frac{1}{5} & 0 < x \leq 1 \\ \frac{3}{10} & 1 < x \leq 3 \\ 0 & \text{elsewhere} \end{cases}$$

(a) Plot this probability function.
(b) Indicate the probability $p(-3 \leq x \leq 1)$.
(c) Is this a continuous distribution? How can one tell?

22. A carnival game uses a large wheel which is spun and which then stops at a number. Suppose that the wheel is so constructed that it can stop at any of the real numbers between 0 and 6, inclusive. Suppose further that each of these numbers has the same likelihood of occuring (i.e., the wheel is ''fair''). Then the number where the wheel stops on any trial is the value of a continuous random variable, with a rule given by:

$$f(x) = \begin{cases} \frac{1}{6} & \text{for } 0 \leq x \leq 6 \\ 0 & \text{otherwise} \end{cases}$$

Graph the distribution of this random variable. Then, use the appropriate areas under this curve to find the following probabilities

(a) $p(x \leq 1.5)$
(b) $p(x \leq 5)$
(c) $p(3 \leq x \leq 4)$
(d) $p(5 \leq x)$

23. In another carnival operation a wheel is used similar to that in problem 23. However, this wheel is biased, with numbers close to 3 much more likely to occur than numbers far from 3. This might be represented by a random variable following the rule:

$$f(x) = \begin{cases} \frac{x(6 - x)}{36} & 0 \leq x \leq 6 \\ 0 & \text{otherwise} \end{cases}$$

(a) Draw a graph of the distribution of this random variable.
(b) On this graph mark off areas corresponding to the probabilities:

$p(x \leq 1.5)$
$p(x \leq 5)$
$p(3 \leq x \leq 4)$
$p(5 \leq x)$.

24. Suppose that you really observed a wheel at a carnival, which might be working in the way described in exercise 22, or in the manner of exercise 23. Describe how you might be able to decide which of these two models of the wheel's behavior seems to provide the best fit to what is actually observed..

25. Consider the following cumulative distribution function for the discrete random variable X:

a	$p(X \leq a)$
20	1.00
18	.98
16	.83
14	.74
12	.63
10	.52
8	.37
6	.21
4	.10
2	.03
0	0

Suppose that 48 observations were made, independently and at random, from the process generating this random variable. Construct the frequency distribution of observations which one should *expect* to obtain.

Chapter 3

A DISCRETE RANDOM VARIABLE: THE BINOMIAL

The main burden of this chapter is to show how the theoretical distribution of a discrete random variable can be constructed. We will develop the distribution known as the "binomial," which will play a very important role in much that follows. However, first of all we will need ways to make elementary probability calculations and to know some "counting rules" that underlie the development to follow.

3.1 / COMPUTING PROBABILITIES

As we have seen, almost any simple problem in probability can be reduced to a problem in counting. Especially for a sample space containing equally probable elementary events, the computation of probability involves two quantities, both of which are *counts of possibilities:* the total number of elementary events, and the number qualifying for a particular event class. The key to solving probability problems is to learn to ask: "How many distinctly different ways can this event happen?" It may be possible simply to list the number of different elementary events that make up the event-class in question, but it is often much more convenient to use a rule for finding this number. Once the probabilities of events are found, the rules of Chapter 1 permit deductions of other probabilities.

There is really no way to become expert in probability computations except by practice in the application of these various counting procedures. Naturally, problems differ widely in the particular principles they involve, but most problems should be approached by these steps:

1. Determine exactly the sample space of elementary events with which this problem deals. Draw a graph or picture such as Table 2.15.1 if necessary.
2. Find out how many elementary events make up the sample space. What are all the distinct outcomes that might conceivably occur? If the elementary events are equally probable, then the probability of any single elementary event is one over the total number of elementary events.
3. Decide on the particular events for which probabilities are to be found. How many elementary events qualify for each event class? If no other counting method is available, *list* the elements of these different events and count them. Be sure to remember that if event A can occur in $n(A)$ ways and an independent event B in $n(B)$ ways, then event $(A \cap B)$ might occur in $n(A)n(B)$ ways.
4. For equally probable elementary events the probability of any event A is simply the ratio of the number of members of A to the total number of elementary events. Even though elementary events are not equally probable, we can still use the fact that elementary events are mutually exclusive to find the probability of any event A by taking the *sum* of the probabilities of all members of that set.

A game of chance such as roulette shows how probabilities may be computed simply by listing elementary events. A standard roulette wheel has 37 equally spaced slots into which a ball may come to rest after the wheel is spun. These slots are numbered from 0 through 36. One half the numbers 1 through 36 are red, the others are black, and the zero is generally green.

In playing roulette, you may bet on any single number, on certain groups of numbers, on colors, etc. One such bet might be on the odd numbers (excluding zero of course). The only elementary events in the sample space are the numbers 0 through 36 with their respective colors. Which numbers qualify for the event "odd"? It is easy to count these numbers: they form the set

$$\{1,3,5,7,9,11,13,15,17,19,21,23,25,27,29,31,33,35\}.$$

Since there are exactly 18 elementary events qualifying as odd numbers, if each slot on the roulette wheel is equally likely to receive the ball, the probability of "odd" is 18/37, slightly less than 1/2.

As a more complicated example that can be solved by listing, let us find the probability that the number contains a "3" as one of the digits. The elementary events qualifying are:

$$\{3,13,23,30,31,32,33,34,35,36\}.$$

Here the probability is 10/37, if the 37 different elementary events are equally likely.

As a final example, consider the probability of a number that is "even and red" on the roulette wheel. These are

$$\{12,14,16,18,30,32,34,36\}$$

and the probability is 8/37.

There is no end to the examples that might be brought in at this point to show this counting principle in operation, but there is little point in spending more time with it. Just remember that almost all simple problems in probability can be solved in this same way. The important thing to understand is, "When in doubt, make a list." This will usually work, although, as we shall see, other counting methods are often more efficient.

3.2 / SEQUENCES OF EVENTS

In many problems, an elementary event may be *the set of outcomes of a series or sequence of observations*. Such sequences can also be considered joint events, but it is convenient to regard an entire sequence as an elementary event. Suppose that any trial of some simple experiment must result in one of K mutually exclusive and exhaustive events, $\{A_1, \cdots, A_K\}$. Now the experiment is *repeated* N times. This leads to a sequence of events: the outcome of the first trial, the outcome of the second, and so on in order through the outcome of the Nth trial. The outcome of the whole series of trials might be the sequence $(A_3, A_1, A_2, \cdots, A_3)$. This denotes that the event A_3 occurred on the first trial, A_1 on the second trial, A_2 on the third, and so on. The place in order gives the trial on which the event occurred, and the symbol occupying that place shows the event occurring for that trial. For example, it might be that the simple experiment is drawing one marble at a time from a box, with replacement after each drawing. The marbles observed may be red, white, or black. For four drawings a sequence of outcomes might be (R, W, B, W), or perhaps (W, W, B, B).

Remember that each possible sequence can be thought of in two ways; it is a joint event from the standpoint of a series of simple experiments, but it is also an elementary event if one thinks of the *experiment itself* as the series of trials with a sample space consisting of n-tuples, or possible sequences of outcomes. Each *whole* or *compound experiment* has as its outcome one and only one sequence. Such experiments producing sequences as outcomes are very important to statistics, and so we will begin our study of counting rules by finding how many different sequences a given series of trials could produce.

3.3 / COUNTING RULE 1: NUMBER OF POSSIBLE SEQUENCES FOR *N* TRIALS

Suppose that a series of N trials were carried out, and that on each trial any of K events might occur. Then the following rule holds:

Counting rule 1: If any one of K mutually exclusive and exhaustive events can occur on each of N trials, then there are K^N different sequences that may result from a set of trials.

As an example of this rule, consider a coin's being tossed. Each toss can result only in a H or a T event ($K = 2$). Now the coin is tossed five times, so that $N = 5$. The total number of *possible* results of tossing the coin five times is $K^N = 2^5 = 32$ sequences. Exactly the same number is obtained for the possible outcomes of tossing five coins simultaneously, if the coins are thought of as numbered and a sequence describes what happens to coin 1, to coin 2, and so on.

As another example, the outcome of tossing two dice is a sequence: what number comes up on the first die, and what number comes up on the second. Here $K = 6$ (six different numbers per die), and $N = 2$. There are exactly $6^2 = 36$ different sequences possible as results of this experiment, as we saw in Section 2.15.

3.4 / COUNTING RULE 2: FOR SEQUENCES

Sometimes the number of possible events in the first trial of a series is different from the number possible in the second, the second different from the third, and so on. It is obvious that if there are different numbers $K_1, K_2, \ldots, K_N$ of events possible on the respective trials, then the total number of sequences will not be given by rule 1. Instead, the following rule holds:

Counting rule 2: If $K_1, \cdots, K_N$ are the numbers of distinct events that can occur on trials $1, \cdots, N$ in a series, then the number of different sequences of N events that can occur is $(K_1)(K_2)\cdots(K_N)$.

For example, suppose that for the first trial you toss a coin (two possible outcomes) and for the second you roll a die (six possible outcomes). Then the total number of different sequences would be $(2)(6) = 12$.

Notice that counting rule 1 is actually a special case of rule 2. If the same number K of events can occur on any trial, then the total number of sequences is K multipled by itself N times, or K^N. We saw this illustrated in Section 2.15.

3.5 / COUNTING RULE 3: PERMUTATIONS

A rule of extreme importance in probability computations concerns the number of ways that objects may be arranged in order. The rule will be given for arrangements of objects, but it is equally applicable to sequences of events:

Counting rule 3: The number of different ways that N distinct things may be arranged in order is $N! = (1)(2)(3)\cdots(N-1)(N)$, (where $0! = 1$). An arrangement in order is called a permutation, so that the total number of permutations of N objects is $N!$ The symbol $N!$ is called "N factorial."

As an illustration of this rule, suppose that a classroom contained exactly 10 seats for 10 students. How many ways could the students be assigned to the chairs? Any of the students could be put into the first chair, making 10 possibilities for chair 1. But, given the occupancy of chair 1, there are only 9 students for chair 2; the total number of ways chair 1 and chair 2 could be filled is $(10)(9) = 90$ ways. Now consider chair 3. With chairs 1 and 2 occupied, 8 students remain, so that there are $(10)(9)(8) = 720$ ways to fill chairs 1, 2, and 3. Finally, when 9 chairs have been filled there remains only 1 student to fill the remaining place, so that there are

$$(10)(9)(8)(7)(6)(5)(4)(3)(2)(1) = (10)! = 3628800$$

ways of arranging the 10 students into the 10 chairs.

Now suppose that there are only N different events that can be observed in N

trials. Imagine that the occurrence of any given event ''uses up'' that event for the sequence, so that each event may occur *once and only once* in the sequence. In this case there must be $N!$ different orders in which these events might occur in sequence. Each sequence is a permutation of the N possible events.

To take a homely example, suppose that a man is observed dressing. At each point in our observation there are three articles of clothing he might put on: his shoes, his pants, or his shirt. However each article can be put on only once. We might observe the following sequences:

(shoes, pants, shirt)
(shoes, shirt, pants)
(shirt, shoes, pants)
(shirt, pants, shoes)
(pants. shoes, shirt)
(pants, shirt, shoes)

The man's putting on any one of these articles of clothing literally uses up that outcome for any sequence of observations, so that there are $N! = (1)(2)(3) = 6$ possible permutation sequences.

The procedure of counting sequences as permutations in order is especially important for sampling without replacement. For example, suppose that a teacher has the names of five students in a hat. She draws the names out at random one at a time, *without* replacement. Then there are exactly 5! or 120 different sequences of names that she might observe. If all sequences of names are equally likely to be drawn, the the probability for any one sequence is 1/120.

What is the probability that the name of any given child will be drawn first? Given that a certain child is drawn first, the order of the remainder of the sequence is still unspecified. There are $(N - 1)! = 4! = 24$ different orders in which the other children can appear. Thus the probability of any given child being first in sequence is $24/120 = .20$ or 1/5.

Suppose that exactly *two* of the names in the hat are girls'. What is the probability that the first two names drawn belong to the girls? Given the first two names are girls', there are still $3! = 6$ ways three boys may be arranged. The girls' names may themselves be ordered in two ways. Thus, the probability of the girls' names being drawn first and second is $(2)(6)/120 = 1/10$.

3.6 / COUNTING RULE 4: ORDERED COMBINATIONS

Sometimes it is necessary to count the number of ways that r objects might be selected from among some N objects in all, $(r \leq N)$. Furthermore, each different *arrangement* of the r objects is considered separately. Then the following rule applies:

Counting rule 4: The number of ways of selecting and arranging r objects from among N distinct objects is $\frac{N!}{(N - r)!}$.

The reasoning underlying this rule becomes clear if a simple example is taken: Consider a classroom teacher, once again, who has 10 students to be assigned to seats. This time, however, imagine that there are only 5 seats. How many different ways could the teacher select 5 students and arrange them into the available seats? Notice that there are 10 ways that the first seat might be filled, 9 ways for the second, and so on, until seat 5 could be filled in 6 ways. Thus there are

$$(10)(9)(8)(7)(6)$$

ways to select students to fill the 5 seats. This number is equivalent to

$$\frac{10!}{5!} = \frac{N!}{(N - r)!}$$

the number of ways that 10 students out of 10 may be selected and arranged, but divided by the number of arrangements of the 5 *un*selected students in the 5 *missing* seats.

As an example of the use of this principle in probability calculations, consider the following: In lotteries it is usual for the first person whose name is drawn to receive a large amount, the second some smaller amount, and so on, until some r prizes are awarded. This means that some r names are drawn in all, and the *order* in which those names are drawn determines the size of the prizes awarded to individuals. Suppose that in a rather small lottery 40 tickets had been sold, each to a different person, and only 3 were to be drawn for first, second, and third prizes. Here $N = 40$ and $r = 3$. How many different assignments of prizes to persons could there be? The answer, by counting principle 4, is

$$\frac{40!}{(40 - 3)!} = \frac{40!}{37!} = (38)(39)(40) = 59{,}280.$$

On how many of these possible sequences of winners would a given person, John Doe, appear as first, second, *or* third prize winner? If he were drawn first, the number of possible selections for second and third prize would be $39!/37! = 1482$. Similarly, there would be 1482 sequences in which he could appear second, and a like number where he could appear third. Thus, the probability that he appears in a sequence of three drawings, winning first, second, or third prize is (by Rule 4, Section 1.5.)

$$p(\text{first, second, or third}) = \frac{3(1482)}{59{,}280} = \frac{3}{40}.$$

3.7 / COUNTING RULE 5: COMBINATIONS

In a very large class of probability problems, we are not interested in the *order* of events, but only in the number of ways that r things could be selected from among N things, *irrespective of order*. We have just seen that the total number of ways of selecting r things from N and ordering them is $\frac{N!}{(N - r)!}$, by rule 4. Each set of r objects has $r!$ possible orderings, by rule 3. A combination of these two facts gives us

Counting rule 5: The total number of ways of selecting r distinct combinations of N objects, irrespective of order, is

$$\frac{N!}{r!(N-r)!} = \binom{N}{r}.$$

The symbol $\binom{N}{r}$ is *not* a symbol for a fraction, but instead denotes the number of combinations of N things, taken r at a time. Sometimes the number of combinations is known as a "binomial coefficient," and occasionally $\binom{N}{r}$ is replaced by the symbols ${}^{N}C_r$ or ${}_{N}C_r$. However, the name and symbol introduced in rule 5 will be used here.

It is helpful to note that

$$\binom{N}{r} = \binom{N}{N-r}.$$

Thus, $\binom{10}{3} = \binom{10}{7}$, $\binom{50}{49} = \binom{50}{1}$, and so on.

As an example of the use of this rule, suppose that a total of 33 persons were candidates for the board of supervisors in some community. Three supervisors are to be elected at large; how many ways could 3 persons be selected from among these 33 candidates? Here, $r = 3$, $N = 33$, so that

$$\binom{N}{r} = \frac{(33)!}{3!\,(30)!} = \frac{(1 \cdot 2 \cdot \cdots \cdot 32 \cdot 33)}{(1 \cdot 2 \cdot 3)(1 \cdot 2 \cdot \cdots \cdot 29 \cdot 30)}.$$

Canceling in numerator and denominator, we get

$$\binom{33}{3} = \frac{(31)(32)(33)}{6} = (31)(16)(11) = 5456.$$

If all sets of three are equally likely to be chosen, the probability for any given set is 1/5456.

Because of their utility in probability calculations, a table of $\binom{N}{r}$ values for various values of N and r is included in Appendix D. Although this table shows values of N only up to 20, and values of r up to 10, other values can be found by the relation given above, and also by the relation known as Pascal's Rule:

$$\binom{N}{r} = \binom{N-1}{r-1} + \binom{N-1}{r}.$$

Still other values may be worked out from the table of factorials also included in the Appendix D, in terms of rule 5. A great many hand calculators have $x!$ as one of their operations, going up to about 50! or 60!. This is a useful feature, since permutation and combination values can be calculated directly using these operations, along with multiplication and division.

These five rules provide the basis for many calculations of probability when the number of elementary events is finite, especially when sampling is without replacement. One of the easiest examples of their application is the calculation of probabilities for poker hands.

The game of poker as discussed here will be highly simplified and not very exciting to play: the player simply deals five cards to himself from a well-shuffled standard deck of fifty-two cards. Nevertheless, the probabilities of the various hands are interesting, and can be computed quite easily.

The particular hands we will examine are the following:

a. one pair, with three different remaining cards.
b. full house (three of a kind, and one pair)
c. flush (all cards of the same suit).

First of all we need to know how many different *hands* may be drawn. A given hand of five cards will be thought of as an event; notice that the order in which the cards appear in a hand is immaterial, so that a number of sequences of cards can correspond to a given hand. By rule 5, since there are 52 different cards in all and only 5 are selected, then there are

$$\binom{52}{5} = \frac{52!}{5!47!}$$

different hands that might be drawn. If all hands are equally likely, then the probability of any given one is $1 \Big/ \binom{52}{5}$, or about 4 in ten million.

There are 13 numbers that a pair of cards may show (counting the picture cards), and each member of a pair must have a suit. Thus by rule 5 there are $13\binom{4}{2}$ different pairs that might be observed. The remaining cards must show three of the twelve remaining numbers and each of the three cards may be of any suit. By rules 1 and 5 there are $\binom{12}{3}$ (4)(4)(4) ways of filling out the hand in this way. Finally, we find that there are

$$13\binom{4}{2}\binom{12}{3}4^3$$

different ways for this event to occur. Thus,

$$p(\text{one pair}) = \frac{13\binom{4}{2}\binom{12}{3}4^3}{\binom{52}{5}}.$$

This number can be worked out by writing out the factorials, canceling in numerator and denominator, and dividing. It is approximately equal to .42. The chances are roughly four in ten of drawing a single pair in five cards, if all possible hands are equally likely to be drawn.

This same scheme can be followed to find the probability of a full house. There are thirteen numbers that the three of a kind may have, and then twelve numbers possible for the pair. The three of a kind must represent three of four suits, and the pair two of four suits. This gives

$$13\binom{4}{3}\ 12\binom{4}{2}$$

different ways to get a full house. The probability is

$$p(\text{full house}) = \frac{13\binom{4}{3}\ 12\binom{4}{2}}{\binom{52}{5}}$$

which is about .0014.

Finally, a flush is a hand of five cards all of which are in the same suit. There are exacly four suit possibilities, and a selection of five out of thirteen numbers that the cards may show. Hence, the number of different flushes is $4\binom{13}{5}$, and

$$p(\text{flush of five cards}) = \frac{4\binom{13}{5}}{\binom{52}{5}}$$

or about .0019.

The probabilities of the various other hands can be worked out in a similar way. The point of this illustration is that it typifies the use of these counting rules for actually figuring probabilties of complicated events, such as particular poker hands. Naturally, a great deal of practice is usually necessary before one can visualize and carry out probability calculations from "scratch" with any facility. Nevertheless, such probability calculations usually depend upon counting how many ways events of a certain kind can occur, if the elementary events are finite in number.

Now we will turn to a special use of these counting rules in finding the distribution of an important discrete random variable.

3.9 / BERNOULLI TRIALS

The very simplest probability distribution is one with only two event classes. For example, a coin is tossed and one of two events, heads or tails, must occur, each with some probability. Or a normal human being is selected at random and his or her sex recorded: the outcome can be only Male or Female. Such an experiment or process that can eventuate in only one of two outcomes is usually called a Bernoulli trial, and we will call the two event classes and their associated probabilities a Bernoulli process.

In general, one of the two events is called a "success" and the other a "failure" or "nonsuccess." These names serve only to tell the events apart, and are not meant to bear any connotation of "goodness" of the event. In the discussion to follow, the symbol p will stand for the probability of a success, and $q = 1 - p$ for the probability of a failure. Thus, in tossing a fair coin, let a head be a success. Then $p = 1/2$, $q = 1 - p = 1/2$. If the coin is biased, so that heads are twice as likely to come up as tails, then $p = 2/3$, $q = 1/3$. In the following, take care to distinguish between p, standing for the probability of a success, and $p(x)$, standing for the probability of some value of a random variable.

3.10 / SAMPLING FROM A BERNOULLI PROCESS

Suppose that some sample space exists fitting a Bernoulli process. Furthermore, suppose that either we sample independently with replacement, or that an infinite number of elementary events exist, so that for each sample observation out of N trials, p and q are unchanged. When the outcome is generated by the same process on every trial, so that p and q remain constant over the trials, the process is said to be **stationary.** If the p and q values change from trial to trial, as may well be the case in some practical situations, then the Bernoulli process is said to be nonstationary. For the time being, we will confine our attention to stationary processes and to independent trials made from such a process.

Now we proceed to make N independent observations. Let N be, say, 5. How many different sequences of five outcomes could be observed? The answer, by rule 1, is $2^5 = 32$. *However, here it is not necessarily true that all sequences will be equally probable.* The probability of a given sequence depends upon p and q, the probabilities of the two events. Fortunately, since trials are independent, one can compute the probability of any sequence by the application of rule 1.14.1.

We want to find the probability of the particular sequence of events:

$$(S, S, F, F, S)$$

where S stands for a success and F for failure. The probability of first observing an S is p. If the second observation is independent of the first, then by rule 1.14.1,

$$\text{probability of } (S, S) = p \cdot p = p^2.$$

The probability of an F on the third trial is q, so that the probability of (S, S) followed by F is p^2q. In the same way the probability of $(S, S, F, F) = p^2q^2$, and that of the entire sequence is $p^2q^2p = p^3q^2$.

The same argument shows that the probability of the sequence (S, F, F, F, F) is pq^4, that of (S, S, S, S, S) is p^5, of (F, S, S, S, F) is p^3q^2, and so on.

Now if we write out all of the possible sequences and their probabilities, an interesting fact emerges: **the probability of any given sequence of *N* independent Bernoulli trials depends only on the number of successes and *p*, the probability of a success.** That is, regardless of the *order* in which successes and failures occur in a sequence, the probability is

$$p^rq^{N-r} \qquad [3.10.1]$$

where r is the number of successes, and $N - r$ is the number of failures. Suppose that in a sequence of 10 trials, exactly 4 successes occur. Then the probability of that particular sequence is

$$p^4q^6.$$

If $p = 2/3$, then the probability can be worked out from

$$\left(\frac{2}{3}\right)^4\left(\frac{1}{3}\right)^6.$$

The same procedure would be followed for any r successes out of N trials for any p.

For example, if we toss a fair coin ($p = 1/2$) six times, what is the probability of observing three heads followed in order by three tails? The answer is

$$p^3q^3 = \left(\frac{1}{2}\right)^3\left(\frac{1}{2}\right)^3 = \frac{1}{64}.$$

This is also the probability of the sequence (H, T, H, T, H, T), of the sequence (H, T, T, T, H, H), and of any other sequence containing exactly three successes or heads.

The probabilities just found are for *particular sequences,* arrangements of r successes and $N - r$ failures in a certain order. What we have found is that if we want to know the probability of a particular sequence of outcomes of independent Bernoulli trials, that sequence will have the same probability as any other sequence with exactly the same number of successes, given N and p.

In most instances, however, one is not especially interested in particular sequences *in order*. We would like to know probabilities of given numbers of successes *regardless* of order in which they occur. For example, when a coin is tossed five times, there are several sequences of outcomes where exactly two heads occur:

(H, H, T, T, T)
(H, T, H, T, T)
(H, T, T, H, T)
(H, T, T, T, H)
(T, H, T, T, H)
(T, H, H, T, T)
(T, H, T, H, T)
(T, T, H, H, T)
(T, T, H, T, H)
(T, T, T, H, H).

Each and every one of these different sequences must have the same probability, $p^2q,^3$, since each shows exactly two successes and three failures. Notice that there are $\binom{N}{r} = \binom{5}{2} = 10$ different such sequences, exactly as counting rule 5 gives for the number of ways 5 things can be taken 2 at a time.

What we want now is the probability that $r = 2$ successes will occur *regardless* of order. This could be paraphrased as "the probability of the sequence (H, H, T, T, T) *or* the sequence (H, T, H, T, T) *or* any other sequence showing exactly 2 successes in 5 trials.'' Such ''or'' statements for mutually exclusive events recall Axiom 3 of 1.4: whenever A and B are mutually exclusive events, then $p(A \cup B) = p(A) + p(B)$. Thus, the probability of 2 successes in any sequence of 5 trials is

$$p(\text{2 successes in 5 trials}) = p^2q^3 + p^2q^3 + \cdots + p^2q^3 = \binom{5}{2} p^2q^3$$

since each of these sequences has the same probability and there are $\binom{5}{2}$ of them.

Generalizing this idea for any r, N, and p, we have the following principle: **in sampling from a stationary Bernoulli process, with the probability of a success equal to p, the probability of observing exactly r successes in N independent trials is**

$$p(r \text{ successes}; N, p) = \binom{N}{r} p^r q^{N-r}. \qquad [3.10.2]$$

In understanding the basis for this rule, the thing to keep in mind is that $p^r q^{N-r}$ is the probability of *any* of the events consisting of a specific sequence showing exactly r successes out of N trials. Then $\binom{N}{r}$ is the number of such sequence events that qualify for the event ''exactly r successes in N trials.'' It is important to notice that there is an exact correspondence between the binomial coefficients $\binom{N}{r}$ and the number of sequences possible in which exactly r successes occur out of N trials.

An experiment carried out in such a way that N independent trials are made from a stationary Bernoulli process is known as **binomial sampling.** In binomial sampling the value of N is predetermined, and it is the value of r, the number of successes, that is left to chance.

For example, imagine that in some very large population of animals 80 percent of the individuals have normal coloration, and only 20 percent are albino (no skin and hair pigmentation). This may be regarded as a Bernoulli process, with ''albino'' being a success and ''normal'' a failure. Suppose that the probabilities are $.20 = p$ and $.80 = q$. A biologist manages to sample this population at random, catching three animals. What is the probability of catching one albino? Here, $N = 3$, $r = 1$, so that

$$p(\text{1 albino in 3 animals}) = \binom{3}{1} (.20)^1(.80)^2$$
$$= .384.$$

If the sampling is random, and if the population is so large that sampling without replacement still permits one to regard the results of the successive trials as independent, then the biologist has about 38 chances in 100 of observing exactly 1 albino in the sample of 3 animals.

3.11 / NUMBER OF SUCCESSES AS A RANDOM VARIABLE: THE BINOMIAL DISTRIBUTION

When samples of N trials are taken from a Bernoulli process, the number of successes is a discrete random variable. Since the various values are counts of successes out of N observations the random variable can take on only the whole values from 0 through N. We have just seen how the probability for any given number of successes can be found. Now we will discuss the distribution pairing each possible number of successes with its probability. This distribution of number of successes in N trials is called the **binomial distribution.** This is the first distribution we have studied that can most readily be described by its mathematical rule.

The general definition of the binomial distribution can be stated as follows:

Any random variable X with probability function given by

$$p(X = r; N,p) = \binom{N}{r} p^r q^{N-r}, \qquad X = 0, 1, \cdots, N, \qquad [3.11.1]$$

is said to have a binomial distribution with parameters N and p.

The unspecified mathematical constants such as N and p that enter into the rules for probability or density functions indicate **parameters.** By changing the particular values of p and N that enter into the rule, one can produce different binomial distributions, each following the same mathematical rule, but differing in the probabilities that particular values of the random variable receive. Thus, what we are calling "the binomial distribution" is actually a whole *family* of binomial distributions. Families of distributions share the same mathematical rule for assigning probabilities or probability densities to values of X; in these rules the parameters are simply symbolized as constants. Actually assigning values to the parameters, such as we did for p and N in the distributions above, gives some particular distribution belonging to the family. Thus, *the* binomial distribution usually refers to the family of distributions having the same rule, and *a* binomial distribution is a particular one of this family found by fixing N and p.

A binomial distribution can be illustrated most simply as follows: Consider a simple experiment repeated independently five times. Each trial must result in only one of two outcomes and the result of five trials is a sequence of outcomes like those in the preceding sections. However, we are not at all interested in the order of the outcomes, but only in the number of successes in the set of trials.

In order to find the probability for each value of the discrete random variable, given N and p, let us begin with the largest value, $X = 5$. By counting rule 5, there must be $\binom{5}{5} = 1$ possible sequence where all of the outcomes are successes. The probability of this sequence is $p^5q^0 = p^5$, so that we have

$$p(X = 5; N = 5,p) = \binom{5}{5} p^5 = p^5.$$

For four successes, we find that by counting rule 5,

$$\binom{5}{4} = \frac{5!}{4!1!} = \frac{(1)(2)(3)(4)(5)}{(1)(2)(3)(4)(1)} = 5,$$

so that four successes can appear in five different sequences. Each sequence has probability p^4q^1. Thus,

$$p(X = 4; N = 5,p) = \binom{5}{4} p^4q^1 = 5p^4q^1.$$

Going on in this way, we find

$$p(X = 3; N = 5,p) = \binom{5}{3} p^3q^2 = 10p^3q^2$$

$$p(X = 2; N = 5,p) = \binom{5}{2} p^2q^3 = 10p^2q^3$$

$$p(X = 1; N = 5,p) = \binom{5}{1} pq^4 = 5pq^4$$

$$p(X = 0; N = 5,p = \binom{5}{0} q^5 = q^5.$$

(Note that $\binom{5}{0} = \frac{5!}{0!5!} = 1$, since 0! is 1 by definition.)

Now, to take a concrete instance of this binomial distribution, let the experiment be that of tossing a fair coin five times. Then $p = 1/2$, and the binomial distribution for the number of heads is

x	$p(x)$		
5	$(1/2)^5$	=	1/32
4	$5(1/2)^4(1/2)$	=	5/32
3	$10(1/2)^3(1/2)^2$	=	10/32
2	$10(1/2)^2(1/2)^3$	=	10/32
1	$5(1/2)(1/2)^4$	=	5/32
0	$(1/2)^5$	=	1/32
			32/32

Notice that the probabilities over all values of X sum to 1.00, just as they must for any probability distribution.

Now consider another example that is similar to this last one, but provides different probability values. Suppose that among American women college students who are undergraduates, only one in ten is married. A sample of five female students is drawn at random. Let X be the number of married students observed. (We will assume the total set of students to be large enough that sampling can be without replacement without affecting the probabilities and that observations are independent.) Hence, $p = .10$ and the distribution is

Table 3.11.2
Binomial distribution with $p = 1$ and $N = 5$.

x	$p(x)$		
5	$(1/10)^5$	=	1/100,000
4	$5(1/10)^4(9/10)$	=	45/100,000
3	$10(1/10)^3(9/10)^2$	=	810/100,000
2	$10(1/10)^2(9/10)^3$	=	7,290/100,000
1	$5(1/10)(9/10)^4$	=	32,805/100,000
0	$(9/10)^5$	=	59,049/100,000
			100,000/100,000

Contrast this distribution with the preceding one: When p was 1/2 the distribution showed the greatest probability for $X = 2$ and $X = 3$, with the probabilities diminishing gradually both toward $X = 0$ and toward $X = 5$. On the other hand, in the second distribution, the most probable value of X is 0, with a steady decrease in probability for the values 1 through 5. The distribution over such values of X is very different in these two situations, even though the probabilities are found by exactly the same *formal* rule. This illustrates once again that the binomial is actually a family of theoretical distributions, each following the same mathematical rule for associating probabilities with values of the random variable, but differing in particular probabilities, depending on the values of p and N.

Almost all theoretical distributions of interest in statistics can be specified by stating the function rule. The way this simplifies the discussion of distributions will be obvious as we go along; indeed, continuous distributions cannot really be discussed at all except in terms of their function rule. Typically, any one of these theoretical distributions will be a member of a family of distributions, each of which follows the same mathematical rule as the others, but which differs according to the values of the constants, or parameters, appearing in the rule.

3.12 / THE BINOMIAL DISTRIBUTION AND THE BINOMIAL EXPANSION

In school algebra you were very likely taught how to expand an expression such as $(a + b)^n$ by the following rule:

$$(a + b)^n = a^n + \frac{n!}{(n-1)!1!}a^{n-1}b + \frac{n!}{(n-2)!2!}a^{n-2}b^2 + \cdots + \frac{n!}{1!(n-1)!}ab^{n-1} + b^n.$$

For example, $(a + b)^3 = a^3 + 3a^2b + 3ab^2 + b^3$ according to this rule. This is the familiar "binomial theorem" for expanding a sum of two terms raised to a power.

Notice that the various probabilities in the binomial distribution are simply terms in such a binomial expansion. Thus, if we take $a = p$, $b = q$, and $n = N$,

$$(p + q)^N = p^N + \binom{N}{N-1}p^{N-1}q + \binom{N}{N-2}p^{N-2}q^2 + \cdots + q^N.$$

Since $p + q$ must equal 1.00, then $(p + q)^N = 1.00$, and the sum of all of the probabilities in a binomial distribution is 1.00.

3.13 / PROBABILITIES OF INTERVALS IN THE BINOMIAL DISTRIBUTION

In Chapter 2 we saw how to find a probability that a value of a random variable lies in an interval, such as $p(2 \leq X \leq 8)$, the probability that the random variable X takes on some value between 2 and 8 inclusive. This idea is easy to extend to a binomial variable.

Consider the binomial distribution shown in Table 3.13.1, with $p = .3$ and $N = 10$.

Table 3.13.1

A binomial distribution with $p = .3$ and $N = 10$.

r	$\binom{N}{r} p^r q^{N-r} = p(X = r)$
10	.00001
9	.00014
8	.00145
7	.00900
6	.03676
5	.10292
4	.20012
3	.26683
2	.23347
1	.12106
0	.02824
	1.00000

First of all we will find the probability $p(1 \leq X \leq 7)$, that X lies between the values 1 and 7 inclusive. This is given by the sum

$p(X = 1)$	.12106
$+ \ p(X = 2)$	.23347
$+ \ p(X = 3)$	.26683
$+ \ p(X = 4)$	.20012
$+ \ p(X = 5)$	.10292
$+ \ p(X = 5)$	.03676
$+ \ p(X = 7)$	.00900
	.97016 $= p(1 \leq X \leq 7)$.

The probability is about .97 that an observed value of X will lie between 1 and 7 inclusive. Thus, if we were drawing random samples of 10 observations from a Bernoulli process where the probability of a success were .3, we should be very likely to observe a number of successes between 1 and 7 inclusive.

By the same token, we can find

$$p(8 \leq X) = p(X = 8) + p(X = 9) + p(X = 10) = .00160.$$

The chances are less than two in a thousand of observing eight or more successes, *if* $p = .30$.

Notice that

$$p(X = 0, \text{ or } 8 \leq X) = .02824 + .00160 = .02984,$$

which is the same as

$$1 - p(1 \leq X \leq 7) = 1 - .97016 = .02984,$$

so that this is also that probability that X falls *outside* the interval bounded by 1 and 7.

Binomial distributions can also be put into cumulative form. Thus, the probability that X falls at or below a certain value a is the probability of the interval $X \leq a$. For this particular distribution, the corresponding cumulative distribution is

Table 3.13.2

The cumulative distribution for Table 3.13.1.

r	$p(x \leq r)$
10	1.00000
9	.99999
8	.99985
7	.99840
6	.98940
5	.95264
4	.84972
3	.64960
2	.38277
1	.14930
0	.02824

In this distribution, we see that about 65 percent of samples of 10 should show 3 or fewer successes, about 85 percent should have 4 or fewer, 99.8 percent should have 7 or fewer. Every sample (100 percent) must have ten or fewer successes, of course, since $N = 10$.

It must be re-emphasized that the binomial distribution is *theoretical*. It shows the probabilities for various numbers of successes out of N trials *if* independent random samplings are carried out from a stationary Bernoulli process and *if* p is the probability of a success. Given a different value of p or of N (or of both) the probabilities will be different. Nevertheless, regardless of N or p, the probabilities are found by the same, binomial, rule.

Table II of Appendix D gives binomial probabilities for $0 \leq r \leq N$ and $1 \leq N \leq 20$, for selected values of p. For p values which are greater than .50, one simply deals with $q = 1 - p$ and with the number of failures, or $N - r$. Thus, for $p = .70$ and $N = 10$, the probability of 6 successes is the same as the probability of 4 failures given $q = .30$. In the table under $N = 10$, $r = 4$, and $p = .30$, this is .2001. Cumulative probabilities such as $p(X \leq r;N,p)$ are found from the sum of the values of $p(x)$ for $x = 0,1, \cdots, r$, given N and p. Probabilities for larger values of N can be found directly by use of the table of factorials provided in Appendix D, Table VIII. Provided that only a few probabilities are needed, these can also be found quickly on a pocket or desk calculator.

A PREVIEW OF A USE OF THE BINOMIAL DISTRIBUTION Section 3.15

Quite often researchers are interested not in the number of successes that occur for some N trial observations, but rather in the *proportion* of successes, r/N. The proportion of successes is also a random variable, taking on fractional (or decimal) values between 0 and 1.00. Such sample proportions will be designated by the capital letter, P, to distinguish them from p, the probability of a success.

The probability of any given proportion P of successes among N cases sampled from a given Bernoulli process is exactly the same as the probability of the number of successes; that is

$$p\left(P = \frac{r}{N}\right) = p(X = r). \qquad [3.14.1]$$

For instance, if $N = 6$ and $r = 4$

$$p(X = 4) = p\left(P = \frac{4}{6}\right) = \binom{6}{4} p^4q^2.$$

The distribution of sample proportions is therefore given by the binomial distribution, the only difference being that each possible value of X becomes a value of $P = X/N$, so that prob. $(P = a) =$ prob. $(X = Na)$ for any particular value a. In the further discussion of the binomial distribution in Chapter 6, care will be taken to specify whether the random variable is regarded as X or P, since the arithmetic is slightly different in the two situations. Nevertheless, the probability of any given P is the same as for the corresponding X, given the sample size N.

3.15 / A PREVIEW OF A USE OF THE BINOMIAL DISTRIBUTION

It is now possible to point ahead to an important use of the binomial distribution. This example will deal with an experiment in psychology where we must use the data to decide whether or not a particular hypothesis actually seems to fit the data we obtain.

The context of the experiment is this: a psychophysical threshold or limen is that value on some physical measurement of a stimulus object at which a human subject is just capable of responding—in somewhat inexact terms, if the stimulus is a point of light in a darkened room, the threshold might be the physical intensity the light would have to have so that the subject would just be able to "see" the stimulus. It has, however, been suspected for some time that subjects may be able to respond in certain ways to stimuli which are actually below their known threshold of awareness. Such stimuli are said to be subliminal; the subject may not really be conscious that he "sees" the light—nevertheless, the subject may be able to respond *as though* capable of seeing the light.

Think of a hypothetical study of this question: "If a human is subjected to a stimulus below the threshold of conscious awareness, can the behavior somehow still be influenced by the presence of the stimulus?" The experimental task is as follows: the subject is seated in a room in front of a square screen divided into four equal parts and instructed that the task is to guess in which part of the screen a small, very faint, spot of light is thrown. He is to do this for many trials, and is told the light will be projected on the screen in a completely "haphazard," "random" manner over the trials. The light projected is made to be so faint that the subject cannot in any conscious sense actually "see" the light. However, unknown to the subject, the spot is always projected into the same one of the four parts of the screen over the various trials. For our computational convenience, suppose that only 10 trials are taken for this one subject.

Our hypothesis goes like this: if the subject really is in no way being influenced by the small "invisible" spot of light, then guesses should be random, haphazard affairs themselves, so that the guess should be right only 1/4 of the time by accident. Thus, under this hypothesis of "only guessing," the sample space of the subject's response to this situation should be distributed in this way:

Class	*p*
right	1/4
wrong	3/4
	4/4 = 1.00

The explicit assumption made is that the various trials for the subject are independent of each other. Now what sampling distribution holds for 10 trials for this subject if this hypothesis is true? The number of correct guesses *could* range from 10 right to 0 right, and the distribution would be a binomial distribution with $p(\text{right}) = 1/4$, $N = 10$. If we apply the rule for the binomial, we find the distribution that appears in Table 13.15.1. This binomial distribution gives all the possible numbers of correct guesses that this subject *could* make in the ten trials, and the probability for each, by guessing in a truly haphazard manner.

Given this theoretical distribution of possible outcomes, we turn to the actual result of the experiment. It is found that the subject guessed correctly on 7 out of the 10 trials. What then is the probability that exactly this result *should* have come up by chance? The binomial probability for 7 correct is 3,240 out of 1,048,576, or about .0031. Thus the probability of getting exactly this number correct by random guessing alone is about 31 chances in 10,000. In other words, if we repeated the experiment, giving the subject 10,000 independent sets of ten trials, about 31 of these repetitions should give us exactly 7 correct guesses. The exact sample result obtained is clearly not very likely to occur if the hypothesis is correct.

However, we should be interested in the probability not only of getting exactly 7 correct, but also in the probability of getting *this many or more* correct, since we are really interested in this result as evidence of whether or not the subject is guessing, or doing something else which gives more correct responses than should simple, haphazard guessing. What is the probability of *7 or more* correct trials? The answer is readily seen if we remember that this is asking for the probability of an interval:

$$p(7 \leq X) = p(7) + p(8) + p(9) + p(10) = .0035(+).$$

Seven or more correct guesses should occur only about 35 times in 10,000 independent replications of this experiment, if guessing alone is responsible for the subject's behavior. Does this unlikely result cast any doubt on the theory that the stimulus has no effect on guessing behavior? The answer is yes. For a theory to be "good," it

Table 3.15.1

A binomial distribution for $p = .25$ and $N = 10$.

x	$p(x)$
10 right	$\binom{10}{10}(1/4)^{10}(3/4)^0 = .0000(+)$
9	$\binom{10}{9}(1/4)^9(3/4)^1 = .0000(+)$
8	$\binom{10}{8}(1/4)^8(3/4)^2 = .0004$
7	$\binom{10}{7}(1/4)^7(3/4)^3 = .0031$
6	$\binom{10}{6}(1/4)^6(3/4)^4 = .0162$
5	$\binom{10}{5}(1/4)^5(3/4)^5 = .0584$
4	$\binom{10}{4}(1/4)^4(3/4)^6 = .1460$
3	$\binom{10}{3}(1/4)^3(3/4)^7 = .2503$
2	$\binom{10}{2}(1/4)^2(3/4)^8 = .2816$
1	$\binom{10}{1}(1/4)^1(3/4)^9 = .1877$
0	$\binom{10}{0}(1/4)^0(3/4)^{10} = .0563$

should forecast results that agree with what we actually obtain. If the subject had come up with 2, 3, or 4 correct responses, then we would have little reason to doubt the ''guessing'' hypothesis, since these results fall among those that are quite probable according to the sampling distribution. However, the results obtained are quite unlikely to occur if the hypothesis is true, and so the evidence does not seem to favor this hypothesis.

On the other hand, are we completely safe in inferring that the subject was *not* just guessing? The answer is, of course, no. Even though this many or more correct responses obtained is improbable if the hypothesis is true, it is still not impossible that the subject was only guessing. We need some way to state just how much of a chance we are taking in saying that the hypothesis is not true, and that the subject is in some way being influenced by the stimulus. *The best measure of the amount of risk we run by abandoning this hypothesis on the evidence is given by the probability of sample results as extreme or more extreme than those actually obtained, if the hypothesis were true.* The evidence apparently diverges considerably from what we should expect (about 1/4 correct) if the subject were guessing, and the probability of such a divergent sample is only about 35 in 10,000. The probability of our being *wrong* in rejecting the guessing-hypothesis equals the probability of the divergent result, or .0035. We *could* be wrong in rejecting the hypothesis that the subject is guessing, but the probability of our being wrong is not very great. The more our sample result departs from what we expect given a hypothetical situation, and the more improbable that departure is, the less credence is given the hypothesis. In a later chapter we shall discuss in detail the rules for (and precautions in) evaluating hypotheses in the light of obtained results. However, this illustration exhibits the general logic underlying all "tests" of hypotheses: from the hypothetical population distribution one obtains a theoretical sampling distribution. Then the obtained results are compared with the sampling distribution probabilities. If the probability of samples such as the one obtained is high, the hypothesis is regarded as tenable. On the other hand, if the probability of such a sample (or one in more extreme disagreement with what is expected) is quite small, then doubt is cast on the hypothesis.

Although this little example should not be taken as a model of sophisticated scientific practice, it does suggest the use of a theoretical distribution of sample results as an aid in making inferences about a hypothetical situation. The binomial distribution is only one of a number of families of distributions that are employed in this general way.

Since we have been rather heavily theoretical in our discussion up to this point, the next section will deal with one method based on the binomial distribution that has some immediate practical use in experimental situations. Although not all of the qualifications involved in the use of such methods can be explained at this time, perhaps it will prove interesting and useful nevertheless. Then, some close relatives of the binomial distribution will be introduced, along with some indications of how these distributions are used as well.

3.16 / THE SIGN TEST FOR TWO GROUPS

We have already seen that the binomial distribution can be used to test a hypothesis about a proportion on the basis of a random sample of N independent observations. This is but one of a very large number of uses of the idea of a binomial variable in applied statistics. Now, a method based on the binomial distribution will be introduced, which is designed for the comparison of two groups of observations.

This simple method based on the binomial distribution is the so-called "sign test." The sign test is used in situations in which *N pairs of matched* observations are made. The first member of each pair is an observation of some type, A, and the second an observation of another type, B. Thus, husband-wife pairs may be observed, in which each husband (type A) is paired with a wife (type B). Now each A individual has a numerical score X_A, and this is paired with a numerical score for the B individual, X_B. The question to be answered is, "Is the distribution of X_A scores identical to the distribution of X_B scores in the long run, if all possible pairs could be observed?" We approach this question by noting the difference $(x_A - x_B)$ for each of the N pairs. If $(x_A - x_B)$ is positive in value, then a "+" or a "success" is recorded. If $(x_A - x_B)$ is negative, the a "$-$" or "failure" is recorded. When $(x_A - x_B) = 0$, a fair coin is tossed, and the result is noted as a success or a failure depending on the outcome of the coin.

If the distribution of X_A values and the distribution of the X_B values really are identical, in the long run for all possible pairs, then it should be true that $p(+) = p(-) = 1/2$. Furthermore, the occurrence of successes and failures should correspond to a Bernoulli process, with N equal to the number of pairs observed, and p

= 1/2. In order to reach a conclusion about the tenability of the hypothesis that $p = 1/2$, one calculates the probability that results as extreme, or more extreme, should occur, given that $p = 1/2$, and given the value of N that is actually used. If this probability is sufficiently small (say .05 or less), then the hypothesis that $p = 1/2$ is rejected, meaning that the hypothesis of equal distributions is rejected as well. If a number of successes or failures equally or more extreme than the number actually obtained does not have a sufficiently small probability, then the hypothesis is not rejected.

For example, suppose that the basic question had been, "Are women actually better drivers than their husbands?" In order to shed light on this question, we take twenty married couples at random, and separately give each wife and each husband a driving test. If a wife scores higher than a husband, this is a "success" or a "+," and if the husband scores higher, this is a "failure," or "−." We decide that we will reject the hypothesis of "no difference in driving ability" only if the number of successes observed should be equaled or exceeded with a probability no larger than .05. That is, we need to find the probability of the number of successes we actually observed, *or more,* in a binomial distribution with $N = 20$ and $p = 1/2$. If this probability is .05 or less, then we will reject the hypotheseis of equal ability.

Suppose that exactly 15 successes were observed. Then the probability of 15 successes or more can be found quite simply in Table II in Appendix D. Here, $N = 20$ and $p = 1/2$, so that we find the column in the table labeled $p = .50$, and the row section for $N = 20$. Adding up the probabilities for 15 or more, we find p (15 or more) $= .0148 + .0046 + .0011 + .0002 = .0207$. Then, since this value is less than the probability of .05 we had decided to employ in making the decision, we reject the hypothesis that husbands and wives are equal in driving ability. The evidence obtained suggests that wives are actually better drivers.

On the other hand, if we were able to observe the entire population of husbands and wives, and give the members of each pair the driving test, it might well be true that there is no long-run difference, or even that husbands turn out to be the better drivers. After all, this is only a sample of twenty from among millions of couples, and by chance it could be that we have drawn a most unrepresentative group. However, we have a bit of insurance: since the probability of fifteen or more successes is less than .05, the chances of our having got such an extreme sample (when husbands and wives are not truly different) is also less than .05 We are thus running a small risk in saying that wives are indeed better than husbands, provided this sample was indeed random.

The sign test is useful in a variety of problems where pairs of observations are to be compared on some characteristic. It can also be extended to situations where the sample size is too large to permit use of Table II; this extension is discussed in Chapter 16.

3.17 / THE GEOMETRIC AND PASCAL DISTRIBUTIONS

A close relative of the binomial distribution is the Pascal distribution, named for the French mathematician-philosopher Blaise Pascal (1623–1662). Whereas a random variable following the binomial rule corresponds to the number of successes r out of a fixed number of trials N, the random variable following the Pascal rule has a different interpretation. Here one is interested in the number of trials, N, necessary in order to achieve a given number of successes r. Thus N is the random variable and r is a constant in a geometric or Pascal distribution. Unlike binomial sampling, in geometric and Pascal sampling r is fixed and N is left to chance.

Consider a stationary and independent Bernoulli process in which the probability of a success on any trial is p. Then for any sequence of trials we might ask, "What is the probability that the first success occurs on the first trial?" or "What is the probability that the first success occurs on the sixth trial?" and so on for a trial of any number. Let us study the first success. Since the first success never *has* to occur at all, short of an indefinitely large number of trials, then the possible trials on which the first success *might* occur are countably infinite in number. The distribution of N, the trial number on which the first success occurs, given trials from a stable Bernoulli process, is known as the **geometric distribution.** In a geometric distribution, the random variable N can take on any value that is one of the counting numbers 1, 2, 3, · · · .This means that, unlike a binomial variable, a geometric variable takes on a countably infinite set of values. However, such a geometric variable is still discrete, since it can take on only whole-number values.

Now, the probability that the first success occurs on the first trial is

$$p(N = 1;p) = p.$$

The probability that the first success occurs on the second trial must be

$$p(N = 2;p) = (1 - p)p,$$

since the first trial must be a failure if the first success occurs on the second trial. In the same way, we can show that

$$p(N = 3;p) = (1 - p)^2 p$$

and so on, until for the probability that the first success occurs on trial $N = n$ we have

$$p(N = n;p) = (1 - p)^{n-1} p.$$

This is the rule for the geometric distribution, in which the random variable is the trial number on which the first success occurs, in trials from a stable, independent, Bernoulli process with probability p.

The Pascal distribution can be thought of as a generalization of the geometric distribution. That is, the random variable in the Pascal distribution is the trial number on which the rth success occurs, where r can be any whole number $r = 1, 2, \cdots$, and where $r \leqslant N$. The geometric distribution is a Pascal distribution in which $r = 1$. For a Pascal distribution, the probability that N equals a given value n depends on the fixed value r, or number of successes, and p, the probability of a success, as follows:

$$p(N = n;r, p) = \frac{(n - 1)!}{(r - 1)!(n - r)!} p^r(1 - p)^{n-r}, \qquad N \geqslant r. \qquad [3.17.1]$$

(Notice that there is no upper limit to N; in principle the rth success might *never* occur. Thus a Pascal variable is discrete, but *countably infinite* in the number of values.)

For an example of how the idea of a Pascal variable can be used, consider the following: A team of investigators are choosing items for inclusion in a test. They believe that one item is moderately easy, so that two out of three schoolchildren of a given age should pass the item. They interpret this to mean that the probability that any given child will pass is 2/3. However, they want to test this belief against the possibility that the item may be harder. Time and expense are factors, so they do not want to sample more children than absolutely necessary in order to arrive at a judgment. Thus the team decides to administer the item to individual children, one at a time, selected at random. They also decide to test only until four children have successfully passed the item. Then, on the basis of the total number of children tested in order to achieve four successes, they will reach a conclusion about the notion that the probability of a given child's passing is 2/3. The more children it is necessary to test in order to achieve four successes, the more doubt will be cast upon 2/3 as the value of p. For reasons that we will elaborate at length later on, the team decides to reject the idea that $p = 2/3$ only if if the number of children required to reach four successes is in some sense "excessive." In particular, $p = 2/3$ will be rejected as the true probability only if N, the number of children actually tested, should be equaled or exceeded with probability of only .05 or less, given that $p = 2/3$ is the true probability of passing the test. Otherwise, $p = 2/3$ will be retained as a tenable hypothesis about the true value.

Now suppose that the investigators begin to test children in the order of their random selection. Sure enough, the fourth child to pass the item is actually the eighth child tested. Thus, $N = 8$, and the fixed number of successes $r = 4$, while the value of p is believed to be 2/3. The critical question then is the value of $p(N \geqslant 8;4, 2/3)$, which is the same as $1 - F(7;4, 2/3)$. Is this, or is it not, .05 or less?

The value of this probability is determined by the Pascal rule. Since

$$F(7;4, 2/3) = p(4) + p(5) + p(6) + p(7),$$

it is necessary to find the probability for each of these values of the random variable. (Remember that one cannot have four successes in fewer than four trials.) Proceeding, we find

$$p(4) = \frac{(4-1)!}{(4-1)!(0)!} p^4 = (2/3)^4 = 16/81,$$

$$p(5) = \frac{(5-1)!}{(4-1)!(1)!} p^4(1-p) = 4(2/3)^4(1/3) = 64/243,$$

$$p(6) = \frac{(6-1)!}{(4-1)!(6-4)!} p^4(1-p)^2 = 10(2/3)^4(1/3)^2 = 160/729,$$

$$p(7) = \frac{(7-1)!}{(4-1)!(7-4)!} p^4(1-p)^3 = 20(2/3)^4(1/3)^3 = 320/2187.$$

Then

$$F(7;4, 2/3) = 16/81 + 64/243 + 160/729 + 320/2187 = 1808/2187,$$

so that $F(7;4, 2/3)$ is about .83. Thus, a number of trials equal to 8 or more can occur with probability of about .17. This is far in excess of the probability of .05 that the investigators had decided upon. On this evidence they decide not to reject the hypothesis that $p = 2/3$.

On the other hand, had it been true that ten trials were required in order to reach four successes, then $F(9;4, 2/3)$ would have been approximately .96. This in turn implies that the probability of ten or more trials required to reach four successes would have been .04, which is less than the criterion agreed upon. In this case, the investigators would have decided to reject the hypothesis that $p = 2/3$. Either it is true that $p = 2/3$, and a rare event has occurred, or 2/3 is not the correct value of p. The investigators have already decided to conclude the latter, should such a rare event occur.

Notice what has been assumed here. First of all, we have assumed that $p = 2/3$ for each and every one of the children tested, so that observation of each child represents a trial in a stationary Bernoulli process. Furthermore, we have assumed that the results for each child, or trial, are independent of those for any other child. Finally, we have assumed that the order of selection of the children was completely random. The failure of any of these assumptions to be true could, of course, make a difference in the final conclusions.

Problems sometimes arise in which interest focuses not on the trial number on which the rth success occurs, but rather on the number of failures that occur before the rth success. However, when $y = N - r$, the number of failures before the rth success, is the random variable, it is customary to refer to the distribution as the **negative binomial distribution,** and this formulation will sometimes be encountered in the statistical literature in place of the Pascal distribution. For all practical purposes these two kinds of distributions are the same, and problems can be solved in terms of either the number of failures before the rth success, $(N - r)$ or the trial number of the rth success, N.

Like the binomial distribution, the Pascal distribution can be used as the basis for a variety of statistical techniques. Ordinarily, the kind of situation producing a random variable following a Pascal rule will be one involving **sequential sampling,** as opposed to simple random sampling. In sequential sampling no predetermined limit is set upon the number of observations. Rather a series of observations are made until sufficient data are accumulated to enable a decision to be made, according to a predetermined criterion. Since Pascal variable is interpreted as the number of trials required in order to reach a given number of successes, this conception lends itself readily to the simplest kinds of sequential experiments.

Other methods of sequential sampling, not depending on the Pascal distribution, are also prominently used in statistics. Unfortunately, space does not permit going further with this topic. However, you should know this approach is a valid alternative to some of the sampling methods and statistical techniques we will be employing. Details may be found in many standard texts, such as Mood, Graybill, and Boes (1974).

3.18 / THE POISSON DISTRIBUTION

Still other relatives of the binomial family of distributions play a large role in theoretical and applied statistics One is the family of Poisson distributions, named after the nineteenth-century French mathematician S. Poisson. A random variable following this rule is referred to as a Poisson variable, and the process generating values of such a random variable is known as a Poisson process. The probability function for a Poisson variable X follows the rule

$$p(x;m) = \begin{cases} \dfrac{e^{-m}m^x}{x!}, & x = 0, 1, 2, 3, \cdots; m > 0; \\ 0, & \text{otherwise}, \end{cases} \qquad [3.18.1]$$

where e is a mathematical constant equal approximately to 2.718, and m is a constant known as the "intensity" of the Poisson process.

Like a Pascal variable, a Poisson variable X can take on only integral or "whole" values—in this case from zero to an indefinitely large value. This random variable thus can assume any of a countably infinite set of values. It is thus a discrete variable.

Although Poisson variables can be given a variety of useful interpretations, perhaps the simplest approach to the study of the Poisson is to regard it as a special case of the binomial, where N is thought of as very large, and p is very small.

Not only does this connection with the binomial distribution permit one interpretation of the Poisson distribution; there are practical consequences as well. In cases in which N is large and p is relatively small, the binomial probabilities may be very laborious to calculate. In this instance, it is a much simpler matter to approximate the exact binomial probabilities through use of Poisson probabilities for the various values of x.

When a Poisson probability for some specific x value is desired, one simply calculates $m = Np$, and then applies the Poisson rule

$$p(x; m) = \frac{e^{-m}m^x}{x!}$$

for the value $X = x$ of interest. Table X in Appendix D gives selected values of e^{-m}, and Table VIII gives various values of factorials. For example, suppose that for $N = 3000$ and $p = .001$, we wished to find the probability for $X = 5$. Then, $m = (3000)(.001) = 3$, and we calculate the probability by taking

$$\begin{aligned} p(5;3) &= \frac{e^{-3}\,3^5}{5!} \\ &= \frac{(.0498)(243)}{120} \\ &= .1008. \end{aligned}$$

This is an exact Poisson probability, $p(5;3)$. It is also an approximation of the binomial probability $p(5;3000, .001)$. The Poisson probability will be approximately equal to the actual binomial probability only for very large N and very small p, of course. Nevertheless, the approximation is good enough to be useful even when N is only moderately large (say $N \geqslant 50$) and p only relatively small ($p \leqslant .2$).

Poisson probabilities and cumulative probabilities can be determined relatively easily in the way just shown, given Table VIII for the values of factorials and Table X for values of e^{-m}. A number of pocket calculators also give values for e^{-m} and for $x!$. Many books on advanced statistics give extensive tables of Poisson probabilities, particularly when they are designed to be used in fields such as the physical sciences or industry. However, since our use of the Poisson will not be extensive, space will not be given to such tables here.

A great many illustrations of Poisson processes occur in the physical and the biological sciences, as well as in everyday life. For example, the degeneration of a radioactive substance can be regarded as a Poisson process. At any given instant the probability is very small that an alpha particle will be emitted, while there are vast numbers of opportunities for such an event to occur. The distribution of bacteria on a petri plate can be viewed as a Poisson process. Each tiny area on the plate can be viewed as a trial, and a bacterium may or may not occur on such an area. The probability of such an occurrence on any given area is very small indeed, but there are very many areas on such a plate. The distribution of misprints in a book can be studied as a Poisson process, as can the occurrence of accidents of a certain kind in a manufacturing plant. The Poisson distribution is thus important in its own right, quite apart from its connection with the binomial distribution.

As a simple example of a problem involving the Poisson distribution consider the following. Suppose it is known that the annual rate of suicides on U.S. college campuses is 5 in 50,000 students, or about .0001. Now a campus of 25,000 students is studied, which is typical in all respects of such campuses, so that its students might be considered a random sample of such students. However, on this campus last year there were 4 suicides. What is the probability of four or more suicides if the true rate is actually .0001?

Here, we take $p = .0001$ and $N = 25{,}000$, so that $m = (.0001)(25{,}000) = 2.5$. Then,

$$\begin{aligned} p(x \geqslant 4;2.5) &= 1 - p(x = 0) - p(x = 1) - p(x = 2) - p(x = 3) \\ &= 1 - \frac{e^{-2.5}(2.5)^0}{0!} - \frac{e^{-2.5}(2.5)}{1!} - \frac{e^{-2.5}(2.5)^2}{2!} - \frac{e^{-2.5}(2.5)^3}{3!} \end{aligned}$$

so that, by Table X (or from any reasonably sophisticated hand calculator) we obtain

$$p(x \geqslant 4;2.5) = 1 - .7576 = .2424.$$

The probability of 4 or more suicides on this campus, if it *is* typical, is about .24. However, if we figure the corresponding probability for six or more suicides, the probability falls to only .04. Thus, for six or more such occurrences, we might begin to have real doubts about the typicality of this campus.

3.19 / THE MULTINOMIAL DISTRIBUTION

The basic rationale underlying the binomial distribution can also be generalized to situations with more than two event classes. This generalization is known as the ''multinomial distribution,'' having the following rule:

Consider K classes, mutually exclusive and exhaustive, and with probabilities p_1, $p_2, \cdots, p_K$. If N observations are made independently and at random, then the probability that exactly n_1 will be of kind 1, n_2 of kind 2, $\cdots$, and n_K of kind K, where $n_1 + n_2 + \cdots + n_K = N$, is given by

$$\frac{N!}{n_1!n_2! \cdots n_K!}(p_1)^{n_1}(p_2)^{n_2} \cdots (p_K)^{n_K}. \qquad [3.19.1]$$

Think once again of colored marbles mixed together in a box, where the following probability distribution holds:

Color	*p*
Black	.40
Red	.30
White	.20
Blue	.10
	1.00

Now suppose that 10 balls were drawn at random and with replacement. The sample shows 2 black, 3 red, 5 white, and 0 blue. What is the probability of a sample distribution such as this? On substituting into the multinomial rule, we have

$$\frac{10!}{(2!(3!)(5!)(0!)}(.4)^2(.3)^3(.2)^5(.10)^0$$

$$= \frac{1 \cdot 2 \cdot 3 \cdot 4 \cdot 5 \cdot 6 \cdot 7 \cdot 8 \cdot 9 \cdot 10}{(1 \cdot 2)(1 \cdot 2 \cdot 3)(1 \cdot 2 \cdot 3 \cdot 4 \cdot 5)(1)}(.4)^2(.3)^3(.2)^5$$

since 0! and $(.1)^0$ are both equal to 1. Working out this number, we find that .0035 is the probability of the *sample* distribution

Black	2
Red	3
White	5
Blue	0
	10

if the probability distribution given above is the true one. Using this multinomial rule, one could work out a probability for each *possible sample distribution.*

Although the multinomial rule is relatively easy to state, a tabulation or graph of this distribution is very complicated: here a sample result is not a single number, as for the binomial, but rather *an entire frequency distributon.* In principle, once a given discrete probability distribution is specified, then the probability of every conceivable sample distribution of size N could be worked out by the multinomial rule. Even though this possibility exists, the multinomial distribution plays a relatively modest role in statistics. Most often, the probability of obtaining an entire frequency distribution is not so much of concern as the probability of one or more indices summarizing the sample distribution. This problem will occupy most of our attention in the sections to come. Nevertheless, some of the methods we will consider in Chapter 15 will be founded directly on the multinomial distribution.

3.20 / THE HYPERGEOMETRIC DISTRIBUTION

Another theoretical probability distribution deserves passing mention for the same reason as the multinomial. The multinomial rule (and the binomial, of course) assumes either that the sampling is done with replacement, or that the sample space is infinite, so that the basic probabilities do not change over the trials made. However, suppose that one were sampling from a finite space *without* replacement; then the probabilities would change for each observation made. By a series of arguments very similar to those used for finding probabilities of poker hands in Section 3.8, we could arrive at a new rule for finding the probabilities of sample results.

This rule describes the hypergeometric distribution, and can be stated as follows:

Given a sample space containing a finite number T of elements, suppose that the elements are divided into K mutually exclusive and exhaustive classes, with T_1 in class 1, T_2 in class 2, $\cdots$, T_K in class K. A sample of N observations is drawn at random without replacement, and is found to contain n_1 of class 1, n_2 of class 2, $\cdots$, n_K of class K. Then the probability of occurrence of such a sample is given by

$$\frac{\binom{T_1}{n_1}\binom{T_2}{n_2}\cdots\binom{T_K}{n_K}}{\binom{T}{N}} \qquad [3.20.1]$$

where $n_1 + n_2 + \cdots + n_K = N$ and $T_1 + T_2 + \cdots + T_K = T$.

For an illustration of the use of the hypergeometric rule, let us return to the problem of drawing marbles at random from a box (Section 3.19), but this time *without* replacement. Suppose there had been 30 balls in the box originally, with the following distribution of colors:

Color	*f*
Black	12
Red	9
White	6
Blue	3
	30

Notice that the relative frequencies are the same as the probabilities in the previous example. Now ten balls are drawn at random without replacement. We want the probability that the *sample* distribution is

Color	*f*
Black	2
Red	3
White	5
Blue	0
	10

Using the hypergeometric rule, we get

$$\frac{\binom{12}{2}\binom{9}{3}\binom{6}{5}\binom{3}{0}}{\binom{30}{10}}$$

which works out to be about .0011. This is not, of course, the same probability as we found for this sample result using the multinomial rule, since the entire sampling scheme is assumed different in this second example. This illustrates that *the sampling scheme adopted makes a real difference in the probability of a given result.*

Ordinarily, this is a practical consideration only when the basic sample space contains a small number of elementary events. When there is a very large number of elementary events in the sample space, the selection and nonreplacement of a particular unit for observation has negligible effect on the probabilities of events for successive samplings. For this reason, the hypergeometric probabilities are very closely approximated by the binomial or multinomial probabilities when T, the total number of elements in the sample space, is extremely large. The distinction between these different distributions becomes practically important only when samples are taken from relatively small sets of potential units for observation. We will have more to say on sampling without replacement from small sample spaces in Section 5.10.

EXERCISES

1. A well-known intelligence test item consists of arranging cartoon pictures in a meaningful order. If there are six pictures in all, into how many orders might a subject arrange these cards?

2. In the test item of exercise 1, only two sequences or orders are scored as correct. If a subject is operating purely by chance, so that each sequence is equally likely to occur, what is the probability of a correct response?

3. Suppose that a child is seated before a toy piano consisting of twelve keys. The child plays four notes. If any note is equally likely to be struck on any attempt, how many different sequences of notes are possible (including repetitions of the same note)? What is the probability of a sequence consisting of the same note struck four times? What is the probability that the sequence consists of at least two different notes? What is the probability of a four-note ascending scale?

4. Each day a cafeteria offers 4 types of salad, 3 types of meat, 8 kinds of vegetables, 2 types of bread, and 6 varieties of dessert. If a person's meal consists of one selection from each of these choices, how many different meals are possible? (Consider two meals to be different if they differ on any item.) Suppose that two people choose their meals independently on the same day. If each is equally likely to select any possible meal, what is the probability that they will come up with exactly the same choices? (**Hint:** Bear in mind how many possible meals there are on which the people could agree.)

5. In arranging a series of interviews, a social scientist wished to assign seven subjects to seven available time periods. If all subjects were available at all times, in how many ways could the assignments of subjects to times be made?

6. Suppose that the researcher of problem 5 actually had 15 time slots available for interviewing the seven subjects. In how many ways could the subjects be assigned to time periods? (Leave answer in symbolic form.)

7. An experimenter has 20 subjects available. It is desirable to form two groups of ten subjects each. In how many ways can the first group be formed? Given that the first group has been selected, how many ways exist for forming the second group? Given the assumption that all assignments of subjects to groups are equally likely, what is the probability of any given assignment? (Leave answers in symbolic form.)

8. In still another experiment, 50 subjects are available. Six groups of 5 subjects each are to be formed. In how many different ways can these assignments of subjects to groups be made? If all assignments are equally probable, what is the probability of any given arrangement of subjects into groups? (Leave answers in symbolic form.)

9. In the situation of exercise 8, exactly 20 of the 50 subjects are black, and the remainder are white. If all assignments of subjects to groups are equally likely, what is the probability of exactly one all-black group, with the remainder of the groups all white? (Remember that there are six possibilities for the all-black group.) What is the probability of exactly two all-black groups? (Leave answers in symbolic form.)

10. In a fraternity house, three boys share a room with a single closet. Each boy can wear each of the other boys' items of clothing, and they share freely. The closet contains 3 pairs of shoes, 7 shirts, 5 pairs of pants, 8 pairs of socks, and 4 coats. If each boy dresses in shoes, shirt, pants, socks, and coat, in how many combinations of clothing may the boys appear together? (Leave the answer in symbolic form.)

11. A college contains three departments, *A*, *B*, and *C*. Department *A* has 10 faculty members, *B* has 15 members, and *C*, 20 members. It is decided to form a college committee consisting of 1/5 of the members of each department. How many possible such committees could be formed? (Leave the answer in symbolic form.)

12. Suppose that two people work in the same office. Given that their birthdays were independent of each other, and that births are equally likely to occur in any month, what is the probability that they have a birthday in the same month?

13. Let there be four people working in an office. Under the same assumptions made in exercise 12, find the probability that no two or more of these people's birthdays fall in the same month. Then find the probability that two or more do have a birthday in the same month. (Leave the answers in symbolic form.)

14. The legislature of a certain state wants to make automobile license plates carry three letters followed by three digits. Will there be enough possible license plates under this system to provide different plates for all of the five million automobiles the state registers? What if the plates contain two letters and three digits? Will there still be enough different plates? How about with three letters and two digits? (Count 0 through 9 as the possible digits, and include all 26 letters of the alphabet.)

15. In the circumstances of exercise 14, where three letters and three digits are used, what is the probability that a car owner will get a plate with three repeated letters and three repeated digits, such as BBB-222, if all combinations are equally probable? What is the probability of three repeated letters and three different digits? Three different letters and three different digits? (Leave answers in symbolic form.)

16. Suppose that in the college mentioned in exercise 11 above, it was decided to form a nine-member committee by selecting faculty members completely at random. What is the probability that such a committee would wind up containing:

(a) Exactly 1/5 of Department *A*?
(b) Exactly 1/5 of Department *B*?
(c) Exactly 1/5 of Department *C*?
(d) Fewer than 1/5 of the members of Department *A*?
(e) *Only* members from Department *C*?

(Again, leave the answers in symbolic form.)

17. A very old machine that manufactures automobile parts is now believed to produce defective parts with a frequency of about 30 in 100. We will assume that this process is stationary and independent. Given that 15 parts produced by this machine are sampled at random, find the probability that one or fewer parts will turn out to be defective. Two or more?

18. In a test of possible side effects of a new medication, a physician matched 20 pairs of persons on their physical characteristics. One member of each pair was given the medicine, and the pair-mate was given a placebo. A "success" was recorded when the member receiving the medication showed more of the side effect than the pair-mate, and a "failure" was recorded otherwise. There were 13 successes and 7 failures. If medication and placebo were equal in their tendency to produce the effect, how likely is a result this much or more deviant from what one should expect?

19. On a multiple-choice examination, each item has exactly 5 options, of which the student must pick one. Only one of the options is correct for each item. If a student is merely guessing at the answer, each option should be equally likely to be chosen. Furthermore, the student's answer on a given item is believed to be independent of the answer on any other. Given that the test has 18 items, and that a student is just guessing on each item, find the probability of the following events:

(a) 3 items correct.
(b) 7 items correct.
(c) 4 items incorrect.
(d) More than 5 items correct.
(e) Fewer than 2 items correct.
(f) Between 2 and 7 items correct (inclusive).

20. In a certain lottery, 40 percent of the tickets were purchased by men, and 60 percent by women. Each person purchased only one ticket. Ten tickets were drawn at random and with replacement. What is the probability that:

(a) Four or more winners were women?
(b) Two or fewer winners were women?
(c) The winners were all of the same sex?
(d) Exactly four men and six women were winners?

21. A Public Health officer in a certain area suspects that 25 percent of children in that area are severely undernourished. When a sample of 20 children is taken at random, and it is found that 9 show severe malnutrition. What is the probability of 9 or more such children in the sample if the true proportion in the population is .25? What would the officer be inclined to conclude?

22. In a test of reaction time, husband and wife pairs were studied. Sixteen pairs chosen at random were given the same reaction-time test, and the individual time noted. In nine of the pairs, the wife showed the faster reaction time, and in seven pairs, the husband showed the shorter time. If husbands and wives tend to be about equal in reaction time, what is the probability of a result this much or more in favor of the wives?

23. In an experiment two matched groups of ten individuals each were employed. Each individual in each group was selected independently and at random. The score of each individual on a certain perceptual test was determined. The results were as indicated in the table on the next page. Before the experiment, the experimenter expected group II to have the larger scores. Use the sign test to examine the hypothesis that the groups represented by these samples are actually equal in the scores one should expect.

Group I	Group II
5	14
8	21
7	23
6	6
21	11
13	5
20	10
17	18
10	21
17	25

24. In an experiment on extrasensory perception (ESP), three shells are used, one of which covers a pea. On each trial the shells are rearranged into a row and the subject's task is to locate the shell under which the pea lies. Suppose that the first, second, and third shells are equally likely to cover the pea on any trial. Suppose also that without ESP a subject is equally likely to guess any of the three shells on a trial. Under these conditions, assuming 10 independent trials, how probable is it that out of 10 trials that the subject guesses the correct shell two or fewer times? Eight or more times?

25. An expert typist tends, on the average, to make just one error for 5 business letters typed, about .2 of an error per letter. What is the probability that the typist will make no errors at all on a letter? (**Hint:** Use the Poisson distribution, letting $m = .2$ and $x = 0$.) What is the probability that the typist will make at least one error?

26. A suicide-prevention unit in a city receives calls for help at a rate of about 2.3 per day. If this is a stationary and independent Bernoulli process, what is the probability of one day without a call? Only one call? Two or more calls in a day? (**Hint:** Poisson.)

27. On a college football team, suppose that a passer has a probability of .60 of a completion on any attempt. If his passing performance reflects an independent and stable random process, what is the probability that in a game he gets his first completion on his fourth attempt? What is the probability of his first completion on his first attempt?

28. Suppose that a certain door-to-door salesman believes that he has probability of .30 of making a sale on any given call. If his sales can be regarded as corresponding to events in a stable and independent Bernoulli process, what is the probability that he makes his fifth sale on his tenth call, given that $p = .30$? What is the probability that he makes his first sale on his fifth call? (Leave your answers in symbolic form.)

29. Students at a very large university campus show the following distribution of religious preferences: Protestant, 32 percent; Catholic, 14 percent; Jewish, 19 percent; other religion, 3 percent; no preference stated, 31 percent. Suppose that a given group of twenty students could be regarded as a random selection of students on this campus. What then is the probability of the following set of religious preferences? (**Hint:** Treat the campus as large enough that sampling can be thought of as with replacement. Leave the answer in symbolic form.)

Preference	Frequency
Protestant	5
Catholic	3
Jewish	4
Other	0
None reported	8

30. An item in a questionnaire asked for a response on a five-point scale extending from "1" for "strongly agree," to "5" for "strongly disagree." If all respondents are actually equally likely to use any of the categories, how probable is the following distribution of 10 responses? (Leave answer in symbolic form.)

Response	*Frequency*
1	0
2	2
3	6
4	2
5	0

31. Suppose that 10 students out of a class of 50 undergraduates are chosen for a behavioral experiment. The total class consists of 15 freshmen, 18 sophomores, 10 juniors, and 7 seniors. If all students are equally likely to be picked, what is the probability of the following? (Leave answers in symbolic form.)

(a) The group chosen consists solely of sophomores.
(b) The group contains exactly 7 seniors and 3 juniors.
(c) The group contains 3 freshmen, 4 sophomores, 2 juniors, and 1 senior.

32. A large metropolitan area reports 2.4 traffic fatalities a day, or .1 per hour, on the average. If traffic fatalities represent a stable independent Poisson process, what is the probability of no traffic fatalities in a given hour? Of 2 or more in an hour? (**Hint:** use $m = .1$)

Chapter 4
CENTRAL TENDENCY AND VARIABILITY

Any frequency distribution is a summarization of data, but for many purposes it is necessary to summarize still further. Rather than compare entire distributions of data with each other or with hypothetical distributions, it is generally more efficient to compare only certain characteristics of distributions. Two such general characteristics of any distribution, whether frequency or probability, obtained or theoretical, are its measures of central tendency and variability. Indices of central tendency are ways of describing the "typical" or the "average" value in the distribution. Indices of variability, on the other hand, describe the "spread" or the extent of difference among the observations making up the distribution.

Perhaps the more basic concern is the description of central tendency. This will be treated first for obtained frequency distributions, and then the ideas will be extended to probability distributions as well. Next, the measurement of the dispersion or spread of a frequency distribution will be taken up, and a parallel treatment will once again be given probability distributions. Finally, an attempt will be made to show how two indices, the mean and standard deviation, form the cornerstone of most statistical inference.

4.1 / THE SUMMATION NOTATION

In this chapter it will be necessary to employ the summation symbol Σ (capital Greek sigma). This is read as "the sum of," and tells us to take the sum of the values represented by the expressions following the symbol. Thus, for example, Σx stands for the sum over all the different values that the variable X can assume. Most simple statistical derivations involve various sums, and the use of this symbol introduces considerable economy of statement into these formulations.

(Strictly speaking, when we wish to indicate summation over a set of values which are labeled as x_1, x_2, and so on up to x_N, this should be indicated by writing

$$\sum_{i=1}^{N} x_i.$$

However, often, when the context is clear, this will be abbreviated to

$$\sum_i x_i, \text{ or, occasionally, } \sum x_i, \text{ or even } \sum x.$$

Nevertheless, the $i = 1$ at the bottom and the N on top of the summation should still be understood. Printed summation signs also vary in size, from very large to very small, depending on the space requirements of the expression that follows the sigma, or the available space for Σ in some formulas. However, the size is irrelevant: it always means to sum the values given by the expression immediately following.)

There are a number of simple rules for the algebraic manipulation of the summation sign. These rules are given and illustrated in Appendix A. The student who has not already encountered summation notation in school algebra or elsewhere is urged to study these rules until thoroughly familiar with them. The exercises following Appendix A should also be done. Actually, the rules themselves are easy, and a little practice at writing out the sums symbolized can familiarize one very quickly with the various ways sums can be manipulated. A little time spent in this way will greatly increase your ability to follow the simple mathematical arguments used in later sections.

4.2 / MEASURES OF CENTRAL TENDENCY

Imagine an obtained distribution of numerical scores. If you were asked to state *one value* that would best "capture" and communicate the distribution as a whole, which value should you choose? One answer is to find that score-value which is a good "bet" about any randomly selected case from this distribution. Such a score may not be exactly correct for any given case, but it should be a good guess about the obtained score for that case. However, there are at least three ways to specify what we mean by a "good bet" about any case: (1) the most frequent (most probable) measurement class, (2) the point exactly midway between the top and bottom halves of the distribution, and (3) the arithmetic average of the distribution. The first of these ways of defining the central tendency leads to the measure known as the **mode;** the second leads to the **median** of the distribution; the third is merely the familiar average, or **mean.**

The mode is the most easily computed and the simplest to interpret of all the measures of central tendency. *The mode of a frequency distribution is merely the midpoint or class name of the most frequent measurement class.* If a case were drawn at random from the distribution, then that case is more likely to fall in the **modal class** than any other. So it is that in the graph of any distribution, the modal class shows the highest "peak" or "hump" in the graph or interval. The graph of a continuous distribution reaches its maximum density at the mode. The mode may be used to describe any distribution, regardless of whether the events are categorized or numerical.

On the other hand, there are some disadvantages to the mode. One is that there may be more than one modal class: it is perfectly possible for two or more measurement classes to show frequencies that are equal to each other and higher than the frequency shown by any other class. In this case, there is ambiguity about which class gives *the* mode of the distribution, as two or more values, or intervals are "most popular." Fortunately, this does not happen very often; even though there are two or more "humps" in the graph of a distribution, the the midpoint of the interval having the highest frequency or probability is still taken as *the* mode. Nevertheless, in such a distribution, where one measurement class is not clearly most popular, the mode loses its effectiveness as a characterization of the distribution as a whole.

Another disadvantage is that the mode is very sensitive to the size and number of class intervals employed when events are numerical; the value of the mode may be made to "jump around" considerably by changing the class intervals for a distribution.

Finally, the mode of a sample distribution forms a very undependable source of information about the mode of the basic probability distribution. For these reasons, our use of the mode will be restricted to situations where: (1) the data are truly nominal scale in nature; (2) only the simplest, most easily computed, measure of central tendency is needed.

The median score in any set of observations or in a distribution is also a good bet about any case in the total set represented. Just as its name implies, the median is the score corresponding to the middle case when all individual cases are arranged in order by scores, or to a score that divide the cases into two intervals having equal frequency. Thus, if you drew a case at random from any set of N observations, and guessed that this case showed the median score, you are just as likely to be guessing too high as too low.

The median for a set of observed data is defined in slightly different ways depending upon whether N is odd or even, and upon whether raw data or a grouped frequency distribution is to be described. For a set of raw scores **when N is odd the median corresponds to the score of individual number $(N + 1)/2$, when all individuals are arranged in order by scores; when N is even, the median is defined as the score-value midway between the scores for individual $(N/2)$ and individual $(N/2) + 1$ in order.** Then either for odd or even N it will be true that exactly as many cases fall above as fall below the median score.

On the other hand, this way of computing a median is not usually applicable if the data have been arranged into a grouped frequency distribution. **For any such grouped distribution, the median is defined as the point at or below which exactly 50 percent of the cases fall.** Consequently the first step in finding the median of a grouped distribution is to construct the *cumulative* frequency distribution. Such a cumulative frequency distribution is illustrated in Table 4.2.1.

The last column in Table 4.2.1 shows the cumulative frequencies for the class intervals. Since, by definition, the median will be that point in the distribution at or below which 50 percent of the cases fall, the cumulative frequency *at* the median score should be $.50N$. Thus, the cumulative frequency is $.50(200) = 100$ at the

Table 4.2.1

A cumulative frequency distribution for a sample of 200 cases.

Class	*f*	*cf*
74–78	10	200
69–73	18	190
64–68	16	172
59–63	16	156
54–58	11	140
49–53	27	129
44–48	17	102
39–43	49	85
34–38	22	36
29–33	6	14
24–28	8	8
	200	

median for this example. Where would such a score fall? It can be seen that it could not fall in any interval below the real limit of 43.5, since only 85 cases fall at or below that point. However, it does fall below the real limit 48.5, since 102 cases fall at or below that point. We have thus located the median as being in the class interval with real limits 43.5 and 48.5.

We must still ascertain the *score* that corresponds to the median, and this is where the process of interpolation comes in. The median score is somewhere in the interval 44–48. We assume that the 17 cases in that interval are evenly scattered over the interval width of 5 units, as in Figure 4.2.1.

The median score, which is greater than or equal to exactly 100 cases, must then exceed not only the 85 cases below 43.5, but also equal or exceed 15 cases above 43.5. In other words, the median is 15/17 of the way *up* the interval from the lower limit. Next, what score is exactly 15/17 of the way between 43.5 and 48.5? Since the difference between these two limits is 5, the class interval size, we take (15/17)5, or 4.4 as the amount that must be added to the lower real limit to find the median score, so that the median must be 43.5 + 4.4 or 47.9.

Figure 4.2.1

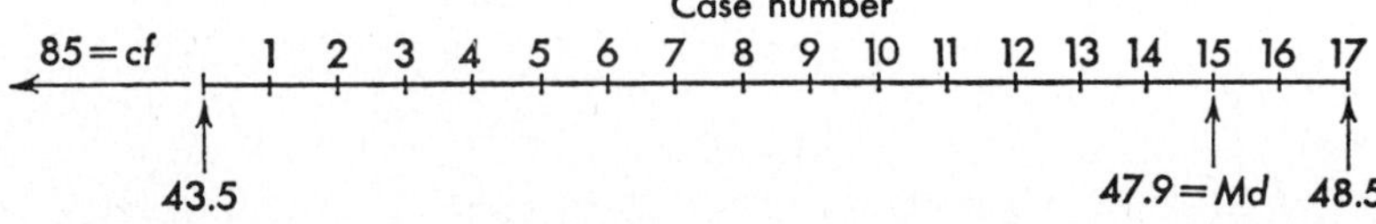

A little computational formula that summarizes the steps just described is given by:

$$\text{median} = \text{lower real limit} + i\,\frac{(.50N - cf \text{ below lower limit})}{f \text{ in interval}} \qquad [4.2.1]$$

where the lower real limit used belongs to the interval containing the median, and the *cf* refers to the cumulative frequency *up to* the lower limit of the interval. For the example, we find

$$\text{median} = 43.5 + 5\,\frac{[.50(200) - 85]}{17} = 47.9.$$

If the median can fall only in some class interval with nonzero frequency, this method of interpolation gives a unique value. However, if the interval frequency happens to be zero, then any point in the interval serves equally well as the distribution median; here one usually takes the midpoint as the median.

Even when the raw data are available, it is often worthwhile to define and compute the median as for a grouped frequency distribution. In the first place, for a sizable N, ordering all of the cases by their score magnitudes can be a considerable chore, and it may be simpler to construct a grouped distribution. Secondly, a troublesome problem arises when two or more cases in the raw data are tied in order at the median position, and here it often makes sense to calculate the median by interpolation as for a grouped distribution.

In principle, a median may be found for any distribution in which the variable represents an interval or even an ordinal scale; it is not applied to a distribution in which the measurement classes are purely categorical, since such classes are unordered.

The median is considerably less sensitive to the distribution's grouping into class intervals than is the mode. Furthermore, when one is making inferences about a large ''population'' of potential observations from a sample, the median is generally more useful and informative than the mode, although the median itself is not ordinarily so useful as the mean, to be discussed next. We will have more to say about characteristics of the median in future sections.

By far the most used and familiar index of central tendency for a set of raw data or a distribution is the *mean,* or *simple arithmetic average*. Surely everyone knows that to take the average of a set of raw scores you simply add them all up and divide by the total number, N:

$$M = \frac{\sum_i x_i}{N}. \qquad [4.2.2^*]$$

(Here, x_i stands for the score of the observation labeled i, and the sum is taken over all of the N different observations i.) Thus, equation 4.2.2* defines the mean for any set of raw data in the form of numerical scores.

Since expressions representing means will occur so frequently in all of the later sections, it is well to point out that we might also represent the arithmetic mean by

$$M = \sum_i \frac{x_i}{N},$$

standing for each value x_i first divided by N and then summed over the individual observations i. The value M represented by either of these expressions is precisely the same; this accords with rule 1 of Appendix A, that the sum of N observations each multiplied by a constant number is the same as the sum itself multiplied by that number. Therefore, in succeeding sections sometimes one and sometimes the other way of expressing the mean will be used, depending on the algebraic and typographical requirements of the particular discussion in which these expressions occur.

Incidentally, it should be mentioned that other texts in statistics often use other symbols for the sample mean. Frequently, the sample mean is shown as $\overline{X}$ or as $\bar{x}$, when X is the variable of interest. Although fine for samples, these symbols do not have clear parallels when one wishes to discuss a population mean. Since we will frequently wish to emphasize such parallels, we will symbolize the sample mean by M.

The definition and computation of the arithmetic mean for raw data is simple enough, but the situation is slightly more complicated when one wishes to find the mean of a grouped distribution of scores. You will recall (Section 2.7) that when a distribution was grouped in class intervals the midpoint x_j of each class interval j was taken to represent the score of each of the cases in the interval. Thus, in an interval 59–73 with midpoint 66 and frequency 16, the sum of the scores of the 16 cases falling into this interval is taken to be 66 summed 16 times, or (66) (16) = $x_j f_j$. Similarly, when all of the scores in any interval are assumed the same, their sum is the midpoint of that particular interval times the frequency for that interval, or $x_j f_j$. Then the sum of *all* of the scores in the distribution is taken to be the sum of the values of x times f over all of the respective intervals

$$M = \frac{1}{N} \sum_{j=1}^{J} x_j f_j \qquad [4.2.3^*]$$

(Note that here x_j is the midpoint of any interval, f_j is the frequency corresponding to that interval, and the sum is taken over *all intervals*.)

Table 4.2.2

Computation of a mean from a grouped distribution

Class	*x*	*f*	*xf*
74–78	76	10	760
69–73	71	18	1278
64–68	66	16	1056
59–63	61	16	976
54–58	56	11	616
49–53	51	27	1377
44–48	46	17	782
39–43	41	49	2009
34–38	36	22	792
29–33	31	6	186
24–28	26	8	208
		200	10,040 = Σxf

$$M = \frac{\sum_j x_j f_j}{N} = \frac{10040}{200} = 50.2$$

For example, consider once again the distribution shown in Tables 4.2.1 and 4.2.2. The frequency of each class interval is multipled by its midpoint, and these are then summed and divided by N to give the distribution mean.

The mean calculated from a distribution with grouped class intervals need not agree exactly with that calculated from raw scores. Information is lost and a certain amount of inaccuracy introduced when scores are grouped and treated as though each corresponded to the midpoint of some interval. The coarser the grouping, in general, the more likely is the distribution mean to differ from the raw-score mean. For most practical work the rule of ten to twenty class intervals gives relatively good agreement, however. Nevertheless, it is useful to think of the mean calculated from any given distribution as the mean of that *particular* distribution, a particular set of groupings with their associated frequencies.

Actually, in modern statistical work it is somewhat unusual to find means calculated from grouped frequency distributions. Any extensive statistical analysis is done by computer these days, and the computer is perfectly capable of working directly with the raw data input, almost regardless of the number of cases involved. Even manual computations of means are much simpler than in former days. Pocket calculators are easy to use for computing means and other statistical indices, and a great many of these calculators have built-in programs for computing means and other statistical indices, where all you have to do is to enter in the raw numbers. These are nice features to look for if you are shopping for a calculator.

There are also two "minor means" which are encountered, though rarely, in statistics. When the influence of extreme values is to be minimized, sometimes one employs the *harmonic mean,* defined by

$$M_H = \frac{N}{\sum_i \left(\frac{1}{x_i}\right)}.$$

The harmonic mean makes a single appearance in this text (expression 12.12.4 and the comment following). The other is the *geometric mean,* defined by

$$M_G = (x_1 x_2 \cdots x_N)^{\frac{1}{N}},$$

the product of all of the x_i values with the Nth root taken. Then the logarithm of M_G is the average logarithm of the x values. (This mean is often used to find averages of ratios. The geometric mean will be used in expression 13.7.4.)

4.3 / THE MEAN AS THE "CENTER OF GRAVITY" OF A DISTRIBUTION

The mean of a distribution parallels the physical idea of a center of gravity, or balance point, of ideal objects arranged in a straight line. For example, imagine an ideal board having zero weight. Along this board are arranged stacks of objects at various positions. The objects have uniform weight and differ from each other only in their positions on the board. The board is marked off in equal units of some kind, and each object is assigned a number according to its position. This is shown in Figure 4.3.1. Now given this idealized situation, at what point would a fulcrum

Figure 4.3.1

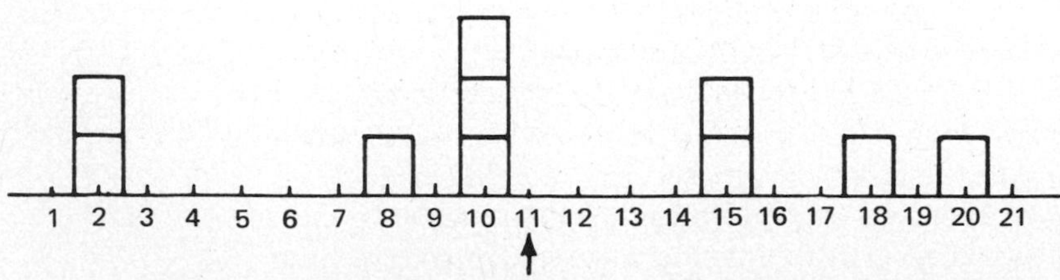

placed under the board create a state of balance? That is, what is the point at which the "push" of objects on one side of the board is exactly equal to the push exerted by objects on the other side? This is found from the mean of the positions of the various objects:

$$M = \frac{2 + 2 + 8 + 10 + 10 + 10 + 15 + 15 + 18 + 20}{10} = 11.$$

Here, the board would exactly balance if a fulcrum were placed at the position marked 11. Note that since there were piles of uniform objects at various positions on the board, this center of gravity was found in exactly the same way as for the mean of a distribution, since the position (midpoint of an interval) was in effect multiplied by the number of objects at that position (the class frequency), and then these values were summed and divided by the number of objects (the total frequency).

In short, the position of any object on the board is analogous to the score of a case, and each case is treated as having equal "weight" in our computations. The arithmetic mean is then like the center of gravity, or balance point. The mean is that score about which deviations in one direction exactly equal deviations in the other.

This property of the mean is bound up in the statement that **the sum of the signed deviations about the mean is zero in any distribution.** A deviation from the mean is simply the signed difference between the score for any case and the mean score:

$$d_i = (x_i - M); \qquad [4.3.1]$$

then it is easy to show that

$$\begin{aligned} \sum_i d_i &= \sum_i X_i - \sum_i M = \sum_i X_i - NM \\ &= NM - NM = 0; \end{aligned} \qquad [4.3.2^*]$$

that is, the sum of the signed deviations about the mean is always zero.

A simple consequence of this fact is that *the mean signed deviation from the mean is zero:*

$$\frac{\sum_i d_i}{N} = \frac{0}{N} = 0. \qquad [4.3.3^*]$$

4.4 / "BEST GUESS" INTERPRETATIONS OF CENTRAL TENDENCY

We have just seen that the tendency for cases in a distribution to differ from the mean in one way is exactly balanced by the tendency to differ in the opposite way.

Suppose once again that you were told to *guess* the score of some case picked at random from a distribution. If you guessed the mean for each case you might be in error to some extent on each trial, since it need not be true that the mean is exactly the same as any obtained score. The *extent* of error for a given case is d, the departure of the true score from the mean. Over all possible cases that might be drawn from the distribution the average signed error would be $\sum_i d_i/N$, the mean signed deviation. But we have just seen that the mean signed deviation is always zero. Hence, the following statement is true: **if the mean is guessed as the score for any case drawn at random from the distribution, on the average the amount of signed error will be zero.** This is a most important interpretation of the mean: the mean is the best guess about the score of any case in the distribution, if one wishes the *average signed error* to be zero.

Now suppose there were some distribution where you had to guess the score of a case picked at random, and you wished to be *absolutely right* with the highest possible probability. Then you should guess the mode rather than the mean; since it is the most frequent score, guessing the mode guarantees the greatest likelihood of hitting the score "on the nose" for a case drawn at random.

On the other hand, it might be that in guessing the score drawn at random you are not interested in being exactly right most often, nor in making signed error zero on the average, but rather wanted to make the smallest absolute error on the average. Here, the sign of the error is unimportant, but the size of the error is what matters. Then you should guess the median for any score. By doing so you would make the *smallest absolute error* on the average. The median is the typical score in this sense: it is closest on the average to all of the scores in the distribution.

There is really no way to say which is the best measure of central tendency in general terms. This depends very much on what one is trying to do and what one wants to communicate in summary form about the distribution. Each of the measures of central tendency is, in its way, a best guess about any score, but the sense of "best" differs with the way error is regarded. If both the size of the errors and their signs are considered, and we want zero error in the long run, then the mean serves as a best guess. If a miss is as good as a mile, and one wants to be exactly right as often as possible, the the mode is indicated. If one wants to come as close as possible on the average, irrespective of sign of error, then the median is a best guess.

From the point of view of purely descriptive statistics, as apart from inferential work, the median is a most servicable measure. Its property of representing the typical (most nearly like) score makes it fit the requirements of simple and effective communication better than the mean in many contexts.

On the other hand, the median is usually inferior to the mean when our purpose is to make inferences beyond the sample. The median has mathematical properties making it difficult to work with, whereas the mean is mathematically tractable. For this and other reasons, mathematical statistics has taken the mean as the focus of most of its inferential methods, and the median is relatively unimportant in inferential statistics. Nevertheless, as a description of a given set of data, the median is extremely useful in communicating the typical score.

4.5 / CENTRAL TENDENCY IN DISCRETE PROBABILITY DISTRIBUTIONS

The ideas of mean, median, and mode apply to distributions of discrete random variables just as they do to frequency distributions. However, as we shall see, in probability distributions measures of central tendency such as the mean often play a much more important role: not only do such measures summarize the distribution; they may also serve as parameters entering into the mathematical rule giving the probability for any value or interval of values of the random variable.

For a discrete random variable, the mode is simply the *most probable* value. For example, in the discrete distribution shown in Table 4.5.1 the mode is 41, since this is the midpoint of the most probable class (recall that each case in any interval of a grouped distribution is ordinarily treated as though it had the value of the midpoint).

A median value for a discrete random variable need not be unique. Any value qualifies if the probability for X less than or equal to that value equals .50, $p(X \leqslant Md) = .5$. The median is any value that evenly divides the distribution. The median of a grouped probability distribution is found in exactly the same way as for a frequency distribution, except that probabilities take the place of frequencies in the computations:

$$Md = \text{lower real limit} + i\left[\frac{.50 - p(X \leqslant \text{lower real limit})}{p(\text{lower limit} \leqslant X \leqslant \text{upper limit})}\right] \qquad [4.5.1]$$

where "lower limit" and "upper limit" refer to the interval containing the median. The example below shows the computation of the median in such a distribution.

Finally, the mean of a discrete distribution is found much as for a grouped frequency distribution: each distinct value of X, or midpoint of an interval, is multiplied by the *probability* that X takes on that value (or that X lies in the interval). Suppose there are J intervals, any one of which might be called j, with midpoint x_j. Then the sum of these products is found:

$$\text{mean} = \sum_{j=1}^{J} x_j\, p(x_j) \qquad [4.5.2^*]$$

summed over all possible values of x_j. Alternatively,

$$\text{mean} = \sum_{x} xp(x), \qquad [4.5.3^*]$$

where the sum is taken over all possible values of x.

In general, **the mean of a discrete random variable is the sum of the products of the different values of X each times the probability that X takes on that value.**

All three central tendency indices are found in Table 4.5.1, an example of a probability distribution grouped into class intervals.

Table 4.5.1

Computation of central tendency measures in a discrete probability distribution.

Class interval	X	*p(X in interval)*	*xp(X in interval)*
74–78	76	.050	3.80
69–73	71	.090	6.39
64–68	66	.080	5.28
59–63	61	.080	4.88
54–58	56	.055	3.08
49–53	51	.135	6.89
44–48	46	.085	3.91
39–43	41	.245	10.85
34–38	36	.110	3.96
29–33	31	.030	.93
24–28	26	.040	1.04
			50.21

Mode = 41 (midpoint of interval with probability = .245)

$$\text{Median} = 43.5 + \frac{5(.50 - .425)}{.085} = 47.91$$

(where .425 is $p(X \leqslant 43.5)$)

Mean = $\Sigma xp(X \text{ in interval}) = 50.21$

4.6 / THE MEAN OF A RANDOM VARIABLE AS THE EXPECTATION

A special term is often used to denote the mean of a probability distribution. This is the "expectation" or the "expected value" of a random variable X. The symbol $E(X)$ simply represents the mean of the probability distribution of X, and if X is discrete,

$$E(X) = \sum_x xp(x) = \text{mean of } X, \qquad [4.6.1^*]$$

the sum being taken over all values that X can assume.

The mean or expectation of a continuous random variable is defined in a way very similar to that for a discrete variable. However, since X may assume any of an infinite set of paricular values, and because, for the reasons outlined in Section 2.18, it is necessary to discuss the *probability density* associated with any particular value of X, the actual definition of the mean is somewhat different. If we let $f(x)$ symbolize the probability density associated with any particular value of X, then

$$E(X) = \int_{-\infty}^{\infty} xf(x)\,dx. \qquad [4.6.2]$$

Here, the integral sign indicates the infinite sum of x times a factor $f(x)\,dx$ for all real number values between the ultimate limits of $-\infty$ and $+\infty$. Notice that, much as in the discrete case, the expectation of the continuous random variable X is actually a sum of products. In the former the products were of values of X each times time probability of that value, but for the continuous case each value of X is weighted by a factor depending on the probability density at that value.

Since the idea of expectation is so pervasive in theoretical statistics, it is very convenient to have available some list of the formal rules for dealing with expectations mathematically. These rules are summarized in Appendix B and should clarify the ways that expectations will be treated algebraically in other sections. As with Appendix A exercises are available. The student is advised to become familiar with these rules, as with the rules of summation, which they greatly resemble. A student who does so should have very little trouble in following the simple derivations in mathematical statistics such as this book contains.

The rules for expectations and their manipulations given in Appendix B are valid either for continuous or for discrete random variables, with a few minor exceptions that need not bother us in an elementary discussion. This is extremely convenient, since it makes it possible to demonstrate certain general features of statistics without having to qualify the result as pertaining to discrete or continuous variables.

4.7 / EXPECTATION AS EXPECTED VALUE

The idea of expectation of a random variable is closely connected with the origin of statistics in games of chance. Gamblers were interested in how much they could "expect" to win in the long run in a game, and in how much they should wager in certain games if the game was to be "fair." Thus, expected value originally meant the expected long-run winnings (or losings) over repeated play; this term has been retained in mathematical statistics to mean the long-run average for any random variable over an indefinite number of samplings.

The use of expectation in a game of chance is easy to illustrate. For example, suppose that someone were setting up a lottery, selling 1000 tickets at $1 per ticket. A prize of $750 will go to the winner of the first draw. Suppose now that you buy a ticket. How *good* is this ticket in the sense of *how much you should expect to gain?* Should you have bought it in the first place? You can think of your chances of winning and losing as represented in a probability distribution where the outcome of any drawing falls into one of two event categories:

Class	*Prob.*
win	1/1000
don't win	999/1000

Translated into the amount of money gained (the random variable X), and with a loss regarded as a negative gain, this distribution becomes

x	$p(x)$
\$749	1/1000
−\$ 1	999/1000

Since this is the distribution of a discrete random variable, the mean of the distribution can be found by the methods of Section 4.5; that is,

$$E(X) = \Sigma xp(x) = 749(1/1000) + (-1)(999/1000)$$
$$= .749 - .999 = -.25$$

This amount, a *minus* 25 cents, is the amount which you can expect to gain by buying the ticket, meaning that if the lottery were run over and over indefinitely and you bought a ticket each time in the long run you should be poorer by a quarter. Should you buy the lottery ticket? Probably not, if you are going to be strictly rational about it; the mean winnings (the expected value) is certainly not in your favor.

On the other hand, suppose that the prize offered were $1000, so that the gain in winning is $999. Now the expected value is

$$E(X) = 999(1/1000) + (-1)(999/1000) = .00.$$

Here the lottery is more worth your while, as there is at least no amount of money to be lost *or* gained in the long run. Such a game is often called "fair." Obviously, truly fair lotteries and other games of chance are hard to find, since their purpose is to make money for the proprietors, not to break even or lose money. If the prize were $2000, you would likely take the opportunity, as in this case the expected value would be exactly the *gain* of one dollar.

In figuring odds in gambling situations one uses the expected value to find what constitutes a fair bet; that is, a bet where the mean of the probability distribution of gains and losses is zero. For instance, it is known that in a particular game the odds are 4 to 1 *against* winning. This means that the probability of winning is 1/5 and that of losing is 4/5. It costs the player exactly $1 to play the game once. How much should the amount he gains by winning be in order to make this a fair game with expectation of zero? The expectation is

$$E(X) = (\text{gain value})p(\text{win}) + (\text{loss value})p(\text{lose})$$
$$= (\text{gain value})(1/5) - \$1(4/5).$$

Setting $E(X)$ equal to zero and solving gives

$$(\text{gain value})(1/5) - \$1(4/5) = 0$$

or

$$\text{gain value} = \$4.$$

In short, one should stand to gain $4 for $1 put up if the game is to be fair. In general, if odds are A to B against winning, the game or bet is fair when B dollars put up gains A dollars.

In betting situations, the random variable is, of course, gains or losses of amounts of money, or of other things having utility value for the person. Nevertheless, the same general idea applies to any random variable; the expectation is the long-run average value that one should observe.

4.8 / THEORETICAL EXPECTATIONS: THE MEAN OF THE BINOMIAL DISTRIBUTION

As an example of how the mean of a theoretical probability distribution may be deduced mathematically, we will consider the binomial distribution once again. Here we will see that the distribution rule alone dictates what the expectation of a binomial variable must be. What we will show is that if X is a binomial variable, then

$$E(x) = Np, \qquad [4.8.1^*]$$

the expectation is the number of observations times the probability of a success.

We start off with the definition of expectation for any discrete random variable

$$E(X) = \sum_x xp(x).$$

For the binomial distribution, the probability that the number of successes X takes on any value r is $\binom{N}{r} p^r q^{N-r}$ for $0 \leq X \leq N$. Thus, any value r multiplied by the probability $p(X = r)$ is

$$r\left[\frac{N!}{r!\,(N - r)!}\,p^r q^{N-r}\right]. \qquad [4.8.2]$$

Now notice that we could factor this expression somewhat, canceling r in the numerator and denominator and bringing an N and a p outside the brackets:

$$Np\left[\frac{(N - 1)!}{(r - 1)!(N - r)!}\,p^{r-1} q^{N-r}\right]. \qquad [4.8.3]$$

For $r = 0$, expression 4.8.2 is equal to zero, and so there is no equivalent expression 4.8.3. On substituting expression 4.8.3 into the expression for $E(X)$ we have

$$E(X) = \sum_{r=1}^{N} Np\left[\frac{(N - 1)!}{(r - 1)!(N - r)!}\,p^{r-1} q^{N-r}\right] \qquad [4.8.4]$$

with the sum going from $r = 1$ to $r = N$, since the term is zero for $r = 0$. By rule 1 in Appendix A, this is the same as

$$E(X) = Np \sum_{r=1}^{N} \frac{(N-1)!}{(r-1)!(N-r)!} p^{r-1} q^{N-r}. \qquad [4.8.5]$$

However, if we wrote out the various terms represented in expression 4.8.5 beyond the summation sign, we would find that each is a binomial probability for a distribution with parameters $N - 1$ and p, and thus their sum must be 1.00. Thus

$$E(X) = Np.$$

We have just proved that the mean of a binomial distribution depends only on the two parameters, N and p. If $N = 10$ and p is 1/2, then the expectation or mean of the distribution is (10) (1/2) = 5. If $N = 25$ and $p = .3$, the mean is (25) (.3) = 7.5, and so on. Notice that the mean *can* be some value that X cannot take on, as in this last example. Nevertheless, the mean or expectation is a perfectly good statement about the "best guess" for any set of N observations, provided that we want our long-run error to be zero in guessing.

In a very similar way, the mean for a Poisson distribution can be found. It turns out that

$$E(X) = m. \qquad [4.8.6]$$

The mean for a Poisson distribution is its "intensity" parameter, m.

Furthermore, in a closely related fashion one can show that the expectation of a Pascal variable N, with probability given by

$$p(N = n; r, p) = \binom{n-1}{r-1} p^r q^{n-r},$$

is

$$E(N) = \frac{r}{p}. \qquad [4.8.7]$$

That is, the expected number of trials required in order to achieve the rth success from a Bernoulli process is r/p. Hence if $p = 1/2$ and $r = 5$, we expect that 10 trials will be required to reach the fifth success, and so on for any other r and p.

4.9 / THE MEAN AS A PARAMETER OF A PROBABILITY DISTRIBUTION

In many important instances of probability distributions, the mean or expectation is a parameter that enters into the function rule assigning a probability or probability-density to each possible value of X. A simple example is the binomial distribution of sample proportions (or, more properly, the distribution of sample proportions P, which can be found from the binomial distribution). Here, it is easy to show that $E(P) = p$, one of the two parameters figuring in the function rule for the binomial distribution.

In the preceding section we showed that, for a binomial distribution with parameters p and N,

$$E(X) = Np.$$

Now suppose the random variable were

$$P = \frac{X}{N}. \qquad [4.9.1^*]$$

By rule 1 for expectations (Appendix B),

$$E(P) = E\left(\frac{X}{N}\right) = \frac{E(X)}{N} = \frac{Np}{N} = p. \qquad [4.9.2^*]$$

For the binomial distribution of proportions P, the expectation is the parameter p.

We have also seen that for a Poisson distribution, the mean of the distribution is the parameter m. For many, although not all, distributions, the mean is one of the parameters of the distribution rule. (One exception is the Pascal distribution, where the mean involves both the parameters r and p.)

For this reason, it will sometimes be convenient to use still another symbol for the mean of a random variable, or of a distribution, especially when we are thinking of the mean in its role as a parameter. In these instances, by analogy with M for a sample, the lowercase Greek letter mu, or μ, will stand for the mean of a random variable or of a probability distribution. That is,

$$E(X) = \mu = \text{mean of the distribution of } X. \qquad [4.9.3^*]$$

In much that follows, small Greek letters will be used to indicate parameters of probability distributions, while Roman letters will stand for sample values. The word "parameter" will always indicate a characteristic of a probability distribution, and the word "statistic" will denote a summary value calculated from a sample. Thus, given some sample space and the random variable X, the mean of the distribution of X is a parameter, μ, and the mean of any given sample of N values of X, or M, is a statistic.

4.10 / RELATIONS BETWEEN CENTRAL TENDENCY MEASURES AND THE "SHAPES" OF DISTRIBUTIONS

In discussions either of obtained frequency or of theoretical probability distributions, it is often expedient to describe the general "shape" of the distribution curve. Although the terms used here can be applied to any distribution, it will be convenient to illustrate them by referring to graphs of continuous distributions.

First of all, a distribution may be described by the number of relative maximum points it exhibits, its "modality." This usually refers to the number of "humps" apparent in the graph of the distribution. Strictly speaking, if the density (or probability or frequency) is greatest at one point, then that value is *the* mode, regardless of whether other relative maxima occur in the distribution or not. Nevertheless, it is common to find a distribution described as **bimodal** or **multimodal** whenever there are two or more pronounced humps in the curve, even though there is only one distinct mode. Thus, a distribution may have no modes (Fig. 4.10.1), may be unimodal (Fig. 4.10.2 and 4.10.4), or may be multimodal (Fig. 4.10.3). Once again, notice that the possibility of multimodal distributions lowers the effectiveness of the mode as a description of central tendency.

Figures 4.10.1–4.10.4

Idealized distributions of various "shapes."

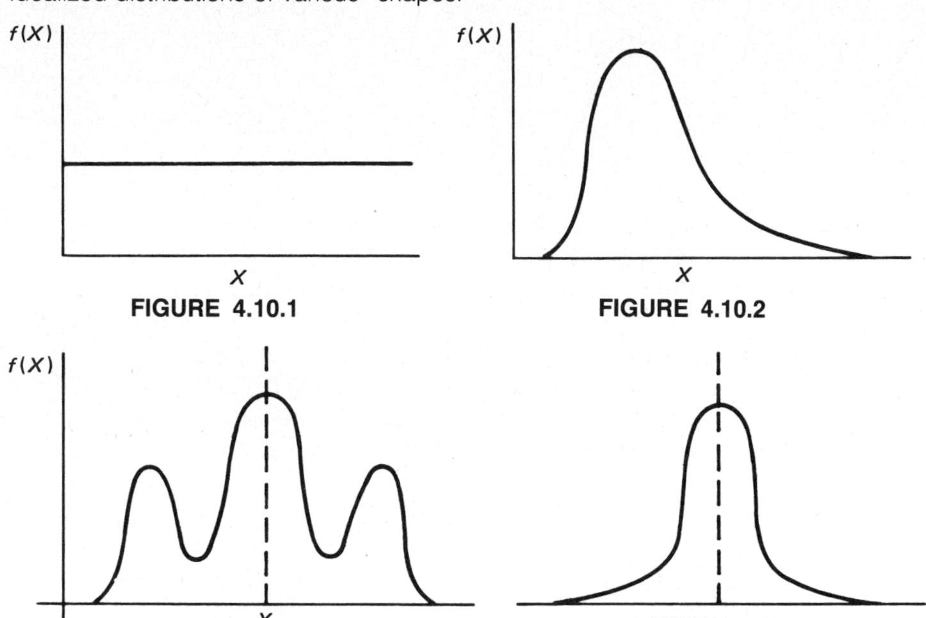

FIGURE 4.10.1

FIGURE 4.10.2

FIGURE 4.10.3

FIGURE 4.10.4

Another characteristic of a distribution is its symmetry, or conversely, its skewness. A distribution is **symmetric** only if it is possible to divide its graph into two ''mirror-image halves,'' as illustrated in Figure 4.10.3 and 4.10.4 Note that in the graph of Figure 4.10.2 there is no point at which the distribution may be divided into two similar parts, as in the other examples. When a distribution is symmetric, it will be true that the mean and the median are equal in value. It is not necessarily true, however, that the mode(s) will equal either the mean or the median; witness the example of Figure 4.10.5. On the other hand, a nonsymmetric distribution is sometimes described as **skewed,** which means that the length of one of the **tails** of the distribution, relative to the central section, is disproportionate to the other. For example, the distribution in Figure 4.10.6 is skewed to the right, or **skewed positively.** In a positively skewed distribution, the bulk of the cases fall into the lower part of the range of scores, and relatively few show extremely high values. This is reflected by the relation

Mean > Median

usually found in a positively skewed distribution.

On the other hand, it is possible to find distributions skewed to the left, or **negatively skewed.** In such a distribution, the long tail of the distribution occurs among the low values of the variable. That is, the bulk of the distribution shows relatively high scores, although there are a few quite low scores (Fig. 4.10.7). Generally in a negatively skewed distribution Median > Mean. Thus, a rough and ready way to

Figures 4.10.5–4.10.7

Relationships of central tendency measures in symmetric and skewed distributions.

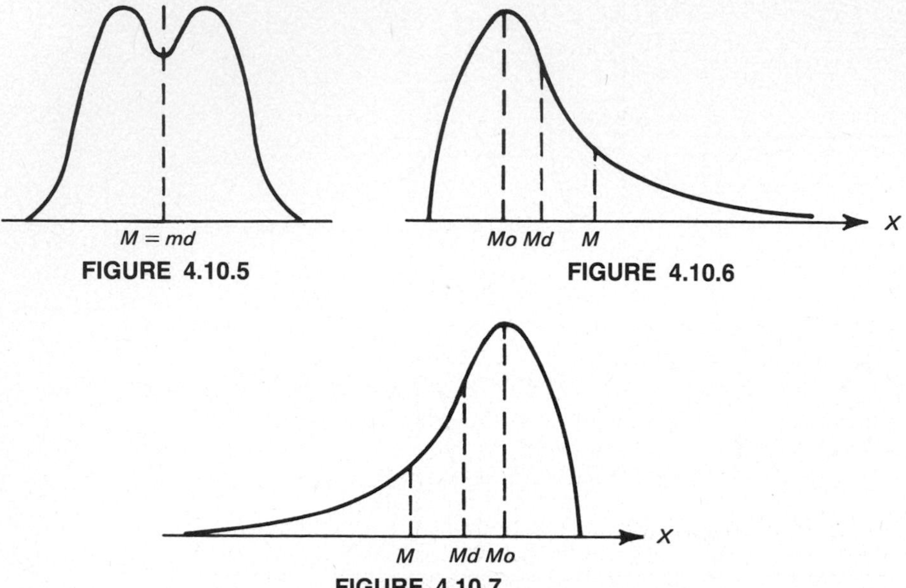

FIGURE 4.10.5

FIGURE 4.10.6

FIGURE 4.10.7

describe the skewness of a distribution is to find the mean and median; if Mean > Median, then you can conclude that the distribution is skewed to the right (positively). If Median > Mean, then you may conclude that negative skewness exists. If a more accurate determination is needed, other indices reflecting skewness are available, although space will not be devoted to these here. A word of warning: if a distribution is symmetric, then Mean = Median, but the fact that Mean = Median does not necessarily imply that the distribution is symmetric.

Describing the skewness of a distribution in terms of measures of central tendency again points up the contrast between the mean and the median as measures of central tendency. The mean is much more affected by the extreme cases in the distribution than is the median. Any alteration of the scores of cases at the extreme ends of a distribution will have no effect at all on the median so long as the rank order of all of the scores is roughly preserved; only when scores near the center of the distribution are altered is there a good chance of altering the median. This is not true of the mean, which is very sensitive to score changes at the extremes of the distribution. The alteration of the score for a single extreme case in a distribution may have a profound effect on the mean. It is evident that the mean follows the skewed tail in the distribution, while the median does so to less extent. *The occurrence of even a few very high or very low cases can seriously distort the impression of the distribution given by the mean, provided that one mistakenly interprets the mean as the typical value*. If you are dealing with a nonsymmetric distribution, and you want to communicate the typical value, you must report the median. On the other hand, in spite of the distribution's shape, the mean always communicates the same thing: the point about which the sum of deviations is zero. *The choice of an index must depend on what the user is trying to get across about the distribution.*

Differences in shape are only rough, qualitative ways of distinguishing among distributions. The only adequate description of a theoretical distribution is its function rule. Distributions that look similar in their graphic form may be very different functions. Conversely, distributions that appear quite different when graphed actually may belong to the same family.

A description in terms of modality and skewness is sometimes useful for giving a general impression of what a distribution is like, but it does not communicate its essential character. Similarly, for obtained frequency distributions, an actual statement of the distribution contains far more information than is ever given by any report of central tendency or shape alone.

4.11 / MEASURES OF DISPERSION IN FREQUENCY DISTRIBUTIONS

An index of central tendency summarizes only one special aspect of a distribution, be it mode, median, or mean. Any distribution has at least one more feature that must be summarized in some way. Distributions exhibit **spread** or **dispersion,** the tendency for observations to depart from central tendency. If central tendency measures are thought of as good bets about observations in a distribution, then measures of spread represent the other side of the question: dispersion reflects the ''poorness'' of central tendency as a description of a randomly selected case, the tendency of observations *not* to be like the average.

The mean is a good bet about the score of any observation sampled at random from a distribution, but *no observed case need be exactly like the mean.* A **deviation** from the mean expresses how ''off'' the mean is as a bet about a particular case, or how much *in error* is the mean as a description of this case:

$$d_i = (x_i - M)$$

where x_i is the value of a particular observation.

In the same way, we could talk about a deviation from the median, if we wished,

$$d'_i = (x_i - Md)$$

or even perhaps from the mode. It is quite obvious that the larger such deviations are, on the whole, from a measure of central tendency, the more do cases differ from each other and the more spread does the distribution show. What we need is an index (or set of indices) to reflect this spread or variability.

First of all, why not simply take the average of the deviations about the mean as our measure of variability:

$$\frac{\sum_i (x_i - M)}{N}$$

or

$$\frac{\Sigma(x - M) \text{ freq}(x)}{N} ?$$

This will not work, however, because in Section 4.3 it was shown that in any frequency distribution the average signed deviation from the mean must be zero.

The device used to get around this difficulty is to take the *square* of each deviation from the mean, and then to find the average of these squared deviations:

$$S^2 = \frac{\sum_i (x_i - M)^2}{N} = \frac{\sum_i d_i^2}{N}. \qquad [4.11.1^*]$$

For any distribution, the index S^2, equal to the average of the squared deviations from the mean, is called the **variance** of the distribution. *The variance reflects the degree of spread, since S^2 will be zero if and only if each and every case in the distribution shows exactly the same score, the mean. The more that the cases tend to differ from each other and the mean, the larger will the variance be.*

The variance is defined in equation 4.11.1 *as though* the raw score of each of N cases were known, and can always be computed by this formula when the raw data have not been grouped into a distribution. However, when data are in a grouped distribution, an equivalent definition of the variance is given by

$$S^2 = \frac{\sum_j (x_j - M)^2 \operatorname{freq}(x_j)}{N} \qquad [4.11.2^*]$$

where x_j is, as usual, the midpoint of an interval j. Here, for each interval, the deviation of the midpoint from the mean is squared, and multiplied by the frequency for that interval. When this has been done for each interval, the average of these products is the variance, S^2. Just as with the mean, the value of S^2 calculated for a grouped distribution need not agree exactly with that based on the raw scores; nevertheless, if a relatively large number of class intervals is used these two values should agree very closely.

4.12 / THE STANDARD DEVIATION

Although the variance is an adequate way of describing the degree of variability in a distribution, it does have one important drawback. The variance is a quantity in squared units of measurement. For instance, if measurements of height are made in inches, then the mean is some number of inches, and a deviation from the mean is a difference in inches. However, the square of a deviation is in *square-inch units,* and thus the variance, being a mean squared deviation, must also be in square inches. Naturally, this is not an insurmountable problem: taking the positive square root of the variance gives an index of variability in the original units. **The square root of the variance for a distribution is called the standard deviation, and is an index of variability in the original measurement units.**

A letter S will be used to denote the standard deviation for a frequency distribution:

$$S = \sqrt{S^2} = \sqrt{\frac{\sum_i (x_i - M)^2}{N}} \qquad [4.12.1^*]$$

or

$$S = \sqrt{\frac{\sum_j (x_j - M)^2 \operatorname{freq}(x_j)}{N}}. \qquad [4.12.2^*]$$

THE COMPUTATION OF THE VARIANCE AND STANDARD DEVIATION
Section 4.13

The variance and standard deviation can be computed from a list of raw scores by formulas 4.11.1 and 4.12.1. This "deviation method" entails finding the mean, subtracting it successively from each score, squaring each result, adding the squared deviations together, and dividing by N to find the variance. Obviously, this can be relatively laborious for a sizable number of cases. However, it is possible to simplify the computations by a few algebraic manipulations of the original formulas. This may be shown as follows: For any single deviation, expanding the square gives

$$(x_i - M)^2 = x_i^2 - 2x_iM + M^2,$$

so that, on averaging these squares, we have

$$\sum_i \frac{(x_i - M)^2}{N} = \sum_i \frac{(x_i^2 - 2x_iM + M^2)}{N}.$$

By rule 4 in Appendix A, the summation may be distributed so that

$$\sum_i \frac{(x_i - M)^2}{N} = \sum_i \frac{x_i^2}{N} - 2\sum_i \frac{x_iM}{N} + \sum_i \frac{M^2}{N}.$$

However, wherever M appears it is a constant over the sum, and so, by rules 1 and 2 in Appendix A,

$$\sum_i \frac{(x_i - M)^2}{N} = \sum_i \frac{x_i^2}{N} - 2M\sum_i \frac{x_i}{N} + \frac{NM^2}{N}$$

or

$$S^2 = \sum_i \frac{x_i^2}{N} - 2M^2 + M^2 = \frac{\sum_i x_i^2}{N} - M^2. \qquad [4.13.1^*]$$

Finally,

$$S = \sqrt{\frac{\sum_i x_i^2}{N} - M^2}. \qquad [4.13.2^*]$$

These last formulas, 4.13.1 and 4.13.2, give a way to calculate the indices S^2 and S with some saving in steps. These formulas will be referred to as the "raw-score computing forms" for the variance and standard deviation. For an example of the use of these forms, study the example shown in Table 4.13.1, based on seven cases. These two methods must always agree exactly with the values obtained by the deviation method, as in the example.

It is also possible to find similar computing forms for grouped distributions. Starting with the definition of the variance of a grouped distribution,

$$S^2 = \frac{\sum_j (x_j - M)^2 \text{ freq}(x_j)}{N}$$

Table 4.13.1

Computation of the variance and the standard deviation from raw data.

Scores	Deviation method		Raw-score method
x_i	$d_i = (x_i - M)$	d_i^2	x_i^2
11	6	36	121
10	5	25	100
9	4	16	81
8	3	9	64
6	1	1	36
−4	−9	81	16
−5	−10	100	25
$35 = \sum_i x_i$	0	$268 = \sum_i d_i^2$	$443 = \sum_i x_i^2$

$M = 5$

$$S^2 = \frac{\sum_i d_i^2}{N} = \frac{268}{7} = 38.29$$

$$S^2 = \frac{\sum_i x_i^2}{N} - M^2 = \frac{443}{7} - 25 = 38.29$$

$$S = \sqrt{38.29} = 6.19$$

we could develop the following computing forms:

$$S^2 = \frac{\sum_j x_j^2 \operatorname{freq}(x_j)}{N} - M^2 = \frac{\sum_j x_j^2 \operatorname{freq}(x_j)}{N} - \left[\frac{\sum_j x_j \operatorname{freq}(x_j)}{N}\right]^2. \qquad [4.13.3^*]$$

Then, as usual,

$$S = \sqrt{S^2}$$

This little derivation is left to you as an exercise.

An example of the computation of the mean and standard deviation from a frequency distribution is given in Table 4.13.2.

Since the operations for squaring and summing a set of numbers are very simple on most pocket calculators, the variance and standard deviation are almost as easy to compute as the mean on these little devices, particularly if only a small set of numbers is involved. In addition, many small calculators are preprogrammed to yield the variance and the standard deviation directly (as well as the mean, of course) once the raw data are entered. Naturally, this is extremely convenient in many statistical problems.

In times past, grouping large sets of data into frequency distributions was the main labor-saving device for computing things such as the variance from such data. These days, when the data set is even moderately large, or when many variances or standard deviations are to be calculated, the work is almost always done on a computer.

Table 4.13.2

Variance and standard deviation from a grouped distribution.

Class	x	f	xf	x^2	x^2f
46–50	48	6	288	2304	13824
41–45	43	8	344	1849	14792
36–40	38	10	380	1444	14440
31–35	33	5	165	1089	5445
26–30	28	3	84	784	2352
21–25	23	1	23	529	529
		33	1284 = Σxf		51382 = $\Sigma x^2 f$

$$M = \frac{1284}{33} = 38.91$$

$$S^2 = \frac{51382}{33} - 1513.99 = 43.04 \qquad S = \sqrt{43.04} = 6.56$$

Statistical packages such as SPSS, referred to in the Introduction, are simple to use, and provide these and a great variety of other statistics, even for very large sets of data.

4.14 / SOME MEANINGS OF THE VARIANCE AND STANDARD DEVIATION

Since the variance and standard deviation will figure very largely in our subsequent work, it is well to gain some intuition about what they represent. One interpretation is provided by the fact that **the variance is directly proportional to the average squared difference between all pairs of observations:**

$$\sum_{(i,j)} \frac{(x_i - x_j)^2}{\binom{N}{2}} = \frac{2N}{N-1} S^2. \qquad [4.14.1]$$

(Here (i,j) indicates summation over all possible *pairs* of scores.) The variance summarizes *how different the various cases are from each other,* just as it reflects how different each case is from the mean. Given some N cases, the more that pairs of cases tend to be unlike in their scores, the larger the variance and standard deviation.

Still another way to think of the variance and standard deviation is by a physical analogy. A deviation from the mean can be identified with a certain amount of *force* exerted by a variety of factors making this case different from others in its group. Think of any score as composed of the mean, plus a deviation from the mean,

$$x_i = M + d_i,$$

where the deviation is the resultant force of these "influences." Picturing a deviation from the mean as we would show a physical force away from a point, we have:

$$\underbrace{M \longrightarrow x_i}_{d_i}.$$

Now suppose that we think of two cases from the larger group, each of which has a deviation from the mean, or d_1 and d_2 respectively. These two cases are independent, so that we can represent their deviations as forces acting at right angles (Fig.

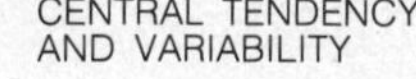

Figures 4.14.1/4.14.2

Physical analogies to the standard deviation

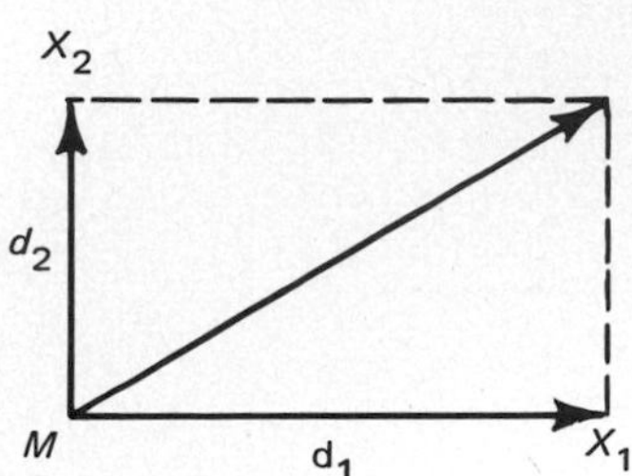

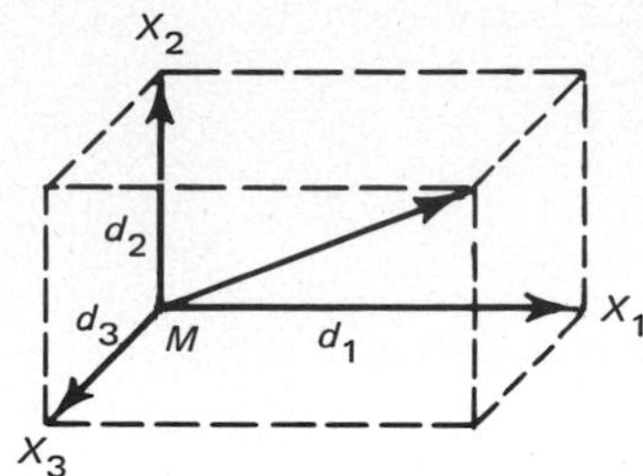

4.14.1). How would you find the *net force* away from the mean for these two cases? The rule of the parallelogram of forces shows the resultant force to be the *diagonal length* in Figure 4.14.1. This diagonal has length

$$\sqrt{\Sigma d^2} = \sqrt{(x_1 - M)^2 + (x_2 - M)^2}$$

by the Pythagorean theorem. If we divide by N before taking the square root, this looks much like the standard deviation.

Suppose that there were three independent cases. When their deviations are interpreted as forces away from the mean, the diagram shown in Figure 4.14.2 holds, and the resultant force is $\sqrt{d_1^2 + d_2^2 + d_3^2}$. The resultant force away from the mean per observation would again be calculated much like a standard deviation. In short, a physical analogy to the standard deviation is a resultant force away from the mean per unit observation. A large standard deviation is analogous to a large "push" away from the mean per observation, due to all the factors making observations heterogeneous. In statistics, "error" is often viewed as such a resultant force away from homogeneity, and in terms of this physical analogy the standard deviation should reflect the net effect of such forces per observation.

The reader acquainted with elementary physics may recognize not only that the mean is the center of gravity of a physical distribution of objects, but also that the variance is the moment of inertia of a distribution of mass. Furthermore, the standard deviation corresponds to the radius of gyration of a mass distribution; this is the real basis for regarding the standard deviation as analogous to a resultant force away from the mean. These physical conceptions and their associated mathematical formulations have influenced the course of theoretical statistics very strongly, and have helped to shape the form of statistical inference as we will encounter it.

4.15 / THE MEAN AS THE "ORIGIN" FOR THE VARIANCE

This question may already have occurred to the reader: "Why is the variance, and hence the standard deviation, always calculated in terms of deviations from the *mean?* Why couldn't one of the other measures of central tendency be used as well?" The answer lies in the fact that **average squared deviation (i.e., the variance) is smallest when calculated from the mean.** That is if the disagreement of any score with the mean is indicated by the square of its difference from the mean, then on the average the mean agrees better with the scores than any other single value one might choose.

This may be demonstrated as follows: Suppose that we chose some arbitrary real number C, and calculated a "pseudo-variance" S_C^2 by subtracting C from each score, squaring, and averaging:

$$S_C^2 = \sum_i \frac{(x_1 - C)^2}{N}.$$

Adding and subtracting M for each score would not change the value of S_C^2 at all. However, if we did so, we could expand each squared deviation as follows:

$$\begin{aligned}(x_i - C)^2 &= (x_i - M + M - C)^2 \\ &= (x_i - M)^2 + 2(x_i - M)(M - C) + (M - C)^2.\end{aligned}$$

Substituting into the expression for S_C^2 and distributing the summation by rule 4, Appendix A, we have

$$S_C^2 = \sum_i \frac{(x_i - M)^2}{N} + 2 \sum_i \frac{(x_i - M)(M - C)}{N} + \sum_i \frac{(M - C)^2}{N}.$$

Notice that $(M - C)$ is the same for every score summed, so that by rules 1 and 2 in Appendix A

$$S_C^2 = \sum_i \frac{(x_i - M)^2}{N} + 2(M - C) \sum_i \frac{(x_i - M)}{N} + (M - C)^2.$$

However, the first term on the right above is simply S^2, the variance about the mean, and the second term is zero, since the average deviation from M is zero. On making these substitutions, we find

$$S_C^2 = S^2 + (M - C)^2.$$

Since $(M - C)^2$ is a squared real number it can be only positive or zero, and so S_C^2 must be greater than or equal to S^2. The value of S_C^2 can be equal to S^2 only when M and C are the same. In short, we have shown that the variance calculated about the mean will always be smaller than about any other point. This method of finding a value that has the property of minimizing the sum of squared deviations is an application of the so-called *principle of least squares*. We will encounter many examples of values determined by this principle in later chapters.

If we are going to use the mean to express the central tendency of a distribution, and if we let the standard deviation indicate the extent of error we stand to make in guessing the mean for any score, then this error-quantity is at its minimum when we guess the mean in place of any other single value for all scores.

On the other hand, if we appraise error by taking the absolute difference (disregarding sign) between a score and a measure of central tendency, then **the average absolute deviation is smallest when the median is used.** This is one of the reasons that squared deviations rather than absolute deviations figure in the indices of variability when the mean is used to express central tendency. When the median is used to express central tendency it is often accompanied by the average absolute deviation from the median to indicate dispersion, rather than by the standard deviation which is more appropriate to the use of the mean. The average absolute deviation is simply

$$A.D. = \sum_i \frac{|x_i - Md|}{N} \qquad [4.15.1]$$

where the vertical bars indicate a disregarding of sign. Analogically speaking this measure is to the median as the standard deviation is to the mean.

4.16 / THE VARIANCE AND OTHER MOMENTS OF A PROBABILITY DISTRIBUTION

The variance and standard deviation of a probability distribution have exactly the same interpretations as do the corresponding indices for a frequency distribution: each is a measure of the variability or spread, the former in squared units and the latter in the original units of the random variable. However, like the mean of a probability distribution, the variance (or standard deviation) often figures as a parameter entering into the function rule for the distribution.

The small Greek letter sigma, σ, is generally used to denote the standard deviation of a random variable, and σ^2 its variance. When the random variable is discrete, the variance is defined by

$$\sigma^2 = \Sigma[x - E(X)]^2 p(x) \qquad [4.16.1^*]$$

which is equivalent to

$$\sigma^2 = E(X^2) - [E(X)]^2, \qquad [4.16.2^*]$$

the expectation of the square of X minus the square of the expectation of X. Then the standard deviation σ of the random variable X is just the square root of the variance, exactly as in a frequency distribution. (See Appendix B.)

By methods almost identical to those used in Section 4.8 to find the mean of the binomial distribution, it can be shown that the variance of a binomial distribution of number of "successes" is

$$\sigma^2 = Npq \qquad [4.16.1^*]$$

Thus, the standard deviation for a binomial distribution is

$$\sigma^2 = Npq. \qquad [4.16.2^*]$$

If the random variable is P, the proportion of successes out of N observations, then the variance becomes

$$\sigma_P^2 = \frac{pq}{N}, \qquad [4.16.3^*]$$

and the standard deviation is

$$\sigma_P = \sqrt{\sigma^2} = \sqrt{\frac{pq}{N}}. \qquad [4.16.4^*]$$

Similarly, the Poisson distribution has a variance

$$\sigma^2 = m. \qquad [4.16.5]$$

(Note this interesting property of a Poisson distribution: its mean must have the same value as its variance.)

Finally, our other prime example of a discrete random variable, the Pascal, has a variance given by

$$\sigma^2 = \frac{r(1 - p)}{p^2}. \qquad [4.16.6]$$

The variance for any continuous distribution has the same form as for any other probability distribution:

$$\sigma^2 = E(X^2) - [E(X)]^2, \qquad [4.16.7^*]$$

the expectation of the square of the variable, minus the square of the expectation. This means that for a random variable X with probability density $f(x)$ at any point $X = x$, the variance is

$$\sigma^2 = \int_{-\infty}^{\infty} x^2 f(x)\, dx - [E(X)]^2. \qquad [4.16.8]$$

A truly mathematical treatment of distributions would introduce not only the mean and the variance, but also a number of other summary characteristics. These are the so-called "moments" of a distribution, which are simply **the expectations of different powers of the random variable.** Thus, the first moment about the origin of a random variable X is

$$E(X) = \text{the mean.}$$

The second moment about the origin is

$$E(X^2),$$

the third is

$$E(X^3),$$

and so on. When the mean is subtracted from X before the power is taken, then the moment is said to be **about the mean;** the variance

$$E[X - E(X)]^2$$

is the second moment about the mean;

$$E[X - E(X)]^3$$

is the third moment about the mean, and so on.

Just as the mean describes the "location" of the distribution on the X axis, and the variance describes its dispersion, so do the higher moments reflect other features of the distribution. For example, the third moment about the mean is used in certain measures of degree of **skewness:** the third moment will be zero for a symmetric distribution, negative for skewness to the left, positive for skewness to the right. The fourth moment indicates the degree of "peakedness" or **kurtosis** of the distribution, and so on. These higher moments have relatively little use in elementary applications of statistics, but they are important for mathematical statisticians in the study of the properties of distributions and in arriving at theoretical distributions fitting observed data. The entire set of moments for a distribution will ordinarily determine the distribution exactly, and distributions are sometimes specified in this way when their general function rules are unknown or difficult to state.

A major use of the mean and standard deviation is in transforming a raw score into a **standardized score,** showing the *relative status* of that score in a distribution. If you are given the information, "John Doe has a score of 60," you really know very little about what this score means. Is this score high, low, middling, or what? However, if you know something about the distribution of scores for the group including John Doe, you can judge the location of the score in the distribution. The score of 60 gives quite a different picture when the mean is 30 and the standard deviation 10 than when the mean is 65 and the standard deviation is 20.

Each value in any distribution can be converted into a standardized score, or z **score,** expressing the deviation from the mean in standard deviation units:

$$z = \frac{x - M}{S}. \qquad [4.17.1^*]$$

The z score tells how many standard deviations away from the mean is x. The two distributions mentioned above give quite different z scores to the score of 60:

$$z_1 = \frac{60 - 30}{10} = 3$$

$$z_2 = \frac{60 - 65}{20} = -.25.$$

The conversion of raw scores to z scores is handy when one wishes to emphasize the *location* or *status* of a score in the distribution, and in future sections we will deal with standardized scores when this aspect of any score is to be discussed.

Changing each of the scores in a distribution to standardized score creates a distribution having a "standard" mean and standard deviation: **the mean of a distribution of standardized scores is always 0, and the standard deviation is always 1.** This is easily shown as follows:

$$\text{mean of } z \text{ scores} = M_z = \sum_i \frac{z_i}{N} = \sum_i \frac{(x_i - M)}{NS}.$$

Since N and S are constant over the summation, this becomes

$$M_z = \frac{1}{NS} \sum_i (x_i - M) = 0 \qquad [4.17.2^*]$$

since the sum of deviations about the mean is always zero.

In a similar fashion,

$$\text{variance of } z \text{ scores} = S_z^2 = \sum_i \frac{z_i^2}{N} = \sum_i \frac{(x_i - M)^2}{NS^2}$$
$$= \frac{S^2}{S^2} = 1, \qquad [4.17.3^*]$$

so that

$$\text{standard deviation of } z \text{ scores} = S_z = 1. \qquad [4.17.4^*]$$

Although it is true that the mean and standard deviation of a distribution of z scores will always be 0 and 1 respectively, *changing the scores in any distribution to z scores does not alter the form of the distribution. The frequency of any given z score is exactly that of the X score corresponding to it in the original distribution.*

Standardized scores have exactly the same format in probability distributions as in frequency distributions. If X is a value of a random variable, then the corresponding standardized score, relative to the probability distribution, is

$$z = \frac{x - E(X)}{\sigma} = \frac{x - \mu}{\sigma}. \qquad [4.17.5^*]$$

Furthermore, the mean of the standardized scores is zero for any probability distribution:

$$E(z) = E\left[\frac{X - E(X)}{\sigma}\right] = \frac{E(X) - E(X)}{\sigma} = 0 \qquad [4.17.6^*]$$

since $E(X)$ and σ are constants over the various possible values of X. The standard deviation of standardized scores is always 1 in a probability distribution:

$$\sigma_z^2 = E(z^2) - [E(z)]^2 = E\left[\frac{X - E(X)}{\sigma}\right]^2 = E\,\frac{[X - E(x)]^2}{\sigma^2} \qquad [4.17.7^*]$$

$$= \frac{\sigma^2}{\sigma^2} = 1$$

and

$$\sigma_z = 1. \qquad [4.17.8^*]$$

Again, the form of the probability distribution is not changed at all by the transformation to z scores, in the sense that the probability (or probability density) of any value of z is simply the probability (or probability density) of the corresponding value of X.

4.18 / TCHEBYCHEFF'S INEQUALITY

There is a very close connection between the size of deviations from the mean and probability, holding for distributions having finite expectation and variance. The following relation is called Tchebycheff's inequality, after the Russian mathematician who first proved this very general principle:

$$\text{prob}(|X - \mu| \geq b) \leq \frac{\sigma^2}{b^2}, \qquad [4.18.1]$$

the probability that a random variable X will differ absolutely from expectation by b or more units ($b > 0$) is *always* less than or equal to the ratio of σ^2 to b^2. Any deviation from expectation of b or more units can be *no more probable* than σ^2/b^2.

This relation can be clarified somewhat by dealing with the deviation in σ units, making the random variable a standardized score. If we let $b = k\sigma$, then the following version of the Tchebycheff inequality is true:

$$\text{prob}\left(\frac{|X - \mu|}{\sigma} \geq k\right) \leq \frac{1}{k^2}; \qquad [4.18.2]$$

the probability that a standardized score drawn at random from the distribution has *absolute* magnitude greater than or equal to some positive number k is *always* less than or equal to $1/k^2$. Thus, given a distribution with some mean and variance, the probability of drawing a case having a standardized score of 2 or more (disregarding sign) must be *at most* 1/4. The probability of a standardized score of 3 or more must be no more than 1/9, the probability of 10 or more can be no more than 1/100, and so on.

Although this principle is very important theoretically, it is not very powerful as a tool in applied problems. The Tchebycheff inequality can be strengthened somewhat if we are willing to make assumptions about the general form of the distribution, however. For example, if we assume that the distribution of the random variable is both *symmetric* and *unimodal,* then the relation becomes

$$p(|z| \geq k) \leq \frac{4}{9}\left(\frac{1}{k^2}\right). \qquad [4.18.3]$$

For such distributions, we can make somewhat ''tighter'' statements about how large the probability of a given amount of deviation may be. For example, how probable is a score value falling three or more standard deviations from the mean in some distribution? By Tchebycheff's inequality, we know that regardless of what the true distribution of scores may be like, a score three or more standard deviations away from the expected value in either direction can be no more probable than 1/9, or about .11. However, if the basic distribution of scores can be assumed to be unimodal–symmetric, then the expression 4.18.3 above tells that a value three or more standard deviations from the mean should be observed with probability *no greater than* (4/9) (1/9), or about .05. Similarly, we could find that at least 1 − 4/9 or 5/9 of such scores fall within one standard deviation to either side of the mean. Furthermore, 1 − (4/9) (1/4) or 8/9 must fall within two standard deviations, and so on.

By adding assumptions about the form of the distribution to the general principle relating standardized values to probabilities we are able to make stronger and stronger probability statements about a sample result's departures from the mean. In order to make very precise statements one has to make even stronger assumptions about the distribution of the random variable, unless dealing with very large samples of cases, as we shall see. This chapter has concluded with the introduction of the Tchebycheff inequality to suggest that the mean and the standard deviation play a key role in the theory of statistical inference. The mean and standard deviation are, of course, useful devices for summarizing data if that is our purpose, although there are situaitons where other measures of central tendency and variability may do equally well or better. The really paramount importance of mean and standard deviation does not emerge until one is interested in making *inferences,* involving the estimation of parameters, or assigning probabilities to sample results. Here, we shall find that most ''classical'' statistical theory is erected around these two indices, together with their combination in standardized scores.

So far in this chapter we have been concerned with measures of central tendency and of variability. Such indices are summary measures of an entire distribution. However, another type of question concerns the relative location of a particular score value within the distribution. Next we will discuss percentiles and percentile ranks, which provide one way to answer this question of location of an individual or a score within any given distribution.

PERCENTILES AND AND PERCENTILE RANKS Section 4.19

In any frequency distribution of numerical scores, the percentile rank of any specific value x is the percent of cases out of the total that fall at or below x in value. In a probability distribution, the percentile rank of a given value x is simply $100F(x)$, or 100 times the cumulative probability associated with $X = x$. Note especially that the percentile rank associated with the median in any distribution must be 50, since 50 percent of all observed values in a frequency distribution must lie at or below the median, and $F(X = Md) = .50$ in a probability distribution.

One use of the idea of percentile ranks is in finding the relative locations of values within given distributions. The value need not actually have occurred in frequency distribution in order for a percentile rank to be determined. One asks where that value would have fallen relative to the other cases if it actually had occurred. This is done by linear interpolation:

$$\frac{(x - \text{value below } x)}{(\text{value above } x - \text{value below } x)}(\text{\%ile rank above } x - \text{\%ile rank below } x) + \text{\%ile rank below} = \text{\%ile rank of } x,$$

where "%ile" is simply an abbreviation for "percentile."

Such a use of percentile ranks is probably already familiar to most students, since this typically is the way that the test performance of an individual is compared to the performance of some reference group, or norm group.

The percent at or below a given value of x is the percentile rank of that value. Often, however, we are interested in the reverse problem: What score x cuts off the bottom G percent of the distribution? That score is then referred to the G percentile of the distribution. In order to find the value corresponding to the Gth percentile, we first look to see if there is any exact value that occurred in the distribution such that the number of cases falling at or below that value is $GN/100$. If so, the value is the Gth percentile. However, it may happen that no value can be located that corresponds exactly to the Gth percentile, among those values of X that actually occurred. In that case, linear interpolation is employed once again:

$$\frac{[(GN/100) - cf \text{ below})(\text{value above} - \text{value below})]}{(cf \text{ above} - cf \text{ below})} + \text{value below} = G\text{th percentile},$$

where "cf below" symbolizes the cumulative frequency nearest below the number $GN/100$ among the values of X that occurred, and "value below" is the actual X value associated with that cumulative frequency. Similarly, "cf above" stands for the cumulative frequency just above $GN/100$ for one of the values of X that occurred, and "value above" is the actual value associated with that cumulative frequency.

A similar set of problems exist when one is trying to find percentile ranks and percentiles in a grouped frequency distribution. Suppose that N is the total number in the distribution, i is the class interval size, and we want to find the percentile rank for some value of X. First of all we locate the class interval in which $X = x$ must fall. That done, we take

$$\text{percentile rank} = \frac{(x - \text{lower limit})}{i}\,\frac{100(\text{frequency in interval})}{N} + \frac{100(\text{cumulative frequency below interval})}{N},$$

using the lower limit of the interval and i, the class interval size, or (upper real limit − lower real limit).

In order to find the Gth percentile value, we find the interval in which this percentile value must fall by taking $NG/100$, and seeing that the interval chosen has a cumulative frequency greater than or equal to $NG/100$, with the interval just below having a cumulative frequency less than $NG/100$. (Remember that the cumulative frequency of an interval is the number of cases in the total distribution falling at or below the *upper* real limit of the interval.) Then we take

$$\frac{(NG/100 - cf \text{ for interval below})}{\text{frequency in interval}}(i) + \text{lower real limit of interval} = G\text{th percentile.}$$

Since we are using linear interpolation to find percentile ranks and percentiles in such a grouped distribution, it follows that we are assuming that within any given class interval, the scores falling into the interval are evenly spaced in value across the extent of the interval.

Occasionally one runs into reports of deciles, quartiles, or other "fractiles" of a distribution. Deciles are simply the values corresponding to the 10th, 20th, 30th, · · · percentiles: those values which divide the distribution into tenths. Similarly, quartiles divide the distribution into fourths and are the values of the 25th, 50th, 75th, and 100th percentiles. Any other arbitrary division of a distribution might be worked out by percentiles as the occasion demanded.

EXERCISES

1. Chooose and calculate the appropriate index of central tendency for the data of exercise 4, Chapter 2. Why must you choose this particular index? If you guess that any individual drawn at random from this group shows this status, how likely are you to be wrong?

2. A social psychologist determined the number of verbal exchanges between pairs of people attempting to solve a puzzle in a given time period. For 45 pairs, the numbers were as follows. Calculate the mean and the median numbers of exchanges observed for this group.

27	31	37	15	31	8	39	33	24
26	23	43	25	47	22	26	42	20
41	7	33	21	33	61	50	27	52
38	32	25	46	54	22	45	63	17
13	21	48	26	40	51	24	53	34

3. A sample of newly divorced women over forty was taken. Each gave the age (in nearest years) at which she had been first married. The results were as follows. Calculate the mean and the median ages.

20	24	20	27	18	22	22	35	26	25	30	18	25	26	30	18
19	25	27	24	26	20	22	28	24	21	24	32	26	34	28	16

4. Calculate the mean and the median from the raw (i.e., ungrouped) data presented in exercise 17, Chapter 2.

5. From the raw data of exercise 7, Chapter 2, calculate the mean income of the group.

6. Using the data of exercise 5, Chapter 2, find the mean, median, and mode for each of the three questionnaire items. What do the relative sizes of mean, median, and mode reflect about the distributions of responses?

7. A teacher was interested in the difference in the time it took students to complete each of two different tests of simple arithmetic. It was believed that practice effects should make the time shorter on the second test. A group of 25 students showed the following differences in time taken on the first and second tests (in minutes). Find the mean difference in time. Does the notion that the second test took less time seem to be supported by the data?

−2.3	−.4	−.6	1.7	1.9
−.3	.7	1.4	.8	−1.3
1.6	1.1	.2	.5	−1.8
−.5	.4	1.5	2.4	.8
.9	−1.5	−1.2	.5	1.2

8. Using the data of exercise 3 above, demonstrate that the sum of the deviations about the mean is indeed zero.

9. A nutritionist was studying the eating patterns of early adolescent boys and girls. Two randomly selected groups were used, the first consisting of 50 thirteen-year-old boys, and the second of 50 thirteen-year-old girls. The data follow in the form of two frequency distributions, showing the calories (in hundreds) consumed daily by the members of each group. Compute the median and the modal number of calories consumed by the boys and by the girls. (Leave the answer in hundreds of calories.)

Class interval	*Boys*	*Girls*
50.00–54.99	1	1
45.00–49.99	2	0
40.00–44.99	4	2
35.00–39.99	16	5
30.00–34.99	12	10
25.00–29.99	7	20
20.00–24.99	5	8
15.00–19.99	2	2
10.00–14.99	1	2
	50	50

10. For the distributions of exercise 9 calculate the mean number of calories consumed by each group. What do these means reveal about the two groups?

11. Suppose that an individual were drawn at random from 150 cases with a distribution like that in Table 2.6.1. Find the expected value for the score of this individual. What is the most likely value for the individual's score?

12. Compute the mean for the distribution of Table 2.4.1. How does this differ from the mean of the distribution in Table 2.6.1, based on the same cases? To what can you attribute the difference, if any?

13. Find the mean, median, and mode for the distribution shown in Table 2.8.2. What do the relative sizes of these three indices reflect about the distribution's form?

14. Choose and calculate the index of central tendency which seems to be most appropriate for the data of exercise 10, Chapter 2. Why did you choose this particular index?

15. Find the expected value for each of the three theoretical distributions described in Section 2.17.

16. Find the expected value of the random variable of exercise 19, Chapter 2.

17. Find the expected value for the random variable of exercise 20, Chapter 2.

18. Calculate the variance and standard deviation for the data of exercise 2 above.

19. What is the value of the standard deviation for the data of exercise 3 above?

20. In the same study referred to in exercise 3 above, another sample of married women over forty, who had never been divorced, was taken. These also reported the age at which they had been married, with the following results. Do the samples of newly divorced and married women differ in mean age of marriage and in the variability of age at marriage? If so, how?

18 27 30 26 28 26 34 22 26 24 23 31 26 29 26
22 20 24 28 32 27 32 20 28 20 26 35 24 29 26

21. Find the variance and standard deviation for the data of exercise 7.

22. What is the standard deviation of the raw income data of exercise 7, Chapter 2?

23. Compare the two groups of exercise 9 on their means and standard deviations. Which group has the higher caloric consumption, on the average? Which is the more variable in terms of calories consumed?

24. Calculate the variance and standard deviation for the distribution of Table 2.8.2.

25. Two fair dice are tossed independently and at random, yielding the number of "spots" on each trial ranging from 2 to 12. Find the variance and standard deviation of the random variable "number of spots coming up on the dice." (Refer to Section 2.15).

26. Find the mean and standard deviations of the following theoretical distributions:

(a) Binomial, with $N = 17$, $p = .35$.
(b) Binomial, with $N = 75$ and $p = .80$.
(c) Poisson, with $m = 1.25$.
(d) Pascal, with $r = 3$ and $p = .40$.
(e) Geometric, with $p = .19$.
(f) Binomial, with $N = 2000$ and $p = .01$.
(g) Poisson, with $m = 20$.

27. Find the variance and standard deviation for each of the three theoretical distributions of Section 2.17.

28. What is the standard deviation of the random variable of exercise 19, Chapter 2?

29. Find the standard deviation of the random variable of exercise 20, Chapter 2.

30. Determine the mean and standard deviation of the random variable, "number of items correct," based on the test described in exercise 19, Chapter 3. Assume that the student in question is only guessing throughout the test.

31. What is the mean and standard deviation of the random variable "number of women winners" as defined by exercise 20, Chapter 3.

32. Convert the ages in exercise 20 above to the corresponding standardized scores. Then show that (within a small rounding error) these new scores must have a mean of zero and a standard deviation of 1.00.

33. Describe the relative position of each of the following ages in the group of divorced women of exercise 3, using standardized scores. Then show where these same ages fall in the married group of exercise 20.

(a) 25
(b) 30
(c) 17

(d) 37
(e) 20

34. Convert the two distributions of exercise 9 into distributions of standardized or z-score values. Determine the following:

(a) What is the percentile rank corresponding to a z-value of -1 among the boys? Among the girls?
(b) An individual consumes 3300 calories per day. Where does this place him among the boys, in standardized score terms? Among the girls?
(c) Between what two standardized score values do the middle 90 percent of the girls lie? Between what two z-values do the middle 90 percent of the boys lie?
(d) Is a caloric intake of 4200 relatively more deviant for a boy or for a girl?

(**Hint:** Any grouped distribution can be converted into a z-score distribution by simply converting the limits of the intervals into z-scores.)

35. A certain probability distribution has a mean of 68 and set a standard deviation of 11. What is the maximum probability that a case drawn at random from this distribution will show a value greater than or equal to 115.5 or less than or equal to 20.5? In this distribution, what is the maximum probability of observing a case which is more than 1.7 standard deviations from the mean? What is the smallest proportion of cases that we should expect to fall in the interval 53.7 to 82.3 in this distribution?

Chapter 5

SAMPLING DISTRIBUTIONS AND POINT ESTIMATION

5.1 / POPULATIONS, PARAMETERS, AND STATISTICS

So far the entire set of elementary events has been called the *sample space,* since this term is useful and current in probability theory. However, in many fields using statistics it is common to find the word **population** used to mean the totality of potential units for observation. These potential units for observation are very often real or hypothetical sets of people, plants, or animals, and *population* provides a very appropriate alternative to *sample space* in such instances. Nevertheless, whenever the term ''population'' is used in the following, we shall mean only the sample space of elementary events from which samples are drawn.

Given a population of potential observations, the particular numerical score assigned to any particular unit observation is a value of a random variable; the distribution of this random variable is the **population distribution.** This distribution will have some mathematical form, with a mean μ, a variance σ^2, and all the other characteristic features of any distribution. If you like, you can usually think of the population distribution as a frequency distribution based upon some large but finite number of cases. However, population distributions are almost always discussed as though they were theoretical probability distributions; the process of random sampling of single units with replacement insures that the long-run relative frequency of any value of the random variable is the same as the probability of that value. Later we shall have occasion to idealize the population distribution and treat it as though the random variable were continuous. This is impossible for real-world observations, but we shall assume that it is ''true enough'' as an approximation to the population state of affairs.

Population values such as μ and σ^2 will be called **parameters of the population** (or sometimes, **true values**). Strictly speaking, a parameter is a value entering as an arbitrary constant in the particular function rule for a probability distribution, although the term is used more loosely to mean any value summarizing the population distribution. Just as parameters are characteristic of populations, so are **statistics** associated with samples.

Some amplification of this idea of a statistic is called for, however. We have already seen that the same sample of data may be used as the basis for a wide variety of statistics. Thus, even in samples from a Bernoulli process, we have examined both the number of successes out of N trials and the proportion of successes, as well as the trial number on which the rth success occurred (as in Pascal sampling). Each of these ways of regarding the same basic set of data provides a useful statistic. However, these are not the only statistics that might have been formed. We might, for example, have taken the product of the proportion of successes and the proportion of failures, or perhaps the logarithm of the proportion of successes. In some contexts we might have been interested in the reciprocal of the proportion of successes. In still other contexts we might have chosen to form a statistic in any of a variety of other ways.

The point is that there is no limit to the number of ways in which statistics can be constructed and associated with samples, even for samples as simple as binomial sequences. Not all of these statistics would be very useful perhaps, but we are perfectly free to define them. **A statistic is simply a function on samples, such that any sample is paired with a value of that statistic.** For samples of numerical data we ordinarily construct and use familiar statistics such as means, variances, medians, percentile ranks and the like because they happen to be simple and useful. However, in some situations we would want to use still other statistics, such as the sum of the logarithms of all of the values, or perhaps the sum of the reciprocals of the values observed, or the difference between the highest and lowest values, and so on for any other way of combining or transforming the values in the sample. Each way of relating samples to new ''summarizing'' values is legitimate.

Moreover, a statistic need not use all of the information in a sample. Certainly the median, like the other percentiles, appears to be based on less of the information in a sample than is the mean or the variance. If we wished, we could use even less of the information in a sample in defining a statistic; thus, we might define our statistic as merely the value of the fifth observation made, for example, and ignore the rest of the sample values. Indeed, we might even let the value of a statistic be a constant, having no relationship to the sample values themselves, so that none of the information in the sample is used in the formation of the statistic.

This point has been stressed in order to bring out a related point: Since there are no ''natural'' statistics among the wide variety possible, and if an unlimited variety of statistics might be associated with any given sample, we need criteria for choosing among such statistics. In the last chapter some descriptive statistics were compared on the basis of what they tell, and what we wish to communicate, about a given sample distribution of values. Now, however, we will examine sample statistics against criteria of what they tell about population distributions. While it is often true that the best way to gain information about a population parameter is through the use of the analogous sample statistic (that is, population mean and sample mean, population variance and sample variance), this is not always or necessarily true.

Thus, in Chapters 8 and 9 we will use some statistics that have no direct parallels in the population. It may be that some other statistic with a different form contains more, or more useful, information about the population from which the sample came than does a familiar statistic. The main business of this chapter is, then, to examine how population distributions determine (or induce) distributions of sample statistics, and then to outline some of the criteria that have been developed for the choice of statistics to be used in inferences about populations.

5.2 / SAMPLING DISTRIBUTIONS

In actual practice, random samples seldom consist of single observations. Almost always some N observations are drawn from the same population. Furthermore, the value of some statistic is associated with the sample. Interest then lies in the distribution of values of this statistic *across all possible samples* of N observations from this population. Accordingly, we must distinguish still another kind of theoretical distribution, called a sampling distribution.

A sampling distribution is a theoretical probability distribution that shows the functional relation between the possible values of a given statistic based on a sample of N cases, and the probability (density) associated with each value, for all possible samples of size N drawn from a particular population.

In general, the sampling distribution of values for a particular sample statistic will not be the same as the distribution of the random variable for the population. However, the sampling distribution always depends in some specifiable way upon the population distribution, provided the probability structure underlying the occurrence of samples is known.

Notice that this definition is not confined to simple random samples, even though in most applications it will be assumed that samples are drawn at random from the population. Nevertheless, some probability structure linking the occurrence of the possible samples with the population must exist and be known if the population distribution is to be related to the sampling distribution of any statistic. For our elementary purposes this probability structure will be that of simple random sampling, in which each possible sample of size N has exactly the same probability of occurrence as any other. However, in more advanced work assumptions other than simple random sampling are sometimes made.

Actually, we have already used sampling distributions in the preceding chapters. For example, a binomial distribution is a sampling distribution. Recall that a binomial distribution is based on a two-category population distribution, or Bernoulli process. A sample of N independent cases is drawn at random from such a distribution, and the number (or proportion) of successes is calculated for each sample. Then the binomial distribution is the sampling distribution showing the relation between each possible sample result and the theoretical probability of occurrence. The binomial distribution is *not* the same as the Bernoulli process unless N is 1; however, given the Bernoulli process and the size of the sample, N, the binomial distribution may be worked out.

Other examples of sampling distributions will now be given. A most important distribution we shall employ is the sampling distribution of the mean. Here, samples of N cases are drawn independently and at random from some population and each observation is measured numerically. For each sample drawn the sample mean M is calculated. **The theoretical distribution that relates the possible values of the sample mean to the probability (density) of each over all possible samples of size N is called the sampling distribution of the mean.**

Furthermore, for each sample of size N drawn, the sample variance S^2 may be found. The theoretical distribution of sample variances in relation to the probability of each is the sampling distribution of the variance. By the same token, *the sampling distribution of any summary characteristic* (mode, median, range, etc.) *of samples of N cases may be found, given the population distribution and the sample size N.*

5.3 / CHARACTERISTICS OF SINGLE-VARIATE SAMPLING DISTRIBUTIONS

A sampling distribution is a theoretical probability distribution, and like any such distribution, is a statement of the functional relation between the values or intervals of values of some random variable and probabilities. Sampling distributions differ from population distributions in that the *random variable is always the value of some statistic based on a sample of N cases,* such as the sample mean, or the sample variance, the sample median, etc. Thus, a plot of sampling distribution, such as that in Figure 5.3.1, always has for the abscissa (or horizontal axis) the different sample statistic values that might occur. Figure 5.3.1, for example, shows a theoretical distribution for sample variances for all possible samples of size 7 drawn from a particular population. Any point on the horizontal axis is a possible value of a sample variance, and the height of the curve on the vertical axis gives the probability density $f(S^2)$, for that particular value.

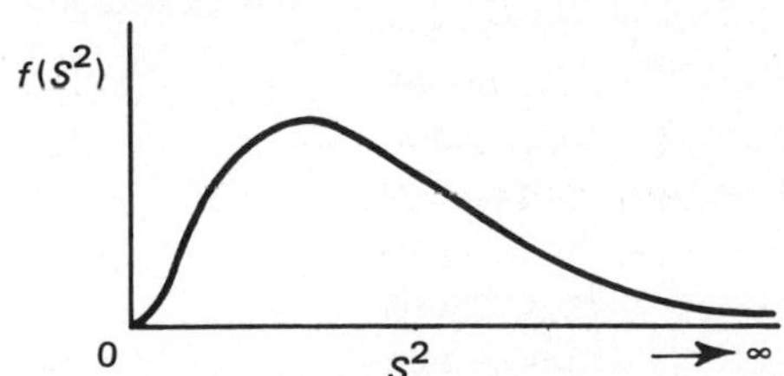

Figure 5.3.1
A theoretical sampling distribution of S^2 for $N = 7$.

Like population distributions, sampling distributions may be either continuous or discrete. The binomial distribution is discrete, although in applied problems it is sometimes treated as though it were continuous. Most of the commonly encountered sampling distributions based on a continuous population distribution will be continuous.

Since a sample statistic is a random variable, the mean and variance of any sampling distribution are defined in the usual way. That is, let G be any sample statistic; then if the sampling distribution of G is discrete, its expectation or mean is

$$E(G) = \mu_G = \sum_g gp(g). \qquad [5.3.1^*]$$

(see Section 4.6). If the variable G is continuous, then

$$E(G) = \mu_G = \int_{-\infty}^{\infty} gf(g)\, dg,$$

just as for any other continuous random variable (Section 4.6).

In the same way, the variance of a sampling distribution for any statistic G can be defined:

$$\sigma_G^2 = E(G - \mu_G)^2 \qquad [5.3.2]$$

or

$$\sigma_G^2 = E(G^2) - [E(G)]^2. \qquad [5.3.3^*]$$

The standard deviation σ_G of the sampling distribution reflects the extent to which sample G values tend to be *unlike* the expectation, or are *in error*. To aid in distinguishing the standard deviation of a sampling distribution from the standard deviation of a population distribution, a standard deviation such as σ_G is usually called the "standard error" of the statistic G:

$$\text{standard error of } G = \sigma_G. \qquad [5.3.4^*]$$

Thus, the standard error of the mean, σ_M is identical to the standard deviation of the distribution of possible sample means, for all possible samples of size N drawn from a specified population. Similar meanings hold for the standard error of the median, the standard error of the standard deviation, the standard error of the range, and so on.

5.4 / SAMPLE STATISTICS AS ESTIMATORS

Some population parameters have obvious parallels in sample statistics. The population mean μ has its sample counterpart in M, the variance σ^2 in the sample variance S^2, the population proportion p in the sample proportion P, and so on. On the other hand, the relationship between population parameters and sample statistics is not necessarily 1 to 1. It is entirely possible for a population distribution to depend upon parameters that are not directly paralleled in one of the common sample statistics, and for sample statistics to be used that are not direct parallels of population parameters.

It is true, however, that a sample of cases drawn from a population contains information about the population distribution and its parameters. Furthermore, a statistic computed from the data in the sample contains some of that information. Some statistics contain more information than others, and some statistics may contain more information about certain parameters than about others.

A central problem of inferential statistics is **point estimation,** the use of the value of some statistic to infer the value of a population parameter. The value of some statistic (or point in the "space" of all possible values) is taken as the "best estimate" of the value of some parameter of the population distribution.

How does one go from a sample statistic to an inference about the population parameter? In particular, which sample statistic does one use, if it is to give an estimate that is in some sense "best"?

The fact that the sample represents only a small subset of observations drawn from a much larger set of potential observations makes it nearly impossible to say that any estimate is exactly like the population value. As a matter of fact they very probably will not be the same, as all sorts of different factors of which we are in ignorance may make the sample a poor representation of the population. Such factors we lump together under the general rubrics *chance* or *random effects*. In the long run such samples should reflect the population characteristics. However, practical action can seldom wait for "in the long run"; things must be decided here and now in the face of limited evidence. We need to know how to use the available evidence in the best possible ways to infer the characteristics of the population.

Various statistics differ in the information they provide about population parameters. They also differ in the extent to which this is "good" information, that can be used to estimate the value of the parameter in question. We are now going to examine some statistics in terms of their properties as estimators.

In the following, it will be well to keep in mind the difference between an "estimator" of some population parameter and an "estimate" of the value of that parameter. An estimator is a formula or method for combining the values occurring in the data. Hence an estimator is a random variable, which takes on values dependent upon the sample data. On the other hand, the particular value that results from the application of that formula is an estimate of the population parameter in question. Viewed as a random variable, with a value arrived at in a certain way from the sample data, the sample mean is an estimator. On the other hand, a particular value of the sample mean, based on a particular sample, is an estimate. Our immediate interest will focus on the properties of statistics as estimators—that is, as random variables with certain characteristics.

5.5 / THE PRINCIPLE OF MAXIMUM LIKELIHOOD

Statisticians have established a number of criteria for choosing among the statistics that might be used as estimators. One of these is the very important **principle of maximum likelihood,** which will be discussed next. Even though a formal presentation of the maximum-likelihood principle is far beyond our scope, some intuitive feel for this idea may help you understand the methods that follow.

Basically, the problem of the person using statistics is, "Given several possible population situations, any one of which might be true, how shall I bet on the basis of the evidence so as to be as confident as possible of being right?" The user of statistical inference knows that any number of things might be true of the population. Fortunately, one can "snoop" on the population to a certain extent by taking a sample, but the evidence gained will almost certainly be faulty. Since this evidence is all there is, the person must use it nevertheless to form an opinion or make a decision. How does one decide on the basis of evidence that is probably erroneous? The principle of maximum likelihood gives a general strategy for such decisions, and may be paraphrased as follows.

Suppose that a random variable X has a distribution that depends only upon some population parameter θ (symbolized by lowercase Greek "theta"). The form of the density function will be assumed known, but not the value of θ. A sample of N independent observations is drawn, producing a set of values $(x_1, x_2, \cdots, x_N)$. Let

$$L(x_1, \cdots, x_N; \theta)$$

represent the likelihood, or probability (density), of this particular sample result *given* θ. For each possible value of θ, the likelihood of the sample result will be different, perhaps. Then, **the principle of maximum likelihood requires us to choose as our estimate that possible value of θ making $L(x_1 \cdots x_N;\theta)$ take on its largest value.**

In effect this principle says that when faced with several parameter values, any of which might be the true one for a population, the best "bet" is that parameter value which *would* have made the sample actually obtained have the highest prior probability. When in doubt, place your bet on that parameter value which would have made the obtained result most likely.

This principle may be illustrated very simply for the binomial distribution. Suppose that a sample is drawn at random from a population of college graduates. Each graduate is classified either as a "liberal arts major" or as a graduate from some other university college. Now three possibilities or hypotheses are entertained about the proportion of college graduates from liberal arts colleges. Hypothesis 1 states that .5 of all graduates are from such colleges, hypothesis 2 states that .4 of the graduates are from such schools, and hypothesis 3 states that .6 are liberal arts graduates.

Now suppose that some 15 college graduates are drawn at random and classified according to "liberal arts degree" versus "other degree." The result is that 9 out of the 15 hold a liberal arts college degree (we are supposing that no one in the population holds more than one degree). What decision about the three available hypotheses should we reach? This is a simple binomial sampling problem, so that the probability of each sample result may be calculated for each of the three hypothetical values of the parameter p: $p = .5$, $p = .4$, or $p = .6$.

If $p = .5$ the prior probability of a sample result such as that obtained would be

$$\binom{15}{9}(.5)^9(.5)^6 = .153.$$

For $p = .4$ the prior probability becomes

$$\binom{15}{9}(.4)^9(.6)^6 = .061.$$

Finally, if p were .6 the prior probability of 9 successes out of 15 would be

$$\binom{15}{9}(.6)^9(.4)^6 = .207.$$

The use of the principle of maximum likelihood to decide among these three possibilities leads to the choice of hypothesis 3, that .6 is the population proportion, since this is the parameter value among the possibilities considered that would have made the obtained sample result most likely, a priori.

This principle has a great deal of use in theoretical statistics, since general methods exist for finding the value of θ that maximizes the likelihood of a sample result. Statistics chosen as estimators because their values substituted for the parameter maximize the likelihood of the sample result are called **maximum-likelihood estimators.** For example, if there were no prior information at all about the number of "liberal arts" graduates, so that any number between 0 and 1.00 might be entertained as a hypothesis about the value of p, the maximum likelihood estimate of p

would be the sample P, or 9/15, since among all possible values of p this value makes the occurrence of the actual result have greatest a priori likelihood (the student may check a few values of p for himself to see that this is true).

The principle of maximum likelihood also provides a fairly routine way of finding estimators having "good" properties. However, the principle of maximum likelihood is introduced here not only because of its importance in estimation, but also because of the general point of view it represents about inference. This point of view is that *true population situations should be those making our empirical results likely; if a theoretical situation makes our obtained data have very low prior likelihood of occurrence, then doubt is cast on the truth of the theoretical situation.* Theoretical propositions are believable to the extent that they accord with actual observation. If a particular result should be very unlikely given a certain theoretical state of affairs, and we do get this result nevertheless, then we are led back to examine our theory. Good theoretical statements accord with observation by giving predictions having high probability of being observed.

Naturally, the results of a single experiment, or even of any number of experiments, cannot prove or disprove a theory. Replications, variants, different ways of measuring the phenomena, must all be brought into play. Even then, proof or disproof is never absolute. Nevertheless, the principle of maximum likelihood is in the spirit of empirical science, and it runs throughout the methods of statistical inference.

5.6 / OTHER DESIRABLE PROPERTIES OF ESTIMATORS

Since there are many ways for devising a sample statistic for estimating a population parameter's value, several criteria are used for judging how effectively a given statistic serves this purpose. As we have just seen, some statistics have the desirable property of being the maximum-likelihood estimator of a particular parameter. In addition, good estimators should be *unbiased,* should be *consistent,* should be *relatively efficient,* and a set of estimators used for estimating a set of parameters should be *sufficient*. A brief description of each of these criteria follows.

Imagine, once again, a population distribution depending only on some parameter θ. Some sample statistic G is to be used as an estimator of θ. Then G is said to be an *unbiased* estimator of θ if the expected value of G is *exactly* θ:

$$E(G) = \theta.$$

That is, when the value of G is an unbaised estimate of θ, then in the long run the average value of G over all possible random samples is exactly the value of θ.

As we shall see in subsequent sections, the sample mean M is an unbaised estimator of the population mean μ. Furthermore, under binomial sampling, the sample proportion P is an unbaised estimator of the population probability p. On the other hand, the sample variance S^2 is an example of a biased estimator, since $E(S^2)$ is not, in general, equal to the population variance σ^2. (A simple correction exists to solve this problem, as we shall see.)

Another desirable property of an estimator is *consistency*. Roughly speaking, this means that *the larger the sample size N, the higher the probability that G comes close to the population value* θ. Statistics that have this property are called *consistent estimators*. The sample mean, the sample variance, and many other common statistics are consistent estimators, as they tend in likelihood to be closer to the population value the larger the sample size. However, not all possible statistics meet this criterion. For example, suppose that we wished to estimate a population mean, and that we chose to use the score of the second case observed in any sample as our estimate of the population mean. Obviously, by the criterion of consistency this is not a good way to make an estimate. For an N of 2 or more, the probability of coming close to the population value does not increase as we increase the sample size. (Notice that this method does give an unbiased estimate, however.)

A third criterion for choosing an estimator is called *relative efficiency*. Imagine the population depending on the parameter θ, once again. This time, however, suppose that there were two statistics, G and H, that might be computed from any sample of size N, and that both G and H are unbiased estimators of θ. Given a constant N, the sampling distribution of G will have some variance of σ_G^2 and the sampling distribution of H will have some variance σ_H^2. Then

$$\frac{\sigma_H^2}{\sigma_G^2} = \text{efficiency of } G \text{ relative to } H.$$

The more efficient estimator has the smaller sampling variance. If G is more efficient then, on the average, the sampling error of G is smaller than it is for H, and thus one tends to make a smaller error in using G rather than H in estimating θ. As we shall see, one of the reasons for preferring the mean to the median is that when the population is of a "normal" type, and both are unbiased estimates of the population mean μ, the mean is relatively more efficient than the median, given the same sample size N.

Still another concept of major importance in the modern theory of statistics is that of *sufficiency*. Once again consider a random variable X which depends only on one parameter θ, and let G be a statistic which is computed from N sample values of X. Then we say that *G is a sufficient estimator* (or a *sufficient statistic*) *if G contains all of the information in the data about the parameter* θ. That is, if G is a sufficient statistic, our estimate of θ cannot be improved by considering any other aspect of the data not already included in G itself.

In some population distributions where there may be more than one parameter required to specify the distribution, then two or more statistics may be required for sufficiency. In these instances one refers to the set of sufficient statistics, rather than to a single sufficient estimator.

Sufficient statistics do not always even exist, and situations can be constructed in which no sufficient set of estimators can be found for a set of parameters. Nevertheless, sets of sufficient estimators, when they do exist, are important, since if one can find a set of sufficient estimators, then it is ordinarily possible to find unbiased and efficient estimators based upon that sufficient set. In particular, when a set of sufficient statistics exist, then the maximum-likelihood estimators will be based upon that set.

In most of the work to follow, the estimators will be the sample mean and the (corrected) sample variance, both of which will fulfill these criteria for good estimators in the particular situations where they will be used. This does not imply, however, that other estimators failing to meet one or more of these criteria are useless. In particular, special situations exist where other methods of estimating central tendency may be better than either mean or median, and variance estimates other than S^2 are called for. Other statistics are useful on occasion, but we will focus most attention on the mean and the corrected variance, since they occupy a central place in the "classical" statistical methods we will be treating.

5.7 / THE SAMPLING DISTRIBUTION OF THE MEAN

By the criteria just listed, the sample mean comes off very well as an estimator of its population counterpart, μ. In general, as an estimate of μ, the sample mean is unbiased, is consistent, and in an important set of circumstances is both efficient relative to other statistics and, taken with S^2, sufficient.

It is simple to show that the sample mean must be an unbiased estimator, or that

$$E(M) = \mu.$$

By definition, for any N independent observations,

$$M = (x_1 + x_2 + \cdots + x_N)/N,$$

so that

$$E(M) = \mathrm{E}(x_1 + x_2 \cdots + x_N)/N.$$

By rule 5 of Appendix B, it must be true that

$$E(M) = [E(x_1) + E(x_2) + \cdots + E(x_N)]/N.$$

For any given observation, $E(x) = \mu$, so that

$$E(M) = N\mu/N = \mu.$$

The expected value of the sample mean is the population mean μ. This same statement can be interpreted in another way: **the mean of the sampling distribution of means is the same as the population mean.**

However, it is not true that the sampling distribution of the mean will be the same as the population distribution; in fact, these distributions will ordinarily be quite different, depending particularly on sample size. In the first place, the variance of the sampling distribution of means will not be the same as σ^2, but instead will be smaller than σ^2 for samples of size 2 or larger. This will be shown next.

Intuitively it seems quite reasonable that the larger the sample size the more confident we may be that the sample mean is a close estimate of μ. When it was said that the mean is a consistent estimator, this was another way of saying that the mean is a better estimate of μ for large than for small samples. Now we can put that intuition on a firm basis by looking into the effect of sample size on the variance and standard deviation of the distribution of sample means.

By definition, the variance of the distribution of means, from samples drawn at random from a population with mean μ and variance σ^2, is

$$\begin{aligned} \sigma^2_M &= E(M^2) - [E(M)]^2 \\ &= E(M^2) - \mu^2. \end{aligned} \qquad [5.7.1^*]$$

Bear in mind that we are assuming that the sample mean is based on N *independent* observations. Let us call any pair of these observations i and j, with scores X_i, and X_j. The square of the sample mean is

$$M^2 = \frac{(x_1 + x_2 + \cdots + x_N)^2}{N^2}$$

$$= \frac{1}{N^2}(x_1^2 + \cdots + x_N^2 + 2\sum_{i<j} x_i x_j),$$

the sum of the squared scores, plus twice the sum of the products of all pairs of scores, all divided by N^2. For any single observation i,

$$E(x_i^2) = \sigma^2 + \mu^2 \qquad [5.7.2]$$

since, from Section 4.16,

$$\sigma^2 = E(x_i^2) - \mu^2. \qquad [5.7.3]$$

For a pair of *independent* observations, i and j, rule 9 in Appendix B gives

$$E(x_i x_j) = E(x_i)E(x_j) = \mu^2. \qquad [5.7.4]$$

Thus, putting these two facts (5.7.2 and 5.7.4) together, we have

$$E(M^2) = \frac{1}{N^2}[E(x_1^2) + E(x_2^2) + \cdots + 2\sum_{i<j} E(x_i x_j)]$$

$$= \frac{N\sigma^2 + N\mu^2 + (N)(N-1)\mu^2}{N^2}$$

$$= \frac{\sigma^2}{N} + \mu^2.$$

Making this substitution in expression 5.7.1 we find

$$\sigma_M^2 = E(M^2) - \mu^2 = \frac{\sigma^2}{N}.$$

The variance of the sampling distribution of means for independent samples of size N is always the population variance divided by the sample size, σ^2/N.

This is a most important fact, and gives direct support to our feeling that large samples produce better estimates of the population mean than do small. When the sample size is only 1, then the variance of the sampling distribution is exactly the same as the population variance. If however, the sample mean is based on two cases, $N = 2$, then the sampling variance is only 1/2 as large as σ^2. Ten cases give a sampling distribution with variance only 1/10 of σ^2, $N = 500$ gives a sampling distribution with variance 1/500 of σ^2, and so on. If the sample size approaches infinity, then σ_M^2 approaches zero. **If the sample is large enough to embrace the entire population, there is no difference between the sample mean and μ.**

In general, the larger the sample size, the more probable it is that the sample mean comes arbitrarily close to the population mean. This fact is often called **the law of large numbers.**

5.8 / STANDARDIZED SCORES CORRESPONDING TO SAMPLE MEANS

The standard error of the mean is

$$\sigma_M = \sqrt{\sigma_M^2} = \frac{\sigma}{\sqrt{N}}, \qquad [5.8.1^*]$$

so that when a sample mean is put into standard form, we have

$$z_M = \frac{M - \mu}{\sigma_M} = \frac{M - \mu}{\sigma/\sqrt{N}}. \qquad [5.8.2^*]$$

The larger the standard score of a mean, relative to μ and the standard error, the *less likely* is one to observe a mean this much or more deviant from μ. Any given degree of departure of M from μ corresponds to a larger absolute standard score value, and hence a less probable class of result, the larger the sample size N.

For example, suppose that in sampling from some population a sample mean was found differing by 10 points from the population mean. Suppose that σ were 5 and the sample size were 2. Then the standard score z_M would be

$$z_M = \frac{10}{5/\sqrt{2}} = 2.8,$$

disregarding sign. The Tchebycheff inequality (Section 4.18) tells us that means this deviant or more so from expectation can occur with probability no greater than about $1/(2.8)^2$ or .13.

Suppose, however, that the sample size had been 200. Here

$$z_M = \frac{10}{5/\sqrt{200}} = 28.28$$

so that a sample M deviating this much or more from expectation could occur only with probability no greater than about $1/(28.28)^2$ or .0013. An extent of deviation of M from the true mean that could occur relatively often for small samples is rare when the sample size is large. Notice that knowing the standard error of the mean is essential if we are going to judge the agreement between a sample and a population mean in terms of the probability of occurrence for a given extent of deviation. For the moment we are assuming that this information is simply given to us.

5.9 / CORRECTING THE BIAS IN THE SAMPLE VARIANCE AS AN ESTIMATOR

It has already been mentioned that the sample variance S^2 is a biased estimate of the population variance σ^2, since

$$E(S^2) \neq \sigma^2. \qquad [5.9.1]$$

This can be demonstrated as follows: the expectation of a sample variance is

$$E(S^2) = E\left(\frac{\sum_i x_i^2}{N} - M^2\right) = E\left(\frac{\sum_i x_i^2}{N}\right) - E(M^2).$$

Let us consider the two terms on the extreme right separately. By rules 5 and 2 for expectations,

$$E\left(\frac{\sum_i x_i^2}{N}\right) = \frac{\sum_i E(x_i^2)}{N}.$$

From the definition of the variance of the population,

$$\sigma^2 = E(x^2) - \mu^2$$

so that

$$E(x_i^2) = \sigma^2 + \mu^2 \qquad [5.9.2]$$

for any observation i. Thus

$$E\left(\frac{\sum_i x_i^2}{N}\right) = \frac{\sum_i (\sigma^2 + \mu^2)}{N} = \sigma^2 + \mu^2. \qquad [5.9.3]$$

Now the variance of the sampling distribution of means is, from 5.7.1,

$$\sigma_M^2 = E(M^2) - \mu^2$$

so that

$$E(M^2) = \sigma_M^2 + \mu^2. \qquad [5.9.4]$$

Putting these two results (5.9.3 and 5.9.4) together, we have

$$E(S^2) = \sigma^2 - \sigma_M^2, \qquad [5.9.5^*]$$

the expectation of the sample variance is the *difference* between the population variance σ^2 and the variance of the sampling distribution of means σ_M^2. In general this difference will not be the same as σ^2, since ordinarily the variance σ_M^2 will not be zero, and so the sample variance is biased as an estimator of the population variance. In particular, the sample variance is, on the average, smaller than the population variance σ^2.

We have just shown in Section 5.7 that

$$\sigma_M^2 = \frac{\sigma^2}{N}.$$

On making this substitution into expression 5.9.5, we have

$$E(S^2) = \sigma^2 - \frac{\sigma^2}{N} = \left(\frac{N-1}{N}\right)\sigma^2, \qquad [5.9.6]$$

so that on the average the sample variance is *too small* by a factor of $\frac{N-1}{N}$.

Since this is true, a way emerges for correcting the variance of a sample to make it an unbiased estimator. **The unbiased estimate of the variance based on any sample of N independent cases is**

$$s^2 = \frac{N}{N-1} S^2. \qquad [5.9.7^*]$$

(In all that follows, we will reserve the symbol S^2 to stand for the uncorrected variance of a sample, as originally defined, and use the lowercase letter s^2 to indicate the corrected variance. Similarly, S will stand for the standard deviation based upon S^2, and s for the standard deviation based upon s^2.)

It is simple to show that s^2 is indeed unbiased:

$$E(s^2) = \frac{N}{N-1} E(S^2) = \frac{N}{(N-1)} \frac{(N-1)}{N} \sigma^2 = \sigma^2. \qquad [5.9.8^*]$$

Quite often it is convenient to calculate the unbiased variance estimate s^2 directly, without the intermediate step of calculating S^2. This is done either by the formula

$$s^2 = \frac{\sum_i x_i^2 - NM^2}{N - 1} \qquad [5.9.9^*]$$

or by

$$s^2 = \frac{\sum_i x_i^2}{N - 1} - \frac{\left(\sum_i x_i\right)^2}{N(N - 1)}. \qquad [5.9.10^*]$$

Incidentally, some modern statistics texts completely abandon the idea of the sample variance S^2 as used here, and introduce only the unbiased estimate s^2 as *the* variance of a sample. However, this is apt to be confusing in some work, and so we will follow the older practice of distinguishing between the sample variance S^2 as a descriptive statistic, and s^2 as the unbiased estimate of σ^2.

Even though we will be using the square root of s^2, or s, to estimate the population σ, it should be noted that s is not *itself* an unbiased estimate of σ, and that

$$E(s) \neq \sigma$$

in general. The correction factor used to make s an unbiased estimate of σ depends upon the form of the population distribution; thus, for the unimodal symmetric "normal" distribution (Chapter 6) an unbiased estimate for large N is provided by

$$\text{unbiased estimate of } \sigma = \left[1 + \frac{1}{4(N - 1)}\right] s. \qquad [5.9.11]$$

Furthermore, special tables exist for correcting the estimate of σ for relatively small samples from such populations (Dixon and Massey, 1957). However, the problem of estimating σ from s is bypassed, in part, by the methods we will use in Chapter 9 and elsewhere, and, if the sample size is reasonably large, the amount of bias in s as an estimator of σ ordinarily is rather small. For these reasons, we will not trouble to correct for the bias in s found from the *unbiased* estimate s^2 of σ^2.

One estimates the standard error of the mean by using the unbiased estimate s^2:

$$\text{estimated } \sigma_M = \frac{\text{estimated } \sigma}{\sqrt{N}} = \sqrt{\frac{s^2}{N}}. \qquad [5.9.12^*]$$

On the other hand, if the sample variance (not the unbiased estimate) is used, then the estimate may be arrived at as follows:

$$\text{estimated } \sigma_M = \sqrt{\frac{s^2}{N}} = \sqrt{\frac{N}{(N - 1)}\frac{S^2}{N}} = \frac{S}{\sqrt{N - 1}}. \qquad [5.9.13^*]$$

5.10 / PARAMETER ESTIMATES BASED ON POOLED SAMPLES.

Sometimes it happens that one has several independent samples where each provides an estimate of the same parameter or set of parameters. The most usual situation occurs when one is estimating μ or σ^2, or both. When this happens there is a real advantage in pooling the sample values to get an unbiased estimate. These pooled estimates are actually weighted averages of the estimates from the different samples.

The big advantage lies in the fact that the sampling error will tend to be smaller for the pooled estimate than that for any single sample's value taken alone.

Suppose that there were two independent samples, based on N_1 and N_2 observations respectively. From each sample a mean M is calculated, each of which estimates the same value μ. Then the pooled estimate of the mean is

$$\text{pooled } M = \text{est. } \mu = \frac{N_1M_1 + N_2M_2}{N_1 + N_2}. \qquad [5.10.1^*]$$

It is easy to see that this must give an unbiased estimate of μ:

$$E\left[\frac{N_1M_1 + N_2M_2}{N_1 + N_2}\right] = \frac{N_1E(M_1) + N_2E(M_2)}{N_1 + N_2}$$

$$= \frac{\mu(N_1 + N_2)}{N_1 + N_2} = \mu.$$

If there were three samples of sizes N_1, N_2, and N_3, giving means M_1, M_2, and M_3, then

$$\text{pooled } M = \text{est. } \mu = \frac{N_1M_1 + N_2M_2 + N_3M_3}{N_1 + N_2 + N_3}.$$

This same idea applies to any number of samples. The fact that the pooled estimate is likely to be better than the single sample estimates taken alone is shown by the standard error of the pooled mean. For two independent samples, each composed of independent observations drawn from the same population,

$$\sigma_M = \frac{\sigma}{\sqrt{N_1 + N_2}}, \qquad [5.10.2^*]$$

which *must* be smaller than the standard error either of M_1 or of M_2. Similarly, for a pooled value of M found from three independent samples,

$$\sigma_M = \frac{\sigma}{\sqrt{N_1 + N_2 + N_3}}, \qquad [5.10.3]$$

and so on for any number of independent samples.

Naturally, these standard errors refer to the situation where either all of the respective samples are drawn from the same population, or where they all come from populations having the same mean and the same population variance σ^2.

In the same general way, it is possible to find pooled estimates of σ^2, for populations with the same variance σ^2, even though the population means may be different. Thus for two independent samples

$$\text{est. } \sigma^2 = \frac{(N_1 - 1)s_1^2 + (N_2 - 1)s_2^2}{(N_1 - 1) + (N_2 - 1)}, \qquad [5.10.4^*]$$

which is an unbiased estimate of the variance σ^2 of each of the populations.

5.11 / SAMPLING FROM RELATIVELY SMALL POPULATIONS

It has been mentioned repeatedly that most sampling problems deal with populations so large that the fact that samples are taken without replacement of single cases can safely be ignored. However, it may happen that the population under study is not only finite, but relatively small, so that the process of sampling without replacement has a real effect on the sampling distribution.

Even in this situation the sample mean is still an unbiased estimate regardless of the size of the population sampled. Hence no change in procedure for mean estimation is needed.

However, for finite populations the sample variance is biased as an estimator of σ^2 in a way somewhat different from the former, infinite-population, situation. When samples are drawn without replacement of individuals, the unbiased estimate of σ^2 is

$$\text{est. } \sigma^2 = \frac{N(T-1)}{(N-1)T} S^2 = \frac{(T-1)}{T} s^2 \quad [5.11.1]$$

where T is the *total number* of elements or individuals in the population.

Another difference from the infinite-population situation comes with the variance of the sampling distribution of means. For a population with T cases in all, from which samples of size N are drawn, the sampling variance of the mean is

$$\sigma_M^2 = \left(\frac{T-N}{T-1}\right)\frac{\sigma^2}{N}. \quad [5.11.2]$$

The variance of the mean tends to be *somewhat smaller* for a fixed value of N when sampling is from a finite population than when it is from an infinite population. Note that here the size of σ_M^2 depends both upon T, the total number in the population, and upon N, the sample size. An unbiased estimate of the variance of the mean is thus given by

$$\text{est. } \sigma_M^2 = \left(\frac{T-N}{T-1}\right)\frac{s^2}{N}. \quad [5.11.3]$$

The square root of this value gives the estimate of σ_M, for sampling from a finite population.

Once this adjustment to the estimated value of σ_M has been made, any of the inferential methods calling for this value may be employed in the usual way.

5.12 / THE IDEA OF INTERVAL ESTIMATION

So far, our discussion of estimation of parameters from statistics has been confined to the subject of "point estimation," making an estimate of a parameter value in terms of the value of a sample statistic. Thus, if a sample of 50 second-grade schoolchildren drawn at random from a large city school system shows a mean arithmetic achievement score of 92.7, we estimate the true value of μ, the mean of all second-grade schoolchildren in that school system, also to be 92.7.

Nevertheless, it is clear that sample mean ordinarily will not be exactly equal to the true population value because of sampling error. It is necessary to qualify our estimate in some way to indicate the general magnitude of this error. Usually this is done by showing a **confidence interval,** which is **an estimated range of values with a given high probability of covering the true population value.** When there is a large degree of sampling error the confidence interval calculated from any sample will be large; the range of values likely to cover the population mean is wide. On the other hand, if sampling error is small the true value is likely to be covered by a small range of values; in this case one can feel confident that the population value has been "trapped" within a small range of values calculated from the sample.

The general idea is to find two values, say a and b, such that the chances are very good that the interval of values from a to b actually does include the population value. Such an interval of values is called a confidence interval.

One rather primitive way to establish an approximate confidence interval for the mean of a population, given a sample mean, is by use of the Tchebycheff inequality (Section 4.18). We known from 4.18.2 that for any probability distribution it must be true that for any positive number k

$$\text{prob}\left\{\left|\frac{(X-\mu)}{\sigma}\right| \geqslant k\right\} \leqslant \frac{1}{k^2}.$$

Since the sampling distribution of the mean is such a probability distribution of the random variable M, it must be true that

$$\text{prob}\left\{\left|\frac{(M-\mu)}{\sigma_M}\right| \geqslant k\right\} \leqslant \frac{1}{k^2}.$$

Then, it must also be true that

$$\text{prob}\left\{-k \leqslant \frac{(M-\mu)}{\sigma_M} \leqslant k\right\} \geqslant 1 - \frac{1}{k^2}$$

which is equivalent to

$$\text{prob}\,(M - k\sigma_M \leqslant \mu \leqslant M + k\sigma_M) \geqslant 1 - \frac{1}{k^2}. \qquad [5.12.1]$$

For any sampling distribution of means, the probability is at least $1 - (1/k^2)$ that the interval of values $M - k\sigma_M$ to $M + k\sigma_M$ covers the true population value μ, where k is any arbitrary positive value.

For example, what is the probability that the true mean μ lies in an interval $M - 3\sigma_M$ to $M + 3\sigma_M$, as found from any sample mean M? Here, $k=3$, so that the probability must be at least $1 - (1/9)$ or .89 that this interval covers the true value of μ.

To go further, suppose that we are sampling from a population where μ is unknown, but where we know that the population standard deviation is 20. There are $N = 50$ cases in each sample. Now we draw a sample, where the mean turns out to be 124. Then, since the standard error of the mean must be

$$\sigma_M = \frac{\sigma}{\sqrt{N}}$$

we find that

$$\sigma_M = \frac{20}{\sqrt{50}} = \frac{20}{7.07} = 2.83.$$

Now let us find the probability that the true mean is covered by the interval of values extending from 3 standard errors below the sample mean to 3 standard errors above. That is, we want the probability that the interval from

$$M - 3\sigma_M = 124 - 3(2.83) = 115.51$$

to

$$M + 3\sigma_M = 124 + 3(2.83) = 132.49$$

covers the value of μ. This we know from expression 5.11.1 above to be greater than or equal to .89. We can then make the following statement: The probability is at least .89 that the interval 115.51 is 132.49 covers the value of the population mean.

Although the best estimate of the population value μ is, of course, the obtained sample mean M, the confidence interval based on any sample can be thought of as a range of "good" estimates to either side of M. Be sure to notice that the probability statement made in connection with a confidence interval *does not refer to the probability of* μ, *but rather to the probability of a sample*. In drawing samples, one may decide beforehand to compute a confidence interval for each one. The actual range of numbers making up the confidence interval on a given occasion will depend on the sample mean, M. For some samples, the confidence interval actually will cover the value of μ; for others, it will not. However, the Tchebycheff inequality lets us say what the minimum probability must be that a given confidence interval covers the value of μ.

Now suppose that instead of drawing samples of 50 observations, we take a sample of 600. Once again, imagine that the population has a standard deviation of 20, and that the actual sample obtained gave a mean of 124. In this circumstance, if we calculate the standard error of the mean we get

$$\sigma_M = \frac{20}{\sqrt{600}} = \frac{20}{24.49} = .82.$$

Now, if we take 3 standard errors to either side of the sample mean we find an interval

$$124 - 3(.82) \text{ to } 124 + 3(.82)$$

or

$$121.54 \text{ to } 126.46.$$

We can state that the probability is at least .89 that the interval 121.54 to 126.46 covers the true value of μ.

Notice in this second example that we have pinned the value of the population mean within a much narrower range of values. This is because the sample size is so much larger in the second example, so that the standard error of the mean is much smaller. The use of the Tchebycheff inequality gives only a very "rough and ready" sort of confidence interval. This method can be improved to a certain extent if we can assume that the sampling distribution of means is unimodal and symmetric. As we shall see in the next chapter, this is an entirely reasonable assumption in the case of means based on samples of at least moderate size ($N \geqslant 20$), almost regardless of the population distribution. Then the following statement is true:

$$\text{prob}\ (M - 3\,\sigma_M \leqslant \mu \leqslant M + 3\,\sigma_M) \geqslant .95. \qquad [5.11.2]$$

That is, the probability is at least .95 that the interval $M - 3\sigma_M$ to $M + 3\sigma_M$ covers the true value of the population mean, for random samples of N independent observations drawn from some population. An approximate "95 percent confidence interval" for the mean can be constructed for moderately large samples, by subtracting and adding $3\sigma_M$ to the obtained value of M. Notice that here we assume nothing about the form of the population distribution, and for the sampling distribution of M we assume only that it is unimodal and symmetric. Later we will employ a principle permitting a much narrower confidence interval to be found.

In this situation, where the population σ is known, an analogy to the idea of finding confidence intervals for the mean is tossing rings of a certain size at a post. The size of the confidence interval is like the diameter of a ring, and random sampling is like random tosses at the post. The predetermined size of the ring determines the chances that the ring will cover the post on a given try, just as the size of the confidence interval governs the probability that the true value μ will be covered by the range of values in the estimated interval.

The width of the confidence interval calculated from any sample depends upon three things: how many standard errors to either side of the sample mean one chooses to go (i.e, the value of k), the population standard deviation, and the size of the sample.

If, instead of going 3 standard errors to each side of M in this last example, we had chosen to take 4 standard errors in either direction, then the following statement would have been true:

$$\text{prob}\ (M - 4\sigma_M \leqslant \mu \leqslant M + 4\sigma_M) \geqslant .97.$$

That is, instead of being at least .95, the probability would be at least .97 that the interval calculated covers the true population mean. This interval would have been 124 − 4(.82) to 124 + 4(.82), or 120.72 to 127.28. This interval is a bit bigger than the interval found before. A larger confidence interval permits one to be more certain that the population value is covered; however, a larger interval introduces more uncertainty about what the population value actually is, since there are more possibilities in the interval.

The larger the standard deviation of the population, the larger will the confidence interval be, other things being equal. The spread of possibilities in the population is reflected in the range of possibilities estimated for the population mean.

Sample size works to reduce the size of the confidence interval; the larger the sample, the smaller the confidence interval, other things being equal. In the extreme situation, where the sample is infinite in size, any confidence interval for the mean must be zero in width. Here, we can be sure that the sample mean is exactly right as an estimate of the population. Even in practical situations, the size of the confidence interval can be made very small by taking a large number of cases.

In the next chapter we will begin to use confidence intervals where the probability of covering the population value can be stated exactly. There, because of other things we will know about the sampling distribution of the mean, we will no longer have to rely on the Tchebycheff inequality in order to make a probability statement.

Even so, the main features of confidence intervals are as they have been introduced here. Any confidence interval consists of a set of values constructed from a sample, which has a known, and generally high, probability of covering the population value of interest.

5.13 / OTHER KINDS OF SAMPLING

The model of simple random sampling is the basis of almost all the discussion to follow. The classical procedures of statistical inference rest upon the sampling scheme in which each and every population element sampled is independent of every other, and is equally likely to be included in any sample.

However, statistical inference is not at all limited to such equally probable samples. Samples may be drawn according to any other probability structure, just so long as the probability of occurrence of any particular sample is known or can be calculated. Samples are drawn by some method consistent with these probabilities, and the probabilities are then taken into account in the treatment of any information gained from the sample. The generic term for the process of drawing samples according to some known probability structure is, simply, **probability sampling.** Simple random sampling is then a special case of probability sampling; the probability structure in simple random sampling dictates equal probability for each possible sample. Many methods exist for treating samples drawn according to other probability structures, and often these methods differ from those we will discuss.

One common scheme is called **stratified sampling.** Here, the population is divided into a number of parts, called "strata." A sample is drawn independently and at random in each part. Given the sizes of the various strata, one can make inferences about the total population represented. Such a scheme is very good for insuring a representative sample, and it may reduce the error in estimation, although it does require special methods over and above those given here.

Another approach to sampling is called *two-stage sampling*. Here, the population may break naturally into relatively large units, and then within these large units one wishes to examine specific individuals or objects. For example, suppose that social workers as a group were being studied. Most social workers are employed by large governmental units such as states, and then within the governmental unit by some agency. Thus, an effective approach to the sampling of social workers might be first to choose a set of states. Then, within the states selected a random sample of agencies could be chosen, with the social workers making up those agencies used as the ultimate sample.

This idea can be extended to sampling in three or more stages, of course. Thus, in the social worker example, one might first sample states, then agencies, then social workers employed by the agencies selected.

Staged sampling of this sort is often the only practical approach when a large and complicated population must be sampled. However, this approach too requires some specialized techniques for statistical inference.

Mention has already been made (Section 3.17) of a sequential approach to sampling. That is, instead of regarding the sample of N independent observations all made at essentially the same time, sampling is done on an individual unit basis. Then the decision to proceed or not to proceed with sampling depends on statistical criteria the investigator has set up ahead of time. Ordinarily these criteria depend on such things as how precise an estimate of the population parameter has been gained, or how much risk remains in making a particular decision in the light of the evidence so far.

The student interested in some of these other sampling strategies might look into Mood, Graybill, and Boes (1974) or Kish (1965). Unfortunately, space does not permit a further discussion of them here.

5.14 / TO WHAT POPULATIONS DO OUR INFERENCES REFER?

Most who use inferential statistics in research rely on the model of simple random sampling. Yet how does one go about getting such a ''truly'' random sample? It is not easy to do, unless, as in all probability sampling, each and every potential member of the population may somehow be listed. Then, by means of a device such as random numbers individuals may be assigned to the sample with approximately equal probabilities.

However, in social and behavioral sciences, interest often lies in experimental or natural effects that, presumably, should apply to very large populations of people or things. Such a listing procedure is simply not possible. Still other experiments may refer to all possible measurements that *might* be made of some phenomenon under various experimental conditions, where estimated true values may be sought from the experimental observation of a few instances. Here, the population is not only infinite, it is hypothetical, since it includes all *future* or *potential* observations of that phenomenon under the different conditions. In sampling from such experimental populations, where there is no possibility of listing the elements for random assignment to the sample, the only recourse of the experimenter is to draw basic experimental units in some more or less random, ''haphazard,'' way, and then *make sure that in the experiment only random factors determine which unit gets which experimental treatment*. In other words, there are two ways in which randomness is important in an experiment: the first is in the selection of the sample as a whole, and the second is in the allotment of individuals to experimental treatments. Each kind of randomness is important for the ''generalizability'' of the experimental results, so that in doing an experiment one takes pains to see that both kinds of randomness are present. However, even given that individual cases are assigned to experimental manipulations at random, the possible inferences are still limited by the fundamental population from which the total sample is drawn.

How does one know the population to which statistical inferences drawn from a sample apply? If random sampling is to be assumed, then the population is defined by the manner in which the sample is drawn. The only population to which the inferences strictly apply is that for which the *units* sampled have equal (and nonzero) likelihood of appearing in any sample. In some contexts a unit may be one thing, and in other contexts quite another: thus, for example, if the units sampled are families inhabiting the same dwelling, it may be that each such family unit has equal probability of appearing in a sample of N such units. However, it would not then necessarily be true that individual persons have equal probability for appearing in any sample. On the other hand, if individual persons are sampled at random, it will not necessarily be true that family units have equal probability of appearing, since families may contain different numbers of people. It is most important to have a clear definition of the unit that is to be sampled, since in random sampling the totality of such possible units with equal probability of appearing in any given sample must make up the population.

It should be obvious that simple random samples from one population may not be random samples of another population. For example, suppose that some one wishes to sample American college students. He or she obtains a directory of college students from a midwestern university and, using a random number table, takes a sample of these students. One is not, however, justified in calling this a random sample of the population of American college students, although it may well be that this a random sample of students at *that* university. *The population is defined not by what is said, but rather by what was done to get the sample*. For any sample, one should always ask the question, "What is the set of potential units that could have appeared in any sample with equal probability?" If there is some well-defined set of units that fits this qualification, then inferences may be made to that population. However, if there is some population whose members could not have been represented in the sample with equal probability, then inferences do not *necessarily* apply to that population when methods based on simple random sampling are used. Any generalization beyond the population actually sampled at random must rest on extrastatistical, scientific, considerations.

From a mathematical-statistical point of view the assumption of random sampling makes it relatively simple to determine the sampling distribution of a particular statistic given some particular population distribution. As we noted above, it is possible to use other probability structures for the selection of samples, and in some situations there are advantages in doing so. The point is that *some* probability structure must be known or assumed to underlie the occurrence of samples if statistical inference is to proceed. This point is belabored only because it is so often overlooked, and statistical inferences are so often made with only the most casual attention to the process by which the sample was generated. The assumption of some probability structure underlying the sampling is a little "price tag" attached to a statistical inference. It is a sad fact that if one knows nothing about the probability of occurrence for particular samples of units for observation, very little of the machinery we are describing here applies. This is why our assumption of random sampling is not to be taken lightly. All the techniques and theory that we will discuss apply to random samples, and do not necessarily hold for any old data collected in any old way. In practical situations, the experimenter may be hard put to show that a given sample is "truly" random, but he must realize that he is acting *as if* this were a random sample from some well-defined population when he applies methods of statistical inference. Unless this assumption is at least reasonable, the probability results of inferential methods mean very little, and these methods might as well be omitted. While there may be plenty of meaning in the data, and this may be worth studying for its own sake, the probability statements we attach to such data depend on random sampling methods.

In the social sciences, studies made "in the field" are often interesting and valuable, even though random sampling may be out of the question in such studies. In addition, the social scientist is often interested in natural experiments, where a set of highly relevant factors or conditions come together in a society or a culture, providing a unique opportunity to observe the responses of individuals or groups of people. To rule such investigations off limits simply because random sampling models do not happen to fit would be the sheerest folly. Such data should be collected and studied with as much scientific and intellectual rigor as we can muster. However, merely prettifying these studies with the ordinary methods of statistical inference adds little or nothing to their meaning. This is a most serious problem in the development of the social sciences, and has not been given the attention that it deserves. One recent and thoughtful treatment of this subject may be found in Cook and Campbell (1979).

Certainly, the application of some statistical method does not somehow magically make a sample random, and the statistical conclusions therefore valid. Inferential methods apply to probability samples, drawn either by simple random sampling or in accordance with some other probability structure. There is no guarantee of their validity in other circumstances.

5.15 / LINEAR COMBINATIONS OF RANDOM VARIABLES, AND APPENDIX C

So far in our discussion of sampling distributions, we have assumed that there is a single variable of interest, and that a sample quantity such as a mean or a variance is to be studied. Then, the population distribution dictates the sampling distribution of that particular statistic.

However, in other circumstances, the sampling process may produce *values of two or more random variables*. Our interest may not be in the usual statistics such as a sample mean at all, but rather in a particular *weighted sum of the values of the random variables*. Thus, each sample may give us the value of one random variable X, and the value of another random variable Y. We might be interested in still another variable W, which is formed by weighting and summing the values of X and Y as they appear in each sample. For example, we might take

$$w = 3x + 4y,$$

or perhaps

$$w = x - y.$$

Over all samples producing values of X and of Y, there will then be a sampling distribution of W as well.

Any weighted sum of two or more variables in known as a *linear combination*. In later chapters we will have occasion to form linear combinations of random variables, and we will need to know something about their distributions over samples. Fortunately, such linear combinations of random variables follow a few simple rules and principles that can be readily summarized. A summary of these is provided in Appendix C. These rules are worthy of study, particularly when references to them are made in later sections. As you will see, the same general principles apply to such linear combinations as to other probability or sampling distributions.

An especially useful notation and concept that is also introduced in Appendix C is that of a *vector*, which is simply an ordered set of values. Thus, as in the example above, when we wish to represent one value $X = x$, and another value $Y = y$, we can actually do so with a pair of values, or a vector, (x,y). This principle extends easily to several variables or values, and gives a simple way to discuss several sets of values at the same time. Some principles involving vectors are given in C.3.

1. A psychologist kept a large population of rats. Each rat was to be observed individually running a maze. If a rat had a tendency to turn right in the maze, it was given a score $X = 1$. On the other hand, if the rat had a tendency to turn left it got a score $X = 2$. If any of the population of rats had equal probability of being a ''right-turner'' or a ''left-turner,'' find the population distribution of X. Find the mean and variance of this distribution.

2. Now suppose that the experimenter of exercise 1 chooses two rats independently and at random to run the maze. On the basis of its performance, each is given a value of X. However, a new variable G is formed by summing the X-values for the two rats, or $G = x_1 + x_2$. Given that the population of rats is as described in exercise 1, find the theoretical sampling distribution of G over all possible pairs of rats. (**Hint:** Simply list all possible pairs of outcomes and the resulting G-values, and find the probabilities.)

3. Using the sampling distribution of G found in exercise 2, find the expected value of G, or $E(G)$. How does this compare with the expected value for the population, or $E(X)$? Show that $E(G)$ in this instance is exactly twice the value of $E(X)$.

4. Again using the sampling distribution of G in exercise 2, find the variance of this sampling distribution, or σ^2_G. How does this value compare with the variance of the population, or σ^2_X? Show that in this instance the variance of the sampling distribution is precisely twice the variance of the population.

5. Using the results of exercises 1 through 4 above, make a verbal comparison of the sampling distribution of G to the population distribution on which it is based. How does this sampling distribution differ from the population distribution on which it is based?

6. Suppose that an experimenter of exercise 1 selected three rats at random and ran each through the maze. Now the score G was formed by taking the X-values for all three rats and adding them up and dividing by 3 or $G = (x_1 + x_2 + x_3)/3$. If the population distribution is as found in exercise 1, what now is the distribution of G?

7. Find the mean and the variance for the sampling distribution of G found in exercise 6. How do these compare with the mean and variance of the population distribution? Show that $E(G) = E(X)$ and $\sigma^2_G = \sigma^2_X/3$ in this instance.

8. See if you can generalize from the results of exercises 1 through 7. That is, if the experimenter selected any number N of rats from the population and formed a statistic G from the sum of their X-values divided by N, what would the mean of the sampling distribution of G turn out to be? The variance σ^2_G? What would the sampling distribution be like relative to this population distribution?

9. An educational researcher has devised a measure of the motivation for schoolchildren to achieve. Children are assigned a score $x = -1$ if they are classified as ''underachievers,'' $x = 0$ for ''parachievers,'' and $x = 1$ for ''overachievers.'' The researcher theorizes that about 1/4 of children in the school system are overachievers, another 1/4 are underachievers, and the rest achieve about at par for their ability. Find the population distribution for the random variable X. What is the value of $E(X)$? What is the value of σ^2_X?

10. Suppose that the researcher of exercise 9 selected two children at random from this school system, and formed the average of their two scores, $M = (x_1 + x_2)/2$. Given the population situation assumed in exercise 9, work out the exact sampling distribution of the statistic M. (**Hint:** Again work out all possible pairs of scores, the corresponding M-value, and the probability of each value.)

11. Work out the expected value and the variance for the sampling distribution found in exercise 10. How do these compare with the expected value and variance for the population, as found in exercise 9? Show that the variance for the sampling distribution is here exactly one-half of the variance for the population.

12. Suppose that the two children selected at random in exercise 10 showed one of the average scores listed listed below. How probable is such an average score, given the specified population?

(a) exactly 0
(b) 1.00
(c) -1.00
(d) .5 or less
(e) $-.5$ or less

13. Imagine now that the researcher of exercise 9 sampled 3 children independently and at random from the population of students as given above, and again computed their average score, M. Find the sampling distribution of M under these conditions. How does this distribution compare in its general "shape" to that to the population distribution? To that of exercise 10?

14. Work out the expected value and the variance of the sampling distribution of M for $N = 3$, as found in exercise 13. Demonstrate that $E(M) = E(X)$, and that $\sigma_M^2 = (\sigma_X^2/3)$.

15. In the sampling distribution found in exercise 13, determine the following probabilities:

(a) $p(M = 1)$
(b) $p(M = 0)$
(c) $p(M \leq 1/3)$
(d) $p(M \leq 0)$
(e) $p(M \leq 2/3)$
(f) $p(M < -1/3)$

16. Let us suppose that instead of the theory about the population stated in exercise 9, there is good reason to believe that 60 percent of the school population consists of underachievers, 30 percent of parachievers, and only 10 percent of overachievers. Repeat exercise 10 under this new assumption. How do the mean and variance of the sampling distribution now relate to the mean and variance of the population? Describe the "form" of this sampling distribution relative to that of the population.

17. A research worker is faced with a large stack of computer cards, each containing data for one individual. Unfortunately, these cards are not labeled, and they could belong either to study I or to study II in this particular research project. As it happens, study I contained 60 percent female subjects while study II contained only 40 percent. The worker draws 15 cards at random and notes the code for sex that each contains. It turns out that there are eight females out of fifteen cards. What should the worker conclude? Why? (**Hint:** Regard "female" as a success in a Bernoulli process.)

18. Consider the table given in exercise 32 of Chapter 1. Suppose that one individual were drawn at random from the 1000 represented in this table. According to the principle of maximum likelihood, what would you guess the occupation of this individual to be, given each of the different patterns possible?

19. A public health officer believes that something between 5 to 20 percent of three-year-old children in a certain area have been exposed to tuberculosis. A random sample of 20 such children is taken, and a test for tuberculosis given to each. Three children show a positive test. Is 5% or 20% a better bet for the proportion actually exposed in this population? What is the *best* bet about the true proportion of children who would give a positive response to the tuberculosis test? Why?

20. An archeologist has assembled a collection of 18 potshards (pieces of pottery) found at the same level of an ancient site. It is already known that only about 20 percent of pottery produced in this area before 3000 B.C. was decorated, although about 40 percent of later pottery had decoration. Four of the potshards in the present group are decorated. What will the archeologist tend to conclude about the age of these potshards? Why? What is being assumed here?

21. There are three typists is a business office. Typist I is not very accurate, and tends to average 1 error per business letter. Typist II is a little better, and makes about 3 errors per five letters. Typist III is quite expert, and makes only one error for each ten letters on the average. The boss reads five letters all typed by the same person, and discovers 2 errors. Who would you guess typed the letters? (**Hint:** Use the rule for the Poisson distribution.)

22. Four interviewers were sent door to door in a neighborhood. On a given try, the occupant of the dwelling might or might not agree to give the interview. One of the interviewers (A) had a past record of 60 percent success in getting interviews, while interviewer B had a record of only 40 percent success. Interviewer C had an even worse record of 30 percent, and D was worst of all with 10 percent. For one of these interviewers, the first success occurred at the third house. If each interviewer was operating true to form, which interviewer do you think this was? Why? (**Hint:** Look at Section 3.17.)

23. One of the interviewers in exercise 22 above had the third successful interview at the fifth house contacted. Again assuming that the true proportion of successes for each interviewer remains constant, and that each interview is independent of every other, which interviewer do you think this was? Why?

24. A candidate for governor of a certain state feels confident in receiving 52 percent of the votes in the impending election. A political scientist takes a random sample of 250 of this state's registered voters, and asks each person whether or not they intend to vote for this candidate. The proportion P who gave an affirmative response was recorded. Find the expected value of P and its sampling variance and standard deviation, *if* the candidate's idea is actually correct. What is the form of this sampling distribution?

25. Suppose that the political scientist of problem 24 found that in the sample of 250 voters, only $P = .40$ expressed the intention of voting for this candidate. If the candidate's theory is right, what z-score does this result represent in the sampling distribution? What is the probability of getting a sample result as deviant as this (in absolute terms) if the expected value of P really is .52? Use Tchebycheff's inequality of a unimodal symmetric distribution.

26. Assume that the researcher of exercise 9 sampled a total of 500 children, and scored each for achievement level. The mean of the scores was taken. What is the expected value of the mean score for any such sample? What is the standard deviation of the sampling distribution of such means? If a sample mean of .25 is obtained, what z-score in the sampling distribution does this represent? Assuming a unimodal symmetric sampling distribution, about how likely is a z-value this much or more deviant from 0 if the researcher's theory about the population is true?

27. Let us assume that the psychologist of exercise 1 took a random sample of 50 rats and ran each one through the maze. The average score for left–right tendency these rats achieved was 1.55. If the population of rats actually has a mean of 1.5 and a variance of .25, where does this result fall (in standardized score terms) in the sampling distribution of means? Again assuming a unimodal symmetric distribution, does this appear to be a relatively likely or a relatively unlikely result, given the population situation specified? Why?

28. Based on the sample result of exercise 27, find an interval of possible values for the population mean that has probability of at least .95 of covering the true value.

29. A personnel manager of an industry wished to see if the introduction of a new training method reduced the hours lost by injury by a group of employees. In the past such employees lost an average of 17.4 hours per year through injury, with a standard deviation of 4.9. A group of 150 employees were given the new training method and then placed on the job. After a year's time, this group had lost only 12.3 hours on the average. If the training method actually made no difference, and these employees truly represent a sample from the larger group with old mean and standard deviation, how likely is a sample deviating this much or more from 17.4? (Assume a unimodal symmetric distribution). How safe is it to say that a trained population will be different from the old group of employees?

30. Form a confidence interval for the mean of the new population of employees in problem 29, such that the probability that this interval covers the true mean is at least .90. Does this interval include the old value of 17.4? What does this say about the new training?

31. Ten observations were drawn at random from some very large population. The values in the sample were as follows. What are the best estimates of the population's mean and variance?

26	25	28	33	42	27	31	37	39	20

32. A sample of 25 employed American men, forty years of age, was drawn at random. Each was asked how many hours he had watched television during the preceding week. The answers are listed below. On the basis of these data estimate the mean and standard deviation of the population of such men, with respect to number of television hours watched that week. According to these estimates, how deviant (in z-score terms) was a man who watched no television at all? What is the maximum percentage of such cases there could possibly be in this population?

9	8	7	14	15
2	11	9	7	5
3	7	10	11	10
6	8	0	14	20
6	10	7	8	9

33. A medical school enrolls 208 students during a year. At random, a sample of 49 students was taken, and it was determined how much each one had spent for books during that school year. The data are given below, in whole dollars spent. On the basis of this sample, estimate the mean and standard deviation for the entire set of students. Estimate the *total* amount spent for books.

512	650	817	791	526	525	602
715	871	592	725	622	551	617
606	540	720	643	681	708	910
545	819	648	677	715	552	619
811	611	814	519	682	644	560
950	646	772	518	754	624	850
632	575	710	731	816	881	939

34. Consider the medical school described in problem 33 once again. Suppose that all possible samples of size 49 were taken from this medical school student body, and that the true mean expenditure for books is 600 dollars a year. Where does the sample mean found in exercise 34 appear to lie in the sampling distribution of means, in terms of its z-score value? (Use the estimate found in exercise 33 in place of the unknown population variance.)

35. Suppose that the same population was sampled three times. On the first occasion 10 observations were made independently and at random, on the second 20 observations were made independently and at random, and on the third occasion 15 observations were similarly made. The results were

Sample 1	*Sample 2*	*Sample 3*
$M_1 = 96$	$M_2 = 105$	$M_3 = 103$
$S_1^2 = 22$	$S_2^2 = 29$	$S_3^2 = 31$

Estimate the mean and the variance of the population. What is the estimated standard error of this estimate of the population mean?

36. A population is known to have a variance of 18.5. Two samples of 90 observations each are made independently and at random from this population. One sample produces a mean of 65 and a variance of 16 and the second independent sample a mean of 70 and a variance of 20. Estimate the population mean, and find the standard error of this estimate.

37. In a certain sampling distribution of means based on 44 cases each, a sample mean of 438 is known to correspond to a z-value of 1.75 in the sampling distribution. If the standard error of the mean is 9, what is the population mean? What is the population variance?

38. In a certain sampling distribution of the mean based on samples of size N_1, a sample mean of 100 corresponds to a z-value of 1.5. However, when samples of size N_2 are taken, a sample mean of 100 corresponds to a z-value of 2.00. How large is N_2 relative to N_1?

39. Two independent samples were drawn from the same population with a standard deviation of 9. The first sample contained 10 cases and the second sample contained 20 cases. The mean of the first sample corresponded to a z-score of 1.8 in its sampling distribution of means, while the second sample mean corresponded to a z-score of .5. How much larger was the first mean than the second? What z-score would correspond to the mean of the pooled samples?

40. A random variable X is generated by a stable and independent Poisson process with intensity 10. Five independent observations of values of X are made, and a mean calculated. Find the mean and standard deviation of the sampling distribution of these means.

Chapter 6
NORMAL POPULATION AND SAMPLING DISTRIBUTIONS

Heretofore, the only theoretical distribution we have considered in any detail has been the discrete binomial distribution. Now we consider a type of distribution having a domain which is *all* of the real numbers. This is the so-called ''normal'' or ''Gaussian'' distribution. The normal distribution is but one of a vast number of mathematical functions one might invent for a distribution; it is purely theoretical. At the very outset let it be clear that, like binomial or other probability functions, the normal distribution is not a fact of nature that one actually observes to be exactly true. Rather, the normal distribution is a theory about what might be true of the relation between intervals of values and probabilities for some variable.

6.1 / NORMAL DISTRIBUTIONS

Like most theoretical functions, a normal distribution is completely specified only by its mathematical rule. Quite often, the distribution is symbolized by a graph of the functional relation generated by that rule, and the general picture that a normal distribution presents is the familiar bell-shaped curve of Figure 6.1.1. The horizontal axis represents all the different values of X, and the vertical axis $f(x)$ their densities. The normal distribution is continuous for all values of X between $-\infty$ and ∞, so that each conceivable nonzero *interval* of real numbers has a probability other than zero. For this reason, the curve is shown as never quite touching the horizontal axis; the tails of the curve show decreasing probability densities as values grow extreme in either direction from the mode, but any interval representing *any* degree of deviation from central tendency is possible in this theoretical distribution. The distribution is absolutely symmetric and unimodal, and mean, median, and mode all have the same value. Bear in mind that since the normal distribution is continuous, the height of the curve shows the probability *density* for each X value. However, as for any continuous distribution, **the area cut off beneath the curve by any interval is a probability,** and the entire area under the curve is 1.00.

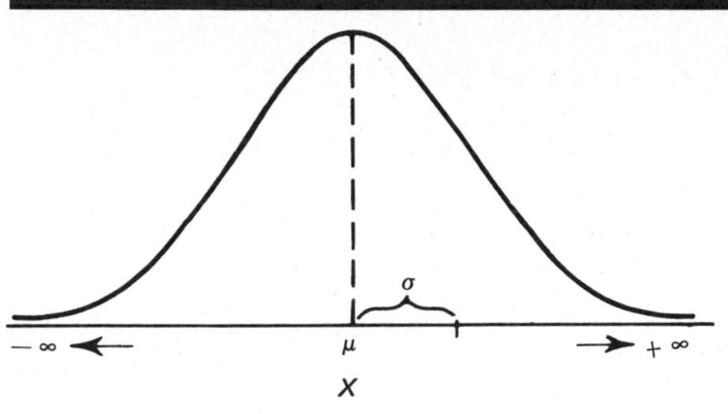

Figure 6.1.1

The graph of a normal distribution.

The mathematical rule for a normal density function is as follows:

$$f(x;\, \mu,\, \sigma^2) = \frac{1}{\sqrt{2\pi\sigma^2}}\, e^{-(x-\mu)^2/2\sigma^2}. \qquad [6.1.1^*]$$

This rule pairs a probability density $f(x)$ with each and every possible value x. This rule looks somewhat forbidding to the mathematically uninitiated, but actually it is not difficult to understand. The π, the e, and the 2, of course, are simply positive numbers acting as mathematical constants. The "working" part of the rule is the exponent

$$-\frac{(x - \mu)^2}{2\sigma^2}$$

where the particular value of the variable X appears, along with two parameters μ and σ^2.

It is important to notice that the precise density value assigned to any x by this rule cannot be found unless the two parameters μ and σ are specified. The parameter μ can be any finite number, and σ can be any finite *positive* number. Thus, like the binomial, the normal distribution rule actually specifies a family of distributions. Although each distribution in the family has a density value paired with each x by this same general rule, the *particular* density that is paired with a given x value differs with different assignments of μ and σ. Thus, normal distributions may differ in their means (Fig. 6.1.2); in their standard deviations (Fig. 6.1.3); or in both means and standard deviations (Fig. 6.1.4) Nevertheless, given the mean and standard deviation of the distribution, the rule for finding the probability density of any value of the variable is the same.

Since the normal distribution is really a family of distributions, statisicians constructing tables of probabilities found by the normal rule find it convenient to think of the variable in terms of standardized or z scores. That is to say, if the random variable is a standardized score, z, so that $\mu = 0$ and $\sigma = 1.00$, then the rule becomes simpler:

$$f(z) = \frac{1}{\sqrt{2\pi}}\, e^{-z^2/2}. \qquad [6.1.2^*]$$

Figures 6.1.2–6.1.4

Normal distributions differing in μ and σ.

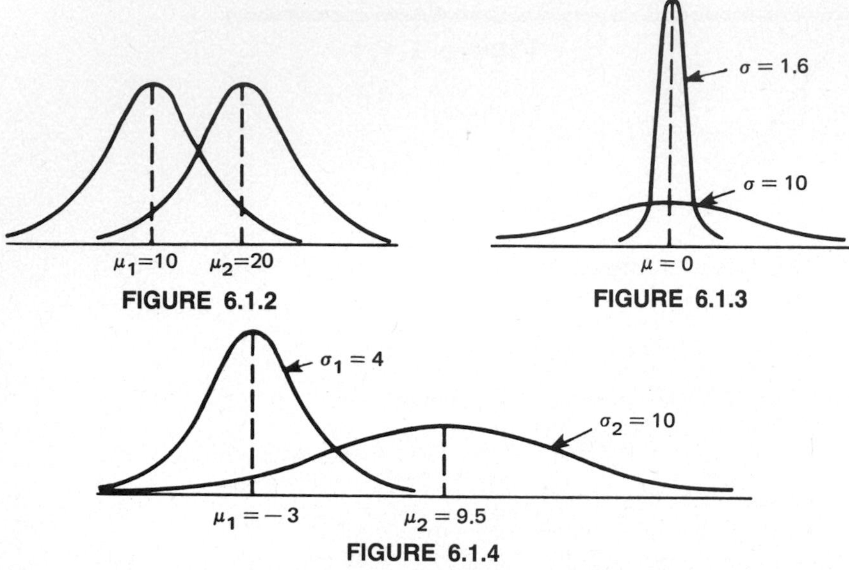

FIGURE 6.1.2

FIGURE 6.1.3

FIGURE 6.1.4

For standardized normal variables, the density depends only on the *absolute* value of z; since both z and $-z$ give the same value z^2, they both have the same density. The higher the z in absolute value, the less the associated density. The standardized-score form of the distribution makes it possible to use one table of densities for any normal distribution, regardless of its particular parameters. Thus, for the density of a score x in a normal distribution, with mean equal to 80 and standard deviation equal to 5, one can look up the density associated with a standardized score $(x - 80)/5$, and this gives the desired number. In the same way, cumulative probabilities are usually found from a table of the standardized normal distribution, such as Table I of Appendix D.

6.2 / CUMULATIVE PROBABILITIES AND AREAS FOR THE NORMAL DISTRIBUTION

In our use of the normal distribution, we will be concerned almost exclusively with cumulative probabilities and with the probabilities of intervals of values. The cumulative probability

$$F(a) = p(X \leq a) \qquad [6.2.1^*]$$

can be thought of as the area under the normal curve in the interval bounded by $-\infty$ and the value a (Fig. 6.2.1).

These cumulative probabilities can be used to find the probability of any interval. For example, suppose that in some normal distribution we want to find the probability that X lies between 5 and 10, given some exact values for μ and σ. This is the probability represented by the area shown in Figure 6.2.2. The cumulative probabil-

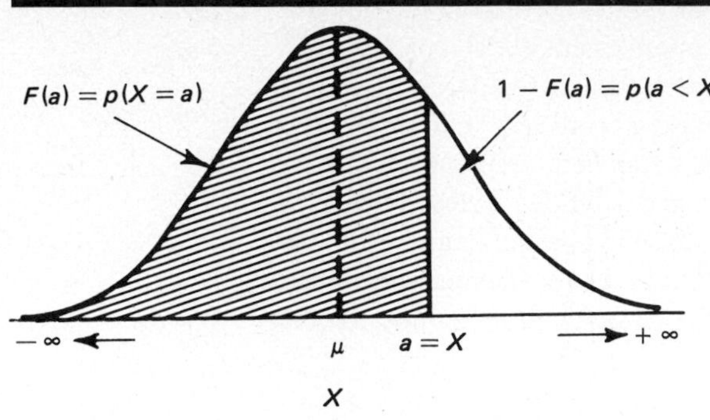

Figure 6.2.1

A cumulative distribution

ity *up to and including* 10, or $F(10)$, minus the cumulative probability up to and including 5, or $F(5)$, gives the probability in the interval:

$$p(5 \leqslant X \leqslant 10) = F(10) - F(5).$$

In the same way, the probability of any other interval with limits a and b can be found from cumulative probabilities:

$$P(a \leqslant X \leqslant b) = F(b) - F(a). \qquad [6.2.2^*]$$

As mentioned above, most tables of the normal distribution give the cumulative probabilities for various *standardized* values. That is, for a given z-score the table provides the cumulative probability *up to and including that standardized score* in a normal distribution; for example,

$$F(2) = p(z \leqslant 2)$$
$$F(-1) = p(z \leqslant -1)$$

and so on.

If the cumulative probability is to be found for a positive z, then Table I in Appendix D can be read directly. For example, suppose that a normal distribution is known to have a mean of 50 and a standard deviation of 5. What is the cumulative probability of a score of 57.5? The corresponding standardized score is

$$z = \frac{57.5 - 50}{5} = 1.5.$$

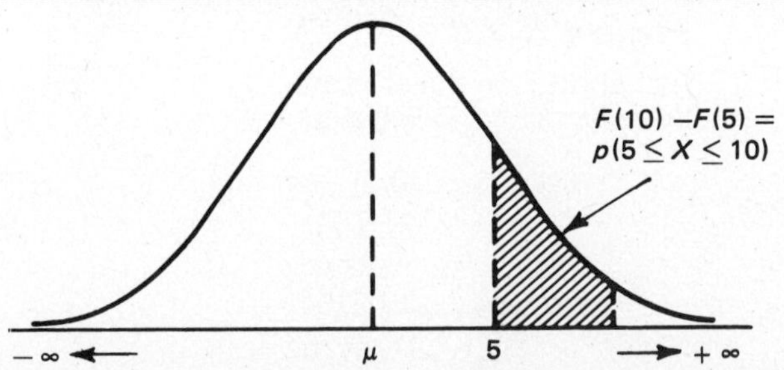

Figure 6.2.2

The probability of an interval in a normal distribution.

A look at Table I shows that for 1.5 in the first column, the corresponding cumulative probability in the column labeled $F(z)$ is .933, approximately. This is the probability of observing a z-value *less than or equal* to 1.5 in a standardized normal distribution. It is thus also the probability of observing a score less than or equal to 57.5 in a normal distribution with a mean of 50 and a standard deviation of 5.

If the z-score is negative, this procedure is changed slightly. Since the normal distribution is symmetric, the density associated with any z is the same as for the corresponding value with a negative sign, or $-z$. However, the cumulative probability for a negative standardized score is 1 minus the cumulative probability for the z-value with a positive sign:

$$F(-z) = 1 - F(z)$$

where z is the positive standardized score of the same magnitude. Figure 6.2.3 will perhaps clarify this point. The shaded area on the left is the cumulative probability for a value $z = -a$. The table gives only the area falling below the point $z = a$, the positive number of the same absolute value as $-a$. However, the shaded area on the *right*, which is $1 - F(a)$, is the same as the shaded area on the *left*, $F(-a)$.

For example, suppose that in a distribution with mean equal to 107 and standard deviation of 70, we want the cumulative probability of the score 100. In standardized form,

$$z = \frac{100 - 107}{70} = -.1.$$

We look in the table for z equal to *positive* .1, which has a cumulative probability of approximately .5389. The cumulative probability of a z equal to $-.1$ must be approximately

$$F(-.1) = 1 - .5389 = .4611.$$

We may also answer questions about the probabilities of various intervals from the table. For example, what proportion of cases in a normal distribution must lie within one standard deviation of the mean? This is the same as the probability of the interval $(-1 \leq z \leq 1)$. The table shows that $F(1) = .8413$, and we know that $F(-1)$ must be equal to $1 - .8413$ or .1587.

Thus,

$$p(-1 \leq z \leq 1) = F(1) - F(-1) = .8413 - .1587 = .6826.$$

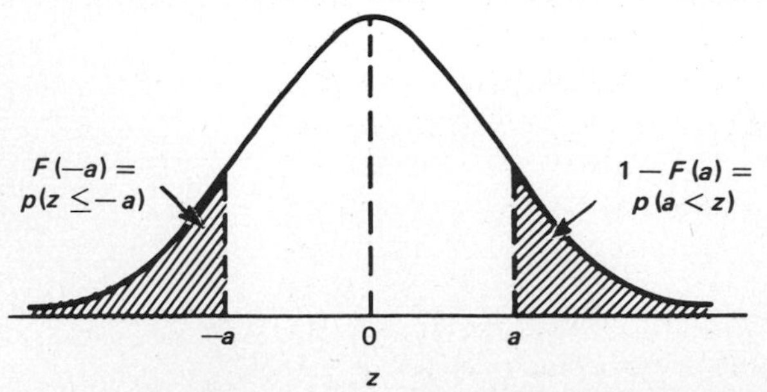

Figure 6.2.3

Probabilities on the extreme tails of a normal distribution.

About 68 percent of all cases in a normal distribution must lie within one standard deviation of the mean.

To find the proportion lying between 1 and 2 standard deviations *above* the mean, we take

$$p(1 \leq z \leq 2) = F(2) - F(1).$$

From the table, these numbers are .9772 and .8413, so that

$$p(1 \leq z \leq 2) = .9772 - .8413 = .1359.$$

About 13.6 percent of cases in a normal distribution lie in the interval between 1σ and 2σ above the mean. By the symmetry of the distribution we know immediately that

$$p(-2 \leq z \leq -1) = .1359$$

as well.

Beyond what number must only 5 percent of all standardized scores fall? That is, we want a number b such that the following statement is true for a normal distribution:

$$p(b < z) = .05.$$

This is equivalent to saying that

$$p(z \leq b) = .95.$$

A look at the table shows that roughly .95 of all observations must have z scores at or below 1.65. Thus,

$$p(1.65 < z) = .05, \text{ approximately.}$$

Similarly, two intervals $(z < -2.58)$ and $(2.58 < z)$ define

$$p(z < -2.58 \text{ or } 2.58 < z) = .005 + .005 = .01.$$

Put in the other way,

$$p(2.58 < |z|) = .01,$$

the probability that z exceeds 2.58 in absolute value is about one in one hundred.

6.3 / THE IMPORTANCE OF THE NORMAL DISTRIBUTION

The normal is by far the most used distribution in inferential statistics. There are at least three very good reasons why this is true.

A population may be assumed to follow a normal law because of what is known or presumed to be true of the measurements themselves. There are two rather broad instances in which the random variable is a measurement of some kind, and the distribution is conceived as normal. The first is when one is considering a hypothetical distribution of "errors," such as errors in reading a dial, in discriminating between stimuli, or in test performance. Any observation can be assumed to represent a "true" component plus an error. Each error has a magnitude, and this number is thought of as a reflection of "pure chance," the resultant of a vast constellation of circumstances operative at the moment. Any factor influencing performance at the moment contributes a tiny amount to the size of the error and to its direction. Furthermore, such errors are appropriately considered to push the observed measurement up or down with equal likelihood, to be independent of each other over samplings, and to "cancel out" in the long run. Thus, in theory, it makes sense that errors of measurement or errors of discrimination follow something like the normal rule.

In other instances, the distribution of *true magnitudes* of some trait may be thought of as normal. For example, human heights form approximately a normal distribution, and so, we believe, does human intelligence. Indeed, there are many examples, especially in biology, where the distribution of measurements of a natural trait seems to follow something closely resembling a normal rule. No obtained distribution is ever exactly normal, of course, but some distributions of measurements have relative frequencies very close to normal probabilities. However, the view that the distribution of almost any trait of living things should be normal, although prominent in the nineteenth century, has been discredited. The normal distribution is not, in any sense, "nature's rule," and all sorts of distributions having little resemblance to the normal distribution occur in all fields.

It may be convenient, on mathematical grounds alone, to assume a normally distributed population. Mathematical statisticians do not devote so much attention to normal distributions just because they think bell-shaped curves are pretty! The truth of the matter is that the normal function has important mathematical properties shared by no other theoretical distribution. Assuming a normal distribution gives the statistician an extremely rich set of mathematical consequences that are useful in developing methods. Very many problems in mathematical statistics either are solved, or can be solved, *only* in terms of a normal population distribution. We will find this true especially when we come to methods for making inferences about a population variance. The normal distribution is the "parent" of several other important theoretical distributions that figure in statistics. In some practical applications the methods developed using normal theory work quite well even when this assumption is not met, despite the fact that the problem can be given a *formal* solution only when a normal population distribution is assumed. In other instances, there just is not any simple way to solve the problem when the normal rule does not hold for the population, at least approximately.

There is a very intimate connection between the size of sample N and the extent to which a sampling distribution approaches the normal form. Many sampling distributions based on large N can be approximated by the normal distribution even though the population distribution itself is definitely not normal. This is the extremely important principle that we will call the **central limit theorem.** The normal distribution is the *limiting form* for large N for a very large variety of sampling distributions. This is one of the most remarkable and useful principles to come out of theoretical statistics.

Thus, you see, it is no accident that the normal distribution is the workhorse of inferential statistics. The assumption of normal population distributions or the use of the normal distribution as an approximation device is not as arbitrary as it sometimes appears; this distribution is part of the very fabric of inferential statistics. These features of the normal distribution will be illustrated in the succeeding sections.

One of the interpretations given to the normal distribution is that it is the limiting form of a binomial distribution as $N \longrightarrow \infty$ for a fixed p.

Imagine samples of N from a Bernoulli distribution with the probabilities of the two categories being p and q, respectively. Then we should expect a binomial distribution of number of successes. Suppose that we do this for all possible samples of size 10. Next we draw all possible samples with $N = 10{,}000$, then repeat the sampling with an N of 10,000,000, and so on. For each sample size, there will be a different binomial distribution we should observe when *all possible* samples of that size have been drawn.

How does the binomial distribution change with increasing sample size? In the first place, the actual range of the possible number of successes grows larger, since the whole numbers 0 through N form a larger set with each increase in N. Second, the expectation, Np, is larger for each increase in N for a fixed p, and the standard deviation $\sqrt{Npq}$ also increases with N. Finally, the probability associated with any given exact value of X tends to *decrease* with *increase* in N.

However, suppose that for any value of N the number of successes X were put into standardized form:

$$z = \frac{x - Np}{\sqrt{Npq}} = \frac{x - E(X)}{\sigma}. \qquad [6.4.1^*]$$

Then regardless of the size of N, the mean of the z-scores for a binomial distribution will be 0, and the standard deviation 1. The probability of any z-score is the same as for the corresponding x. Any z-score interval can be given a probability in the particular binomial distribution, and the same interval can also be given a probability for a normal distribution. We can compare these two probabilties for any interval; if the two probabilities are quite close over all intervals, the normal distribution gives a good approximation to the binomial probabilities. On making this comparison of binomial and normal interval probabilities, we should find that the larger the sample size N the better is the fit between the two kinds of probabilities. **As N grows infinitely large, the normal and binomial probabilities become identical for any interval.**

For example, imagine that $p = .5$, and $N = 5$. Here the expectation is (5)(.5), or 2.5, and the standard deviation is $\sqrt{5(.5)(.5)} = \sqrt{1.25}$ or about 1.12. The binomal distribution for these values of p and N is given in Table 6.4.1.

Table 6.4.1

A binomial distribution with $p = .5$ and $N = 5$.

x	$p(x)$
5	.0312
4	.1563
3	.3125
2	.3125
1	.1563
0	.0312

For the moment, let us pretend that this is actually a continuous distribution, and that the possible numbers of successes are midpoints of class intervals. Furthermore, let us turn each X into its standardized value z, by taking

$$z = \frac{x - E(X)}{\sigma} = \frac{x - 2.5}{1.12}.$$

Then the distribution is

Table 6.4.2 The normal approximation to a binomial with $p = .5$ and $N = 5$.

z real limits	*z midpoint*	*p(x)*	*Normal*
(1.79 to 2.68)	2.23	.0312	.0367
(.89 to 1.79)	1.34	.1563	.1500
(.00 to .89)	.45	.3125	.3132
(−.89 to .00)	−.45	.3125	.3132
(−1.79 to −.89)	−1.34	.1563	.1500
(−2.68 to −1.79)	−2.23	.0312	.0367

Here, the z intervals were found by converting each real limit for the distribution of successes (regarded as continuous) into its equivalent z-score: thus, $1.79 = (4.5 - 2.5)/1.12$, and so on. Note that the probabilities are the same for these intervals as for the original X-values.

The last column gives the probabilities that these same intervals would have in a normal distribution, found by using Table I in Appendix D. Thus, the interval .89 to 1.79 has probability

$$F(1.79) - F(.89) = .96327 - .81327 = .1500$$

and the other probabilities are found in the same way. The top and bottom intervals are given the probabilities

$$1 - F(1.79) = .0367$$

and

$$F(-1.79) = .0367$$

in order to make the probabilities over all intervals sum to 1.00.

Observe how closely the normal probabilities approximate their binomial counterparts: each normal probability is correct to two decimal places as an approximation to a binomial probability. The difference in the two probabilities is smallest for the middle intervals, and is larger for the extremes. Thus, even when N is only 5, the normal distribution gives a respectable approximation to binomial probabilities for $p = .5$.

Now suppose that for this same example, N had been 15. Here the expectation is 7.5, and the standard deviation is $\sqrt{(15)(.25)} = 1.94$. The distribution, with z-score intervals and both binomial and normal probabilities, is shown in Table 6.4.3.

For this distribution the normal approximation gives an even better fit to the exact binomial probabilities; the average absolute difference in probability over the inter-

Table 6.4.3
The normal approximation to a binomial with $p = .5$ and $N = 15$.

x	z intervals	p(x)	Normal probabilities
15	3.608 to 4.124	.00003	.0002
14	3.092 to 3.608	.0005	.0012
13	2.577 to 3.092	.0032	.0036
12	2.061 to 2.577	.0139	.0147
11	1.546 to 2.061	.0416	.0409
10	1.030 to 1.546	.0916	.0909
9	.515 to 1.030	.1527	.1501
8	.000 to .515	.1964	.1984
7	−.515 to .000	.1964	.1984
6	−1.030 to −.515	.1527	.1501
5	−1.546 to −1.030	.0916	.0909
4	−2.061 to −1.546	.0416	.0409
3	−2.577 to −2.061	.0139	.0147
2	−3.092 to −2.577	.0032	.0036
1	−3.608 to −3.092	.0005	.0012
0	−4.124 to −3.608	.00003	.0002

vals is about .001, whereas when N was only 5, the average absolute difference was about .004. In general, as N is made larger, the fit between normal and binomial probabilities grows increasingly good. In the limit, when N approaches infinite size, the binomial probabilities are exactly the same as the normal probabilities for any interval, and thus one can say that the normal distribution is the limit to the binomial. This is true regardless of the value of p.

On the other hand, for any finite N, the more p departs from .5, the less well does the normal distribution approximate the binomial. When p is not exactly equal to .5, the binomial distribution is somewhat skewed for any finite sample size N, and for this reason the normal probabilities will tend to fit less well than for $p = .5$, which always gives a symmetric distribution.

In any practical use of the normal approximation to binomial probabilities, it is important to remember that here we regard the binomial distribution as though it were continuous, and actually find the normal probability associated with an intrval with real lower limit of $(x - .5)$ and real upper limit of $(x + .5)$, where x is any given number of successes. Thus, in order to find the normal approximation to the binomial probability of x, we take

$$\text{prob}\left(\frac{x - Np - .5}{\sqrt{Npq}} \leqslant z \leqslant \frac{x - Np + .5}{\sqrt{Npq}}\right) \qquad [6.4.2]$$

by use of the standardized normal tables. Similarly, in terms of sample $P = x/N$, we have

$$\text{prob}\left(\frac{P - p - .5/N}{\sqrt{pq/N}} \leqslant z \leqslant \frac{P - p + .5/N}{\sqrt{pq/N}}\right). \qquad [6.4.3]$$

This adjustment, by which the probability of an interval of values is taken in place of the exact value of x or P, gives rise to the so-called **correction for continuity.**

That is, in later sections when the normal distribution is to be used to approximate a binomial probability, we will deal with the z-value

$$z = \frac{x - Np - .5}{\sqrt{Npq}} \qquad [6.4.4^*]$$

when x is larger than Np, and with

$$z = \frac{x - Np + .5}{\sqrt{Npq}} \qquad [6.4.5^*]$$

when x is less than Np. This will allow for the fact that the normal distribution is being used in order to approximate the probability in a discrete distribution. Only when the sample size N is relatively large does this correction become unimportant enough to ignore.

As noted above, the normal distribution need not be particularly good as an approximation to the binomial if either p or q is quite small, and if either p or q is extremely small the normal approximation may not be satisfactory even when N is quite large. In these circumstances another theoretical distribution, the Poisson, provides a better approximation to binomial probabilities. This approximation of binomial probabilities by use of Poisson probabilities was outlined in Chapter 3.

6.5 / THE THEORY OF THE NORMAL DISTRIBUTION OF ERROR

The fact that the limiting form of the binomial is the normal distribution actually provides a rationale for thinking of random error as distributed in this normal way. Consider an object measured over and over again independently and in exactly the same way. Imagine that the value Y obtained on any occasion is a sum of two independent parts:

$$Y = T + e.$$

That is, the obtained score Y is a sum of a constant *true* part T plus a random and independent *error* component. However, the error portion can also be thought of as a sum:

$$e = g[e_1 + e_2 + e_3 + \cdots + e_N]$$

Here, e_1 is a random variable that can take on only two values:

$$e_1 = 1, \text{ when factor 1 is operating}$$

$$e_1 = -1, \text{ when factor 1 is not operating.}$$

The g is merely a constant, reflecting the "weight" of the error in Y. Similarly, the other random errors are attributable to different factors, and take on only the values 1 and -1. Now imagine a vast number of influences at work at the moment of any measurement. Each of these factors operates independently of each of the others, and whether any factor exerts an influence at any given moment is purely a chance matter. If you want to be a little anthropomorphic in your thinking about this, visualize old Dame Fortune tossing a vast number of coins on any occasion, and from the result of each deciding on the pattern of factors that will operate. When coin one comes up heads, e_1 gets value 1, and if tails, value -1, and the same principle determines the value for every other error portion of the observed score.

Now under this conception, the number of factors operating at the moment one observes Y is a number of successes in N independent trials of a Bernoulli experiment. If this number of successes is X, then

$$e = gX - g[N - X] = g[2X - N].$$

The value of e is exactly determined by X, the number of "influences" in operation at the moment, and the probability associated with any value of e must be the same as for the corresponding value of X. **If N is very large, then the distribution of e must approach a normal distribution.** Furthermore, if any factor is equally likely to operate or not operate at a given moment, so that $p = 1/2$, then

$$E(e) = g[2E(X) - N],$$

so that, since $E(X) = Np = N/2$,

$$E(e) = 0.$$

In the long run, over all possible measurement occasions the errors all "cancel out." This makes it true that

$$E(Y) = E(T) + E(e) = T,$$

The long-run expectation of a measurement Y is the true value T, provided that error really behaves in this random way as an additive component of any score.

This is a highly simplified version of the argument for the normal distribution of errors in measurement. Much more sophisticated rationales can be invented, but they all partake of this general idea. Moreover, the same kind of reasoning is sometimes used to explain why distributions of natural traits, such as height, weight, and size of head, follow a more or less normal rule. Here, the mean of some population is thought of as the "true" value, or the "norm." However, associated with each individual is some departure from the norm, or error, representing the culmination of all the billions of chance factors that operate on him, quite independently of other individuals. Then by regarding these factors as generating a binomial distribution, we can deduce that the whole population should take on a form like the hypothetical normal distribution. However, this is only a theory about how errors might operate and there is no reason at all why errors must behave in the simple additive way assumed here. If they do not, then the distribution need not be normal in form at all.

6.6 / TWO IMPORTANT PROPERTIES OF NORMAL POPULATION DISTRIBUTIONS

As suggested earlier in this chapter, the normal distribution has mathematical properties that are most important for theoretical statistics. For the moment, we are going to discuss only two of these general properties, both of which will be useful to know in later sections. The first has to do with the sampling distribution of means for samples drawn from a normally distributed population. The second deals with the independence of sample mean and sample variance.

Suppose that samples of N independent observations are being drawn from a population with mean μ and standard deviation σ. Furthermore, suppose that it is known that this population shows a normal distribution of the random variable of interest, X. Can we then say anything special about the sampling distribution of means M drawn from this normal population?

As it happens, we can say a great deal:

Given random samples of N independent observations each drawn from a normal population, then the distribution of the sample means is normal, irrespective of the size of N.

In other words, when one can assume that the population itself is normally distributed with respect to the random variable X, the problem of the sampling distribution of M is solved: this sampling distribution will itself be normal, with expectation equal to μ and standard error equal $\sigma/\sqrt{N}$. *This is true regardless of the size of the sample, or N* (so long as N is at least 1, of course).

A great deal of use is made of this principle in statistical inference, especially in those situations where one is forced to deal with small samples. As we shall see in subsequent chapters a normal population distribution is almost always assumed when the sample size is small, since this assumption permits one to say exactly what the sampling distribution of M must be like.

Any sample consisting of N independent obsrvations of the same random variable provides both a sample mean M and a sample variance. The sample mean estimates μ and the unbiased sample variance, s^2, estimates σ^2. These two values (M, s^2) obtained from any sample can be thought of as a joint event. But are these two estimates independent? Or, does the value found for one of these statistics somehow depend on the value found for the other?

The answer to these questions is provided by the following key principle:

Given random and independent observations, the sample mean M and the sample variance (either S^2 or s^2) are independent if and only if the population distribution is normal.

The information contained in the sample mean in no way dictates the value of the sample variance, and vice versa, when a normal population is sampled. Furthermore, **unless the population actually is normal, these two sample statistics are not independent across samples.**

This is a most important principle, since a great many problems concerned with a population mean can be solved only if one knows something about the value of the population variance. At least an estimate of the population variance is required before particular sorts of inferences about the value of μ can be made. Unless the estimate of the population variance is statistically independent of the estimate of μ made from the sample, no simple way to make these inferences may exist. This question will be considered in more detail in Chapters 10 and 11. For the moment, suffice it to say that this principle is one of the main reasons for statisticians' assuming normal population distributions.

As we have just seen, there are some very good reasons for the user of statistical methods to be interested in populations having a normal distribution of the random variable under study. Sometimes there are compelling reasons to believe that the population really would show such a distribution, could all of the observations actually be made. On other occasions, no very good reasons can be advanced why the population should *not* be normal, and so one assumes this to be the case, even though evidence to support this assumption may be lacking.

Nevertheless, it is quite common for us to be concerned with populations where the distribution should *definitely not be normal*. We may know this either from empirical evidence about the distribution, or because some theoretical issue makes it impossible for scores to represent this kind of random variable. Simple illustrations are distributions of intelligence scores among graduate engineers, and of measures of socially nonconforming behavior among normal adults. In principle both these distributions should be extremely skewed, though for quite different reasons: in the first, there is a selection in the education process making it rather unlikely for anyone in the middle or low intelligence ranges to qualify as an engineer; in the second, social nonconformity is something that most people must show in very small degree. The assumption of a normal distribution does not make sense in such instances.

Very often an inference must nonetheless be made about the mean of such a population. To do this effectively the experimenter needs to know the sampling distribution of the mean, and to know this exactly, one has to be able to specify the particular form of the population distribution. However, if we had enough evidence to permit this, we would likely have an extremely good estimate of the population mean in the first place, and would not need any other statistical methods!

The way out of this apparent impasse is provided by the **central limit theorem,** which can be given an approximate statement as follows:

If a population has a finite variance σ^2 and mean μ, then the distribution of sample means from samples of N independent observations approaches a normal distribution with variance σ^2/N and mean μ as the sample size N increases. When N is very large, the sampling distribution of M is approximately normal.

The central limit theorem may be the only example in the universe where one almost gets something for nothing! *Absolutely nothing is said in this theorem about the form of the population distribution.* Regardless of the population distribution, if sample size N is large enough, the normal distribution is a good approximation to the sampling distribution of the mean. This is the heart of the theorem, since, as we have already seen, the sampling distribution will have a variance σ^2/N and mean μ for any N.

The sense of the central limit theorem is illustrated by Figures 6.7.1 and 6.7.2. The solid curve in Figure 6.7.1 is a very skewed population distribution in z-score form, and the broken curve is a standardized normal distribution. The other figures show the standardized form of the sampling distribution of means for samples of size

Figures 6.7.1/6.7.2

A negatively skewed population distribution (Fig. 6.7.1, left) and a sampling distribution of M for $N = 2$ (Fig. 6.7.2 right), with comparable normal distributions

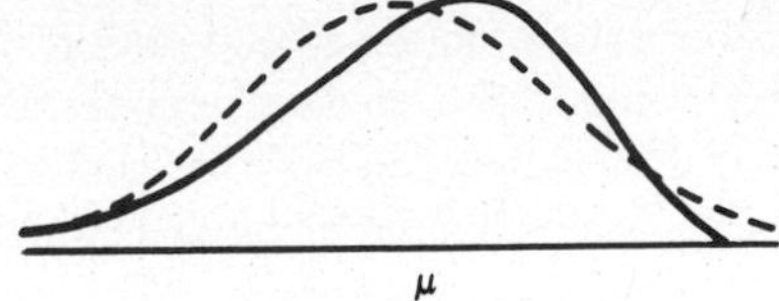

2, 4, and 10 respectively from this population together with the corresponding standardized normal distributions. Notice that it is *not* the number of samples drawn, but rather the *size* of each sample, that is the effective part of the central limit theorem. A vast number of samples of any size N may be drawn; the larger N is for each sample, the more nearly is the distribution of means normal.

Even with relatively small sample sizes it is obvious that each increase in sample size gives a sampling distribution more nearly symmetric and tending more toward the normal distribution with the same mean and variance. This symmetry increases with increasing sample size until, in the limit, the normal distribution is reached.

It must be emphasized that in most instances the tendency for the sampling distribution of M to be like the normal distribution is very strong, even for samples of moderate size. Naturally, the more similar to a normal distribution the original population distribution, the more nearly will the sampling distribution of M be like the normal distribution for any given sample size. However, even extremely skewed or other nonnormal distributions may yield sampling distributions of M that can be approximated quite well by the normal distribution, for samples of at least moderate size. In the examples shown in Figures 6.7.3 and 6.7.4, the correspondence between the exact probabilities of intervals in the sampling distribution and intervals in the normal distribution is fairly good even for samples of only 10 observations; for a rough approximation, the normal distribution probabilities might be useful even here in some statistical work. In a great many instances in social research, a sample size of 30 or more is considered large enough to permit a satisfactory use of normal probabilities to approximate the unknown exact probabilities associated with the

Figures 6.7.3/6.7.4

Sampling distributions of M for $N = 4$ (Fig. 6.7.3 left) and $N = 10$ (Fig. 6.7.4 right), with comparable normal distributions.

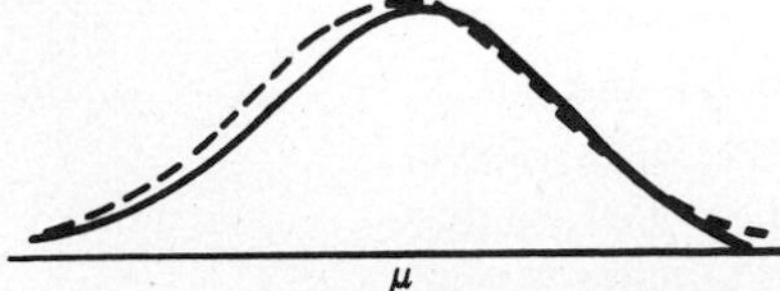

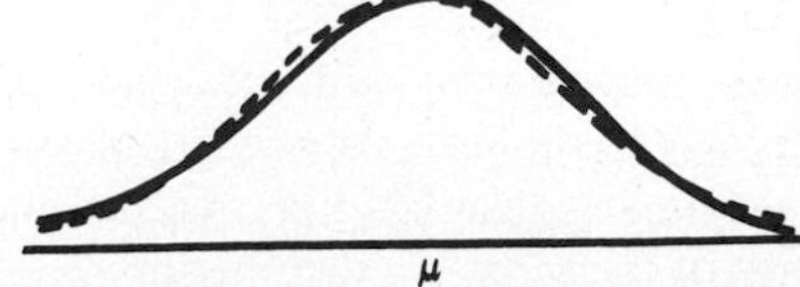

sampling distribution of M. Thus, even though the central limit theorem is actually a statement about what happens *in the limit* as N approaches an infinite value, the principle at work is so strong that in many instances the theorem is practically useful even for moderately large samples.

The mathematical proof of this theorem is extremely advanced, and many eminent mathematicians over the centuries contributed to its development before it was finally proved in full generality. However, some intuitive feel for why it should be true can be gained from the following example. Here we will actually work out the sampling distribution of the mean for a special and very simple population distribution.

Imagine a random variable with the following distribution:

x	$p(x)$
5	1/3
4	1/3
3	1/3

The μ of this little distribution is 4. However, instead of dealing directly with this random variable, let us consider the deviation d from μ, or

$$d = (X - \mu).$$

The distribution of d has a form identical to the distribution of X itself, although the mean d is zero. Since X can assume only three values, each equally probable, there are three possible deviation values of d, also equally probable.

Suppose that sample observations were taken two at a time, independently and at random. The deviation value of the first observation made is d_1 and that of the second is d_2. Corresponding to each joint event (d_1, d_2) there is a mean deviation $(d_1 + d_2)/2$, which is equal to $M - \mu$. The following table shows the mean deviation value that is produced by each possible joint event:

d_2			
1	0	.5	1
0	−.5	0	.5
−1	−1	−.5	0
	−1	0	1 d_1

Since X_1 and X_2 are independent, then d_1 and d_2 are also independent.

The probability associated with each cell in the table is thus $p(d_1)p(d_2)$ or 1/9.

Now let us find the probabilities of the various values of mean d. The value −1 can occur in only one way, and so its probability is 1/9. On the other hand, −.5 can occur in two mutually exclusive ways, giving a probability of 2/9. In this same way we can find the other probabilities, and form the distribution of the mean d and the mean X values, as follows:

M	Mean d	$p(M) = p(\text{mean } d)$
5.0	1.0	1/9
4.5	0.5	2/9
4.0	0	3/9
3.5	−0.5	2/9
3.0	−1.0	1/9

where M = mean $d + \mu$. Now notice that whereas in the original distribution there was no distinct mode, in this sampling distribution there *is* a distinct mode at 4, with exact symmetry about this point. What causes the sampling distribution to differ from the population distribution in this way? Look at the table of mean deviations corresponding to joint events once again: there are simply more possible ways for a sample to show a small than an extreme deviation from μ. There are more ways for a sample joint event to occur where the deviations tend to "cancel out" than where deviations tend to cumulate in the same direction.

Exactly the same idea could be illustrated for any discrete population distribution, whatever its form. The small and "middling" deviations of M from μ always have a numerical advantage over the more extreme deviations. This superiority in number of possibilities for a small mean deviation increases as N is made larger. Regardless of how skewed or otherwise irregular a population distribution is, by increasing sample size one can make the advantage given to small deviations so big that it will overcome any initial advantage given to extreme deviations by the original population distribution. The numerical advantage given to relatively small deviations from μ effectively swamps any initial advantage given to other deviations by the form of the population distribution, as N becomes large.

Perhaps this simple example gives you some feel for why it is that the sampling distribution of the mean approaches a unimodal symmetric form. Of course, this is a far cry from showing that the sampling distribution must approach a *normal* distribution, which is the heart of the central limit theorem. Nevertheless, the basic operation of chance embodied in this theorem is of this general nature: for large N it is relatively much easier to get a small deviation of M from μ than a large one.

6.8 / CONFIDENCE INTERVALS FOR THE MEAN

When the idea of a confidence interval was introduced in the last chapter, it was illustrated in terms of Tchebycheff's inequality. This was done without our having to specify the exact form of the sampling distribution of a statistic such as M. On the other hand, these confidence intervals were associated with quite general statements such as "the probability is at least .89" or the "probability is at least .95" that the obtained interval actually covers the population mean value.

Now we are ready to construct confidence intervals for the mean, where more precise probability statements may be made. This can be done because two principles can be applied: the sampling distribution of M is normal when the population is normal, and, when N is large, the sampling distribution of M is normal without respect to the population distribution.

In any normal distribution of sample means, with population mean μ and standard deviation μ_M, the following statement is true: **Over all samples of N, the probability is .95 for the event**

$$-1.96\mu_M \leqslant M - \mu \leqslant +1.96\sigma_M \;. \qquad [6.8.1]$$

That is, very nearly 95 percent of all possible sample means from the population in question must lie within 1.96 standard errors to either side of the true mean. You can check this for yourself from Table I.

However, this is still the same event even if we alter the inequality by changing all the signs and adding M to each term: **Over all sample values of M, the probability is .95 for the event**

$$M - 1.96\sigma_M \leqslant \mu \leqslant M + 1.96\sigma_M \;. \qquad [6.8.2^*]$$

That is, over all possible samples, the probability is about .95 that the range between $M - 1.96\sigma_M$ and $M + 1.96\sigma_M$ will include the true mean, μ.

This range of values between $M \pm 1.96\sigma_M$ is called the **95 percent confidence interval for μ.** The two boundaries of the interval, or $M + 1.96\sigma_M$ and $M - 1.96\sigma_M$, are called **the 95 percent confidence limits.**

Be sure to notice that μ is not a random variable and the probability statement is really not about μ, but about *samples*. Before any sample mean is seen, one may decide to compute the 95 percent confidence interval for each and every sample. The actual range of numbers obtained for a sample will depend on the value of M. In short, over all possible samples, there will be many possible 95 percent confidence intervals. Some of these confidence intervals will represent the event ''covers the true mean,'' and others will not. If one such confidence interval were sampled at random, then the probability is .95 that it covers the true mean.

The 99 percent confidence interval for the mean is given by

$$M - 2.58\sigma_M \leqslant \mu \leqslant M + 2.58\sigma_M. \qquad [6.8.3^*]$$

Notice that the 99 percent confidence interval is larger than the 95 percent interval. More possible values are included, and thus one can assert with more confidence that the true mean is covered. A 100 percent confidence interval would approach

$$M - \infty\sigma_M < \mu < M + \infty\sigma_M.$$

Here, one can be completely confident that the true value is covered, since the range includes all possible values of μ. On the other hand, the 0 percent interval is simply *M;* for a continuous sampling distribution of M, the probability is, in effect, zero that M exactly equals μ. Any other confidence interval may be found by finding the z values in a normal distribution cutting off the desired percent of cases in the middle of the distribution. These new z values then take the place of the values of $\pm$ 1.96 and $\pm$ 2.58 used above.

It is interesting to note that, in principle, the 95 percent, or 99 percent, or any other confidence interval for a mean might be defined in many other ways. For example, for a normal sampling distribution of M it is also true that the probability is approximately .95 that

$$-1.75\sigma_M \leqslant M - \mu \leqslant 2.33\sigma_M.$$

This implies that a 95 percent confidence interval might possibly be found with limits

$$M - 1.75\sigma_M \text{ and } M + 2.33\sigma_M.$$

If many different ways of defining the 95 percent confidence interval are possible, what is the advantage of defining this confidence interval in terms of the *middle* portion of a normal distribution? In the particular case of the mean, this way of defining the confidence interval gives the *shortest* possible range of values such that the probability statement holds. Naturally, there is an advantage in pinning the population parameter within the narrowest possible range with a given probability, and this dictates the form of the confidence interval we use.

This method for finding a confidence interval applies either when the population is normally distributed, or when the sample size is sufficiently large to make the central limit theorem apply. For many practical purposes, a sample size of 30 or more cases seems to be large enough to permit this method to be used.

A more serious drawback is the assumption that the population value of σ is known. Strictly speaking, these confidence intervals do not have the correct probabilities unless a known and constant value of σ is used in figuring the value of σ_M for each sample. However, this too is not an insurmountable problem. When σ is unknown, and the sample size N is large, confidence intervals that are approximately correct may still be found. We do this by using estimated σ or $s/\sqrt{N}$, in place of the unknown value of σ_M. That is, if the sample size is large (in this instance certainly thirty or more) the 95 percent confidence interval is found by taking the limits

$$M - (1.96)\,(s/\sqrt{N}) \text{ and } M + (1.96)\,(s/\sqrt{N}) \qquad [6.8.4^*]$$

and interpreting this interval in the usual way. Although a full rationale for our doing this will not be developed until a later chapter, this is the way that confidence intervals for the mean are usually calculated.

6.9 / SAMPLE SIZE AND THE ACCURACY OF ESTIMATION OF THE MEAN

The width of any confidence interval for the mean μ depends upon σ_M, the standard error, and anything that makes σ_M proportionately smaller reduces the width of the interval. Thus, any increase in sample size, operating to reduce σ_M, makes the confidence interval shorter.

A practical result of this relation between the standard error of the mean and sample size is that *the population mean may be estimated within any desired degree of precision, given a large enough sample size*. Especially if the desired accuracy can be stated in population σ units, the required N is easy to find. For instance, how many cases should one sample in order to make the probability be .99 that the sample mean lies within $.1\sigma$ of the true mean? That is, the experimenter wants

$$\text{prob}(|M - \mu| \leq .1\sigma) = .99.$$

Assuming that the sampling distribution is nearly normal, which it should be if sample size is large enough, this is equivalent to requiring that the 99 percent confidence interval should have limits

$$M - .1\sigma$$

and

$$M + .1\sigma.$$

However, this is the same as saying that

$$.1\sigma = 2.58\sigma_M$$

so that

$$.1\sigma = 2.58\frac{\sigma}{\sqrt{N}}.$$

Solving for N, we find

$$\sqrt{N} = 25.8$$

$$N = 665.64.$$

In short, if the experimenter makes 666 independent observations, he or she can be sure that the probability of the estimate's being wrong by more than $.1\sigma$ is only one in one hundred. Notice that we do not have to say exactly what σ is in order to specify the desired accuracy in σ units, and to find the required sample size.

Obviously, the question of adequate sample size could be settled this way in many social science studies. Once we have some idea of the desirable accuracy of estimation, stated in population standard deviation units, then we can set the sample size so as to give that degree of accuracy of estimation a high probability. Although this is a very sensible and easy way to determine sample size, it is not used as often as it should be in social science research. The main reason is that social scientists have usually not paid much attention to the question of the accuracy of estimation that is desirable, or necessary, to attain. Although this is a complicated scientific question, other disciplines routinely concern themselves with such issues of accuracy in estimation. Perhaps as the social sciences continue to develop and mature, these questions will become as routine as they are in other sciences.

6.10 / USE OF A CONFIDENCE INTERVAL IN A QUESTION OF INFERENCE

In order to see a simple example of how a confidence interval might be used in a problem concerning a population mean, imagine the following: A gerontologist is engaged in a study of the eating habits of the elderly. In particular this scientist is concerned that, due to inflation and other factors, the elderly may not be eating as well today as in years past. A study had been done ten years ago, in which it was found that healthy women over age seventy, living independently and not in a nursing or retirement home, had an average daily intake of 2032 calories. Now the gerontologist wishes to estimate the current daily calorie intake of such a population. In particular, the question to be asked is, ''Is the average daily calorie intake of this population the same as it was ten years ago?''

A sample of 100 women over the age of seventy is drawn, where each woman sampled meets the criteria listed above. The daily caloric intake of each individual sampled is measured. The sample mean turns out to be 1847, with a standard deviation s equal to 310. On the face of it, this sample suggests that elderly women are consuming fewer daily calories on the average than they did ten years ago.

This is, after all, just a sample result, and it could be true that the population of elderly women still has the same, or even a higher, average calorie intake than before. Can the investigator say with assurance that the present calorie intake is not the same? This question may be answered by calculating confidence limits for the population mean. Here, since

$$s = 310$$

the estimated standard error of the mean is

$$\text{est } \sigma_M = 310/\sqrt{100} = 31.$$

Then the approximate ninety-five percent confidence interval has limits given by

$$1847 - (1.96)(31) \text{ and } 1847 + (1.96)(31)$$

or

$$1786.24 \text{ to } 1907.76.$$

Now notice that the interval between these two values *does not* contain the former population value of 2032. Since the probability is .95 that the true population value of the mean is covered in this interval, then the probability is only .05 or less that the sample represents a population with a mean outside this range of values. In other words, the investigator rejects the hypothesis that the new population mean is the same as the old. However, this could be the wrong conclusion. What is the probability of the investigtor's being wrong in this conclusion? It is no more than .05, the probability of the sample confidence interval's not covering the true value of μ.

This exemplifies one use to which a confidence interval for a mean may be applied: checking on a hypothetical population situation in the light of sample evidence. Such "hypothesis testing" forms an important part of statistical inference. The construction of a confidence interval is one way that a hypothesis about an exact value of some population's mean may be tested in the light of a sample. It the confidence interval does not cover the hypothetical value of the mean, then that hypothesis may be rejected. If the K percent confidence interval is used, one says that the hypothesis is rejected at the $(100 - K)$ percent level. We will have a great deal more to say about procedures and conventions for testing hypotheses in subsequent chapters. For the moment, let it suffice that this is one highly useful application of the idea of a confidence interval.

6.11 / CONFIDENCE INTERVALS FOR PROPORTIONS

It was seen in Section 6.4 that as N is made larger, the binomial distribution can be approximated quite well by the normal distribution. In consequence, it is possible to construct confidence intervals for population proportions in much the same way as confidence intervals for the mean, given a large N.

When a sample of N independent observations is taken, the sample proportion P is found by taking $P = x/N$, where x is the number of "successes," or events of a particular kind. When N is reasonably large (say $N \geqslant 20$ when P is about .5, but $N \geqslant 50$ for $p \leqslant .4$), the distribution of values of P should be close to a normal distribution, where

$$E(P) = p$$

and

$$\sigma_P = \sqrt{pq/N}.$$

We would like to be able to form a confidence interval that is highly likely to cover the value of p, the population proportion. At first glance it appears that this could be done simply by taking $z = 1.96$ or 2.58 and substituting into

$$P - z\sqrt{\frac{pq}{N}} \text{ and } P + z\sqrt{\frac{pq}{N}}.$$

However, there is a hitch in doing this: the value of p that we are trying to estimate is, of course, unknown. Nevertheless, the possible values of p corresponding to these limits must satisfy the relation

$$\frac{N(P - p)^2}{(p - p^2)} = z^2.$$

Simply solving this quadratic equation for p by methods familiar from high school algebra yields the following expression for the confidence interval:

$$\frac{N}{N + z^2}\left[P + \frac{z^2}{2N} \pm z\sqrt{\frac{PQ}{N} + \frac{z^2}{4N^2}}\right] \qquad [6.11.1]$$

where $Q = 1 - P$ and z is the standard score in a normal distribution cutting off the *upper* $(K/2)$ proportion of cases. These limits define the 100 $(1-K)$ percent confidence interval for p.

If N is very large, (certainly $N \geqslant 100$), this confidence interval may be replaced by the simpler approximation

$$P - z\sqrt{\frac{PQ}{N - 1}} \leqslant p \leqslant P + z\sqrt{\frac{PQ}{N - 1}}. \qquad [6.11.2]$$

For example, suppose that we were interested in the question of how many full professors at state universities prefer teaching activities, and how many prefer research and writing. We decide to take a random sample sample from among all such professors, and ask each his or her preference. Here, quite arbitrarily, let us say that

"success" = prefers teaching
"failure" = prefers research

We will not allow a response of indifference or "no preference," so that by eliminating such responses from our sample, we are restricting the population of interest to "professors who have a definite preference." Now suppose that we are successful in drawing a random sample of full professors at state universities, providing $N = 132$ usable responses.

From this sample, the proportion of successes, or P, turns out to be .62. In order to find the 95 percent confidence interval for p, the population proportion, we use 6.11.1 and take

$$\frac{132}{132 + (1.96)^2}\left[.62 + \frac{(1.96)^2}{2(132)} \pm 1.96\sqrt{\frac{(.62)(.38)}{132} + \frac{(1.96)^2}{4(132)^2}}\right]$$

which works out to provide the confidence limits of approximately

.535 and .698

then the probability is about .95 that in the population of college professors the proportion preferring teaching to research is covered by the range of values

.535 to .698.

Can one say it is reasonable to suppose that just as many professors prefer teaching as prefer research, on the basis of this evidence? If the preferences for the two activities were exactly equal, then the population proportion would be .50. Notice, however, that this value is not covered by the 95 percent confidence interval just constructed. Thus, on the basis of this evidence, the hypothesis of equal preference may be rejected at the .05, or 5 percent, level.

EXERCISES

1. A teacher believes that the class scores on a final examination should be approximately normally distributed, provided the class is large enough. If this belief is correct, what proportion of the class should fall at or below the following z-values?

(a) -1.2 **(d)** -1.78
(b) .96 **(e)** $-.43$
(c) 1.88 **(f)** 2.15

2. Another teacher wishes to curve class grades in order to make them correspond to values in a normal distribution with a mean of 100 and a standard deviation of 10. In order to do this, one first gives each score a cumulative relative frequency. These are then equated with corresponding z-values in a normal distribution, and thus into the scores desired. Carry out this process with the following set of scores:

18, 21, 23, 25, 30, 31, 36, 38, 40, 42, 45, 46, 48, 50, 52, 57, 60, 75

3. A mean from a sample of 36 cases has a value of 100. How probable is a sample mean of 100 or more when the population being sampled is normal with the following parameters?

(a) Mean 103, standard deviation 10
(b) Mean 99, standard deviation 4
(c) Mean 80, standard deviation 50
(d) Mean 98, standard deviation 24
(e) Mean 110, standard deviation 80

4. A normal distribution has a mean of 500 and a standard deviation of 10. It is desired to divide this distribution into five intervals of values such that the probability of each interval will be exactly .20. List the desired intervals, starting with the highest.

5. A study was concerned with the age at which American mothers have their first children. Twenty years ago the age at first delivery was known to be 23.6. However, it is also known that this distribution is not symmetric, since mothers tend to have their first children at early rather than at late ages. A sample of 200 first births was drawn at random from records for the past year, and the age of each mother recorded. This sample yielded a mean of 24.1 years, with a standard deviation S of 5.6. How likely is a sample mean this much or more deviant from expectation (in either direction) if the population mean is truly 23.6? Is it reasonable to conclude that the average age has changed? What does the presumed nonnormal distribution of this population do to our inference in this situation?

6. Under each of the conditions outlined for a population in exercise 3, how probable is it that a sample mean based upon 36 independent observations will fall in the interval of values 95–105?

7. Using the method illustrated in Section 6.4 in the text, compare the binomial probabilities for $N = 8$, $p = .5$ with the corresponding normal probabilities. Then compare $N = 8$, $p = .4$ with normal probabilities. Is the latter appreciably the poorer?

8. Prior to a presidential primary in a certain state, a candidate claimed approximately 45 percent of the votes of the party. A newspaper took a random sample of 500 registered voters in that party and found that only 37 percent indicated they would vote for the candidate. If the true proportion of voters favoring the candidate is .45, how probable is a sample result showing 37 percent or fewer in favor? Would you say that there is good reason to doubt the candidate's assertion?

9. A study found that a large group of children judging the radius of a circle made errors which were approximately normal in their distribution, with a mean of .1 inch, and a standard deviation of .03 inch. A new study focussed on the extent to which these judgments might change if the circle were embedded in a larger circle. To examine this, ten children were drawn at random and asked to judge the radius of such a circle. They showed a mean error of −.2 inch. If embedding the circle in a larger figure is having no effect, and these children can be thought of simply as a sample from the former population, how likely is a sample mean differing this much from expectation (in either direction)?

10. A known population has an average height of 68.2 inches. The population distribution is normal, and it is known that the middle fifty percent of the population have heights between 66.5 and 69.9 inches. If a sample of 25 individuals were drawn independently and at random from this population, how likely is it that their mean height would exceed 72 inches? How likely is it that their mean height would be less than 60 inches? Beyond what mean height should only about 5 percent of sample means fall?

11. A random sample of 3000 income tax returns was taken from a given tax year. Each was examined for the number of exemptions claimed for that year. The mean was found to be 3.78, with a standard deviation of .97. Find the 99 percent confidence interval for the true mean based on these data.

12. A standardized test was designed so that it should produce an approximately normal distribution for normal adults. The testmaker stated that the middle fifty percent of the scores for all such adults should lie between 294 and 306. What is the mean and standard deviation for this population?

13. Suppose that a sample of 40 adults is taken at random from the population described in exercise 12. Find the probability that the mean of this sample will fall between the limits 303 and 304, and between the limits 304 and 305. What are the odds that the mean will fall into the first interval as opposed to the second?

14. In a study of reading, subjects were asked to identify 100 words flashed on a screen under varying background conditions. Subjects often confused these words with other words with similar appearance. A sample of 269 subjects showed an average of 7.91 such confusions out of the 100 words presented to each. The standard deviation S for these subjects was .47. Find the 99 percent confidence limits for the average number of confusions in a population of such subjects.

15. Suppose that a sample of 20 normal adults is taken at random, and each person given the test described in exercise 12. What is the probability that this sample will have a mean of 294 or less? What is the probability of a mean of 306 or more?

16. As a part of an experiment, four distinct random samples are drawn from the same population of subjects. The 99 percent confidence interval for the mean is found for each sample. What is the probability that every single confidence interval found will cover the population mean? What is the probability that at least one confidence interval will fail to do so? (**Hint:** Regard the event "covers the population mean" as a "success" in a Bernoulli process.)

17. In a very large experiment, 200 random samples were taken independently from the same population. The experimenter believed that the population mean was actually 67.9 units. When the 90 percent confidence interval was found for each of the 200 samples, some failed to cover the value of 67.9. If the population mean actually was 67.9, how many such "non-covering" intervals should have been expected?

18. For the situation described in exercise 17, show (in symbols) the probability that all 200 intervals cover the population value. Is this probability large or small? Is it therefore likely that or unlikely that one or more intervals failed to cover the true value?

19. A dress manufacturer suspected that about 25 percent of the bolts of cloth that were being purchased from a certain wholesaler contained defects. Accordingly, a random sample of 75 from among a large number of bolts obtained from this source were examined. It was found that 15 had defects. Find the 95 percent confidence limits for the true proportion of defective bolts.

20. In exercise 19 above, how likely is the manufacturer to obtain 15 or fewer defective bolts if the true proportion is actually .25? Is the manufacturer running a large risk in concluding that 25 percent or more are defective on the basis of this evidence?

21. In the past the city council of a large municipal government has proposed numerous bonding issues, only to have them defeated. Consistently, 54 percent of the voters are against any such issue. A new bonding issue is about to come up for vote. A random poll of 1000 voters shows that 51 percent favor this issue, and 49 percent are against. Form the 99 percent confidence interval for the true proportion of voters favoring this issue. Does this include the value .46? Does the council have any reason to believe that 50 percent or more of voters might favor this issue?

22. Prior to the further development of a small electric automobile, a company conducted a survey to see how many middle-income families would be interested in such a car, if it were offered for sale. The results showed that 37 percent of 5000 such families expressed an interest. Form the 99 percent confidence interval for the true percentage of such families, given that the sample was drawn at random from among all middle-income families in the United States. How likely was the company to have received responses at least this favorable if the true percentage were only 30?

23. In a study of the effects of a medication upon the body temperature of normal adults a scientist wishes to be 95 percent sure that the estimates made from a sample are within .01F degrees of the population mean. The population under study is believed to have a standard deviation in body temperature of .07°F. How many subjects should be used in the sample if these conditions are to be met?

24. Suppose that the experimenter in exercise 23 did not know the standard deviation of the population. Nevertheless, it was desired to estimate mean body temperature to within 1/4 of a standard deviation, with 99 percent confidence. How many subjects were needed?

25. The public health officer of exercise 19, Chapter 5, drew a random sample of 900 children from the area under study. Some 135 children showed a positive test for tuberculosis. What is the probability that no more children than this should show this reaction if the true population proportion is .18? How much risk is the officer running in stating that the true proportion is less than .18?

26. The psychologist of exercise 27, Chapter 5, took a sample of 127 rats and ran each through the maze. The average score was 1.52. If, as the psychologist suspects, the average score of the population of rats is truly 1.5, how probable is a sample mean this far or farther from expectation in either direction?

27. For exercise 29 of Chapter 5, find the 90 percent confidence interval for the mean of the population of employees under the new training (using 4.9 as the standard deviation once again). How do these exact limits differ from those found in exercise 30 of Chapter 5.?

28. Use the data of exercise 31, Chapter 5, and find the 95 percent confidence limits for the mean, given that the population is normal with a standard deviation of 7.

29. An experimenter wishes to be 95 percent certain that the mean of a sample will fall within K standard deviations of the true mean. Make a table showing how the size of K changes with different sample sizes. Use the following values for N: 1, 5, 10, 20, 30, 40, 50, and 100.

30. Four independent random samples were taken from the same population. The means and variances of the samples are given below. Establish the 95 percent confidence interval based upon the mean and variance from the pooled samples.

Samples			
1	2	3	4
$N_1 = 83$	$N_2 = 46$	$N_3 = 74$	$N_4 = 88$
$M_1 = 106.2$	$M_2 = 112.9$	$M_3 = 107.8$	$M_4 = 109.3$
$S_1^2 = 10.11$	$S_2^2 = 15.26$	$S_3^2 = 13.19$	$S_4^2 = 14.19$

Is this a "better" confidence interval than that figured from any single one of the samples? Explain why.

31. For exercise 33, Chapter 5, find the 99 percent confidence interval for the mean amount spent for books.

32. For exercise 32, Chapter 5, what are the odds that a 40-year-old American watched between 6 and 10 hours of television during the week, as opposed to some other number of hours? (Use the unbiased estimate of the variance figured from the sample here.)

33. For the data of exercise 35, Chapter 5, find a pooled estimate of the population mean and standard deviation. How probable is a mean value this large or larger if the mean of the population is actually 102?

34. Imagine that the archeologist of exercise 20, Chapter 5, had assembled 720 potshards from the same level of some dig, where each potshard appeared to represent a different pot. Suppose that 27 percent of these potshards showed decoration. What is the probability of this few or fewer such potshards if this site is later than 3000 B.C.?

35. In Chapter 2, Section 2.17 discusses three discrete random variables following very simple rules. Suppose that a sample of 50 observations were made in turn from each of these three distributions. A sample mean is computed for each. For which of these theoretical distributions would you anticipate that the sampling distribution of means would be most nearly normal? Why? Which sampling distribution would have the smallest standard error?

Chapter 7
HYPOTHESIS TESTING

The title of this chapter might well be "How to decide how to decide," or "How much evidence is enough?" The social scientist or anyone else who samples from a population is trying to decide, or at least form an opinion on, something about the population. Quite often the investigator wants to decide if some hypothetical population situation appears reasonable in the light of the sample evidence. Sometimes the problem is to judge which of several possible population situations is best supported by the evidence at hand. In either instance, one is trying to make up his or her mind from evidence. However, as we shall see, there are many possible ways of making this kind of decision, depending on what information in the sample is actually used, the various possibly true population situations being compared, and especially the risk one is willing to take of being wrong in the decision.

Granting for the moment that we know the information we need from the sample and how to extract it, how do we judge the tenability of a particular hypothesis about the population? It seems reasonable that a tenable or "good" hypothesis about the population should provide us with a "good" expectation about the sample result in terms of the relevant statistic. That is, the hypothesis gives a good fit to the data if, as a consequence of that hypothesis' being true, we can deduce an expected value of the statistic that agrees quite well with the value actually observed; if the hypothesis is true, there is some value that we should expect the appropriate sample statistic to show. If the hypothesis is a good one, our obtained sample result should fall into a region of values relatively close to the expected value the hypothesis dictates for our statistic.

A substantial deviation between the obtained value of the statistic and its expected value under the hypothesis implies one of two things: either the hypothesis is right, and the difference in value between the statistic and its expectation is the product of chance, or the hypothesis is wrong and we were not led to expect the right thing of our sample statistic. In particular, if the sample value falls so far from expectation that it lies in an interval of values very improbable given the hypothesis, but this sort of sample result would be made probable if some alternative hypothesis were true, then doubt is cast on the original hypothesis and we reject it in favor of the alternative. Obviously, however, these notions of "disagree with expectation," "improbable," and "probable" are relative matters; when do we say that a sample result disagrees "enough" with expectation under some hypothesis to warrant rejection of the hypothesis itself?

What we need is a decision rule, a guide giving the conclusion we will reach depending on how the data turn out, and which can be formulated even before the data themselves are seen. However, there is literally no end to the number of decision-rules that one might formulate for a particular problem. Some of these ways of deciding might be very good in a particular circumstance, but no so good in others, and so we need to decide how to decide on the basis of what we wish to accomplish in a specific situation. The branch of mathematics known as *decision theory* treats of this problem of choosing a decision rule, and we are going to apply some of its elementary principles here. We will find that there is no rigid formula supplying the right way to decide in all situations, and, indeed, we often lack the information necessary for choosing an optimal decision rule. For this reason social scientists and others using statistical inference most often fall back on accepted conventions for evaluating evidence, and these conventions may have very little to do with the principles of decision theory. Nevertheless, the application of principles from decision theory to the problem of statistical inference does shed some light on the ideas underlying tests of statistical hypotheses.

Although we will speak of deciding about some single hypothesis, in practice the experimenter always behaves as if the choice lies between *two* hypotheses. A great many issues connected with the use of statistical inference can be understood only if this decision-making task is presupposed.

The process of comparing two hypotheses in the light of sample evidence is usually called a "statistical test" or a "significance test." The competing hypotheses are stated in terms of the form, or of one or more paremeters, of the distribution for the population. Then the statistic containing the relevant information is chosen, having a sampling distribution dictated by the population distribution specified in the particular hypotheses. The hypothesis that dictates the particular sampling distribution against which the obtained sample value is compared is said to be "tested." The significance test is based on the sampling distribution of the statistic given that the particular hypothesis is true. If the observed value of the statistic falls into an interval representing a kind and degree of deviation from expectation which is improbable given the hypothesis, but which would be relatively probable given the other, alternative, hypothesis, then the first hypothesis is said to be rejected. The actual decision to entertain or to reject any hypothesis is thus based on whether or not the sample statistic falls into a particular region of values in the sampling distribution dictated by that hypothesis.

As we look into this problem of the choice of a decision rule, we will be using the mean M and the proportion P as examples of statistics on which decisions are based. However, the issues raised apply to the testing of any hypothesis. Before we can consider these ideas more closely, some standard terminology must be given.

7.1 / STATISTICAL HYPOTHESES

A statistical hypothesis is usually a statement about one or more population distributions, and specifically about one or more parameters of such population distributions. It is always a statement about the population, not about the sample. The statement is called a hypothesis because it refers to a situation that *might* be true. Statistical hypotheses are almost never equivalent to the hypotheses of science, which are usually statements about phenomena or their underlying bases. Quite commonly, statistical hypotheses grow out of or are implied by scientific hypotheses, but the two are seldom identical. The statistical hypothesis is usually a concrete discription of one or more summary aspects of one or more populations; there is no implication of *why* populations have these characteristics.

We shall use the following scheme to indicate a statistical hypothesis: a letter H followed by a statement about parameters, the form of the distribution, or both, for one or more specific populations. For example, one hypothesis about a population could be written

H: the population in question is normally distributed with $\mu = 48$ and $\sigma = 13$, and another
H: the population in question has a Bernoulli (two-class) distribution with $p = 5$.

Hypotheses that completely specify a population distribution are known as **simple hypotheses.** In general, the sampling distribution of any statistic also is completely specified given a simple hypothesis and N, the sample size.

We will also encounter hypotheses such as:

H: the population is normal with $\mu = 48$.

Here, the exact population distribution is not specified, since no requirement was put on σ, the population standard deviation. When the population distribution is not determined completely, the hypothesis is known as **composite.**

Hypotheses may also be classified by whether they specify *exact* parameter values, or merely *a range* or *interval* of such values. For example, the hypothesis

$$H: \mu = 100$$

would be an **exact hypothesis,** although

$$H: \mu \geqslant 100$$

would be **inexact.**

The hypothesis actually to be tested will be given the symbol H_0. Such a hypothesis is commonly referred to as the *null hypothesis,* meaning that this is the hypothesis that is assumed true in generating the sampling distribution to be used in the test. The other hypothesis, which is assumed to be true when the null hypothesis is false, is referred to as the *alternative hypothesis,* and is often symbolized by H_1.

Both the null and the alternative hypothesis should be stated before any statistical test of significance is attempted.

Incidentally, there is an impression in some quarters that the term "null hypothesis" refers to the fact that in experimental work the parameter value specified in H_0 is very often zero. Thus, in many experiments the hypothetical situation "no experimental effect" is represented by a statement that some mean or difference between means is exactly zero. However, as we have seen, the tested hypothesis can specify any of the possible values for one or more parameters, and this use of the word "null" is only incidental. It is far better for the student to think of the null hypothesis H_0 as simply designating that hypothesis actually being tested, the one which, if true, determines the sampling distribution referred to in the test.

In applied situations very seldom will a hypothesis specify the sampling distribution completely. Simple hypotheses are just not of interest or are not available in most practical situations. On the other hand, it *is* necessary that the sampling distribution be completely specified before any precise probability statement can be made about the sample results. For this reason assumptions usually are made, which, taken together with the hypothesis itself, determine the relevant statistic and its sampling distribution and justify a test of the hypothesis. These assumptions differ from hypotheses in that they are rarely or never tested against the sample data. They are assertions that are simply assumed true, or the evidence is collected in such a way that they must be true (e.g., random sampling).

These assumptions are necessary for the formal justification of many of the methods to be discussed. It is not true, however, that these assumptions are always or even ever realized in practice. The results of any data analysis leading to statistical inference can always be prefaced by "If such and such assumptions are true, then. . . ." The effects of the violation of these assumptions on the conclusions reached can be very serious in some circumstances, and only minor in others. In later sections the importance of the most common assumptions will be discussed. Nevertheless, these assumptions should always be kept in mind, along with the conditional character of any result subject to the assumptions being true.

7.2 / TESTING A HYPOTHESIS IN THE LIGHT OF SAMPLE EVIDENCE

Every test of a hypothesis involves the following features:

1. The hypothesis to be tested is stated, together with an alternative hypothesis.
2. Additional assumptions are made, permitting one to specify the sample statistic that is most relevant and appropriate to the test, as well as the sampling distribution of that statistic.
3. Given the sampling distribution of the test statistic when the hypothesis to be tested is true, a **region of rejection** is decided upon. This is an interval of possible values deviant from expectation if the hypothesis were true, but which more or less accord with expectation if the alternative were true. This region of rejection contains values relatively improbable of occurrence if the first hypothesis were true, but relatively probable given the alternative. The risk one is willing to take in rejecting the tested hypothesis *falsely* determines the size

of the region of rejection, and the alternative hypothesis determines its location.

4. The sample itself is obtained. If the computed value of the test statistic falls into the region of rejection, then doubt is cast on the hypothesis, and it is said to be rejected in favor of the alternative. If the result falls out of the region of rejection, then the hypothesis is not rejected, and the experimenter may choose either to accept the hypothesis or to suspend judgment, depending on the circumstances. A sample result falling into the region of rejection is said to be **statistically significant,** or **to depart significantly from expectation** under the hypothesis.

As an example, suppose that an experimenter entertains the hypothesis that the mean of some population is 75:

$$H_0: \mu = 75.$$

This is actually a composite hypothesis, since nothing whatever is said about the form of the population distribution nor about other parameters such as the standard deviation. The experimenter is, however, prepared to assume that whatever is true about the mean the population has a normal distribution, and that the standard deviation for the population is 10; for the moment, let us imagine that both of these assumptions are reasonable in the light of what the experimenter already knows about this problem. In effect, the addition of these two assumptions to the original hypothesis makes this hypothesis simple:

$$H_0: \mu = 75, \sigma = 10, \text{ normal distribution.}$$

Now, what of the alternative hypothesis? It might be that the alternate hypothesis is

$$H_1: \mu \neq 75.$$

Note that this alternative happens to be inexact, and really asserts that some (unspecified) value of μ other than 75 is true. However, even if H_1 is true and H_0 false, the experimenter will still assume that the distribution is normal with a standard deviation of 10.

It has already been decided that 25 independent observations will be made. Given this sample size N, the information in the hypothesis H_0, and the assumptions, the sampling distribution of the mean is known completely: this must be a normal distribution with a mean of 75, and a standard error of $10/\sqrt{25}$ or 2, if the hypothesis H_0 is true.

Now in accordance with H_1, the experimenter decides to reject this hypothesis H_0 only if the sample mean falls among the extremely deviant ones in either direction from the value 75; in fact, H_0 will be rejected only if the probability is .05 or less for the occurrence a sample mean as deviant or more so from 75. Thus, the region of rejection for this hypothesis contains only 5 percent of all possible sample results when the hypothesis H_0 is actually true. Such sample values depart widely from expectation and are relatively improbable given that H_0 is true; on the other hand, samples falling into these regions are relatively more probable if a situation covered by H_1 is true. These regions of rejection in the sampling distribution are shown in Figure 7.2.1.

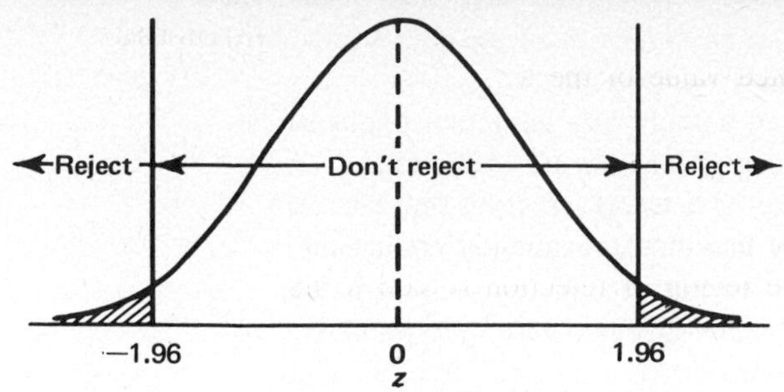

Figure 7.2.1
Possible rejection regions for a normal sampling distribution.

Tables of the normal distribution show that a z score of 1.96 has a cumulative probability $F(1.96)$ of about .975. Hence only .025 of all sample means should lie *above* 1.96 standard errors from μ. Similarly, it must be true that about .025 of all sample means will lie below -1.96 standard errors from μ. In short, the region of rejection includes all sample means such that, in this sampling distribution,

$$1.96 \leq z_M$$

or

$$z_M \leq -1.96$$

giving a total probability of .025 + .025 or .05 for the combined intervals. A value of the sample mean that lies exactly on the boundary of a region of rejection is called a **critical value** of M; the critical z_M values here are -1.96 and 1.96, so that the critical values of M, given $\sigma_M = 2$, are

$$-1.96\sigma_M + \mu = -1.96(2) + 75 = 71.08$$

and

$$1.96\sigma_M + \mu = 1.96(2) + 75 = 78.92.$$

Now the sample is drawn and turns out to have a mean of 79. The corresponding z_M score is

$$z = \frac{79 - \mu}{\sigma_M} = \frac{79 - 75}{2} = 2.$$

Notice that this z_M score *does* fall into the region of rejection, since it is greater than a critical value of 1.96. Then the experimenter says that the simple result is **significant beyond the 5 percent level.** Less than 5 percent of all samples should show results this deviant (or more so) from expectation under H_0, if H_0 is actually true. Since the experimenter has decided in advance that samples falling into this region show sufficient departure from expectation to be called *improbable* results, then doubt is cast on the truth of the hypothesis, and H_0 is said to be rejected.

In its essentials this simple problem exemplifies all significance tests. Naturally, the details vary with the particular problem. Hypotheses may be about variances, or any other characteristic feature of one or more populations. Other statistics may be appropriate to a test of the particular hypothesis. Other assumptions may be made to permit the specification of the sampling distribution of the relevant statistic. Other regions of rejection may be chosen in the sampling distribution. *But once these specifications are made, the experimenter knows, even before drawing the sample, how the decision will be made in the light of what the sample shows.*

Here the region of rejection was simply specified without any particular explanation. In this step, however, lies a key problem of significance tests. In the choice of the region of rejection the experimenter is deciding how to decide from the data. It is perfectly obvious that there are very many ways that the experimenter could have chosen a rejection region. It could have been decided to reject the hypothesis if the sample result fell 1.65 standard errors below the mean specified by the hypothesis, or if the sample result fell between 3 and 4 standard errors above the hypothetical mean. Each of these intervals has small probability in a normal sampling distribution. There is no law against using any decision rule for rejection, although on intuitive grounds alone the one actually used here seems somehow more sensible than the others suggested. We now examine this question in more detail.

7.3 / A PROBLEM IN DECISION THEORY

Next to breathing, perhaps the most common human activity is decision making. Every moment of existence is filled with choices that the human being must make. From the time that the caveman pondered ''Shall I leave the campfire to go out into the dark for a moment?,'' to the investor asking, ''Shall I buy this common stock while its value is low?,'' life has remained full of choices among actions. Each of these decisions carries with it an element of risk. There might have been a tiger waiting in the dark for the caveman, even though our ancestor hoped that he might catch a small animal for his own family to eat. The investor might buy the stock and see it streak upward in value; or see it hit rock bottom immediately after purchase.

Decision theory, which actually grew out of problems of economic decision making, has established some general principles which extend to decision situations of many kinds. Any number of choices of action are usually open to the person who must make a decision, and once a decision to act has been made, some event or chain of events is going to occur. Sometimes the outcome of a choice of action is almost certain: if I decide to place my hand in a roaring fire, the outcome of a burn is almost sure to occur. However, in most instances the decision maker does not know exactly what the outcome will be be. Such decisions are called ''risky'' for this reason. In most decision settings there are various ''states of the world'' that might or might not be true (e.g., the tiger in the bushes, the health of the stock market). Then the outcome of any decision depends both on what is decided and what turn out to be true.

In statistical inference these possible states of the world are formulated as hypotheses. We try to form an advance judgment on which state of the world obtains by drawing a sample. Then on the basis of that evidence we try to make a decision about the hypothesis under study. Since our knowledge is necessarily imperfect, these decisions too are risky.

In order to illustrate this point, along with some of the principles of decision theory, a very simple example will next be used. Even this example, homely as it is, illustrates the problems of how to decide how to decide in the presence of evidence.

Suppose that you were placed in the following decision situation. You are seated before a table, on which rests a bag full of paper currency. You are told that there are exactly fifty bills in all in the bag. Furthermore, one of two situations is possible: either the bag contains eighty percent $2 bills, with the remainder being $1 bills, or the bag contains forty percent $2 bills, with the remainder in ones. Your task is to guess which bag is actually on the table. If the guess is right, you will get the bag of money; should your guess be wrong, then you will get nothing.

In order to make your decision a little easier, the experimenter is willing to allow you to extract ten bills from the bag at random, with replacement after each draw. You are expected to tell the experimenter ahead of time how you will decide after having seen the sample of bills.

The choice open to you is equivalent to a decision between the exact hypotheses

$$H_0: p = .80$$
$$H_1: p = .40.$$

That is, either the true proportion in the bag for $2 bills is .80 or it is .40.

Being a good statistician, you know that drawing a $2 bill from the bag is a "success," and the distribution of P, the proportion of successes among ten draws, is binomial, with $N = 10$. However, the actual sampling distribution depends on p, which in turns depends on which bag is actually on the table. The binomial distribution under each of the two hypotheses is shown in Table 7.3.1.

Table 7.3.1

Binomial distributions of P for $p = .8$ and $p = .4$, with $N = 10$.

	$p(P)$	
P	if $p = .8$	if $p = .4$
1.0	.107	.0001(+)
.9	.268	.002
.8	.302	.011
.7	.201	.042
.6	.088	.111
.5	.026	.200
.4	.006	.251
.3	.001	.215
.2	.000(+)	.121
.1	.0000(+)	.040
.0	.0000(+)	.006

Now how should you decide on the basis of the sample evidence? You are going to need a decision rule, telling you how to proceed on the basis of the number of $2 bills in the sample. There are any number of decision rules that you might use, but for the sake of simplicity we are going to evaluate the efficacy of only three rules that you might possibly use:

Decision rule 1: If P is greater than or equal to .8, choose H_0; otherwise H_1.
Decision rule 2: If P is greater than or equal to .6, choose H_0; otherwise H_1.
Decision rule 3: If P falls between .2 and .8 inclusive, choose H_0; otherwise H_1.

Regardless of which decision rule is selected there are two ways that you could be right, and two ways of making an error. This is diagrammed below:

		True situation H_0	H_1
Decision	H_0	correct	error
	H_1	error	correct

If you decide on H_0, and H_1 is actually true, an error is made. Furthermore, you make an error by deciding on H_1 and H_0 is actually true. The other possibilities lead, of course, to correct decisions.

Given the sampling distributions under each of the two hypotheses, and any decision rule, we can find the probabilities of these two kinds of error.

Suppose that you have no idea which hypothesis is true, and all you have to go by is the occurrence of a P value leading you reject one or the other of the hypotheses. Suppose that decision rule 1 were adopted. This rule requires that the occurrence of any P of .8 or more lead automatically to a decision that H_0 is true. What is the probability of observing such a sample result *if* H_1 is true? Notice that whenever this happens, there will be a wrong decision and so this probability is that of *one kind* of error. The binomial distribution for $p = .4$ shows a probability of about .013 for P greater than or equal to .8, and so the probability is .013 of *wrongly* deciding that H_0 is true.

In the same way, we can find the probability of the error made in choosing H_1 when H_0 is true. By decision rule 1, H_1 is chosen whenever P is less than or equal to .7; in the distribution under H_0, this interval of values has a probability of about .323. Thus, the probabilities of error and correct decisions, under decision rule 1, are as shown in Table 7.3.2.

Table 7.3.2

		True situation H_0	H_1
Decision	H_0	.677	*.013*
	H_1	*.323*	.987

The probability of a correct decision is $1 - p(\text{error})$ for either of the possibly true situations (the two error probabilities appear in italics). Notice that you are far more likely to make an erroneous judgment by using this rule when H_0 is true than for a true H_1.

Exactly the same procedure gives the erroneous and correct decision probabilities under rule 2, as shown in Table 7.3.3.

Table 7.3.3

	H_0	H_1
H_0	.966	.166
H_1	.034	.834

By this second rule, the probability of error is relatively smaller when H_0 is true than it is for rule 1. However, look what happens to the probability of the other error This illustrates a general principle in the choice of decision rules: *Any change in a decision rule that makes the probability of one kind of error smaller will ordinarily make the other error probability larger* (other things, such as sample size, being equal). 

Before trying to choose between rules 1 and 2, let us write down the probabilities for decision rule 3 (See Table 7.3.4).

Table 7.3.4

	H_0	H_1
H_0	.625	.952
H_1	.375	.048

Even on the face of it this decision rule does not look sensible. The probability of error is large when H_0 is true, and the experimenter is almost sure to make an error if H_1 is the true situation! This illustrates that not all decision rules are reasonable if the experimenter has any concern at all with making an error; here regardless of what is true the experimenter has a larger chance of making an error using this rule than in using either rules 1 or 2. Rules such as this are called **inadmissible** by decision theorists. We need confine our attention only to the two relatively ''good'' rules 1 and 2.

There is a real problem in deciding between the rules 1 and 2; rule 1 is good for making error probability small when H_1 is true, but risky when H_0 represents the true state of affairs. On the other hand, rule 2 makes for small chance of error when H_0 is true, but gives relatively large probability of error when H_1 holds. Obviously, any choice between these two rules must have something to do with the relative importance of the two kinds of errors. It might be that making an error is a minor matter when H_0 is true, but very serious given H_1. In this case rule 1 is preferable. On the other hand, were error very serious given H_0, rule 1 could be disastrous. In short, any rational way of choosing among rules must involve some notion of the loss involved in making an error. Thus, before completing this example, we need to examine the role that possible losses to the decision maker can play in the choice of a decision rule.

7.4 / EXPECTED LOSS AS A CRITERION FOR CHOOSING A DECISION RULE

In the everyday world, a decision followed by an outcome always involves some payoff to the decision maker. That is, the combination of a decision and the outcome of that decision carries some positive or negative value for the person affected. The exact nature of these values need not concern us here; they may reflect safety, dollars, time, prestige, self-respect, pleasure, or any of a variety of other things of importance to a human being. We will simply assume that it is possible to assign a value to any combination of what was decided and what turned out to be true. When the choice is between H_0 and H_1, then that value can be represented by

$$u(H_0|H_1),$$

standing for the value of choosing H_0 when H_1 turns out to be true. Similar meaning attaches to $u(H_1|H_0)$ and so on, for the other combinations of decision and outcome.

For our purposes we need consider these values only as *losses*. Each loss value is some positive number or zero given to an action-outcome combination. If the combination represents the best available action under the true circumstances, then the decision is *right* and the loss value is zero. On the other hand, if the action is less desirable than the best available action, then this is an *error* and the loss value is positive. (A more technical term for such a value is *opportunity loss*.)

Now let us return to your choice between the two hypotheses of Section 7.3. In your present decision-making situation, when H_0 is true the money you receive from the bag is $90. On the other hand, when H_1 is true the amount you receive is only $70. Then the possible losses associated with the four possible action-outcome possibilities would be

Table 7.4.1

		True situation H_0	H_1
Choose	H_0	0	$70
	H_1	$90	0

Now let us examine the *expected loss,* given that you use a particlar decision rule, and given that H_0 is actually true. This value is given by

$$E(u|H_0) = u(H_0|H_0)p(\text{decide } H_0|H_0) + u(H_1|H_0)\, p(\text{decide } H_1|H_0).$$

Similarly, we can find $E(u|H_1)$ by taking

$$E(u|H_1) = u(H_0|H_1)p(\text{decide } H_0|H_1) + u(H_1|H_1)p(\text{decide } H_1|H_1).$$

These expected losses can be computed for each of the three decision rules we have been considering. For decision rule 1 the probabilities are as shown in Table 7.3.2, and the u values as given in Table 7.4.1. Thus,

$$E(u|H_0) = (0)(.677) + (90)(.323) = 29.07$$

and

$$E(u|H_1) = (70)(.013) + (0)(.987) = .91.$$

In other words, when the bag of money on the table actually represents H_0, then the expected loss to you in using decision rule 1 is just over twenty-nine dollars. If, on the other hand, the bag represents H_1, then the expected loss is only ninety-one cents. Clearly, decision rule 1 is much better when H_1 is actually true than when H_0 happens to be true.

Now in the same way the expected losses under H_0 and H_1 can be calculated for decision rule 2, and for decision rule 3. These expected losses are summarized in Table 7.4.2.

Table 7.4.2

		True	
		H_0	H_1
Decision rule	1	29.07	.91
	2	3.06	11.62
	3	33.75	66.64

Decision rule 1 gives a small loss when H_1 is true, but a high loss under H_0. On the contrary, decision rule 2 gives a relatively small loss when H_0 is true, but a larger loss under H_1. The table again shows the absurdity of rule 3, which gives the highest loss of the three rules regardless of which hypothesis is true.

Can we choose between rules 1 and 2, however, on the basis of expected loss? A device within decision theory for choosing between rules is known as the ''minimax'' principle. Applying this principle leads us to choose that rule showing the *minimum maximum-expected-loss* over all possible true situations. For expected-loss tables such as 7.4.2, a minimax decision rule is one having a largest entry in its row that is smaller than the largest entry in any other row of the table. Notice that in this situation, rule 2 has 11.62 as the largest value in its row; this is smaller than the largest value for rule 1, or 29.07. Thus, applying the minimax principle, we find that rule 2 is the one to use. Using rule 2, the largest that we expect long-run loss to be is smaller than for either rule 1 or rule 3.

This minimax criterion for choosing among decision rules is historically important in the theory of games and in decision theory, and it does give a way to compare decision rules on the basis of their expected losses. However, this criterion is no longer taken very seriously by statisticians. Among other things, it is unduly conservative, especially in scientific decision making, since it focuses only on the extreme potential loss in making errors, and not on the very large gains one can make through increased knowledge despite the risk of being wrong on occasion. An alternative to the minimax criterion will be discussed next.

One thing that has been left completely out of this example so far is the fact that you may know something, or at least believe something, about the bag on the table before a sample is even drawn. That is, you might be thought of as having a subjective probability, say $\mathcal{P}(H_0)$, that the bag on the table actually does represent H_0, and another subjective probability, $\mathcal{P}(H_1)$, that the bag represents H_1. The experimenter might even have gone so far as to create these subjective probabilities for you, by saying right at the outset that ''the chances are about one in four that H_0 is true, and the chances are about three in four that H_1 is true.'' Then, another way to choose among the decision rules exists.

We can define *subjective expected loss* for rule 1 by

$$SE(u|\text{rule 1}) = E(u|H_0)\mathcal{P}(H_0) + E(u|H_1)\mathcal{P}(H_1).$$

In the same way, the subjective expected loss for any other rule can be defined, given the values of $\mathcal{P}(H_0)$ and $\mathcal{P}(H_1)$.

Suppose that you choose the decision rule having the lowest subjective expected loss. It turns out that, since we have established that $\mathcal{P}(H_0) = 1/4$ and $\mathcal{P}(H_1) = 3/4$,

$$SE(u|\text{rule 1}) = 29.07(1/4) + (.91)(3/4) = 7.95$$

$$SE(u|\text{rule 2}) = 3.06(1/4) + (11.62)(3/4) = 9.48$$

$$SE(u|\text{rule 3}) = (33.75)(1/4) + (66.64)(3/4) = 58.42$$

If you seek to minimize subjective expected loss in this experiment, then decision rule 1 should be chosen. Given that $\mathcal{P}(H_0) = 1/4$ and $\mathcal{P}(H_1) = 3/4$, then on the average decision rule 1 is the best way to decide.

This illustrates how decision rules may also be compared by their subjective expected losses; the lower the subjective expected loss, the better the rule. In this case, given the loss values, and given the personal probabilities, rule 1 is clearly superior to rule 2. Even though error has a fairly high probability by this rule when H_0 is true, your own weighing of what you expect to be true of the situation tends to discount the likelihood of such errors.

Your little problem in decision theory is now solved: provided that you can attach loss values to the action-outcome combinations, and provided that personal probabilities are associated with the competing hypotheses, a subjective expected loss can be calculated for each possible decision rule, and the best one adopted. *This does not guarantee that you will make the right decision, but only that you will have chosen a good way to make the decision*. Essentially, that is the basic concern of decision theory.

7.5 / FACTORS IN SCIENTIFIC DECISION MAKING

It requires no great logical leap to see that your problem in trying to decide how much money is in a bag is very like the dilemma of the business executive trying to decide which of the two lots of merchandise to buy, or that of the scientist confronted with two competing hypotheses about human behavior. In all such situations, the decision maker will rely on a sample to make a judgment between the two possible states of affairs. There will be a variety of decision rules that might be used, and a choice needs to be made among these possible ways of deciding. Just as some of the criteria of decision theory could be applied in the money-bag problem to help in the choice of a decision rule, so these same principles can be extended to scientific or practical decision making.

However, particularly in the case of scientific decision making, the choice of a decision rule is not nearly so simple as it may be in a business situation. In business, it is frequently true that a given action must be followed by one of a fairly small and specifiable set of outcomes. The decision maker does not know what the particular outcome of an action will be, but at least one can state the possible outcomes. The actual outcome may depend upon economic conditions in general, quality of product, competition, and so on. Furthermore, each of these possible outcomes to any action can be assigned a profit or loss value, again in principle.

On the other hand, the possible outcomes of actions taken by the scientist are far more difficult to specify. It is clear that some definite courses of action are usually open to the experimenter: to decide for or against a hypothesis, to publish or withhold findings, to collect or not to collect more data, to pursue or abandon the line of investigation, to ask for funds for further research, or to forget the whole matter. By stretching a point, we might say that finding out what the true situation is constitutes an outcome following any action. However, this outcome may not apply to the experimenter at all, but rather may serve as an outcome for the science or for humanity.

It is important to recognize that the scientist as a mere human being may be capable of foreseeing only the short-term monetary, prestige, or other outcome possibilities of the actions, which are insignificant compared to the long-range impact of the action on knowledge and human welfare.

For the sake of argument, let us nonetheless suppose that the set of actions and outcomes could be specified for the scientific decision maker. Then, in principle, criteria such as minimax expected loss or minimum subjective expected loss could be applied in the choice of a scientific decision rule, provided that values could be assigned to the action-outcome combinations.

In the sections to follow various procedures will be given that are in common use in such research. Most such procedures use a rule for deciding when a result is significant that *might* be justified by a decision-making argument very similar to that just given. The experimenter wants to avoid errors in inference, presumably because such errors lead to losses, at least in time and effort. Decision rules differ in the expected losses they give, and it behooves the experimenter to choose a decision rule that minimizes expected loss. The act of deciding on a region of rejection in the test of a hypothesis is the choice of a decision rule, and should be subject to the same considerations.

Unfortunately, the cost of erroneous decisions is almost never considered in social science research. Indeed, in most instances in such research it seems very unlikely that a numerical value could ever be assigned to the loss incurred in an erroneous decision. Just how bad is it to be wrong in a scientific inference? This problem is not quite so difficult in many other fields using statistical inference, especially in research on applied problems. In business decisions it is often possible to give a value in dollars to the outcomes that might result from various decisions, and statistical decision theory then provides guides for choosing an effective decision rule. Even in some applied social science areas, losses involved in errors might be reckoned in the same general way. Thus, in studying methods for selecting people for various jobs the cost of an error may be a calculable thing, and the decision rule may be chosen on that basis. Similarly, this possibility may exist for studies of diagnostic methods, of training effectiveness, and so on. By and large, however, most social or behavioral scientists would not know how to assign loss values to decision errors. Even in a relatively clear-cut situation such as diagnosis or selection, the costs to the people involved or even to the community may be extremely important. For example, consider a diagnostic method for potential suicides. Errors of the two different types (false positives and false negatives) have enormously different consequences for the person and for society. Such costs are most difficult to assess, and the effort required is seldom expended. Even so, the decision rules obtained by omitting such incalculable costs could be disastrous. Such an omission is sometimes called the "accountant's error," although that term fails to convey the importance of the problem. In still other circumstances the loss values of errors may be so small that it really does not make much difference how or what the experimenter decides.

In the same way, even if we admit the existence of personal probabilities or $\mathcal{P}$ values for the experimenter, typically one is not in a position to assess objectively what these values should be, even though this is theoretically possible. Perhaps in the future a standard part of statistical analysis will be the statement of prior personal probabilities before the collection of the data. Personal probabilities can be found by the odds the experimenter is willing to accept in betting on the truth of a hypothesis, and it might be possible to make this a standard part of statistical practice. This "Bayesian" approach to inference is already common is some applied fields.

In short, we use much of the terminology of statistical decision theory without its main feature, the choice of a decision rule having optimal properties for a given purpose. Instead, we use conventional decision rules, completely ignoring questions of the loss involved in errors and the degree of prior certainty of the experimenter. These conventional rules can be justified by decision theory in some contexts, but they are surely not appropriate to every situation.

Must the scientist make a decision about what is true from a given set of data? Naturally, choosing to suspend judgment and wait for more evidence is a decision to adopt a course of action. However, why cannot this time-honored strategy of the scientist be used more often than apparently it is? As we have seen, one is usually cut off from the possibility of choosing an appropriate decision rule, using the minimax principle or minimum subjective-expected-loss. There is no recourse except to fall back on some conventional rule. No convention can possibly be ideal or even sensible for all the problems to which it may be applied, as we shall illustrate below.

It is entirely possible that the course of action represented by "no decision, suspend judgment" will be best in a given situation, especially if the penalty involved in waiting to decide is small compared to the loss involved in an incorrect decision.

Why, then, do social scientists bother with significance tests at all? *Regardless of what one is going to do with the information—change an opinion, adopt a course of action, or what not—one needs to know relatively how probable is a result like that obtained, given a hypothetical true situation. Basically, a significance test gives this information, and that is all.* The conventions about significance level and regions of rejection can be regarded as ways of defining "improbable." The occurrence of a significant result in terms of these conventions is really a signal: "Here is a direction and degree of deviation which falls among those relatively unlikely to occur given that the tested hypothesis is true, but which is relatively more likely given the truth of some other hypothesis." On deciding to reject the original hypothesis on the basis of a significant result, then at least one knows the probability of error in doing so. This does not mean that one must decide against the hypothesis simply because some conventional level of significance was met. Other options, such as suspending judgment, may actually be better actions under the circumstances regardless of the result of the conventional signficance test. Even more emphatically, the occurrence of a nonsignificant result does not mean that you must accept the hypothesis as true. As we shall see, one often has not the foggiest idea of the error probability in saying that the tested hypothesis is true; here, making a decision to accept the tested hypothesis is absurd in the light of the unknown, and perhaps very large, probability of such an error. On occasion, however, the probability of such an error can be assessed, and here one can feel safe in asserting that the tested hypothesis is true when this error probability is small.

To sum up: it is perfectly all right to discuss conventional hypothesis testing in the decision-theoretic language, but the user is speaking as though the conventional rule were a good way to decide, which it may not be. *There is no God-given rule about when and how to make up your mind in general.*

7.6 / TYPE I AND TYPE II ERRORS

As we have noted already, in any choice between two hypotheses, there are two ways of being correct in the decision, and two ways of making an error. Ordinarily, these action-outcome possibilities and their associated probabilities are symbolized as follows:

		True	
		H_0	H_1
Decide	H_0	$1 - \alpha$	β
	H_1	α	$1 - \beta$

The probability of making an error of the type "decide H_1 and H_0 is true" is symbolized by α, the lowercase Greek letter alpha. Then the probability of being correct, given that H_0 is true, is symbolized by $1 - \alpha$, since given that H_0 is true, we must have decided either on H_0 or H_1.

In a similar way, the probability of making an error of the type "decide H_0 and H_1 is true" is symbolized by the lowercase Greek letter beta, or β. Then the probability of making a correct decision, given that H_1 is true must be $1 - \beta$.

The time has also come to dignify with names the two kinds of errors we have been discussing. **Type I error is that made when H_0 (the tested hypothesis) is falsely rejected.** One makes a Type I error whenever the sample result falls into the rejection region even though H_0 is true. Thus the probability α gives the risk run of making a Type I error:

$$\alpha = \text{probability of Type I error.} \qquad [7.6.1^*]$$

The errors of Type II are those made by not rejecting H_0 when it is false. When a sample result does not fall into the rejection region, even though some H_1 is true, we are led to make a Type II error. Thus for a given true alternative H_1, β is the probability of Type II error

$$\beta = \text{probability of Type II error.} \qquad [7.6.2^*]$$

7.7 / CONVENTIONAL DECISION RULES

In social science research, given some hypothesis H_0 to be tested, the region of rejection is usually found as follows:

CONVENTION: Set α, the probability of falsely rejecting H_0, equal to some small value. Then, in accordance with the alternative H_1, choose a region of rejection such that the probability of observing a sample value in that region is equal to α when H_0 is true. The obtained result is significant beyond the α level if the sample statistic falls within that region.

Ordinarily, the values of α used are .05 or .01, although, on occasion, larger or smaller values are employed. Even though only one hypothesis, H_0, may be exactly specified, and this determines the sampling distribution employed in the test, in choosing the region of rejection one acts as though there were two hypotheses, H_0 and H_1. The alternate hypothesis H_1 dictates what portion or **tail** of the sampling distribution contains the rejection region for H_0. In some problems, the region of rejection is contained in only one tail of the distribution, so that only extreme deviations in a given direction from expectation lead to rejection of H_0. In other problems, big deviations of either sign are candidates for the rejection region, so that the region of rejection lies in both tails of the sampling distribution.

In all that follows, these conventional rules will be used for finding a rejection region. Some hypothesis will be designated as H_0, and an arbitrary α value will be chosen, which, taken together with H_1, determines the sample values that lead to a possible rejection of H_0.

Although the conventional rules usually involve α values such as .05 and .01, the choice of α is not always totally arbitrary. The choice of α may also be dictated by a balance that the experimenter strikes between the two kinds of errors possible in drawing conclusions from a test of a hypothesis.

The conventions about the size of the α probability of Type I error actually grew out of a particular sort of experimental setting. Here it is known in advance that one kind of error is extremely important and is to be avoided. In this kind of experiment these conventional procedures do make sense when viewed from the decision-making point of view. Furthermore, designation of the hypothesis H_0 as the "null" hypothesis and the arbitrary setting of the level of α can best be understood within this context.

As an example of an experimental setting where Type I error is clearly to be avoided, imagine that one is testing a new medicine, with the goal of deciding if the medicine is safe for the normal adult population. By "safe" we will mean that the medicine fails to produce a particular set of undesirable reactions on all but a very few normal adults. Now in this instance, deciding that the medicine is safe when actually it tends to produce reactions in a relatively large proportion of adults is certainly an error to be avoided. Such an error might be called "abhorrent" to the experimenter and the interests he or she represents. Therefore, the hypothesis "medicine unsafe" or its statistical equivalent is cast in the role of the null hypothesis, H_0, and the value of α chosen to be extremely small, so that the abhorrent Type I error is very unlikely to be committed. A great deal of evidence against the null hypothesis is required before H_0 is to be rejected. The experimenter has complete control over Type I error, and regardless of any other feature this study of the medicine may have, one can be confident of taking very little risk of asserting that H_1, or "medicine safe," is true when actually H_0, or "medicine unsafe," is true.

On the other hand, the experimenter always has some control over the values of

β under the various possibly true alternatives. As we shall see, by the choice of an appropriate sample size and by exercise of control over the size of σ^2 in the population considered, the value of β for any possibly true alternative to H_0 may be made as small as desired. The essential point is that the experimenter is absolutely free to set α at any level and that the conventional levels are dictated by the notion that Type I errors are bad and must be avoided. The experimenter does not have this same freedom with respect to Type II errors; low risk of Type II error must be bought in terms of sample size and other features of the test procedure. The value of β can always be made small against any given alternative, but only at some cost to the experimeter. An inappropriately small value of α makes it more difficult than otherwise to avoid Type II error.

Within contexts such as the test of a new medication, where Type I error is abhorrent, setting α extremely small is manifestly appropriate. Here, considerations of Type II error are actually secondary. In some instances in a social science as well, Type I error is clearly to be avoided, and from the outset the experimenter wants to be sure that this kind of error is very improbable. The designation of one hypothesis as H_0 should rest on which kind of error is to have the small probability α.

On the other hand, in some social science research it is very hard to see exactly why the particular hypothesis tested, or H_0, should be the one we are loath to abandon, and why Type I errors necessarily have this drastic character. Granting that scientific discretion is commendable, the mistaken conclusion that "something really happened" is not *necessarily* worse than overlooking a real experimental phenomenon. In some situations, perhaps, we should be far more attentive to Type II errors, and less attentive to setting α at one of the conventional levels. Furthermore, if the conventional α levels are to be used, a little more thought might be given to deciding exactly what *is* the null hypothesis we want to be so careful not to reject falsely.

7.8 / THE POWER OF A STATISTICAL TEST

Given only the one exact hypothesis H_0, and the required value of α, one may still think of an inference involving two hypotheses:

H_0, the hypothesis actually being tested
H_1, the hypothesis, against which H_0 is to be compared.

Once any true H_1 is specified, then we can determine β, the probability of an error in decision given the true H_1. Similarly, one may compute the probability $1 - \beta$, which is the probability of being right in rejecting H_0 given that H_1 is true. This probability, $1 - \beta$, is often called the **power** of the statistical test. It is literally the probability of finding out that H_0 is wrong, given the decision rule and the true value under H_1.

The idea of power can be discussed most simply in terms of the choice between two exact hypotheses. Admittedly, practical problems very seldom present us with this sort of choice, and so any example will appear somewhat contrived. Nevertheless, any power calculation always begins as though this were the choice available to the experimenter.

For example, a psychologist working in industry is assigned to study the possibility of using only blind persons in a certain job requiring unusually fine hearing and touch discrimination, but no use of sight at all. Heretofore, only sighted persons have been used, and past experience has shown that for the sighted population the average performance score is 138, with a standard deviation of 20. If the blind perform exactly like the sighted, then their population should have this same mean and standard deviation. On the other hand, if the mean performance of blind persons is at least 142, then a considerable benefit will accrue to the company by employing them. From what is already known about the requirements of the job, and blind performance in other settings, there is good reason to believe that one of these two situations will prove true.

Accordingly, the experimenter frames two hypotheses, both of which deal with a population of blind persons who might be placed on this job:

$$H_0 : \mu = 138$$

$$H_1 : \mu = 142.$$

It is assumed that the population distribution in either situation will have a standard deviation of 20.

Now a sample of 100 blind persons is to be drawn at random and put into this job situation on an experimental basis. The psychologist feels that this sample is large enough to permit using the normal approximation to the sampling distribution of the mean. Note that here $\sigma_M = 2$.

The following conventional rule is used:

If the sample result falls among the highest 5 percent of means in a normal distribution given H_0, then reject H_0; otherwise, reject H_1.

This region of rejection for H_0 is shown in Figure 7.8.1.

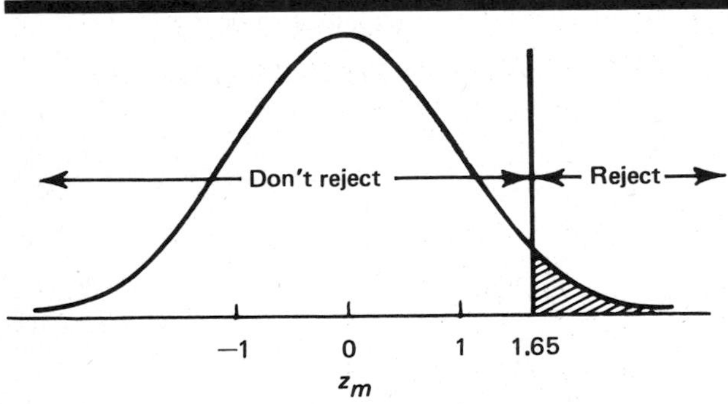

Figure 7.8.1
A region of rejection on the right tail of a normal distribution.

In this instance, there are once again two kinds of errors that can be made. The probabilities for each kind are labeled α and β respectively:

$$\alpha = \text{probability (reject } H_0 | \mu = 138)$$

$$\beta = \text{probability (accept } H_0 | \mu = 142).$$

In this situation with two *exact* hypotheses, it does make sense to talk about accepting or deciding in favor of H_0, since both probabilities can be known. In essence, the conventional rule says to fix α at .05, and the region of rejection chosen does just that. Given that H_0 is true, then the region of rejection must be bounded by a z_M score such that

$$F(z_M) = .95, \text{ or } 1 - F(z_M) = .05.$$

The normal tables show that this z_M score is 1.65. In terms of a sample mean

$$z_M = \frac{M - 138}{\sigma_M} = \frac{M - 138}{2}.$$

Thus, the *critical* value of M, forming the boundary of the rejection region, is

$$M = 138 + 1.65\sigma_M = 138 + 3.30 = 141.30.$$

However, notice that assigning the value of .05 to the α probability of error automatically fixes β as well.

What would the z_M score for this critical value of 141.30 be if H_1 were true? This is found from

$$z_M = \frac{141.3 - 142}{2} = -.35.$$

In a normal distribution, $F(-.35) = .36$, approximately, and so we can see that $\beta = .36$. The two error probabilities are then

$$\alpha = .05$$

$$\beta = .36.$$

Thus by the rule used, H_0 is rejected when a sample exceeds 141.3. This gives a probability of error β shown by the shaded region in Figure 7.8.2, if H_1 is true.

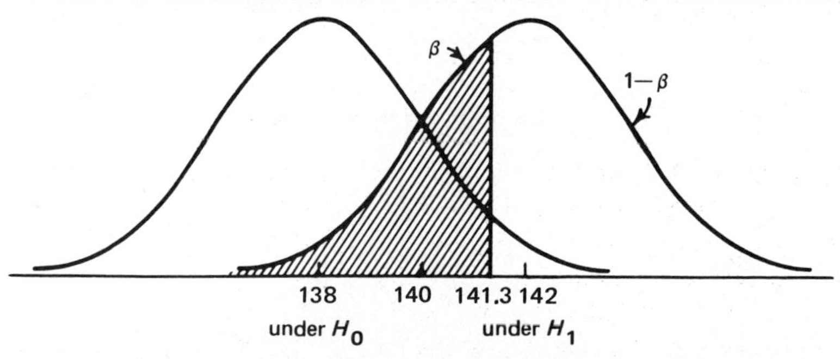

Figure 7.8.2

Type II error (β) and power $1 - \beta$ when H_1 is true.

We see that here the β probability of making an error is much greater than α. Our experimenter is rather unlikely to decide on H_1 when H_0 is true, but has considerable chance of concluding H_0 when H_1 is true.

The unshaded area under the curve to the right is $1 - \beta$, or the power of the test, given H_1. In this instance the power is about .64, so that the chances of correctly rejecting H_0 are about six in ten. Notice that *the power of a test of H_0 cannot be found until some true situation H_1 is specified, as in this example.*

What factors determine the power of a statistical test in a situation such as we have just examined? The power of a test always depends upon four things:

1. The particular alternative hypothesis H_1 that is to be assumed true if H_0 is false.
2. The value of α chosen by the experimenter and the selection of a region of rejection corresponding to α.
3. The size of the sample.
4. The variability of the population under study.

We shall examine each of these factors in turn.

7.9 / POWER OF TESTS AGAINST VARIOUS TRUE ALTERNATIVES

The power of a test of H_0 is not unlike the power of a microscope. It reflects the ability of a decision rule to detect from evidence that the true situation differs from a hypothetical one. Just as a high-powered microscope lets us distinguish gaps in an apparently solid material that we would miss with low power or the naked eye, so does a high-powered test of H_0 almost insure us of detecting when H_0 is false. Pursuing the analogy further, any microscope will reveal "gaps" with more clarity the larger these gaps are; the larger the departure of H_0 from the true situation H_1, the more powerful is the test of H_0, other things being equal.

For example, suppose that the hypothesis to be tested is

$$H_0: \mu = \mu_0$$

but that the true hypothesis is

$$H_1: \mu = \mu_1.$$

Here, μ_0 and μ_1 symbolize two different possible numerical values for μ. It will be assumed that under either hypothesis the sampling distribution of the mean is normal with σ known. Suppose now that the α probability of error is set at .05, and the rejection region is on the *right tail* of the distribution. The decision rule is such that the critical value of M (the smallest M leading to the rejection of H_0) is $\mu_0 + 1.65\sigma_M$. The power of the test can then be figured for any possible true H_1.

First of all, suppose that μ_1 were actually $\mu_0 + \sigma_M$. Then in the distribution under H_1, the critical value of M has a z-score given by

$$z_M = \frac{M - \mu_1}{\sigma_M}$$

$$= \frac{(\mu_0 + 1.65\sigma_M) - (\mu_0 + 1\sigma_M)}{\sigma_M} = .65.$$

The probability of a sample's falling into the region of rejection for H_0 is found from Table I to be .26. This probability is $1 - \beta$, the power of the test.

This may be clarified by Figure 7.9.1. The region of rejection for H_0 is the segment of the horizontal line to the right of $M = \mu_0 + 1.65\sigma_M$. The shaded portion under the curve for $\mu = \mu_0 + 1\sigma_M$ denotes β, the probability of *failing* to reject H_0, and the power is the unshaded part of the curve.

If another alternative hypothesis were true, say,

$$H_1: \mu_1 = \mu_0 + 3\sigma_M$$

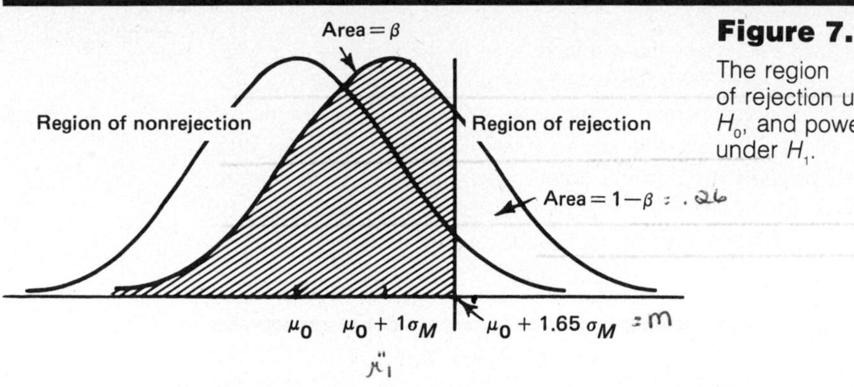

Figure 7.9.1

The region of rejection under H_0, and power under H_1.

then the power would be much larger. Here, relative to the sampling distribution based on the *true* mean or $\mu_0 + 3\sigma_M$, the critical value of M would correspond to a z_M-score of

$$z_M = \frac{(\mu_0 + 1.65\sigma_M) - (\mu_0 + 3\sigma_M)}{\sigma_M} = -1.35.$$

Table I shows us that above a z-score of -1.35 lie .91 of sample means in a normal sampling distribution, so that the power is .91.

The power of the test for any true value of μ_1 can be found in the same way. Often, to show the relation of power to the true value μ_1, so called **power functions** or **power curves** are plotted. One such curve is given in Figure 7.9.2, where the horizontal axis gives the possible values of true μ_1 in terms of μ_0 and σ_M, and the vertical axis the value of $1 - \beta$, the power for that alternative. Notice that for this particular decision rule, the power curve rises for increasing values of μ_1, and approaches 1.00 for very large values. On the other hand, for this decision rule, when true μ_1 is less than μ_0 the power is very small, and approaches 0 for decreasing μ_1 values. In any statistical test where the region of rejection is in the direction of the true value covered by H_1, the greater the discrepancy between the tested hypothesis and the true situation, the greater the power.

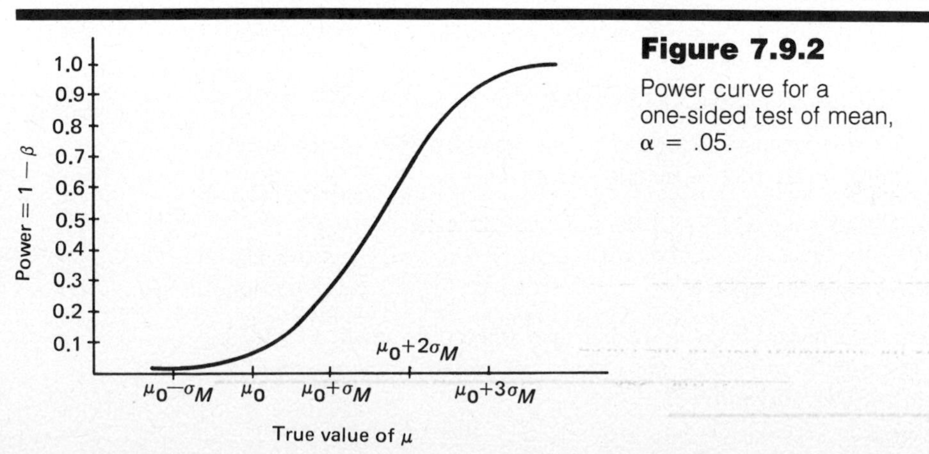

Figure 7.9.2

Power curve for a one-sided test of mean, $\alpha = .05$.

7.10 / POWER AND THE SIZE OF α

Since β will ordinarily be small for large α, as we have seen in the discussion above, it follows that setting α larger makes for relatively more powerful tests of H_0. For example, the two power curves given below for the same decision rule show that if α is set at .10 rather than at .05, the test with α = .10 will be more powerful than that for α = .05 over all possibly true values under alternative H_1 (note Figure 7.10.1). Making the probability of error in rejecting H_0 larger has the effect of making the test more powerful, other things being equal.

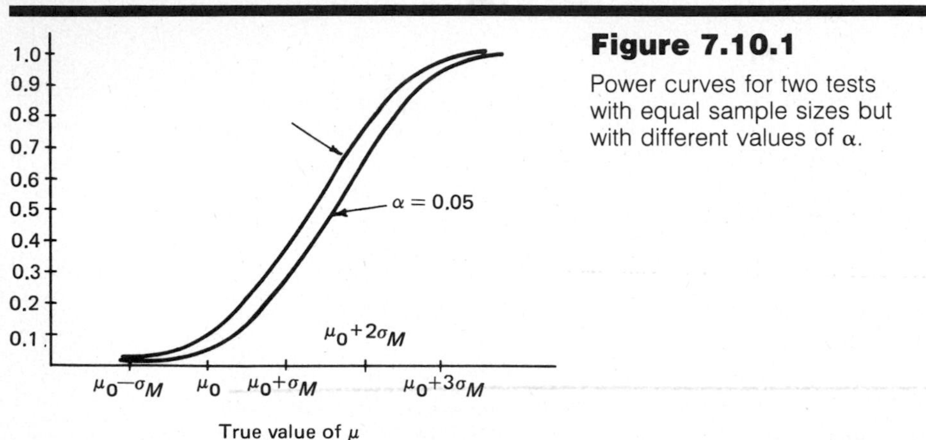

Figure 7.10.1

Power curves for two tests with equal sample sizes but with different values of α.

In principle, if it is very costly to make the mistake of overlooking a true departure from H_0, but not very costly to reject H_0 falsely, one could (and perhaps should) make the test more powerful by setting the value of α at .10, .20, or more. This is not ordinarily done in social science research, however. There are at least two reasons why α is seldom taken to be greater than .05: In the first place, as observed in Section 7.7 the problem of relative losses incurred by making errors is seldom solved in such research; hence conventions about the size of α are adopted. The other important reason is that given some fixed α the power of the test can be increased either by increasing sample size or by reducing the standard error of the test statistic in some other way.

7.11 / THE EFFECT OF SAMPLE SIZE ON POWER

Given a population with true standard deviation σ, the standard error of the mean depends inversely upon the square root of sample size N. That is,

$$\sigma_M = \frac{\sigma}{\sqrt{N}}.$$

When N is large, then the standard error is smaller than when N is small. Provided that $1 - \beta > \alpha$ to begin with, *increasing the sample size increases the power of a given test of H_0 against a true alternative H_1*.

For example, suppose that

$$H_0: \mu = 50$$
$$H_1: \mu = 60$$

where H_1 is true. Let us assume that true $\sigma = 20$. If samples of size 25 are taken, then

$$\sigma_M = \frac{20}{5} = 4.$$

Now the α probability for this test is fixed at .01, making the critical value of M be

$$M = \mu_0 + 2.33\sigma_M = 50 + (2.33)(4) = 59.32.$$

In the true sampling distribution (under H_1) this amounts to a z-score of

$$z_M = \frac{59.32 - 60}{4} = \frac{-.68}{4} = -.17$$

so that .57 of all sample means should fall into the rejection region for H_0. The power here is thus .57.

Now let the sample size be increased to 100. This changes the standard error of the mean to

$$\sigma_M = \frac{20}{10} = 2$$

and the critical value of M to

$$M = 50 + (2.33)(2) = 54.66.$$

The corresponding z-score when $\mu = 60$ is

$$z_M = \frac{54.66 - 60}{2} = -2.67$$

making the power now in excess of .99. With this sample size we would be almost certain to detect correctly that the H_0 is false when this particular H_1 is true. With only 25 cases, we are quite likely not to do so.

The disadvantages of an arbitrary setting of α can thus be offset, in part, by the choice of a large sample size. Other things being equal and regardless of the size chosen for α, the test may be made powerful against any given alternative H_1 in the direction of the rejection region, provided that sample N can be made very large. (See Figure 7.11.1.)

Once again, however, it is not always feasible to obtain very large samples. In much social science research, samples of substantial size are costly for the experimenter, if not in money, then in effort. Our ability to attain power through large samples not only partly offsets the failure to choose a decision rule according to cost of error in such research; large samples may not really be necessary in some research using them, especially if the error thereby made improbable is actually not very important. The matter of sample size and power will be considered once again in Chapter 8.

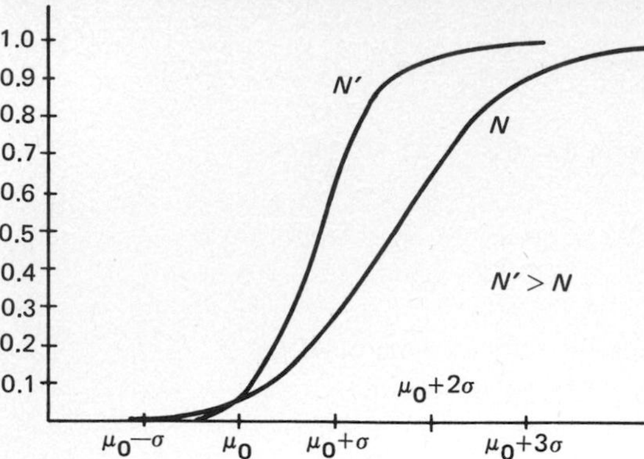

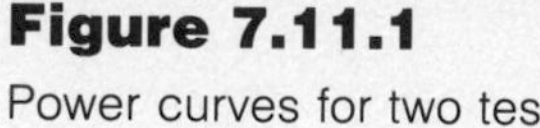

Figure 7.11.1

Power curves for two tests with $\alpha = .05$ but with different sample sizes, $N' > N$.

7.12 / POWER AND "ERROR VARIANCE"

Even given a fixed sample size, the experimenter has one more device for attaining power in tests of hypotheses. *Anything that makes σ, the population standard deviation, small will increase power, other things being equal.* This is one of the reasons for the careful control of conditions in good experimentation. By making conditions constant, the experimenter rules out many of the factors that contribute to variation in the observations. Statistically this amounts to a relative reduction in the size of σ for some experimental population. By ruling out some of the error variance from the observations one is *decreasing* the standard error of the mean, and thus *increasing* the power of the test against whatever hypothesis H_1 is true. Experiments in which the variability attributable to experimental or sampling error is small are said to be **precise;** the result of such precision is that the experimenter is quite likely to be able to detect when something of interest is happening. The application of experimental controls is like restricting inferences to populations with smaller values of σ^2 than otherwise, and thus control over error variation through careful experimentation implies powerful statistical tests. It follows that controlled experiments in which there is little ''natural'' variation in the materials observed can attain statistical power with relatively few observations, while those involving extremely variable material may require many observations to attain the same degree of power.

7.13 / ONE-TAILED REJECTION REGIONS

The primary notions of errors in inference and the power of a statistical test have just been illustrated for an extremely artifical situation, in which a decision must be made between two exact alternatives. Such situations are almost nonexistent in social science research. Instead, the experimenter is far more likely to be called on to evaluate inexact hypotheses, each of which encompasses a whole range of possibly true values. What relevance, then, does the discussion of the exact two-alternative case have to what researchers actually do? The answer is that researchers make inferences as though deciding between two exact alternatives, even though actual interest lies in judging between inexact hypotheses. Thus, the mechanism we have been using for decisions between exact hypotheses is exactly the same as for any other set of alternatives.

In some examples two inexact hypotheses are compared, having the form

$$H_0: \mu \leq \mu_0$$

$$H_1: \mu > \mu_0.$$

Here the entire range of possible values for the parameter under study (in this case, the population mean) was divided into two parts, that above and that below (or equal to) an exact value μ_0. The interest of the experimenter is in placing the true value of μ either above or below μ_0.

In this instance the appropriate region of rejection for H_0 consists of values of M relatively much larger than μ_0. Such values have a rather small probability of representing true means covered by the hypothesis H_0, but are more likely to represent true means in the range of H_1. Such a rejection region consisting of sample values in a particular direction from the expectation given by the exact value included in H_0 are called **directional** or **one-tailed** rejection regions. For the particular hypotheses compared in Section 7.8, the rejection region was one-tailed, since only the right or "high-value" tail of the sampling distribution under H_0 was considered in deciding between the hypotheses.

For other questions, the two inexact hypotheses are of the form

$$H_0: \mu \geq \mu_0$$

$$H_1: \mu < \mu_0.$$

Once again the region of rejection is one-tailed, but this time the lower or left tail of the sampling distribution contains the region of rejection for H_0. The choice of the particular rejection region thus depends both on α and the alternative hypothesis H_1.

Tests of hypotheses using one-tailed rejection regions are also called **directional.** The direction or sign of the value of the statistic (such as z_M) is important in directional tests since the sample result must show not only an extreme departure from expectation under H_0 but also a departure in the right direction to be considered strong evidence against H_0 and for H_1.

Directional hypotheses are implied when the basic question involves terms like "more than," "better than," "increased," "declined." The essential question to be answered by the data has a clear implication of a difference or change in a specific direction. An example will probably clarify this point. As a fairly plausible situation, imagine a psychologist who has constructed a new test of verbal fluency.

This test has been very carefully standardized on a population of American adults. The standardization has been so carried out that the mean score is 100 and the standard deviation of the test is 15. Furthermore, the distribution of scores in the standardization population is approximately normal.

The psychologist is now interested in the application of the test to residents of England. From knowledge of the skills required by the test, there is some reason to suspect that English adults may, on the average, score somewhat higher than American adults. So, in order to arrive at some idea if this may be the case, and especially to see if the test may require some scoring modification for administration in England, it is decided to try the test on a sample of English adults. Basically, the question to be answered is "Do English adults tend to score higher on this test than

Americans?'' We will suppose that there is no reason to question the standard deviation of scores among the English, nor the general form of the distribution of scores.

The answer to this question is tantamount to a decision between two *inexact* hypotheses:

$$H_0: \mu \leq 100$$

$$H_1: \mu > 100$$

i.e., the English population has either a mean score less than or equal to the American mean, or a mean greater than the American mean.

In choosing the decision rule to be used, the experimenter decides to set α equal to .01. A result greater than 100 tends to favor H_1, and so the region of rejection includes all z-scores greater than or equal to 2.33, since this value cuts off the highest 1 percent of sample means in a normal sampling distribution.

What, however, is the hypothesis actually being tested? As written, H_0 is inexact, since it states a whole region of possible values for μ. One exact value is specified, however: this is $\mu = 100$. Actually, then, the hypothesis tested is $\mu = 100$ against some unspecified alternative greater than 100. In effect, the decision rule can be put in the following form:

		True situation	
		$\mu = 100$	$\mu > 100$
Decide	$\mu = 100$	.99	(β)?
	$\mu > 100$	.01	$(1 - \beta)$?

The α probability of error can be specified in advance as .01, but β and the power are unknown, depending as they do upon the true situation. The experimenter has no real interest in the hypothesis that $\mu = 100$, and may even feel extremely confident that the true mean is not precisely 100. Nevertheless, this is a useful dummy hypothesis, in the sense that *if one can reject $\mu = 100$ with $\alpha = .01$, then one can reject any other hypothesis that $\mu < 100$ with $\alpha < .01$*. In other words, by this decision rule, if the researcher can be confident of not making an error in rejecting the hypothesis actually tested, then there is even more confidence in rejecting any other hypothesis covered by H_0.

But what does this do to β? Given some true mean μ_1 covered by H_1, *the power of the test of $\mu = 100$ is less than the power for any other hypothesis covered by H_0 with fixed α*. If the test is of any other exact hypothesis embodied in H_0, such as $\mu = 90$, then neither α nor β exceed those for $\mu = 100$. Testing the exact hypothesis with given α and β probabilities can be regarded as testing *all* hypotheses covered by H_0, with *at most* α and β probabilities of error (the β depending on the *true* mean, of course). This is illustrated below and in Figure 7.13.1.

		True situation			
		$\mu = 100$	$\mu < 100$	$\mu = \mu_1 > 100$	$\mu > \mu_1$
Decide	H_0	$1 - \alpha$	$> 1 - \alpha$	β	$< \beta$
	H_1	α	$< \alpha$	$1 - \beta$	$> 1 - \beta$

Figure 7.13.1

Representation of α and β for some other hypothesis covered by H_0, in a one-tailed test.

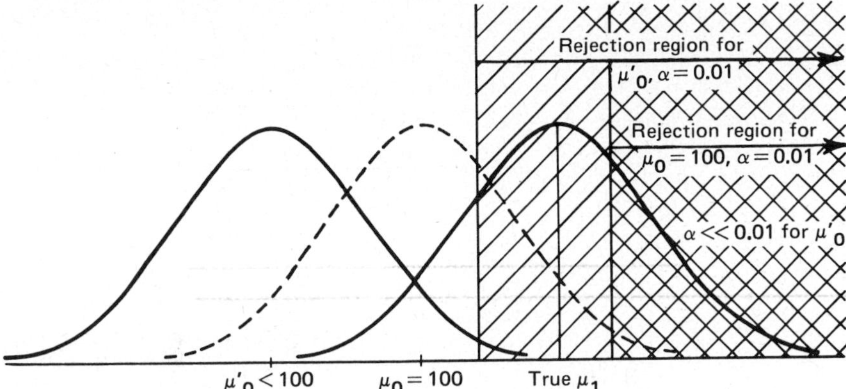

In other words, by testing the exact value $\mu = 100$, and using a one-tailed rejection region corresponding to H_1, we can be sure that the probability of wrongly rejecting any of the values covered by H_0 is *no more* than the .05 or .01 we pick for α. Furthermore, we can be sure that the power of the test is *at least* as great for the other hypotheses under H_0 as it is for 100, given some true alternative value greater than 100.

7.14 / TWO-TAILED REJECTION REGIONS

Many times an experimenter goes into a problem without a clearly defined notion of the direction of difference to expect if H_0 is false. We ask "Did something happen?" "Is there a difference?" or "Was there a change?" without any specification of expected direction. Next we will examine techniques for nondirectional hypothesis testing.

Imagine a study carried out on the "optical dominance" of human subjects. There is interest in whether or not the dominant eye and the dominant hand of a subject tend to be on the same or different sides. Subjects are to be tested for both kinds of dominance, and then classified as "same side" or "different side" in this respect. We will use the letters "S" and "D" to denote these two classes of subjects.

The experimenter knows that in 70 percent of subjects in this population the right hand is dominant. It is also known that the right eye is dominant for 70 percent of subjects. From such past knowledge about the relative frequency of each kind of dominance the experimenter reasons that if there actually is no tendency for eye dominance to be associated with hand dominance, then in a particular population of subjects one should expect 58 percent S and 42 percent D. (Why?) However, little is known about what to expect if there is some connection between the two kinds of dominance. To try to answer this question, our experimenter draws a random sample of 100 subjects, each with a full set of eyes and hands, and classifies them.

The question at hand may be put into the form of two hypotheses, one exact and one inexact:

$$H_0: p = .58$$

$$H_1: p \neq .58.$$

The first hypothesis represents the possibility of no connection between the two kinds of dominance, and the inexact alternative is simply a statement that H_0 is not true, since H_1 does not specify an exact value of population p.

The α chosen is .05. The experimenter then is faced with the choice of a rejection region for the hypothesis H_0. Either a very high percentage of S subjects in the sample or a very low percentage would tend to discount the credibility of the hypothesis that $p = .58$, and would lend support to H_1. Thus the rejection region is chosen so that the H_0 will be rejected when extreme departures from expectation of *either* sign occur. This calls for a rejection region on *both* tails of the sampling distribution of sample P when $p = .58$. Since the sample size is relatively large the binomial distribution of sample P may be approximated by a normal distribution (Section 6.4) and the two rejection regions may be diagrammed as shown in Figure 7.14.1.

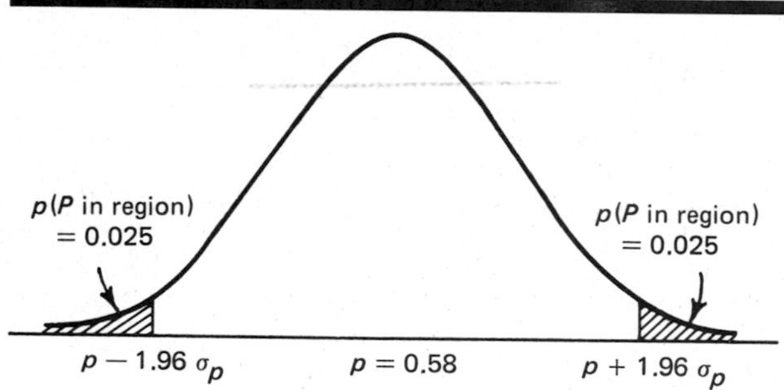

Figure 7.14.1
A two-tailed rejection region, for $\alpha = .05$.

Since the total probability of Type I error has been set at .05, the rejection region on the upper tail of the distribution will contain *the highest 2.5 percent,* and the lower tail *the lowest 2.5 percent of sample proportions,* given that $p = .05$. In short, each region should contain exactly $\alpha/2$ proportion of all samples under H_0. Consequently, either a very large or very small sample P will lead to a rejection of H_0, and the total probability of Type I error is .05. The z-score cutting off the upper rejection region is 1.96, and that for the lower is -1.96. Any sample giving a z beyond these two limits will lead to a rejection of H_0.

The basic sampling distribution is binomial, and the standard error of P is

$$\sigma_P = \sqrt{\frac{pq}{N}} = \sqrt{\frac{(.58)(.42)}{100}}$$

$$= .049$$

when H_0 is true (4.16.4). Using the normal approximation to the binomial we find

$$z = \frac{P - p}{\sigma_P}.$$

Now suppose that the proportion of S subjects is .69. Then

$$z = \frac{.69 - .58}{.049} = 2.24.$$

This value exceeds the critical z score of 1.96, and so the result is said to be significant beyond the 5 percent level. By concluding that eye and hand dominance do tend to be related the experimenter runs a risk of less than .05 of being wrong.

Suppose, however, that the sample result had come out to be only .53. Then

$$z = \frac{.53 - .58}{.049} = -1.02$$

making the result nonsignificant. Here, the experimenter should most likely suspend judgment pending further evidence, and not be willing to assert that H_0 is true: the risk in that assertion is unknown. Indeed, the best guess about the true value of p is .53, not .58, and given enough sample observations, a P value of .53 might be enough to warrant rejection. Thus, for a nonsignificant result such as this the experimenter cannot wisely accept H_0 as true. The best available choice usually is to suspend judgment, and look for more evidence.

One additional refinement of this test must be pointed out here. The actual sampling distribution of concern here is the binomial, which is, of course, discrete. On the other hand, we are approximating this discrete distribution by a continuous normal distribution. In order that the fit between binomial probabilities and the probabilities for intervals under a normal curve be as good as possible, the "correction for continuity" (6.4.4) is generally made. Such a correction consists in regarding a given P value as simply the midpoint in an interval of values with limits $P - .5/N$ and $P + .5/N$. Then the z-value is computed, not from the difference between P and p, but for the difference between the limiting value of the interval and p. That is, we take

$$z = \frac{P - p - (.5/N)}{\sigma_P}$$

if P happens to be greater in value than p, or we take

$$z = \frac{P - p + (.5/N)}{\sigma_P}$$

if P happens to be smaller in value than p. For the example above, the correction for continuity amounts to .5/100 or .005. This has the result of making the z-value in the test 2.14 rather than the 2.24 found previously.

Obviously, the correction for continuity can make a difference in the conclusions reached from a test, although it happened in this example not to do so. In general, the correction really should be used, especially when the sample size N is relatively small.

Although the example just concluded dealt with a hypothesis about a proportion, much the same procedure applies to two-tailed tests of means. An exact and an inexact hypothesis are opposed:

$$H_0: \mu = \mu_0$$

$$H_1: \mu \neq \mu_0.$$

The exact hypothesis is tested by forming two regions of rejection, each region containing exactly $\alpha/2$ proportion of sample results when H_0 is true, and lying in the higher and lower tails of the distribution. A sample result falling into either of these regions of rejection is said to be significant at the α level.

7.15 / RELATIVE MERITS OF ONE- AND TWO-TAILED TESTS

In deciding whether a hypothesis should be tested with a one- or two-tailed rejection region, the primary guide to the experimenter must be the original question. Is one looking for a directional difference between populations, or a difference only in kind or degree? By and large, most significance tests done in social science research are nondirectional, simply because research questions tend to be framed this way. However, there are situations where one-tailed tests are clearly indicated by the question posed.

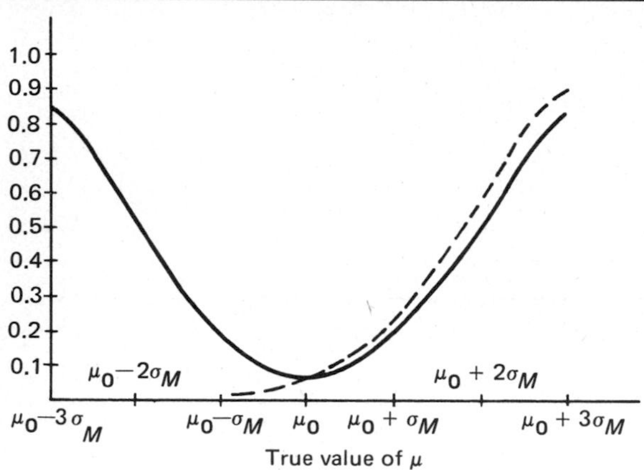

Figure 7.15.1

Power curves for a two-tailed and a one-tailed test of a mean, $\alpha = .05$.

The powers of one- and two-tailed tests of the same hypothesis will be different, given the same α level and the same true alternative. If a one-tailed test is used, and the true alternative is in the direction of the rejection region, then the one-tailed test is more powerful than the two-tailed over all such possibly true values of μ. In a way, we get a little statistical credit in the one-tailed test for asking a more searching question. On the other hand, if the true alternative happens to be on the tail opposite the rejection region in a one-tailed test, the power is very low; in fact, the power will always be less than α in the test of a mean. If you will, we are penalized for framing a stupid question. Power curves for one- and two-tailed tests for means are compared in Figure 7.15.1.

In many circumstances calling for one-tailed tests the form of the question or of the sampling distribution makes it clear that the only alternative of logical or practical consequence must lie in a certain direction. For example, when we turn to the problem of simultaneously testing many means for equality (Chapter 10) it will turn out that the only rejection region making sense lies on one tail of the particular sampling distribution employed. In other circumstances the question involves considerations of "which is better" or "which is more," where the discovery of a difference from H_0 in one direction may have real consequences for practical action, although a mean in another direction from H_0 may indicate nothing. For example, consider a treatment for some disease. The cure rate for the disease is known, and we want to see if the treatment *improves* the cure rate. We have not the slightest practical interest in a possible decrease in cures; this, like no change, leads to non-adoption of the treatment. The thing we want to be sure to detect is whether or not the new treatment is really better than what we have. Thus, we can safely ignore the low power for detecting a poorer treatment, so long as we have high power for detecting a really better treatment. In many such problems where different practical actions depend on the sign of the deviation from expectation, a one-tailed test is clearly called for. Otherwise, the two-tailed test is safest for the experimenter asking only "What happens?"

7.16 / INTERVAL ESTIMATION AND HYPOTHESIS TESTING

We have already examined the concept of a confidence interval for a population parameter such as μ and p. These are, of course, constructed by finding sample confidence limits a certain number of standard errors to either side of the sample mean or proportion. Then we are justified in saying that the probability has some value such as .95 or .99 that the true value of μ or p is covered by the interval between those limits. By using confidence intervals, we are able to "shade our bet" about the value of μ or p estimated from M or P.

However, it is also true that confidence limits have a close connection with the procedures we use in two-tailed tests of hypotheses about μ or p. That is, if one forms a 95 percent confidence interval for the value of μ, based on a sample of N cases drawn independently and at random, then ninety-five out of one hundred such confidence intervals will cover the true value of μ. Any hypothetical value that you might have for μ and that is covered by the confidence interval *could not* be rejected at the $\alpha = .05$ level. That is, if μ_0 is the value you have in a null hypothesis

$$H_0: \mu = \mu_0$$

and it is true that

$$M - 1.96\sigma_M \leq \mu_0 \leq M + 1.96\sigma_M,$$

then it must also be true that

$$-1.96 \leq \frac{M - \mu_0}{\sigma_M} \leq 1.96,$$

meaning that the hypothesis cannot be rejected for $\alpha = .05$ in a two-tailed test.

This fact leads to another interpretation and use for a confidence interval: a 100 $(1 - \alpha)$ percent confidence interval contains all of the values of μ (or of p) that *cannot* be rejected at the 100 (α) percent level of significance. A value not covered by the confidence interval would be significant beyond the 100 (α) level. Therefore, the reporting of a confidence interval along with a point estimate of a population's μ or p-value immediately establishes a range of hypotheses that would not be rejected at a particular level. Any hypothesis lying outside of that range of values would prove to be significant if tested (two tails).

In some research settings, a very good case can be made that the purpose of data collection is neither to permit an immediate decision to be made nor to obtain point estimates of parameters. Instead, the purpose of data collection is to modify the experimenter's *degrees of belief* in the various situations that may exist. The experimenter starts out with various hypotheses about the true situation. Experimentation is never begun in a state of total ignorance, however. For various theoretical and empirical reasons there are grounds to believe in some of these hypotheses more strongly than others, *before* the data are seen. Then the data alter these beliefs, so that some may be strengthened, some weakened, and still others unchanged. In short, some statisticians and researchers would argue that scientific investigation does not usually deal with decision making, but rather with alteration in personal probabilities. As evidence accumulates, some beliefs become so strong that these propositions are regarded as proved, and become part of the body of empirical science. Other beliefs fall away as evidence fails to support them.

The idea of personal probability has already been introduced in the discussion of decision rules. Let us suppose once again that it is possible to assign numerical values to various hypotheses, each standing for the experimenter's personal probability that a given proposition is true.

For example, in the situation described in Section 7.3, the true proportion is either

$$H_0\!:p = .80$$

or

$$H_1\!:p = .40,$$

so that we may regard the personal probability for any other H to be zero. Suppose that in the light of what you already know there is good reason to believe much more strongly in H_0 than in H_1. Imagine the personal probabilities assigned to these hypotheses are

$$\mathcal{P}(H_0) = .75$$

$$\mathcal{P}(H_1) = .25.$$

In other words, you are willing to give odds of three to one that H_0 is true. These personal probabilities of the experimenter are *prior,* since they reflect the degrees of belief in the two alternatives before the evidence. After the data are observed, something should happen to these personal probabilities; the degree of belief in each of the hypotheses should change somewhat.

The personal probability for H_0, given the evidence y, is designated by $\mathcal{P}(H_0|y)$, and that for H_1 by $\mathcal{P}(H_1|y)$. These are *posterior* values, since they *do* depend upon what is observed.

How should evidence affect the degree of belief in a given hypothesis? The key is given by Bayes' theorem, discussed in Section 1.13. If personal probabilities operate like ordinary, relative-frequency, probabilities, then by Bayes' theorem,

$$\mathcal{P}(H_0|y) = \frac{p(y|H_0)\mathcal{P}(H_0)}{p(y|H_0)\mathcal{P}(H_0) + p(y|H_1)\mathcal{P}(H_1)}.$$

The $p(y|H_0)$ and $p(y|H_1)$ are the usual probabilities dealt with in statistical inference: the conditional probability of evidence given that the particular hypothesis is true. Notice that, other things being equal, the more likely is y given H_0, and the less likely is y given H_1, the larger is $\mathcal{P}(H_0|y)$. Under this conception, evidence made likely by one hypothesis and not by the other always strengthens belief in the first, and weakens belief in the second.

However, the effective *strength* of the evidence also depends upon the *prior personal probabilities*. The amount that evidence changes degree of belief in H_0 and H_1 depends on the *prior* $\mathcal{P}(H_0)$ and $\mathcal{P}(H_1)$ values. For example, suppose that the experimenter observes a sample proportion P of .6 among 10 cases. From the sampling distributions given in Section 7.3, we can see that

$$p(.6|H_0) = .088$$

$$p(.6|H_1) = .111.$$

Applying Bayes' theorem, we find

$$\begin{aligned}\mathcal{P}(H_0|.6) &= \frac{p(.6|H_0)\mathcal{P}(H_0)}{p(.6|H_0)\mathcal{P}(H_0) + p(.6|H_1)\mathcal{P}(H_1)} \\ &= \frac{.066}{.066 + .028} \\ &= .70 \\ \mathcal{P}(H_1|.6) &= .30.\end{aligned}$$

This evidence lowers the personal probability for H_0 somewhat, since it is less probable under H_0 than H_1. However, the posterior personal probability for H_0 is *still* much greater than for H_1, reflecting the original discrepancy in the degree of belief the experimenter holds for each.

On the other hand, suppose that the sample evidence is a P of .3. Here,

$$p(.3|H_0) = .001; \; p(.3|H_1) = .215.$$

Notice that this value of P is *very* improbable if H_0 were true. Now,

$$\begin{aligned}\mathcal{P}(H_0|.3) &= \frac{(.001)(.75)}{(.001)(.75) + (.215)(.25)} \\ &= \frac{.00075}{.00075 + .05375} = .01 \\ \mathcal{P}(H_1|.3) &= 1 - (H_0|.3) = .99.\end{aligned}$$

In spite of the original weighting given the evidence by the prior personal probabilities, this time the P obtained is so unlikely given H_0 that the experimenter is almost certain that H_1 is true in the light of the evidence.

This way of evaluating evidence actually uses the raw materials of ordinary tests of hypotheses; as we have seen, the net result of any significance test is a statement of the likelihood of the sample result given that H_0 is true. However, a new ingredient is added: the prior personal probabilities of the experimenter. Intuitively, this approach has much to recommend it. In many ways it seems a far better description of what the scientist actually does with the statistical result than is the usual discussion of a significance test, burdened as it is with irrelevant decision terminology. In this conception, scientists are viewed as changing their minds in the light of evidence, but they are not thought of as using some phony decision rule.

On the other hand, mathematical statisticians are not always enthusiastic about this approach. The general idea is very old, and at one time was carried to extreme and unreasonable lengths. Even metaphysical questions were treated by this general method. Much of modern theoretical statistics grew out of attempts to get away from the idea of prior probabilities, and it is not surprising that some statisticians look at such uses of Bayes' theorem with a jaundiced eye.

This is not an ideal place to go further into this approach to inferential statistics. However, the student who would like to become acquainted with this approach is urged to look into the books by Phillips (1975), by Winkler (1974), or by Novick and Jackson (1974). Each gives a relatively simple introduction to statistics as seen from a Bayesian point of view.

7.18 / SIGNIFICANCE TESTS AND COMMON SENSE

Stripped of the language of decision theory and of concern with personal probabilities, all that a significant result implies is that one has observed something relatively unlikely given the hypothetical situation, but relatively more likely given some alternative situation. Everything else is a matter of what one does with this information. Statistical significance is a statement about the likelihood of the observed result, nothing else. It does not guarantee that something important, or even meaningful, has been found.

By shifting our point of view slightly, we can regard the statistical significance or nonsignificance of a result as a measure of the "surprisal value" of that result. That is, the surprisal value of a result can be thought of as the reciprocal of the probability (or likelihood) of the result given H_0. If a result is relatively likely given H_0, the surprisal value of this result is quite small. On the other hand, if the result falls among unlikely events given H_0, then it has great surprisal value. An impossible result should lead to infinite surprise. Then, since we tend to favor those hypotheses that lead to unsurprising results, H_0 is rejected only when the surprisal value of our result, given H_0, is quite high.

The simple matter of parsimony should lead one to be cautious in rejecting H_0. If trivial deviations from expectation are to lead us to rejection of a given hypothesis, our ideas about what is true might flit all over the place. The conservatism of the conventional rules for hypothesis testing act as a kind of brake on our tendency to follow trivial or ephemeral tendencies in the data.

These are very good arguments indeed for the traditional procedures of hypothesis testing. However, these arguments to not apply with equal force to all situations, nor do they engage the full conceptual machinery of decision making.

Any social scientist who is seriously concerned with making an important decision from data should pay attention to the possible losses the various outcomes represent. There is no guarantee that the use of any of the conventionalized rules for deciding significance is appropriate for this particular purpose. On the other hand, so long as the real costs of decision errors in scientific research remain as obscure as they are at present, then these conventions may not be a bad thing. In a young science, still mapping out its area of study, the error of pursuing a "phantom" result *is* costly, perhaps even more so than failing to recognize a real experimental effect when we see one. If these conventions are actually used to determine which lines of research are pursued and which are abandoned, then at least we tend to follow up the big departures from expectation that our data show. 

However, conventions about significant results should not be turned into canons of good scientific practice. Even more emphatically, a convention must not be made a superstition. It is interesting to speculate how many of the early discoveries in physical science would have been statistically significant in the experiments where they were first observed. Even in the crude and poorly controlled experiment, some departures from expectation stand out simply because they are interesting and suggest things to us that we might not be able to explain. These are matters that warrant looking into further regardless of what the conventional rule says to decide. Statistics cannot do the scientist's basic job—looking and wondering and looking again.

It is a grave error to evaluate the "goodness" of an experiment only in terms of the significance level of its results. Regardless of the scientist's convictions about personal probability, even a little bit of evidence coming out of a careful experiment is far more persuasive than a great deal from a sloppy one. To what population of subjects do the conclusions refer? Has a significant result been "bought" by restricting the population so much that the result fails to have any generality at all? In an effort to achieve a large sample N has the experimenter sacrificed all claim that the sample is random? Finally, how potent is the finding in a *predictive* sense: how much do these results permit us to reduce our uncertainty about the status of a given individual in this situation? This is by no means the same as statistical significance! As we shall see in later chapters, it is entirely possible for a highly significant result to contribute little to our ability to predict behavior, and for a nonsignificant result to mask an important gain in predictive ability.

It is very easy for research persons, particularly young persons, to become overconcerned with statistical method. Sometimes the problem itself seems almost secondary to some elegant method of data analysis. Easy access to computers has made it possible even for the novice in research to apply highly sophisticated methods quite routinely. Highly analyzed and significant results are often confused with good results. But overemphasizing the role of statistical analysis and statistical significance in research is like confusing the paintbrush with the painting. This form of statistical inference is a valuable tool in research, but it is never the arbiter of good research.

The remainder of this book is about statistical inference, and largely about significance tests. Since a text in statistics must necessarily deal mostly with statistics, there just is not room to discuss all the *if's, and's,* and *but's* that accompany the use of this tool in research. The skills of doing good research are acquired slowly, and often painfully, although a good share both of native curiosity and of common sense helps matters along. Statistical texts can display the tools, but they cannot give a guide to their use in every conceivable situation. If the user has some idea of how the tool works perhaps its uses and limitations in a given situation will become clearer. Expert help is available and should always be sought. But if there is ever a conflict between the use of a statistical technique and common sense, then common sense comes first. Careful observation is the main business of empirical science, and statistical methods are useful only so long as they help, not hinder, the systematic exploration of data and the cumulation and coordination of results.

EXERCISES

1. A study was concerned with the tendency of alcoholic middle-aged men who relapse after having undergone a certain new course of treatment. In the past, under an old treatment, 39 percent of such men relapsed into alcoholism within six months. It was felt that the six-month relapse rate may be different following the new treatment. A random sample of 150 male alcoholics was thus given the new treatment. After its conclusion, the number that relapsed within six months was noted. Formulate the null and the alternative hypotheses for this example. Sketch the relative locations of the rejection and nonrejection regions in a test of the null hypothesis.

2. In the study of exercise 1, interest actually focused on the possibility that the new treatment was better, as there is little practical value in a treatment that increases relapse rate. Given this interest, state the null and alternative hypotheses, and sketch the rejection and nonrejection regions that appear to be appropriate.

3. There was concern in a public school system that the method of teaching reading then in use might be inferior to other methods. A standardized test was available, giving national norms on reading achievement. For fifth-graders, this test showed a national average of 172 with a standard deviation of 16. A random sample was taken consisting of 250 fifth-graders taught be the method in question, and each was given the test. Frame the null and the alternative hypotheses for this problem, and indicate the regions of rejection and nonrejection that would be appropriate here. Let the probability of incorrectly rejecting the null hypothesis be equal to .01. in this instance.

4. Actually test the hypothesis stated in exercise 3, given that the random sample of students showed a mean of 170. What conclusions would you draw?

5. Suppose that a child transferring into a school must be placed either in a sixth-grade-level or a seventh-grade-level mathematics class. A standardized test of arithmetic achievement is used. It has been found that people who succeed in the seventh-grade-level class have an average score of 148 on entering, with a standard deviation of 20. On the other hand, those who belong in the lower class have an average entering score of 132, with a standard deviation of 15. Frame this as a decision problem, showing the different correct decisions and errors that can occur.

6. Assume that the two distributions of scores described in exercise 5 are both normal. Then work out the probabilities of erroneous and correct decisions about this child if the following rules are applied:

(a) Assign to 6th grade if score $\leqslant$ 132; 7th grade otherwise.
(b) Assign to 7th grade if score $\geqslant$ 148; 6th otherwise.
(c) Assign to 6th if score $\leqslant$ 139; otherwise 7th.
(d) Assign to 6th if score $\leqslant$ 140; otherwise 7th.
(e) Assign to 6th if score $\leqslant$ 141; otherwise 7th.

If the losses associated with each kind of error are equal, which is the best decision rule?

7. Let us assume that the losses associated with errors in exercises 5 and 6 are known. Let A represent the loss associated with placing a child in a grade which is too high, and let B

represent the loss involved in the other kind of error. Decide which of the five rules listed above should be used if the errors have the loss values

(a) $A = 10, B = 10$
(b) $A = 20, B = 10$
(c) $A = 10, B = 20$
(d) $A = 10, B = 12$

If only one of the last three rules may be used, which one is best given each of the loss values indicated above?

8. A social science study is to be based on the use of telephone interviews. A new city is about to be included in the study. If 40 percent of the persons contacted agree to be interviewed, it will be worthwhile to go on with the study in this city. On the other hand, if 20 percent (or fewer) will be interviewed, it is not worth proceeding. Based on past experience, it is believed that one of these situations (20 percent or 40 percent) is likely to be true. Therefore, a random selection of 10 residential calls are made. Depending on the results, the study will either be completed or abandoned for this city. Find the probabilities of errors and of correct decisions under the following rules:

(a) Continue if 6 or more calls are successful, and not otherwise.
(b) Continue if 2 or more calls are successful; not otherwise.
(c) Continue if 4 or more calls are successful; not otherwise.
(d) Continue if 3 or more calls are successful; not otherwise.
(**Hint:** Use Table II, Appendix D.)

9. Let A symbolize the loss involved in the study above if erroneously continued. Let B stand for the loss if the study is wrongly abandoned. Then find the best decision rule to use under each pair of these values for A and B.

(a) $A = \$1000$, $B = \$1000$
(b) $A = \$1500$, $B = \$1000$
(c) $A = \$1000$, $B = \$5000$
(d) $A = \$10{,}000$ $B = \$1000$

10. In the past, about 60 percent of cities such as that in exercises 8 and 9 have given a satisfactory number of interviews, while about 40 percent have not. The study director thus tends to believe that the odds are about 3 to 2 that this city will prove productive. Under these circumstances, and given equal loss values for the two kinds of errors, which decision rule should the study director prefer?

11. A certain random variable is normally distributed and has a standard deviation of 4.2. Twenty-six observations were made independently and at random, and yielded a sample mean value of 31 for this random variable. Test the hypothesis that the mean of the random variable is 28.6 against the alternative hypothesis that the mean has some other value. (Use the .05 level for significance.)

12. A psychological test was standardized for the population of tenth-grade students in such a way that the mean must be 500 and the standard deviation 100. A sample of 90 twelfth-grade students was selected independently and at random, and each given the test. The sample mean turned out to be 506.7. On this basis, can one say that the population distribution for twelfth-grade students would differ from that for tenth-graders?

13. For a particular task given to subjects in an experiment, the researcher theorized that about 25 percent of this population of subjects should be able to complete the task within the allotted time. In order to test out this hunch, he took twenty subjects chosen at random, and gave each the task independently. Of this group, 45 percent actually did finish within the time allotted. Test the hypothesis that the true proportion is .25 or less against the alternative hypothesis that the proportion is greater than .25.

14. Suppose that in exercise 13 above the experimenter had taken 200 subjects in order to test the hypothesis, and that the sample proportion had come out to be .28. Test the hypothesis that the true proportion is .25 against the alternative that the true proportion is not .25.

15. Are young minority children judged to be more or less cooperative than nonminority peers, when the teacher doing the rating is nonminority? In order to answer this question, a random sample of 123 nonminority first-grade teachers were asked to rate the cooperation of each of their students. Each had about the same number of minority and nonminority students. The score then assigned to the teacher was the average difference in ratings (i.e. minority average − nonminority) given to the two groups of children in that classroom. Over the entire sample of teachers the mean score was +.23, with a standard deviation of 1.27. Test the hypothesis that there is no difference in the way teachers tend to rate these two groups of children, using a two-tailed test and $\alpha = .05$.

16. A nine-hole golf course was supposed to have a par of 30, but over a long period of time the population of golfers who played this course had a mean of 38.2, with a standard deviation of 3.3. A designer was called in to extend this course to 18 holes, on the understanding that the last nine would have the same difficulty level as the first nine. After the course was finished a random sample of 121 golfers played the course, and produced an average score of 42.6 on the last nine holes. Given that this sample was drawn from the population who played the old nine holes, test the hypothesis that the two sets of nine holes are truly equal in difficulty. Use an α value of .05.

17. Some eighty rats selected at random were taught to run a maze. All of them finally succeeded in learning the maze, and the average number of trials to perfect performance was 15.91. However, long experience with a population of rats trained to run a similar maze shows that the average number of trials to success is 15, with a standard deviation of 2. Would you say that the new maze appears to be harder for rats to learn than the older, more extensively used maze?

18. A teacher wished to study the change in student attitude toward the federal government produced in a political science course. At the beginning of the class, therefore, a specially constructed attitude test was given, and then repeated at the end of the course. The student's score was the difference between scores on the second and the first test. A total of 368 students took the tests on both occasions. The mean change score was 2.3 points, and the sample standard deviation S was 10.5. Assuming that the students in the course constitute a random sample, would you say that there was a significant change in attitude? (**Hint:** Since the sample is large, use the unbiased estimate of the population variance as taken from the sample.)

19. Given a normal distribution with a standard deviation of 10, suppose that exactly one of the following two hypotheses must be true:

$$H_0: \mu = 100, \quad H_1: \mu = 105$$

If the probability of Type I error is to be .10, and the probability of Type II is also to be .10, what sample size should be used, and what is the critical value of the sample mean leading one to choose H_0 or H_1 as true?

20. Given the conditions of exercise 19 suppose that Type I error probability is to be fixed at .01, and Type II error probability at .10. Now what is the sample size that should be used, and what is the critical value of M?

21. In exercise 11 what is the power of the test against the alternative hypothesis that the mean is 32? Against the hypothesis that the mean is 25? (Use .05 as the probability of Type I error.)

22. In exercise 12 what is the power of the test against the hypothesis that the mean is 510? Against the hypothesis that the mean is 480? Against the hypothesis that the mean is 520? (Use .05 as the probability of Type I error.)

23. In exercise 17, what is the power of the test against the hypothesis that the mean is 16? That the mean is 20?

24. In exercise 17, if the experimenter had wished to have a power of .95 against the alternative hypothesis that the true mean is 16, what number of subjects would have been sufficient? (Use .05 as the probability of Type I error.)

25. Sketch the power function for a test of hypothesis about a mean which is two-tailed for $\alpha = .10$ and for $\alpha = .05$. (Assume a normal population distribution and, for simplicity, let $\sigma/\sqrt{N} = 1.00$. Divide the X axis of the plot into units of .5 or less.) What does this plot show about the relationship of power to the probability of Type I error?

26. For exercise 29 of Chapter 5, test the hypothesis that the new training method is no better than the old, against the alternative that the new method does cut down on injuries. Use the stated value of 4.9 for the standard deviation of the population, and take $\alpha = .01$.

27. Imagine that an experimenter draws a random sample and uses it to test a hypothesis. The α value is set at .05. Then another, independent random sample is drawn and the same hypothesis is tested once again, with the same α level. What is the probability that a Type I error is made on both tests? On neither test? On at least one test?

28. Suppose that the experimenter of exercise 27 takes a whole series of independent random samples, J in all. With each sample the same null hypothesis is tested. What is the probability of no Type I errors in this set of J tests? What is the probability of at least one Type I error? Work out this probability of at least one Type I error for $\alpha = .05$ and for $J = 3, 4, 5,$ and 6. What happens to the probability of at least one Type I error as the total number of tests is increased?

29. Suppose that an experimenter were trying to decide among three hypotheses:

$$H_0\colon \mu = 200, \quad H_1\colon \mu = 210, \quad H_2\colon \mu = 220$$

One and only one of these hypotheses must be true. Drawing a sample of data, it is determined that if H_0 is true, such a result has a likelihood of .15; if H_1 is true the sample result has a likelihood of .23; and if H_2 is true the sample result has a likelihood of .20. Prior to the experiment, the experimenter believes that the probability that H_0 is true is 1/2, and that H_1 and H_2 each has probability of only .25. Describe the experimenter's probabilities for these three hypotheses *after* the sample results are in.

30. Suppose that an experimenter is entertaining three hypotheses, one and only one of which must be true:

$$H_0\colon \mu = 200, \quad H_1\colon \mu = 210, \quad H_2\colon \mu = 220$$

It is known that the population is normal, and that the population standard deviation is 20. A sample of 25 cases is to be drawn and the sample mean computed. Decision Rule A is formulated: If $M \leq 205$, accept H_0; if $205 < M < 215$, accept H_1; if $215 \leq M$ accept H_2. Find the probabilities of being correct and of being in error under this rule, and display these probabilities in a table similar to that in Section 7.3.

31. Suppose that the experimenter of problem 30 above formulates a new rule, Decision Rule B: If $M \leqslant 207$ accept H_0; if $207 < M < 214$ accept H_1; if $M \geqslant 214$ accept H_2. Calculate the probabilities of correct decisions and of errors under Decision Rule B.

32. The experimenter in exercises 29, 30, and 31 learns that losses are connected with erroneous decision in the following way:

		True state		
		H_0	H_1	H_2
Decision	Accept H_0	0	5	20
	Accept H_1	10	0	10
	Accept H_2	20	5	0

Under the criterion of making the expected loss as small as possible, is Decision Rule A or Rule B the better rule? Why?

Chapter 8
INFERENCES ABOUT POPULATION MEANS

This chapter deals with ways to make inferences about means. The procedures for point and interval estimation and for hypothesis testing described in the past three chapters will be applied, first to questions about the mean of a single population, and then to questions about the difference in the means of two populations. A new distribution, that of the test statistic *t*, will be the basis for most of these methods.

8.1 / LARGE-SAMPLE PROBLEMS WITH UNKNOWN POPULATION σ^2

In some of the examples of hypothesis testing up to this point we have actually "fudged" a bit on the usual situation: we have assumed that σ^2 is somehow known, so that the standard error of the mean is also known exactly. In these examples the author did not explain how σ^2 became known, largely because he could not think up a good reason. Now we must face the cold facts of the matter: for inferences about the population mean, σ^2 is seldom known. Instead, we must use the only substitute available for σ^2, which is our unbiased estimate s^2, calculated from the sample.

Notice that this problem does not exist for hypotheses about a population proportion p, since the existence of an exact hypothesis about p specifies what the value of the standard error of P, the sample proportion, must be. Therefore, the special techniques of this chapter apply only to inferences about means, and not to inferences about proportions.

From what we have already seen of the relation between sample size and accuracy of estimation, it makes sense that for large samples s^2 should be a very good estimate of σ^2. *In general, for large samples, there is rather little risk of a sizable error when one uses* s *in place of* σ *in estimating the standard error of the mean.*

Hence, when the sample size is quite large, tests of hypotheses about a single mean are carried out in the same way as when σ is known, except that the standard error of the mean is estimated from the sample:

$$\text{est. } \sigma_M = \frac{s}{\sqrt{N}} = \frac{S}{\sqrt{N-1}} \cdot \qquad [8.1.1^*]$$

The standardized score corresponding to the sample mean is then referred to the normal distribution. This step is justified by the central limit theorem when N is large, regardless of the population distribution's form.

For example, consider the following problem. A small rodent characteristically shows hoarding behavior for certain kinds of foodstuffs when the environmental temperature drops to a certain point. Numerous previous experiments have shown that in a fixed period of time, and given a fixed food supply, the mean amount of food hoarded by an animal is 9 grams. The experimenter is currently interested in possible effects that early food deprivation may have upon such hoarding behavior in the animal as an adult. So, the experimenter takes a random sample of 175 infant animals and keeps them on survival rations for a fixed period while they are at a certain age, and on regular rations thereafter. When the animals are adults each one is placed in an experimental situation where the lowered temperature condition is introduced. The amount of food each hoards is recorded, and a score is assigned to each animal.

What is the null hypothesis implied here? The basic experimental question is "Does the experimental treatment (deprivation) tend to affect the amount of food hoarded?" The experimenter has no special reason to expect either an increase or a decrease in amount, but is interested only in finding out if a difference from normal behavior occurs. This question may be put into the form of a null and an alternative hypothesis:

$$H_0: \mu_0 = 9 \text{ grams}$$

$$H_1: \mu \neq 9 \text{ grams.}$$

Suppose that the conventional level chosen for α is .01, so that the experimenter will say that the result is significant only if the sample mean falls among either the upper .005 or the lower .005 of all possible results, given H_0. Reference to Table I shows that .005 is the probability of a z-score in a normal distribution falling at or below -2.58, and the probability is likewise .005 for a z equal to or exceeding $+2.58$. Accordingly, the sample result will be significant only if

$$\frac{M = E(M)}{\text{est. } \sigma_M} \qquad [8.1.2]$$

equals or exceeds 2.58 in absolute magnitude (disregarding sign). When the null hypothesis is true, $E(M) = 9$, and for a sample this large the value of the standard error of the mean should be reasonably close to $\frac{s}{\sqrt{N}}$ or $\frac{S}{\sqrt{N-1}}$, the value of the sample estimate.

Everything is now set for a significance test except for the sample results. The sample shows a mean of 8.8 grams of food hoarded, with a standard deviation, S, of 2.30. The estimated standard error of the mean is thus

$$\text{est. } \sigma_M = \frac{2.30}{\sqrt{175 - 1}} = \frac{2.30}{13.19} = .17.$$

The standardized score of the mean is found to be

$$\frac{8.80 - 9.00}{.17} = -1.18.$$

This result does not qualify for the region of rejection for $\alpha = .01$. Since the experimenter feels able to reject H_0 only if the α probability of error is no more than .01, then H_0 will not be rejected on the basis of this sample. On the other hand, the risk run in accepting H_0 is unknown, so that "suspend judgment, pending more evidence" seems the appropriate choice here.

Confidence intervals may also be found by the methods of Chapter 6. However, either when σ^2 is unknown, or when the population distribution has unknown form, a normal sampling distribution is assumed only for large samples. Just as in significance tests, the estimated standard error of the mean can be used in place of σ_M in finding confidence limits when the sample is relatively large (say $N \geq 100$).

For example, the experimenter studying hoarding behavior computes the approximate 99 percent confidence limits in the following way:

$$M - 2.58 \text{ (est. } \sigma_M)$$

and

$$M + 2.58 \text{ (est. } \sigma_M)$$

so that for this problem, the numerical confidence limits are

$$8.8 - 2.58(.17) \text{ or } 8.36$$

and

$$8.8 + 2.58(.17) \text{ or } 9.24.$$

The experimenter can say that the probability is approximately .99 that the true value of μ is covered by an interval such as that between 8.36 and 9.24.

8.2 / THE DISTRIBUTION OF *t*

In inferences about μ, the ratio we would like to evaluate and refer to a normal sampling distribution is the standardized score

$$z_M = \frac{M - E(M)}{\sigma_M}. \qquad [8.2.1]$$

However, when we have only an *estimate* of σ_M, then the ratio we really compute and use is not a normal standardized score at all, although it has much the same form. The ratio actually used is

$$t = \frac{M - E(M)}{\text{est. } \sigma_M} = \frac{M - E(M)}{s/\sqrt{N}}. \qquad [8.2.2^*]$$

There is an extremely important difference between the two ratios, z_M and t. For z_M, the numerator $(M - E(M))$ is a random variable, the value of which depends upon the particular sample drawn from a given population situation; on the other hand, the denominator is a constant, σ_M, which is the same regardless of the particular sample of size N we observe. Now contrast this ratio with the ratio t: just as before, the numerator of t is a random variable, but the denominator is also a random variable, since the particular value of s—and hence the estimate of σ_M—is a sample quantity. Over several different samples, the same value of M must give us precisely the same value of z_M; however, over different samples, the same value of M will give us different t values. Similar intervals of t and z_M values should therefore have different probabilities of occurrence. For this reason it is risky to use the ratio t as though it were z_M unless the sample size is very large.

The solution to the problem of the nonequivalence of t and z_M rests on the study of t itself as a random variable. That is, suppose that the t ratio were computed for every conceivable sample of N independent observations drawn from some normal population distribution with true mean μ. Each sample would have some t value,

$$t = \frac{M - E(M)}{\text{est. } \sigma_M} = \frac{M - \mu}{s/\sqrt{N}}. \qquad [8.2.3^*]$$

Over the different samples the value of t would vary, of course, and the different possible values would each have some probability-density. A random variable such as t is an example of a **test statistic**, so called to distinguish it from an ordinary descriptive statistic or estimator, such as M or s^2. The t value depends on other sample statistics, but is not itself an estimate of a population value. Nevertheless, such test statistics have sampling distributions just as ordinary sample statistics do, and these sampling distributions have been studied extensively.

In order to find the exact distribution of t, one must assume that the basic population distribution is normal. There are two reasons for this assumption. The first reason is that it permits us to specify the distribution of the numerator of t, or $(M - \mu)$, without regard to sample size. The second is that only for a normal distribution will the basic random variables in numerator and denominator, sample M and s, be statistically independent; this is a use of the important fact mentioned in Section 6.6. Unless M and s are independent, the sampling distribution of t is extremely difficult to specify exactly. On the other hand, for the special case of normal populations, the distribution of the ratio t is quite well known. In order to learn what this distribution is like, let us take a look at the rule for the density function associated with this random variable.

The density function for t is given by the rule:

$$f(t;\, \nu) = G(\nu)\left[1 + \frac{t^2}{\nu}\right]^{-(\nu+1)/2} \qquad \begin{matrix} -\infty < t < \infty \\ 0 < \nu \end{matrix}. \qquad [8.2.4]$$

Here, $G(\nu)$ stands for a constant number which depends *only* on the parameter ν (lowercase Greek nu), and how this number is found need not really concern us.

The "working part" of the rule involves only ν and the value of t. This looks very different from the normal distribution function rule in Section 6.1. As with the normal function rule, however, a quick look at Figure 8.2.1, the graph corresponding to this mathematical expression, tells us much about the distribution of t (for $\nu > 1$).

First of all, notice that the distribution of sample t values must be symmetric, since a positive and a negative value having the same absolute size must be assigned the same probability density by this rule. Second, the largest possible density value is assigned to $t = 0$. Thus $t = 0$ is the distribution mode. Observe also that the distribution is unimodal and "bell-shaped." If we inferred from the symmetry and unimodality of this distribution that the mean of t is also 0, we should be quite correct. In short, the t distribution is a unimodal, symmetric, bell-shaped distribution having a graphic form much like a normal distribution, even though the two function rules are quite dissimilar. Loosely speaking, the curve for a t distribution differs from the standardized normal in being "plumper" in extreme regions and "flatter" in the central region, as Figure 8.2.1 shows. (Note that both t and the standardized normal distribution have a mean of zero, $\nu > 1$.)

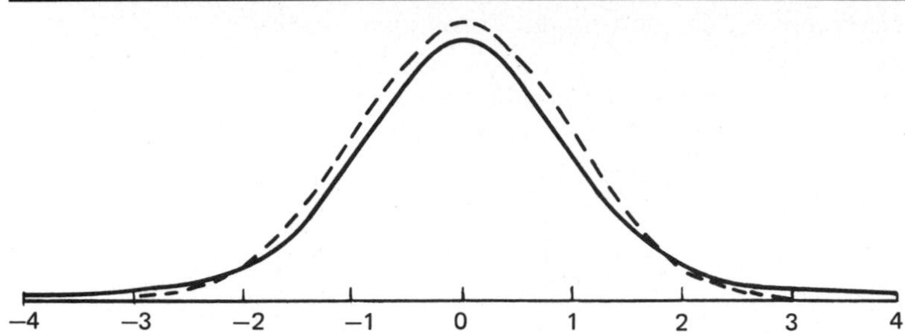

Figure 8.2.1

Distribution of t with $\nu = 4$, and standardized normal distribution

The most important feature of the t distribution will appear if we return for a look at the function rule. Notice that the only unspecified constants in the rule are those represented by ν and $G(\nu)$, which depends only on ν. This is a one-parameter distribution: the single parameter is ν, called *the degrees of freedom. Ordinarily, in most applications of the t distribution to problems involving a single sample, ν is equal to $N - 1$, one less than the number of independent observations in the sample.* For samples of N independent observations from any normal population distribution, the exact distribution of sample t values depends only on the degrees of freedom, $N - 1$. Remember, however, that the value of $E(M)$ or μ must be specified when a t ratio is computed, although the true value of σ need not be known.

In principle, the value of ν can be any positive number, and it just happens that $\nu = N - 1$ is the value for the degrees of freedom for the particular t distributions we will use first. Later we will encounter problems calling for t distributions with other numbers of degrees of freedom. Like most theoretical distributions, the t distribution is actually a family of distributions, with general form determined by the function rule, but with particular probabilities dictated by the parameter. For $\nu > 2$ the variance of the t distribution is $\nu/(\nu - 2)$, so that the smaller the value of ν the larger the variance. As ν becomes large the variance of the t distribution approaches 1.00, which is the variance of the standardized normal distribution.

Incidentally, the random variable t is often called "Student's t," and the distribution of t, "Student's distribution." This name comes from the statistician W. S. Gosset, who was the first to use this distribution in an important problem, and who first published his results in 1908 under under the pen name "Student." Distributions of the general "Student" form have a number of important applications in statistics.

8.3 / THE t AND THE STANDARDIZED NORMAL DISTRIBUTIONS

As we have seen, the "shape" of the t distribution is not unlike that of the normal distributon. Just as for the standardized normal, the mean of the distribution of t is 0 for $v > 1$ although the variance of t is greater than 1.00 for finite $v > 2$. Given any extreme interval of fixed size on either tail of the t distribution, the probability associated with this interval in the t distribution is larger than that for the corresponding normal distribution of z_M. The smaller the value of v, the larger is this discrepancy between t and normal probabilities at the extreme ends of each distribution. This reflects and partly explains the danger of using a t ratio as though it were a z ratio except when v is quite large: for small v extreme values of t are relatively more likely than comparable values of z_M. A small sample size corresponds to a small value of v, or $N - 1$, and thus there is serious danger of underestimating the probability of an extreme deviation from expectation when sample size is small. This is apparent in the illustration (Figure 8.2.1) showing the distribution of t together with the standardized normal function.

On the other hand, what happens to the distribution of t as v becomes large (sample size grows large) is suggested both by Figure 8.2.1 and by the variance of a t distribution. *As sample size N grows large, the distribution of t approaches the standardized normal distribution. For large numbers of degrees of freedom, the exact probabilities of intervals in the t distribution can be approximated closely by normal probabilities.*

The practical result of this convergence of the t and the normal probabilities is that the t ratio *can* be treated as a z_M ratio, provided that the sample size is substantial. The normal probabilities are quite close to—though not identical with—the exact t probabilities for large v. On the other hand, when sample size is small the normal probabilities cannot safely be used, and instead one uses a special table based on the t distribution.

How large is "large enough" to permit use of the normal tables? If the population distribution is truly normal, even forty or so cases permit a fairly accurate use of the normal tables in confidence intervals or tests for a mean. If really good accuracy is desired in determining interval probabilities, the t distribution should be used even when the sample size is around 100 cases. Beyond this number of cases, the normal probabilities are extremely close to the exact t probabilities. For example, in the "hoarding" experiment just discussed, use of t rather than z values would have given confidence limits of about $M \pm 2.6$ (est σ_M) instead of $M \pm 2.58$(est σ_M), a very slight difference.

Recall that the stipulation is made that the population distribution be *normal* when the t distribution is used, even when the normal approximations are substituted for the exact t-distribution probabilities. Clearly, this requirement limits the usefulness of the t distribution, since this is an assumption often hard to justify in practical situations. Fortunately, when sample size is fairly large, and provided that the parent distribution is roughly unimodal and symmetric, the t distribution still gives an adequate approximation to the exact (and often unknown) probabilities under these circumstances. However, you should insist on a relatively *larger* sample size the *less* confident you are that the normal rule holds for the population, especially when one-tailed tests of hypotheses are made, since a very skewed population distribution can make the t probabilities for one-tailed tests considerably in error. Once again, it is wise to plan on somewhat larger samples when one is considering a one-tailed test using the t distribution and the population is not assumed normal.

8.4 / TABLES OF THE t DISTRIBUTION

Unlike the table of the standardized normal function, which suffices for all possible normal distributions, tables of the t distribution must show many different distributions each depending on the value of ν, the degrees of freedom. Consequently, tables of t are usually given only in abbreviated form.

Table III in Appendix D shows selected percentage points of the distribution of t, in terms of the value of ν. Different ν values appear along the left-hand margin of the table. The top margin gives values of Q, which is $1 - p(t \leq a)$, one minus the cumulative probability that t is less than or equal to a specific value a, for a distribution within the given value for ν. A cell of the table then shows the value of t cutting off the upper Q proportion of cases in a distribution for ν degrees of freedom.

This sounds rather complicated, but an example will clarify matters considerably: suppose that $N = 10$, and we want to know the value *beyond which* only 10 percent of all sample t values should lie. That is, for the distribution of t shown in Figure 8.4.1 we want the t value that cuts off the shaded area in the curve, the upper 10 percent:

First of all, since $N = 10$, $\nu = N - 1 = 9$. So, we enter the table for the row marked 9. Now since we want the upper 10 percent, we find the column for which $Q = .1$. The corresponding cell in the table is the value of t we are looking for, $t = 1.383$. We can say that in a t distribution with 9 degrees of freedom, the probability is .10 that a t value *equals or exceeds* 1.383. Since the distribution is symmetric, we also know that the probability is .10 that a t value *equals or falls below* -1.383. If we wanted to know the probability that t equals or exceeds 1.383 in absolute value, then this must be $.10 + .10 = .20$, or $2Q$.

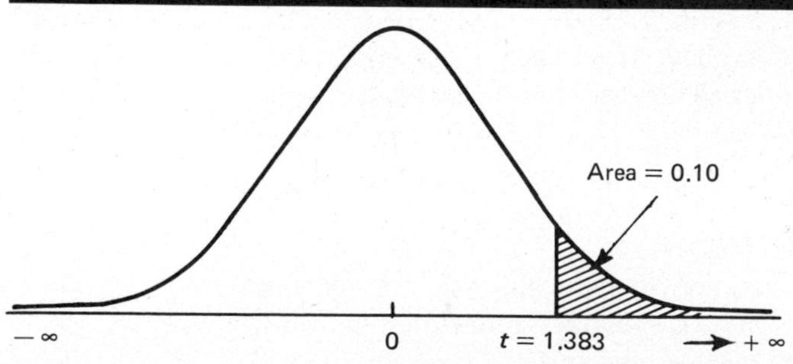

Figure 8.4.1
The upper 10 percent area in a t distribution with 9 degrees of freedom.

Suppose that in a sample of 21 cases, we get a t value of 1.98. We want to see if this value falls into the upper .05 of all values in the distribution. We enter row $\nu = 21 - 1 = 20$, and column $Q = .05$; the t value in the cell is 1.725. Our obtained value of t is larger than this, and so the obtained value does fall among the top 5 percent of all such values. On the other hand, suppose that the obtained t had been -3. Does this fall either in the top .001 or the bottom .001 of all such sample values? Again with $\nu = 20$, but this time with $Q = .001$, we find a t value of 3.552. This means that at or above 3.552 lie .001 of all sample values, and also at or below -3.552 lie .001 of all sample values. Hence our sample value does not fall into either of these intervals; we can say that the sample value does not fall into the rejection region for $\alpha = .002$.

The very last row in Table III, marked ∞, shows the z scores that cut off various areas in a normal distribution curve. If you trace down any given column, you find that as v gets larger the t value bounding the area specified by the column comes closer and closer to this normal deviate value, until finally, for an infinite sample size, the required value of t is the same as that for z.

For one-tailed tests of hypotheses, the column Q values are used to find the t value which exactly bounds the rejection region. If the region of rejection is on the upper tail of the distribution, then Q is the probability of a sample value's falling into the region greater than or equal to the tabled value of t. If the region is on the lower tail, the t value in the table is given a negative sign, and Q is the probability of a sample's falling at or below the negative t value. If a two-tailed region is to be used, then the total α probability of error is $2Q$, and the number in the table shows the absolute value of t that bounds the rejection region on *either* tail.

8.5 / THE CONCEPT OF DEGREES OF FREEDOM

Before we proceed to the uses of the t distribution, it is well to examine the notion of degrees of freedom. The degrees of freedom parameter reflects the fact that a t ratio involves a sample standard deviation as the basis for estimating σ_M. Recall the basic definitions of the sample variance and the sample standard deviation:

$$S^2 = \frac{\Sigma(x - M)^2}{N}$$

and

$$S = \sqrt{\frac{\Sigma(x - M)^2}{N}}.$$

The sample variance and standard deviation are both based upon a sum of squared deviations from the sample mean. However, recall another fact of importance about deviations from a mean: in 4.3 it was shown that

$$\sum_i (x_i - M) = 0;$$

that is, the sum of deviations about the mean must be zero.

These two facts have an important consequence: Suppose that you are told that $N = 4$ in some sample, and that you are to guess the four deviations from the mean M. For the first deviation you can guess any number, and suppose you say

$$d_1 = 6.$$

Similarly, quite at will, you could assign values to two more deviations, say

$$d_2 = -9$$

$$d_3 = -7.$$

However, when you come to the fourth deviation value, you are *no longer free* to guess any number you please. The value of d_4 *must* be

$$d_4 = 0 - d_1 - d_2 - d_3$$

or

$$d_4 = 0 - 6 + 9 + 7 = 10.$$

In short, given the values of any $N - 1$ deviations from the mean, which could be any set of numbers, the value of the last deviation is completely determined. Thus we say that there are $N - 1$ degrees of freedom for a sample variance, reflecting the fact that only $N - 1$ deviations are "free" to be any number, but that given these free values, the last deviation is completely determined. It is not the sample size per se that dictates the distribution of t, but rather the number of degrees of freedom in the variance (and standard deviation) estimate.

Perhaps this little argument will give you some intuition about the concept of degrees of freedom, and why it is important. However, it must be said that the concept is a good bit more sophisticated than it may appear here, and that in future sections a better explanation will indeed be required. For the moment, however, this interpretation will have to suffice.

8.6 / SIGNIFICANCE TESTS AND CONFIDENCE LIMITS FOR MEANS USING THE t DISTRIBUTION

For the moment you can relax; there is really nothing new to learn! When the null hypothesis concerns a single mean, then the test is carried out just as before, except that the table of t (Appendix D, Table III) is used instead of the normal table (Appendix D, Table I). The α level is chosen, and the value (or values) of t corresponding to the region of rejection can be determined from the t table. The number of degrees of freedom used is simply $v = N - 1$. The t ratio is

$$t = \frac{M - E(M)}{\text{est. } \sigma_M} \qquad [8.6.1]$$

where

$$\text{est. } \sigma_M = S/\sqrt{N - 1} \qquad [8.6.2]$$

and

$$E(M) = \text{expected value of } M \text{ if } H_0 \text{ is true.}$$

Then the value of t which is obtained from the sample is compared with values in the rejection region specified by Table III. If the obtained t ratio falls into the rejection region chosen, the sample result is said to be significant beyond the α level.

Naturally, all the considerations hitherto discussed, especially the assumed normal distribution of the population, should be faced before the sample size and rejection region are decided upon. If large samples are available then the assumption of a normal population is relatively unimportant; on the other hand, this matter should be given some serious thought if you are limited to a very small sample size.

The t distribution may also be used to establish confidence limits for the mean. For some fixed percentage representing the confidence level, $100(1 - \alpha)$ percent, the sample confidence limits depend upon three things: the sample value of M, the estimated standard error, or est. σ_M, and the number of degrees of freedom, v. For some specified value of ν, then the $100(1 - \alpha)$ percent confidence limits are found from

$$M - t_{(\alpha/2;\nu)}\text{ (est. }\sigma_M)$$
$$M + t_{(\alpha/2;\nu)}\text{ (est. }\sigma_M). \qquad [8.6.3^*]$$

Here, $t_{(\alpha/2;\nu)}$ represents the value of t that bounds the upper $\alpha/2$ proportion of cases in a t distribution with ν degrees of freedom. In Table III this is the value listed for $Q = \alpha/2$ and ν. Thus, if one wants the 99 percent confidence limits, the value of $\alpha = .01$, and one looks in the table for $Q = .005$.

For example, imagine a study using 8 independent observations drawn from a normal population. The sample mean is 49 and the estimated standard error of the mean is 3.70. Now we want to find the 95 percent confidence limits. First of all, $\alpha = .05$, so that $Q = .025$. The value of ν is $N - 1$, or 7. The table shows a t value of 2.365 for $Q = .025$ and $\nu = 7$, so that $t_{(\alpha/2\nu)} = 2.365$. The confidence limits are

$$49 - (2.365)(3.70) = 40.25$$

and

$$49 + (2.365)(3.70) = 57.75.$$

Over all random samples, the probability is .95 that the true value of μ is covered by an interval such as that between 40.25 and 57.75, the confidence interval calculated for this sample.

8.7 / QUESTIONS ABOUT DIFFERENCES BETWEEN POPULATION MEANS

Examples of hypotheses about single means often sound rather ''phony'' in their experimental contexts, and the reason is not hard to find. In most experimental work it is not true that the experimenter knows about one particular population in advance and then draws a single sample for the purpose of comparing some experimental population to the known population. Rather, it is far more common to draw two samples, to only one of which the experimental treatment is applied; the other sample is given no treatment, and stands as a control group for comparison with the treated group. In other situations, two different treatments may be compared. The treatments may be something done by the experimenter, or nature may have already made the groups different in some respect. The advantages of this method over the single-sample procedure are obvious; the experimenter can exercise the same experimental controls on both samples, making sure that insofar as possible they are treated in exactly the same way, with the only systematic experimental difference being in the fact that something was done to representatives of one sample which was not done to members of the other. Then, if a very large difference appears between the two samples, this difference should be a product of the experimental treatments and not just a peculiarity introduced by the way in which the data were gathered.

The best available estimate of the difference between the means of two populations is the obtained difference between the means of the two samples. As always, this estimate of the population difference is in error to some unknown extent, and although the obtained difference between the sample means is the best guess the experimenter can make, there is absolutely no guarantee that this estimate is exactly correct. It could well be true that the difference the experimenter observes has no real connection with the treatment administered, and is purely a chance result.

What is needed is a way of applying statistical inference to differences between means of samples representing two populations. First, large sample distributions of *differences* between sample means will be studied. Then, the application of the t distribution to small sample differences will be introduced.

Suppose that we wished to test a hypothesis that two populations have means which differ by some specified amount, say 20 points. This is tested against the hypothesis that the population means do not differ by that amount. In our more formal notation:

$$H_0: \mu_1 - \mu_2 = 20$$

$$H_1: \mu_1 - \mu_2 \neq 20.$$

We draw a sample size N_1 from population 1, and an *independent* sample of size N_2 from population 2, and consider the difference between their means, $M_1 - M_2$. Now suppose that we kept on drawing pairs of independent samples of these sizes from these populations. For each pair of samples drawn, the difference $M_1 - M_2$ is recorded. What is the distribution of such sample *differences* that we should expect in the long run? In other words, what is the sampling distribution of the difference between two means?

In the first place,

$$E(M_1 - M_2) = E(M_1) - E(M_2) = \mu_1 - \mu_2, \qquad [8.7.1^*]$$

by Rule 4 of Appendix B. Thus, the *expected difference between the two sample means is the difference between the population means.*

Furthermore, since these are two independent samples, of N_1 and N_2 independent cases respectively, then

$$\text{var}(M_1 - M_2) = \sigma^2_{M_1} + \sigma^2_{M_2}. \qquad [8.7.2^*]$$

by C2.6 of Appendix C.

Hence, the standard error of the difference, $\sigma_{\text{diff.}}$, is

$$\sigma_{\text{diff.}} = \sqrt{\sigma^2_{M_1} + \sigma^2_{M_2}} = \sqrt{\frac{\sigma_1^2}{N_1} + \frac{\sigma_2^2}{N_2}} \qquad [8.7.3^*]$$

provided that samples 1 and 2 are completely independent. Then, for two *large* samples drawn from populations having different variances σ_1^2 and σ_2^2, the standard error of the difference is estimated by

$$\text{est. } \sigma_{\text{diff.}} = \sqrt{(\text{est. } \sigma_{M_1})^2 + (\text{est. } \sigma_{M_2})^2} \qquad [8.7.4^*]$$

Notice that there is no requirement at all that the samples be of equal size. Regardless of the sample sizes, the expectation of the difference between two means is always the difference between their expectations, and the variance of the difference between two *independent* means is the *sum* of the separate sampling variances.

Furthermore, these statements about the mean and the standard error of a difference between means are true regardless of the form of the parent distributions. However, the form of the sampling distribution can also be specified under either of two conditions:

If the distribution for each of the two populations is normal then the distribution of differences between sample means is normal.

This is a direct consequence of Principle C2.8, Appendix C. When we can assume both populations normal, the form of the sampling distribution is known to be *exactly* normal.

On the other hand, one or both of the original distributions may not be normal; in this case the central limit theorem comes to our aid:

As both N_1 and N_2 grow infintely large, the sampling distribution of the difference between means approaches a normal distribution, regardless of the form of the original distributions.

In short, when we are dealing with two very large samples, then the question of the form of the original distributions becomes irrelevant, and we can approximate the sampling distribution of the difference between means by a normal distribution.

Thus, when sample size is large, the test statistic for a difference between means is

$$z_{\text{diff.}} = \frac{(M_1 - M_2) - E(M_1 - M_2)}{\text{est. } \sigma_{\text{diff.}}}. \qquad [8.7.5^*]$$

This value may be referred to a normal distribution. The expected difference depends upon the hypothesis tested, and the estimated $\sigma_{\text{diff.}}$ is found directly from the estimate σ_M^2 for each sample by 8.7.4.

The exact hypothesis actually tested is of the form

$$H_0: \mu_1 - \mu_2 = k,$$

where k is any difference of interest. Quite often, the experimenter is interested only in $k = 0$, but it is entirely possible to test any other meaningful difference value. The alternative hypothesis may be directional,

$$H_1: \mu_1 - \mu_2 > k$$

or

$$H_1: \mu_1 - \mu_2 < k,$$

or nondirectional,

$$H_1: \mu_1 - \mu_2 \neq k,$$

depending on the form of the original question.

When both samples are large, confidence limits are found exactly for a single mean, except that $(M_1 - M_2)$ and est. $\sigma_{\text{diff.}}$ are substituted for M and est. σ_M respectively. Thus, 95 percent confidence limits for a difference with large samples are

$$M_1 - M_2 - 1.96 \text{ (est. } \sigma_{\text{diff.}})$$
$$M_1 - M_2 + 1.96 \text{ (est. } \sigma_{\text{diff.}}). \qquad [8.7.6^*]$$

8.8 / AN EXAMPLE OF A LARGE-SAMPLE SIGNIFICANCE TEST FOR A DIFFERENCE BETWEEN MEANS

An experimenter working in the area of motivational factors in perception was interested in the effects of deprivation upon the perceived size of objects. Among the studies carried out was one done with orphans, who were compared with nonorphaned children on the basis of the judged size of parental figures viewed at a distance. Each child was seated at a viewing apparatus in which cutout figures appeared. Each figure was actually of the same size and at the same distance from the viewer, although the children were unaware all the figures had the same size. A device was provided on which the child could actually judge the apparent sizes of the different figures in numerical terms. Several of the figures in the set viewed were obviously parents, whereas others were more or less neutral, such as milkmen, postmen, nurses, and so on. Each child was given a score, which was itself a difference in average judged size of parental and nonparental figures.

Now two independent randomly selected groups were used. Sample 1 was a group of orphaned children without foster parents. Sample 2 was a group of children having a normal family with both parents. Both populations of children sampled showed the same age level, sex distribution, educational level, and so forth.

The question asked by the experimenter was, "Do deprived children tend to judge the parental figures relatively larger than do the nondeprived?" In terms of a null and alternative hypothesis,

$$H_0: \mu_1 - \mu_2 \leq 0$$
$$H_1: \mu_1 - \mu_2 > 0.$$

The α level for significance decided upon was .05. The actual results were

Sample 1	*Sample 2*
$M_1 = 1.82$	$M_2 = 1.61$
$s_1 = .7$	$s_2 = .9$
$N_1 = 125$	$N_2 = 150$

These sample sizes are rather large, and the experimenter felt safe in using the normal approximation to the sampling distribution, even though she had no idea about the distribution form for the two populations sampled. The ratio used was

$$\frac{(M_1 - M_2) - E(M_1 - M_2)}{\text{est. } \sigma_{\text{diff.}}}$$

In this problem, $E(M_1 - M_2) = 0$, under the hypothesis tested. It was obviously necessary for the experimenter to estimate the standard error of the difference, since both σ_1 and σ_2 were unknown to her. This estimate was found by first estimating $\sigma^2_{M_1}$ and $\sigma^2_{M_2}$:

$$\text{est. } \sigma^2_{M_1} = \frac{s_1^2}{N_1} = \frac{.49}{125} = .004$$

$$\text{est. } \sigma^2_{M_2} = \frac{s_2^2}{N_2} = \frac{.81}{150} = .005.$$

Then,

$$\text{est. } \sigma_{\text{diff.}} = \sqrt{\text{est. } \sigma^2_{M_1} + \text{est. } \sigma^2_{M_2}} = \sqrt{.004 + .005} = .095.$$

On making these substitutions, the experimenter found

$$t = \frac{1.82 - 1.61}{.095} = 2.21.$$

The rejection region implied by the alternative hypothesis is on the *upper* tail of the sampling distribution. For a normal distribution the upper 5 percent is bounded by $z = 1.65$. Thus, the result is significant; deviations this far from zero have a probability of less than .05 of occurring by chance alone when the true difference is zero.

The experimenter may conclude that a difference exists between these two populations, *if* an α value less than .05 is a small enough probability of error to warrant this decision. However, the experimenter does not necessarily conclude that parental deprivation causes an increase in perceived size. The statistical conclusion suggests that it *might* be safe to assert that a particular direction of numerical difference exists between the mean scores of the two populations of children, but the statistical result is absolutely noncommittal about the reason for this difference, if such exists. The experimenter here takes the step of advancing a reason at her own peril. The statistical test as a mathematical tool is absolutely neutral about what these numbers measure, the level of measurement, what was or was not represented in the experiment, and, most of all the cause of the experimenter's particular finding. As always, the test takes the numerical values as given, and cranks out a conclusion about the conditional probability of such numbers, given certain statistical conditions.

For this example; the 95 percent limits are

$$.21 - 1.96(.095)$$

$$.21 + 1.96(.095)$$

or .024 and .396. Notice that since the value $\mu_1 - \mu_2 = 0$ does not fall within these values this value can be rejected as a hypothesis beyond the .05 level (two-tailed).

8.9 / USING THE *t* DISTRIBUTION TO TEST HYPOTHESES ABOUT DIFFERENCES

Given the assumption that both populations sampled have normal distributions, any hypothesis about a difference can be tested using the t distribution, regardless of sample size. Here, for the null hypothesis

$$H_0: \mu_1 - \mu_2 = k$$

the test is

$$t = \frac{(M_1 - M_2) - k}{\text{est. } \sigma_{\text{diff}}}. \qquad [8.9.1^*]$$

However, one additional assumption becomes necessary: *in order to use the t distribution for tests based on two (or more) samples, one must assume that the standard deviations of both (or all) populations are equal.* The basis for this assumption will be discussed in the next chapter.

Given these assumptions, then the distribution of t for a difference has the same form as for a single mean, except that the degrees of freedom are

$$\nu = N_1 - 1 + N_2 - 1 = N_1 + N_2 - 2.$$

When samples are drawn from populations with equal variance, then the estimated standard error of a difference takes a somewhat different form. First of all, when $\sigma_1 = \sigma_2 = \sigma$,

$$\sigma_{\text{diff.}} = \sqrt{\frac{\sigma^2}{N_1} + \frac{\sigma^2}{N_2}} = \sqrt{\sigma^2\left(\frac{1}{N_1} + \frac{1}{N_2}\right)}. \qquad [8.9.2^*]$$

Now, as we showed in Section 5.10, when one has two or more estimates of the same parameter σ^2 the *pooled* estimate is actually better than either one taken separately. From 5.10.4 it follows that

$$\text{est.}\sigma^2 = \frac{(N_1 - 1)s_1^2 + (N_2 - 1)s_2^2}{N_1 + N_2 - 2} \qquad [8.9.3]$$

is our best estimate of σ^2 based on the two samples. Hence

$$\begin{aligned} \text{est. } \sigma_{\text{diff.}} &= \sqrt{\text{est. } \sigma^2\left(\frac{1}{N_1} + \frac{1}{N_2}\right)} \\ &= \sqrt{\left(\frac{(N_1 - 1)s_1^2 + (N_2 - 1)s_2^2}{N_1 + N_2 - 2}\right)\left(\frac{N_1 + N_2}{N_1 N_2}\right)}. \end{aligned} \qquad [8.9.4^*]$$

This estimate of the standard error of the difference ordinarily forms the denominator of the t ratio when the t distribution is used for hypotheses about a difference.

Let us imagine that two random samples of subjects are being compared on the basis of their scores on a motor learning task. The subjects are allotted to two experimental groups, with 5 in the first group and 7 in the other. In group 1 a subject is rewarded for each correct move made, and in the second each incorrect move is punished. The score is the number of errors made in a fixed set of trials. The experimenter wishes to find evidence for the question, "Does the kind of motivation employed, reward or punishment, affect the performance?" This question implies the null and alternative hypotheses:

$$H_0: \mu_1 - \mu_2 = 0$$

$$H_1: \mu_1 - \mu_2 \neq 0.$$

The experimenter is willing to assume that the population distributions of scores are normal, and that the population variances are equal. The probability of Type I error decided upon is .01. Since this is a two-tailed test, a glance at Table III shows that for $N_1 + N_2 - 2$ or $5 + 7 - 2 = 10$ degrees of freedom, and for $2Q = .01$, the required t value is 3.169. Thus an obtained t ratio equaling or exceeding 3.169 in absolute value is grounds for rejecting the hypothesis of no difference between population means.

The sample results are

$$M_1 = 18 \qquad M_2 = 20$$
$$s_1^2 = 7.00 \qquad s_2^2 = 5.83$$

The estimated standard error of the difference is found by the pooling procedure given by 8.9.4

$$\begin{aligned} \text{est. } \sigma_{\text{diff.}} &= \sqrt{\text{est. } \sigma^2 \left(\frac{1}{N_1} + \frac{1}{N_2}\right)} \\ &= \sqrt{\frac{(4)(7) + (6)(5.83)}{10}\left(\frac{12}{35}\right)} \\ &= \sqrt{2.16} = 1.47. \end{aligned}$$

Thus, the t ratio is

$$t = \frac{(M_1 - M_2) - E(M_1 - M_2)}{\text{est. } \sigma_{\text{diff.}}} = \frac{-2}{1.47} = -1.36.$$

This value comes nowhere close to that required for rejection, and thus if α must be no more than .01 the experimenter does not reject the null hypothesis. The best choice may be to suspend judgment, pending more evidence.

Confidence intervals are found just as for a single small-sample mean: the limits are

$$\begin{aligned} &(M_1 - M_2) - t_{(\alpha/2;\nu)} (\text{est. } \sigma_{\text{diff.}}) \\ &(M_1 - M_2) + t_{(\alpha/2;\nu)} (\text{est. } \sigma_{\text{diff.}}). \end{aligned} \qquad [8.9.5^*]$$

For this example, the 99 percent limits are

$$-2 - (3.169)(1.47)$$

and

$$-2 + (3.169)(1.47)$$

or approximately -6.66 to 2.66. The probability is .99 that true $\mu_1 - \mu_2$ is covered by an interval such as this. Once again, notice that this interval *does* contain the value 0, indicating the hypothesis entertained above is not rejected.

8.10 / THE IMPORTANCE OF THE ASSUMPTIONS IN A *t* TEST OF A DIFFERENCE BETWEEN MEANS

The rationale for a t test of a difference between the means of two groups rests upon two assumptions: the populations each have a normal distribution, and each population has the same variance, σ^2. On the other hand, in practical situations these assumptions are sometimes violated with rather small effect on the conclusions.

The first assumption, that of a normal distribution in the populations, is apparently the less important of the two. So long as the sample size is even moderate for each group quite severe departures from normality seem to make little practical difference in the conclusions reached. Naturally, the results are more accurate the more nearly unimodal and symmetric the population distributions are, and thus if one suspects radical departures from a generally normal form then he should plan on larger samples. Furthermore, the departure from normality can make more difference in a one-tailed than in a two-tailed result, and once again some special thought should be given to sample size when one-tailed tests are contemplated for such populations. By and large, however, this assumption may be violated almost with impunity provided that sample size is not extremely small.

The assumption of homogeneity of variance is more important. In older work it was often suggested that a separate test for homogeneity of variance be carried out before the t test itself, in order to see if this assumption were at all reasonable. However, the most modern authorities suggest that this is not really worth the trouble involved. In circumstances where they are needed most (small samples), the tests for homogeneity are poorest. Furthermore, for samples of equal size relatively big differences in the population variances seem to have relatively small consequences for the conclusions derived from a t test. Even so, when the variances are quite unequal the use of different sample sizes can have serious effects on the conclusions. The moral should be plain: given the usual freedom about sample size in experimental work, *when in doubt use samples of the same size.*

However, sometimes it is not possible to obtain an equal number in each group. Then one way out of this problem is by the use of a correction in the value for degrees of freedom. This is useful when one cannot assume equal population variances and samples are of different size. In this situation, the t ratio is calculated as in Section 8.7, where the separate standard errors are computed from each sample and the pooled estimate is not made. Then the corrected number of degrees of freedom is found from

$$\nu = \frac{(\text{est. } \sigma^2_{M_1} + \text{est. } \sigma^2_{M_2})^2}{(\text{est. } \sigma^2_{M_1})^2/(N_1 + 1) + (\text{est. } \sigma^2_{M_2})^2/(N_2 + 1)} - 2. \qquad [8.10.1]$$

This need not result in a whole value for ν, in which case the use of the nearest whole value for ν is sufficiently accurate for most purposes. When somewhat greater accuracy is desired, the approximate formula for critical values of t given in Section 12.12 is useful. When both samples are quite large, then both the assumptions of normality and of homogeneous variances become relatively unimportant, and the method of Section 8.7 can be used.

8.11 / THE POWER OF t TESTS

The idea of the power of a statistical test was discussed in the preceding chapter only in terms of the normal distribution. Nevertheless, the same general considerations apply to the power of tests based on the t distribution. Thus, the power of a t test increases with sample size, increases with the discrepancy between the null hypothesis value and the true value of a mean or a difference, increases with any reduction in the true value of σ, and increases with any increase in the size of α, given a true value covered by H_1.

Unfortunately, the actual determination of the power for a t test against any given true alternative is more complicated than for the normal distribution. The reason is that when the null hypothesis is false, each t ratio computed involves $E(M)$ or $E(M_1 - M_2)$, which is the exact value given by the null (and false) hypothesis. If the true value of the expectation could be calculated into each t ratio, then the distribution would follow the t function tabled in the appendix. However, when H_0 is false, each t value involves a false expectation; this results in a somewhat different distribution, called the **noncentral *t* distribution.** The probabilities of the various t's cannot be known unless one more parameter, δ, is specified beside ν. This is the so-called noncentrality parameter, defined by

$$\delta = \left| \frac{\mu - \mu_0}{\sigma_M} \right|. \qquad [8.11.1]$$

The parameter δ expresses the absolute difference between the true expectation μ and that given by the null hypothesis, or μ_0, in terms of σ_M. For a hypothesis about a difference, tested by independent samples,

$$\delta = \left| \frac{(\mu_1 - \mu_2) - (\mu_{0_1} - \mu_{0_2})}{\sigma_{\text{diff.}}} \right|. \qquad [8.11.2]$$

Unlike the regular, central, t distribution, the distribution of noncentral t is rather difficult to show in a concise table, depending as it does on the two parameters ν and δ. However, this distribution is the basis for the power graphs for t that appear in some advanced texts in statistics. The power for a t test can also be approximated. One simple way to approximate the power of a t test is as follows: first, some alternative value of the population mean, or μ_1, must be specified. Then the power of a t test of $\mu_0 = k$ is approximated from

$$\text{power} = 1 - \text{prob.} \left\{ z \leq \left(\frac{t' - \delta}{\sqrt{1 + \frac{(t')^2}{2\nu}}} \right) \right\} \qquad [8.11.3]$$

where δ is estimated by taking

$$\text{est. } \delta = |(\mu_1 - \mu_0)/\text{est. } \sigma_M|$$

and where t' represents the upper critical value of t for ν degrees of freedom, in a two-tailed test where α has been specified. After calculating the ratio to the right of the inequality, this value is referred to a normal distribution, where the probability of a z-value less than or equal to this number is found. This probability is the approximate value of β. Then the power is found by subtracting this probability from 1.00.

This same procedure can be applied to estimating the power of t for a hypothesis about a difference as well. Here we take

$$\text{est. } \delta = \left| \frac{\text{difference under } H_0 - \text{difference under } H_1}{\text{est. } \sigma_{\text{diff.}}} \right|. \qquad [8.11.4]$$

The critical value of t, or t', must be found for $N_1 + N_2 - 2$ degrees of freedom, of course.

As an example of such a power calculation involving a difference between means, suppose that there were two independent samples, each containing 35 independent observations drawn at random.

The hypothesis to be tested is

$$H_0: \mu_1 - \mu_2 = 0$$

against the alternative

$$H_1: \mu_1 - \mu_2 \neq 0.$$

The .01 level is to be used for α.

What is the power of this test if the alternative hypothesis, $\mu_1 - \mu_2 = 4$, turns out to be true? Suppose that it is found that $s_1^2 = 19$ and $s_2^2 = 17$. Then

$$\text{est. } \sigma_{\text{diff.}} = \sqrt{\frac{(34)(19) + (34)(17)}{34 + 34}\left(\frac{2}{35}\right)} = 1.01.$$

Then we estimate δ by taking

$$\delta = \left|\frac{4}{1.01}\right| = 3.96.$$

The critical value of t in this test is read from Table III to be about

$$t' = 2.66,$$

since 68 degrees of freedom is closest to the tabled value for 60. Then

$$\begin{aligned} \text{power} &= 1 - \text{prob}\left\{z \leqslant \left(\frac{2.66 - 3.96}{\sqrt{1 + \dfrac{(2.66)^2}{2(68)}}}\right)\right\} \\ &= 1 - \text{prob}\{z \leqslant -1.26\} \\ &= .896. \end{aligned}$$

The power of this test, when the true difference is 4, should be about .90. The power against any other true alternative can be evaluated in this same way, of course.

The big difficulty with this or any similar power calculation is in knowing the value of σ_M or of $\sigma_{\text{diff.}}$. Of course, when sample size is fairly large, one can use estimates of these values, as we have done here. On the other hand, it is common to want a power calculation before a sample is drawn, as an aid in determining sample size itself. Power calculations or the use of power tables become difficult under these circumstances, if one does not already have a value for the standard error. An alternative approach will now be explored.

8.12 / STRENGTH OF ASSOCIATION

When an experimenter assigns subjects to two experimental groups, and proceeds to give each a different treatment, he or she is generally looking for evidence of a statistical relation. That is, the question asked is basically of the form, "If I do something different to these two groups in one way, is this reflected in a difference between the groups in another way?" The manipulation that is under the control of the experimenter is called the *independent variable*. The performance of the two groups that is not under the experimenter's control is called the *dependent variable*. For notational convenience, let us called the independent variable (which may be either a number, or a qualitative label) the variable X.

We symbolize the dependent variable, which usually is a number, by Y. The question then becomes, "Is X related to Y?"

Evidence for a statistical relation exists when different X treatments lead to different *expectations* about Y. When the expectation of Y is different, depending on which X treatment the group in question receives, then we can say that a *statistical relation,* or *association,* exists between X and Y. On the other hand, if we are led to expect the same Y value, regardless of which X any group receives, then we say that there is *no association*.

Now suppose that you were asked to guess the Y score of an individual drawn at random from one of the two groups given different X treatments. Obviously, you would not have a very good idea about what to guess. However, as we have seen before, when in doubt it is probably wisest to guess the mean of all of the Y values. Then at least your error will be zero on the average.

If you are guessing the mean of all Y scores for any single individual, how "off" should your guess tend to be? Recall that the measure of spread about a mean is the variance. Let us designate this variance about the mean Y score for all of the cases as σ_Y^2. We can say that this value, σ_Y^2, is a measure of our *uncertainty* about the Y scores in general.

Now suppose that you were asked to guess a Y score once again, but this time you are given the X group to which that individual belongs. Then what should you guess? The best guess is again a mean, but this time it is *the mean Y value of all individuals within that same X grouping*. Very likely you will still not be exactly right in your guess about the score of an individual in this X group, and the spread about the mean Y value, given X, will be a reflection of how "off" your guess will tend to be. Let us designate this variance about the mean Y score within an X group by $\sigma_{Y|X}^2$, the variance of Y *given* X. Then $\sigma_{Y|X}^2$ can be said to reflect your uncertainty about the Y scores even given the information about X.

In general, we will say that *the strength of a statistical relation is reflected in the extent to which knowing X reduces our uncertainty about Y*. That is, since our uncertainty about Y not knowing X is proportional to σ_Y^2, and our uncertainty about Y X is proportional to $\sigma_{Y|X}^2$, then the *reduction in uncertainty* about Y afforded by X should be based on

$$\sigma_Y^2 - \sigma_{Y|X}^2.$$

We can say that a statistical relation exists when the variability of Y given X is smaller than the variability of Y in general.

It is convenient to turn this reduction in uncertainty into a **relative reduction** by dividing by σ_Y^2, giving

$$\omega^2 = \frac{\sigma_Y^2 - \sigma_{Y|X}^2}{\sigma_Y^2}. \qquad [8.12.1^*]$$

The relative reduction in uncertainty about Y given by X is shown by the index ω^2 (Greek omega, squared). Sometimes the value ω^2 is called **the proportion of variance in Y accounted for by X**. Viewed either as a relative reduction in uncertainty, or as a proportion of variance accounted for, the index ω^2 represents the strength of association between independent and dependent variables. (The index ω^2 is almost identical to two other indices to be introduced later, *the intraclass correlation* and the *correlation ratio,* usually represented by the symbols ρ_I and η^2 respectively. However, since these indices were developed for and are used in somewhat different contexts, it seems better to use the relatively neutral symbol ω^2 here, to avoid later confusion.)

This index reflects the predictive power afforded by a relationship: when ω^2 is zero, then X does not aid us at all in predicting the value of Y. On the other hand, when ω^2 is 1.00, this tells us that X lets us know Y exactly. All intermediate values of the index represent different degrees of predictive ability. Notice that for any precise functional relation, $\omega^2 = 1.00$, since there can be only one Y for each possible X. A value less than unity tells us that precise prediction is not possible, although X nevertheless gives *some* information about Y unless $\omega^2 = 0$.

Let us return to the problem of guessing scores drawn from two different X groups or populations. Suppose that you are equally likely to have to guess a Y value from group 1 as from group 2. Furthermore, suppose that the value of the variance within group 1, σ_1^2, is exactly the same as the variance within group 2, or σ_2^2. Let us also designate the mean of group 1 by μ_1 and the mean of group 2 by μ_2. Then it will be true that

$$\sigma_Y^2 = \sigma_{Y|X}^2 + \frac{(\mu_1 - \mu_2)^2}{4} \qquad [8.12.2]$$

where μ_1 is the mean of population 1, μ_2 that of population 2, and

$$\frac{\mu_1 + \mu_2}{2} = \mu,$$

the mean of the marginal distribution (Appendix C).

On substituting into 8.12.1 we find

$$\omega^2 = \frac{(\mu_1 - \mu_2)^2}{4\sigma_Y^2}. \qquad [8.12.3^*]$$

For two treatment populations with equal variances, the strength of the statistical association between treatment and dependent variable varies directly with the squared difference between the population means, relative to the unconditional, total variance of Y. This connection between the strength of association index ω^2 and the difference between population means provides another approach to estimating the power of a t test. This will be examined in the next section.

8.13 / STRENGTH OF ASSOCIATION, POWER, AND SAMPLE SIZE

As we have just seen, there is a very close connection between the value of ω^2 in the populations and the difference between the two populations means. Thus, it is possible to rewrite the usual hypothesis about a difference between two means

$$H_0: \mu_1 - \mu_2 = 0$$

in terms of a hypothesis about strength of association:

$$H_0 : \omega^2 = 0$$

Furthermore, the alternative hypothesis

$$H_1 : \mu_1 - \mu_2 \neq 0$$

is equivalent to

$$H_1 : \omega^2 > 0.$$

In addition, the parameter δ, necessary for us to specify before we can calculate the power of a t test, can be stated in terms of ω^2. That is,

$$\delta = \sqrt{\frac{2N\omega^2}{1 - \omega^2}}. \qquad [8.13.1]$$

This can be a distinct advantage, since we seldom know values such as σ_Y^2 and $\sigma_{Y|X}^2$, especially in advance of a sample, when power calculations are often most useful. However we can preset the power of a t test in terms of the ω^2 value we want to be sure to detect as a significant result.

If you wish to calculate the power of a t test, given that a certain value of ω^2 is true, the methods of the preceding section give a reasonable approximation. However, the value of δ is found from 8.13.1, rather than from 8.11.4.

An even more important situation occurs when we wish to choose a sample size that will give our test a certain power, given that a particular ω^2 value is true in the population. Actually, a rough and ready way exists to determine sample size, once power against a certain ω^2 value is specified. This method will now be described.

Suppose that two experimental groups of equal size are to be chosen, in order to test the hypothesis that the means of two populations are equal (or, alternatively, that ω^2 is equal to zero). A two-tailed test is to be carried out, with δ equal to one of the conventional values such as .05 or .01.

Suppose further that we want to be very sure that if the true strength of association is at least some value of ω^2, we will get a significant result some sizable proportion (or $1 - \beta$) of the time. In other words, we want to set the power at some given level, given that some specific value of ω^2 is true.

Then a first approximation to the total size of sample required *in each group* can be obtained from

$$N \geqslant \frac{[z_{(1-\beta)} - z_{(\alpha/2)}]^2(1 - \omega^2)}{2\omega^2}, \qquad [8.13.2]$$

where $z_{(1-\beta)}$ cuts off the lower $(1 - \beta)$ proportion of a normal distribution, and $z_{(\alpha/2)}$ cuts off the lower $\alpha/2$ proportion. This gives a rough approximation to N, the total number of cases required. If N is a decimal number, the next largest whole number is taken. Furthermore, since this estimate of N will tend to be slightly small, it is wise to add one more case to each group, or two more cases to the total size of the sample.

The ease of applying this rule can be illustrated by an example. Imagine that we plan to test the difference between the means of two independent groups by use of t, where the α value of .01 will be employed. We would like, say, the probability to be about .90 of rejecting H_0 of no difference if the true value of ω^2 is at least .25. Then, from the normal table we find

$$z_{(1-\beta)} = z_{(.90)} = 1.28$$

$$z_{(\alpha/2)} = z_{(.005)} = -2.58$$

and

$$N \geqslant \frac{(1.28 + 2.58)^2(.75)}{2(.25)}$$

or

$$N \geqslant 22.35.$$

Since the estimate is 22.35, we then know that 23 cases in each group should be in the right ballpark. However, 24 cases in each group will be used. This should make the power be about .90 when the true value of ω^2 is .25 or more.

The required sample size can be estimated in this way for any other combination of α, power, and ω^2. For a one-tailed test, z_α may be substituted for $z_{(\alpha/2)}$, of course. Such a rough and ready procedure should be adequate for most practical purposes. However, if greater precision in finding N is required, one can explore various powers obtained for a given ω^2 by use of 8.11.3 above, or one can look into power tables such as those given in Winer (1971) and other books on experimental design.

8.14 / STRENGTH OF ASSOCIATION AND SIGNIFICANCE

When the difference $\mu_1 - \mu_2$ is zero, then ω^2 must be zero. In the usual t test for a difference, the hypothesis of no difference between means is equivalent to the hypothesis that $\omega^2 = 0$. On the other hand, when there is any difference at all between population means, the value of ω^2 must be greater than 0.

A true difference is ''big'' in the sense of predictive power only if the square of that difference is large relative to σ_Y^2. However, in significance tests such as t, we compare the difference we get with an estimate of $\sigma_{\text{diff.}}$. The standard error of the difference can be made almost as small as we choose if we are given a free choice of sample size. Unless sample size is specified, there is no *necessary* connection between significance and the true strength of associaton.

Virtually any study can be made to show significant results if one uses enough subjects, regardless of how small ω^2 may be. There is surely nothing on earth that is completely independent of anything else. The strength of an association may approach zero, but it should seldom or never be exactly zero. If one applies a large enough sample of the study of any relation, trivial or meaningless as it may be, sooner or later a signficant result will almost certainly be achieved. Such a result may be a valid finding, but only in the sense that one can say with assurance that some association is not exactly zero. The degree to which such a finding enhances our knowledge is debatable. If the criterion of strength of association is applied to such a result, it becomes obvious that little or nothing is actually contributed to our ability to predict one thing from another.

This kind of problem occurs when people pay too much attention to the significance test and too little to the degree of statistical association the finding represents. This clutters up the literature with findings that are often not worth pursuing, and which serve only to obscure the really important predictive relations that occasionally appear. The serious scientist should ask not only, "Is there any association between X and Y?" but also, "How much does my finding suggest about the power to predict Y from X?" Much too much emphasis is paid to the former, at the expense of the latter, question.

Can sample size be too large? In one sense, even posing this question sounds like heresy! Social scientists are usually trained to think that large samples are good things, and there is much to support this notion. As we have seen, the most elegant features within theoretical statistics are the limit theorems, each implying a connection between sample size and the goodness of inference.

On the other hand, samples are expensive in terms of time, effort, and money. The experimenter purchases information through the sample, and it may be that too high a price is paid for the information received.

It seems reasonable that sample size, and the consequent investment in the experiment or study, can never really be discussed apart from what the experimenter is trying to do and the stakes involved. So long as the experimenter's primary purpose is in precise estimation, then the larger the sample the better. When we want to come as close as possible to the true parameter values we can always do better by increasing sample size.

This is not, however, the purpose of many studies. These studies are, in the strict sense, exploratory, where the main relationships in some area are to be mapped out. The study serves as a guide for directions that will be pursued in other, perhaps more refined, explorations of the problem. Of main concern are statistical relations that are relatively large and that give considerable promise that a more or less precise relationship is there to be discovered and refined. The experimenter does not want to waste time, effort, and funds by concluding that an association exists when the degree of prediction afforded by that association is negligible.

When this is the situation it is advisable to look into the effects of sample size on the probability of finding a significant result given a *weak* association. Trivial associations may well show up as signficant results when the sample size is very large. If the experimenter wants significance to reflect a sizable association in the data, and also wants to be sure not to be led by a signficant result into some blind alley, then attention should be paid to both aspects of sample size. Is the sample size *large* enough to give confidence that the big associations will indeed show up, while being *small* enough so that trivial associations will be excluded from significance?

8.15 / ESTIMATING THE STRENGTH OF A STATISTICAL ASSOCIATION FROM DATA

It is quite possible to estimate the amount of statistical association implied by any obtained difference between two independent means. The ingredients for this kind of estimation are essentially those used in a t test.

In the first place, a serviceable statistic for *the proportion of variance accounted for in this sample* may be had by taking

$$\frac{t^2}{t^2 + N_1 + N_2 - 2} = \text{proportion of variance in } Y \text{ accounted for by } X. \qquad [8.15.1]$$

However, such sample values tend to be somewhat biased as representations of ω^2, the proportion of variance accounted for in the population.

(A number of ways have been proposed for estimating the strength of a statistical association from obtained differences between means. For reasons to be elaborated later, none of these methods is entirely satisfactory. The method to be introduced here is thus only one of the ways that may be encountered in the statistical literature, but it seems to have as much to recommend it as any other.)

For samples from two populations, each of which has the same true variance, $\sigma^2_{Y|X}$, a rough estimate of ω^2 is provided by

$$\text{est. } \omega^2 = \frac{t^2 - 1}{t^2 + N_1 + N_2 - 1}. \qquad [8.15.2]$$

(A more general form for estimating ω^2 will be given in Chapter 10.) Notice that if t^2 is less than 1.00, then this estimate is negative, although ω^2 cannot assume negative values. In this situation the estimate of ω^2 is set equal to zero.

Let us consider an example using this estimate. Imagine a study involving two groups of 30 cases each. Subjects are assigned at random to these two groups, and each set of subjects is given a different treatment. The results are

Group 1	*Group 2*
$M_1 = 65.5$	$M_2 = 69$
$s_1^2 = 20.69$	$s_2^2 = 28.96$
$N_1 = 30$	$N_2 = 30$

First of all the t ratio is computed in the usual way (Section 8.9):

$$\text{est. } \sigma^2 = \frac{(29)(20.69 + 28.96)}{58} = 24.83$$

and

$$\text{est. } \sigma_{\text{diff.}} = \sqrt{\frac{24.83(2)}{30}} = 1.29.$$

Thus,

$$t = \frac{65.5 - 69}{1.29} = -2.71.$$

For a two-tailed test with 58 degrees of freedom, this value is significant beyond the .01 level. Thus, we are fairly safe in concluding that some association exists.

What do we estimate the true degree of association to be? Substituting into 8.15.2, we find

$$\text{est. } \omega^2 = \frac{(2.71)^2 - 1}{(2.71)^2 + (60 - 1)} = .096.$$

Our rough estimate is that X (the treatment administered) accounts for about 10 percent of the variance of Y (the obtained score). By way of contrast, we find that the proportion of variance accounted for *in this sample* is slightly larger, or .11.

Suppose, however, that the groups had contained only 10 cases each, and that the results had been:

Group 1	*Group 2*
$M_1 = 65.5$	$M_2 = 69$
$s_1^2 = 5.55$	$s_2^2 = 7.78$
$N_1 = 10$	$N_2 = 10$

Here

$$\text{est. } \sigma^2 = \frac{9(5.55 + 7.78)}{18} = 6.67$$

$$\text{est. } \sigma_{\text{diff.}} = \sqrt{6.67\left(\frac{2}{10}\right)} = 1.15$$

so that

$$t = \frac{-3.5}{1.15} = -3.04.$$

For 18 degrees of freedom, this value is also significant beyond the .01 level (two-tailed), and once again we can assert with confidence that some association exists.

Again, we estimate the degree of association represented by this finding:

$$\text{est. } \omega^2 = \frac{(3.04)^2 - 1}{(3.04)^2 + 19} = .29.$$

Here, our rough estimate is that X accounts for about 29 percent of the variance in Y, Even though the difference between the sample means is the same in these two examples, and both results are significant beyond the .01 level, the second experiment gives a much higher estimate of the true association than the first.

8.16 / PAIRED OBSERVATIONS

Sometimes it happens that subjects are actually sampled in pairs. Even though each subject is experimentally different in one respect (nominally, the independent variable) from the pair-mate and each has some distinct dependent variable score, the scores of the members of a pair are not necessarily independent. For instance, one may be comparing scores of husbands and wives; a husband is ''naturally'' matched with his wife, and it makes sense that knowing the husband's score gives us some information about his wife's, and vice versa. Or individuals may be matched on some basis by the experimenter, and within each matched pair the members are assigned at random to experimental treatments. This matching of pairs is one form of experimental control, since each member of each experimental group must be identical (or nearly so) to the pair-mate in the other group with respect to the matching factor or factors, and thus the factor or factors used to match pairs is less likely to be responsible for any observed difference in the groups than if two unmatched groups are used.

Given two groups matched in this pairwise way, either by the experimenter or otherwise, it is still true that the difference between the means is an unbiased estimate of the population difference (in two matched populations):

$$E(M_1 - M_2) = \mu_1 - \mu_2. \qquad [8.16.1]$$

However, the matching, and the consequent *dependence* within the pairs, changes the standard error of the difference. Thus, for matched pairs

$$\sigma^2_{\text{diff.}} = \sigma^2_{M_1} + \sigma^2_{M_2} - 2\ \text{cov.}(M_1, M_2). \qquad [8.16.2^*]$$

The first of these terms is just $\sigma^2_{M_1}$, and the second is $\sigma^2_{M_2}$. However, what of the third term? Let us denote this last term above as cov.(M_1,M_2), the **covariance** of the means. (The covariance of two random variables is defined in Appendix B, following Rule 9.)

From Rule 9 Appendix B we find that the value of the covariance must be zero when the variables are independent. On the other hand, when variables are dependent the expectation is *not* ordinarily zero. In general, for groups matched by pairs, this covariance is a positive number, and thus the variance and standard error of a difference between means will usually be *less* for matched than for unmatched groups. This fact accords with the experimenter's purpose in matching in the first place: to remove one or more sources of variability, and thus to lower the sampling error.

The unknown value of cov.(M_1,M_2) could be something of a problem, but actually it is quite easy to bypass this difficulty altogether. Instead of regarding this as two samples, we simply think of the data coming from one sample of *pairs*. Associated with each pair i is a difference

$$D_i = (y_{i1} - y_{i2}), \qquad [8.16.3]$$

where y_{i1} is the score of the member of pair i who is in group 1, and y_{i2} is the score of the member of pair i who is in group 2. Then an ordinary t test for a *single* mean is carried out using the scores D_i. That is,

$$M_D = \frac{\sum_i D_i}{N} \qquad [8.16.4]$$

and

$$s_D^2 = \frac{\sum_i D_i^2}{N-1} - \frac{N(M_D)^2}{N-1}. \qquad [8.16.5]$$

Then

$$\text{est. } \sigma_{M_D} = \frac{s_D}{\sqrt{N}} \qquad [8.16.6]$$

and t is found from

$$t = \frac{M_D - E(M_D)}{\text{est. } \sigma_{M_D}} \qquad [8.16.7^*]$$

with $N - 1$ degrees of freedom. *Be sure to notice that here N stands for the number of differences, which is the number of pairs.* *

This method turns the test of a hypothesis about two matched populations into the test of an exact hypothesis about a population of pairs. Any of the usual hypotheses about a single mean may be tested. Although the value stated in the hypothesis usually amounts to $E(M_D) = \mu_D = 0$, any other exact value could be used. The test may be either one- or two-tailed.

Matching pairs for an experiment or sampling a population of pairs is very common. Some caution must be exercised in this matching process, however. In the first place, it can happen that the factor on which subject pairs are matched is such that the means are *negatively* related. Thus, for example, suppose that one had an effective measure of the dominance of personality of an individual. It just might be that highly dominant women tend to marry men with low dominance, and vice versa, so that among husband-wife pairs, dominance scores are negatively related. Then, if our interest is basically that of comparing men and women generally on such scores, it would be a mistake to match, since the negative relationship would lead to a larger, rather than a smaller, standard error of the difference than would a comparison of unmatched groups.

Furthermore, such matching may be less efficient than the comparison of unmatched random groups, unless the factor used in matching introduces a relatively strong positive relationship between the means. While a positive relationship, reflected in a positive covariance term, does reduce the standard error of the difference, this procedure also *halves* the number of degrees of freedom. Dealing with a sample of N pairs gives only half the number of degrees of freedom available when we deal with two independent groups of N cases each. Thus, if the factor entering into the matching is only slightly relevant to the differences between the groups, or is even irrelevant to such differences, matching is not a desirable procedure. The experimenter should have quite good reasons for matching before adopting this procedure in preference to the simple comparison of two randomly selected groups.

8.17 / COMPARING MORE THAN TWO MEANS

In the social sciences it is quite common for an investigator to be interested in an independent variable X that is represented by more than two groupings. Then the dependent variable Y is to be compared among the several X groups. The question is, as always, "Is a difference in X status reflected in a difference in Y means?"

For example, a study might involve the marital status of women between the ages of 30 and 40 as the independent variable X. Three groupings are used, "never been married," "married," and "divorced or widowed." The dependent variable Y might be a measure such as "satisfaction with amount of personal independence." The idea is to see how these three marital groupings of women vary in their averages on this measure.

Obviously, a simple difference cannot be calculated for all three groups considered simultaneously. However, each pair of groups could be compared, and their difference tested by means of a statistic such as t. There would then be three such differences to be examined.

There is a serious problem in this approach, however. Suppose that the t test between groups 1 and 2 were carried out, using $\alpha = .05$. Then the probability of a Type I error should be .05, or about 5 in one hundred pairs of samples. Similarly the difference between groups 1 and 3 could be tested at the same α level, and so could the difference between groups 2 and 3.

In making t tests for all three differences, a Type I error *might* be made for none, for one, for two, or even for all three of the tests. What is the probability that we would be making *at least one* Type I error? If each test can be thought of as an independent trial of a Bernoulli process where $p = \alpha = .05$, then the binomial distribution shows that the probability of making *no* Type I errors in the three tests is

$$\text{prob(no errors)} = \binom{3}{0}(.05)^0(.95)^3 = .86.$$

Hence the probability of making at least one Type I error in these three tests is

$$\text{prob(one or more)} = 1.00 - .86 = .14.$$

In other words, in making three t tests, the probability of making at least one Type I error is almost three times as large as the probability of such an error for a single test, or $\alpha = .05$, even assuming that the tests are independent of each other, as we have done. In general, if we make some J independent *t tests, using* $\alpha = .05$ for each, then the probability of making at least one Type I error is

$$\text{probability one or more Type I errors} = 1 - (1 - \alpha)^J. \quad [8.17.1]$$

When J is large, this can be a large probability. For example, suppose that we have eight groups, and we examine all twenty-eight differences among them by means of t tests, for $\alpha = .05$. Given that the tests are independent of each other, there will be a probability of .76 for one or more Type I errors. For twenty groups this probability is almost one.

The problem is complicated further by the fact that tests of all differences among a set of means cannot be independent of each other, for reasons we will examine later. Thus, the actual probability of one or more Type I errors might be even worse than the values we calculate.

There is a principle from probability theory that gives us an upper bound for the probability that one or more such tests will lead to a Type I error. This is one of the so-called "Bonferroni inequalities" (Feller, 1968).

Essentially this says that if there are some J confidence intervals constructed simultaneously, we may think of the event C_j as meaning "the jth confidence interval covers the true mean." Then the event that all confidence intervals simultaneously cover their true means can be symbolized by $(C_1 \cap C_2 \cap \cdots \cap C_J)$ with probability $p(C_1 \cap C_2 \cap \cdots \cap C_J)$. The Bonferroni inequality says that if $(1 - \alpha)$ is the probability for each of the events C_j,

$$p(C_1 \cap C_2 \cap \cdots \cap C_J) \geq 1 - J\alpha.$$

This also implies that the probability that at least one interval *fails* to cover the true mean is

$$1 - p(C \cap C_2 \cap \cdots \cap C_J) \leq J\alpha.$$

(If $J\alpha \geq 1.00$, then the upper bound is set equal to 1.00, of course.) In terms of tests, this is equivalent to saying that the maximum probability for at least one Type I error is $J\alpha$. There is also a practical implication: If we have a series of confidence intervals or corresponding tests, and if we want the probability of making at least one Type I error to be *no larger* than α, then we can set the level of each test at α/J. This will make the maximum probability of a Type I error among all the tests be exactly α. We will have more to say about this problem in Chapter 12. For the moment the subject of t tests will be left behind while we explore ways to compare two or more means by one, overall, test.

EXERCISES

1. In a study of leadership, a sample of 186 U.S. military officers was taken, and each was given a test designed to measure "inner direction" of actions. Previously, this test had been given to a large group of business executives, who produced a mean value of 83.4. The random sample of military officers had a mean score of 82.8, with S = standard deviation 6.5. Does the population of military officers appear to differ from the business population in this respect? Use $\alpha = .05$.

2. A standardized test for college entry has a mean of 500 and a standard deviation of 10, with a roughly normal distribution of scores when applied to high school seniors across the United States. A coaching service promises greatly improved scores on this test. Suppose that the scores of 223 seniors who had used this service were examined, and found to have a mean of 493 with S = standard deviation 18.9. If this group of coached students actually is a random sample of high school seniors, what can one say about the claim the coaching service makes? Is it reasonable to assume that the students who were coached represent a random sample? Why or why not?

3. Over a long period of time a public library system in a city has charged 5 cents a day for overdue books. At the beginning of this year it was found that the population of library cardholders owed an average of 17.34 cents. In an effort to reduce the total debt, the library mailed out notices to everyone with a record of any charges unpaid. Since that time, a random sample of cardholders has been taken. This sample of 392 individuals showed a mean of 17.02 cents due, with S = standard deviation 2.53. Do the notices seem to be having the desired effect? Use $\alpha = .01$.

4. A random sample of 175 American women were asked to record their body temperatures twice a day for a full month. From their records an average value was found for each woman. The mean of these values was 98.7 with a standard deviation S of .95. Test the hypothesis that the mean body temperature of such American women is 98.6 against the alternative that the mean is some other value.

5. Find the 99 percent confidence interval for the mean in exercise 4.

6. In a study of truth in advertising, a government agency opened 500 boxes selected at random of a well-known brand of raisin bran. For each box the actual number of raisins was counted. The mean number of raisins was 32.4, with a standard deviation $S = 4.1$. Evaluate the company's claim that each box contains 34 raisins on the average, against the alternative of fewer raisins than claimed.

7. Find the 95 percent confidence interval for the mean in exercise 6.

8. Suppose that the body weight at birth of normal children (single births) within the United States is approximately normally distributed and has a mean of 115.2 ounces. A pediatrician believes that the birth weights of normal children born of mothers who are habitual smokers may be lower on the average than for the population as a whole. In order to test this hypothesis, records were taken of the birth weights of a random sample of 20 children from mothers who are heavy smokers. The mean of this sample is 114.0 with $S = 4.3$. Evaluate the pediatrician's hunch.

9. Reevaluate the data of exercise 8 on the assumption that a sample of 80 children had been used.

10. For the results of exercise 8 find the 99 percent confidence interval for the mean birth weight of normal children from smoking mothers.

11. Suppose that in a certain large community the number of hours that a TV set is turned on in a given home during a given week is approximately normally distributed. A sample of 26 homes was selected, and careful logs were kept of how many hours per week the TV set was on. The mean number of hours per week in the sample turned out to be 36.1 with a standard deviation S of 3.3 hours. Find the 95 percent confidence interval for the mean number of hours that TV sets are played in the homes of this community.

12. For the data of exercise 11, test the hypothesis that the true mean number of hours is 35. Test the hypothesis that the mean number of hours is 30.

13. The same government agency referred to in exercise 6 has decided to compare two well-known brands of raisin bran with respect to the numbers of raisins each contain on the average. Some 100 boxes of Brand A were taken at random, and the same number of boxes of Brand B were randomly selected. On the average the Brand A boxes contained 38.7 raisins, with $S = 3.9$, and Brand B contained an average of 36 raisins with $S = 4$. Test the hypothesis that the two brands are actually identical in the average number of raisins that their boxes contain. Let H_1 be "not H_0."

14. For exercise 13 find the 99 percent confidence interval for the difference in average number of raisins for Brands A and B.

15. As editor of a journal in psychology you tend to believe that the contributors to that journal now use shorter sentences on the average than they did a few years ago. In order to test this hunch, you take a random sample of 150 sentences from journal articles published ten years ago and take a random sample of 150 sentences from articles published within the last two years. The first sample shows a mean length of 127 type spaces per sentence, whereas the second sample shows a mean length of 113 type spaces. The first standard deviation $S = 41$, and the second standard deviation $S = 45$. Should you conclude that the recent articles do tend to have shorter sentences?

16. Find the 95 percent confidence interval for difference in sentence length from exercise 15.

17. In an experiment, subjects were assigned at random between two conditions, five to each. Their scores turned out as follows:

Condition A	*Condition B*
128	123
115	115
120	130
110	135
103	113

Can one say that there is a significant difference between these two conditions? What must one assume in carrying out this test?

18. Find the 99 percent confidence interval for the difference between Conditions A and B in exercise 17. On the evidence of this confidence interval, could one reject the hypothesis that the true mean of Condition B is five points higher than that of condition A?

19. In an experiment, the null hypothesis is that two means will be equal. The variance of each population is believed to be equal to 16. If $\alpha = .05$, two-tailed, and the test is to have a power of .90 against the alternative that $M_1 - M_2 = 3$, about how many cases should one take in each experimental group?

20. Suppose that two brands of gasoline were being compared for mileage. Samples of each brand were taken and used in identical cars under identical conditions. Nine tests were made of Brand I and six tests of Brand II. The following miles per gallon were found.

Brand I	*Brand II*
16	13
18	15
15	11
23	17
17	12
14	13
19	
21	
16	

Are the two brands significantly different? What must be assumed here in order to carry out the test?

21. In a study of the effects of author attribution upon the perceived political tone of otherwise neutral quotations, fifty unfamiliar quotations on a variety of subjects were chosen. A sample of 40 college students was drawn, and assigned at random between two groups, 20 to each. Group I received the quotations attributed to authors such as Lenin, Marx, and Mao. Group II received the same quotations but attributed to authors such as Jefferson, Franklin Roosevelt, and Lincoln. Each student rated each statement on a scale running from 1, for "democratic" to 5 for "totalitarian." Total ratings were then calculated for each student, and the mean and standard deviation figured for each group. The results were as follows:

Group I	*Group II*
$M_1 = 155.0$	$M_2 = 130.00$
$S_1 = 44.6$	$S_2 = 30.5$

Was there an effect of author attribution on the perception of the quotations? Use $\alpha = .05$.

22. For the data of exercise 21 above, find the 95 percent confidence interval for the true difference in the means.

23. Consider the data of exercise 9, Chapter 4, as two independent random samples of 50 cases each. Test the hypothesis that boys and girls differ on the average by 1000 calories, against the alternative that the difference is greater than this in the direction of the boys. Use $\alpha = .01$.

24. Calculate the 99 percent confidence limits for the difference between the means of exercise 23. Does this interval include the value 1000? What does this reflect about the significance test carried out in exercise 23?

25. An experimenter was interested in dieting and weight losses among men and among women. It was believed that in the first two weeks of a standard dieting program, women would tend to lose more weight than men. As a check on this notion, a random sample of 15 husband-wife pairs were put on the same strenuous diet. Their weight losses after two weeks showed the following:

Pair	Husbands	Wives
1	5.0 lbs.	2.7 lbs.
2	3.3	4.4
3	4.3	3.5
4	6.1	3.7
5	2.5	5.6
6	1.9	5.1
7	3.2	3.8
8	4.1	3.5
9	4.5	5.6
10	2.7	4.2
11	7.0	6.3
12	1.5	4.4
13	3.7	3.9
14	5.2	5.1
15	1.9	3.4

Did wives lose significantly more ($\alpha = .05$) than husbands? What are we assuming here?

26. In the comparison of two brands of gasoline, mentioned in problem 20, the investigator felt that the type of automobile in which the gasoline was tried would make a difference in the mileage. Therefore it was decided to take ten different makes of automobile, and to draw a pair at random from within each make. One member of the pair of automobiles was assigned gasoline Brand I and the other member gasoline Brand II. Then the mileages produced by the *pairs* of cars was as follows:

Pair	Brand I	Brand II
1	19	16
2	24	22
3	21	20
4	23	15
5	14	13
6	16	16
7	15	14
8	17	18
9	16	15
10	19	20

Do the two brands appear to be significantly different ($\alpha = .05$)?

27. Find the 95 percent confidence limits for the difference in mileage between the two types of gasoline, given the data of exercise 26.

28. An experiment was concerned with the possible effect of a small lesion in a particular area of a hamster's brain upon the activity level of the animal. A total of thirty hamsters were selected at random, and randomly divided into two groups of 15 animals each. Group I was used as a control group, and given no lesion, while Group II was given a lesion in the designated area. Since 3 animals died in group II, the actual number turned out to be 12 for that group. After full recovery by the remaining hamsters, each was given access to a running wheel, and the number of revolutions per fixed unit of time recorded. The following data show activity in hundreds of revolutions per unit time. Did the lesion significantly reduce the activity of the hamsters ($\alpha = .05$)?

Group I	*Group II*
$M_1 = 9.26$	$M_2 = 5.14$
$S_1 = 1.45$	$S_2 = 2.81$

29. Find the 95 percent confidence limits for the mean difference in wheel revolutions for the two populations of hamsters in exercise 28.

30. What is the estimated value of ω^2 from the data of exercise 28? What should we be inclined to say about the likely strength of association between presence or absence of a lesion and the activity of a hamster?

Chapter 9

THE CHI-SQUARE AND THE *F* DISTRIBUTIONS

The essential ideas of inferential statistics are most easily discussed in terms of inferences about means or proportions, and so attention has been focused almost exclusively on these matters in the preceding chapters. However, population distributions can be compared in terms of variability as well as central tendency, and it is important to have inferential methods for the variance at our disposal.

The three basic sampling distributions used so far (the binomial, the normal, and the *t* distribution) no longer apply directly when the variance of a population is under study. Rather, we must turn to two new theoretical distributions. The first of these is called the **chi-square distribution,** or the distribution of the random variable χ^2 (small Greek chi, squared). We will use this distribution first in making inferences about a single population variance, although it has many other applications. The second distribution we will consider is usually called the ***F* distribution,** or the distribution of the random variable F (after Sir Ronald Fisher, who developed the main applications of this distribution). The study of this theoretical distribution grows out of the problem of comparing two population variances. The uses of both of these distributions extend far beyond the problems for which they were originally developed, since, like the normal distribution, they provide good approximations to a large class of other sampling distributions that are not easy to determine exactly. These five theoretical distributions, the three already studied plus the two to be introduced in this chapter, make up the arsenal of theoretical functions from which the statistician draws most heavily; almost all the elementary methods of statistical inference rest on one or more of these theoretical distributions. Furthermore, these five theoretical functions have very close connections with each other, and after we conclude our discussion of the chi-square and F distributions, some of these relationships will be pointed out.

Finally, brief mention will be made of three distributions that are close relatives of chi-square or F distributions. These are the *exponential,* the *gamma,* and the *beta distributions*. While not so important in applied statistics as the five ''major'' distributions mentioned above, they do have some special interpretations that are worth knowing about.

9.1 / THE CHI-SQUARE DISTRIBUTION

Suppose that there exists a population having a normal distribution of scores Y. The mean of this distribution is $E(Y) = \mu$, and the variance is

$$E(Y - \mu)^2 = \sigma^2.$$

Now cases are sampled from this distribution *one* at a time, $N = 1$. For each case sampled the **squared standardized score**

$$z^2 = \frac{(y - \mu)^2}{\sigma^2}$$

is computed. Let us call this squared standardized score $\chi^2_{(1)}$ so that

$$\chi^2_{(1)} = z^2.$$

Now we will look into the sampling distribution of this variable $\chi^2_{(1)}$.

First of all, what is the range of values that χ^2 might take on? The original normal variable Y ranges over all real numbers, and this is also the range of the standardized variable z. However, $\chi^2_{(1)}$ is always a squared quantity, and so its range must be all the *nonnegative* real numbers, from zero to a positive infinity. We can also infer something about the form of this distribution of $\chi^2_{(1)}$; the bulk of the cases (about 68 percent) in a normal distribution of standard scores must lie between -1 and 1. Given a z between -1 and 1, the corresponding $\chi^2_{(1)}$ value lies between 0 and 1, so that the bulk of this sampling distribution will fall in the interval between 0 and 1. This implies that the form of the distribution of $\chi^2_{(1)}$ will be very skewed, with a high probability for a value in the interval from 0 to 1, and relatively low probability in the interval with lower bound 1 and approaching positive ∞ as its upper bound. The graph of the distribution of $\chi^2_{(1)}$ is represented in Figure 9.1.1.

Figure 9.1.1 pictures **the chi-square distribution with one degree of freedom. The distribution of the random variable $\chi^2_{(1)}$ where**

$$\chi^2_{(1)} = \frac{(y - \mu)^2}{\sigma^2} \qquad [9.1.1]$$

and Y is normally distributed with mean μ and variance σ^2 is a chi-square distribution with 1 degree of freedom.

Now let us go a little further. Suppose that samples of *two* cases are drawn independently and at random from a normal distribution. We find the squared standardized score corresponding to each observation:

$$z_1^2 = \frac{(y_1 - \mu)^2}{\sigma^2}$$

$$z_2^2 = \frac{(y_2 - \mu)^2}{\sigma^2}.$$

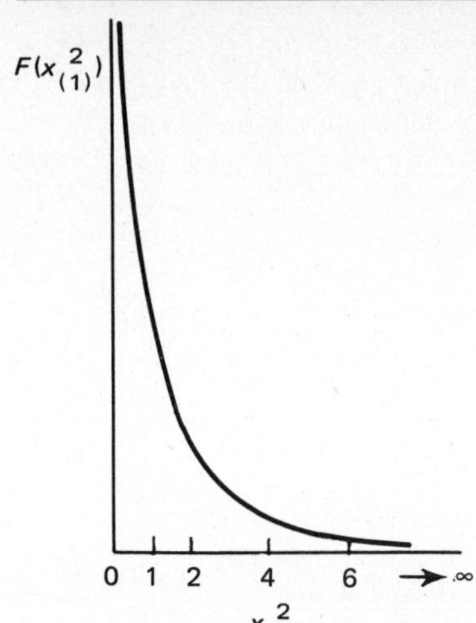

Figure 9.1.1

The distribution of χ^2 for $\nu = 1$.

If the *sum* of these two squared standardized scores is found over repeated independent samplings, the resulting random variable is designated as $\chi^2_{(2)}$:

$$\chi^2_{(2)} = \frac{(y_1 - \mu)^2}{\sigma^2} + \frac{(y_2 - \mu)^2}{\sigma^2} = z_1^2 + z_2^2. \qquad [9.1.2]$$

If we look into the distribution of the random variable $\chi^2_{(2)}$ we find that the range of possible values extends over all nonnegative real numbers. However, since the random variable is based on *two* independent observations, the distribution is somewhat less skewed than for $\chi^2_{(1)}$; the probability is not so high that the sum of two squared standardized scores should fall between 0 and 1. This is illustrated in Figure 9.1.2. This illustrates a **chi-square distribution with two degrees of freedom.**

Finally, suppose that we took N independent observations at random from a normal distribution with mean μ and variance σ^2 and defined the random variable

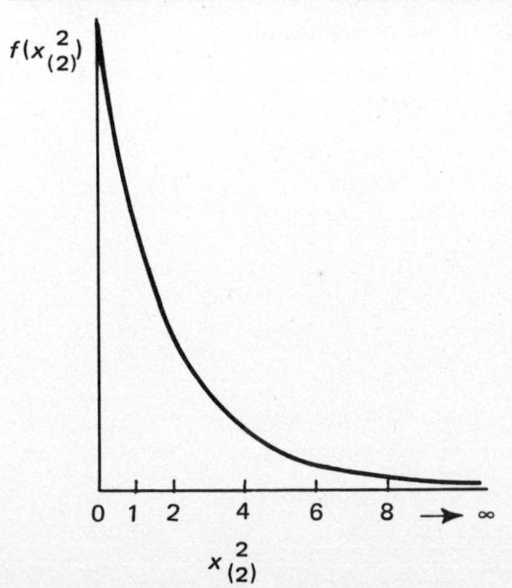

Figure 9.1.2

The distribution of χ^2 for $\nu = 2$.

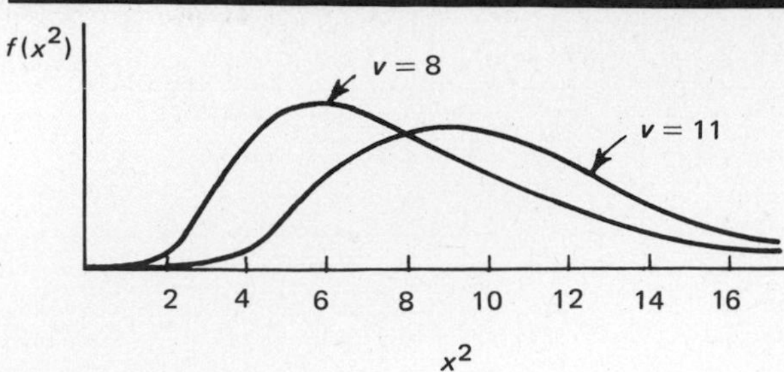

Figure 9.1.3

The general form of the distribution of χ^2 for larger numbers of degrees of freedom.

$$\chi^2_{(N)} = \frac{\sum_{i}^{N} (y_i - \mu)^2}{\sigma^2} = \sum_i z_i^2. \qquad [9.1.3]$$

The distribution of this random variable will have a form that depends upon the number of independent observations taken at one time. **In general, for N independent observations from a normal population, the sum of the squared standardized scores for the observations has a chi-square distribution with N degrees of freedom.** Notice that the standardized scores must be relative to the *population mean* and the *population standard deviation*.

The function rule assigning a probability density to each possible value of χ^2 is given by

$$f(\chi^2;\nu) = h(\nu)e^{-\chi^2/2}(\chi^2)^{(\nu/2)-1}, \qquad \text{for } \chi^2 \geq 0, \; \nu > 0. \qquad [9.1.4]$$

As shown in Figures 9.1.1 through 9.1.3, the plot of this density function always presents the picture of a very positively skewed distribution, at least for relatively small values of ν, with a distinct mode at the point $\nu - 2$ for $\nu > 2$. However, as ν is increased, the form of the distribution appears to be less skewed to the right. In this function rule the value $h(\nu)$ is a constant depending only on the parameter ν. Consequently, there is only one value other than χ^2 that must be specified in order to find the density: this is the parameter ν. Like the t distribution, the distribution of

χ^2 depends only on the degrees of freedom, the parameter ν. The family of chi-square distributions all follow this general rule, but the exact form of the distribution depends on the number of degrees of freedom, ν. In principle, the value of ν can be any positive number, but in the applications we will make of this distribution in this chapter, ν will depend only on the sample size, N.

The mean of a chi-square distribution is simply the value of the parameter ν,

$$E(\chi^2_{(\nu)}) = \nu$$

and the variance is

$$\text{Var}(\chi^2_{(\nu)}) = 2\nu.$$

When several independent random variables each have a chi-square distribution, then the distribution of the *sum* of the variables is also known. This is a most important and useful property of variables of this kind, and can be stated more precisely as follows:

If a random variable $\chi^2_{(\nu_1)}$ has a chi-square distribution with ν_1 degrees of freedom, and an independent random variable $\chi^2_{(\nu_2)}$ chi-square distribution with ν_2 degrees of freedom, then the new random variable formed from the sum of these variables

$$\chi^2_{(\nu_1 + \nu_2)} = \chi^2_{(\nu_1)} + \chi^2_{(\nu_2)} \qquad [9.1.5]$$

has a chi-square distribution with $\nu_1 + \nu_2$ degrees of freedom.

In short, the new random variable formed by taking the sum of two independent chi-square variables is itself distributed as χ^2, with degrees of freedom equal to the sum of those for the original distributions.

9.2 / TABLES OF THE CHI-SQUARE DISTRIBUTION

Like the t distribution, the particular distribution of χ^2 depends on the parameter ν, and it is difficult to give tables of the distribution for all values of ν that one might need. Thus, Table IV in Appendix D, like Table III, is a condensed table, showing values of χ^2 that correspond to percentage points in various distributions specified by ν. The rows of Table IV list various degrees of freedom ν and the column headings are probabilities Q, just as in the t table. The numbers in the body of the table give the values of χ^2 such that in a distribution with ν degrees of freedom, *the probability of a sample chi-square value this large or larger is* Q (Fig. 9.2.1).

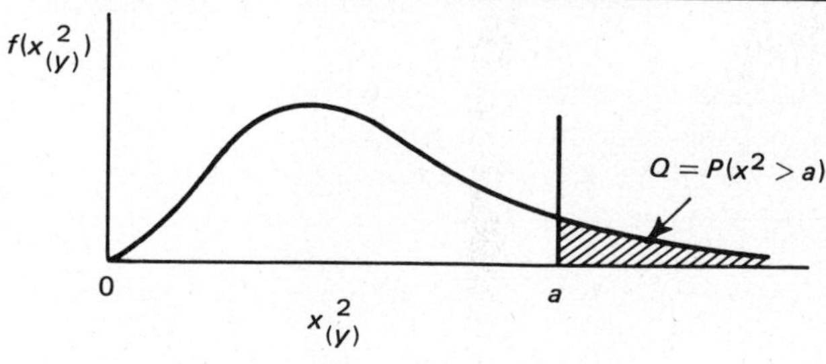

Figure 9.2.1

A right-tail area in a chi-square distribution with ν degrees of freedom.

For example, suppose that we are dealing with a chi-square distribution with 5 degrees of freedom. Find the row ν equal to 5, and the column headed .05. The value in this row and column is approximately 11.071, showing that for 5 degrees of freedom, random samples showing a chi-square value of 11.071 *or more* should in the long run occur about five times in a hundred. As another example, look at the row labeled $\nu = 2$ and the column headed .10. The entry in this combination of row and column indicates that in a distribution with 2 degrees of freedom, chi-square values of 4.605 or more should occur with probability of about 1 in 10, under random sampling. Finally, look at the row for $\nu = 24$, and the column for .001; the cell entry indicates that chi-square values of 51.179 or more occur with probability of only about .001 in a distribution with 24 degrees of freedom.

9.3 / THE DISTRIBUTION OF THE SAMPLE VARIANCE FROM A NORMAL POPULATION

In this section, the sampling distribution of the sample variance will be studied. It will be assumed that the population actually sampled is normal, and the results in this section apply, strictly speaking, only when this assumption is true.

Recall that in Section 4.11 the sample variance was defined by

$$S^2 = \frac{\sum_{i=1}^{N} (y_i - M)^2}{N}.$$

Then, since

$$\frac{NS^2}{\sigma^2} = \sum_{i=1}^{N} \frac{(y_i - M)^2}{\sigma^2}$$

it can be shown that for N independent observations from a normal population,

$$\frac{NS^2}{\sigma^2} = \chi^2_{(N-1)}, \qquad [9.3.1^*]$$

the ratio NS^2/σ^2 is a chi-square variable with $N - 1$ degrees of freedom.

If, instead of S^2, we use the unbiased estimate s^2, we find that again for N independent observations from a normal population,

$$\frac{(N-1)s^2}{\sigma^2} = \chi^2_{(N-1)} \qquad [9.3.2^*]$$

meaning that the ratio of $(N - 1)s^2$ to σ^2 is a random variable distributed as chi-square with $N - 1$ degrees of freedom. This fact is the basis for inferences about the variance of a single normally distributed population.

This same idea applies to the sampling distribution of the estimate of a variance based upon the *pooling* of independent estimates, provided that the basic population sampled is normal. Thus, it is also true that when the unbiased pooled estimate of σ^2 is made from two independent samples of N_1 and N_2 cases,

$$\frac{(N_1 + N_2 - 2)\text{ est. }\sigma^2}{\sigma^2} = \chi^2_{(N_1+N_2-2)}. \qquad [9.3.3^*]$$

The pooled unbiased sample estimate of σ^2, multiplied by the degrees of freedom and divided by the true value of σ^2, is a chi-square variable with $\nu = N_1 + N_2 - 2$.

9.4 / TESTING EXACT HYPOTHESES ABOUT A SINGLE VARIANCE

Just as for the mean, when the population is normal, it is possible to test exact hypotheses about a single population variance (and, of course, a standard deviation). The exact hypothesis tested is

$$H_0: \sigma^2 = \sigma_0^2$$

where σ_0^2 is some specific positive number. The alternative hypothesis may be either directional or nondirectional, depending, as always, on the original question.

As usual, some value of α is decided upon, and a region of rejection adopted depending both on α and the alternative H_1. The test statistic itself is

$$\chi^2_{(N-1)} = \frac{(N-1)s^2}{\sigma_0^2}. \quad [9.4.1^*]$$

The value of this test statistic, computed from the s^2 actually obtained and the σ_0^2 dictated by the null hypothesis, is referred to the distribution of χ^2 for $N - 1$ degrees of freedom.

For example, one might ask this question about height: "It is well known that men and women in the United States differ in terms of their mean height; is it true, however, that women show less variability in height than do men?" Now let us assume that from the past records of the Selective Service System we actually *know* the mean and standard deviation of height for American men between the ages of twenty and twenty-five years. However, such evidence is lacking for women. Assume that the standard deviation of height for the population of men twenty to twenty-five years is known to be 2.5 inches. For this same age range, we want to ask if the population of women shows this same standard deviation, or if women are *less* variable, with their distribution having a smaller σ. The null and alternative hypotheses can be framed as

$$H_0: \sigma^2 \geqslant 6.25 \text{ (or } [2.5]^2)$$
$$H_1: \sigma^2 < 6.25.$$

Imagine that we plan to draw a sample of 30 women at random, each between the ages of twenty and twenty-five years, and measure the height of each. The test statistic will be

$$\chi^2_{(29)} = \frac{(29)s^2}{6.25}.$$

What, however, is the region of rejection? Here, *small* values of χ^2 tend to favor H_1, that women actually are less variable than men. Hence, we want to use a region of rejection on the left (or small-value) tail of the chi-square distribution with 29 degrees of freedom. If $\alpha = .01$, then from Table IV in Appendix D the value of χ^2 leading to rejection of H_0 should be less than the value given by the row with $\nu = 29$, and the column for $Q = .99$. This value is 14.257.

Now the actual value of s^2 obtained turns out to be 4.55, so that

$$\chi^2_{(29)} = \frac{(29)(4.55)}{6.25} = 21.11.$$

This value is larger than the critical value decided upon, and we cannot reject H_0 if α is to be .01.

This example illustrates that, as with the t and the normal distribution, either or both tails of the chi-square distribution can be used in testing a hypothesis about a variance. Had the alternative hypothesis in this problem been

$$H_1: \sigma^2 \neq 6.25,$$

then the rejection region would lie in both tails of the chi-square distribution. The rejection region on the lower tail of the distribution would be bounded by a chi-square value corresponding to $\nu = 29$ and $Q = .995$, which is 13.121; the rejection region on the upper tail would be bounded by the value for $\nu = 29$ and $Q = .005$, which is 52.336. Any obtained χ^2 value falling *below* 13.121 or *above* 52.336 would let one reject H_0 beyond the .01 level.

9.5 / CONFIDENCE INTERVALS FOR THE VARIANCE AND STANDARD DEVIATION

Finding confidence intervals for the variance is quite simple, provided that the normal distribution rule holds for the population.

Suppose that we had a sample of some N independent observations, and we wanted the 95 percent confidence limits for σ^2. For samples from a normal distribution, it must be true that

$$\text{prob}\left[\chi^2_{(N-1;.975)} \leqslant \frac{(N-1)s^2}{\sigma^2} \leqslant \chi^2_{(N-1;.025)}\right] = .95 \qquad [9.5.1]$$

where $\chi^2_{(N-1;.975)}$ is the value in a chi-square distribution with $N - 1$ degrees of freedom cutting off the upper .975 of samples values, and $\chi^2_{(N-1;.025)}$ the value cutting off the upper .025 of sample values. This inequality can be manipulated to show that it is also true that

$$\text{prob}\left[\frac{(N-1)s^2}{\chi^2_{(N-1;.025)}} \leqslant \sigma^2 \leqslant \frac{(N-1)s^2}{\chi^2_{(N-1;.975)}}\right] = .95. \qquad [9.5.2]$$

That is, the probability is .95 that the true value of σ^2 will be covered by an interval with limits found by

$$\frac{(N-1)s^2}{\chi^2_{(N-1;.025)}}$$

and [9.5.3]

$$\frac{(N-1)s^2}{\chi^2_{(N-1;.975)}}.$$

Suppose, for example, that a sample of 15 cases is drawn from a normal distribution. We want the 95 percent confidence limits for σ^2. The value of s^2 is 10; this is our best single estimate of σ^2, of course. For $\nu = 14$ and for $Q = .025$, $\chi^2_{(14;.025)}$ is found from Table IV to be 26.12, and the value of $\chi^2_{(14;.975)}$ is 5.63. Thus the confidence limits for σ^2 are

$$\frac{(14)(10)}{26.12} \text{ or } 5.36$$

and

$$\frac{(14)(10)}{5.63} \text{ or } 24.87.$$

We can say that the probability is .95 that an interval such as 5.36 to 24.87 covers the true value of σ^2.

Unfortunately, because of the skewness of a chi-square distribution, confidence intervals for σ^2 are not necessarily "shortest" in the same sense that holds for confidence intervals for the mean. Thus, these intervals are not as useful as they might be, particularly when our interest is in finding the best possible estimates of σ^2. More advanced methods do exist for finding the shortest possible confidence interval for σ^2, and tables for this purpose are given in Tate and Klett (1959).

9.6 / THE NORMAL APPROXIMATION TO THE CHI-SQUARE DISTRIBUTION

As the number of degrees of freedom grows infinitely large, the distribution of χ^2 approaches the normal distribution. You should hardly find this principle surprising by now! Actually the same mechanism is at work here as in the central limit theorem: for any ν greater than 1, the chi-square variable is equivalent to a *sum* of ν independent random variables. Thus, like the mean, M, given enough summed terms the sampling distribution approaches the normal form in spite of the fact that each component of the sum does not have a normal distribution when sampled alone.

This fact is of more than theoretical interest when very large samples are used. For very large ν, the probability for any interval of values of $\chi^2_{(\nu)}$ can be found from the normal standardized scores

$$z = \frac{\chi^2_{(\nu)} - \nu}{\sqrt{2\nu}} \qquad [9.6.1]$$

since the mean and variance of a chi-square distribution are ν and 2ν respectively. However, this approximation is not good unless ν is extremely large. A somewhat better approximation procedure is to find the value of $\chi^2_{(\nu)}$ that cuts off the *upper Q* proportion of cases by taking

$$\chi^2_{(\nu;Q)} = \frac{1}{2}\{z_Q + \sqrt{2\nu - 1}\}^2. \qquad [9.6.2]$$

This can be used to find chi-square values for ν greater than 100, which are not given in Table IV. The necessary z_Q values are listed at the bottom of Table IV, and can also be found from Table I.

9.7 / THE IMPORTANCE OF THE NORMALITY ASSUMPTION IN INFERENCE ABOUT σ^2

In the preceding chapter it was pointed out that although the rationale for the t test demands the assumption of a normal population distribution of scores, in practice the t test may be applied when the parent distribution is not normal, provided that sample N is at least moderately large. *This is not the case for inferences about the variance, however.* One runs a considerable risk of error in using the chi-square distribution either to test a hypothesis about a variance or to find confidence limits unless the population distribution is normal, or approximately so. The effect of the violation of the normality assumption is usually minor for large N, but can be quite serious when inferences are made about variances, even for moderate N. Indeed, for some population forms, the effect of using the chi-square statistics may actually grow more serious for larger samples. Thus, the assumption of a normal distribution is important in inferences about the variance.

It is rather rare to find a problem in research that centers on the value of a single population variance. More common is the situation where the variances of two populations are compared for equality.

Imagine two distinct populations, each showing a normal distribution of the variable Y. The means of the two populations may be different, but each population shows the same variance, σ^2. We draw two independent random samples: the first sample, from population 1, contains N_1 cases, and that from population 2 consists of N_2 cases.

For each possible pair of samples, one from population 1, and one from population 2, we take the *ratio* of s_1^2 to s_2^2 and call this ratio the random variable F:

$$F = \frac{s_1^2}{s_2^2} = \frac{\text{est. } \sigma_1^2}{\text{est. } \sigma_2^2}. \qquad [9.8.1^*]$$

The sampling distribution of this ratio of two independent unbiased estimates of the population variance, each based on a sample of independent observations from a normal distribution, is called the F distribution.

Notice that when we put the two variance estimates in ratio, this is actually the ratio of two independent chi-square variables, each divided by its degrees of freedom:

$$F = \frac{[\chi^2_{(\nu_1)}/\nu_1]}{[\chi^2_{(\nu_2)}/\nu_2]} \qquad [9.8.2]$$

since

$$\frac{s_1^2}{s_2^2} = \frac{(s_1^2/\sigma_1^2)}{(s_2^2/\sigma_2^2)}$$

provided that $\sigma_1^2 = \sigma_2^2$ (that is to say, the hypothesis of the equality of the population variances is true).

Showing F as a ratio of two independent chi-square variables each divided by its ν is actually a way of defining the F variable:

A random variable formed from the ratio of two independent chi-square variables, each divided by its degrees of freedom, is said to be an F ratio, and to follow the rule for the F distribution.

When this definition is satisfied, meaning that both parent populations are normal, have the same variance, and that the samples are drawn are independent, then the theoretical distribution of F values can be found.

Mathematically, the F distribution is rather complicated. However, for our purposes it will suffice to remember that the density function for F depends only upon two parameters, ν_1 and ν_2, which can be thought of as *the degrees of freedom associated with the numerator and the denominator of the F ratio*. The range for F is *nonnegative real numbers*. The expectation of F is $\nu_2/(\nu_2 - 2)$ for $\nu_2 > 2$. In general form, the distribution for any fixed ν_1 and ν_2 is nonsymmetric, although the particular "shape" of the function curve varies considerably with changes in ν_1 and ν_2. However, it is worth noting that for the values of ν_1 and ν_2 that will usually concern us (that is, integral values, with $\nu_1 < \nu_2$) the F distribution is skewed positively, and is unimodal for $\nu_1 > 2$. On the other hand, the distribution can take on other forms when other conditions are set for ν_1 and ν_2.

Before we can apply this theoretical distribution to an actual problem of comparing two variances, the use of F tables must be discussed.

9.9 / THE USE OF *F* TABLES

Since the distribution of F depends upon two parameters ν_1 and ν_2, it is even more difficult to present tables of F distributions than those of χ^2 or t. Tables of F are usually encountered only in drastically condensed form.

Such tables give only those values of F that cut off the *upper* proportion Q in an F distribution with ν_1 and ν_2 degrees of freedom. The only values of Q given are those that are commonly used as α in a test of significance (that is, the .25, .10, .05, .025, and the .01 values). Table V in Appendix D shows F values required for significance at the $\alpha = Q$ level, given ν_1 and ν_2. The columns of each table give values of ν_1, the degrees of freedom for the numerator, and the rows give values of ν_2, the degrees of freedom for the denominator. Each separate table represents one value of Q. The entries in the body of the table are the values of F required for significance at this level.

The use of Table V can be illustrated in the following way: Suppose that two independent samples are drawn, containing $N_1 = 10$ and $N_2 = 6$ cases, respectively. The degrees of freedom associated with the two variances are $\nu_1 = 10 - 1 = 9$ and $\nu_2 = 6 - 1 = 5$. For the ratio

$$F = \frac{s_1^2}{s_2^2}$$

the degrees of freedom must be 9 for the numerator, and 5 for the denominator. Now suppose that the obtained $F = 7.00$. Does this fall into the upper .05 of values in an F distribution, with 9 and 5 degrees of freedom? We turn to Table V and find the page for .05. Then we look at the column for $\nu_1 = 9$ and the row for $\nu_2 = 5$. The tabled value is 4.77. Our obtained F value of 7.00 exceeds 4.77, and thus our sample result falls among the upper 5 percent in an F distribution.

As another example, suppose that $\nu_1 = 1$, and $\nu_2 = 45$. For the problem at hand, the rejection region contains the *upper* .01 proportion in an F distribution. What value must our sample result equal or exceed in order to be significant? Table V shows that for the .01 level, the required F for this number of degrees of freedom is between 7.31 (the F for $\nu_2 = 40$) and 7.08 (the F for $\nu_2 = 60$). It is difficult to find the exact value of F required, but we are reasonably sure that the required F is somewhere around 7.3.

In both of these illustrations we have dealt only with the upper tail of the F distribution. This would be appropriate if we were doing a one-tailed test, where $H_1: \sigma_1^2 > \sigma_2^2$. However, it is possible, although somewhat more difficult, to find F values corresponding to two-tailed rejection regions.

Let $F_{(\nu_1;\nu_2)}$ stand for an F ratio with ν_1 and ν_2 degrees of freedom, and let $F_{(\nu_2;\nu_1)}$ be an F ratio with ν_2 degrees of freedom in the *numerator,* and ν_1 degrees of freedom in the *denominator*. The numbers of degrees of freedom in numerator and denominator are simply reversed for these two F ratios. Then it is true that, for any positive number C,

$$p[C \leqslant F_{(\nu_1;\nu_2)}] = p\left[F_{(\nu_2;\nu_1)} \leqslant \frac{1}{C}\right] \qquad [9.9.1]$$

the probability that $F_{(\nu_1;\nu_2)}$ is greater than or equal to some number C is the same as the probability that the reciprocal of $F_{(\nu_1;\nu_2)}$ *is less than or equal to the reciprocal* of C. Practically, this means that the *value required for F on the lower tail of some particular distribution can always be found by finding the corresponding value required on the upper tail of a distribution with numerator and denominator degrees of freedom reversed, and then taking the reciprocal.*

9.10 / USING THE F DISTRIBUTION TO TEST HYPOTHESES ABOUT TWO VARIANCES

In an investigation of the effect of stress on children's performance of a reasoning test, it is felt that competition with peers represents one form of stressful situation to a child. However, the experimenter suspects that competition has different effects on different children. The conjecture is that bright children might be stimulated to do even better than otherwise by the competitive atmosphere, but that the relatively dull child will appear even more at a disadvantage. One implication of this notion is that if groups of children are sampled from a population having a normal distribution of ability, but the groups are given different amounts of "competitive stress," the group subjected to the greater stress should show a relatively *greater variance* among the scores.

Letting population 1 stand for the potential group of children tested under stress, and 2 for the control, or nonstressed, population, we frame the following hypotheses:

$$H_0: \sigma_1^2 \leqslant \sigma_2^2$$
$$H_1: \sigma_1^2 > \sigma_2^2.$$

The rejection region decided upon is one-tailed, reflecting the experimenter's prior "hunch" about effects of stress. The α level to be used is .05.

Thirty-two children are selected at random, and assigned at random to two experimental groups of sixteen cases each. Group 1 is given the stress treatment, and group 2 is not. The results are

$$s_1^2 = 5.8, s_2^2 = 1.7.$$

If the exact null hypothesis is true, so that both of these values are estimates of the same population value σ^2, the ratio

$$F = \frac{s_1^2}{s_2^2}$$

should be distributed as the F distribution with $N_1 - 1 = 15$ and $N_2 - 1 = 15$ degrees of freedom. The value required for significance at the .05 level, one-tailed, is found from Table V to be 2.40. However, the obtained value of F is

$$F = \frac{5.8}{1.7} = 3.41,$$

which exceeds the value required. On this evidence we can reject the null hypothesis of equal variances at the .05 level. We are fairly safe in saying that the experimental increase in stress seems to increase the variability of scores.

Had the alternative hypothesis been two-tailed, then we would have had to consider the required value of F on both tails of the distribution, of course. In this instance, the procedure given in 9.9 would be used.

Like the chi-square distribution, the use of F depends on the assumption that the population distributions are themselves normal. The failure of this assumption to hold true can make a difference in the conclusions reached from an F test of the equality of variances, *unless* the sample sizes are rather large. (About thirty in each sample should be relatively safe, however.)

9.11 / RELATIONSHIPS AMONG THE THEORETICAL DISTRIBUTIONS

Now that the major sampling distributions have been introduced, some of the connections among these theoretical distributions can be examined in more detail. Over and over again, the binomial, normal, t, chi-square, and F distributions have proved their utility in the solution of problems in statistical inference. Remember, however, that none of these distributions is empirical in the sense that someone has taken a vast number of samples and found that the sample values actually do occur with exactly the relative frequencies given by the function rule. Rather, it follows mathematically (read "logically") that if we are drawing random samples from certain kinds of populations, various sample statistics *must* have distributions given by the several function rules. Like any other theory, the theory of sampling distributions deals with "if-then" statements. This is why the assumptions we have introduced are important; if we wish to apply the theory of statistics to making inferences from samples, then we cannot expect the theory necessarily to provide us with correct results unless the conditions specified in the theory hold true. As we have seen, from a practical standpoint these assumptions may be violated to some extent in our use of these theoretical distributions as approximations, especially for large samples. However, these assumptions are quite important for the *mathematical* justification of our methods, in spite of the possible applicability of the methods to situations where the assumptions are not met.

Apart from the general requirement of simple random sampling of independent observations, the most usual assumption made in deriving sampling distributions is that the population distribution is normal. The chi-square, the t, and the F distributions all rest on this assumption. The normal distribution is, in a real sense, the "parent" distribution to these others. This, as mentioned in Chapter 6, is one of the main reasons for the importance of the normal distribution: the normal function rule not only provides probabilities that are often excellent approximations to other probability (density) functions, but it also has highly convenient mathematical properties for deriving other distribution functions based on normal populations.

The chi-square distribution rests directly upon the assumption that the population is normal. As you will recall from Section 9.1, the chi-square variable is basically a sum of squares of independent *normal* variables, each with mean 0 and with variance 1. At the elementary level, the problem of the distribution of the sample variance can be solved explicitly only for normal populations; this sampling distribution depends on the distribution of χ^2, which in turn rests on the assumption of a normal distribution of single observations Furthermore, in the limit the distribution of χ^2 approaches a normal form.

There are close connections in theory between the F distribution and both the normal and chi-square distributions. Basically, the F variable is a ratio of two independent chi-square variables, each divided by its degrees of freedom. Since a chi-square variable is itself defined in terms of the normal distribution, then the F distribution also rests on the assumption of two (or more) normal populations.

In general, a t variable is a standardized normal variable z_M in ratio to the square root of a chi-square variable divided by ν. Let us look at t^2 for a single mean in the light of this definition:

$$t^2 = \frac{z_M^2}{\chi^2/(N-1)}. \qquad [9.11.1]$$

The numerator of t^2 is, by definition, a chi-square variable with 1 degree of freedom, and the denominator is a chi-square variable divided by its degrees of freedom, $\nu = N - 1$. Furthermore, these two chi-square variables are independent, by the principle stating that for a normal population, s^2 is independent of M (Section 6.6). Thus, for a single mean, t^2 qualifies as an F ratio, with 1 and $N - 1$ degrees of freedom. In general,

$$t^2_{(\nu)} = F_{(1, \nu_2)} \qquad \nu = \nu_2; \qquad [9.11.2]$$

the square of t with ν degrees of freedom is an F variable with 1 and ν degrees of freedom.

You can check this for yourself by examining the column for 1 and ν_2 degrees of freedom in the F table. In the table of F values required for significance at the .05 level, each of the entries in this column is simply the square of the entry in the t table for $\nu = \nu_2$ and $t_{.025}$. Similarly, in the table for $\alpha = .01$, each entry in the column for 1 and ν_2 degrees of freedom is the square of the corresponding entry for $t_{.005}$ in the distribution of t.

This relation between t^2 and F lets us illustrate the importance of the assumption that the two populations have the same true σ^2 when a difference between means is tested. For n cases in each of two independent groups, the value of z^2 represented by the numerator of t^2 must be

$$z^2_{\text{diff.}} = \frac{(M_1 - M_2)^2}{\sigma^2_{\text{diff.}}}. \qquad [9.11.3]$$

When $\mu_1 = \mu_2$, this is a chi-square variable with one degree of freedom. Furthermore, the square of the denominator term in the t ratio must correspond to

$$\frac{\text{est. } \sigma^2_{\text{diff.}}}{\sigma^2_{\text{diff.}}} \qquad [9.11.4]$$

for this to be a chi-square variable divided by its degrees of freedom. When each population has the same true variance, σ^2, the denominator term we compute is equivalent to such a variable, and the value t^2 is then the square of the ratio we actually find:

$$t^2 = \frac{(M_1 - M_2)^2}{\text{est. } \sigma^2_{\text{diff.}}} = \frac{(\chi^2_{(1)}/1)}{(\chi^2_{(\nu_2)}/\nu_2)} = F. \quad [9.11.5]$$

The ratio one calculates, not knowing the true $\sigma^2_{\text{diff.}}$, actually is distributed exactly as t (or its square distributed as F) when the true variances are equal. However, if the values of the two variances are not the same, the ratio of the estimated $\sigma^2_{\text{diff.}}$ to the true $\sigma^2_{\text{diff.}}$ is equivalent to

$$\frac{\text{est. } \sigma^2_{\text{diff.}}}{\text{true } \sigma^2_{\text{diff.}}} = \frac{\chi^2_a\sigma^2_1 + \chi^2_b\sigma^2_2}{(n-1)(\sigma^2_1 + \sigma^2_2)} \quad [9.11.6]$$

where χ^2_a and χ^2_b symbolize two possibly different values of a chi-square variable with $n - 1$ degrees of freedom. This ratio is not necessarily distributed as a chi-square variable divided by its degrees of freedom. Thus, when variances are unequal for the two populations, the ratio we compute is not really distributed *exactly* like the random variable t, since the square of the ratio we compute cannot be equivalent to an F ratio. For this reason, a correction procedure, such as that in Section 8.10, is required when the variances are unequal and sample size is small.

The important relationships among these theoretical distributions are summarized in Table 9.11.1, showing how the distribution represented in the column depends for its derivation upon the distribution represented in the row.

Table 9.11.1

Connections among the theoretical distributions.

Distribution	*Chi-square*	*F*	*t*
Normal	Parent, and limiting form as $\nu \to \infty$. Defined as sum of normal and independent z^2 values.	Parent, making values in numerator and denominator independent χ^2/ν values.	Parent, and limiting form as $\nu \to \infty$. Numerator is normal z.
Chi-square		Variables in numerator and denominator are independent χ^2/ν.	Denominator is $\sqrt{\chi^2/\nu}$.
F			$t^2_{(\nu)} = F_{(1,\nu)}$

All four of these theoretical distributions will figure in the next three chapters as we work though the rationale for the analysis of variance and comparisons among means. Having some idea of the interrelations of these distributions will be of help in understanding how the F distribution can be used to test a hypothesis about *several* means.

9.12 / THE EXPONENTIAL, GAMMA, AND BETA DISTRIBUTIONS

Our introduction to the more important continuous distributions will conclude with a brief discription of three families of distributions which are very closely related to the chi-square and the F distributions. In addition, these distributions have special uses in their own right, some of which are of potential value in the social and behavioral sciences.

The exponential distribution has close conceptual ties to the geometric distribution of Section 3.17, and to the Poisson distribution of Section 3.18. Suppose that a random process follows a Poisson law producing "successes" at the rate of m per unit of time. The time is recorded until the *first* (or the *next*) success occurs. Then the distribution of time until the first success occurs follows the rule of *the exponential distribution:*

$$f(x; m) = \begin{cases} e^{-mx}m & \text{for } x \geqslant 0,\ m > 0 \\ 0 & \text{otherwise.} \end{cases}$$

An exponential distribution of the random variable X has a mean $E(X) = 1/m$, and a variance of $1/m^2$.

Thus, for example, a pigeon has been trained to peck at a bar in order to secure a reward of grain. The trained pigeon has been observed to peck at a stable rate of 5 times a minute. What then is the average time that will elapse before the pigeon will give its first peck after having been placed in front of the bar? The answer is $E(X) = 1/5 = .2$ minutes, or 12 seconds. These times to the first peck will have a variance of $1/25 = .04$, or a standard deviation of .2 minute. The probability density for a given time such as $X = 6$ seconds, or .1 minute, is $e^{-.5}(.5)$, and so on for any other value of X. (The random variable does not necessarily have to represent only time until the first success; any other continuous dimension such as the distance to the first "success" may also be used with this model.)

A generalization of the exponential distribution is the *gamma distribution*. One interpretation of a gamma variable is as follows: Again suppose that a Poisson process produces successes at the stable rate of m per unit of time. Then if X is the elapsed time until the rth success, the random variable X follows the gamma law

$$f(x; r, m) = \begin{cases} \dfrac{e^{-mx}(mx)^{r-1}m}{(r-1)!} & \text{for } x \geqslant 0,\ r > 0,\ m > 0 \\ 0 & \text{otherwise.} \end{cases}$$

Here, when the value of the parameter r is an integer, $(r-1)!$ is an ordinary factorial. However, for noninteger values, $(r-1)!$ stands for a value of the so called "gamma function," studied in the advanced calculus. (Here, as well, X may represent time, distance, or any other continuous dimension relevant to the context, while m is a rate of "success" events per unit interval on this dimension.) Then the expected value of X for the gamma distribution with parameters r and m is $E(X) = r/m$, and the variance of X is $\text{Var}(X) = r/(m^2)$.

For example, the police department in a very large city knows from past records that murders tend to be reported at the rate of 2.6 per week. On a certain January 1, the police department will begin keeping the records for that year. How many weeks should it take them to record exactly fifty murders? Notice that the time it could take to reach this number could be zero weeks (i.e., a very bad early New Year's morning), or it could be some large, or even infinite number of weeks (miraculously, there are no more murders in this city). However, for any starting point, the number of weeks until fifty murders occur should follow a gamma distribution, with $m = 2.6$ and with $r = 50$. The expected number of weeks should be

$$E(X) = 50/2.6 = 19.23 \text{ weeks}$$

and the variance should be

$$E(X) = 50/(2.6)^2 = 7.40.$$

The murder rate in that year could be compared with past years, or with any hypothetical rate, in terms of the time it takes to reach some arbitrary number of cases reported.

A random variable following a *beta distribution* has a probability density for any value $X = x$ given by

$$f(x;r,N) = \begin{cases} \dfrac{(N-1)!}{(r-1)!(N-r-1)!} x^{r-1}(1-x)^{N-r-1} & \text{for } 0 \leq x \leq 1,\ 0 < \text{r} < \text{N} \\ 0 & \text{otherwise.} \end{cases}$$

Once again, for integer values of r and N, where $0 < r < N$, the factorial expressions are found in the usual way. However, for noninteger r or N, each factorial is replaced by a value from the gamma function, as mentioned above. The expected value of a beta variable X is given by $E(X) = r/N$, and the variance is $\text{Var}(X) = r(N - r)/ N^2(N + 1)$.

One of the uses of the beta distribution is when two independent Poisson-like processes are occurring, each producing successes at the same rate m. Then if A stands for the time taken to the rth success in the first process, and B the time taken to the $(N - r)$th success in the second process, the variable $X = A/(A + B)$ has a beta distribution with parameters r and N. Thus, a psychologist might be interested in comparing two groups performing the same task, to see if they come up with errors ("successes") at the same rate. Suppose that the time to the *tenth error* is recorded for each group. This time is called A for the first group, and B for the second. Then, $X = A/(A + B)$ and is distributed as a beta variable with $r = 10$, $N - r = 10$, if the occurrence of an error in either group is the outcome of a stable, independent, Poisson process with the same rate m.

Formally, any chi-square variable may also be considered a gamma variable where $r = \nu/2$ and $m = 1/2$. Furthermore, for an F variable with ν_1 and ν_2 degrees of freedom, if one takes

$$X = \frac{F}{F + (\nu_2/\nu_1)}$$

then X is a beta variable with $r = \nu_1/2$ and $(N - r) = \nu_2/2$. Thus, all of the theoretical continuous variables we have studied in this chapter are related.

Any intermediate text in mathematical statistics will give more details on the gamma and beta distributions, and tables are to be found for each type of distribution in the *Biometrika Tables for Statisticians* (1968).

EXERCISES

1. A normal distribution has a certain mean μ and a certain variance σ^2. What is the probability that a value drawn from this distribution represents a z value of 1.15 or more? What is the probability that $z = -1.15$ or less? What is the probability that *either* $z \geq 1.15$ *or* $z \leq -1.15$? What then is the probability that $z^2 \geq (1.15)^2$? Show by use of Table IV of Appendix D that this is exactly the probability that a value of $\chi^2 \geq 1.3230$. Why should this be true?

2. Using the same reasoning as in exercise 1, find the probability $p(\chi^2 \geq h)$ where h represents in turn each of the values given below, in a chi-square distribution with one degree of freedom.

(a) $h = 1.00$
(b) $h = 2.00$
(c) $h = 3.00$
(d) $h = .50$
(e) $h = .33$

3. A normal distribution has mean $\mu = 100$ and variance $\sigma^2 = 10$. Five values are taken independently and at random from this distribution. The z-value is computed for each value drawn, and then the sum of the squared z-values found. What should one expect this sum to be? What is the approximate probability that this sum is greater than 15? Greater than 11? Less than 1.6?

4. A sample of seven observations drawn independently and at random from a normal distribution gave the following results:

22, 2, 0, 30, 28, 26, 32

Test the hypothesis that the population standard deviation is 10 against the alternative that it has some other value. Use $\alpha = .05$. Then test the same hypothesis against the alternative $\sigma^2 < 10$.

5. For the data of exercise 4 find the 95 percent confidence limits for the population standard deviation.

6. A study was concerned with the effects of anxiety upon test performance. It was theorized that very anxious people might not only show a lower average performance relative to their nonanxious peers, but also a higher degree of variability in performance. A random sample of 22 college students showing normal general ability but high test-anxiety was taken. Each such student was given a simple mathematics test. In general, the scores of college students on this test follow an approximately normal distribution with a mean of 55 and a standard deviation of 6. The sample of anxious students had a mean of 51 and a standard deviation $s = 10$. Test both of the hypotheses implied here, using $\alpha = .05$ for each.

7. For exercise 6 establish both a 95 percent confidence interval for the mean of the population of anxious students, and a 95 percent confidence interval for their variance.

8. Suppose that a sample of 20 independent observations is drawn at random from a normal population with standard deviation of 30. What is the probability that the sample value S will turn out to be less than or equal to 25?

9. A certain achievement test is standardized in such a way that a score value of 80 lies one standard deviation below the mean score value of 100 in the normally distributed population. A sample of 30 scores is drawn at random and independently from this population. What is the approximate probability that the sample standard deviation S will be at least 1.2 times larger than the population standard deviation?

10. In exercise 9 above, what is the expected value of the ratio of the sample variance S^2 to the population variance σ^2? What is the *most likely* value of that ratio? (**Hint:** Look carefully at Section 9.1, following 9.1.4.)

11. Six observations are drawn independently and at random from a normal population. The values obtained are

106, 98, 97, 103, 101, 105

Find the 95 percent confidence interval for the population variance.

12. Assume that the sample given in exercise 32, Chapter 5 represents a normally-distributed population. Then test the hypothesis that the variance of this population is 20, against the alternative that the variance is greater than 20, using $\alpha = .05$.

13. From exercise 14, Chapter 6, test the hypothesis that the population standard deviation is .50, as opposed to the hypothesis that $\sigma \neq .50$. Assume that the population in question is normal. (**Hint:** See Section 9.6.)

14. Find the 95 percent confidence interval for σ^2 in exercise 13.

15. In a study of the size of the vocabularies of six-year-old normal children, the experimenter entertained the notion that size of vocabulary should increase from ages five to six, but that the variability of these two age groups should be about the same in this respect. Previous studies had shown that the standard deviation for five-year-olds was about 24 words, and that size of vocabulary is approximately normally distributed. A sample of 25 six-year-old children showed that $S^2 = 945$. Does the standard deviation of the population of six-year-olds seem to be equal to about 24?

16. An experimenter drew random samples from two normal distributions, with different means, but with the same variance. The first sample contained 18 independent observations, and the second contained 11. If the two sample variances turned out to be $S_1^2 = 92$ for the first sample, and $S_2^2 = 86$ for the second, what are the 99 percent confidence limits for the variance of either population?

17. Suppose that the experimenter of exercise 15 had taken both a sample of 25 six-year-old children, and a similar random sample of nine-year-olds, also consisting of 25 children. The six-year-olds again showed $S^2 = 945$, while the nine-year-old children had $S^2 = 919$. Is it reasonable to conclude that six-year-olds are more variable than the nine-year-olds? Use $\alpha = .05$.

18. The test carried out in exercise 17, Chapter 8 involves the assumption that the two populations have exactly the same variance. Test this assumption, using $\alpha = .01$.

19. Using the two samples of exercise 21, Chapter 8, test the hypothesis that the two populations have the same variance, against the alternative that the variance for population I is greater. Again use $\alpha = .05$.

20. Use the data of exercise 20, Chapter 8 to test the hypothesis that the two brands of gasoline have the same variance in terms of mileage, against the hypothesis that Brand I is the more variable. The .05 level for α may be used.

21. Two normally distributed populations are being compared in order to see if they have the same values for their variances. The alternative hypothesis is that Population I has the larger variance. A sample of 13 independent observations are drawn from Population I and a separate random sample of 9 independent observations are made of Population II. Sample I shows that $S_{\text{I}}^2 = 141$, while Sample II shows $S_{\text{II}}^2 = 123$. What can one conclude about the variances of the two populations?

22. Suppose that independent random samples each consisting of 16 cases were drawn at random from two normal populations. The first sample produced a sample standard deviation $S = 38.2$ and the second sample a standard deviation $S = 89.2$. The null hypothesis was $H_0: \sigma_{\mathrm{I}}^2 = \sigma_{\mathrm{II}}^2$ and the alternative was "not H_0." Test the null hypothesis.

23. Is it possible to test a hypothesis of the form $H_0: \sigma_{\mathrm{I}}^2 = k\sigma_{\mathrm{II}}^2$. Explain why it should or should not be possible, and, if possible, how one would go about doing so.

24. An airport has an information counter at which people tend to arrive and ask questions at the rate of four per minute. The process seems to be stable throughout the day, and the arrivals and departures at the counter may be assumed to be independent. A random sample of 100 different times and days is taken. On each occasion the time is recorded between the departure of a person at the counter and the arrival of the next person with a question. The mean of this sample of 100 occasions is taken. What is the expected value and the standard error of such a mean? (**Hint:** Use the exponential distribution.)

25. Suppose that in exercise 24 above the time is recorded until the next 10 people have asked for information. Once again the mean of 100 occasions is taken. What is the expected value and the standard error of such means? (**Hint:** Use the gamma distribution.)

Chapter 10
THE GENERAL LINEAR MODEL AND THE ANALYSIS OF VARIANCE

In Chapter 8 and elsewhere we have discussed the idea of a statistical relation between some independent variable X and a dependent variable Y. The variable X may stand for some set of qualitatively different groupings, or treatments, under the control of the experimenter. The variable Y, however, is not under the direct control of the experimenter, although its value may depend in some way on the X value.

In other contexts, the independent variable X may be quantitative, representing groups or treatments that differ in amount or degree of the thing that the values of X symbolize. Such an independent variable is still thought of as under the direct control of the experimenter, while the Y value is thought of as free to vary in some way as the X values are varied.

Although our discussion so far has been confined to the situation where there are only two independent-variable groups, most studies employ a number of qualitatively different groups, or a number of different values of the variable X. Thus, we need ways to extend the ideas underlying the comparison to two groups to the situation where any number of different groups may be compared on the basis of their Y, or dependent-variable, values.

In order to do this, we must place the notion of a statistical relation between X and Y on a more concrete basis. This calls for a *mathematical model* of how the variable X influences or is associated with the variable Y. This model should be simple, but it should also be general enough to cover both the situations of qualitative and of quantitative groupings in terms of the independent variable. It should also allow expansion to many independent variables and to variables of mixed types. The model must also distinguish between systematic connections between X and Y, and chance or ''error'' variations in Y.

There are many different kinds of mathematical formulations, or models, that one could invent in order to represent the relationship between any independent variable X and any dependent variable Y. However, the simplest and most flexible of such models is known as a *linear model*. In essence, *any linear model of data states that an observed value of the dependent variable is equal to a weighted sum of values associated with one or more independent variables, plus a term standing for error.* For the moment we will consider only the most elementary version of this linear model, involving only one independent variable X and one dependent variable Y. However, the concept is broad enough to cover much more complicated situations, as we shall see.

Now we will formulate a simple linear model as it applies to three experimental groups. There is nothing special about the number "3" here; we could do the same thing for 2 or for any number of groups. The use of three groups simply lets us show some features most clearly.

Suppose that the independent X represents *three qualitatively different groups or experimental treatments*. In all, these groups contain some N subjects or observations. Any subject will be in exactly one of the groups, and will exhibit some value of the dependent variable, or y_i. Then a linear model showing how y_i depends on the X variable would be

$$y_i = a_0x_0 + a_1x_{i1} + a_2x_{i2} + a_3x_{i3} + e_i. \qquad [10.1.1]$$

Here, x_0 is a constant value that enters into the y_i score for each subject. The symbol x_{1i} stands for the value of X as it applies to observation i depending on whether or not that observation is in group 1. The value of x_{i2} pertains to whether or not observation i is in group 2, and similarly for x_{i3}. The symbols a_0, a_1, a_2, and a_3 stand for weights that the x values receive in making up the y_i score value. Finally, the term e_i represents the error component of the observed y_i score.

Now you may well ask how we talk about the "values" of x_{i1}, x_{i2}, and x_{i3} when it was clearly stated that the three treatments or groups are qualitatively different. Here, the values of x are defined in such a way as to indicate the membership of an individual subject or observation in a qualitatively distinct group, as follows:

$$x_0 = 1, \text{ for all } i$$

$$x_{i1} \begin{cases} = 1, \text{ if } i \text{ is in group 1} \\ = 0, \text{ otherwise} \end{cases}$$

$$x_{i2} \begin{cases} = 1, \text{ if } i \text{ is in group 2} \\ = 0, \text{ otherwise} \end{cases}$$

$$x_{i3} \begin{cases} = 1, \text{ if } i \text{ is in group 3} \\ = 0, \text{ otherwise.} \end{cases}$$

This device, known as the use of "indicator variables" or "dummy variables," lets us sort out the subjects into the qualitative groupings, according to the x values assigned to particular individuals i.

Then, according to this linear model, the y value of any individual i in group 1 is found from

$$y_i = a_0 + a_1 + e_i.$$

For a subject in group 2, the linear model states that

$$y_i = a_0 + a_2 + e_i,$$

and for group 3,

$$y_i = a_0 + a_3 + e_i.$$

The linear model thus lets us state in detail the make up of any y_i value, in terms of the overall level of the Y values, the "effect" of the treatment group, and error.

In another study, the three groups may be *quantitatively* different with respect to the independent variable. Thus, group 1 might receive 3 hours of practice on some task, group 2, 6 hours, and group 3, 9 hours. Then the linear model for an individual in group 1 would be

$$y_i = a_0 + a_1(3) + e_i.$$

For group 2 the model would be

$$y_i = a_0 + a_2(6) + e_i$$

and for group 3,

$$y_i = a_0 + a_3(9) + e_i.$$

In other words, now the X values are no longer merely indicators, as before, but numerical values standing for amount or degree of treatment applied.

In the next section we will examine a broad version of such a model, known as the general linear model, and point out some of the variations that will prove useful in the remainder of this text.

10.2 / A GENERAL LINEAR MODEL

In an experiment of other research study, there are many different things that are going to influence the outcome that we observe for any individual subject. Thus, suppose that we wished to investigate the reading achievement score of fifth-grade boys (variable Y). Then an immediate decision has to be made that some things are going to be held constant. We have already decided to study human children, and not animals; we have decided to study children in the fifth grade, not children of all ages; we have decided to study boys, not boys and girls. Furthermore, when we decide to use the elementary school down the street, we may tacitly have limited the observations to one socio-economic and ethnic mix, and a particular staff of fifth-grade teachers. When we decide to give all the tests next Tuesday, we have also imposed certain constant constraints. In short, when a group of observations are made, many things must be limited or held constant all of which will go into determining a general level that the obtained values of Y will exhibit. Our model needs to reflect this, and so we will enter a constant value x_0 to stand for this general level we have established by defining our sample as we have. This value receives some weight a_0 in any observation.

However, a great many things are still going to make fifth-grade boys differ in their reading achievement. Some of these things may be under our control, as when we can ask the different teachers in the school to use different reading texts of our choice. Still other things that may create differences in reading achievement are things we know about and can measure, although they may not be under our direct control. Proper nutrition, or its lack, conceivably affects reading achievement, as does general ability, and so on for a host of other things. All of these known factors or variables in the study, that can operate to make the Y values different, may be thought of as the study's systematic features. Each of these features can be symbolized as a variable, such as X_j. Perhaps we can identify a set of some J of these variables, $(X_1, \cdots, X_J)$ which we can either control or observe systematically for any individual.

Finally there are the things that make boys different from each other in reading achievement, but which we can neither control nor measure, or perhaps even adequately explain. Such factors, peculiar to a particular individual in a particular time and place, we identify as ''error'' in our study. This too must have a place in our underlying model.

Without going further toward specifying the exact nature of the constant, the systematic, and the error components that underlie the Y values we observe for particular individuals, we can represent these things in a model:

$$y_i = a_0x_0 + a_1x_{i1} + a_2x_{i2} + \cdots + a_Jx_{iJ} + e_i \qquad [10.2.1^*]$$

Here, x_0 stands for the general level of the group of observations, and a_0 is the weight that the general level carries in determining the score of individual i. Then x_{ij} symbolizes the value of some variable x_1 as it relates to individual i, and a_1 represents the weight or ''effect'' of that variable in determining y_i. Similarly, for all the other variables that we can identify and observe systematically in the experiment, each variable X_j has some influence x_{ij} on the value y_i. The weight of that influence is given by a_j. Finally, there is the component of the y_i value that entails all of the factors that we either do not or cannot know about. This is error, or e_i. In short, we can envision the actual y_i score for some individual i as being a weighted sum of values of a number of variables X_j, plus a term standing for error.

This is a *general linear model*. It is general in the sense that we really have not specified the precise nature of the X variables, exactly how many there are, nor the weights which they receive. Rather, these specifics of the model may be tailored to fit a given experimental or research situation. Such a conception is enormously flexible in lending itself to all sorts of data-collection situations. By adapting and interpreting the linear model in various ways, we are able to use it as a basis for describing situations which, on their surface, appear quite disparate, but which nevertheless fit under this general conception of how an observed score or value is composed of the weighted contibutions of many factors.

Most of the techniques to be discussed in this and subsequent chapters presuppose that such a linear model underlies the data. However, there are many variations that can be applied to this general linear model to fit the particular kind of situation or experiment that is being modeled. As suggested in the previous section, two variations have to do with whether the x values in the model are simply ''indicators'' or ''markers'' showing qualitatively different treatments or groups, or whether any x stands for a numerical value (other than 0 or 1) representing an amount of the experimental treatment in question.

When the x values in a linear model are indicators, taking on only the values of 0 and 1, the model is sometimes said to be an *experimental design model*, or *an analysis of variance model*. We will use the latter term, since the statistical techniques known collectively as "the analysis of variance" will provide our chief use of such models.

More generally, when the x variables in the linear model can take on any real number values, this is called *a multiple regression model*. As we shall see in Chapter 14, experimental design models such as we will study in this chapter are actually special cases of the multiple regression model. Various kinds of *problems in correlation and problems in regression,* as they will be outlined in Chapters 13 and 14, are also special instances of this overarching model.

In every single instance in which we will be using a linear model, or special case of the general linear model, we will have to identify the values or weights represented by a_0 through a_J. In other words, one of the key problems will be deciding on how much weight each variable actually does have in determining variations among the Y values we observe. In order to find the values of these weights, we will always rely on *the criterion of least squares,* a simple version of which we first encountered in Chapter 4. Although it will not be possible for us to illustrate this principle each time it occurs, or to show how it applies in complex situations, you should remember that this is the main avenue for turning the formal statement of the relation between Y and the various X variables into a numerical form that we can actually observe and use. (A detailed and moderately simple treatment of linear models and their variations may be found in Ward and Jennings (1973). This would make good material for study following Chapter 14 of this book, for the student seriously interested in getting an overview of methods for the analysis of experiments.)

With this preamble, we will now return to the immediate problem, which is to describe the relationship between the qualitative group to which an individual belongs and the Y value that one should expect that individual to show.

10.3 / ANALYSIS OF VARIANCE MODELS

For the time being, we are going to confine our attention to analysis of variance models, or experimental design models, as special cases of the general linear model. Recall that for an experimental design model, the x_{ij} values entering into the model of 10.2.1 are indicator variables, taking on only the value "1" if individual i is in group j, and the value "0" otherwise. Then the pattern of "1" and "0" values as they apply to the set of observations is called *the structural matrix* (or simply the "structure") of the experiment being described.

For example, this time imagine a group of 8 subjects, assigned by twos to four groups, each representing a qualitatively different treatment. Then the structure of the experiment would be shown as in Table 10.3.1, where the rows stand for the subjects, and the columns contain the x_{ij} values.

Any display of the x_{ij} values for the experiment, such as that shown in Table 10.3.1, is the structural matrix. Any designed experiment can be represented by such a matrix of "1" and "0" values.

Now applying the general linear model to this specific situation, we have, for $J = 4$ groups,

$$y_i = a_0x_0 + a_1x_{i1} + a_2x_{i2} + a_3x_{i3} + a_4x_{i4} + e_i. \qquad [10.3.1]$$

	subjects	X_0	X_{i1}	X_{i2}	X_{i3}	X_{i4}
	1	1	1	0	0	0
	2	1	1	0	0	0
groups	3	1	0	1	0	0
	4	1	0	1	0	0
	5	1	0	0	1	0
	6	1	0	0	1	0
	7	1	0	0	0	1
	8	1	0	0	0	1

Table 10.3.1

The structure of an experiment involving eight subjects in four groups

Thus we find that the eight observation values y_i should be made up as follows:

$$
\begin{aligned}
y_1 &= a_0 + a_1 + e_1\\
y_2 &= a_0 + a_1 + e_2\\
y_3 &= a_0 + a_2 + e_3\\
y_4 &= a_0 + a_2 + e_4\\
y_5 &= a_0 + a_3 + e_5\\
y_6 &= a_0 + a_3 + e_6\\
y_7 &= a_0 + a_4 + e_7\\
y_8 &= a_0 + a_4 + e_8
\end{aligned}
$$

All observations share the constant a_0. Only observations 1 and 2 share the "effect" a_1, only observations 3 and 4 the effect a_2, and so forth. An error term, e_i, is then associated with any subject or observation i.

Naturally, we could have constructed such a model for any number of subjects arranged into any number of groups. Regardless of how many treatment groups, however, the score of each subject is assumed to consist of a constant part common to all subjects, an effect of the group to which the subject belongs, and an error term unique to that subject in that group.

10.4 / LEAST SQUARES AND THE IDEA OF AN EFFECT

In the linear model describing the relationship of variable X to variable Y, the various constant values a_0, a_1, a_2, and so on were described as weights or "effects." The term "effect" is used especially when an analysis of variance model is being discussed. What are these effects, and how are they estimated?

Obviously, the a_j weights in such a linear model could be anything, provided that they represent some reasonable theory of how the X variables relate to Y. However, we want to employ a principle that will provide the "best weights" or "best-fitting weights" to a given data situation. That is, if we wish to use our linear model to describe the Y values that occur on the basis of the X values that entered into the experiment, or to predict the Y value for an individual on the basis of information about the X values that individual shows, we want weights a_j that will provide the best such description or prediction. As mentioned above, the most usual way of achieving a good fit between data and such unknown constants in statistics is the so-called "principle of least squares" (or equivalently, the principle of minimum variance). This principle has already been used in Section 4.15, where it was shown that taking deviations about the sample mean gives the lowest value of the average squared deviation, or variance. We will now invoke this principle in order to find the best-fitting a_j values for our analysis of variance model.

Since e_i is a part of the linear model underlying any y_i observation, then it must be true that for any i:

$$e_i = y_i - a_0 x_0 - a_1 x_{i1} - \cdots - a_J x_{iJ}. \qquad [10.4.1]$$

Then we choose values for $a_0, a_1, \cdots, a_J$ so as to minimize the variance of these e_i values. That is, we minimize

$$\text{variance of errors} = \sum_i^N \frac{e_i^2}{N} - (M_e)^2. \qquad [10.4.2]$$

We will also place two more restrictions or "side conditions" on our model: on the average, across all observations, we would like the value of e to be zero. That is, we set $M_e = 0$. We also require the average of the a_j values to be zero over the J groups. (The first of these restrictions makes the problem of minimizing the variance of the errors equivalent to minimizing the sum of their squares.) Then when minimization methods from the differential calculus are applied, it turns out that the sum of the squared errors is a minimum when

$$\begin{aligned} a_0 &= M_Y \\ a_1 &= M_1 - M_Y \\ a_2 &= M_2 - M_Y \\ &\cdots \\ a_J &= M_J - M_Y. \end{aligned} \qquad [10.4.3]$$

In other words, in an analysis of variance model, where N observations are arranged among J different groups, if the sum of the squared errors is to be minimized, the value of a_0 must be the mean of the Y values. Furthermore, any effect a_j must be the difference between the mean of the Y values in group j and the mean of all of the Y values.

Now we have a model for describing a set of data, which must fit the data in the least squares sense:

$$y_i = M_Y + (M_j - M_Y) + e_i. \qquad [10.4.4^*]$$

Note that this implies that the best prediction one can make about the score of any individual i in group j is M_j, the mean score for that group.

10.5 / POPULATION EFFECTS

We have just seen the results of applying the least squares criterion to a linear analysis of variance model, for a particular set of data containing N cases arranged among J experimental or treatment groups. However, we are usually not as interested in describing a particular set of data so much as we are in large populations of potential observations in the different experimental treatments. Fortunately, the model and the least squares argument apply just as well to vast populations of potential observations as they do to a small set of N cases.

Suppose that we now think of a large population of potential observations within experimental treatment j. Each observation i in treatment population j will have a dependent variable score y_{ij}. For the population, the model equivalent to 10.3.1 is, for each i and j,

$$y_{ij} = \alpha_0 + \alpha_j + e_{ij} \qquad [10.5.1*]$$

(Here, j subscripts have been added to both y and e to emphasize that we are speaking about individual i in population j; this will help us keep up with individuals and populations later on. Furthermore, the a values in expression 10.3.1 have been replaced with Greek letters to emphasize that this is a model of the population.)

Then by the criterion of least squares it must be true that

$$\alpha_0 = \mu$$

and

$$\alpha_j = \mu_j - \mu.$$

The constant α_j is called the effect of treatment j. The effect α_j is simply the difference between the mean of the population j, or μ_j, and the general mean over all treatment populations, μ.

To get an idea of what these effects actually represent, consider the grand population, formed by pooling the J treatment populations. Now let the proportionate representation of any population j in the grand population be $p(x_j)$, so that the probability that any observation drawn at random from the grand population belongs to population j is $p(x_j)$. If the J treatment populations are equally represented in the grand population, then for every j the value of $p(x_j) = 1/J$.

Now under these conditions, the mean μ of the grand population would be the weighted average of the several population means:

$$\mu = \sum_j p(x_j)\mu_j.$$

In the special case where $p(x_j) = 1/J$ for all j, the grand mean would be

$$\mu = \frac{\sum_j \mu_j}{J}.$$

Then, as we have seen, the effect of treatment j is defined as the deviation of the mean of population j, μ_j, from the grand popolation mean, μ:

$$\textbf{effect of treatment } j = (\mu_j - \mu) \qquad [10.5.2*]$$

or $$\alpha_j = (\mu_j - \mu).$$

This symbol α_j (*not* to be confused with the alpha standing for the probability of Type I error) will stand for the effect of any single treatment j.

Since the grand population mean μ is also the weighted sum of all of the treatment population means, it follows that the weighted sum of all of the effects must be zero:

$$\sum_j p(x_j)\alpha_j = \sum_j p(x_j)(\mu_j - \mu) = \mu - \mu = 0. \qquad [10.5.3^*]$$

In the special case where $p(x_j) = 1/J$ for all treatment populations j, then the sum of the effects is zero:

$$\frac{\sum_j \alpha_j}{J} = 0 \text{ implies that } \sum_j \alpha_j = 0. \qquad [10.5.4^*]$$

Now suppose that there is absolutely no effect associated with any treatment. This means that

$$\alpha_j = 0$$

for each and every treatment population j. This is equivalent to the statement that

$$\mu_1 = \mu_2 = \cdots = \mu_J = \mu,$$

where the index numbers 1, 2, $\cdots$, J designate the various treatments. *The complete absence of effects is equivalent to the absolute equality of all of the population means.*

Notice that when there are no treatment effects,

$$\sum_j p(x_j)\alpha_j^2 = 0; \qquad [10.5.5^*]$$

that is, the weighted sum of the α_j^2 must be zero, since each and every α_j is zero when no treatment effects exist.

The expected score value over all individuals i in population j is

$$E(y_{ij}) = \mu + \alpha_j + E(e_{ij}),$$

or

$$\mu_j = \mu_j + E(e_{ij}).$$

Thus, for any population j, the expectation of e_{ij} over all individuals is zero:

$$E(e_{ij}) = 0. \qquad [10.5.6^*]$$

10.6. / MODEL I: FIXED EFFECTS

It is important to distinguish between two different sampling situations to which the analysis of variance applies. These differ both in the way that experimental treatments are selected and in the kinds of inferences one makes from the analysis. The formal statistical models applying to these two sampling situations have become known as "Model I" or the "fixed effects model," and "Model II," the "random effects model" (or, sometimes, "components of variance model"). The situations calling for these two different models will be compared briefly.

Imagine a situation where several experimental treatments are to be administered.

Suppose that there are J different such treatments, and that each treatment is to be administered to one and only one experimental group. Each of the J groups consists of individuals chosen at random and independently and assigned at random to the groups. For example, four different tranquilizing drugs are to be compared for the effect each has on driving skill. Some N subjects are chosen at random and independently and allotted at random to four nonoverlapping groups. The individuals in the first group are given drug 1, those in the second drug 2, and so on. Then the groups are to be compared on the dependent variable Y, a score on driving skill. Or, perhaps, six different models of teaching second-grade arithmetic are known. School children from a specific population are sampled and allotted at random to six different groups, each group representing one of the six instructional methods. The groups are then compared on their average achievement after a year's instruction. In both examples, members of some small set of treatments are to be compared, and each treatment of interest is actually used in the experiment.

Experiments to which Model I applies are distinguished by the fact that inferences are to be made only about differences among the J treatments actually administered, and about no other treatments that might have been included. In advance of the actual experiment, the experimenter decides to see if differences in effect exist among some fairly small set of treatments or treatment combinations. There is interest in these treatments or combinations, and *no* others. Each treatment of immediate interest to the experimenter is actually included in the experiment, and the set of treatments or treatment combinations applied exhausts the set of treatments about which the experimenter wants to make inferences. The effect of any treatment is "fixed," in the sense that it must appear in any complete repetition of the experiment on new subjects.

As an example to which the fixed effects model does not apply, consider an experiment involving a projective test. This test consists of ten different stimulus cards given in turn to the subject, who gives verbal responses to each. Among the things recorded about the behavior of a subject is the mean time between the presentation of a card and the first response. The experiment is designed to study the effect of the order of presentation of the cards to a subject upon average first response time. However, there are 10! or 3,628,800 different possible orders of presentation of the cards. The experimenter takes a random sample of, say, twenty such orders, and tests a different group of randomly selected subjects under each. Here, the experimenter is not really interested in the twenty orders actually administered so much as in the possible effects of order *in general*. Thus, the fixed effects model is inappropriate for this problem.

The methods to be discussed in this chapter, as well as those in Chapter 12 and most of those in Chapter 14, will assume the fixed-effects model. Other models will apply in Chapters 11 and 13, however.

10.7 / DATA GENERATED BY A FIXED-EFFECTS MODEL

To gain some feel for this linear model, imagine three samples consisting of three observations each. Suppose that these three samples represent identical population distributions, and that there is *no* variability (that is, no error) within any of the populations. If the mean of each of the populations were $\mu = 40$, then our sample results should look like this:

Sample 1	*Sample 2*	*Sample 3*
40	40	40
40	40	40
40	40	40

There should be no differences either between or within samples if this were the true situation. When this is true the linear model becomes simply

$$y_{ij} = \mu,$$

since $\alpha_j = 0$, $e_{ij} = 0$ for all i and j.

Now suppose that the three samples are given different treatments, and that treatment effects exist, but that there is once again no variability within a treatment population (again, no error). Our results might look like this:

Sample 1	*Sample 2*	*Sample 3*
40 − 2 = 38	40 + 6 = 46	40 − 4 = 36
40 − 2 = 38	40 + 6 = 46	40 − 4 = 36
40 − 2 = 38	40 + 6 = 46	40 − 4 = 36

Here there are differences between observations in different treatments, but there are no differences within a treatment sample. The linear model here is

$$y_{ij} = \mu + \alpha_j,$$

since $\alpha_j \neq 0$ while $e_{ij} = 0$, for any i and j.

In actuality there is always variability in a population, so that there is sampling error. The actual data we might obtain would undoubtedly look something like this:

Sample 1	*Sample 2*	*Sample 3*	
40 − 2 + 5 = 43	40 + 6 − 5 = 41	40 − 4 + 3 = 39	
40 − 2 + 2 = 40	40 + 6 + 1 = 47	40 − 4 − 2 = 34	
140 − 2 − 3 = 35	40 + 6 + 8 = 54	40 − 4 + 1 = 37	
$M_1 = 39.3$	$M_2 = 47.3$	$M_3 = 36.7$	$M = 41.1$

Here, a random error component has been added to the value of μ and the value of α_j in the formation of each score. The linear model in this situation is

$$y_{ij} = \mu + \alpha_j + e_{ij}.$$

Notice that not only do differences exist between observations in different treatments, but also between observations in the same treatment.

If we estimate the effect of treatment 1 by taking

$$\text{est. } \alpha_1 = M_1 - M = 39.3 - 41.1 = -1.8$$

it happens in this example that we are almost right, since the data were fabricated so that $\alpha_1 = -2$. Likewise, our estimate of α_2 is in error by .2 and our estimate of α_3 in error by −.4. Although these errors may seem rather slight in this example, there is no guarantee in any given experiment that they will not be very large. Thus we need to evaluate how much of the apparent effect of any experimental treatment is, in fact, due to error before we can decide that something systematic is actually occurring.

This example should suggest that evidence for experimental effects has something to do with the differences *between* the different groups relative to the differences that exist *within* each group. Next, we will turn to the problem of separating the variability among observations into two parts: the part that should reflect both experimental effects and sampling error, and the part that should reflect sampling error alone.

10.8 / THE PARTITION OF THE SUM OF SQUARES FOR ANY SET OF *J* DISTINCT SAMPLES

In this section we are going to leave the study of population effects for a while, and show how the variability in any set of *J* experimentally different samples may be partitioned into two distinct parts. Actually, we will do this in terms of the sum of squared deviations about the grand mean for the samples, rather than the sample variance itself.

Any score y_{ij} in sample j exhibits some deviation from the grand sample mean of all scores, M:

$$\text{deviation} = (y_{ij} - M).$$

This deviation can be thought of as composed of two parts,

$$(y_{ij} - M) = (y_{ij} - M_j) + (M_j - M), \tag{10.8.1}$$

the first part being the deviation of y_{ij} from the mean of group j, and the second being the deviation of the group mean from the grand mean.

Now suppose that we square the deviation from M for each score in the entire sample, and sum these squared deviations across all individuals i in all sample groups j:

$$\begin{aligned}\sum_j \sum_i (y_{ij} - M)^2 &= \sum_j \sum_i [(y_{ij} - M_j) + (M_j - M)]^2 \\ &= \sum_j \sum_i (y_{ij} - M_j)^2 + \sum_j \sum_i (M_j - M)^2 \\ &\quad + 2 \sum_j \sum_i (y_{ij} - M_j)(M_j - M).\end{aligned} \tag{10.8.2}$$

Now look at the last term on the right in equation 10.8.2 above:

$$\begin{aligned}2 \sum_j \sum_i (y_{ij} - M_j)(M_j - M) &= 2 \sum_j (M_j - M) \sum_i (y_{ij} - M_j) \\ &= 0\end{aligned} \tag{10.8.3}$$

since the value represented by the term $(M_j - M)$ is the same for all i in group j, and the sum of $(y_{ij} - M_j)$ must be zero when taken over all i in any group j.

Furthermore,

$$\sum_j \sum_i (M_j - M)^2 = \sum_j n_j(M_j - M)^2 \tag{10.8.4}$$

since, once again, $(M_j - M)$ is a constant for each individual i figuring in the sum. Putting these results together, we have

$$\sum_j \sum_i (y_{ij} - M)^2 = \sum_j \sum_i (y_{ij} - M_j)^2 + \sum_j n_j(M_j - M)^2. \quad [10.8.5^*]$$

This identity is usually called the *partition of the sum of squares,* and is true for any set of J distinct samples. Verbally, this fact can be stated as follows: the total sum of squared deviations from the grand mean can always be separated into two parts, the sum of squared deviations within groups, and the weighted sum of squared deviations of group means from the grand mean. It is convenient to call these two parts

$$\text{SS within} = \sum_j \sum_i (y_{ij} - M_j)^2 \quad [10.8.6^*]$$

for **sum of squares within groups,** and

$$\text{SS between} = \sum_j n_j(M_j - M)^2 \quad [10.8.7^*]$$

for **sum of squares between groups.** Thus, it is a true statement that

$$\text{SS total} = \text{SS within} + \text{SS between}. \quad [10.8.8^*]$$

The meaning of this partition of the sum of squares into two parts can easily be put into common sense terms: Individual observations in any sample will differ from each other, or show variability. These obtained differences among individuals can be due to two things. Some pairs of individuals are in different treatment groups, and their differences are due either to the different treatments, or to chance variation, or to both. The sum of squares between groups reflects the contribution of different treatments, as well as chance, to intergroup differences. On the other hand, individuals in the *same* treatment groups can differ only because of chance variation, since each individual within the group received exactly the same treatment. The sum of squares within groups reflects these intragroup differences due only to chance variation. Thus, in any sample two kinds of variability can be isolated: the sum of squares between groups, reflecting variability due to treatments *and* chance, and the sum of squares within groups, reflecting chance variation alone.

10.9 / ASSUMPTIONS UNDERLYING INFERENCES ABOUT TREATMENT EFFECTS

The partition of the sum of squares is possible for any set of J distinct samples, and no special assumptions about populations or sampling are necessary in its derivation. However, before we can use sample data to make inferences about the existence of population effects, several assumptions must be made. These are as follows.:

1. For each treatment population j, the distribution of e_{ij} is assumed normal.
2. For each population j, the distribution of e_{ij} has a variance σ_e^2, which is assumed to be the same for each treatment population.
3. The errors associated with any pair of observations are assumed to be independent. A consequence of this assumption is that if h and i stand for any pair of observations, and j and k for any pair of treatments, then

   $$E(e_{ij}e_{hj}) = 0$$

   and

   $$E(e_{ij}e_{hk}) = 0.$$

In short, we are going to regard our observations as independently drawn from normal treatment populations each having the same variance, and with error components independent across all pairs of observations.

10.10 / THE MEAN SQUARE BETWEEN GROUPS

The next question is how to use the partition of the sum of squares in making inferences about the existence of treatment effects. First of all, we will examine the expectation of the sum of squares between groups. Let us define

$$M_{ej} = \frac{\sum_i e_{ij}}{n_j} \text{ and } M_e = \frac{\sum_j \sum_i e_{ij}}{N} = \frac{\sum_j n_j M_{ej}}{N}.$$

Then for any group j, a simple substitution from 10.5.1 above shows that

$$M_j = \frac{\sum_i y_{ij}}{n_j} = \mu + \alpha_j + M_{ej} \qquad [10.10.1]$$

and

$$M = \frac{\sum_j \sum_i y_{ij}}{N} = \mu + M_e. \qquad [10.10.2]$$

since

$$\sum_j \frac{n_j \alpha_j}{N} = \sum_j p(x_j)\alpha_j = 0.$$

Furthermore, let M_j be the mean of the sample given the particular treatment j, and let M stand for the mean over all individuals in all treatments for the experiment. Then over all possible samples of n_j individuals from population j,

$$E(M_j) = \mu + \alpha_j + E(M_{ej}),$$

That is, the expected value of M_j depends on μ, on the effect α_j, and on the expected mean error over the n_j observations in group j. Since

$$E(M_j) = \mu_j,$$

it must be true that

$$E(M_{ej}) = 0.$$

Thus, the deviation of any sample group mean from the grand sample mean is actually

$$(M_j - M) = \alpha_j + (M_{ej} - M_e).$$

From this it follows that

$$\text{SS between} = \sum_j n_j(M_j - M)^2 = \sum_j n_j[\alpha_j + (M_{ej} - M_e)]^2. \qquad [10.10.3]$$

On taking the expectation of the SS between, we find

$$E(\text{SS between}) = E\sum_j n_j[\alpha_j + (M_{ej} - M_e)]^2 \qquad [10.10.4]$$
$$= \sum_j n_j\alpha_j^2 + E\sum_j n_j(M_{ej} - M_e)^2,$$

since, within any j, $E(M_{ej}) = 0$, and $E(M_e) = 0$. Then, since $E(M_{ej}^2) = \sigma_e^2/n_j$ and $E(M_e^2) = \sigma_e^2/N$,

$$E(\text{SS between}) = \sum_j n_j\alpha_j^2 + (J - 1)\sigma_e^2. \qquad [10.10.5^*]$$

Ordinarily, we deal with the *mean square between,*

$$\text{MS between} = \frac{\text{SS between}}{J - 1}. \qquad [10|10.6^*]$$

Then

$$E(MS \text{ between}) = \sigma_e^2 + \frac{\sum_j n_j\alpha_j^2}{J - 1}. \qquad [10.10.7^*]$$

The mean square between groups is an unbiased estimate of σ_e^2, the error variance, plus a term that can be zero only when there are no treatment effects at all. When the hypothesis of no treatment effects is absolutely true, then,

$$E(\text{MS between}) = \sigma_e^2. \qquad [10.10.8^*]$$

If any true treatment effects at all exist, then,

$$E(\text{MS between}) > \sigma_e^2. \qquad [10.10.9^*]$$

Accordingly, we can see that the mean square between groups gives one piece of the evidence needed to adjudge the existence of treatment effects. The sample value of MS between should be an unbiased estimate of error variance alone when no treatment effects exist. On the other hand, the value of MS between must be an estimate of σ_e^2 plus a positive quantity when any treatment effects exist.

Each sum of squares in the analysis of variance is associated with some number of degrees of freedom. What is the number of degrees of freedom for MS between? There are really only J different sample values that go into the computation of MS between: these are the J values of M_j. *Thus, there are $J - 1$ degrees of freedom for MS between.*

Naturally, MS between is always a sample quantity, and thus it must have a sampling distribution. However, it is easy to see what this sampling distribution must be: when there are no treatments effects, MS between is an unbiased estimate of σ_e^2. For this estimate, as for other estimates of σ^2 for normal populations (Section 9.3),

$$\frac{(\text{est. }\sigma_e^2)}{\sigma_e^2} = \frac{\chi^2_{(\nu)}}{\nu}.$$

The ratio of MS between to σ_e^2 must be a chi-square variable divided by degrees of freedom, *when* there are no treatment effects *and* the parent populations are normal (assumption 1, Section 10.9).

As yet we have no idea of the value of σ_e^2, so that the sampling distribution of MS between cannot be used directly to provide a test of the hypothesis of no treatment effects. Now, however, let us investigate the sampling distribution of MS within.

10.11 / THE MEAN SQUARE WITHIN GROUPS

What population value is estimated by the mean square within groups? Under the fixed-effects model it is obvious that the treatments administered cannot be responsible for differences that occur among observations within any given group. This kind of within-groups variation should be a reflection of random error alone. Keeping this in mind, let us find the expectation of the sum of squares within groups.

$$E(\text{SS within}) = E\left[\sum_j \sum_i (y_{ij} - M_j)^2\right].$$

For any given sample j,

$$E\frac{\left[\sum_i (y_{ij} - M_j)^2\right]}{n_j - 1} = \sigma_e^2 \qquad [10.11.1]$$

since for any sample j this value is an unbiased estimate of the population error variance, σ_e^2. Thus,

$$\begin{aligned} E(\text{SS within}) &= \sum_j E \sum_i (y_{ij} - M_j)^2 \\ &= \sum_j (n_j - 1)\sigma_e^2 \qquad [10.11.2^*] \\ &= (N - J)\sigma_e^2. \end{aligned}$$

We define

$$\text{Mean square within} = \text{MS within} = \frac{\text{SS within}}{N - J}.$$

Then it must be true that

$$E(\text{MS within}) = \sigma_e^2, \qquad [10.11.3^*]$$

the expected value of the mean square within is simply error variance alone. **The degrees of freedom for MS within is**

$$\sum_j (n_j - 1) = N - J. \qquad [10.11.4^*]$$

Surely you can anticipate the turn the argument takes now! We have MS between, which estimates σ_e^2 when there are no treatment effects, but a value greater than σ_e^2 when effects exist. Moreover, we have another estimate of σ_e^2 given by MS within, which does *not* depend on the presence or absence of effects: Two variance estimates which *ought* to be the same under the null hypothesis suggest the F distribution (Section 9.8), and this is what we use to test the hypothesis.

10.12 / THE *F* TEST IN THE ANALYSIS OF VARIANCE

The usual hypothesis tested using the analysis of variance is

$$H_0: \mu_1 = \cdots = \mu_j = \cdots = \mu_J,$$

the hypothesis that all treatment population means are equal. The alternative is just

H_1: not H_0,

implying that some of the population means are different from others. As we have seen, these two hypotheses are equivalent to the hypothesis of no effects and its contrary:

$H_0: \alpha_j = 0$, for all j

$H_1: \alpha_j \neq 0$, for some j.

The argument in Sections 10.8 through 10.11 has shown that *when H_0 is true,*

$E(\text{MS between}) = \sigma_e^2$

and

$E(\text{MS within}) = \sigma_e^2$,

both the mean square between and the mean square within are unbiased estimates of the same value, σ_e^2. On the other hand, *when the null hypothesis is false,* then

$E(\text{MS within}) < E(\text{MS between})$.

Since these mean squares divided by σ_e^2 are each distributed as chi-square variables divided by their respective degrees of freedom when H_0 is true, it follows that their ratio should be distributed as F, provided that MS between and MS within are *independent* estimates of σ_e^2. From the principle of Section 6.6 the following can be proved: **For *J* samples of independent observations, each drawn from a normal population distribution, MS between and MS within are statistically independent.** For each sample, the mean M_j is independent of the variance estimate s_j^2, provided that the population distribution is normal. By an extension of the principle given in Section 6.6, MS between, based on the J values M_j, must be independent of MS within, based on the several s_j^2 values; each piece of information making up MS between is independent of the information making up MS within, given normal parent distributions.

Finally, we have all the justification needed in order to say that the ratio

$$\frac{(\text{MS between}/\sigma_e^2)}{(\text{MS within}/\sigma_e^2)} = \frac{\text{MS between}}{\text{MS within}} \qquad [10.12.1^*]$$

is distributed as F with $J - 1$ and $N - J$ degrees of freedom, *when the null hypothesis is true*. This statistic is the ratio of two independent chi-square variables, each divided by its degrees of freedom, and thus is exactly distributed as F when H_0 is true.

The F ratio used in the analysis of variance always provides a *one-tailed* test of H_0 in terms of the sampling distribution of F. Evidence for H_1 must show up as an F ratio greater than 1.00, and an F ratio less than 1.00 can signify nothing except sampling error (or perhaps nonrandomness of the samples or failure of the assumptions). Therefore, for the analysis of variance, the F ratio obtained can be compared directly with the one-tailed values given in Table V. An α level is chosen in advance, and this value determines the section of the table one uses to determine the significance of the obtained F ratio.

10.13 / COMPUTATIONAL FORMS FOR THE SIMPLE ANALYSIS OF VARIANCE

Although the argument given above dealt with sums of squares defined as follows:

$$\text{SS total} = \sum_j \sum_i (y_{ij} - M)^2$$

$$\text{SS within} = \sum_j \sum_i (y_{ij} - M_j)^2$$

$$\text{SS between} = \sum_j n_j(M_j - M)^2$$

most users of the analysis of variance find it more convenient to work with equivalent, but computationally simpler, versions of these sample values. These computational forms will be given below.

First of all, the total sum of squares can be shown to be equal to

$$\text{SS total} = \sum_j \sum_i y_{ij}^2 - \frac{\left(\sum_j \sum_i y_{ij}\right)^2}{N}. \qquad [10.13.1^*]$$

It is easy to show that this is true:

$$\text{SS total} = \sum_j \sum_i (y_{ij} - M)^2 = \sum_j \sum_i (y_{ij}^2 - 2y_{ij}M + M^2)$$

$$= \sum_j \sum_i y_{ij}^2 - 2M \sum_j \sum_i y_{ij} + \sum_j \sum_i M^2,$$

by rules 3 and 4 in Appendix A. This last expression reduces further to

$$\sum_j \sum_i y_{ij}^2 - 2M(NM) + NM^2$$

or

$$\sum_j \sum_i y_{ij}^2 - NM^2,$$

by the definition of the sample grand mean, $M = \sum_j \sum_i y_{ij}/N$. Making one last substitution for M gives the computing formula, 10.13.1.

The computing formula for the sum of squares between groups can be worked out in a similar way:

$$\text{SS between} = \sum_j n_j(M_j - M)^2 = \sum_j n_j (M_j^2 - 2MM_j + M^2)$$

$$= \sum_j n_j M_j^2 - 2M \sum_j n_j M_j + M^2 \sum_j n_j$$

$$= \sum_j \frac{\left(\sum_i y_{ij}\right)^2}{n_j} - 2NM^2 + NM^2$$

or

$$\text{SS between} = \sum_j \frac{\left(\sum_i y_{ij}\right)^2}{n_j} - \frac{\left(\sum_j \sum_i y_{ij}\right)^2}{N}. \qquad [10.13.2]$$

Finally, the computing formula for the sum of squares within groups is found by

$$\text{SS within} = \text{SS total} - \text{SS between}$$

$$= \sum_j \sum_i y_{ij}^2 - \frac{\left(\sum_j \sum_i y_{ij}\right)^2}{N} - \sum_j \frac{\left(\sum_i y_{ij}\right)^2}{n_j} + \frac{\left(\sum_j \sum_i y_{ij}\right)^2}{N} \quad [10.13.3^*]$$

$$= \sum_j \sum_i y_{ij}^2 - \sum_j \frac{\left(\sum_i y_{ij}\right)^2}{n_j}.$$

Ordinarily, the simplest computational procedure is to calculate both the sum of squares total and the sum of squares between directly, and then to subtract SS between from SS total in order to find the SS within.

10.14 / A COMPUTATIONAL OUTLINE FOR THE ONE-WAY ANALYSIS OF VARIANCE

It is natural for the beginner in statistics to be a little staggered by all of the arithmetic that the analysis of variance involves. However, take heart! With a bit of organization and with the aid of a good calculator simple analyses can be done quite quickly. The important thing is to form a clear mental picture of the different sample quantities you will need to compute, and how they combine. Below is an outline of the steps to follow:

1. Start with a listing of the raw scores separated by columns into the treatment groups to which they belong.
2. Square each score (y_{ij}^2) and then add these squared scores over all individuals in all groups. The result is $\sum_j \sum_i y_{ij}^2$. Call this quantity A.
3. Now sum the *raw* scores over all individuals in all groups to find $\sum_j \sum_i y_{ij}$. Call the resulting value B.
4. Now for a single group, say group j, sum all of the raw scores in that group and square the sum, to find $\left(\sum_i y_{ij}\right)^2$. Divide by the number in that group: $\left(\sum_i y_{ij}\right)^2 / n_j$.
5. Repeat step 4 for each group, and then sum the results across the several groups to find $\sum_j \frac{\left(\sum_i y_{ij}\right)^2}{n_j}$. Call this quantity C.
6. The **sum of squares total** is found from $A - B^2/N$.
7. The **sum of squares between** is $C - B^2/N$.
8. The **sum of squares within** is

   $$\text{SS total} - \text{SS between} = A - C.$$

9. Divide SS between by $J - 1$ to give **MS between.**
10. Divide SS within by $N - J$ to give **MS within.**

11. Divide MS between by MS within to find the F **ratio.**
12. Carry out the test by referring the F ratio to a table of the F distribution with $J - 1$ and $N - J$ degrees of freedom.

In actual research settings analyses of variance are now most often done by computer, by use of one of the statistical packages, such as SPSS. Nevertheless, it is a good idea to have the computational routine in mind, at least for the simple analysis of variance, as an aid in understanding what this technique actually provides.

10.15 / THE ANALYSIS OF VARIANCE SUMMARY TABLE

The results of an analysis of variance are often (though not invariably) displayed in a table similar to Table 10.15.1.

In practice, the column labeled SS in Table 10.15.1 contains the actual values of the sums of squares computed from the data. In the df column appear the numbers of degrees of freedom associated with each sum of squares; these numbers of degrees of freedom must sum to $N - 1$. The MS column contains the values of the mean squares, each formed by dividing the sum of squares by its degrees of freedom. Finally, the F statistic is formed from the ratio of the mean square between groups to the mean square within groups.

The student does well to form the habit of arranging the results of an analysis of variance in this way. Not only is it a good way to display the results for maximum clarity, but it also forms a convenient device for organizing and remembering the computational steps.

Table 10.15.1

Standard layout for the results of a simple analysis of variance.

Source	*SS*	*df*	*MS*	*F*
Treatments (between groups)	$\sum_j \frac{\left(\sum_i y_{ij}\right)^2}{n_j} - \frac{\left(\sum_j \sum_i y_{ij}\right)^2}{N}$	$J - 1$	$\frac{\text{SS between}}{J - 1}$	$\frac{\text{MS between}}{\text{MS within}}$
Error (within groups)	$\sum_j \sum_i y_{ij}^2 - \sum_j \frac{\left(\sum_i y_{ij}\right)^2}{n_j}$	$N - J$	$\frac{\text{SS within}}{N - J}$	
Totals	$\sum_j \sum_i y_{ij}^2 - \frac{\left(\sum_j \sum_i y_{ij}\right)^2}{N}$	$N - 1$		

An experiment was carried out to study the effect of a small lesion introduced into a particular structure in a rat's brain on its ability to perform in a discrimination problem. The particular structure studied is bilaterally symmetric, so that the lesion could be introduced into the structure on the right side of the brain, the left side, both sides, or neither side (a control group). Four groups of randomly selected rats were formed, and given the various treatments. Originally the control group contained 7 rats and each of the experimental groups 14 rats, but due either to death or postoperative incapacity only the following numbers were actually observed in the discrimination situation. The experimenter was prepared to assume that loss of rats during the study was a purely random event, which had no bearing on the results. The final data were as shown in Table 10.16.1.

Now the simple analysis of variance will be illustrated for these data. Given this listing of the raw scores according to treatment groups, we first square and sum over all individuals in all groups (step 2 above):

$$A = \sum_j \sum_i y_{ij}^2 = 20^2 + 18^2 + \cdots + 18^2 + 25^2 = 24424.$$

Next, the raw scores over all observations are summed, and the result squared (step 3):

$$B = \left(\sum_j \sum_i y_{ij}\right) = (20 + 18 + \cdots + 18 + 25) = 970$$

$$\frac{B^2}{N} = \frac{(970)^2}{40} = 23522.5.$$

Using steps 4 and 5 above we find

$$C = \sum_j \frac{\left(\sum_i y_{ij}\right)^2}{n_j} = \frac{(159)^2}{7} + \frac{(266)^2}{11} + \frac{(322)^2}{13} + \frac{(223)^2}{9} = 23545.1.$$

Table 10.16.1

Group			
I	*II*	*III*	*IV*
20	24	20	27
18	22	22	35
26	25	30	18
19	25	27	24
26	20	22	28
24	21	24	32
26	34	28	16
159	18	21	18
	32	23	25
	23	25	223
	22	18	
	266	30	
		32	
		322	

Then (steps 6, 7, and 8),

$$\text{SS total} = A - \frac{B^2}{N} = 24424 - 23522.5 = 901.5$$

$$\text{SS between} = C - \frac{B^2}{N} = 23545.1 - 23522.5 = 22.6$$

$$\text{SS within} = 901.5 - 22.6 = 878.9.$$

Steps 9, 10, and 11 are represented in the following summary table:

Source	*SS*	*df*	*MS*	*F*
Treatments (between groups)	22.6	3	7.5	$\frac{7.5}{24.4}$
Error (within groups)	878.9	36	24.4	
Totals	901.5	39		

Ordinarily, at this point, the obtained F ratio would be compared with the value shown in Table V for 3 and 36 degrees of freedom and the specified α level. For $\alpha = .05$, the required F is 2.84, with 40 degrees of freedom used as the value nearest to 36. However, this step is really not necessary for this example, since the obtained F value is less than one, and the null hypothesis cannot be rejected. There is not enough evidence to warrant the conclusion that mean differences, or effects, truly exist among these treatment populations.

10.17 / THE F TEST AND THE t TEST

When only two independent groups are being compared in the experiment, and a nondirectional alternative hypothesis is being considered, it makes no difference whether the analysis of variance or the t test shown in Section 8.9 is used. As noted in Section 9.11, the square of a variable distributed as t with $N - 2$ degrees of freedom will be distributed as F with 1 and $N - 2$ degrees of freedom. A simple analysis of variance for two groups always yields an F ratio that is the same as the *square* of the t ratio calculated as in 8.9 for the same data. If the obtained F value is significant for any α, the corresponding value of $t = \sqrt{F}$ will be significant at the same α level in a two-tailed test. However, if the alternative hypothesis is directional, the sign of the difference between the two means must be considered; in this situation if F is significant at the α level, the one-tailed t test will show significance at the $\alpha/2$ level, provided that the sign of the obtained difference is appropriate to the alternative hypothesis.

This direct parallel between the F test in the analysis of variance and the t test for a difference in means holds only for the case of two groups, with an important exception that will be discussed in Chapter 12. One is never really justified in carrying out all the $\binom{J}{2}$ different t tests for differences among J groups, and then regarding this as some kind of substitute for the analysis of variance. One reason is the inflation in the apparent Type I error rate, as discussed in Section 8.17. Another is that such t tests carried out on all pairs of means must necessarily extract redundant, overlapping, information from the data, and as a result a complicated pattern of dependency must exist among the tests. In most instances the analysis of variance, testing a single meaningful hypothesis at a known rate of Type I error, is to be preferred.

10.18 / THE IMPORTANCE OF THE ASSUMPTIONS IN THE FIXED-EFFECTS MODEL

In the development of the fixed-effects model for the analysis of variance a number of assumptions were made. These assumptions help to provide the theoretical justification for the analysis and the F test. On the other hand, it is sometimes necessary to analyze data when these assumptions clearly are not met; indeed, it seldom stands to reason that they are exactly true. In this section we will examine the consequences of the application of the analysis of variance and the F test when these assumptions are not met.

In the first place, note that the inferences made in this chapter are about *means*. The models to be described in the next chapter provide inferences about *variances*, and *the remarks made in this section apply only to the present, fixed-effects, model.*

The first assumption listed in Section 10.9 specifies a normal distribution of errors, e_{ij}, for any treatment population j. This is equivalent to the assumption that each population j has a normal distribution of scores, y_{ij}. What are the consequences for the conclusions reached from the analysis when this assumption is not true? It can be shown that, other things being equal, *inferences made about means that are valid in the case of normal populations are also valid even when the forms of the population distributions depart considerably from normal, provided that the n in each sample is relatively large*. There is no hard-and-fast rule about how large a sample needs to be in order to be safe in this respect. A common sense approach is to apply the analysis even with small samples unless there is good reason to doubt the normal assumption.

Consequently, we need not worry unduly about the normality assumption so long as we are dealing with relatively large samples. In circumstances where the assumption of normality appears more or less unreasonable, the experimenter might do well to take a somewhat larger number of observations than otherwise. The more severely the population distributions are thought to depart from normal form, the relatively larger should the n per sample be.

The second assumption listed in Section 10.9 states that the error variance, σ_e^2, must have the same value for all treatment populations. Ordinarily, other things being equal, *this assumption of homogeneous variances can be violated without serious risk, provided that the number of cases in each sample is the same*. On the other hand, *when different numbers of cases appear in the various samples, violation of the assumption of homogeneous variances can have very serious consequences for the validity of the final inference*. The moral is again plain: whenever possible, an experiment should be planned so that the number of cases in each experimental group is the same, unless the assumption of equal population variances is eminently reasonable in the experimental context.

Since the analysis of variance is based on the assumption of equal variances, it may seem quite sensible to carry out a test for homogeneous variances on the sample data and then use the result of that test to decide if the analysis of variance is legitimate. Such tests for the homogeneity of several variances exist, and some statistics books advocate these procedures. However, the standard tests for equality of several variances are extremely sensitive to any departure from normality in the populations. The statistician says that these tests with outcomes that depend heavily on incidental assumptions are not "robust." It could easily turn out that one would refrain from carrying out the analysis of variance because variances were apparently unequal, when a test of equality of *means* would actually be quite justifiable. Consequently, a test for homogeneity of variance before the analysis of variance has rather limited practical utility, and modern opinion holds that the analysis of variance can and should be carried on without a preliminary test of variances, especially in situations where the number of cases in the various samples can be made equal (see Box, 1953, 1954).

The third assumption in Section 10.9 requires statistical independence among the error components, e_{ij}. The assumption of independent errors is most important for the justification of the F test in the analysis of variance, and, unfortunately, violations of this assumption have important consequences for the results of the analysis. *If this assumption is not met, very serious errors in inference can be made.* In general, great care should be taken to see that data treated by the fixed-effects analysis of variance are based on independent observations, both within and across groups (that is, each observation in no way relates to any of the other observations). This is most likely to present a problem in studies where repeated observations are made of the same experimental subjects, perhaps with each subject being observed under each of the experimental treatments. In some experiments of this sort there is good reason to believe that the performance of the subject on one occasion has a systematic effect on subsequent performances under the same or another experimental condition. In the fixed-effects model, such systematic connections or dependencies among observations amount to a lack of statistical independence among errors, in violation of the assumption (a more appropriate model for handling this situation will be examined in Chapter 11). For this reason some authors suggest that data based on repeated observations should never be treated under the fixed-effects model for analysis of variance. However, this seems to be a point of experimental technique on which the statistician must tread very lightly. The circumstance that observations were repeated does not, ipso facto, imply that observations must be regarded as statistically dependent; a shrewd experimenter can sometimes get subjects into a stable state very early by preexperimental warm-up techniques, and in such situations it may be quite reasonable to assume that repeated observations are statistically independent. In other kinds of experiments, still other grounds may exist for assuming independence of repeated observations. The point, as always, is that statistics is limited in its ability to legislate experimental practice. Statistical assumptions must not be turned into prohibitions against particular kinds of experiments, although these assumptions must be borne in mind by the experimenter exercising thought and care in matching the particular experimental situation with an appropriate form of analysis. More will be said about the problem of repeated observations in the next chapter.

Since it is true that

$$E(M_j) = \mu_j \text{ and } E(M) = \mu,$$

an unbiased estimate of the magnitude of any fixed effect μ_j can be obtained by taking

$$\text{est. } \mu_j = M_j - M.$$

In this way the experimenter can gain an idea about the magnitudes of the various effects that appear to be present in the data.

Furthermore, following an analysis of variance it is often of interest to judge the extent to which the experimental treatments actually are accounting for variance in the dependent variable. A perfectly good *descriptive* statistic for this purpose is immediately at hand: *the ratio of the SS between to the SS total is an index of the proportion of variance accounted for in this sample*. This index can be symbolized by $\eta^2_{y \cdot x}$ (small Greek eta, squared), and is referred to as the "correlation ratio" in some statistical writing.

However, it is also useful to have an estimate of the population strength of association between the independent variable X and the dependent population variable Y. The population index ω^2 defined in Section 8.12 may be extended readily to the situation where there are J different populations;

$$\omega^2 = \frac{\sigma^2_Y - \sigma^2_{Y|X}}{\sigma^2_Y} = \frac{\sum_j \alpha_j^2 p(x_j)}{\sigma^2_Y}. \qquad [10.19.1]$$

Then a rough sample estimate of this population value can be had by taking

$$\text{est. } \omega^2 = \frac{\text{SS between} - (J - 1) \text{ MS within}}{\text{SS total} + \text{MS within}}. \qquad [10.19.2]$$

When F turns out to be less than 1.00, then this estimate will be negative. In that case, the estimated value of ω^2 is set at zero, of course.

Determining the power of an F test shares the difficulty we have already encountered in the case of t: When the null hypothesis of equal population means is true, then the ratio we compute and call "F" actually does have the F distribution given in Table V. On the other hand, when H_0 is not true, then the ratio we employ does not have an F distribution at all. Rather, it follows a sampling distribution known as "noncentral F." Although "central F" has two parameters, ν_1 and ν_2, noncentral F has three parameters, ν_1, ν_2, and δ^2. Under our usual assumptions, when H_0 is not true the value of δ^2 can be expressed as

$$\delta^2 = \frac{\sum_j n_j \alpha_j^2}{\sigma^2_e}. \qquad [10.19.3]$$

If one finds it easier to think of the population situation in terms of true proportion of variance accounted for, or ω^2, rather than in terms of δ^2, the second index may be converted into the first by the relation

$$\omega^2 = \frac{\delta^2}{N + \delta^2} \qquad [10.19.4]$$

when our assumptions of Section 10.5 are true.

Determination of the power of F thus depends on which noncentral F distribution actually applies when H_0 is not true, and that in turn depends on what we specify as the value of δ^2. In some problems it is possible to decide on such a value to investigate, and then standard power charts, such as those given in Winer (1971), may be used.

10.20 / THE TWO-WAY ANALYSIS OF VARIANCE

We now have enough background to discuss how the simple one-way analysis of variance can be extended to cover a more complicated experimental setup in which there are two different sets of treatments. Suppose that an experimenter is interested in ''level of aspiration'' as the dependent variable in an experiment. An experimental task has been developed consisting of a difficult game apparently involving motor skill, yielding a numerical score that can be attached to a person's performance. But this appearance is deceptive: unknown to the subject, the game is actually under the control of the experimenter, so that each subject is made to obtain exactly the same score. After a fixed number of trials, during which the subject unknowingly receives the preassigned score, the individual is asked to predict what the score will be on the next group of trials. However, before this prediction, the subject is given ''information'' about how the score compares with some fictitious norm group. In one experimental condition the subject is told that the first performance is *above average* for the norm group, in the second that it is *average,* and in the third that it is *below average*. There are thus three possible experimental ''standings'' that might be given to any subject. (Of course, after the experiment, each subject is fully informed of this little ruse by the experimenter.)

However, *two different norm groups* are used in the information given subjects. One-half of the subjects are told that they are being compared with college men, and the other half are told that they are being compared with professional athletes. Hence, there are two additional experimental treatments: ''college norms'' and ''professional athlete norms.''

The dependent score value y is based on the report the subject makes about anticipated performance in the next group of trials. Since each subject has obtained the same score, this anticipated score on the next set of trials is treated as equivalent to a level of aspiration that the subject has set. Each subject is tested privately, and no communication is allowed between subjects until the entire experiment is completed.

In this example interest is focused on two distinct experimental factors: the information about standing that the subject is given, and the norm groups used for comparison. Either or both of these experimental factors might possibly influence the value that a subject anticipates as the next score (the level of aspiration). A random sample of sixty male college students is selected and assigned at random to each of the six possible treatment *combinations*. This can be diagrammed as follows:

Norm information	Athlete	10	10	10
	College	10	10	10
		Above	*Average*	*Below*
			Standing	

Table 10.20.1

Assignment of subjects to treatment combinations, by numbers in each combination.

This experiment represents an instance where two different sets of experimental treatments are crossed, or given in every combination. Here there are six distinct sample groups, each group being given a particular combination of two kinds of treatments. *Three* questions are of interest:

1. Are there systematic effects due to experimental set alone (averaged over the norm groups)?
2. Are there systematic effects due to norm information alone (averaged over the experimental sets)?
3. Are there systematic effects due neither to norm information alone nor to experimental set alone, but attributable only to the *combination* of a particular norm group with a particular experimental set?

Notice that this study could be viewed as two separate experiments carried out on the same set of subjects: There are three groups of twenty subjects each, differing only in experimental set or ''standing''; exactly the same set of ''norm-group'' conditions are represented in each experimental ''standing'' group. On the other hand, looking at Table 10.20.1 by rows rather than by columns, we see that there are also two samples of thirty subjects each, differing systematically by ''norm group.'' Each ''norm-group'' sample has exactly the same representation of the other experimental conditions within it.

Question 3 above cannot, however, be answered by the comparison of ''norm groups'' alone or by the comparison of experimental ''set'' groups alone. This is a question of ''interaction,'' *the unique effects of combinations of treatments*. This is the important new feature of the two-way (or higher) analysis of variance: we will be able to examine *main effects* of the separate experimental variables or factors just as in the one-way analysis, as well as *interaction effects,* differences apparently due only to the unique combinations of treatments. Before we go further into the topic of experiments where two or more experimental factors are to be studied, some important terms must be introduced.

In the experiment described above, two experimental factors are under study, the first being the ''standing information'' given the subject, and the second being the ''norm information.'' Each subject is to get a combination of a category of ''standing information'' and a category of ''norm information.'' In experiments such as this, where two experimental factors are present and each category or level of one factor occurs with each level of the other, the two factors are said to be **completely crossed.** All possible combinations of levels of two (or more) experimental factors occur in a completely crossed experiment. Furthermore, if all possible combinations of factor levels occur an equal number of times, the experiment is said to be **balanced.** Note that the experiment described above is thus completely crossed and balanced.

The remainder of this chapter will be devoted to experiments where there are two completely crossed and balanced factors, and then a different random sample of subjects is assigned to each particular *combination* of the other two factors: that is, each combination of treatments is administered to a separate and distinct set of randomly selected subjects. This arrangement is often called a *factorial design*. However, this is but one of the simplest kinds of experimental designs, and it is entirely possible to design and analyze a wide variety of experiments where different factors are not completely crossed or balanced in this way.

The discussion of the two-way analysis of variance will also be limited here to *replicated experiments*. For our purposes this means that **within each treatment combination there are at least two independent observations made under identical experimental circumstances.** The requirement that the experiment be replicated is introduced here so that an error sum of squares will be available, permitting the study of tests both for treatment effects and for interaction. If there were only one observation for each treatment combination, we would not be able to test separately for interaction effects, since in this situation there is no direct way to estimate error variance apart from interaction effects. Occasionally, experiments are carried out where only one observation is made per treatment combination; under the fixed effects model this makes it necessary to know or to assume that no interaction effects exist if a test for main treatment effect is to be carried out. This assumption is often very questionable, and most circumstances requiring a nonreplicated experiment will fit into one of the models to be discussed in Chapter 11. For this reason, our discussion for the fixed-effect model will be confined to replicated experiments of the factorial type outlined above.

An *orthogonal design* for an experiment can be defined as a way of collecting observations that will permit one to estimate and test for the various treatment effects and for interaction effects separately. The potential information in the experiment can be ''pulled apart'' for study in an orthogonal design. Any factorial layout can be regarded as an orthogonal design provided that: (1) the observations within a given treatment *combination* are sampled at random and independently from a normal population, and (2) the number of observations in *each possible combination* of treatments is the same. Thus, the usual procedure in setting up an experiment to be analyzed by the two-way (or higher) analysis of variance is to assign subjects at random and independently to each combination of treatments so as to have an equal number in each combination. This means that in a table representing the experimental groups, such as Table 10.20.1, the cells of the table all contain the same number of observations. Let us call the number of ''row'' treatments R, and the number of ''column'' treatments C. For experiments of this sort where each cell in an $R \times C$ data table contains the same number n of cases, each *row* will contain Cn cases, and each *column* Rn cases. If at all possible, experiments should be set up in this way, not only to insure orthogonality, but also to minimize the effect of nonhomogeneous population variances should they exist (see Section 10.18).

It is also possible to design two-factor experiments that will be orthogonal even though the number of cases in the various cells differ, provided that **proportionality** holds within the cells of any given row or column. This means that for any treatment combination jk,

$$n_{jk} = \frac{n_j n_k}{N}.$$

Even so, the beginning experimenter is advised to seek either an advanced text, or preferably, expert help, whenever the cells in table of results have varying numbers.

10.21 / THE FIXED-EFFECTS MODEL FOR A TWO-WAY EXPERIMENT

The version of the general linear model applicable to an experiment with two completely crossed factors (i.e., independent variables), in an $R \times C$ table, can be expressed like this:

$$y_{ijk} = a_0 + \sum_{j=1}^{C} a_j x_{ij} + \sum_{k=1}^{R} b_k w_{ik} + \sum_{j=1}^{C} \sum_{k=1}^{R} c_{jk}(x_{ij} w_{ik}) + e_{ijk} \qquad [10.21.1]$$

Here, there are two experimental factors corresponding to the independent variables X and W; the particular values x_{ij} and w_{ik} are again indicators taking on the value 1 when individual i is in group j or in group k respectively, and value 0 otherwise. Notice that the product $(x_{ij} w_{ik})$ will be 1 only when i is in *both* group j and group k; it will be zero otherwise. The a, b, c values are simply constants standing for the effects. The e_{ijk} term represents the error associated with individual i in treatment-combination j and k.

Then in terms of a population of individuals i, under the least squares criterion this model becomes

$$y_{ijk} = \mu + \alpha_j + \beta_k + \gamma_{jk} + e_{ijk} \qquad [10.21.2^*]$$

where

$$\alpha_j = \mu_j - \mu \text{ and } \beta_k = \mu_k - \mu.$$

The new features of equation 10.21.2 are the inclusion of a term representing the effect of row treatment k, or β_k, and a term standing for the interaction effect, γ_{jk} (small Greek gamma). The interaction effect is the experimental effect created by the combination of treatments j and k over and above any effects associated with treatments j and k considered separately:

$$\begin{aligned} \gamma_{jk} &= \mu_{jk} - \mu - \alpha_j - \beta_k \\ &= \mu_{jk} - \mu_j - \mu_k + \mu. \end{aligned}$$

The interaction effect μ_{jk} is thus equal to the mean of the population given both of the treatments j and k, minus the mean of the treatment population j, minus the mean of the population given treatment k, plus the grand mean.

Just as in the simple one-way analysis of variance under the fixed-effects model, in the two-way situation it is assumed that the experimental treatments and treatment combinations are fixed, and that the only inferences to be made are about those treatments and treatment combinations actually represented in the experiment.

The following equalities are true of the effects:

$$\sum_j \alpha_j = 0$$

$$\sum_k \beta_k = 0$$

$$\sum_j \gamma_{jk} = 0$$

$$\sum_k \gamma_{jk} = 0.$$

The effects, being deviations from a grand mean μ, sum to zero over all the different "levels" of a given kind of treatment. However,

$$\sum_j \alpha_j^2 > 0$$

unless $\alpha_j = 0$ for each j;

$$\sum_k \beta_k^2 > 0$$

unless $\beta_k = 0$ for each k;

$$\sum_j \sum_k \gamma_{jk}^2 > 0$$

unless $\gamma_{jk} = 0$ for each and every combination j,k.

10.22 / THE PARTITION OF THE SUMS OF SQUARES FOR THE TWO-WAY ANALYSIS OF VARIANCE

Once more we start off by looking at how the total sum of squares can be partitioned for a set of data. For any individual i in any treatment combination jk, the deviation of the score y_{ijk} from the sample grand mean M can be written as

$$y_{ijk} - M = (y_{ijk} - M_{jk}) + (M_j - M) + (M_k - M) + (M_{jk} - M_j - M_k + M). \quad [10.22.1]$$

If the deviation for each score is squared, and these squares are summed over all individuals in all combinations j and k, we have

$$\sum_j \sum_k \sum_i (y_{ijk} - M)^2 = \sum_j \sum_k \sum_i (y_{ijk} - M_{jk})^2 + \sum_j n_j(M_j - M)^2 + \sum_k n_k(M_k - M)^2 + \sum_j \sum_k n_{jk}(M_{jk} - M_j - M_k + M)^2. \quad [10.22.2]$$

Now let us examine the various individual terms on the right of expression 10.22.2. We call

$$\sum_j \sum_k \sum_i (y_{ijk} - M_{jk})^2 = \text{SS error} \quad [10.22.3^*]$$

the **sum of squares for error,** since it is based on deviations from a cell mean for individuals treated in exactly the same way; the only possible contribution to this sum of squares should be error variation.

Next, consider

$$\sum_j n_j(M_j - M)^2 = \text{SS columns}, \quad [10.22.4^*]$$

which is the **sum of squares between columns.** Here the deviations of column treatment means from the grand mean make up this sum of squares. This sum of squares reflects two things: the treatment effects of the columns *and* error. Notice that this sum of squares is identical to the sum of squares between groups found in the one-way analysis, if the different experimental groups are regarded as columns in the table.

The third term is

$$\sum_k n_k(M_k - M)^2 = \text{SS rows}, \quad [10.22.5^*]$$

which is the **sum of squares between rows.** It is based upon deviations of the row means from the grand mean, and thus reflects both row-treatment effects and error. This is the same as the sum of squares between groups if data were regarded as coming only from experimental groups corresponding to the rows.

Finally, the fourth term

$$\sum_j \sum_k n_{jk}(M_{jk} - M_j - M_k + M)^2 = \text{SS interaction} \qquad [10.22.6^*]$$

the **sum of squares for interaction.** This sum of squares involves only *interaction effects and error*.

The partition of the sum of squares for a two-way analysis can be written in the following schematic form:

$$\text{SS total} = \text{SS error} + \text{SS columns} + \text{SS rows} + \text{SS interaction}. \qquad [10.22.7^*]$$

Whereas in the one-way analysis the total sum of squares can be broken into only two parts, a sum of squares between groups and a sum of squares within groups (error), in the two-way analysis with replication the total sum of squares can be broken into *four* distinct parts. The principle generalizes to experimental layouts with any number of treatment and treatment combinations, but we shall stop with the two-way situation.

10.23 / THE MEAN SQUARES AND THEIR EXPECTATIONS

Before we turn to an examination of the sampling distribution of the various mean squares, the assumptions we must make to determine these sampling distributions will be stated:

1. The errors e_{ijk} are normally distributed with expectation of zero for each treatment-combination population jk.
2. The errors e_{ijk} have exactly the same variance σ_e^2 for each treatment-combination population.
3. The errors e_{ijk} are independent, both within each treatment combination and across treatment combinations.

You will note that these are essentially the same assumptions made for the one-way model, except that now we deal with treatment-combination populations, the entire set of potential observations to be made under any combination of treatments.

We begin by finding the expectation of the *error* sum of squares:

$$E \sum_j \sum_k \sum_i (y_{ijk} - M_{jk})^2 = RC(n - 1)\sigma_e^2. \qquad [10.23.1]$$

(Since the number n_{jk} in each cell is assumed to be the same, hereafter n will be written to signify this number. Remember that the R represents the number of rows and the C the number of columns; the number of observations in any row is Cn, and in any column is Rn.) Let

$$\text{MS error} = \frac{\text{SS error}}{RC(n - 1)}. \qquad [10.23.2^*]$$

Then,

$$E(\text{MS error}) = \frac{E(\text{SS error})}{RC(n-1)} = \frac{RC(n-1)\sigma_e^2}{RC(n-1)} = \sigma_e^2. \qquad [10.23.3^*]$$

The expected value of the mean square error is simply the error variance σ_e^2.

Now look at the mean square between columns. Since there are C columns, the mean square between columns is found from

$$\text{MS columns} = \frac{\text{SS columns}}{C-1} \qquad [10.23.4^*]$$

in exactly the same way as for the MS between in a one-way analysis of variance. Then since within any column $\sum_k \beta_k = 0$ and $\sum_k \gamma_{jk} = 0$, it follows that

$$E(\text{MS columns}) = \sigma_e^2 + \frac{Rn \sum_j \alpha_j^2}{C-1}. \qquad [10.23.5^*]$$

When the hypothesis of no column effects is true,

$$E(\text{MS columns}) = \sigma_e^2, \qquad [10.23.6]$$

but when the hypothesis is false,

$$E(\text{MS columns}) > \sigma_e^2. \qquad [10.23.7]$$

The mean square between columns and the mean square error are independent and unbiased estimates of the same variance σ_e^2 when the null hypothesis of no column effects is true. This hypothesis can be tested by the F ratio,

$$F = \frac{\text{MS columns}}{\text{MS error}} \qquad [10.23.8^*]$$

with $C - 1$ and $RC(n - 1)$ degrees of freedom. The rationale is precisely the same as that given in Sections 10.10 through 10.12.

In the same way we examine the expectation of the mean square for rows:

$$E(\text{MS rows}) = \sigma_e^2 + \frac{Cn \sum_k \beta_k^2}{R-1}. \qquad [10.23.9^*]$$

The expectation of the mean square rows can be exactly σ_e^2 only when the hypothesis of no row effects is true; otherwise,

$$E(\text{MS rows}) > \sigma_e^2. \qquad [10.23.10]$$

The mean square between rows is an unbiased estimate of σ_e^2 when the null hypothesis of no row effects is true, and it is independent of the mean square error. The hypothesis of no row effects is then tested by the ratio

$$F = \frac{\text{MS rows}}{\text{MS error}} \qquad [10.23.11]$$

with $R - 1$ and $RC(n - 1)$ degrees of freedom.

If we define

$$\text{MS interaction} = \frac{\text{SS interaction}}{(R-1)(C-1)} \qquad [10.23.12^*]$$

then

$$E(\text{MS interaction}) = \frac{E(\text{SS interaction})}{(R-1)(C-1)} = \sigma_e^2 + \frac{\sum_j \sum_k n\gamma_{jk}^2}{(R-1)(C-1)}. \qquad [10.23.13^*]$$

When there are no interaction effects *at all*, then

$$E(\text{MS interaction}) = \sigma_e^2. \qquad [10.23.14]$$

The mean square for interaction is also an unbiased estimate of the error variance σ_e^2. Otherwise,

$$E(\text{MS interaction}) > \sigma_e^2. \qquad [10.23.15]$$

The mean square for interaction is independent of the mean square for error, and so the hypothesis of no interaction effects may be tested by

$$F = \frac{\text{MS interaction}}{\text{MS error}} \qquad [10.23.16^*]$$

with $(R-1)(C-1)$ and $RC(n-1)$ degrees of freedom.

Thus we see that it is possible to make separate tests of the hypothesis of no row effects, the hypothesis of no column effects, and the hypothesis of no interaction effects, all from the same data. Furthermore, under the fixed-effects model, and given an orthogonal experimental design, estimates of the three different kinds of effects are independent of each other.

10.24 / COMPUTING FORMS FOR THE TWO-WAY ANALYSIS WITH REPLICATIONS

In carrying out an analysis of variance the following computing forms are generally used. These sums of squares are algebraically equivalent to those given in Section 10.22.

$$\text{SS total} = \sum_j \sum_k \sum_i y_{ijk}^2 - \frac{\left(\sum_j \sum_k \sum_i y_{ijk}\right)^2}{N} \qquad [10.24.1]$$

$$\text{SS rows} = \frac{\sum_k \left(\sum_j \sum_i y_{ijk}\right)^2}{Cn} - \frac{\left(\sum_j \sum_k \sum_i y_{ijk}\right)^2}{N} \qquad [10.24.2]$$

$$\text{SS columns} = \frac{\sum_j \left(\sum_k \sum_i y_{ijk}\right)^2}{Rn} - \frac{\left(\sum_j \sum_k \sum_i y_{ijk}\right)^2}{N} \qquad [10.24.3]$$

$$\text{SS error} = \sum_j \sum_k \sum_i y_{ijk}^2 - \frac{\sum_j \sum_k \left(\sum_i y_{ijk}\right)^2}{n} \qquad [10.24.4]$$

$$\text{SS interaction} = \frac{\sum_j \sum_k \left(\sum_i y_{ijk}\right)^2}{n} - \frac{\sum_k \left(\sum_j \sum_i y_{ijk}\right)^2}{Cn} - \frac{\sum_j \left(\sum_k \sum_i y_{ijk}\right)^2}{Rn} + \frac{\left(\sum_j \sum_k \sum_i y_{ijk}\right)^2}{N} \quad [10.24.5]$$

$$= \text{SS total} - \text{SS rows} - \text{SS columns} - \text{SS error}.$$

Notice that the sum of squares for columns is calculated just as for a one-way analysis of data arranged into columns. Furthermore, the sum of squares for rows is identical to the sum of squares between groups when the data are arranged into a table where the experimental groups are designated by rows. The sum of squares total is also calculated in exactly the same way as for a one-way analysis. The only new features here are the computations for error and for interaction. Generally, the error term is calculated directly, and then the interaction term is found by subtracting the sum of squares, rows, columns, and error all from the sum of squares total.

10.25 / A COMPUTATIONAL OUTLINE FOR THE TWO-WAY ANALYSIS UNDER THE FIXED-EFFECTS MODEL

1. Arrange the data into an $R \times C$ table, in which the R rows represent the R different treatments of one kind, and the C columns the C different treatments of the other kind. Each cell in the table should contain the same number n of observations. There are $N = RCn$ distinct observations in all.
2. Square each raw score and sum over all individuals in all cells to find $\sum_j \sum_k \sum_i y_{ijk}^2$. Call this quantity A.
3. Sum the raw scores in a given *cell jk* to find $\sum_i y_{ijk}$. Do this for *each cell,* and reserve these values for use in later steps.
4. Now sum the resulting values (step 3) over *all cells* to find $\sum_j \sum_k \sum_i y_{ijk}$. Call this quantity B. Find the **sum of squares total** by $A - \dfrac{B^2}{N}$.
5. Next take the RC different values found in step 3 and sum the cell totals for *a given row across columns* to find $\sum_j \sum_i y_{ijk}$. The result for any row k will be designated by D_k.
6. Having carried out step 5 for *each row,* square each of the D_k, sum over all of the various rows to find $\sum_k D_k^2$. Divide this quantity by Cn, the number of observations per row. Then

$$\frac{\sum_k D_k^2}{Cn} - \frac{B^2}{N}$$

is the **sum of squares for rows.**
7. Now return to the quantities found in step 3. This time sum the cell totals for

a given column across rows to find $\sum_k \sum_i y_{ijk}$ and call this value for column j, G_j.

8. Having carried out step 7 for each column, square each of the G_j and sum across the various columns to find $\sum_j G_j^2$. Divide this quantity by Rn, the number of observations per column. Then

$$\frac{\sum_j G_j^2}{Rn} - \frac{B^2}{N}$$

is the **sum of squares for columns.**

9. Once again return to the cell totals found in step 3. For a given cell jk call the total H_{jk}. Now square H_{jk} for each cell and sum across *all cells* to find $\sum_j \sum_k H_{jk}^2$. Divide this by n, the number of observations per cell. Then

$$A - \frac{\sum_j \sum_k H_{jk}^2}{n}$$

is the **sum of squares for error.**

10. Find the **sum of squares for interaction** by taking

SS total − SS rows − SS columns − SS error

or

$$\frac{\sum_j \sum_k H_{jk}^2}{n} - \frac{\sum_k D_k^2}{Cn} - \frac{\sum_j G_j^2}{Rn} + \frac{B^2}{N}.$$

11. Enter these sums of squares in the summary table.
12. Divide the SS rows by $R - 1$ to find MS rows.
13. Divide the SS columns by $C - 1$ to find MS columns.
14. Divide the SS interaction by $(R - 1)(C - 1)$ to find MS interaction.
15. Divide the SS error by $RC(n - 1)$ to find MS error.
16. The hypothesis of no row effects is tested by

$$F = \frac{\text{MS rows}}{\text{MS error}}$$

with $R - 1$ and $RC(n - 1)$ degrees of freedom.

17. The hypothesis of no column effects is tested by

$$F = \frac{\text{MS columns}}{\text{MS error}}$$

with $C - 1$ and $RC(n - 1)$ degrees of freedom.

18. The hypothesis of no interaction is tested by

$$F = \frac{\text{MS interaction}}{\text{MS error}}$$

with $(R - 1)(C - 1)$ and $RC(n - 1)$ degrees of freedom.

Table 10.25.1

Source	SS	df	MS	F
Rows	$\dfrac{\sum_k\left(\sum_j\sum_i y_{ijk}\right)^2}{Cn} - \dfrac{\left(\sum_j\sum_k\sum_i y_{ijk}\right)^2}{N}$	$R - 1$	$\dfrac{\text{SS rows}}{R - 1}$	$\dfrac{\text{MS rows}}{\text{MS error}}$
Columns	$\dfrac{\sum_j\left(\sum_k\sum_i y_{ijk}\right)^2}{Rn} - \dfrac{\left(\sum_j\sum_k\sum_i y_{ijk}\right)^2}{N}$	$C - 1$	$\dfrac{\text{SS col.}}{C - 1}$	$\dfrac{\text{MS col.}}{\text{MS error}}$
Interaction	$\dfrac{\sum_j\sum_k\left(\sum_i y_{ijk}\right)^2}{n} - \dfrac{\sum_k\left(\sum_j\sum_i y_{ijk}\right)^2}{Cn} - \dfrac{\sum_j\left(\sum_k\sum_i y_{ijk}\right)^2}{Rn} + \dfrac{\left(\sum_j\sum_k\sum_i y_{ijk}\right)^2}{N}$	$(R - 1)(C - 1)$	$\dfrac{\text{SS int.}}{(R - 1)(C - 1)}$	$\dfrac{\text{MS int.}}{\text{MS error}}$
Error (within cells)	$\sum_j\sum_k\sum_i y_{ijk} - \dfrac{\sum_j\sum_k\left(\sum_i y_{ijk}\right)^2}{n}$	$RC(n - 1)$	$\dfrac{\text{SS error}}{RC(n - 1)}$	—
Totals	$\sum_j\sum_k\sum_i y_{ijk}^2 - \dfrac{\left(\sum_j\sum_k\sum_i y_{ijk}\right)^2}{N}$	$RCn - 1$	—	—

For the fixed-effects model, the results of a two-way analysis of variance are displayed in Table 10.25.1, with the algebraic expressions and symbols replaced by the corresponding values obtained.

10.26 / AN EXAMPLE

Suppose that the experiment on level of aspiration, outlined in Section 10.20, had actually been carried out, and the data shown in Table 10.26.1 obtained.

We wish to examine three null hypotheses: (1) there is no effect of the standing given the subject, corresponding to the hypothesis of no column effects; (2) the actual norm group given the subjects has no effect, corresponding to the hypothesis of no row effects; and (3) the norm-group-standing combination has no unique effect, corresponding to the hypothesis of no row-column interaction. The α level chosen for each of these three tests will be .05.

Following the computational outline given in Section 10.25 we first find the square of each of the scores, and sum:

$$A = \sum_j\sum_k\sum_i y_k^2 = (52)^2 + (48)^2 + \cdots + (22)^2 + (17)^2 = 66872.$$

Table 10.26.1

Results for the level of aspiration experiment.

Norms	Standing: Above	Average	Below
College men	52	28	15
	48	35	14
	43	34	23
	50	32	21
	43	34	14
	44	27	20
	46	31	21
	46	27	16
	43	29	20
	49	25	14
	464	302	178
Professional athletes	38	43	23
	42	34	25
	42	33	18
	35	42	26
	33	41	18
	38	37	26
	39	37	20
	34	40	19
	33	36	22
	34	35	17
	368	378	214

The sum of the scores in each cell (step 3) is given in Table 10.26.1. Taking the sum of the cell sums gives the total sum,

$$B = \sum_j \sum_k \sum_i y_{ijk} = 464 + 302 + \cdots + 214 = 1904.$$

Hence the total sum of squares is

$$A - \frac{B^2}{N} = 66872 - \frac{(1904)^2}{60} = 6451.7.$$

Now the cell totals are summed for each row:

$$D_1 = \sum_j \sum_i y_{ij1} = 464 + 302 + 178 = 944$$

$$D_2 = \sum_j \sum_i y_{ij2} = 368 + 378 + 214 = 960.$$

The sum of squares for rows is found from

$$\frac{\sum_k D_k^2}{Cn} - \frac{B^2}{N} = \frac{(944)^2 + (960)^2}{30} - \frac{(1904)^2}{60}$$

$$= 4.3.$$

In a similar way, we find the sum of squares for columns by first summing cell totals for each column

$$G_1 = \sum_k \sum_i y_{i1k} = 464 + 368 = 832$$

$$G_2 = \sum_k \sum_i y_{i2k} = 302 + 378 = 680$$

$$G_3 = \sum_k \sum_i y_{i3k} = 178 + 214 = 392.$$

The sum of squares for columns is found from

$$\frac{\sum_j G_j^2}{Rn} - \frac{B^2}{N} = \frac{(832)^2 + (680)^2 + (392)^2}{20} - \frac{(1904)^2}{60}$$
$$= 4994.1.$$

Next, the sum of squares for error will be calculated. We begin by squaring and summing the *cell totals:*

$$\sum_j \sum_k H_{jk}^2 = (464)^2 + (302)^2 + \cdots + (214)^2 = 662288.$$

The sum of squares for error is

$$A - \frac{\sum_j \sum_k H_{jk}^2}{n} = 66872 - \frac{662288}{10} = 643.2.$$

The only remaining value to be calculated is the sum of squares for interaction; this is done by subtraction, as follows:

$$\begin{aligned} \text{SS total} - \text{SS rows} - \text{SS cols.} - \text{SS error} & \\ &= 6451.7 - 4.3 - 4994.1 - 643.2 \\ &= \text{SS interaction} = 810.1. \end{aligned}$$

Table 10.26.2 is the summary tablᴦ for this analysis of variance.

Table 10.26.2

Summary of analysis for level of aspiration study.

Source	*SS*	*df*	*MS*	*F*
Rows (norm groups	4.3	1	4.3	.36
Columns (standings)	4994.1	2	2497.05	209.8
Interaction	810.1	2	405.1	34.0
Error (within cells)	643.2	54	11.9	
Totals	6451.7	59		

The hypothesis of no row effects cannot be rejected, since the F value is less than unity. For the hypothesis of no column effects, an F of approximately 3.15 is required for rejection at the 5 percent level; the obtained F of 209 far exceeds this, and so we may conclude with considerable confidence that column effects exist. In the same way, the F for interaction effects greatly exceeds that required for rejecting the null hypothesis, and so there seems to be reliable evidence for such interaction effects.

Our conclusions from this analysis of variance are:

1. There is apparently little or no effect of norm group alone on level of aspiration.

2. The experimental standing does seem to affect level of aspiration when considered over the different norm groups.
3. There is apparently an interaction between norm group and standing, meaning that the magnitude and direction of the effects of standing differ for different norm groups.

In short, the standing one is told makes a difference in the aspiration level, but the kind and extent of difference that it makes depends upon the norm group to which the individual is being compared.

The different column effects can be estimated from the column means and the overall mean:

$$\text{est. } \alpha_1 = 41.6 - 31.7 = 9.9$$

$$\text{est. } \alpha_2 = 34.0 - 31.7 = 2.3$$

$$\text{est. } \alpha_3 = 19.6 - 31.7 = -12.1$$

(Because of rounding error these do not quite total zero, as they should.) In a similar way, interaction effects may be estimated from the means of cells, the rows, and the columns:

$$\text{est. } \gamma_{11} = 46.4 - 31.5 - 41.6 + 31.7 = 5.0$$

$$\text{est. } \gamma_{21} = 30.2 - 31.5 - 34.0 + 31.7 = -3.6$$

and so on. The estimated total effect of the "above" treatment on a subject in the college norm group is thus

$$\text{est. } (\alpha_1 + \gamma_{11}) = 9.9 + 5.0 = 14.9$$

Note that for group 1 ("above" standing) with unspecified norm group, the best guess we can make about the effect of this treatment on any individual is 9.9 units. However, if we are told that this individual also belongs to the group given the "college men" norms, our best bet is 14.9 as the amount of effect. In the same way, the effect of any column treatment j within a row-treatment population k is estimated to be $\alpha_j + \gamma_{jk}$, since there is little evidence for row effects.

10.27 / THE INTERPRETATION OF INTERACTION EFFECTS

The presence or absence of interaction effects, as inferred from the F test for interaction, can have a very important bearing on how one interprets and uses the results of an experiment. When the presence of column effects is inferred, this implies that the populations represented by the columns have means that differ; the amount and direction of difference between sample means for any pair of columns provides an estimate of the corresponding difference between population means. When interaction effects are absent, differences among the means representing different column-treatment populations have the same size and sign, even though the populations are conceived as receiving still another treatment represented by one of the rows. This suggests that the difference between a pair of column means in the data is our best bet about the difference to be expected between a pair of individuals given different column treatments, quite irrespective of the particular row treatment that might have been administered. On the other hand, when interaction effects exist, *varying differences* exist between the means of populations representing different column treatments, depending on the particular row treatment that is applied. It is still true that differences between column means in the data provide an estimate of the difference we should expect between individuals given the particular treatments, but only on the average, over all of the different row treatments that might have been applied. When a particular row treatment is specified, it may be that quite another size and direction of difference should be expected between individuals given different column treatments. In short, interaction effects lead to a qualification on the estimate one makes of the differences attributable to different treatments; when interaction effects exist, the best estimate one can make of a difference attributable to one factor depends on the particular level of the other factor.

For example, suppose that an experimenter is comparing two methods of instruction in golf. Let us represent these two methods as the column treatments in the data table. The other factor considered is the sex of the student; the study employs a group of 50 boys and a group of 50 girls, with 25 subjects in each group taught by Method I and the remainder by Method II. After a fixed period of instruction by one or the other method, each member of the sample is given a proficiency test. Suppose that the sample means for the four subgroups turn out as follows:

		Method		
		I	*II*	
Sex	*Girls*	55	65	60
	Boys	75	45	60
		65	55	

For a small enough estimated error variance, such data would lead to the conclusion that no difference exists between boys and girls in terms of performance on the proficiency test, but that *both column effects and interaction effects do exist*. Now suppose that you want to decide which method to use for the instruction of an individual student. If *low* scores indicate good performance, but you don't know or don't wish to specify the sex of the student, then Method II clearly is called for, since the best estimate is that Method I gives a higher mean than Method II over both sexes. However, suppose that you know that the individual to be instructed is a *girl:* in this case, you do *much* better to choose Method I, since you have evidence that *within the population of girls, mean II is higher than mean I*.

Significant interaction effects usually reflect a situation very like this: overall estimates of differences due to one factor are fine as predictors of average differences over *all possible levels of the other factor,* but it will not necessarily be true that these are good estimates of the differences to be expected when information about the category on the other factor is given. Significant interaction serves as a warning: treatment differences *do* exist, but to specify exactly *how* the treatments differ, and especially to make good individual predictions, one must look *within* levels of the *other* factor. The presence of interaction effects is a signal that in any predictive use of the experimental results, effects attributed to particular treatments representing one factor are best qualified by specifying the level of the *other* factor. This is extremely important if one is going to try to use estimated effects in forecasting the result of applying a treatment to an individual; when interaction effects are present, the best forecast can be made only if the individual's status on *both* factors is known.

For these reasons, the presence of interaction effects can be most important to the interpretation of the experiment. The estimated effects of any given treatment are not "best bets" about any randomly selected individual when interaction effects are present; the best prediction entails knowing the other treatment or treatments administered.

10.28 / PROPORTION OF VARIANCE ACCOUNTED FOR IN TWO-WAY EXPERIMENTS

Just as in a one-way experimental situation, following an analysis of variance one may wish a descriptive statistic for the proportion of variance accounted for by each factor, and by their interaction. Simply by taking each of these sums of squares and dividing them by the total sum of squares shows how much variance each type of effect accounts for in this sample. Thus, in the preceding example, Table 10.26.2, the sample proportions are .77 for the column factor (standings), .13 for interaction, but only .00067 for the row factor. The systematic features of this experiment thus account for a total of about 90 percent of the variance in the Y values observed.

The strength of association measure ω^2 for a one-way experimental design may be extended in a straightforward way to the situation where there are two independent variables. When the experimental design has an equal number of observations in each combination of two independent variables X and W, then the definition of $\omega^2_{Y|X}$, the proportion of variance accounted for by X alone in the population is

$$\omega^2_{Y|X} = \frac{\left(\sum_j \alpha_j^2\right) \Big/ C}{\sigma_Y^2}$$

Similarly, we can define

$$\omega^2_{Y|W} = \frac{\left(\sum_k \beta_k^2\right) \Big/ R}{\sigma_Y^2},$$

and

$$\omega^2_{Y|XW} = \frac{\left(\sum_j \sum_k \gamma_{jk}^2\right) \Big/ RC}{\sigma_Y^2}.$$

This last index is the proportion of variance accounted for uniquely by the combination of *both* X and W.

Given these definitions, and our results about the expectations of mean squares for the two-way analysis of variance (Section 10.23), we can form rough estimates of these values of ω^2 by taking

$$\text{est. } \omega^2_{Y|X} = \frac{\text{SS columns} - (C - 1)\text{ MS error}}{\text{MS error} + \text{SS total}}$$

$$\text{est. } \omega^2_{Y|W} = \frac{\text{SS rows} - (R - 1)\text{ MS error}}{\text{MS error} + \text{SS total}}$$

$$\text{est. } \omega^2_{Y|XW} = \frac{\text{SS interaction} - (R - 1)(C - 1)\text{ MS error}}{\text{MS error} + \text{SS total}}.$$

For the example in Section 10.26, these estimated values are

$$\text{est. } \omega^2_{Y|X} = \frac{4994.1 - (2)(11.9)}{11.9 + 6451.7} = .77$$

$$\text{est. } \omega^2_{Y|XW} = \frac{810.1 - (2)(11.9)}{11.9 + 6451.7} = .12.$$

Since the F ratio shows a value less than 1.00 in the test for row differences, the estimate of $\omega^2_{Y|W}$ is set equal to zero. These estimates also suggest that a strong association exists between the treatments symbolized by X and the dependent variable Y. Knowing X alone tends to reduce our ''uncertainty'' about Y by about 77 percent. There is apparently a further accounting for around 12 percent of the variance of Y if one knows *both* of the categories represented by the X and W treatment combination.

10.29 / TWO-WAY (OR HIGHER) ANALYSIS OF VARIANCE WITH FIXED EFFECTS

The list of assumptions in Section 10.23 is an almost exact parallel to the list in Section 10.9. Similar assumptions are made for more complex experiments requiring a higher-order analysis, provided that the fixed-effects model is appropriate.

As you may have anticipated, the same relaxation of assumptions is possible in the two- or multiway analysis as in the one-way analysis. For experiments with a relatively large number of observations per cell, the requirement of a normal distribution of errors seems to be rather unimportant. In an experiment where it is suspected that the parent distributions of dependent-variable values are very unlike a normal distribution, perhaps a correspondingly large number of observations per cell should be used.

When the data table represents an equal number of observations in each cell, the requirement of equal error variance in each treatment combination population may also be violated without serious risk, at least in terms of Type I error. However, in some circumstances the power of the F test may be affected. On the whole, there are two good reasons for planning experiments with equal n per cell: the experimental design will thus be orthogonal (Section 10.20) and the possible consequences of nonhomogeneous variances on the probability of Type I error will be minimized.

Regardless of the simplicity or complexity of the experiment, however, the error portions entering into the respective observations should be independent if the fixed-effects model is to apply.

10.30 / THE ANALYSIS OF VARIANCE AS A SUMMARIZATION OF DATA

It may appear that the main use of the analysis of variance, particularly for two-factor or multifactor experiments, is in generating a number of F tests on the same set of data, and that the partition of the sum of squares is only a means to this end. However, this is really a very narrow view of the role of this form of analysis in experimentation. The really important feature of the analysis of variance is that it permits the separation of all of the potential information in the data into distinct and nonoverlapping portions, each reflecting only certain aspects of the experiment. For example, in the simple, one-way analysis of variance, the mean square between groups reflects both the systematic differences among observations that are attributable to the experimental manipulations, as well as the chance, unsystematic differences attributable to all of the other circumstances of the experiment. On the other hand, the mean square within the groups reflects only these latter, unsystematic, features. Under the fixed-effects model, these two statistics are independent, completely nonoverlapping, ways of summarizing the data. The information contained in one is nonredundant with the information contained in the other. Estimates of the effects of the treatments are independent of estimates of error variability. The mechanics of the analysis of variance allow the experimenter to arrange and summarize data in these nonredundant ways, in order to decide if effects exist and to estimate how large or important those effects may be.

Similarly, for a two-factor experiment we arrive at a mean square for one treatment factor, and a separate mean square for the other. These two mean squares reflect quite nonredundant aspects of the experiment, even though they were each based on the same basic data: the first sum of squares reflects only the effects attributable to the first experimental factor (plus error), and the second those effects attributable to the other (plus error). Under the statistical assumptions we make these two mean squares are independent of each other. Furthermore, the mean squares for interaction and for error are independent of each other and of the treatment mean squares. The analysis of variance lets the experimenter "pull apart" the factors that contribute to variation in his experiment, and identify them exclusively with particular summary statistics. For experiments of the orthogonal, balanced type considered here, the analysis of variance is a routine method for finding the statistics that reflect particular, meaningful, aspects of the data. The real importance of the analysis of variance lies in the fact that it routinely provides such succinct overall "packaging" of the data.

For the moment, however, let us consider the several F tests obtained from a two- or multifactor experiment. The sums of squares and mean squares for columns, for rows, for interaction, and for error are all, under the assumptions made, independent of each other. However, are the three or more F tests themselves independent? Does the level of significance shown by any one of the tests in any way predicate the level of significance shown by the others? Unfortunately, it can be shown that although the mean squares are independent such F tests are *not* independent. Some connection exists among the various F values and significance levels. This is due to the fact that each of the F ratios involves the same mean square for error in the denominator; the presence of this same value in each of the ratios creates some statistical dependency among them. The practical result is much the same as for the multiple t tests discussed in Section 8.17. Thus, if three F tests are carried out, and these tests actually are independent, then the probability is $1 - (.95)^3$ or about .14 that *at least one* of the tests will show significance at the .05 level. However, for the usual situation where the tests are not independent, one has no ready way to calculate the exact number that should be expected by chance.

For really complicated analyses of variance, the problem becomes much more serious, since a fairly large number of F tests may be carried out, and the probability may be quite large that one or more tests gives spuriously significant results. The matter is further complicated by the fact that the F tests are not independent, and the number to be expected by chance is quite difficult to calculate exactly. For this reason, when large numbers of F tests are performed, it is wise to set the α value for each test at a smaller value than usual in order to compensate for this inflation of error rate. Furthermore, the experimenter should not pay too much attention to isolated results that happen to be significant. Rather, the pattern and interpretability of results, as well as the strength of association represented by the findings, form a more reasonable basis for the overall evaluation of the experiment. When the number of degrees of freedom for the mean square error is very large, then the various F tests may be regarded as approximately independent, and the number of significant results at the .05 level to be expected should be close to 5 percent of the tests performed. Even here, however, the importance of a particular result is very difficult to interpret on the basis of significance level alone. A great deal of thought must go into the interpretation of a complicated experiment, quite over and above the information provided by the significance tests.

10.31 / ANALYZING EXPERIMENTS WITH MORE THAN TWO EXPERIMENTAL FACTORS

The essential features of any fixed-effects analysis of variance have now been discussed. In experiments involving three or more different experimental factors, the total sum of squares is partitioned into even more parts, but the basic ideas of the partition, the mean squares, and the F tests are the same. In a three-factor experiment, not only are there mean squares representing the interactions of particular pairs of the experimental factors, but also a mean square representing the simultaneous interaction of all three of the factors. The higher the order of the experimental design, the larger becomes the number of possible interactions representing every combination of two or more factors. Each and every significant interaction represents a new qualification on the meaning of the results. Moreover, as suggested above, if many significance tests are carried out on the same data, the probability of at least one spuriously significant result may be very large, and this probability, as well as the number of Type I errors to be expected by chance, cannot be determined routinely. For these reasons, very complicated experiments with many factors are somewhat uneconomical to perform, since they require a large number of observations as a rule, and a complete analysis of the data yields so many statistical results that the experiment as a whole is often very difficult to interpret in a statistical light. In planning an experiment, it is a temptation to throw in many experimental treatments, especially if the data are inexpensive and the experimenter is adventuresome. However, this is *not* always good policy if the investigator is interested in finding meaning in the results; other things being equal, the simpler the experiment the better will be its execution, and the more likely will one be able to decide what actually happened and what the results actually mean.

The fixed-effects analysis of variance is one of the statistical techniques encountered most in current social and behavioral research. Its advantages are many: the general technique is extremely flexible and applies to a wide variety of experimental arrangements. Indeed, the availability of a statistical technique such as the analysis of variance has done much to stimulate inquiry into the logic and economics of the *planning* of experiments. Statistically, the F test in the fixed effects analysis of variance is relatively *robust;* as we have seen, the failure of at least two of the underlying statistical assumptions does not necessarily disqualify the application of this method in practical situations. Computationally the analysis is relatively simple and routine and provides a condensation of the main statistical results of an experiment into an easily understood form.

However, the application of the analysis of variance never transforms a sloppy experiment into a good one, no matter how elegant the experimental design appears on paper, or how neat and informative the summary table appears to the reader. Furthermore, experiments should be planned so as to capture the phenomena under study in the clearest, most easily understood form, and this does not necessarily mean that one of the "textbook" experimental designs or a treatment by the analysis of variance will best clarify matters. The experimental *problem* must come first in planning, and not the requirements of some particular form of analysis, even though, ideally, both should be considered together from the outset. If it should come to a choice between preserving the essential character of the experimental problem, or using a relatively elegant technique such as the analysis of variance, then the problem should come first.

EXERCISES

1. Following the pattern of Section 10.7 construct the scores for an experiment with 4 samples of 4 observations each where $\mu = 100$, $\mu_1 = 95$, $\mu_2 = 104$, $\mu_3 = 98$, and $\mu_4 = 103$. Let the error terms be as follows:

I	*II*	*III*	*IV*
−5	−2	0	3
1	5	−4	−6
0	6	−1	4
2	−2	3	−7

From these "data," estimate μ and the separate effects, and compare them with the true values.

2. By use of the artificial data of problem 1, show it is indeed true that

$$\text{SS between} = \sum_j n_j[\alpha_j + M_{ej} - M_e]^2$$

3. Three independent random samples were taken of U.S. males aged thirteen years. Sample I was drawn from among those boys who are otherwise normal, but who have less than average motor coordination for their age level. Sample II consisted of boys with average motor coordination, while Sample III consisted consisted of boys with greater than average motor coordination. There were 20 boys in each sample. Each boy then filled out a questionnaire that yielded a rating for him on interest in athletic participation. The data are shown below. Can we say that motor coordination is related to interest in athletic participation?

Group					
I		*II*		*III*	
7	8	15	19	6	20
3	3	17	4	24	17
9	4	15	4	9	7
12	7	6	12	11	19
3	2	4	19	8	24
10	3	19	20	12	12
13	10	20	5	15	15
14	5	20	7	8	7
8	2	4	4	25	16
11	7	6	8	18	23

4. In an effort to curb inflation, the U.S. government imposed certain restrictions on the use of credit cards. In order to check on the effects that this might be having on the use of a certain card to purchase gasoline, an investigator sampled 25 large service stations honoring this card in five different regions of the country. The average daily number of charges using this card at that station was taken for six months ago, and then the present daily average number of charges was subtracted for that station. The data for each station, arranged by region are shown below. Does decrease in the use of the credit card seem to vary across the different regions? Use $\alpha = .05$.

Region				
A	*B*	*C*	*D*	*E*
6.3	9.3	1.5	−1.7	2.7
2.1	−2.1	2.7	−2.0	3.9
5.8	6.6	4.4	3.6	3.2
7.9	5.2	3.1	3.8	.6
−4.2	−3.8	4.6	4.5	1.7

5. Carry out an analysis of variance on the data of exercise 17, Chapter 8. How do the results of this analysis compare with those found using the t test for a difference between means?

6. For exercise 3, above, estimate the effects associated with the three "coordination" groups.

7. In a study of the effects of sleep deprivation upon the ability of human subjects to solve intellectual problems, sixty subjects were distributed at random among six groups. Each person was first given a score on a test of reasoning. Then each individual was deprived of a certain number of hours of sleep, after which the reasoning test was administered once again. The dependent variable was the difference between the first and the second administration of the test. These difference scores are given on page 371. Can one say that sleep loss affects the scores on this test?

Hours deprived					
8	*12*	*16*	*20*	*24*	*28*
20	32	70	95	83	100
19	25	65	90	75	96
18	24	59	84	75	95
11	20	55	80	71	94
10	19	50	76	62	80
10	18	47	75	62	78
7	15	40	60	50	78
6	14	38	59	48	72
5	10	31	40	40	72
5	8	27	38	39	71

8. Estimate the effects associated with the sleep-deprivation groups of exercise 7. What would you estimate the strength of association to be between sleep deprivation and loss in score?

9. Carry out an analysis of variance on the following data, corresponding to two randomly chosen and independent experimental groups of ten cases each.

I		*II*	
19	16	18	21
20	20	19	19
24	20	19	23
20	18	24	17
22	24	18	18

10. Test the significance of the difference between the means of the two groups in exercise 9 by using the t test. How does this compare with the results of exercise 9?

11. Carry out a one-way analysis of variance for the following data based on four independent groups:

A	*B*	*C*	*D*
29	19	31	33
41	13	37	47
27	21	23	33
17	23	21	25
33	17	31	37

12. For the data of exercise 11 estimate the various treatment effects and the strength of association between the experimental and dependent variables.

13. The following experiment utilized three groups of differing numbers of cases. Test the hypothesis of the equality of all of the population means.

I	*II*	*III*
.5	.6	.8
.4	.6	.9
.6	.5	.8
1.0	1.0	1.1
.6	.9	1.2
.5	.4	1.1
1.1	.7	

14. Estimate the true effects in exercise 13.

15. In the United States, engineering schools tend to be found in large state universities, medium-sized state universities, private universities, and technical institutes. A large proportion of the engineers graduating this past spring obtained very good jobs. You wish to see if the starting salaries tend to relate to type of school attended. A random sample of 28 newly graduated engineers is taken, seven from each type of school. The results in terms of starting salary (in thousands of dollars) are as follows:

Medium state u.	*Large state u.*	*Technical*	*Private u.*
$M_1 = 17.85$	$M_2 = 22.29$	$M_3 = 18.75$	$M_4 = 24.00$
$S_1^2 = 35.84$	$S_2^2 = 18.20$	$S_3^2 = 16.24$	$S_4^2 = 35.42$

Do significant differences appear to exist among the four types of schools with reference to starting salaries? (**Hint:** Look at expression 10.8.5 in the text, and recall the basic definition of S^2. Then look at expression 10.8.6. How can you find the grand mean given the four group means?)

16. Use the results of exercise 15 to estimate the four fixed effects, and the strength of association.

17. Clergymen of the same large denomination from six large geographical areas of the United States were sampled at random. Each clergyman sampled was given an attitude test, providing a score on the "liberalism" of his attitude toward modern life. The following data resulted:

S.E.	*S.W.*	*N.E.*	*N.W.*	*Midwest*	*Far West*
27	29	34	44	32	45
43	49	43	36	28	50
40	27	30	30	54	30
30	46	44	28	50	33
42	26	32	42	46	35
29	48	42		36	47
30	28	41		41	
41	30	33			
28	47	31			
	50	40			

Test the hypothesis that the means of these populations are equal. What is the estimate of ω^2, the proportion of the variance in "liberalism" accounted for by geographical region?

18. Consider the following sets of data:

Group 1	*Group 2*	*Group 3*	*Group 4*
1.69	1.82	1.71	1.69
1.53	1.93	1.82	1.82
1.91	1.94	1.75	1.86
1.82	1.60	1.64	1.90
1.57	1.78	1.52	1.39
1.77	1.85	1.73	1.56
1.94	1.98	1.86	1.74
1.60	1.72	1.68	1.83
1.74	1.83	1.54	1.47
1.74	1.75	1.75	1.64

We wish to carry out an analysis of variance on these data, and test for equality of means ($\alpha = .05$). However, we will simplify our computations by subtracting 1.00 from each number and multiplying by 100. Complete the analysis and carry out the F-test. Should the transformation of the numbers ($X' = 100\,(X - 1)$) affect the results of our F test? Does it affect the values of M.S. between and M.S. within? How?

19. Estimate ω^2 for the data given in exercise 18 above, as well as the effects associated with the four groups.

20. In the construction of a projective test, 40 more or less ambiguous pictures of two or more human figures were used. In each picture, the sex of at least one of the figures was only vaguely suggested. In a study of the influence of the introduction of extra cues into the pictures, one set of 40 was retouched so that the vague figure looked slightly more like a woman, in another set each was retouched to make the figure look slightly more like a man. A third set of the original pictures was used as a control. The forty pictures were administered to a group of 18 male college students and an independent group of 18 female college students. Six members of each group saw the pictures with female cues, six the pictures with male cues, and six the original pictures. Each subject was scored according to the number of pictures in which the indistinct figure was interpreted as a female. The results follow:

	Female cues		*Male cues*		*No cues*	
Female subjects	29	36	14	5	22	25
	35	33	8	7	20	30
	28	38	10	16	23	32
Male subjects	25	35	3	5	18	7
	31	32	8	9	15	11
	26	34	4	6	8	10

Complete the analysis of variance.

21. For exercise 20 above, estimate the column effects, the row effects, and the interaction effects. What must be true for each of these sets of effects?

22. Suppose that in the experiment of exercise 20, everything had turned out as shown, *except* that the distinction between males and females (the row categories) was not made. Instead, imagine that these data were presented only for the three column groups, and a one way analysis had been carried out. What would happen to the error sum of squares in such an analysis, relative to the error sum of squares in exercise 20. What would this new error sum of squares actually include?

23. In a study of post-meningitic and post-encephalitic brain damage, each of 36 subjects was given a battery of tests, providing a composite score for each. Low scores on this composite measure presumably represented a considerable degree of residual brain damage. The subjects were divided into 3 groups according to type of initial infection, and into 3 crossed groups according to time since apparent physical recovery from the illness. The data follow:

	1–2 years		*3–5 years*		*7–10 years*	
Post-encephalitic	76 75	73 62	69 72	53 55	59 41	43 57
Post-meningitic	81 83	89 75	82 91	70 74	68 75	50 47
Post-operative control	75 65	84 63	85 76	79 87	98 82	100 79

Do there seem to be significant differences in performance among the post-encephalitic, post-meningitic, and control groups? Among the groups according to time since recovery? Is there apparent interaction between type of illness and time since recovery? (Use $\alpha = .05$ here.) State, verbally, your conclusions from these data.

24. Estimate the main effects and the interaction effects for exercise 23.

25. An experiment was carried out on the relation between the size and the wall color of a room used for a standardized interview, and the measured anxiety level of the respondent. The following results were obtained:

Room size		Red	Yellow	Green	Blue
	Small	160 155 170	134 139 144	104 175 96	86 71 112
	Medium	175 152 167	150 156 159	83 89 79	110 87 100
	Large	180 154 141	170 133 128	84 86 83	105 93 85

Room color

Complete the analysis of variance on these data.

26. Estimate the magnitudes of all of the main and interaction effects in exercise 25.

27. Suppose that α level of .05 is used for each of the tests carried out in exercise 25. If each test is assumed independent of every other, what is the probability of a Type I error on at least one of these tests? Suppose that $\alpha = .01$. Now what is the probability of a Type I error on at least one such test? If a number of F tests are to be made following an analysis of variance, should a relatively large or a relatively small value of α be preferred for each? Why?

28. Suppose that, given the information about effects in exercise 25, one was told that an interview of a subject was to be held in some unspecified room. What anxiety level should be predicted? If the room is specified as being green, what anxiety level should be predicted? If the information is that the room is small and red what should one predict?

29. In an experiment involving two factors, each at three levels, each combination of factors contained only one subject. Subjects had been assigned to the nine possible combinations totally at random, however. The data were as follows:

		Factor A		
		A_1	A_2	A_3
Factor B	B_2	37	36	40
	B_2	25	35	28
	B_3	20	22	30

Based on previous studies, the experimenter involved felt quite confident that no interaction effects should exist involving factors A and B. Given this assumption, complete the analysis.

30. Suppose that the experimenter in exercise 29 is wrong about the interaction between factors A and B. How might this affect the results here, relative to a larger study in which interaction was also examined?

31. In a certain experiment, two sets of experimental factors were used. The first (columns) contained 3 levels, and the second (rows) 4 levels. Twelve independent subjects assigned at random to the treatment combinations provided the following data. Test for significant row and column effects. What must one assume in doing so?

	I	*II*	*III*
A	92	40	24
B	−13	98	16
C	12	−8	64
D	82	83	46

32. Within the analysis of variance, could one test a specific hypothesis about the total population mean, such as $H_0: \mu = k$? If so, what form would the F test take for this hypothesis? Does this have anything to do with the fact that the total number of degrees of freedom is usually taken to be $N - 1$ instead of N?

33. At what point in the analysis of variance does it become critical that the population variances are homogeneous?

Chapter 11

THE ANALYSIS OF VARIANCE: MODELS II AND III, RANDOM EFFECTS AND MIXED MODELS

11.1 / RANDOMLY SELECTED TREATMENTS

The previous chapter was concerned exclusively with the variation of the general linear model appropriate for the fixed-effects analysis of variance model. Now we will turn to still another model for the analysis of variance. This is known variously as the *random-effects model, components of variance model,* or Model II. This, too, is a variation on the general linear model, but where the effects associated with the treatments or treatment combinations are no longer conceived as fixed. Rather, different effects may appear over different repetitions of the same experiment. Such effects are thus unlike fixed effects, which are thought of as constants, irrespective of how many times the experiment is carried out.

The random-effects model is designed especially for experiments in which inferences are to be drawn about an entire set of distinct treatments or factor levels, including some not actually observed. For such experiments, many more treatment categories or factor levels are possible than can be actually represented in a given experiment. However, our interest lies in the whole range of treatment possibilities, even though what we use in a given experimental setup is only a random sample of the potential treatments or levels we might have observed. Before the experiment is carried out, a random sample of treatments is drawn from among all possible such treatments, and then inferences are made about all such treatments on the basis of this limited sample.

For example, suppose that in a behavioral experiment we suspect that the personality of the experimenter may be having an effect on the results. In principle there are a vast number of people, each presumably having a distinct personality, who might possibly serve as the person who conducts the experiment. Obviously, it is reasonable to narrow this population of potential experimenters down to people having the requisite technical skills to conduct the experiment (for example, English-speaking behavioral scientists), but even this leaves a great many people from which to choose. To put it mildly, trying out each such person in our experiment would complicate matters! So, instead, we draw a random sample from among such qualified persons. Suppose that five different persons are chosen; each of the five persons conducts the experiment on a different sample of n subjects assigned at random. Each experimenter constitutes an experimental treatment given to one group of subjects; in all other respects subjects are treated exactly the same. Since the experimental treatments employed are themselves a random sample, this experiment fits the random-effects, rather than the fixed-effects, model: an inference is to be made about experimenter effects in general from observation of only five experimenters sampled at random.

Parenthetically, the student is warned that designation of the fixed-effects model as Model I, of the random-effects model as Model II, and the later mixed model as Model III, though convenient for the organization adopted here, is not uniform across authors. It is probably better for the student to learn to think of these three models as "fixed effects," "random effects" (or "components of variance"), and "mixed" from the very outset, rather than relying on any number designation alone.

11.2 / RANDOM EFFECTS AND MODEL II

Imagine an independent variable X, representing an experimental manipulation that can be broken down into a very large number of qualitative categories or levels. Thus, in the preceding example, use of any given experimenter represents a qualitatively different category from the use of another experimenter. Here there are thousands of experimenters, who might be used, and hence thousands of levels of independent variable X. Now let us designate one such level of X as g. Associated with any g will be a whole population of potential observations i, each having a y_{ig} value for the dependent variable. The linear model showing the makeup of any value y_{ig} in any such population g would be just as in the preceding chapter:

$$y_{ig} = \mu + \alpha_g + e_{ig}, \qquad [11.2.1]$$

where μ is the grand mean of all the populations, α_g is the effect of being in population g, and e_{ig} is the error component of y_{ig}. Over all of the different populations that can be symbolized by g, there will be a whole distribution of values α_g. This distribution of effects will have a mean μ_α, and a variance σ^2_α, over all possible populations g. Let us assume further that each population g is equally represented in the grand population. Then, since we are still dealing here with effects α, even when the number of groups is very large it will be true that $\mu_\alpha = 0$. However, only when $\alpha = 0$ for every single population will we know the value of σ^2_α, since then it must be true that $\sigma^2_\alpha = 0$.

Now in the actual experiment, a sample of N subjects will be assigned at random to a sample of J different treatment groups. Note that *only* J of the many possible treatments will be used. Before the experiment, a random sample of treatments is drawn. One of the selected treatments is designated randomly as group 1, another as group 2, and so on up to group J. Let us again call any one of these treatment groups j.

Now consider an individual i who is a member of treatment group j. What is the model for the dependent variable value y_{ij}? It is given by the linear model

$$y_{ij} = \mu + a_j + e_{ij}. \qquad [11.2.2^*]$$

Here

$$a_j = \sum_g \alpha_g v_{gj}$$

where v_{gj} is an indicator variable defined as follows:

$$v_{gj} = \begin{cases} 1 \text{ if treatment } g \text{ is selected and labeled as group } j \\ 0 \text{ otherwise.} \end{cases}$$

The effect a_j of being in treatment group j is thus the same as the effect of being in exactly one of the treatment populations. However, the exact value of a_j depends on *which* treatment was selected and labeled j in the sample. For a group labeled j, the value of a_j will thus vary over repetitions of the experiment, depending on which treatment is selected and called j. Thus a_j is a random variable, and we call values such as a_j random effects. The a_j values have a distribution with a mean of zero, and a variance equal to σ_A^2, where $\sigma_A^2 = \sigma_\alpha^2$.

The assumptions to be made in deriving a test of the hypothesis of no treatment effects are as follows:

1. The possible values a_j represent a random variable having a distribution with a mean of zero, and a variance σ_A^2.
2. For any treatment j, the errors e_{ij} are normally distributed with a mean of zero and a variance σ_e^2, which is the same for each possible treatment j.
3. The J values of the random variable a_j occurring in the experiment are all independent of each other.
4. The values of the random variable e_{ij} are all independent.
5. Each pair of random variables a_j and e_{ij} are independent.

You will note that here we are making assumptions about two different kinds of distributions. First of all there is a distribution of possible values for the effects a_j that might appear in the experiment. For the experiment proper, a random sample is taken from among all possible such effects values when a random sample of treatments is drawn. For the time being, no special assumptions need be made about the distribution of effects from which the sample of effects is drawn, other than that the mean is zero and that the variance has some (finite) value σ_A^2. Later, when we discuss this model more generally, we will assume that the distribution of effects is normal. The second kind of distribution is that of errors within particular treatments. Just as in Model I, we assume errors to be normally and independently distributed, with the same variance irrespective of the particular treatment under which the observation is made. Notice that we are, at present, making no assumptions about the number of potential treatments from which a sample of J is taken for the experiment, except that this number is greater than J. We are, however, assuming that within any given treatment population an infinite number of potential units for observation exist.

In this section the mean squares for the one-way analysis will be examined. *The discussion will be restricted to the situation where exactly the same number of observations n are made under each treatment.* Although the treatments may have different numbers of sample observations in the one-way case for the fixed-effects model, some difficulties arise in Model II unless the numbers of observations are equal.

The partition of the sum of squares for the random effects model is carried out in exactly the same way as for Model I. Furthermore, the mean squares are found just as before:

$$\text{MS between} = \frac{\text{SS between}}{J - 1} = \frac{\sum_j n(M_j - M)^2}{J - 1}$$

and

$$\text{MS within} = \frac{\text{SS within}}{N - J} = \frac{\sum_j \sum_i (y_{ij} - M_j)^2}{N - J}.$$

Bear in mind that for any random sample of J treatments or factor levels, there is a corresponding random sample of effects a_j. Let the sample mean of these effects be denoted by

$$M_a = \frac{\sum_j a_j}{J}.$$

Although the mean of the effects over all of the *possible* treatments must be zero, the value of M_a need not be zero, since this is but the mean of a sample of effects actually occurring in the experiment.

In the same way, sample means for error may be defined:

$$M_{ej} = \frac{\sum_i^n e_{ij}}{n}$$

and

$$M_e = \frac{\sum_j \sum_i e_{ij}}{N} = \frac{\sum_j M_{ej}}{J}.$$

Then the deviation of any mean M_j from the grand mean of the sample, M, can be rewritten as

$$(M_j - M) = (a_j - M_a) + (M_{ej} - M_e),$$

the deviation of effect a_j from the mean of all of the effects in the sample, plus the deviation of the average error in group j from the grand mean of all the errors in the sample.

Because of this makeup of the deviation of M_j from M, it can be shown that

$$E(\text{MS between}) = n\sigma_A^2 + \sigma_e^2. \qquad [11.3.1^*]$$

The expected mean square between is the weighted sum of two variances, that of the population of treatment effects and that of the error.

Now we turn to the mean square within treatments, where

$$E(\text{MS within}) = \sum_j \frac{\sigma_e^2}{J} = \sigma_e^2. \quad [11.3.2^*]$$

Thus the mean square within is always an unbiased estimate of error variance alone.

In Model II, if there are no treatment effects at all, either for those represented in the experiment or for any other possible treatment in the set sampled, then it must be true that $\sigma_A^2 = 0$. Hence the null hypothesis of no treatment effects for Model II is usually written

$$H_0: \sigma_A^2 = 0. \quad [11.3.3^*]$$

Now suppose that this null hypothesis were true. In this case

$$E(\text{MS between}) = \sigma_e^2$$

$$E(\text{MS within}) = \sigma_e^2$$

so that *both* mean squares are unbiased estimates of error variance alone.

The mean square between can be shown to be independent of the mean square within under the assumptions of Section 11.2, so that when H_0 is true the ratio

$$F = \frac{\text{MS between}/\sigma_e^2}{\text{MS within}/\sigma_e^2} = \frac{\text{MS between}}{\text{MS within}} \quad [11.3.4^*]$$

can be referred to the F distribution with $J - 1$ and $N - J$ degrees of freedom, as a test of this null hypothesis. Significant values of F lead to a rejection of $H_0: \sigma_A^2 =$ *0 in favor of* $H_1: \sigma_A^2 > 0$.

11.4 / AN EXAMPLE

Suppose that the experiment described in the introduction to this chapter was actually carried out. Five experimenters chosen at random conduct the same experiment, each on a different set of eight subjects randomly sampled and assigned at random among the experimental groups. The data are as shown in Table 11.4.1.

The hypothesis that $\sigma_A^2 = 0$ (that there are no experimenter effects) is to be tested using $\alpha = .01$.

The computations are:

$$\text{SS total} = 1455.94 - \frac{(240.8)^2}{40} = 6.32$$

$$\text{SS between} = \frac{(44.0)^2 + \cdots + (49.3)^2}{8} - \frac{(240.8)^2}{40}$$

$$= 1453.09 - 1449.62 = 3.48$$

$$\text{SS within} = \text{total} - \text{between} = 6.32 - 3.48 = 2.85.$$

Table 11.4.1

Results of using five different experimenters, chosen at random

Experimeter				
1	2	3	4	5
5.8	6.0	6.3	6.4	5.7
5.1	6.1	5.5	6.4	5.9
5.7	6.6	5.7	6.5	6.5
5.9	6.5	6.0	6.1	6.3
5.6	5.9	6.1	6.6	6.2
5.4	5.9	6.2	5.9	6.4
5.3	6.4	5.8	6.7	6.0
5.2	6.3	5.6	6.0	6.3
44.0	49.7	47.2	50.6	49.3

$\sum_j \sum_i y_{ij} = 240.8$

Table 11.4.2

Source	SS	df	MS	E(MS)	F
Between (experimenters)	3.48	4	.870	$8\sigma_A^2 + \sigma_e^2$	10.72
Within	2.84	35	.081	σ_e^2	
Total	6.32	39			

Table 11.4.2 is the summary table for this analysis. The F value required for rejection at the .01 level for 4 and 35 degrees of freedom is between 4.02 and 3.83 (that is, between the values for 30 and 40 degrees of freedom denominator). Accordingly, the hypothesis of no experimenter effects may be rejected. There is sufficient evidence to say that experimenter effects exist and contribute to the variance of Y.

11.5 / ESTIMATION OF VARIANCE COMPONENTS IN A ONE-WAY ANALYSIS

Instead of estimating effects directly by taking differences of the treatment means from the grand mean, as in the fixed-effects model, in Model II we will estimate σ_A^2, the true variance due to treatments. For the foregoing experiment, this is the true variance attributable to experimenters.

Since the expectation of the mean square between is

$$E(\text{MS between}) = n\sigma_A^2 + \sigma_e^2,$$

an unbiased estimate of σ_A^2 may be found by taking

$$\frac{\text{MS between} - \text{MS within}}{n} = \text{est. } \sigma_A^2. \qquad [11.5.1]$$

(Note: est. $\sigma_A^2 = 0$ when MS within $\geqslant$ MS between.)

For the example, this estimate is

$$\text{est. } \sigma_A^2 = \frac{.870 - .081}{8} = .099.$$

The variance of y_{ij} over the population of all possible potential observations is

$$\sigma_Y^2 = E(y_{ij} - \mu)^2 = \sigma_A^2 + \sigma_e^2$$

so that the true variance consists of two independent parts or components: the variance due to treatments, and that due to error alone. The best estimate of the total variance σ_Y^2 is given by

$$\text{est. } \sigma_Y^2 = \text{est. } \sigma_A^2 + \text{est. } \sigma_e^2 = \frac{\text{MS between} + (n-1)\text{ MS within}}{n}. \qquad [11.5.2]$$

For the example,

$$\text{est. } \sigma_Y^2 = .099 + .081$$
$$= .18.$$

This fact that the total variance must consist of two components allows one to make a somewhat more informative use of the estimates of σ_A^2 and σ_e^2. We can take the ratio of the estimated σ_A^2 to the estimated total variance to find the estimated proportion of variance accounted for by the treatments,

$$\text{est. proportion of variance accounted for} = \frac{\text{est. } \sigma_A^2}{\text{est. } \sigma_Y^2} \qquad [11.5.3]$$

$$\text{est. proportion of variance accounted for by experimenters} = \frac{(.099)}{.180} = .55.$$

Here we estimate that over one-half of the variance among observations seems to be due to experimenter differences. This would be a most important finding in a real experiment, as it would suggest rather strongly that different experimenters' repetitions of this experiment would not necessarily be comparable.

One way of expressing the idea that a factor accounts for a given amount of variance is by the index known as *the population intraclass correlation coefficient:*

$$\rho_I = \frac{\sigma_A^2}{\sigma_A^2 + \sigma_e^2}. \qquad [11.5.4^*]$$

The intraclass correlation coefficient for the grand population will be zero only when σ_A^2 is zero, and will reach unity only when $\sigma_e^2 = 0$, given that $\sigma_Y^2 > 0$. Notice that this index is simply another way of expressing the proportion of variance attributable to the factor A. The population index ρ_I is identical to ω^2 in its general form and meaning, although ρ_I applies to the random-effects and ω^2 to the fixed-effects model so that slightly different estimation methods apply in the two situations.

Quite often the intraclass correlation is used to express the fact that observations in the same category are related, or tend on the average to be more like each other than observations in different categories. The larger the value of ρ_I, the more similar do observations in the same treatment category tend to be, relative to observations in different categories. For example, in a study of the similarity in intelligence of twins, a random sample of sets of twins might be taken, each pair of twins constituting a natural "treatment" with $n = 2$. An estimated intraclass correlation greater than zero would indicate that some of the variability in intelligence is accounted for by variation among sets of twins, so that pairs of twins tend to be more alike in this respect than are pairs of nontwins. The value of ρ_I is thus a measure of the homogeneity of observations within classes, relative to between classes.

Still another special application of this idea of an intraclass correlation occurs in the study of the reliability of repeated measurements of individuals. The reliability of a single measurement y_{ij} for an individual j can be defined in a way equivalent to ρ_I for a population of such individuals. The estimate of ρ_I given by 11.5.3 is then an estimate of the true reliability for such measurements. A modern discussion of these methods for estimating reliability in terms of the random-effects analysis of variance is given in Winer (1971), and in Guilford and Fruchter (1978).

Under Model II, it is also possible to estimate intervals for the proportion of variance accounted for by a factor. Here, however, we must assume that the basic distribution of effects is normal. Given this assumption, the required limits can be found from the distribution of the quantity θ, where

$$\theta = \frac{\sigma_A^2}{\sigma_e^2} \qquad [11.5.5]$$

and

$$\text{true proportion of variance accounted for} = \frac{\theta}{1 + \theta} = \rho_I. \qquad [11.5.6]$$

The required confidence interval is found as follows: Let F' be the value in an F distribution with $J - 1$ and $N - J$ degrees of freedom, cutting off the *upper* $\alpha/2$ proportion in the distribution, and let F'' be the value cutting off the *lower* $\alpha/2$. For this distribution, it must be true that

$$\text{prob. } (F'' \leqslant F \leqslant F') = 1 - \alpha.$$

Furthermore, regardless of the true value of σ_A^2, it will be true that the ratio

$$\frac{\text{MS between}/(\sigma_e^2 + n\sigma_A^2)}{\text{MS within}/(\sigma_e^2)} \qquad [11.5.7^*]$$

is distributed as an F variable with $J - 1$ and $N - J$ degrees of freedom, since numerator and denominator *are* independent chi-square variables divided by degrees of freedom when Model II is true, and the basic distribution of effects is normal. This ratio is equivalent to

$$\left(\frac{\text{MS between}}{\text{MS within}}\right)\left(\frac{1}{1 + n\theta}\right), \qquad [11.5.8]$$

so that

$$\text{prob}\left(F'' \leqslant \frac{\text{MS between}}{\text{MS within}}\,\frac{1}{1 + n\theta} \leqslant F'\right) = 1 - \alpha.$$

By algebraic operation on this inequality, it can be shown that the $100(1 - \alpha)$ percent confidence interval for θ is

$$\frac{1}{n}\left[\frac{\text{MS between}}{(\text{MS within})\, F'} - 1\right] \leqslant \theta \leqslant \frac{1}{n}\left[\frac{\text{MS between}}{(\text{MS within})\, F''} - 1\right]. \qquad [11.5.9^*]$$

This may be turned into a confidence interval for ρ_I very easily by the relation given by 11.5.6.

The value of F' can be found directly from the F tables for $\alpha/2$, and the degrees of freedom $J - 1$, and $N - J$. However, the value of F'' must be found by the methods of Section 9.9 for values on the *lower* tail of an F distribution. These confidence intervals, like those for σ^2 given in Section 9.5, need not be optimal.

As an example let us find the 95 percent confidence interval for θ and ρ_I for the data of Section 11.4. Here,

$$\frac{\text{MS between}}{\text{MS within}} = 10.72.$$

The first thing we need is the value of F', that value cutting off the *upper* 2.5 percent in an F distribution with 4 and 35 degrees of freedom. Table V shows this to be about 3.2. Next, we must find F'', the value on the lower tail of this same distribution cutting off the lower 2.5 percent. From the same table we take the value cutting off the *upper* 2.5 percent in a distribution *wtih 35 and 4 degrees of freedom:* this is about 8.4. Thus,

$$F'' = \frac{1}{8.4} = .119.$$

The approximate 95 percent confidence interval for θ is then

$$\frac{1}{8}\left[\frac{10.72}{3.2} - 1\right] \leq \theta \leq \frac{1}{8}\left[\frac{10.72}{.119} - 1\right]$$

$$.29 \leq \theta \leq 11.14.$$

The corresponding interval for ρ_I is

$$\frac{.29}{1 + .29} \leq \rho_I \leq \frac{11.14}{1 + 11.14}$$

or

$$.22 \leq \rho_I \leq .92.$$

11.6 / TESTING OTHER HYPOTHESES, AND CALCULATING POWER UNDER MODEL II

The hypothesis tested in the example above was that there exist no effects of the individual treatments, so that $\sigma_A^2 = 0$. The theory is not limited to this situation, however. When the distribution of effects can be assumed to be normal it is possible to test many other hypotheses about the ratio $\theta = \sigma_A^2/\sigma_e^2$ or, equivalently, about ρ_I.

It was noted in 11.5.7 that when there is a value of $\theta > 0$ which characterizes the experimental population, the random variable

$$\frac{\text{MS between}}{(\text{MS within})(1 + n\theta)} \qquad [11.6.1]$$

actually is distributed as F with $J - 1$ and $J(n - 1)$ degrees of freedom. On the other hand, the ordinary ratio (MS between/MS within) is not distributed as F in this instance.

In the general case, the hypothesis tested is

$$H_0: \theta \leqslant \theta_0 \text{ versus } H_1: \theta_0 < \theta$$

where θ_0 is any hypothetical value, $0 \leqslant \theta_0$.

This general hypothesis is tested as follows: after α is chosen, we determine the F value cutting off the upper α proportion of cases in a distribution with $J - 1$ and $N - J$ degrees of freedom. Call this value F_α. Now we find the F ratio for the sample in the usual way (Section 11.3). The hypothesis is rejected in favor of H_1 when

$$\frac{\text{MS between}}{\text{MS within}} \geqslant (1 + n\theta_0)F_\alpha. \qquad [11.6.2^*]$$

These facts permit us to find the power of F in a much simpler way under Model II than under Model I. The problem in finding the power of the F test when a given value of θ is true is that of finding the probability

$$\Pr\left(\frac{\text{MS between}}{\text{MS within}} > F_\alpha; \theta, \nu_1, \nu_2\right).$$

That is, the power of the test corresponds to the probability of rejecting the null hypothesis at the α level, when there are ν_1 and ν_2 degrees of freedom, and the particular value of θ is indeed true.

Now since the ratio in 11.6.1 above is distributed as F, it follows that

$$\begin{aligned}\Pr\left[\frac{\text{MS between}}{\text{MS within}} > F_\alpha\right] &= \Pr[F(1 + n\theta) > F_\alpha] \\ &= \Pr[F > F_\alpha/(1 + n\theta)].\end{aligned} \qquad [11.6.3^*]$$

This last probability can be calculated from the distributon of F, and thus the power can be determined. Bear in mind, however, that this method of determining power depends upon the assumption that effects are sampled randomly from a normal distribution of such effects. This method may also require fairly extensive tables of the distribution of F, and such tables are not always easy to find. [One reasonably extensive set of tables for F is found in Dixon and Massey (1957)].

11.7 / IMPORTANCE OF THE ASSUMPTIONS IN MODEL II

By now it should surely be clear that the actual *arithmetic* of the simple analysis of variance is the same regardless of the model adopted. The analysis of variance is, strictly speaking, only a way of arranging this arithmetic. However, the inferences made from the sample values are really quite different, depending on the model invoked. All the inferences made under Model I concern means (and differences between means). On the other hand, the inferences made using Model II deal with variances: that is, *the basic inference has to do with the variance of the population of effects actually sampled by the experimenter*. This distinction between the two models has an influence on the importance of the assumptions made in each.

In the first place, the assumption of normality can be quite important in Model II. Provided that one is concerned only with a test of the hypothesis $\sigma_A^2 = 0$, then slight departures from normality among the error distributions should have only minor consequences for the conclusions reached using a reasonably large sample size. Here, no assumption at all about the distribution of the α values need be made. On the other hand, for tests of the more general hypothesis given in Section 11.6 it is most important that the distributions *both* of effects and of errors be normal in form. In the same way, interval estimates for θ and for ρ_I depend heavily upon the assumptions of normal distributions of the α and the e values for their validity.

The assumption of equality of variances has a somewhat different status in the random effects than in the fixed-effects model. If the random-effects model applies, and if the effects α_j are independent of the errors, e_{ij}, we need assume only one distribution of errors with variance σ_e^2, so that in a sense, the error variance for all possible observations given some treatment j must be the same as for any other treatment. However, the important assumptions are those involving independence: the errors *must* be independent both of the particular treatment effects and of each other. The simple random-effects model is not really applicable to data where the *errors* in observations must be related to the effect of the treatment applied, or to each other. [In another model (11.18), we will modify the assumptions so as to permit one special kind of dependency to exist between errors.]

Although statistical dependency among the error components of scores creates a special problem in any analysis of variance, it is not true that the *observations themselves* must be completely independent for Model II to apply. We have just seen that the intraclass correlation coefficient expresses the degree of similarity or dependency among observations given the same treatment. When σ_A^2 is greater than zero some statistical relation must exist between pairs of score values within the possible treatment groups. As we saw in Section 11.6 hypotheses about the degree of relatedness can be tested via Model II. However, the form of relatedness that makes the simple Model II analysis inapplicable is most often trend or serial relation, implying non-independence of the error terms.

11.8 / TWO-FACTOR EXPERIMENTS WITH SAMPLING OF LEVELS

Now we turn to the analysis of experiments involving two factors, each of which is represented in the experiment only by a sample of its levels.

For example, suppose that a projective test involves ten cards administered individually to a subject. The subject must respond by giving as many free associations to each card as possible. These responses are scored in a number of ways, but the total number of responses is an important index of overall performance. The developer of this test has some idea that the *order* in which cards are presented has a bearing on the total number of responses given, and so would like to see if this factor of order is an important one in accounting for variation in test performance. If it should turn out that order is important, an order of presentation for the cards will be sought that will be optimal in evoking responses. Furthermore, this investigator has worked out a standard set of instructions for test administrators, in the hope that administrator effects on performance are thereby made negligible. However, it is desirable to see if this is true.

Here are two factors, order of presentation and test administrator, that may account for variance in total response to the test. Obviously, neither factor can be represented at all levels in any experiment, since there are exactly 3,628,800 different ways of presenting the ten cards, and a very large number of persons who might be trained to administer the test. Hence, the experimenter decides to conduct a study in which each of these factors will be sampled. From a single set of data the question of the relative contribution to variance of each of the factors can be answered, as well as the secondary question of possible interaction between test administrator and order of presentation.

This hypothetical experiment will be developed as an example in Section 11.11. For the time being we will turn our attention to extending Model II to cover such situations.

11.9 / MODEL II FOR TWO-FACTOR EXPERIMENTS

The two different factors in the experiment will be designated A and B. A random sample of C different levels of A will be drawn, and shown as columns in the data table, and a random sample of R different levels of B will appear as the rows in the table. Within each combination shown by a cell of the data table, n observations are to be made at random.

The score of individual i in column j and row k of the table will be thought of as a sum:

$$y_{ijk} = \mu + a_j + b_k + c_{jk} + e_{ijk}.$$

Here, a_j is the random variable standing for the effect of the sample treatment appearing in the data table as column j, b_k is the random variable indicating the effect of the sample treatment in row k, and c_{jk} is the random interaction effect associated with cell jk. The term e_{ijk} is a random variable standing for the error effect of the observation of individual i under the joint conditions indicated by column j and row k. Observe that in this linear model all components of a score except the grand mean μ are values of *random variables,* the sampled effects, plus error.

The assumptions made are:

1. The a_j are normally distributed random variables with mean zero and variance σ_A^2.
2. The b_k are normally distributed with mean zero and variance σ_B^2.
3. Each c_{jk} has a normal distribution with mean zero and variance σ_{AB}^2.
4. The e_{ijk} are normally distributed with mean zero and variance σ_e^2.
5. The a_j, the b_k, the c_{jk}, and the e_{ijk} are pair wise independent.

11.10 / THE MEAN SQUARES

The computations for the two-way analysis under Model II are exactly the same as for Model I (Section 10.25). Thus, the total sum of squares is partitioned into a sum of squares for rows, a sum of squares for columns, a sum of squares for interaction, and an error sum of squares.

On examining the mean squares, however, we will find that the really important differences between Models I and II appear. You may recall that in Section 10.21 it was pointed out that for the fixed-effects model both row and column effects sum to zero. Also the following things are true of the interaction effects γ_{jk}:

$$\sum_j \gamma_{jk} = 0, \sum_k \gamma_{jk} = 0, \text{ and } \sum_j \sum_k \gamma_{jk} = 0.$$

That is, in Model I, over the column treatments the interaction effects sum to zero, over the row treatments interaction effects sum to zero, and over both rows and columns these effects sum to zero. The net result is that in the sum of squares for, say, rows, only row effects and error can possibly be included:

$$(M_k - M)^2 = (\beta_k + M_{ek} - M_e)^2.$$

For the fixed-effects model, any interaction effect cannot possibly contribute to the sum of squares for rows because this sum of squares is itself based on the data *summed* over columns, automatically making the interaction effects sum to zero. For the same reason, the sum of squares for columns does not include any of the interaction effects.

On the other hand, in Model II the set of R row treatments represents a *sample* from a large set of possible such treatments, and the C column treatments another sample from a large set. Here it is no longer true that the a_j must sum to zero over the sample columns, or that the b_k must sum to zero over the sample rows. Nevertheless, the column effects are removed when the grand mean is subtracted from the mean of any row. Similarly, the row effects are removed when the grand mean is subtracted from the mean of any column.

However, interaction effects are defined only so that

$$\underset{j}{E}(c_{jk}) = 0$$

$$\underset{k}{E}(c_{jk}) = 0$$

$$\underset{jk}{E}(c_{jk}) = 0.$$

That is, the expected value of the interaction term c_{jk} is zero over *all possible* treatments that might have been selected for column j with a fixed row treatment k. Furthermore, the expected value of the interaction term c_{jk} is zero over all possible treatments that might have been selected for row k with a fixed column treatment j. Note that these requirements hold only when one considers *all possible* row or column treatments that might have been selected. Unlike the situation in the fixed-effects model, *here there is no requirement at all that the interaction effects sum to zero over the particular set of R row treatments or over the particular set of C column treatments that just happened to appear in the sample*. For this reason, no sum of interaction effects must be zero in any given set of data. The important consequence of this fact is that *the sum of squares for rows reflects deviations due not only to row treatment effects, but also to interaction effects*. In the same way *the sum of squares for columns includes both column and interaction effects*. This has a most important bearing on estimates and tests made under Model II, as we shall see.

It then will always be true that under Model II,

$$E(\text{MS between rows}) = \sigma_e^2 + n\sigma_{AB}^2 + Cn\sigma_B^2. \qquad [11.10.1^*]$$

Unlike the expectation for fixed effects, *here the mean square for rows estimates a weighted sum of the error variance, the variance due to row treatments, and the interaction variance,* σ^2_{AB}.

A parallel situation holds for factor A, represented by the columns:

$$E(\text{MS columns}) = \sigma^2_e + n\sigma^2_{AB} + Rn\sigma^2_A. \quad [11.10.2^*]$$

Turning to the expected value for the mean square interaction, we find that

$$E(\text{MS interaction}) = \sigma^2_e + n\sigma^2_{AB}. \quad [11.10.3^*]$$

The expectation of the mean square error is, as always,

$$E(\text{MS error}) = \sigma^2_e. \quad [11.10.4^*]$$

11.11 / HYPOTHESIS TESTING IN THE TWO-WAY ANALYSIS UNDER MODEL II

We have just seen that the expected values both for the mean square for rows and the mean square for columns are different in Model II from those in Model I. The practical implication of this fact is that the hypotheses of no row and no column effects are tested in a different way for this model than for Model I. Consider the test of the hypothesis

$$H_0: \sigma^2_A = 0.$$

When this hypothesis is true, then

$$E(\text{MS columns}) = \sigma^2_e + n\sigma^2_{AB} \quad [11.11.1]$$

which is *not* the same as the expectation of mean square error, but rather that of *mean square interaction:*

$$E(\text{MS interaction}) = \sigma^2_e + n\sigma^2_{AB}. \quad [11.11.2]$$

These two mean squares (MS columns and MS interaction) can be shown to be independent, and when each is divided by $\sigma^2_e + n\sigma^2_{AB}$, each is distributed as χ^2 divided by degrees of freedom. Thus, the hypothesis of no column effects is usually tested by

$$F = \frac{\text{MS columns}}{\text{MS interaction}} \quad [11.11.3^*]$$

with $C - 1$ and $(R - 1)(C - 1)$ degrees of freedom.

In the same way, the hypothesis

$$H: \sigma^2_B = 0$$

is tested by a comparison of MS rows with the interaction MS, since if this hypothesis is true,

$$E(\text{MS rows}) = \sigma^2_e + n\sigma^2_{AB}, \quad [11.11.4]$$

which is the same as the expected value of the mean square for interaction. The hypothesis is tested by the ratio

$$F = \frac{\text{MS between rows}}{\text{MS interaction}} \qquad [11.11.5^*]$$

with $R - 1$ and $(R - 1)(C - 1)$ degrees of freedom. The hypothesis of no interaction is tested just as for Model I:

$$H_0: \sigma^2_{AB} = 0$$

tested by

$$F = \frac{\text{MS interaction}}{\text{MS error}} \qquad [11.11.6^*]$$

with $(R - 1)(C - 1)$ and $RC(n - 1)$ degrees of freedom.

It should be emphasized that the test of row and column effects against interaction is really appropriate *only* when the factors have been randomly sampled, as in Model II (and in some instances of the mixed models introduced in sections to follow). When the experiment qualifies for Model I, the ratio of the mean square for a factor to the interaction mean square is *not necessarily* distributed as F, even when the null hypothesis is true. This is one of the major practical distinctions between the fixed-effects and the random-effects models.

There is one situation, however, when the main effects are not tested against interaction in Model II. This occurs when the experimenter has decided that the interaction variance σ^2_{AB} is zero. There is either some theoretical or some empirical reason to believe that interaction effects should not occur, or there is *strong* evidence for no interaction in the data. Of course, there is a difficult problem involved in deciding when there really *is* strong evidence that interaction effects do not exist. When the mean square for interaction is equal to or less than the mean square for error, we are ordinarily quite justified in reaching this conclusion. However, a simple rule that applies to a broad class of situations is that due to Paull (1950): **when each of the mean squares for error and for interaction has a number of degrees of freedom greater than six, then pool these into a combined estimate of error when the F ratio for interaction is less than 2.00.**

When there is sufficient evidence that no interaction effects exist, the two sums of squares, error and interaction, are pooled as follows:

$$\text{pooled MS error} = \frac{\text{SS interaction} + \text{SS error}}{(R - 1)(C - 1) + RC(n - 1)} \qquad [11.11.7]$$

When $\sigma^2_{AB} = 0$, the expectation of this pooled MS error is σ^2_e. The expectation of MS rows is also σ^2_e under the hypothesis of no row effects, so that the test may be carried out by use of the ratio

$$F = \frac{\text{MS rows}}{\text{pooled MS error}} \qquad [11.11.8^*]$$

with $R - 1$ and $RCn - R - C + 1$ degrees of freedom. In the same way a test of no column effects is carried out by

$$F = \frac{\text{MS columns}}{\text{pooled MS error}} \qquad [11.11.9^*]$$

with $C - 1$ and $RCn - R - C + 1$ degrees of freedom. This procedure is *not* recommended, however, unless the evidence for no interaction effects is strong enough to satisfy a criterion such as Paull's. (An even simpler rule to remember and apply is this: test the interactions using some relatively large value of α, such as .25. If the interactions still aren't significant, pool.)

11.12 / AN EXAMPLE

As an example, suppose that developer of the projective test mentioned in Section 11.8 actually carried out the experiment. Recall that this study was designed to test both for the presence of effects associated with order of presentation and with particular administrators of the test. Four qualified persons were selected at random and trained to administer the test. Also, four orders of presentation of the test were selected at random. Each adminstrator gave the test to a different pair of randomly selected normal adults under each one of the selected order conditions, so that a total of thirty-two different test performances were observed in all. The dependent variable was the total number of responses a subject gave to the test cards. The experimenter was prepared to assume normal distributions of order effects, of administrator effects, of interaction effects, and of error. The data were as shown in Table 11.12.1. The α level chosen for the test of each hypothesis was .05.

Table 11.12.1

Results of the projective test study, with four orders and four administrators.

	Administrator				
Order	*1*	*2*	*3*	*4*	*Totals*
I	26	30	25	28	
	25	33	23	30	220
II	26	25	27	27	
	24	33	17	26	205
III	33	26	30	31	
	27	32	24	26	229
IV	36	37	37	39	
	28	42	33	25	277
Totals	225	258	216	232	931

The analysis proceeds in the usual manner for two-way designs (Section 10.25):

$$\text{SS total} = \sum_j \sum_k \sum_i y_{ijk}^2 - \frac{\left(\sum_j \sum_k \sum_i y_{ijk}\right)^2}{N}$$

$$= 27965 - \frac{(931)^2}{32}$$

$$= 878.7$$

$$\text{SS rows} = \frac{\sum_k\left(\sum_j\sum_i y_{ijk}\right)^2}{Cn} - \frac{\left(\sum_j\sum_k\sum_i y_{ijk}\right)^2}{N}$$

$$= \frac{(220)^2 + \cdots + (277)^2}{8} - \frac{(931)^2}{32} = 363.0$$

$$\text{SS columns} = \frac{\sum_j\left(\sum_k\sum_i y_{ijk}\right)^2}{Rn} - \frac{\left(\sum_j\sum_k\sum_i y_{ijk}\right)^2}{N}$$

$$= \frac{(225)^2 + \cdots + (232)^2}{8} - \frac{(931)^2}{32} = 122.3$$

$$\text{SS error} = \sum_j\sum_k\sum_i y_{ijk}^2 - \frac{\sum_j\sum_k\left(\sum_i y_{ijk}\right)^2}{n}$$

$$= 27965 - \frac{(51)^2 + (50)^2 + \cdots + (64)^2}{2}$$

$$= 310.5$$

$$\text{SS interaction} = \text{SS total} - \text{SS rows} - \text{SS columns} - \text{SS error}$$

$$= 878.7 - 363.1 - 122.3 - 310.5 = 82.8.$$

Table 11.12.2 summarizes the analysis.

Table 11.12.2

Source	*SS*	*df*	*MS*	*E(MS)*	*F*
Rows (orders)	363.1	3	121.0	$\sigma_e^2 + 2\sigma_{AB}^2 + 8\sigma_B^2$	$\left(\frac{121.0}{15.7} = 7.7\right)$
Columns (admin.)	122.3	3	40.8	$\sigma_e^2 + 2\sigma_{AB}^2 + 8\sigma_A^2$	$\left(\frac{40.8}{15.7} = 2.6\right)$
Interaction	82.8	9	9.2	$\sigma_e^2 + 2\sigma_{AB}^2$	
Error	310.5	16	19.4	σ_e^2	
Totals	878.7	31			

Ordinarily, the tests both for row and for column effects would be carried out by dividing the mean square for rows or columns by mean square for interaction. However, it is immediately apparent that here the mean square for error is larger than mean square for interaction, so that the estimate of σ_{AB}^2 is zero. Notice that the Paull criterion is also satisfied. Thus, we find

$$\text{pooled MS error} = \frac{82.8 + 310.5}{9 + 16} = 15.7,$$

which has 9 + 16 or 25 degrees of freedom.

We test for row effects by finding

$$F = \frac{121.0}{15.7} = 7.7.$$

For 3 and 25 degrees of freedom, this value exceeds that required for the .05 level, and so the hypothesis of no row effects is rejected. The same procedure is carried out for columns, and here

$$F = \frac{\text{MS columns}}{\text{pooled MS error}} = 2.6$$

for 3 and 25 degrees of freedom. This fails to reach the value of 2.99 required to reject the null hypothesis. (These F values are enclosed in parenthesis in the summary table to show that they are not obtained in the usual way. In reports of such data, the F tests would usually be accompanied by an explanatory footnote to the summary table.)

The conclusions from the experiment are, then, as follows:

1. Order does seem to have an effect on the total number of responses given in the test.
2. There is insufficient evidence to determine if true administrator differences exist.
3. There is virtually no evidence for administrator-order interaction.

11.13 / POINT ESTIMATION OF VARIANCE COMPONENTS

The four expressions in the E(MS) column of the summary table allow us to estimate the components of variance, as follows:

$$\text{est. } \sigma_B^2 = \frac{\text{MS rows} - \text{MS interaction}}{Cn} \quad [11.13.1]$$

$$\text{est. } \sigma_A^2 = \frac{\text{MS columns} - \text{MS interaction}}{Rn} \quad [11.13.2]$$

$$\text{est. } \sigma_{AB}^2 = \frac{\text{MS interaction} - \text{MS error}}{n} \quad [11.13.3]$$

$$\text{est. } \sigma_e^2 = \text{MS error.} \quad [11.13.4]$$

If any of the first three estimates turn out to be negative in sign, then that component is estimated to be zero.

In the example above, the estimate of σ_{AB}^2 is zero, and the pooled MS error was taken to represent σ_e^2. Thus, to be consistent, one should estimate the other component as follows:

$$\text{est. } \sigma_B^2 = \frac{\text{MS rows} - \text{pooled MS error}}{Cn}$$

$$= \frac{121.0 - 15.7}{8} = 13.6$$

$$\text{est. } \sigma_A^2 = \frac{40.8 - 15.7}{8} = 3.14$$

$$\text{est. } \sigma_e^2 = 15.7.$$

The proportion of variance accounted for by factor B (rows) is estimated from

$$\frac{\text{est. } \sigma_B^2}{\text{est. } (\sigma_e^2 + \sigma_{AB}^2 + \sigma_A^2 + \sigma_B^2)} = \frac{13.16}{15.7 + 0 + 13.16 + 3.14} = .41.$$

An estimated 41 percent of the variance in total response is attributable to order of presentation of this projective test.

For factor A (columns) the estimated proportion of variance accounted for is

$$\frac{3.14}{15.7 + 0 + 13.16 + 3.14} = .098$$

so that if factor A accounts for any variance at all, our best guess is that this is less than 10 percent of the total variance. Approximate confidence intervals for the variance components σ_A^2, σ_B^2, and so forth, can be found by methods outlined in Scheffé (1959, pp. 231–235).

11.14 / MODEL III: A MIXED MODEL

Multifactor experiments involving Model II are relatively rare in social and behavioral research. It is far more common to encounter experiments where one or more factors have fixed levels and the remaining factors are sampled. This situation calls for a third model of data, in which each individual observation results in a score that is a sum of *both* fixed and random effects. Obviously, mixed models such as this apply only to experiments where two or more factors are under study.

As an example of an experiment fitting a mixed model, suppose that a study is concerned with the muscular tension induced in subjects by three different varieties of task. The subject performs using pencil and paper with the preferred hand, meanwhile holding the bulb of a sensitive pressure recording gauge in the other. The mean reading on this gauge during the performance provides the dependent variable score. Three separate kinds of tasks are administered: in one type the subject solves relatively complicated problems in arithmetic; in a second type a short, imaginative, composition is written; and in the third a careful tracing of a line drawing must be made.

The subjects in this experiment are to be children in the fifth grade in public schools in a large city. Since there may be sex differences in terms of the three tasks involved, for this study the experimenter decided to use only male subjects, on the supposition that males are more likely to exhibit strong emotional responses to the three tasks given. Furthermore, different school settings may differ in their emphasis on these sorts of skills. Thus, it was decided to sample from among the large number of fifth-grade classrooms in the city, and select at random six classrooms to be observed, with six boys then selected at random from each classroom. From a given classroom, two boys each are given one of the three tasks, for a total of six observations per classroom, or thirty-six observations in all. Thus "classrooms" represents a random-effects factor, while "tasks" is a fixed-effects factor.

Let the factor having fixed levels be labeled A, and represented by the columns of the table, and let the randomly sampled factor be B, and shown by the rows of the data table. Now it is assumed that

$$y_{ijk} = \mu + \alpha_j + b_k + c_{jk} + e_{ijk} \qquad [11.14.1]$$

where α_j is the fixed effect of the treatment indicated by the column j, b_k is the random variable associated with the kth row, c_{jk} the random interaction effect operating in the cell jk, and e_{ijk} is the random error associated with observation i in the cell jk. We make the following assumptions:

1. The b_k and the c_{jk} are jointly normal, each with mean of zero and with variances σ_B^2 and σ_{AB}^2 respectively.
2. The e_{ijk} are normally distributed, with mean zero and variance σ_e^2.
3. The e_{ijk} are independent of the b_k and the c_{jk}.
4. All error terms e_{ijk} are independent of each other.

11.15 / THE EXPECTED MEAN SQUARES IN A MIXED MODEL

Within this mixed model, the sum of the α_j over all of the C fixed treatments must, as usual, be equal to zero:

$$\sum_j \alpha_j = 0.$$

Furthermore, within any row k the sum of the interaction effects, c_{jk}, will also be equal to zero. Thus,

$$c_k = \sum_j \frac{c_{jk}}{C} = 0.$$

From this it follows that the mean of the interaction effects for the entire experiment must be zero:

$$M_c = \sum_k \sum_j \frac{c_{jk}}{RC} = 0.$$

On the other hand, it is not true that within any column the sum of the row effects will be zero. The b_k effects represent a sampled factor, and within any column the average of the row effects, or M_b, will be the same as for any other column, but will not necessarily be equal to zero. For the same reasons, the average of the interaction effects within a column, or c_j, need not be equal to zero, but may differ from column to column. There will also be an average error within any row, or M_{ek}, an average error within any column, or M_{ej}, and an average error overall, or M_e.

The partition of the sum of squares and the calculation of the mean squares proceed exactly as for any two-way analysis of variance. However, in the mixed model, the squared difference between M_j, the mean of a particular column j, and M, the general mean, can be thought of as having the following composition:

$$\begin{aligned}(M_j - M)^2 &= (\alpha_j + M_b + c_j + M_{ej} - M_b - M_e)^2 \\ &= (\alpha_j + c_j + M_{ej} - M_e)^2.\end{aligned}$$

Notice that **any deviation of a column mean from the grand mean includes not only the column effect and a mean error deviation, but also an average interaction effect** c_j. This is because there is no requirement at all that interaction effects must sum to zero within columns when the row effects, and thus the interaction effects, are only a random sample of such effects. Thus column means, formed by summing observations across rows, include not only the fixed column effects, but also some of the random interaction effects as well.

For this reason, given the mean square for columns calculated in the usual way, it can be shown that the expectation for this mean square is

$$E(\text{MS columns}) = \sigma_e^2 + n\sigma_{AB}^2 + \frac{Rn\left(\sum_j \alpha_j^2\right)}{C - 1}. \qquad [11.15.1^*]$$

In Model III, the mean square for columns (the fixed-effects factor) is an estimate of a weighted sum of the error variance, the sum of the squared effects associated with the column factor itself, *and* the interaction variance.

Now let us examine the composition of the mean square for rows (the sampled factor). This mean square is based on the squared deviation of each row mean, M_k, from the general sample mean M. Applying the model once again we find that

$$(M_k - M)^2 = (b_k + M_{ek} - M_b - M_e)^2.$$

Here, the deviation depends only on row effects and upon error deviation; the interaction effects do not appear in any such deviation making up the mean square for rows. Remember that within any given row the sum of the c_{jk} values taken over columns must be zero. Because the columns factor is *not* sampled, and each possible column treatment is actually represented in the experiment, interaction effects within rows *do* cancel out and sum to zero. As a result it can be shown that

$$E(\text{MS rows}) = \sigma_e^2 + Cn\sigma_B^2, \qquad [11.15.2^*]$$

the mean square for rows estimates a weighted sum of the error variance and the variance attributable to the sampled factor B.

The interaction variance is based on the squared difference between the deviation from M for the mean of each cell jk in the data table, and the corresponding row mean's and column mean's deviations:

$$[(M_{jk} - M) - (M_k - M) - (M_j - M)]^2$$
$$= (c_{jk} - c_j + M_{ejk} - M_{ek} - M_{ej} + M_e)^2.$$

Observe that any such deviation depends only on interaction effects and on error. Then it can be shown that

$$E(\text{MS interaction}) = \sigma_e^2 + n\sigma_{AB}^2. \qquad [11.15.3^*]$$

That is, the interaction mean square is an estimate of a weighted sum of the error variance and the interaction variance.

When the hypothesis of no column effects is true, so that $\alpha_j = 0$ for each of the column treatments j, then

$$E(\text{MS columns}) = \sigma_e^2 + n\sigma_{AB}^2. \qquad [11.15.4^*]$$

Notice that this is exactly the same as the expectation for the MS interaction. Under our assumptions the ratio

$$F = \frac{\text{MS columns}}{\text{MS interaction}} \qquad [11.15.5^*]$$

with $C - 1$ and $(R - 1)(C - 1)$ degrees of freedom provides an appropriate test of the hypothesis of no column effects. *Be careful to notice that it is the fixed-effects factor that is tested against interaction in this mixed model.*

On the other hand, when the hypothesis of no row effects is true,

$$E(\text{MS rows}) = \sigma_e^2,$$

which is the same as

$$E(\text{MS error}) = \sigma_e^2.$$

Then the test for the existence of row effects (that is a test of $H_0{:}\sigma_B^2 = 0$) is given by

$$F = \frac{\text{MS rows}}{\text{MS error}} \qquad [11.15.6^*]$$

with $R - 1$ and $RC(n - 1)$ degrees of freedom.

Thus, we see that there are two distinctly different kinds of F tests employed in the analysis of variance for this mixed model: the hypothesis that all the fixed effects are zero is tested by comparing the mean square for that factor against the mean square for interaction. On the other hand, the hypothesis associated with the random-effects factor is tested by comparing its associated mean square with the mean square for error. *The important procedural differences among these three models thus have to do with the way in which hypotheses are tested, and especially with the denominators used in the various F ratios.*

The test for interaction in a two-way analysis is the same for all three models:

$$F = \frac{\text{MS interaction}}{\text{MS error}} \qquad [11.15.7^*]$$

with $(R - 1)(C - 1)$ and $RC(n - 1)$ degrees of freedom.

When there is good reason to believe that interaction effects do not exist, there is some advantage in pooling the interaction and the error sums of squares to get a new estimate of the error variance. This is done exactly as outlined in Section 11.11.

11.16 / AN EXAMPLE FITTING MODEL III

Let us suppose that the experiment outlined in Section 11.14 had actually been carried out, and that the data were as shown in Table 11.16.1. In this table the two values appearing under each combination of classroom and task represent the independent performances of the two subjects in that combination.

Table 11.16.1

Results of the experiment on handpressure.

	Tasks			
Classrooms	*I*	*II*	*III*	*Total*
1	7.8	11.1	11.7	
	8.7	12.0	10.0	
	16.5	23.1	21.7	61.3
2	8.0	11.3	9.8	
	9.2	10.6	11.9	
	17.2	21.9	21.7	60.8
3	4.0	9.8	11.7	
	6.9	10.1	12.6	
	10.9	19.9	24.3	55.1
4	10.3	11.4	7.9	
	9.4	10.5	8.1	
	19.7	21.9	16.0	57.6
5	9.3	13.0	8.3	
	10.6	11.7	7.9	
	19.9	24.7	16.2	60.8
6	9.5	12.2	8.6	
	9.8	12.3	10.5	
	19.3	24.5	19.1	62.9
Totals	103.5	136.0	119.0	358.5

The .05 level is chosen for α, and the analysis proceeds as follows:

$$\text{SS total} = (7.8)^2 + \cdots + (10.5)^2 - \frac{(358.5)^2}{36} = 123.57$$

$$\text{SS columns} = \frac{(103.5)^2 + (136.0)^2 + (119.0)^2}{12} - \frac{(358.5)^2}{36} = 44.04$$

$$\text{SS rows} = \frac{(61.3)^2 + \cdots + (62.9)^2}{6} - \frac{(358.5)^2}{36} = 6.80$$

$$\text{SS error} = (7.8)^2 + \cdots + (10.5)^2 - \frac{(16.5)^2 + \cdots + (19.1)^2}{2}$$

$$= 14.54$$

$$\text{SS interaction} = 123.57 - 44.04 - 6.80 - 14.54 = 58.19.$$

Table 11.16.2 is the summary table for this analysis.

First of all, for 10 and 18 degrees of freedom, the F test for interaction exceeds the required F of 2.41, and the hypothesis of no interaction *is* rejected.

For 2 and 10 degrees of freedom an F value of 4.10 is required for significance, so that the hypothesis of no column effects is *not* rejected. Notice that the F ratio

Table 11.16.2

Source	*SS*	*df*	*MS*	*E(MS)*	*F*
Columns (tasks)	44.04	2	22.02	$\sigma_e^2 + 2\sigma_{AB}^2 + \dfrac{12\sum_j \alpha_j^2}{2}$	3.78
Rows (class)	6.80	5	1.36	$\sigma_e^2 + 6\sigma_B^2$	1.68
Interaction (tasks by class)	58.19	10	5.82	$\sigma_e^2 + 2\sigma_{AB}^2$	7.19
Error	14.54	18	.81	σ_e^2	
Total	123.57	35			

for columns is formed by dividing the column mean square by the mean square for interaction. If this experiment had been incorrectly analyzed under Model I, quite a different conclusion would have been reached; this illustrates the importance of using the proper model in testing hypotheses.

For 5 and 18 degrees of freedom an F value of 2.77 is required in order to reject the null hypothesis at the 5 percent level. The F ratio formed by the mean square for rows over the mean square error gives a value smaller than this, so that the hypothesis of no subject effects is also not rejected.

In summary, there is insufficient evidence to permit us to conclude either that there are task effects or that class effects exist in this experiment. There is, however, fairly strong evidence for the presence of interaction effects. There is something about the combination of a particular classroom with a particular task that accounts for variance in the data. Thus, within classrooms, task differences apparently exist, but these tend to be different for different classrooms. Perhaps the capacity of a task to produce tension in a given boy depends on his classroom background.

11.17 / RANDOMIZATION AND CONTROLS

The basic statistical tools in experimentation are randomization and control. As distinct from informal observation, every experiment worthy of the name is conducted under specific controls. All of the physical and situational factors that the experimenter can manage are kept as constant as possible in the course of the experiment. Efforts are made to have the units that are the basis for observations in the experiment, whether they be humans, animals, plants, or what not, be just as homogeneous as possible. Differences among the units might affect the outcome or the interpretability of the experiment. Often, the mere effect of giving a subject a treatment of any kind is checked by the use of control groups, participating in the experimental setting but receiving no treatment. Controls are exercised in hundreds of ways in any well-planned experiment. Nevertheless, there are countless factors that might affect the outcome of the experiment, but which the experimenter does not know to control, or know how to control.

Randomization enters into the experiment in at least two different ways. The actual individuals observed in any sample may differ in any number of ways from the "typical" individual in the population as a whole. It is entirely possible to obtain a sample of N individuals who are "peculiar" as compared with the population they represent, and the conclusions reached from this sample need not be indicative of what the population as a whole tends to be like. By sampling at random from the population, the experimenter is able to identify the peculiarities of the sample with random error, and to allow for the possibility of an atypical sample in the conclusions.

Furthermore, given some sample it will always be true that factors other than the ones manipulated by the experimenter will contribute to the observed differences between subjects in the particular situation. If it should happen that some extraneous factor operates unevenly over several treatment groups or over different subjects, this can create spurious differences, or mask true effects in the data. Such a factor is a "nuisance factor," playing a role analogous to "noise" in communication. Randomization of subjects over treatments is one device for "scattering" the effects of these nuisance factors through the data. Often, when particular nuisance factors are known, levels of these factors are scattered at random throughout the design. In order to randomize an experiment, the researcher uses some scheme such as a table of random numbers to allot individual subjects to experimental groups or nuisance levels to subgroups in a purely random, unsystematic, "chancy" way. By randomization, the possibility of "pileups" of nuisance effects in particular treatment groups is identified with random error, and the experimenter can rest assured that over all repetitions of the experiment under the same conditions, true effects will eventually emerge if they exist.

Every experiment is randomized to some extent. Ordinarily, a study having one or more experimental factors and certain constant controls where every other factor contributing to variance is randomized is called a **fully randomized** design: usually a random sample is drawn, and then subjects are allotted to treatments or treatment combinations purely at random. In addition, levels of one or more nuisance factors may be assigned at random. The experiments used as examples in Chapter 10 were of this type. Furthermore, many other experiments fitting either the fixed-effects or the random-effects models can also be thought of in this way if nature or some other agency is conceived as carrying out the random assignment of individuals to groups.

On the other hand, sometimes it is advantageous to the experimenter *not* to randomize the effects of particular nuisance factors, but rather to represent them systematically in the experiment. These nuisance factors are treated as though they were experimental factors, when actually the purpose of representing them in the experiment in the first place is to control these factors and thereby reduce error variance. It should be obvious from our discussion of the indices ω^2 and ρ_I that deliberate introduction of any systematic effects that account for a portion of the variance σ_Y^2 must also *reduce* the error variance σ_e^2. That is, the true variance of any dependent variable is thought of as a sum of components:

$$\sigma_Y^2 = \text{(the variance attributable to systematic features of the experiment)} + \text{(the variance attributable to unsystematically represented factors)}.$$

The more *relevant* factors that can be introduced into the experiment in a systematic way, the smaller will be the variance considered as error. Furthermore, the smaller the error variance, then the more precise will be the experiment, in that confidence intervals will be smaller and true effects more likely to be detected if present.

A large part of experimental design deals with this strategy of deliberately introducing nuisance factors into experiments so as remove them from our estimates of error variance. Two of the most common and simple of these strategies will be examined next.

11.18 / RANDOMIZED BLOCKS AND REPEATED MEASURES DESIGNS

Our discussion of mixed models for the analysis of variance will conclude with a brief look at two experimental situations where such mixed models are especially appropriate. Each of these experimental designs is widely used, and the ideas underlying each can be extended to quite complex situations, although we can, of course, be concerned here only with their simple, two-way, form.

The first situation ordinarily calling for a mixed model is the *randomized blocks design*. In this design there is one factor A that is actually the experimental variable of interest. However, there is another factor B that is primarily of a nuisance character. *The experimenter knows that factor B accounts for some variance in Y*. Each level or category of the nuisance factor consists of a group of subjects who are "matched," each showing the same status on this nuisance factor. Then *within* each such matched group, all the different levels of factor A are represented.

It often happens that there are very many levels to the nuisance factor, or many combinations of factors, and the experimenter must sample from among these levels for a representation in the experiment. A sample of K nuisance levels is drawn, with each to contain some nJ subjects. Each group representing a qualitatively different category on the nuisance factor or factors is called **a matched group,** or a **block** of subjects. Then within each block, subjects are allotted at random to the J different treatments of interest. Such a matched group design, with sampling of the nuisance factor levels, is often called a **randomized blocks design** (especially when there are J observations per block, or one subject per treatment within a matched group).

Levels on the matching or "blocking" factor may represent qualitative differences of any kind, or combinations of factor levels. *The important point is that the factor or factors underlying the groupings must actually contribute to variance*. If the factors leading to the matching of subjects have nothing at all to do with the dependent variable, then this is not an economical way to design an experiment, since formation of matched groups is troublesome and costly, and the experimenter gets nothing in return (or may actually lose something) for the extra effort.

In the simplest version of a randomized blocks design, K different groups are sampled each representing a level of the nuisance factor, B. Each block or level of B contains exactly J individuals. Then the individuals within a block are *randomly* assigned to the J different levels of factor A. Thus only one individual receives each treatment-block, or AB, combination.

A randomized blocks experiment such as this is usually analyzed as **a mixed model design without replication.** The analysis of variance proceeds just as for any two-way analysis, except that the partition of the sum of squares contains only three terms:

$$\text{SS total} = \text{SS blocks} + \text{SS treatments} + \text{SS interaction}.$$

Notice that a separate SS error cannot be calculated here, since there is only one observation per cell. Table 11.18.1 is the summary table for the analysis. Since there is only one individual per cell in this instance ($n = 1$) the only F test to be made is that for treatments, which involves the MS columns divided by the MS interaction. The blocks (rows or matched groups effects) cannot be tested in the usual way unless at least 2 observations are made in each block and treatment combination.

Table 11.18.1

Summary table for a randomized blocks design, $n = 1$.

Source	*SS*	*df*	*MS*	*E(MS)*	*F*
Blocks		$K - 1$	$\frac{\text{SS rows}}{K - 1}$	$\sigma_e^2 + J\sigma_B^2$	
Within blocks		$[K(J - 1)]$			
Treatments		$J - 1$	$\frac{\text{SS columns}}{J - 1}$	$\sigma_e^2 + \sigma_{AB}^2 + \frac{K \sum_j \sigma_j^2}{J - 1}$	$\frac{\text{MS treatments}}{\text{MS blocks by treatment}}$
Blocks by treatments		$(J - 1)(K - 1)$	$\frac{\text{SS int.}}{(J - 1)(K - 1)}$	$\sigma_e^2 + \sigma_{AB}^2$	
Total		$JK - 1$			

Matched group designs of the randomized blocks type are extremely common in any number of research areas and especially in agriculture (hence the name, each "block" being a separate set of plots of ground). It is a highly efficient way of proceeding, provided that one can be reasonably sure that the factors on which individuals are matched actually are associated with the dependent variable; presumably this kind of information is available to the experimenter from knowledge of the research area or from pilot studies for the experiment. (The student may already have recognized that the matched pairs design, analyzed by use of the t test in Chapter 8 is an example of a randomized blocks design with $J = 2$.)

The logical extension of the idea of randomized blocks designs is to make an individual subject stand for a block, and to give each of J different treatments to each subject. Ordinarily, the treatments are assigned in some random order to a given subject, and K subjects are used in all. Then every subject is observed under J treatments, and every treatment is tried on K subjects, for a total of JK observations in all.

In randomized blocks, it is reasonable to assume that even when two different subjects are members of the same block, the error portions of their y scores will be independent. The analysis outlined above rests on this assumption. However, when repeated measures are taken of the same subject over different treatments, it may be true that a dependency, or *correlation,* exists between the error portions of such scores (see Appendix B for a definition of correlation). Thus, the assumptions underlying the randomized blocks model are relaxed a bit to allow for this possibility. For repeated measures, where each subject i receives each treatment, we will assume that

$$E(e_{ij}, e_{im}) = \rho\sigma_e^2, \quad \text{when } j = m$$

$$E(e_{ij}, e_{im}) = \sigma_e^2, \quad \text{when } j = m.$$

Here j and m are two of the J treatments. When j and m are different, we will assume that the errors for the same individual i have some relationship or correlation ρ (lowercase Greek rho). When j and m are the same, then e_{ij} and e_{im} are the same, and $\rho = 1.00$. Furthermore in this simplest version of the repeated measurements model, we assume that the value of this correlation term ρ is the same for any individual i and for any two different treatments j and m. All of the other assumptions are the same as for any other mixed model.

Under the new assumption of errors that may be dependent within a subject, the expected mean squares become as follows:

$$E(\text{MS between subjects}) = \sigma_e^2[1 + (J - 1)\rho] + J\sigma^2_{sub.} \qquad [11.18.1]$$

$$E(\text{MS between treatments}) = \sigma_e^2[1 - \rho] + \sum_j \frac{K\alpha_j^2}{J - 1} + \sigma^2_{sub.\ by\ A} \qquad [11.18.2]$$

$$E(\text{MS subjects by treatments}) = \sigma_e^2[1 - \rho] + \sigma^2_{sub.\ by\ A} \qquad [11.18.3]$$

Here, $\sigma^2_{sub.\ by\ A}$ represents subject-treatment interaction variance. In such an analysis, MS between subjects is calculated just as for MS between for any factor B in a two-way analysis, except that this time the factor B is ''subjects.'' The MS between treatments is found as for any fixed factor A, and MS subjects by treatments is computed as for interaction. However, in the summary table for a subjects by treatments, or repeated measures design, it is customary first to break the total sum of squares into two major divisions, SS between subjects, and SS within subjects, where

SS within subjects = SS between treatments + SS subjects by treatments

The summary table for such an analysis takes the form shown in Table 11.18.2.

Table 11.18.2

Summary table for the analysis of variance for a repeated measures design.

Source	*SS*	*df*	*MS*	*E(MS)*	*F*
Between subjects (rows)		$K - 1$	$\frac{\text{SS sub.}}{K - 1}$	$\sigma_e^2\,[1 + (J - 1)\rho] + J\sigma^2_{sub.}$	
Within subjects		$K(J - 1)$			
Between treatments (columns)		$J - 1$	$\frac{\text{SS treat.}}{J - 1}$	$\sigma_e^2\,(1 - \rho) + \sum_j \frac{K\alpha_j^2}{J - 1} + \sigma^2_{sub.byA}$	$\frac{\text{MS treat.}}{\text{MS sub. by treat}}$.
Sub. by treat.		$(J - 1)(K - 1)$	$\frac{\text{SS sub.} \times \text{treat.}}{(J - 1)(K - 1)}$	$\sigma_e^2\,(1 - \rho) + \sigma^2_{sub.byA}$	

This way of arranging the arithmetic really has no bearing on the numerical results, but it does help in keeping up with which F tests are appropriate in such a set-up, particularly in the more complex designs.

Given the expected mean squares above, it can be seen that under Model III the only meaningful F test here must be

$$F = \frac{\text{MS between treatments}}{\text{MS subjects by treatments}} \qquad [11.18.4]$$

with $J - 1$ and $(K - 1)(J - 1)$ degrees of freedom. This is the only F ratio where numerator and denominator have the same expected values when the null hypothesis, $H_0{:}\alpha_j = 0$ for all j, is true. However, note that in this simple subjects by treatments design, there is no way to test for subject effects. This creates no big problem, since "subjects" is ordinarily introduced as a nuisance factor anyway.

Please do not carry away the impression that all the problems introduced by repeated observations are solved when "subjects" appears as a factor in the experimental analysis, and the mixed or random effects model is employed. This is not necessarily true. It is true that the analysis of variance for these models can be applied when a *particular form* of dependency exists among the observations (strictly speaking, when the dependency involves *equal correlations* among pairs of observations within subjects). However, we have no good reason to believe that most repeated observation data have this character, and we have ample reason to believe that in some such data the error components of observations are dependent in other ways, particularly when the material is such that learning or other serial changes occur. Although we will not go into this topic further here, any standard text on experimental design such as Winer (1971) or Kirk (1968) explores repeated measures designs in detail.

11.19 / THE GENERAL PROBLEM OF EXPERIMENTAL DESIGN

The factorial design mentioned in the last chapter and the randomized blocks design just discussed are but two of the many ways that the experimenter may choose to design a study. The literature of experimental design is full of ways for collecting and analyzing data for particular experimental situations. Many of these designs in common use are found in references such as Edwards (1973), Kirk (1968), and particularly in Winer (1971). A more advanced treatment of the subject is given in Cochran and Cox (1957). An informative, nonmathematical introduction to design is given in D. R. Cox (1958).

From a broad point of view, the problem of choosing a design for an experiment is a problem in economics. The experimenter has some question or questions to answer. The following points must be assured.

1. The actual data collected will contain all the information needed to make inferences, and this information can be extracted from the data.
2. The important hypotheses can be tested validly and separately.
3. The level of precision reached in estimation, and the power of the statistical tests, will be satisfactory for the purpose.

Consideration number 1 involves the actual selection of treatments and treatment combinations that the experimenter will observe. What are the factors of interest? How many levels of a factor will be observed? Will these levels be sampled or regarded as fixed? Which factors should be crossed in the design. Consideration 2 is is closely related to the first: must all combinations of factors to be observed, or is there interest in only some treatment combinations? If only part of the possible set of treatment combinations can be observed, is it possible to make separate inferences about the various factors and combinations? Consideration 3 involves the choice of sample size and of experimental controls. Is the sample size contemplated large enough to give the precision of estimation or power (or both) that the experimenter feels is necessary to have for the inferences?

However, attendant to each of these considerations are parallel considerations of cost.

4. The more different treatments administered, necessitated by the more questions asked of the data, the more the experiment will cost in time, subjects, effort, and other expenses.
5. The more kinds of information the experimenter wants to gain the larger the set of assumptions that may be required to obtain valid inferences.
6. The more hypotheses that can be tested validly and separately, the greater the number of treatment combinations necessary, and the larger may be the required number of subjects. Furthermore, the clarity of the statistical findings may be lessened, and the experiment as a whole may be harder to interpret.
7. The experimenter can increase precision and power by larger samples, or by exercising additional controls in the experiment, either as constant control, or by a matching procedure. Each possibility has its real costs in time, effort, and perhaps, money.

Other things being equal, the experimenter would like to get by with as few subjects as possible. Even ''approximately'' random samples are extremely difficult to obtain, and often the sheer cost of the experiment in time and effort goes up with each slight increase in sample size. Furthermore, the experiment may not be carried out as carefully with a large number as with only a few subjects. Throwing in lots of ill-considered treatments just to see what will happen can be an expensive pastime for the serious experimenter. The number of treatment combinations can increase very quickly when many factors are added, and some of these combinations may be of no interest at all to the experimenter or add little or nothing to precision, so that a high price may be paid for discovering ''garbage'' effects in the data. If a large sample is out of the question, power and precision can be bought at the price of constant or matching controls. However, remember that constant control reduces the generality of the conclusions the experimenter can reach from a set of data, and matching may be extremely hard to carry out.

All in all, there are several things the experimenter wants from the experiment and several ways to get them. Each desideratum has its price, however, and the experimenter must somehow decide if the gain in designing the experiment in a particular way is offset by the loss that may be incurred in so doing. This is why the problem of experimental design has strong economic overtones.

Texts in experimental design present ways of laying out the experiment so as to get "the most for the least" in a given situation. Various designs emphasize one aspect of another of the considerations and costs involved in getting and analyzing experimental data. Texts in design can give only a few standard types or layouts that study and experience have shown optimal in one or more ways, with, nevertheless, some price paid for using each design. Obviously, the best design for every conceivable experiment does not exist "canned" somewhere in a book, and experienced researchers and statisticians often come up with novel ways of designing an experiment for a special purpose. On the other hand, a study of the standard experimental designs is very instructive for any experimenter, if only to appreciate the strengths and weaknesses of different ways of laying out an experiment.

We can go no further into the question of experimental design here. However, a principle for the beginner to remember is, *keep it simple!* Concentrate on how well and carefully you can carry out a few *meaningful* manipulations on a few subjects, chosen, *if possible,* randomly from some well-defined population and *randomized* among treatments. The experiment and its meaning is the thing to keep in mind, and not some fancy way of setting up the experiment that has no real connection with the basic problem or the economics of the situation. Then, when the novice finally knows his way around in his experimental area, the refinements of design are open for exploration.

EXERCISES

1. A large chain of retail stores was used in a study of the effects of management style on employee satisfaction. A random sample of eight stores having different managers was taken. Then, within each store a random sample of ten employees was each given a questionnaire, which provided a score on "employee satisfaction." Do store managers appear to account for variation in employee satisfaction within this chain of stores? The data were as follows:

			Manager				
A	*B*	*C*	*D*	*E*	*F*	*G*	*H*
20	16	19	26	23	27	20	19
21	17	18	25	21	23	20	18
19	16	18	20	22	23	20	21
18	15	20	25	21	25	21	21
20	17	16	26	23	28	23	19
23	19	20	21	22	29	22	19
20	18	17	21	22	30	22	18
21	16	17	25	24	26	21	20
19	14	18	25	20	25	21	18
20	16	18	27	21	28	21	19

2. Estimate the proportion of variance that seems to be attributable to managers in exercise 1. Does it seem reasonable to suppose that managers alone could account for this much variance in employee satisfaction? Do stores differ in other ways besides managers, and could these other things be accounting for some of the differences in satisfaction?

3. An interstate highway runs across a large state. The highway patrol of the state monitors this highway and, among other things, gives tickets to speeders. However, does the patrol tend to cover some short sections more thoroughly than others? In order to find evidence for this question, the highway was divided into five-mile stretches, and a random sample of seven of these stretches was taken. Then, for each stretch, 10 days were chosen at random, and the number of speeding tickets given on each day recorded. These data are given below.

Highway stretches						
A	*B*	*C*	*D*	*E*	*F*	*G*
2	13	7	17	9	4	12
0	5	8	4	3	5	4
5	10	6	9	4	6	5
5	4	6	5	6	16	4
15	7	16	5	7	10	6
9	5	2	6	4	4	5
3	6	7	8	13	5	7
21	8	4	5	5	6	5
4	7	5	6	5	8	6
2	10	2	3	10	1	9

Would you say that the highway patrol's coverage is spread evenly along this highway, or does it tend to vary over different stretches?

4. Estimate the proportion of variance in tickets given that is accounted for by highway stretch. What does this result mean?

5. As part of an experiment on psychogalvanic reactions of dogs to unexpected visual stimuli, "artificial" human faces were constructed by combining elements of actual human features. There were six positions in which such elements might appear in a given face, and six possible elements in each position. Thus there were 6^6 or 46,656 different faces possible. As part of a pretest of the main experiment, eight of these possible stimulus faces were chosen at random, and each presented to a different group of two dogs. The results were as follows:

Stimuli							
1	*2*	*3*	*4*	*5*	*6*	*7*	*8*
18	12	20	15	16	19	17	22
19	14	21	13	18	20	19	25

Test the hypothesis of no true between-stimulus variance.

6. From the results of exercise 5 what proportion of the total variance in dogs' responses would you judge to be due to stimulus differences?

7. A random sample of five supermarkets was selected from a very large chain, and within each supermarket the number of items purchased on a single day by each of 10 customers selected at random was noted. The entries in the table on page 408 are the numbers of items purchased, arranged by store.

Test the hypothesis of no variance in number of items purchased attributable to stores. (Use $\alpha = .05$.)

Stores					
1	2	3	4	5	6
13	11	18	45	26	18
16	14	27	48	29	36
19	17	51	27	32	14
16	17	57	33	32	14
19	25	9	18	38	28
22	26	18	27	41	31
19	24	19	36	18	14
19	26	28	40	21	25
13	17	30	28	12	28
6	23	34	10	25	30

8. In exercise 7 test the hypothesis that variance between stores accounts for 50 or more percent of the total variance in number of items purchased.

9. In a certain one-way random-effects design, the mean square between treatments was 28.9 and the mean square within was 3.7. The mean square between had 10 degrees of freedom, and the mean square within had 33 degrees of freedom. Find the 95 percent confidence limits for ρ_I, variance accounted for.

10. What is the power of the test described in exercise 9 if the alternative hypothesis that $\rho_I = .50$ is true? Use $\alpha = .05$.

11. A one-way, random-effects analysis of variance with 6 and 21 degrees of freedom gave an F-value just significant at the .01 level for the hypothesis that $\sigma_A^2 = 0$. Test the hypothesis that $\rho_I \leq .10$ against the alternative that $\rho_I > .10$.

12. Suppose that in exercise 29, Chapter 10 both factor A and factor B had been sampled to produce the three actual levels used for each. How would this fact change the analysis? How would it change the assumptions that need to be made in order to carry out the tests?

13. An investigator was interested in the possible effects of the measured body-build or "somatotype" of a man, together with the ethnic background of that person, on his tendency to be overweight. Thus, a sample of 5 different somatotypes was chosen, along with a sample of 4 different ethnic-geographical groupings. Two men were found in each somatotype and

		Ethnic-geographic group			
		A	B	C	D
Somatotype	I	4.21 3.38	4.37 3.29	2.91 3.35	3.22 2.46
	II	3.58 1.82	3.58 1.37	2.31 2.96	3.02 2.11
	III	3.30 1.89	3.01 2.34	2.01 1.85	2.17 1.22
	IV	3.37 2.21	2.49 3.86	3.36 3.10	2.78 2.79
	V	3.56 2.58	3.70 3.48	3.44 2.78	3.74 2.35

ethnic combination, and then a typical week's calorie consumption of each was found. The table on page 408 shows the average daily consumption in thousands of calories for each man. Does there seem to be a relation between somatotype, ethnic-geographic background, or their combination, and average daily calorie consumption? (Use $\alpha = .05$.)

14. For the data of exercise 13, see if you can estimate the proportion of variance accounted for by each of the two main factors, and by their interaction.

15. In the data table for exercise 1 above, suppose that the first five responses under each manager represent employees with five or more years of experience in the store. The second five responses for each manager represent employees with fewer than five years' experience. Test the hypothesis of no interaction between manager and the experience of the employee. Test the hypothesis of no difference in satisfaction between the experienced and less experienced employees. Does the format of the second test depend on the outcome of the first? How? (Use $\alpha = .05$ throughout.)

16. An automobile company was interested in the comparative efficiency of three different sales approaches to one of their products. They selected a random sample of 10 different large cities, and then assigned the various selling approaches at random to three agencies within the same city. The results in terms of sales volume over a fixed period for each agency were as follows:

		Approach		
		A	*B*	*C*
	1	38	27	28
	2	47	45	48
	3	40	24	29
	4	32	23	33
	5	41	34	26
City	*6*	39	23	31
	7	38	29	34
	8	42	30	25
	9	45	31	26
	10	41	27	34

Does there seem to be a significant difference ($\alpha = .05$) among these three sales approaches? Would you test for city differences here? Explain.

17. Suppose that the data given in exercise 25, Chapter 10 had represented two factors, each with randomly selected levels. Carry out the appropriate analysis of variance and F tests.

18. In a pilot study preliminary to a larger experiment there was interest in the proportion of variance attributable to a particular experimental factor. Four levels of this factor were selected at random, and three subjects were assigned at random to each. The data turned out as follows:

Level 1	*Level 2*	*Level 3*	*Level 4*
3.9	1.9	5.3	4.5
4.7	2.8	3.0	5.4
2.4	3.2	4.5	4.7

Establish 95 percent confidence limits for the true proportion of variance of the dependent variable accounted for by the experimental factor.

19. In a given experiment, twelve randomly selected pairs of children, each pair matched with respect to age, sex, and intelligence, were used. The members of each pair were randomly assigned to two fixed experimental conditions. The data in terms of the dependent variable are given below:

Pair	*Condition I*	*Condition II*
1	18	23
2	19	34
3	27	26
4	25	23
5	28	29
6	15	30
7	17	14
8	29	41
9	36	37
10	25	24
11	46	45
12	31	32

Carry out an analysis of variance and F test on these data.

20. Analyze the data of exercise 19 via a t test for matched pairs, and compare the result with that found in the preceding exercise. Do the two methods agree as they should?

21. In an experiment on paired-associate learning, eight randomly chosen subjects were presented with three different lists of 35 pairs of words to learn. Each subject was successively given the three lists in some randomly chosen order. The score for a subject was the number of pairs correctly recalled on the first trial. Do these three lists seem to be significantly different ($\alpha = .05$) in their difficulty for subjects?

		List A	*List B*	*List C*
Subject	1	22	15	18
	2	15	9	12
	3	16	13	10
	4	19	9	10
	5	20	12	13
	6	17	14	12
	7	14	13	10
	8	17	19	18

22. Exercise 5, Chapter 2 gives data for three questionnaire items rated by 30 subjects. Suppose that the subjects had been chosen at random. Are there significant differences among the three items? (Use $\alpha = .05$).

23. Below is an analysis of variance summary table. Carry out F-tests ($\alpha = .05$ throughout) and interpret the findings under

(a) A fixed-effects model.
(b) A random-effects model (both variables).
(c) Factor A fixed, Factor B random.

Source	*SS*	*d.f.*
Factor *A*	386.0	5
Factor *B*	1427.2	9
Interaction	2699.6	45
Error	3109.3	120
Total	7622.1	179

24. Suppose that in exercise 7, Chapter 10 each row of the data represents a single subject, where each subject was tested under each sleep deprivation condition. Then complete the appropriate analysis of variance for this situation.

25. For the data shown in Table 11.16.1, how might one estimate the average squared effects of the tasks? How could you estimate the interaction variance? The variance for classes? How might these estimates be used to find the proportion of total variance each apparently accounts for in the experiment?

26. In a study of human memory, subjects were given lists of fifty words to memorize. *A priori,* however, the experimenter felt that the four lists chosen (regarded as fixed treatments in the experiment) might differ considerably in difficulty. Therefore each of ten subjects chosen at random was given all four lists in a randomly assigned order. The data, in terms of words correctly recalled, turned out as follows. Are the lists significantly different from each other?

Subjects	*Lists* I	*II*	*III*	*IV*
1	15	21	24	20
2	27	35	32	34
3	26	28	30	30
4	38	41	40	39
5	14	22	19	23
6	29	33	30	31
7	32	37	34	35
8	45	45	40	42
9	28	38	35	32
10	33	34	33	29

27. Suppose that the four lists used in exercise 26 above had been a random sample from a very large set of lists that might be used in such a study. What difference, if any, would this have made in the analysis?

28. Write out the expected mean squares for exercise 26, on the assumption that errors within subjects have a constant degree of dependence.

29. Write out the expected mean squares for exercise 26 on the assumption that the "Lists" factor was randomly sampled, and that there is no dependency of any kind between errors.

30. For exercise 23, estimate the proportion of variance accounted for by factors A, B, and $A \times B$ interaction, on the assumption of a completely random model.

31. For exercise 3, suppose that the rows of the table represent days chosen at random, so that the seven stretches of highway were all observed on each day. Reanalyze the data accordingly.

Chapter 12
COMPARISONS AMONG MEANS

Experimentation is almost never conducted in the hermetically sealed atmosphere that statistical reports often suggest. Most emphatically, good research requires all the care and control that the experimenter can provide in the actual experiment. But research hypotheses are not entertained or dismissed in the cavalier way that a statistics book may imply. These books set down the formal rules of the game, a set of prearranged signals from one scientist to another that something worth looking into may have been found, but that the results contain sampling error. Research reports are written in this conventionalized language, and no doubt will continue to be until something better is devised. Nevertheless, the formal rules of the game should not be confused with the actual work of the scientist.

The novice in the use of statistics often feels that statistical tests somehow cut one off from the close examination of data. There is something that seems so final about an analysis of variance summary table, for example, that one is tempted to regard this as some ultimate distillation of all meaning in the experiment. Nothing could be further from the truth. Statistical summaries and tests are certainly useful, but they are not *ends* of experimentation. The important part of data analysis often begins when the experimenter asks "What accounts for the results I obtain?" and turns to the detailed exploration of the data.

Even so simple a step as comparing the sums of squares for the various main and interaction terms to the total sum of squares can be informative. Thus, a ratio such as (SS between/SS total) can always be interpreted as the proportion of variance in Y accounted for by the treatment factor *in this experiment*. Furthermore, in a multifactor experiment, the total of all the systematic sums of squares, relative to the total sum of squares, gives an index of the total predictability in this situation. No one need be hesitant to explore any or all of these ratios in an attempt to understand what is going on. Although, as we have seen, these may be somewhat biased as estimates of the proportion of variance accounted for in the population, such comparisons of the separate "systematic" sums of squares to the total sum of squares are perfectly good ways to describe which variables seem to be "pulling the load" and which do not, relative to the dependent variable. Later, we will see that even these simple ratios of sums of squares can be given a direct interpretation of indices of predictive relationship.

Considered by themselves, all that the F tests in an analysis of variance can tell you is that *something* seems to have happened. If the F is significant, then some effects presumably exist that can be expected to occur again under similar circumstances; if the test is not significant, something notable still may have happened, but if treatment effects exist they are at least partially obscured by other variations. Other than this, a F test alone tells almost nothing. If all experimenters were satisfied with the general statement that something did (did not) happen, then science would progress by slow stages indeed.

The material in this chapter deals with two devices for analyzing data in more detail than that provided by the ordinary analysis of variance. Both methods are designed for comparing means or groups of means in a variety of ways. First, the technique of **planned comparisons** or **contrasts** will be introduced. Here, instead of planning to analyze the data to see if any overall experimental effects exist, the researcher at the outset has a number of particular questions to be answered separately. This technique of planned comparisons is used *instead of* the ordinary analysis of variance and F test. Indeed, as will be shown, the usual analysis of variance actually summarizes the evidence for many possible such individual questions that might be asked of the data.

Next, the technique of **incidental** or **post hoc comparisons** (or **contrasts**) will be discussed. Here, differences among means combined in any number of ways can be evaluated for signficance, *after* the overall F test has shown significance. This procedure of post hoc comparisons is an important supplement to the usual analysis of variance, and is very useful for the further exploration of data after the initial analysis has suggested the existence of real effects.

Both techniques, planned and post hoc comparisons, apply only to fixed-effects or mixed-model experiments. The inferences to be made concern particular population means combined in particular ways, and these methods make most sense when applied to treatment groups representing *fixed-level, nonsampled, factors*.

12.1 / ASKING SPECIFIC QUESTIONS OF DATA

In Section 5.15 the idea of linear combinations of random variables was introduced, and principles for specifying the sampling distribution for a particular linear combination of random variables are given in Appendix C. Our concern with linear combinations is justified by the fact that the evidence pertaining to particular experimental questions can often be found from various ways of combining the data—specifically, by linear combinations of means. In this section we will see how special linear combinations are useful to the experimenter.

Consider a study of the influence of the manner of persuasion upon the tendency of persons to change their attitude toward some institution or group. The particular attitude studied is that toward a minority group, and the experimental modes of persuasion used were:

1. a motion picture favorable to the minority group;
2. a lecture on the same topic, also favorable to the minority group;
3. a combination of the motion picture and the lecture.

Subjects were to be assigned at random to the different experimental groups, each having first been given a preliminary attitude test. Following the experimental treatment, each subject was to be given the test once again; the dependent variable here is the *change* in attitude score. It was recognized, however, that the mere repetition of a test may have some influence on an individual's changing score, and so a randomly selected control group was also to be used, in which the subjects were to be given no special experimental treatment.

However, during the design of the experiment the question arose, "Would the subjects perhaps change as a result of seeing *any* movie or hearing *any* lecture?" Thus it was decided to introduce two more control groups, the first to be shown a movie completely unrelated to the minority group, and the second given a lecture by the same person, but on a quite different topic.

The final experimental design can be diagrammed as follows:

Experimental groups			*Control groups*		
I	*II*	*III*	*IV*	*V*	*VI*
Movie	Lecture	Movie and lecture	Nothing	Neutral movie	Neutral lecture

A total sample of thirty subjects was drawn at random, and subjects were assigned at random to the six conditions, with five subjects per condition.

In this study, the experimenter was not interested in the overall existence of treatment effects. Rather, from the outset interest lay in answering the following specific questions.

1. Do the experimental groups as a whole tend to differ from the control groups?
2. Is the effect of the experimental lecture-movie combination different from the average effect of either experimental movie or experimental lecture separately?
3. Is the effect of the experimental lecture different from the effect of the experimental movie?
4. Among the control groups, is there any effect of the neutral movie or lecture as compared with the group receiving no treatment?

In other words, the experimenter entertained a number of particular questions before the collection of the data, and wished to analyze the data to answer those questions. An overall analysis of variance and F test would give indication of the existence of *any* systematic effects. However, the main interest lay in the particular differences among population means corresponding to answers to these questions.

Notice that the evidence pertaining to each question comes from the various sample means combined in some special way. For example, the evidence for question 3 involves only the difference between the means for groups I and II. On the other hand, the evidence for question 2 involves both the difference between mean I and mean III *and* the difference between mean II and mean III. Still other combinations of means pertain to the other questions asked. Can one, however, attach a significance level to each of these comparisons among means, permitting a statement about such differences among the *population* means?

This is a problem of planned comparisons among means, and the technique used will now be outlined. Let it be emphasized that this procedure applies only when the experimenter has specific questions to be asked *before* the data are analyzed, and this method may be used *instead* of the ordinary analysis of variance and F test.

12.2 / PLANNED COMPARISONS

The basic theory underlying planned comparisons has been briefly outlined in Section C.2, and Appendix C. There it is noted that given normally distributed variables sampled independently and at random, values from any linear combination of those random variables will also be normally distributed (C.2.8). Furthermore, if the mean and variance for each variable are known, then the mean and variance of the sampling distribution of the linear combination are also known.

These principles were used in 8.7 to establish the rationale for the t test of independent differences. Now we will go far beyond the simple two-sample case and apply these principles for any linear combination of means.

First, we need to define a **population comparison or comparison among population means:** given the means of J distinct populations, $\mu_1, \cdots, \mu_J$, a comparison among those means is any linear combination or weighted sum, with weights c_j not all equal to zero:

$$\psi = c_1\mu_1 + c_2\mu_2 + \cdots + c_J\mu_J = \sum_{j=1}^{J} c_j\mu_j.$$

We will use the symbol ψ (small Greek psi) to stand for the value of some particular *population* comparison. The weights for a comparison ψ are *some set of real numbers* $(c_1, \cdots, c_J)$ *not all zero.*

In addition, we impose the requirement that the *sum* of the weights c_j equals zero

$$\sum_{j=1}^{J} c_j = 0. \qquad [12.2.1^*]$$

In the following we will simply assume that this requirement is met, and defer giving the reason for this requirement until Section 12.6.

A **sample comparison** is defined exactly as for a population comparison, except that **sample means** are weighted and summed:

$$\hat{\psi} = c_1M_1 + \cdots + c_JM_J = \sum_j c_jM_j \qquad [12.2.2]$$

(the caret or "hat" over the psi will always indicate that this comparison is a sample value). Once again, we will require that $\sum_j c_j = 0$. Be sure to notice that the symbol for a comparison, either ψ or $\hat{\psi}$, stands for a single number, since it equals a *weighted sum of numbers*.

Some of the statistical properties of comparisons will now be viewed in the light of the principles summarized in Appendix C. Since each sample comparison is a linear combination, it is true from C.2.2 that

$$E(\hat{\psi}) = E\left(\sum_j c_j M_j\right) = \sum_j c_j E(M_j) = \psi. \qquad [12.2.3]$$

Any sample comparison is an unbiased estimate of the population comparison involving the same weights c_j. If we wish to evaluate the weighted sum of population means, our best estimate is the weighted sum of sample means. In the example above, each question asked by the experimenter can be answered by some particular comparison among population means, as reflected in the same weighted sum of the sample means.

Note especially that when all the population means are equal, so that $E(M_j)$ is the same for all j, by C.2.3, $E(\hat{\psi}) = 0$.

Suppose that each of the sample means M_j is based on a different sample of n_j cases drawn at random from a population with true mean μ_j. Furthermore, suppose that each population has the same variance, σ_e^2, so that the sampling distribution of M_j values has a variance σ_e^2/n_j.

By the principle of C.2.6 and C.2.7 in Appendix C, the sampling variance of a comparison $\hat{\psi}$ based on independent means must be

$$\text{var}(\hat{\psi}) = \sum_j c_j^2 \text{ var.}(M_j) = \sigma_e^2 \sum_j \frac{c_j^2}{n_j}. \qquad [12.2.4]$$

Thus, for example, suppose that the weights in a sample comparison of five means were $(-2, -1, 3, 1, -1)$, so that

$$\hat{\psi} = (-2)M_1 + (-1)M_2 + (3)M_3 + (1)M_4 + (-1)M_5.$$

Furthermore, suppose that each mean is based on 10 cases, so that $n = 10$ for all j. Provided that σ_e^2 is the variance of each population, then the sampling distribution of values of $\hat{\psi}$ must have a variance given by

$$\begin{aligned}\text{var}(\hat{\psi}) &= \sigma_e^2\left[\frac{(-2)^2 + (-1)^2 + (3)^2 + (1)^2 + (-1)^2}{10}\right]\\ &= \sigma_e^2\left(\frac{16}{10}\right)\\ &= \frac{8\sigma_e^2}{5}.\end{aligned}$$

Obviously, in no practical situation will we know the value of σ_e^2. However, *the value of* σ_e^2 *can be estimated from the data in exactly the same way that it is estimated in the analysis of variance:*

$$\text{est. } \sigma_e^2 = \text{MS error}.$$

(Here, σ_e^2 is the same as σ_e^2 in the expected mean squares for analysis of variance, and mean square error is the same as mean square within groups in the one-way analysis.) Thus

$$\text{est. var}(\hat{\psi}) = (\text{MS error}) \sum_j \frac{c_j^2}{n_j}. \qquad [12.2.5^*]$$

Now we have the three essential ingredients for statistical inference: an unbiased estimate $\hat{\psi}$ of some population value ψ, an estimate of the sampling variance of $\hat{\psi}$, and the form of the sampling distribution of $\hat{\psi}$ (normal if populations are normal; see C.2.8.)

Next, the question of a test will be undertaken.

12.3 / INTERVAL ESTIMATES AND TESTS FOR PLANNED COMPARISONS

Given the value of the planned sample comparison $\hat{\psi}$, a confidence interval for the population comparison value ψ can be found from

$$\hat{\psi} - t_{(\alpha/2;\nu)} \sqrt{\text{est. var}(\hat{\psi})} \leq \psi \leq \hat{\psi} + t_{(\alpha/2;\nu)} \sqrt{\text{est. var}(\hat{\psi})}. \qquad [12.3.1]$$

For the 100 $(1 - \alpha)$ percent confidence interval, the value of $t_{(\alpha/2;\nu)}$ represents the value cutting off the *upper* $\alpha/2$ proportion of sample values in a distribution of t with ν degrees of freedom. The number of degrees of freedom is the same as the degrees of freedom for the mean square error used to estimate σ_e^2. Such a confidence interval for ψ rests on the usual assumptions for a fixed-effects analysis of variance: normally distributed populations, each with variance σ_e^2, and independent errors. As in the overall F test for such an analysis, the assumption of normality is relatively innocuous for reasonably large samples.

On the other hand, violation of the assumption of homogeneity of error variances can have rather serious consequences in some instances, although such violations have their minimum effect when n is constant over all groups (*cf.* Section 10.18).

This confidence interval can be used to test any hypothetical value of ψ at all. If some hypothetical value of ψ fails to be covered by the confidence interval, then that hypothesis may be rejected beyond the α level (two-tailed).

In most instances, the only hypothesis of interest in planned comparisons is

$$H_0: \psi = 0.$$

Then a test for this hypothesis is given by

$$t = \frac{\hat{\psi}}{\sqrt{\text{est. var}(\hat{\psi})}}, \qquad [12.3.2^*]$$

distributed as t with degrees of freedom as for mean square error ($N - J$ in the one-factor experiment). If the hypothesis is to be tested against a directional alternative, then the sign of the t is considered, just as in the ordinary test for two means.

Notice that the t test for a difference between two independent means (Section 8.9) is merely a test of a comparison

$$\hat{\psi} = (1)M_1 + (-1)M_2$$

where the two weights are 1 and -1 respectively. However, the t test can be used to test any particular comparison, regardless of the number of groups involved.

When the hypothesis to be tested is nondirectional, the t test for any planned comparison may be replaced by the equivalent F test:

$$F = \frac{(\hat{\psi})^2}{\text{MS error}\left(\sum_j c_j^2/n_j\right)} \quad [12.3.3.*]$$

with 1 and $N - J$ degrees of freedom. (Recall, once again, that t^2 is distributed as an F variable.)

Although this discussion has referred to comparisons planned among a set of J means for data in a one-factor experiment, exactly the same ideas apply when comparisons are planned in a two-factor (or higher) experiment fitting the fixed-effects or mixed model. Thus, when the comparisons are to be made among the J column means in a two-factor Model I experiment with K rows and n per cell in the data table,

$$\text{est. var}(\hat{\psi}) = (\text{MS error})\left(\sum_j \frac{c_j^2}{Kn}\right) \quad [12.3.4]$$

since each mean is based on Kn observations. The t or F value is computed just as in 12.3.2 or 12.3.3, with $JK(n - 1)$ degrees of freedom for the denominator. The test has this form, however, only when the fixed-effects model applies.

When the second factor, represented by the rows, is sampled, so that the mixed model applies, then any comparison among the column means has an estimated variance given by

$$\text{est. var}(\hat{\psi}) = (\text{MS interaction})\left(\sum_j \frac{c_j^2}{Kn}\right) \quad [12.3.5*]$$

and the number of degrees of freedom for t or the denominator of F is $(J - 1)$ times $(K - 1)$.

If, for some reason, cell means are to be compared in a two-way table, with n per cell, the estimated variance for the comparison is given by

$$\text{est. var}(\hat{\psi}) = (\text{MS error})\left(\frac{\sum_j \sum_k c_{jk}^2}{n}\right) \quad [12.3.6]$$

where the mean square error is the mean square within cells, and the degrees of freedom are found from $(JK)(n - 1)$.

12.4 / INDEPENDENCE OF PLANNED COMPARISONS

A single planned comparison among means is usually tested by a simple t ratio, as we have seen. However, seldom does an experimenter have interest in only one comparison on the data. Usually there are sets of questions to be answered, each corresponding to some comparison among means. This brings up the very critical problem of the independence of comparisons.

Just as the possible answers to some questions may depend logically upon answers to others, so the values of some comparisons made on a given set of means may depend upon the values of other comparisons. This fact has serious consequences for estimates and tests of several comparisons, since the questions involved in the respective comparisons cannot be given truly separate and unrelated answers unless the comparisons are statistically independent of each other. Fortunately, a simple method exists for determining whether or not two comparisons are independent, given normal population distributions with equal variances σ_e^2. **The determination of the independence of two comparisons depends only on the weights each involve, and in no way on the means actually observed.** One can plan comparisons that will be independent *before* the data are collected.

The solution to the problem of independent comparisons rests on still another general principle having to do with linear combinations of variables, C.2.9 of Appendix C. This principle is another instance of the extraordinary utility of the assumption of normal distributions in statistical theory.

Suppose that there were J independent samples, each representing a normal population. Each population is assumed to have the same variance σ_e^2. Also for the moment assume that within each of the samples, there will be the same number n of observations. Any group j will, of course, have a mean M_j. Now imagine *two* comparisons among these same means, the first, $\hat{\psi}_1$, with weights symbolized by c_{1j}, and the second, $\hat{\psi}_2$, with weights symbolized by c_{2j}. From C.2.9 of Appendix C, we know that the values $\hat{\psi}_1$ and $\hat{\psi}_2$ are themselves independent *provided* that

$$\sum_j c_{1j}c_{2j} = 0. \qquad [12.4.1]$$

In other words, **given the means of J independent samples, each of size n, one can decide if two different linear combinations of those means are independent simply by seeing if the products of the weights assigned to each mean sum to zero:**

$$c_{11}c_{21} + \cdots + c_{1J}c_{2J} = 0.$$

This principle is necessarily true only for samples from normal populations with equal variances.

Two comparisons satisfying this condition are said to be **orthogonal comparisons;** orthogonality of two comparisons is equivalent to the statistical independence of sample comparisons only when the populations are normal with equal σ_e^2, however. When two comparisons are statistically independent, the information each provides is actually nonredundant and unrelated to the information provided by the other. The estimate $\hat{\psi}_1$ is unrelated to the estimate $\hat{\psi}_2$. Thus, seeing if comparisons are orthogonal lets the experimenter judge whether or not they give unrelated, nonoverlapping pieces of information about the experiment.

When the J distinct samples have different sizes, symbolized by n_j, the criterion for orthogonality for two comparisons among the means becomes

$$\sum_j \frac{c_{1j}c_{2j}}{n_j} = 0,$$

so that the product of comparison weights for each sample is weighted inversely by n_j before the sum is taken. If this weighted sum of products is zero, then the comparisons may be regarded as orthogonal.

In the following, the terms "independent" and "orthogonal" will be used interchangeably. However, this is proper only because normal distributions with homogeneous variances are assumed throughout the discussion.

12.5 / AN ILLUSTRATION OF INDEPENDENT AND NONINDEPENDENT PLANNED COMPARISONS

Four treatment groups were observed in an experiment, with a sample of six randomly assigned subjects in each. The means of the four groups were

I	*II*	*III*	*IV*
17	24	27	16.

The mean square error, found by the usual method for a one-way analysis of variance, was 5.6, with 24 − 4, or 20, degrees of freedom. Before the data were seen, the experimenter expressed interest basically in the following questions:

1. Does mean I differ from the average of means II, III, and IV?
2. Does mean II differ from the average of means III and IV?
3. Does mean III differ from mean IV?
4. Does the average of I and II differ from average of III and IV?

It is assumed that each population is normally distributed, with the same variance in each. Now, each of the questions can be put into the form of comparisons among means by using the following sets of weights. In Table 12.5.1, the rows represent the various questions (comparisons) and the columns the various samples. In the cells the numbers are the weights c_j to be assigned to a given sample mean for a given comparison.

Table 12.5.1

Weights corresponding to comparisons among four means.

	Means			
Question	*I*	*II*	*III*	*IV*
1	1	$-\frac{1}{3}$	$-\frac{1}{3}$	$-\frac{1}{3}$
2	0	1	$-\frac{1}{2}$	$-\frac{1}{2}$
3	0	0	1	−1
4	$\frac{1}{2}$	$\frac{1}{2}$	$-\frac{1}{2}$	$-\frac{1}{2}$

There is certainly nothing mysterious about the way these weights were chosen. The first comparison is designed to investigate a difference between mean I and the *average* of means II, III, and IV. This calls for a weighting of mean I by unity, and subtracting the average of the three other means, which is tantamount to weighting each one by a *negative* 1/3 in the comparison with mean I. Similarly, mean I does not figure in the second question, and so gets weight 0, while mean II is contrasted with the *average* for means III and IV. The other weights are found in similar ways. Note that the sum of weights for each row in the table is zero, as it should be for comparisons among means.

Are these four comparisons orthogonal, and thus independent of each other? Consider the weights in rows 1 and 2; the sum of their products across columns is

$$\sum_j c_{1_j}c_{2_j} = (1)(0) + \left(\frac{-1}{3}\right)(1) + \left(\frac{-1}{3}\right)\left(\frac{-1}{2}\right) + \left(\frac{-1}{3}\right)\left(\frac{-1}{2}\right) = 0,$$

so that comparison 1 and comparison 2 are orthogonal and hence independent. In the same way, by computing the sum of the products of weights we see that comparison 2 and comparison 3, and comparison 1 and comparison 3 are also independent.

Now look at comparison 1 and comparison 4. Here the sum of the products of weights is

$$(1)\left(\frac{1}{2}\right) + \left(\frac{-1}{3}\right)\left(\frac{1}{2}\right) + \left(\frac{-1}{3}\right)\left(\frac{-1}{2}\right) + \left(\frac{-1}{3}\right)\left(\frac{-1}{2}\right) = \frac{2}{3}$$

so that comparisons 1 and 4 are not orthogonal. Neither are comparisons 2 and 4. On the other hand, comparisons 3 and 4 are orthogonal.

The computation and tests for the first three comparisons will next be illustrated. Presumably, our experimenter would not test comparison 4 separately because of its nonindependence of the others. The information gained from comparison 4 is redundant, depending on the outcomes of the first three comparisons.

The value of comparison 1 is

$$\hat{\psi}_1 = (1)(17) + \left(\frac{-1}{3}\right)(24 + 27 + 16) = -5.3.$$

The estimated variance of this comparison is, from 12.2.5,

$$\begin{aligned}\text{est. var.}(\hat{\psi}_1) &= \frac{5.6}{6}\left[(1)^2 + \left(\frac{-1}{3}\right)^2 + \left(\frac{-1}{3}\right)^2 + \left(\frac{-1}{3}\right)^2\right]. \\ &= .93(1.33) = 1.24.\end{aligned}$$

Under the hypothesis that $\psi_1 = 0$, the t ratio is, from 12.3.2,

$$t = \frac{-5.3}{\sqrt{1.24}} = -4.8.$$

For a nondirectional test, with 20 degrees of freedom, this result is significant beyond the 1 percent level, so that we reject the hypothesis that the true comparison value is zero. Our experimenter can assert confidently that population mean I does differ from the average of means II, III, and IV.

In a similar fashion we find

$$\hat{\psi}_2 = (1)24 + \left(\frac{-1}{2}\right)(27) + \left(\frac{-1}{2}\right)(16) = 2.5$$

with

$$\begin{aligned}\text{est. var}(\hat{\psi}_2) &= \frac{5.6}{6}\left[(1)^2 + \left(\frac{-1}{2}\right)^2 + \left(\frac{-1}{2}\right)^2\right] \\ &= 1.40.\end{aligned}$$

Then

$$t = \frac{2.5}{\sqrt{1.40}} = 2.11.$$

This is just significant at the 5 percent level for a nondirectional test; mean II does differ significantly from means III and IV. Finally,

$$\hat{\psi}_3 = 27 - 16 = 11,$$

with

$$\text{est. var}(\hat{\psi}_3) = \frac{5.6}{6}(1 + 1) = 1.87.$$

Here,

$$t = \frac{11}{\sqrt{1.87}} = 8.04,$$

which is very significant. We can say with confidence that population mean III is different from population mean IV.

12.6 / THE INDEPENDENCE OF SAMPLE COMPARISONS AND THE GRAND MEAN

Comparisons in which the sum of the weights is zero (that is, contrasts) represent weighted differences among sets of means. Such contrasts are usually of most interest to the experimenter. In principle, any set of weights (not all zero) could be used in carrying out a comparison among several sample means; however, as we have seen one usually uses only sets of weights such that $\sum_j c_j = 0$ for each comparison. Now one reason for this requirement can be shown: *by insisting that the weights applied in any comparison sum to zero, we make each comparison value be independent of the value of the grand sample mean.*

The mean of any sample is a linear combination of random variables (Appendix C, comment following C.2.8). Given N individuals divided among J sample groups, the grand mean over all individuals in all groups is a linear combination

$$M = \sum_j \sum_i \frac{y_{ij}}{N} \qquad [12.6.1]$$

where each score gets a weight of $1/N$. Furthermore, again in terms of the individual scores, any sample comparison can be written variously as

$$\hat{\psi} = \sum_j c_j M_j = \sum_j c_j \sum_i \frac{y_{ij}}{n_j} = \sum_j \sum_i \left(\frac{c_j}{n_j}\right) y_{ij}. \qquad [12.6.2]$$

Each *individual* score y_{ij} actually gets a weight (c_j/n_j), depending on the group j to which it belongs.

In most instances, we want to ask questions about combinations of population means that will be unrelated to any consideration of what the overall mean of the combined populations is estimated to be. Thus, we choose weights such that any comparison to be made is orthogonal to the linear combination standing for the grand mean. In terms of the basic scores, it should be true that

$$\sum_j \sum_i \left(\frac{1}{N}\right)\left(\frac{c_j}{n_j}\right) = 0, \qquad [12.6.3]$$

which in turn implies that

$$\sum_j c_j = 0.$$

(This is the basis for the comments in Section C.4 of Appendix C about the unit vector, since the weights applied to the data to find the grand mean are simply $1/N$ times the unit vector. Then, by C.3.9 and the following, orthogonality with the unit vector implies orthogonality with the mean.)

The comparison for the mean can also be the basis of a test of a hypothesis about the population mean, such as $H_0: \mu = k$. One takes $\hat{\psi}_M = M$, and then $F_{(1,N-J)} = (\hat{\psi}_M - k)^2/(MSW/N) = t^2_{(N-J)}$.

12.7. / THE NUMBER OF POSSIBLE INDEPENDENT COMPARISONS

It stands to reason that given any finite amount of data, only a finite set of questions may be asked of those data if one is to get nonredundant, nonoverlapping, answers. There is just so much information in any given set of data; once this information has been gained, asking further questions leads to answers that depend upon the answers already learned.

This idea of the amount of information in a set of data has a statistical parallel in the number of possible independent (orthogonal) comparisons to be made among J means:

Given J independent sample means, there can be no more than $J - 1$ comparisons, each comparison being independent both of the grand mean and of each of the others.

In other words, the experimenter can frame no more than $J - 1$ different comparisons standing for questions about the sample means, if these comparisons to be completely independent of each other and the grand mean. This does not say that many different sets of $J - 1$ mutually independent comparisons cannot be found for any set of data. It does say that once a set of $J - 1$ comparisons has been found where the comparisons are independent of each other and the grand mean, it is impossible to find one more comparison which is also independent both of the grand mean and all of the rest. Thus the number of questions the experimenter may ask of the data as planned comparisons is limited, if the statistical answers are to be regarded as independent.

On the other hand, there are a great many *sets* of orthogonal comparisons that may be applied to any given set of data. In fact, there is no limit at all to the number of ways in which one might choose weights for the various comparisons, and such that a set of $J - 1$ such comparisons could be formed. The point is that in any set of such comparisons, that set may consist of no more than $J - 1$ which are mutually orthogonal.

This was the difficulty in the example of Section 12.5. There were only four sample means, and four comparisons were planned; it is impossible to find four mutually orthogonal comparisons *whatever* the weights used, if they all are to be orthogonal to the grand mean in addition. This principle also illuminates the dependencies among multiple t tests; given J samples, there will be $(J)(J - 1)/2$ possible t tests between pairs of means. If J is greater than two, this will always be more than the number of possible independent comparisons. This, then, is another reason why making ordinary t tests on all the differences among J means is not a very good idea (Section 8.17). Not only is the nominal α value for such tests not accurate; the pieces of information contributed by the different tests are partially redundant.

On the other hand, must the experimenter be limited only to inferences corresponding to orthogonal comparisons in the data? Fortunately, the answer is "no," since there is a proper way of inspecting all differences or comparisons of interest, orthogonal or not, in a set of data, where the significance level attached to each comparison is quite meaningful. However, this method properly belongs under post hoc rather than planned comparisons, and will be discussed in Section 12.12.

12.8 / PLANNED COMPARISONS AND THE ANALYSIS OF VARIANCE

There is a very intimate connection between analysis of variance and the technique of planned comparisons. **Each and every degree of freedom associated with treatments in any fixed-effects analysis of variance corresponds to some possible comparison of means.** The number of degrees of freedom for the mean square between is the number of possible *independent* comparisons to be made on the means. Any analysis of variance is equivalent to a breakdown of the data into *sets* of orthogonal comparisons (an idea we will develop at length in Chapter 14.)

In order to show how this is true, we need to define the **sum of squares for a comparison:** the sum of squares for any comparison $\hat{\psi}_g$ is

$$\text{SS}\ (\hat{\psi}_g) = \frac{(\hat{\psi}_g)^2}{w_g} \qquad [12.8.1^*]$$

where

$$w_g = \sum_j \frac{c_j^2}{n_j}. \qquad [12.8.2^*]$$

For any comparison $\hat{\psi}_g$ this sum of squares has *one* degree of freedom. It follows that

$$\text{MS}\ (\hat{\psi}_g) = \text{SS}\ (\hat{\psi}_g). \qquad [12.8.3]$$

Notice that the F ratio for a comparison $\hat{\psi}_g$ could be written as

$$F_g = \frac{\text{MS}\ (\hat{\psi}_g)}{\text{MS error}}, \qquad [12.8.4^*]$$

with 1 and $N - J$ degrees of freedom.

Extending this idea further, suppose that there were $J - 1$ independent comparisons on the data, and that the SS were calculated for each of them. Then

$$\text{SS (all } \hat{\psi}_g) = \text{SS } (\hat{\psi}_1) + \cdots + \text{SS } (\hat{\psi}_g) + \cdots + \text{SS } (\hat{\psi}_{J-1}) \qquad [12.8.5^*]$$

with $J - 1$ degrees of freedom.

It can be shown that **for any set of $J - 1$ independent sample comparisons on any set of J means,**

$$\text{SS (all } \hat{\psi}_g) = \sum_g \text{SS } (\hat{\psi}_g) = \text{SS between groups.} \qquad [12.8.6^*]$$

The total of the sum of squares for any $J - 1$ independent comparisons on J means is always equal to the sum of squares between the J groups.

Furthermore, by 12.8.4 and 12.8.6 above, it must then be true that the average F value over the $J - 1$ comparisons is the overall F value:

$$F_{\text{overall}} = \left(\sum_{g=1}^{J-1} F_g \right) \Big/ (J - 1) = \frac{\text{MS between}}{\text{MS within}}. \qquad [12.8.7^*]$$

As we have seen, the F test in the fixed-effects analysis of variance relates to the hypothesis that all of the J populations sampled have equal means. When this hypothesis is true, it implies that any comparison made among the population means will have a value of zero (see C.2.3). Hence the overall F test is an ''omnibus'' test that all possible comparisons among population means are zero. Alternatively, each of an independent set of these comparisons may be tested separately if the experimenter has definite questions to ask about the data to begin with. If there are no such questions, then the overall F test still permits asking ''did *anything* happen?'' On the other hand, a much more powerful test of a particular hypothesis may be possible by testing one or more comparisons separately than by the overall F test. The important thing to remember is that for each degree of freedom in the sum of squares between groups or treatments, there is a potential prior question to be asked of the data. The F test gives evidence to let us judge if all of a set of $J - 1$ such orthogonal comparisons are simultaneously zero in the populations. For this reason, if planned orthogonal comparisons are tested separately, the overall F test is not carried out, and vice versa.

The same idea extends to higher-order analyses of variance as well. Suppose that in a two-factor experiment, there are R rows and C columns in the data matrix. Then the SS rows is the sum of the $R - 1$ separate and independent sums of squares corresponding to orthogonal comparisons that might be made on row means. When the experimental design itself is orthogonal, each and every one of these $R - 1$ comparisons is itself orthogonal to any of the $C - 1$ independent comparisons that might be made on the column means. Thus, for an orthogonal design, the SS rows and SS columns are independent, since the values making up each sum are independent. For such a design, and given normal population distributions with equal variances, the mean square for rows and the mean square for columns are completely independent values. Furthermore, the $(R - 1)(C - 1)$ potential orthogonal comparisons summarized as mean square interaction are independent of those made on rows, of those made on columns, and of each other. A considerable part of the statistician's ability to formulate different kinds of experimental designs for different purposes comes from the general principle showing which potential comparisons are independent and which are not in a given way that data might be collected.

Quite often it happens that no set of prior questions exists that would use up all the possible $J - 1$ independent comparisons in the data. The experimenter may have only one or two questions that particularly need to be answered, and is content to lump all other comparisons into a single test, corresponding to "are there any other effects?" Then a combination of planned comparisons and analysis of variance techniques is possible.

Suppose that out of $J - 1$ possible comparisons we decide that only two are of overriding interest. Let us call these comparisons $\hat{\psi}_1$ and $\hat{\psi}_2$. We calculate SS $(\hat{\psi}_1)$ and SS $(\hat{\psi}_2)$ and carry out the tests for each. Now what of all the remaining $J - 1 - 2$ independent comparisons we might make? A very simple way exists for finding an SS value for those remaining. First of all, by the analysis of variance SS between groups may be calculated in the ordinary way. Then

$$\text{SS (all } \hat{\psi} \text{ independent of } \hat{\psi}_1 \text{ and } \hat{\psi}_2) = \text{SS between} - \text{SS}(\hat{\psi}_1) - \text{SS}(\hat{\psi}_2). \qquad [12.9.1^*]$$

In other words, one simply subtracts the sum of squares for comparisons 1 and 2 from the sum of squares between groups to find the sum of squares for *all remaining* comparisons independent of the first two. Instead of only one F test, the experimenter now makes three tests: one for comparison 1, one for comparison 2, and one for all remaining comparisons, given by

$$F = \frac{\text{SS between} - \text{SS}(\hat{\psi}_1) - \text{SS}(\hat{\psi}_2)}{(J - 3)\text{ MS error}} \qquad [12.9.2^*]$$

with $(J - 3)$ and $(N - J)$ degrees of freedom.

Any subset of some total possible set of independent comparisons may be planned for in this way: the comparisons of special interest are checked for independence and then tested. All remaining comparisons independent of those tested are embodied in the difference between the SS between and the sum of the SS for the special comparisons tested. The F test for "other comparisons" has degrees of freedom equal to $J - 1$ reduced by one for each comparison tested separately. If this F value is significant, then these comparisons of secondary interest can be examined individually by post hoc methods.

12.10 / AN EXAMPLE USING PLANNED COMPARISONS

The experiment outlined in Section 12.1 was carried out with a total of 30 subjects, assigned at random into groups of 5. The weights for the comparisons representing the four basic questions are shown in Table 12.10.1. A check using the criterion for orthogonality (Section 12.4) shows that these comparisons can be regarded as independent.

It was decided to test each of these comparisons at the .05 level. The data turned out as shown in Table 12.10.2 The numbers in the table are changes in attitude score for each person.

The numeral computations for the ordinary one-way analysis of variance are first carried out; not only does this give the MS error needed for the comparisons tests, but also the SS between, which is useful in testing any remaining comparisons.

Table 12.10.1

Comparison weights for the experiment of Section 12.1.

	Treatments			*Controls*		
Comparison	*Movie*	*Lecture*	*Movie & lecture*	*Nothing*	*Neutral movie*	*Neutral lecture*
1	$\frac{1}{3}$	$\frac{1}{3}$	$\frac{1}{3}$	$-\frac{1}{3}$	$-\frac{1}{3}$	$-\frac{1}{3}$
2	$\frac{1}{2}$	$\frac{1}{2}$	−1	0	0	0
3	1	−1	0	0	0	0
4	0	0	0	1	$-\frac{1}{2}$	$-\frac{1}{2}$

Table 12.10.2

Changes in attitude scores, following indicated treatments.

	Movie	*Lecture*	*Movie & lecture*	*Nothing*	*Neutral movie*	*Neutral lecture*	
	6	3	7	−6	5	−1	
	10	6	9	0	−5	3	
	1	−1	4	−5	3	2	
	6	5	9	2	−4	−1	
	4	2	3	2	5	−6	
Totals	27	15	32	−7	4	−3	68
Means	5.4	3	6.4	−1.4	.8	−.6	

$$\text{SS total} = 720 - \frac{(68)^2}{30} = 565.9$$

$$\text{SS between} = \frac{(27)^2 + \cdots + (-3)^2}{5} - \frac{(68)^2}{30} = 256.3$$

$$\text{SS error} = 309.6$$

$$\text{MS error} = \frac{309.6}{24} = 12.9.$$

Now for comparison 1:

$$\hat{\psi}_1 = \left(\frac{1}{3}\right)(5.4) + \left(\frac{1}{3}\right)(3) + \left(\frac{1}{3}\right)6.4 - \left(\frac{1}{3}\right)(-1.4) - \left(\frac{1}{3}\right)(.8) - \left(\frac{1}{3}\right)(-.6)$$

$$= \frac{16}{3} = 5.3$$

$$w_1 = \frac{1}{5}\left[\frac{1}{9} + \frac{1}{9} + \frac{1}{9} + \frac{1}{9} + \frac{1}{9} + \frac{1}{9}\right] = \frac{2}{15}$$

$$\text{SS}(\hat{\psi}_1) = \frac{15(5.3)^2}{2} = 210.7.$$

The test for comparison 1 is given by

$$F = \frac{210.7}{12.9} = 16.3$$

which is significant far beyond the 5 percent level for 1 and 24 degrees of freedom. There does seem to be a reliable difference between experimental and control groups, in general.

For comparison 2,

$$\hat{\psi}_2 = \left(\frac{1}{2}\right)(5.4) + \left(\frac{1}{2}\right)(3) - (1)(6.4) = -2.2$$

$$w_2 = \frac{1}{5}\left[\left(\frac{1}{2}\right)^2 + \left(\frac{1}{2}\right)^2 + 1\right] = .3$$

$$\text{SS }(\hat{\psi}_2) = \frac{(2.2)^2}{.3} = 16.1,$$

so that the F test for this comparison is

$$F = \frac{16.1}{12.9} = 1.2.$$

This is not significant. There is not enough evidence to say that the combined movie-lecture effect is different from the average of their separate effects.

Comparison 3 gives

$$\hat{\psi}_3 = (1)(5.4) - 1(3) = 2.4$$

with

$$w_3 = \frac{1}{5}(1 + 1) = .4$$

so that

$$\text{SS }(\hat{\psi}_3) = \frac{(2.4)^2}{.4} = 14.4.$$

The F test for comparison 3 is then

$$F = \frac{14.4}{12.9} = 1.1,$$

again not significant. There is not enough evidence to say that a movie-lecture difference exists in the populations.

The value for comparison 4 is

$$\hat{\psi}_4 = (1)(-1.4) + \left(\frac{-1}{2}\right)(.8) + \left(\frac{-1}{2}\right)(-.6) = -1.5$$

with

$$w_4 = \frac{1}{5}\left[1 + \frac{1}{4} + \frac{1}{4}\right] = .3.$$

The sum of squares

$$\text{SS}\ (\hat{\psi}_4) = \frac{(-1.5)^2}{.3} = 7.5$$

is less than MS error, so that the F test is definitely not significant.

Only four of the five possible independent comparisons have been made. The sum of squares for the fifth comparison can be found from

$$\text{SS between} - \text{SS}\ (\hat{\psi}_1) - \text{SS}\ (\hat{\psi}_2) - \text{SS}\ (\hat{\psi}_3) - \text{SS}\ (\hat{\psi}_4)$$
$$= 256.3 - 210.7 - 16.1 - 14.4 - 7.5 = 7.6$$

This last comparison might be tested, but we can see that since its sum of squares is less than MS error, it cannot be significant.

The results of this analysis can be put into tabular form as shown in Table 12.10.3.

Table 12.10.3

Summary table showing four planned comparisons.

Source	*SS*	*df*	*MS*	*F*
Between groups	256.3	5		
Comparison:				
1	210.7	1	210.7	16.3
2	16.1	1	16.1	1.2
3	14.4	1	14.4	1.1
4	7.5	1	7.5	—
Remainder	7.6	1	7.6	—
Error (within groups)	309.6	24	12.9	—
Totals	565.9	29		

As always, however, we should bear in mind that for any set of several significance tests, the probability that at least one of tests will give significant results can be relatively large, even though the nominal α level used for each separate test is conventionally small. A related problem in the evaluation of a set of results is the now familiar principle that the several F or t ratios, even for a set of independent comparisons, need not be independent.

The overall meaning of a set of statistical results such as ours is rather difficult to assess. This gives at least one reason to prefer a single overall F test to a series of $J - 1$ tests for planned comparisons. That is, when a single F test is carried out, then the probability of a Type I error is α, but when a series of $J - 1$ F tests is conducted, then the probability of a Type I error in at least one of these tests is appreciably larger than α. We have, however, at least arranged the data into portions pertaining to hypotheses we believe important and meaningful for the overall interpretation of the experiment. This was the contribution of the analysis by planned comparisons.

12.11 / THE CHOICE OF THE PLANNED COMPARISONS

Given any J independent means, there are any number of ways to choose the $J - 1$ independent comparisons to be made among these means. The important thing is that the experimenter have definite prior questions to be answered, and that these questions can be framed as orthogonal, or independent, comparisons. The data that are actually collected must then be adequate to provide answers to these questions in terms of orthogonal comparisons. For example, in the experiment just analyzed, the experimenter might have had the following initial concerns:

1. Is there any effect of showing the experimental movie, as opposed to the initial lecture or nothing?
2. Is there any effect of the experimental lecture, as opposed to the experimental movie or nothing?
3. Is the effect of the experimental movie the same whether or not it is accompanied by a lecture?
4. Does the neutral movie have the same effect as the neutral lecture?

These four quetions can be embodied in four comparisons quite different from those employed above (Table 12.11.1). A check shows that each of these comparisons is independent of each of the others. Notice, however, that some of these comparisons are *not* independent of some of the first set.

Table 12.11.1

Alternative comparisons for the experiment of Section 12.1.

Comparisons	*Movie*	*Lecture*	*Movie & lecture*	*Nothing*	*Neutral movie*	*Neutral lecture*
1	$\frac{1}{2}$	$-\frac{1}{2}$	$\frac{1}{2}$	$-\frac{1}{2}$	0	0
2	$-\frac{1}{2}$	$\frac{1}{2}$	$\frac{1}{2}$	$-\frac{1}{2}$	0	0
3	$-\frac{1}{2}$	$-\frac{1}{2}$	$\frac{1}{2}$	$\frac{1}{2}$	0	0
4	0	0	0	0	1	-1

For any given experiment, there will undoubtedly be many interesting questions that could be framed as planned comparisons. Only the particular experimenter can decide, of course, which comparisons are of interest. As a rule, one should not try to think up a sensible question to correspond to each of the $J - 1$ degrees of freedom associated with J means just because it can, in principle, be done. Rather, the technique of planned comparisons should be used only when there are a few important specific questions to ask of the data that will clarify the whole experiment if answered. Once these questions have been decided upon, they must be phrased so that the answer will be evident from some way of weighting and combining means. Usually, such questions resolve themselves into differences between groups of means, just as, in the last set of comparisons, the answer to question 1 depends upon the difference between the averages of two groups of means:

$$\frac{(\text{movie}) + (\text{movie \& lecture})}{2} - \frac{(\text{lecture}) + (\text{nothing})}{2}.$$

This difference weights the (movie) mean and the (movie & lecture) means each by 1/2, and the (lecture) and (nothing) group means each by $-1/2$, so that this difference gives us our weights for that comparison. The same is true for each of the other comparisons discussed here.

(If a few comparisons have already been planned, and there is interest in finding other comparisons that would be orthogonal to these, then the Gram-Schmidt method outlined in C.4 of Appendix C may be used. Thus, suppose that two orthogonal comparisons are planned, out of $J - 1$ that are possible. Then, in the terminology of Appendix C, the weights already being used form three of the vectors in the orthogonal basis $\mathbf{V}$: the two sets of weights for the planned comparison plus the unit vector $\mathbf{i}$. Using any arbitrary vector $\mathbf{x}$ to stand for a member of $\mathbf{X}$ and carrying out the Gram-Schmidt procedure will give either a new orthogonal vector of weights, or will give the zero vector. If the zero vector is found, a new selection for $\mathbf{x}$ is made and the process repeated. Any nonzero vector $\mathbf{v}$ found from this process will be orthogonal to the vectors corresponding to the original comparison weights and to the unit vector. Going on in this way, one can find a complete set of $J - 1$ orthogonal comparison weights.

For example, let us find the fifth orthogonal comparison for the example shown in Table 12.11.1. Arbitrarily, we choose an initial vector, say $\mathbf{x}' = [1,0,0,0,0,0]$. Then, reverting to the notation of C.4, Appendix C, and eliminating denominators, we have

$$\begin{bmatrix}1\\0\\0\\0\\0\\0\end{bmatrix} - \frac{1}{6}\begin{bmatrix}1\\1\\1\\1\\1\\1\end{bmatrix} - \frac{1}{4}\begin{bmatrix}1\\-1\\1\\-1\\0\\0\end{bmatrix} - \frac{(-1)}{4}\begin{bmatrix}-1\\1\\1\\-1\\0\\0\end{bmatrix} - \frac{(-1)}{4}\begin{bmatrix}-1\\-1\\1\\1\\0\\0\end{bmatrix} - \frac{(0)}{2}\begin{bmatrix}0\\0\\0\\0\\1\\-1\end{bmatrix} = \begin{bmatrix}1/12\\1/12\\1/12\\1/12\\-2/12\\-2/12\end{bmatrix} \text{ or } \begin{bmatrix}1\\1\\1\\1\\-2\\-2\end{bmatrix}$$

Thus, the set of weights $(1,1,1,1,-2,-2)$ provides a fifth orthogonal comparison among the six means. If this first trial vector of weights, $\mathbf{x}$, had given us only zero weights, we would have chosen another trial vector $\mathbf{x}$ and kept on going until a set of nonzero weights was found.)

12.12 / INCIDENTAL OR POST HOC COMPARISONS IN DATA

Even though tests for planned comparisons form a useful technique in experimentation, it is far more common for the experimenter to have no special questions to begin with. Initial concern is to establish only that some real effects or comparison differences do exist in the data. Given a significant over-all test, the task is then to explore the data to find the source of these effects, and to try to explain their meaning.

There are a number of methods that have been devised for testing the significance of post hoc comparisons, only three of which will be discussed here. The first is the method due to Scheffé (1959), which has advantages of simplicity, applicability to groups of unequal sizes, and suitability for any comparison. This method is also known to be relatively insensitive to departures from normality and homogeneity of variance.

The idea of a comparison here is exactly the same as defined in Section 12.2. A sample comparison $\hat{\psi}$ is a linear combination of sample means

$$\hat{\psi} = \sum_j c_j M_j$$

where $\sum_j c_j = 0$.

After the overall F has been found significant, then *any* comparison ψ may be made. *Unlike planned comparisons, there is no requirement that such post hoc comparisons be independent.* Any and all comparisons of interest may be made. Often the experimenter may be interested in examining all pairs of means, but any comparison $\hat{\psi}$ is legitimate under this method.

Given any comparison g made on the data after a significant F has been found for the relevant factor, the significance of the comparison value $\hat{\psi}_g$ may be found by use of the following confidence interval:

$$\hat{\psi}_g - \mathbf{S}\sqrt{V(\hat{\psi}_g)} \leq \psi_g \leq \hat{\psi}_g + \mathbf{S}\sqrt{V(\hat{\psi}_g)} \qquad [12.12.1^*]$$

where

$$\sqrt{V(\hat{\psi}_g)} = \sqrt{(\text{MS error})w_g} = \sqrt{\text{est. var}(\hat{\psi}_g)} \qquad [12.12.2^*]$$

and

$$\mathbf{S} = \sqrt{(J-1)F_\alpha}. \qquad [12.12.3^*]$$

(Take care to note that boldface $\mathbf{S}$ as used here is *not* the sample standard deviation.) The w_g is defined in Section 12.8, and F_α is the value required for significance at the α level, with $J - 1$ and $N - J$ degrees of freedom (that is, the F required for significance at the α level for an *overall* test of the J means). For any α, this gives the $100(1 - \alpha)$ percent confidence interval for ψ_g, the true value of the comparison. When the confidence interval fails to cover zero, the comparison is said to be significant, and identified as one possible contributor to the overall significance of F.

(Actually, all one really needs to do to determine if the confidence interval covers the value of zero in the procedure given above is to calculate $\mathbf{S}\sqrt{V(\hat{\psi}_g)}$. If this is smaller than the absolute value of $\hat{\psi}_g$, then the confidence interval does not cover zero, and the comparison is significant. Furthermore, it is possible to calculate the F value for the $\hat{\psi}_g$ just as for a planned comparison. Then this F ratio is significant only if it exceeds $\mathbf{S}^2$. All three ways if proceeding—by confidence interval, by absolute difference, and by F ratio—are equivalent.)

The meaning of this confidence interval for post hoc comparisons requires some special comment. If we consider all possible comparisons ψ_g that might be carried out on the true means of the J groups, then *the probability is* $1 - \alpha$ *that the statement 12.12.1 is true simultaneously for all* ψ_g. That is, if we could work out all possible comparisons on the data, and for each comparison calculate a 95 percent confidence interval, then the chances are 95 in 100 that every one of these confidence intervals would contain the true value for that comparison. There is only a 5 percent chance that one or more confidence intervals will not cover the corresponding true comparison value.

Note that one says that a comparison is significant at the α level if the confidence interval for that comparison does not cover zero. Any confidence interval for a comparison ψ_g that includes zero within its limits is said to be nonsignificant.

If the overall F test is significant at the α level, then some possible comparison $\hat{\psi}_g$ must be significant at or beyond the same level. Indeed, a significant F test can be interpreted as evidence that at least one true comparison value among all those possible is not zero. This does not mean that just because the overall F was significant you will necessarily find the significant comparisons, but only that they exist to be found. Hence our interpretation of a significant F as a signal, ''Something's here—start looking.''

This statement is not to be interpreted to mean that post hoc comparisons are somehow illegal or immoral if the original F test is not significant at the required α level. It means only that the probability statements that one makes about such comparisons are not necessarily true when F does not reach that level, and that going through the procedure may be something of a waste of time if one really is interested in accurate probability statements. Nevertheless, you can investigate comparisons among means whenever and wherever you like, and such comparisons may be quite suggestive about what is going on in the data, or about new questions calling for new experiments. What one cannot do is to attach an unequivocal probability statement to such post hoc comparisons, unless the conditions underlying the method have been met. It is not correct to assert that some post hoc comparison is significant at the α level by this method unless the overall F is significant at the α level, but you can look at anything or say anything else you wish.

The Scheffé method can be applied to any comparisons among the means, and to as many comparisons as it is desired to test, orthogonal or not. A very common problem is the investigation of all comparisons corresponding to differences between sample means. Unfortunately, for tests of all differences between pairs of means, the Scheffé method may be unnecessarily conservative. There are, however, several methods available for the post hoc testing of all differences between means. Three of the best-known of these are the Newman-Keuls test, the Duncan multiple range procedure, and the Tukey ''honestly significant difference,'' or HSD, method. All three of these methods rely on a statistic known as the ''studentized range,'' to be described below. Space does not permit a discussion of all of these methods, and we will go further only into the Tukey HSD method since it is simple, widely used, and flexible in application. The reader is directed to Winer (1971) for a full discussion of the Newman-Keuls test, and to Edwards (1972) for information on the Duncan multiple range procedures.

The *studentized range statistic* may be defined as follows: suppose that J independent samples were employed, each yielding a mean. These mean are arranged in order of magnitude, and thc highest and lowest mean noted.

The *range* of these J means is then, of course, given by the difference between the largest and the smallest mean. The studentized range statistic q is then

$$q = \frac{M_{max} - M_{min}}{\sqrt{\text{MS error}/n}} \qquad [12.12.4]$$

where MS error is found from a preceding analysis of variance. The n here is the sample size, if all samples are of equal size. However, if the samples have different sizes, an approximate q statistic can be found by using $2n_1n_2/(n_1 + n_2)$ in place of n, where n_1 is the size of the sample with the largest mean, or M_{max}, and n_2 is the sample size for M_{min}. Alternatively, n may be set equal to the harmonic mean of all the sample sizes. (See Section 4.2.)

The statistic q has a sampling distribution which depends upon two parameters: k, the number of means actually covered by the range making up the numerator of q, and ν, the number of degrees of freedom associated with MS error. Table XI in Appendix D gives the .05 and the .01 points in a studentized range distribution, where k is shown by the column entry, and the degrees of freedom by the row. Thus, for $k = 3$ and 10 degrees of freedom, the .05 value is 3.15, meaning that in a studentized range distribution with these parameters, .05 of all q values fall at or above 3.15.

In the Tukey HSD test, we always take $k = J$, the entire number of groups. Having decided on the value of α to use, we next look under the correct degrees of freedom to find the q value that cuts off the upper required α proportion of the distribution. Let us call this value q_α. Then,

$$\text{HSD} = \text{honestly significant difference} = q_\alpha \sqrt{\frac{\text{MS error}}{n}}. \qquad [12.12.5^*]$$

(Note the comment on n following 12.12.4) Then we examine all pairs of means and note their differences. If the absolute difference between any pair of means equals or exceeds the HSD, then we reject the hypothesis of the equality of means of the populations represented by the samples. Over all *experiments* the probability is α that a difference will exceed the HSD. This method will be illustrated in the following section, along with the application of the Scheffé method to post hoc comparisons other than simple differences.

One additional option exists for post hoc tests among pairs of means (or for any other set of tests for that matter). In Section 8.17, the Bonferroni inequality was mentioned, and it was suggested that by making the Type I probability for each of a number of tests (say K in all) be equal to α/K, then the probability of a Type I error in one or more tests could be made to be *no larger* than α. There is evidence that when only a very few means are to be compared, so that K *is* small, this strategy is an improvement on the Scheffé method as applied to pairs. However, when K is large the method becomes intolerably conservative, at least for the conventional values of α.

Since this procedure usually requires t values for percentage points not often listed in textbook tables, the following approximation is useful. In order to find the approximate value of t cutting off the upper α proportion in a distribution with ν degrees of freedom, take

$$t_{(\alpha;\nu)} = z_\alpha + [(z_\alpha^3 + z_\alpha)/4(\nu - 2)], \qquad [12.12.6]$$

where z_α is the corresponding value in a standardized normal distribution. This should be close enough for most practical purposes.

12.13 / AN EXAMPLE OF POST HOC COMPARISONS FOLLOWING A ONE-WAY ANALYSIS OF VARIANCE

In a psychological study subjects were assigned randomly to five different groups, representing five different experimental treatments. Twelve subjects were used in each group. The means of the various groups were

I	*II*	*III*	*IV*	*V*
63	82	80	75	70

and the analysis of variance is summarized in Table 12.13.1. For an α of .05, the required F for 4 and 55 degrees of freedom is approximately 2.53, so that this obtained value is significant. Hence, post hoc comparisons may be tested for significance. The .05 level will be used in these tests as well.

Table 12.13.1

Source	*SS*	*df*	*MS*	*F*
Between groups	2856	4	714	4.0
Error (within groups)	9801	55	178.2	
Totals	12652			

First the Tukey HSD method will be illustrated for the differences between all pairs of means. For this purpose it is convenient to arrange the means in order, as in Table 12.3.2 so that the difference between any pair is immediately apparent.

Table 12.13.2

All pairwise differences for five means.

	I 63	*V* 70	*IV* 75	*III* 80	*II* 82
I 63	0	7	12	17	19
V 70		0	5	10	12
IV 75			0	5	7
III 80				0	2
II 82					0

The numbers in the body of Table 12.13.2 represent the difference between the mean shown in the column and the mean shown in the row. Since both columns and rows list the means in order of magnitude, then all that need be shown are the differences above the diagonal; the same differences would appear with a negative sign below the diagonal.

In order to apply the Tukey HSD criterion to these differences, we first note that $k = 5$, the total number of groups, and that MS error in Table 12.13.1 has 55 degrees of freedom. Using 60 as the closest degrees of freedom in Table XI and using $\alpha = .05$, we find the value of $q_{.05}$ to be 3.98. We already know that MS error $= 178.2$, and that $n = 12$. Thus, we take

$$\text{HSD} = 3.98\sqrt{178.2/12} = 15.34.$$

Now we look at Table 12.13.2 once again, and see that two differences exceed this HSD value: the difference between groups I and II, and between groups I and III. Hence, we can say that these two differences are significant beyond the .05 level.

If we apply the Scheffé confidence interval method to this set of pairwise differences, we find that it takes a difference of 17.33 or more to be called significant at the .05 level. Thus, only the difference between groups I and II is significant under the Scheffé method, whereas we found two significant differences using the Tukey HSD method. This illustrates the point made earlier about the conservatism of the Scheffé method when applied to pairwise comparisons.

By way of comparison, the Bonferroni inequality implies that we should divide $\alpha = .05$ by the number of pairs, or 10, if we wish the probability to be no more than .05 that at least one Type I error occurs. Then, using a (two-tailed) significance level of .005, and approximating the required t value of about 2.92 by use of 12.12.6, we find that in order to be significant a difference must equal or exceed (in absolute value)

$$2.92\sqrt{\frac{2\text{MS error}}{12}} = 2.92\sqrt{\frac{178.2}{6}} = 15.91.$$

Two differences actually do exceed this value, and are significant. In making this test for all ten differences, we know that the probability of a Type I error in at least one test is no larger than .05. Notice that exactly the same pair of differences turn out significant in this method as in the Tukey HSD.

Many other types of comparisons may be made by the Scheffé method however. Suppose that we compare the average of II and III with the average of I, IV, and V. Here the value of the comparison is

$$\begin{aligned}\hat{\psi}_g &= \frac{82 + 80}{2} - \frac{63 + 75 + 70}{3} \\ &= \left(\frac{1}{2}\right)(82) + \left(\frac{1}{2}\right)(80) - \left(\frac{1}{3}\right)63 - \left(\frac{1}{3}\right)75 - \left(\frac{1}{3}\right)(70) = 11.67.\end{aligned}$$

Furthermore,

$$w_g = \frac{1}{12}\left(\frac{1}{4} + \frac{1}{4} + \frac{1}{9} + \frac{1}{9} + \frac{1}{9}\right) = \frac{5}{72} = .069,$$

and

$$\sqrt{V(\hat{\psi}_g)} = \sqrt{(178.2)(.069)} = 3.51.$$

Since $J - 1 = 4$,

$$S = \sqrt{(4)F_{(.05)}} = \sqrt{4(2.53)} = 3.18.$$

Therefore, the confidence interval is

$$\hat{\psi}_g - (3.18)(3.51) \leqslant \psi_g \leqslant \hat{\psi}_g + (3.18)(3.51)$$

or

$$11.67 - 11.16 \leqslant \hat{\psi}_g \leqslant 11.67 + 11.16.$$

This interval does *not* include zero, and so this comparison also is significant beyond the .05 level.

Even more complicated questions can be asked by way of post hoc comparisons if we desire. For example, one question might be, "Is the difference between group I and group II three times the difference between groups IV and III?" This corresponds to the comparison

$$\hat{\psi}_g = (M_1) - (M_2) - 3(M_4) + 3(M_3)$$
$$= (63) - (82) - 3(75) + 3(80) = -4.$$

Here,

$$w_g = \frac{1}{12}(1 + 1 + 9 + 9) = \frac{20}{12} = 1.67$$

and

$$\sqrt{V(\hat{\psi}_g)} = \sqrt{(178.2)(1.67)} = 17.25.$$

Once again we use $S = 3.18$ and $\alpha = .05$. Then the confidence interval is

$$-4 - (3.18)(17.25) \leq \psi_g \leq -4 + (3.18)(17.25)$$

or

$$-58.86 \leq \psi_g \leq 50.86.$$

This interval does include zero, and so using $\alpha = .05$ we cannot reject the hypothesis that the difference between population means I and II actually is three times the difference between population means IV and III.

The mere fact that one can find a significant comparison does not insure that the comparison is a meaningful one. It is definitely not profitable to work out every conceivable comparison among the means and test each for significance, in hopes that something of meaning will emerge. Just the reverse procedure should be used: inspecting the data, the experimenter comes to tentative conclusions about where the large and interpretable effects lie. These tentative conclusions are then tested.

12.14 / PLANNED OR POST HOC COMPARISONS?

It is obvious that in any given experiment it is always possible either to plan comparisons to be tested in lieu of the overall F test, or to perform post hoc comparisons should the overall F be significant. What are the arguments for and against these two ways of proceeding?

An important point in favor of planned comparisons is this: consider any true comparison ψ among J means ($J > 2$), such that $\psi \neq 0$. **The probability of a test's detecting that ψ is not zero is greater with a planned than with an unplanned comparison on the same sample means.** Thus, for any particular comparison, the test is more powerful when planned than when post hoc. **Put differently, the confidence interval for any given comparison is shorter when that comparison is planned than when it is post hoc.**

The practical implication is that the *importance* of the comparison should dictate whether or not it is tested by the planned or the post hoc procedure. If the experimental question represented by the comparison is an important one for the interpretation of the experiment, and it is essential that a Type II error not be made in the accompanying test, then a planned comparison should be carried out. On the other hand, if the question is a minor one, and a considerable risk of overlooking a true nonzero value can be tolerated, then the post hoc method suffices.

However, we have also seen that the number and variety of planned comparisons to be made and tested is limited by the independence requirement. No such requirement holds for the post hoc method. The data may be only partly explored by the planned method, but fully by post hoc methods.

In summary, then, let it be said that the planned comparisons method is best suited for situations where a few overriding concerns dictate the interpretation of the whole experiment. Here one must have the most powerful tests possible for resolving these issues. The post hoc method is suited for trying out hunches gained during the data analysis and for inferring the sources of the significant overall F test. We will return to the subject of planned and post hoc comparisons in Chapter 14 where their applicability to still another kind of experiment will be pointed out.

EXERCISES

1. A highway department was concerned with developing an effective reflectorized sign for use at certain critical locations. Originally they planned to look at all combinations of three background colors (red, blue, and black) with three possible letter colors (white, yellow, and green). However, three of these combinations proved to have insufficient contrast and they were dropped from the study. Visual stimuli were made up in the remaining six combinations, and flashed on a screen for a subject, who reacted to the stimuli as quickly as possible. Ten subjects saw all stimuli, and each was given a total reaction time score under each condition. The average scores across subjects for each color combination are shown below (in units of .1 second). The mean square used for error (actually subjects by treatments) was 109.87, with 45 degrees of freedom. Set up the weights for comparisons representing the following questions:

(a) Do white letters produce different reaction times than yellow or green letters?
(b) Given white letters, does a red background give a different reaction time than a blue or black background?
(c) Given white letters, is a blue background different from a black?
(d) Given yellow letters, is a blue background different from black?
(e) Is combination 6 different from combinations four or five?

Combination	*Background*	*Letters*	*Mean*
1	red	white	13
2	blue	white	15
3	black	white	10
4	blue	yellow	20
5	black	yellow	18
6	red	green	35

2. Check on the orthogonality of each pair of comparison weights found in exercise 1 above. To which of these comparisons would a sixth comparison corresponding to the following also be orthogonal?

Comparison question 6: Is the average of combinations 2 and 3 different from the average of combinations 4 and 5?

3. Test the significance of each of the planned comparisons outlined in exercise 1. Use $\alpha = .05$.

4. Suppose that the experimenters of exercise 1 decided that instead of planned comparisons, they would carry out an F test for all possible differences between the means of the six combinations. In view of the results for exercise 3 above, find the value of this F statistic. Is this significant for $\alpha = .05$?

5. Consider the data of exercise 18, Chapter 10 once again. Design and carry out a set of orthogonal comparisons, beginning with the comparison of mean for Group 1 with the average of Groups 2, 3, and 4. (You may choose to transform these scores as called for in the original exercise, if you wish.)

6. Using the data of exercise 18, Chapter 10, demonstrate the relation between the comparison sums of squares and the sum of squares between groups.

7. The data of exercise 25, Chapter 10 contain 12 cells, where a cell mean can be calculated for each. See if using these twelve means you can work out two sets of comparison weights for the rows. Then try to work out three sets of comparison weights for the columns. How many orthogonal comparisons could still be made? What do these represent? See if you can work out the weights for at least one of these.

8. For exercise 23 of Chapter 10, there are three groups represented by the rows. Work out the three t tests between pairs of rows for these data. Does the average of these three t values equal the F found for rows in this exercise? Does this say something about the orthogonality (and independence) of these t tests?

9. In a study of the effects of reward upon learning in small children, an experimenter used 20 children divided equally into four groups. Each child was given a puzzle which could be solved only if the steps were learned in order. For Group I, the child was rewarded for every correct move until the puzzle was learned. In Group II, 75 percent of a child's moves were rewarded on a random schedule. In Group III, only 25 percent of correct moves were rewarded, and in Group IV no moves were rewarded. The table below shows the number of trials it took each child to learn the correct sequence of moves. The experimenter entertained the following hypotheses:

(a) Constant reward will produce faster learning than the average of the other conditions.
(b) Frequent reward will produce faster learning than the average of infrequent or no reward.
(c) Infrequent reward will produce faster learning than no reward.

Test each of the experimenter's hypotheses, using $\alpha = .05$.

	Group		
Constant reward	*Frequent*	*Infrequent*	*Never*
12	9	15	17
13	10	16	18
11	9	17	12
12	13	16	17
12	14	16	19

10. Instead of treating the questions of exercise 9 through planned comparisons, see if they can be answered in terms of post hoc comparisons, following an F test. Again use $\alpha = .05$.

11. If each of the comparisons outlined in exercise 9 above is tested for significance with $\alpha = .05$, what is the probability of a Type I error for at least one of these tests? If each were tested using, say, $\alpha = .017$, what is the probability that one or more will represent a Type I error?

12. Consider a set of seven groups, each containing 10 subjects. Planned comparisons are to be made with the following weights:

			Groups			
I	*II*	*III*	*IV*	*V*	*VI*	*VII*
4	4	4	−3	−3	−3	−3
1	1	−2	0	0	0	0
0	0	0	1	1	−1	−1

See if you can complete the set of orthogonal comparisons. (**Hint:** Look for sets of two or more groups which have, so far, received only the same signs.)

13. Explain why there are only $J - 1$ rather than J orthogonal comparisons among J independent groups. What does this have to do with the requirement that the sum of each group of comparison weights must be zero?

14. Consider four groups, and let there be two orthogonal comparisons as follows. Now pick a number and multiply the weights for comparison 1 by this number. Then pick another number and multiply the weights for comparison 2 by that number. Are the comparisons based on the new weights still orthogonal? Why should this be?

	I	*II*	*III*	*IV*
ψ_1	1	1	−1	−1
ψ_2	1	−1	−1	1

15. Suppose that in exercise 12 the first two comparisons shown were to be tested for significance as planned comparisons. The hypotheses to be tested are $H_0\!:\psi_1 \geq 0$, versus $H_1\!:\psi_1 < 0$, and $H_0\!:\psi_2 \leq 0$ versus $H_1\!:\psi_2 > 0$. An α level of .05 is to be used in each test. Explain the procedures you would use.

16. Consider the following set of means, each corresponding to one fixed experimental treatment on a separate group of 10 independent subjects:

I	*II*	*III*	*IV*	*V*
86	95	92	80	104

If the mean square error is 40, test the significance of the planned comparisons having the following weights:

I	*II*	*III*	*IV*	*V*
1	−1	0	0	0
0	0	0	1	−1
−1	−1	0	1	1
1	1	−4	1	1

17. Suppose that in exercise 16 above there had been prior interest only in the first two comparisons individually. Make an analysis of variance summary table showing the resulting mean squares and tests of significance.

18. Given the overall significance level established for exercise 7, Chapter 10, carry out post hoc tests for the difference between all pairs of means, using the Tukey HSD.

19. In the following table, the means for a three by three factorial experiment with *n* observations per cell are symbolized. If interactions among the cell means are to be independent of comparisons among row or among column means, see if you can work out at least 3 of the 4 orthogonal comparisons underlying the interaction sum of squares, and show that they are orthogonal to row and column interactions.

M_{11}	M_{12}	M_{13}
M_{21}	M_{22}	M_{23}
M_{31}	M_{32}	M_{33}

20. For exercise 25, Chapter 10 treat the following comparisons among columns as planned, and test for significance:

I	*II*	*III*	*IV*
1	−1	−1	1
1	0	0	−1
0	1	−1	0

21. For exercise 25, Chapter 10 suppose that the comparisons listed in exercise 20 above had been applied to the "room color" treatments. Carry out a post hoc test for these comparisons, using $\alpha = .05$.

22. For exercise 20, Chapter 10 carry out a test of the planned comparisons corresponding to "male versus female cues" and to "cues versus no cues."

23. Construct the 95 percent confidence interval for the two comparisons made in problem 22 above.

24. For exercise 20, Chapter 10, carry out a test of the comparison corresponding to "same-sex subjects and cues versus different-sex subjects and cues." To which sum of squares does this comparison contribute?

25. For exercise 25, Chapter 10, carry out and test post hoc comparisons on all differences of between-column means, using the Tukey method, and $\alpha = .05$.

26. For exercise 25, Chapter 10 devise and carry out a test of two planned comparisons among rows, other than those in exercise 20 above.

27. Suppose that planned comparisons between lists I and II and between lists III and IV had been planned for exercise 26, Chapter 11. Find the 95 percent confidence limits for these comparisons.

28. If it is appropriate to do so, carry out post hoc comparisons between all pairs of "list" means for exercise 26, Chapter 11.

29. Why are comparison methods limited to designs corresponding to fixed or mixed models?

30. Below is an analysis of a variance summary table.

Source	*SS*	*df*
Between rows		(4)
Comparison 1	203	1
Comparison 2	25	1
Other row comparisons	250	2
Between columns		(6)
Comparison 1	78	1
Other column comparisons	507	5
Interaction		(24)
Comparison 1	215	1
Other comparisons		23
Error	3405	105
Total	5660	139

Assuming a fixed-effects model, complete the table and carry out the tests indicated ($\alpha = .01$).

31. See if you can suggest circumstances under which the theory of comparisons might extend to statistical analyses other than the comparison of means in the analysis of variance.

Chapter 13
PROBLEMS IN CORRELATION AND REGRESSION

So far in applying the different variations of the general linear model we have confined our attention to situations where the independent variable X represents a set of essentially qualitative distinctions. These distinctions have sometimes represented preformed groups of subjects, and at other times they have stood for different levels of some treatment actually administered by the investigator. The model adopted in these instances has treated the variable X strictly as an indicator, taking on only the values 0 and 1 depending on which of a set of groups is being specified.

Now we are going to consider situations in which the independent variable X takes on *any real number value*. Such a variable may, for example, represent the amount of some treatment that a subject received in an experiment, or it may be the score that the subject earned on some test. In short, we are now going to deal with quantitative independent or predictor variables as well as a quantitative dependent variable Y.

One distinction that we will find useful concerns the sampling scheme employed for obtaining the values of the independent variable X. In some situations, the experimenter exercises no control whatsoever over the values of X that occur in the study, nor over the values of Y. Rather, in this sampling situation each individual included simply "brings along" a value of X and a value of Y. We will call this approach "a problem in correlation."

On the other hand, it may be that the investigator forms groups, and then arranges that a given group will receive a given, predetermined, value of X. This might be the situation in an experiment in which different amounts of medication are given to different groups of patients, Here, the values of X are not sampled, as in a problem in correlation, but rather are deliberately preselected. This sampling scheme will be called "a problem in regression."

In this chapter we will usually assume that the data are collected as for a problem in correlation, and defer consideration of the other sampling situation until Chapter 14. Although, both theoretically and mechanically, the methods employed in these two situations are virtually identical, the two approaches do differ in the actual sampling methods used, the breadth of inferences made from the results, and in some of the assumptions ordinarily made about the population sampled.

13.1 / SIMPLE LINEAR RELATIONS

Problems in correlation and in regression are both concerned with three main questions:

1. Does a statistical relation affording some predictability appear between the random variables X and Y?
2. How strong is the apparent degree of the statistical relation, in the sense of possible predictive ability the relation affords?
3. Can a simple rule be formulated for predicting Y from X, and if so, how good is this rule?

The ordinary techniques we have studied heretofore apply to the first two of these questions, but the third is a new feature. In this chapter we are going to study the possibility of applying *the linear model as a rule for the prediction of Y from X*. We are going to act *as though* the true relation actually were a function, and, using a function rule, make predictions or "bets" about Y values from knowledge of X values. Then we are going to evaluate the *goodness* of this prediction rule in terms of how well one actually would do by predicting according to the rule. If the statistical relation actually is a function then some rule exists that affords perfect prediction; for the usual statistical relation, no rule permits perfect prediction, but some function rule may nevertheless provide a good "fit" to the relation under study. Proceeding in this fashion gives us two important advantages: quite often we are able to achieve a fair degree of predictive ability by adopting a particular function rule, even though the true relation itself is not really a precise function. Second, by studying our errors using this rule, we gain information about how the rule might be made better and how the general form of the relation might be specified more adequately.

The model that we will use to describe the relation between X and Y will be linear, just as in Chapters 10 through 12. However, the model we will first employ will be very simple:

$$y_i = a_0 + bx_i + e_i, \tag{13.1.1}$$

where, as before, a_0 is a constant that enters into the value of y_i for any individual i, b is a constant weight that applies to the value x_i, and e_i stands for error. Note that if there were no errors, then

$$y'_i = a_0 + bx_i, \tag{13.1.2}$$

which is strictly linear in the sense that a plot of all of the (x_i, y'_i) pairs of values would fall along a straight line.

Naturally, there is no law that says the relationship between values such as x_i and values such as y_i must be linear. The best description of how the x_i are related to y_i

values in a given set of data might call for a very different mathematical function. Why, then, do we emphasize such linear rules or models?

The reasons for starting with linear rules for prediction are several: linear functions are the simplest to discuss and understand; such rules are often good approximations to other, much more complicated, rules; and we will find that in certain circumstances the only prediction rule that *can* apply is linear. However, do not jump to the conclusion that just because we deal first with linear prediction this is the only important way to predict, or that all real relationships must be more or less like linear functions. In the next chapter we will find that there are many other, nonlinear, function rules that might also be applied to a given problem.

Before we can turn to inferential methods appropriate to problems in correlation, we need some terminology for describing a linear relation between variables X and Y in any fixed set of N observations, or sample of N cases. Thus, the succeeding few sections will deal with the descriptive statistics of correlation and regression.

13.2 / THE DESCRIPTIVE STATISTICS OF CORRELATION AND REGRESSION

Just as we discussed how one could find and interpret the mean and variance as descriptions of particular aspects of a set of data, we will now turn to the problem of finding a linear rule that "fits" a given set of data as well as possible. For the moment, our interest is only in a specific set of data, the scores for some particular set of N observed individuals.

Imagine this kind of situation: a teacher of a large introductory college course is interested in the possible relationship between the high school preparation in mathematics that a student has and success in the course. In a particular semester the teacher has a class of 91 students, and at the outset each student is asked the number of mathematics courses taken in high school (four years). The teacher weights these courses in a routine way and assigns scores running from 2 through 8 to the students. Let us call these "mathematics scores" X.

The teacher, however, files these reports away and does not look at them until after the final examination in the course has been given. The actual raw scores on this examination will be called the variable Y. After both scores for each student are known, the teacher asks this question: "To what extent is there a linear relation between the X and the Y scores?" In other words, how well does a simple linear rule allow one to predict the Y score of a student drawn at random from this group, given the information about the X score? The problem is to find the best possible linear rule for predicting from these data, and then to evaluate the goodness of such a rule.

Actually, the teacher is not especially interested in predicting the raw Y score of a student so much as in the *relative* performance of the student in terms of Y. That is, the teacher would like to be able to predict the standard score z_Y, given by

$$z_Y = \frac{y - M_Y}{S_Y}.$$

This prediction is to be based on the standard score z_X, where

$$z_X = \frac{x - M_X}{S_X}.$$

Since a linear rule is to be used for this prediction, this means a rule of the form

$$z'_Y = A + Bz_X \qquad [13.2.1]$$

where B and A are constants. (Here, and in the following, we will use capital letters for the unknown constants in a rule involving z-scores, and lowercase letters for rules involving raw score values.) The predicted score is labeled as z'_Y to indicate that it *need not* be the same as z_Y, the true standard score for any given individual. Several individuals may have the same z_X standard score, but quite different z_Y scores; by use of the rule, however, *only one z'_Y or predicted standard score will be given for each z_X value.*

The problem is shown graphically in Figure 13.2.1. Here, the horizontal axis represents possible values for z_X, the vertical axis the possible values for z_Y, and any point within the plane defined by these two axes represents a pair of z-scores (z_X,z_Y) that might be associated with any individual observation. Points in the functional relation $z'_Y = A + Bz_X$ lie along the straight line in the figure. For the particular value of z_X represented in the figure, the linear rule affords a *predicted* value z'_Y; this *need not* correspond to the actual value z_Y corresponding to any individual showing the particular value of z_X shown in the figure. The extent of "miss" or error between the predicted value z'_Y and z_Y for an individual is represented by the vertical distance between the two points (z_X,z_Y) and (z_X,z'_Y). We would like our prediction rule to be such that, across all individuals, the fit between predicted and actual standard scores on Y is as good as possible. The reader may be puzzled by the use of the term "prediction" in this content; the teacher actually *has* the two scores for each of the 91 students. Why not merely look at the standardized Y score for any student? Actually the methods to be developed here will apply to situations where the user wants to go beyond the immediate data, and to forecast the Y or z_Y score for an individual for which this information is not already available. However, the basis for these methods is best seen if one deals only with one intact group of N cases, each having two scores, X and Y. For the moment, "prediction" consists of drawing one case at random from this particular group, noting the z_Y, and then finding a predicted value of z'_Y by use of the linear rule.

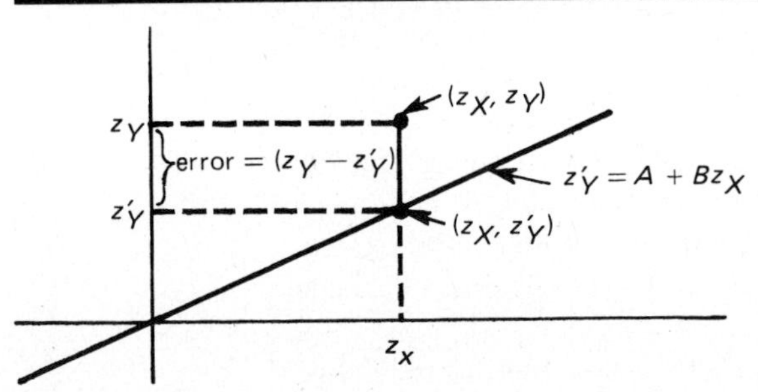

Figure 13.2.1

Plot of a linear regression equation for the prediction of the standard score on Y from the standard score on X.

The first problem is to find constants A and B that will make the linear rule give the "best possible" predictions. These constants are found by the method of least squares, which we have already encountered in Chapters 4 and 10. In this context, applying the criterion of least squares means that we want to minimize the sum of squared errors in prediction. Thus, for any individual case i we will make some prediction, z'_{Y}; this need *not* be the same as the true value z_{Y_i} for that individual, and so some error will exist,

$$e_i = (z'_{Y_i} - z_{Yi}).$$

The least-squares criterion requires that we choose A and B in such a way that the average squared error over individual predictions be as small as possible. Thus, given N individuals i, we want to choose A and B so as to make

$$\frac{\sum_i (z'_{Yi} - z_{Yi})^2}{N} \qquad [13.2.2]$$

have its minimum possible value. (Please understand that in the following we are able to use elementary algebra to find the values of A and B only because the problem is a very simple one. In actual practice, the methods of the differential calculus would be used in this or any more complicated situation in order to find a least-squares solution.)

Now, first of all we will show that by the least-squares criterion, *if we are predicting standard scores, the value of the constant A must be zero, so that the best linear rule is actually*

$$z'_{Yi} = Bz_{Xi}. \qquad [13.2.3]$$

This can be shown as follows: Substituting 13.2.1 into 13.2.2 and rearranging terms we have

$$\frac{\sum_i (z'_{Yi} - z_{Yi})^2}{N} = \frac{\sum_i [(Bz_{Xi} - z_{Yi}) + A]^2}{N}. \qquad [13.2.4]$$

On carrying out the square for each i and summing, we have

$$\frac{\sum_i [(Bz_{Xi} - z_{Yi}) + A]^2}{N} = \frac{\sum_i (Bz_{Xi} - z_{Yi})^2}{N} + 2A\frac{\sum_i (Bz_{Xi} - z_{Yi})}{N} + \frac{\sum_i A^2}{N}$$

$$= \frac{\sum_i (Bz_{Xi} - z_{Yi})^2}{N} + A^2 \qquad [13.2.5]$$

since A and B are constants, and the mean of each set of z-scores must be zero.

Now, assuming B fixed, for what value of A can the expression on the right in 13.2.5 be at its smallest? The first term is a mean of squared numbers and hence must be positive, and A^2 must be positive as well; it follows that this entire expression can be at its smallest value *only* when A is zero. Thus, the value of A dictated by the least-squares criterion is zero.

Next, we will show that by the least-squares criterion the value of B for predicting z-scores must be

$$B = \frac{\sum_i z_{Xi}z_{Yi}}{N} = r_{xy}\,. \qquad [13.2.6]$$

This value of B is actually the **correlation coefficient** (or Pearson product-moment correlation coefficient) r_{xy}, about which we will have much to say. By definition,

$$r_{xy} = \frac{\sum_i z_{Xi} z_{Yi}}{N}. \qquad [13.2.7^*]$$

Thus, by the criterion of least squares, our prediction rule must be

$$z'_{Yi} = r_{xy} z_{Xi}. \qquad [13.2.8^*]$$

We can show as follows how the least-squares criterion dictates that $B = r_{xy}$ for the prediction of standard scores. Since we know that A must be zero, this time we start by substituting expression 13.2.3 into 13.2.2.

$$\frac{\sum_i (z'_{Yi} - z_{Yi})^2}{N} = \frac{\sum_i (Bz_{Xi} - z_{Yi})^2}{N}. \qquad [13.2.9]$$

Expanding the square on the right hand of 13.2.9 and summing, we have

$$\begin{aligned} \frac{\sum_i (z'_{Yi} - z_{Yi})^2}{N} &= \frac{B^2 \sum_i z^2_{Xi}}{N} - \frac{2B \sum_i z_{Xi} z_{Yi}}{N} + \frac{\sum_i z^2_{Yi}}{N} \\ &= B^2 - 2Br_{xy} + 1 \end{aligned} \qquad [13.2.10]$$

since the variance of standard scores is always 1. Now suppose that B differed from r by some amount c, either a positive or negative number:

$$B = r + c.$$

Substituting $r + c$ for B in 13.2.10 above, we find

$$\begin{aligned} \frac{\sum_i (z'_{Yi} - z_{Yi})^2}{N} &= (r + c)^2 - 2(r + c)r + 1 \\ &= r^2 + 2rc + c^2 - 2r^2 - 2rc + 1 \\ &= (1 - r^2) + c^2. \end{aligned} \qquad [13.2.11]$$

When $B = r_{xy}$, so that c is zero, the mean squared error must be at its smallest,

$$\frac{\sum_i (z'_{Yi} - z_{Yi})^2}{N} = (1 - r^2_{xy}), \qquad [13.2.12^*]$$

and for any $c \neq 0$, the value must be $(1 - r^2_{xy})$ *plus a positive number,* c^2. *Hence taking* $B = r_{xy}$ *gives the least squared error in linear prediction, on the average, for standard scores.*

Expression 13.2.8 is called the *prediction equation* (or *regression equation*) for z_Y predicted from z_X. The constant B as found from 13.2.6 is called the *standardized regression coefficient* or *standardized regression weight* for predicting z_Y from z_X. Thus in the special case where standard scores z_Y are to be predicted from standard scores z_X, the value of $B = r_{xy}$. However, as we will see, when raw scores are to be predicted, a somewhat different form of regression weight will be needed.

The notion of the mean squared error (13.2.2) is an important one in its own right, and we will give it a special symbol and name. Let

$$S^2_{z_Y \cdot z_X} = \frac{\sum_i (z'_{Yi} - z_{Yi})^2}{N} = 1 - r^2_{xy} \qquad [13.2.13^*]$$

be called the **sample variance of estimate for standard scores.** This variance of estimate reflects the *poorness* of the linear rule for prediction of standard scores, the extent to which squared error is, on the average, large.

Most often, however, this index is discussed in terms of its positive square root

$$S_{z_Y \cdot z_X} = \sqrt{1 - r^2_{xy}}, \qquad [13.2.14^*]$$

which is called the **sample standard error of estimate for predicting standard scores.**

Obviously, there is a close connection between the size of the standard error of estimate in a sample and the value of r in the regression equation: **the larger the absolute value of r_{xy}, the smaller is the standard error of estimate.** Now we turn to some interpretations of the index r_{xy}.

13.3 / SOME PROPERTIES OF THE CORRELATION COEFFICIENT IN A SAMPLE

The connection between the variance of estimate and the correlation coefficient shows us at once that r_{xy} can take on values *only* between -1 and 1. Notice that the variance of estimate, being a weighted sum of squares, can be only a positive number or zero. If r_{xy} were less than -1, or greater than $+1$, then the variance of estimate could not be positive. Hence, $-1 \leq r_{xy} \leq 1$.

What does it mean when r_{xy} is exactly zero? When this is true, one predicts $z'_Y = 0$, corresponding to the mean Y, *regardless* of the value of X; for any X, the mean of Y is the best linear prediction when the correlation is zero. Furthermore,

$$S^2_{z_Y \cdot z_X} = 1$$

for $r_{xy} = 0$. This means that when the correlation coefficient is zero the variance of estimate for standard scores is exactly the same as the variance of the standard scores z_Y with X *unspecified*. Thus, when r_{xy} is zero, predicting by the linear rule *does not* reduce the variability of z_Y below the variability present when z_X is unknown. In short, the fact that r_{xy} is zero implies that if a predictive statistical relation exists for the set of data, it is not linear, and the linear rule gives no predictive power.

On the other hand, when r_{xy} is either $+1$ or -1, the variance of error in prediction is zero, so that each prediction is exactly right. These ultimate limits for r_{xy} can occur only when X and Y *are* functionally related, *and* follow a linear rule.

All intermediate values of r_{xy} indicate that some prediction is possible using the linear rule, but that this prediction is not perfect, and some error in prediction exists. Any value of r_{xy} between 0 and 1 in absolute magnitude indicates either that the relationship is not functional or that if it is a function, the rule is not exactly linear, although a linear rule does afford some predictability.

In the regression equation for standard scores the correlation coefficient plays the role of converting a standard score in X into a predicted standard score in Y. Rather loosely, the correlation coefficient can be said to be "the rate of exchange," the value of a "standard deviation's worth" of X in terms of *predicted* standard deviation units of Y.

Finally, the fact that we are able to define the correlation coefficient in terms of standard scores shows that r_{xy} is a *dimensionless index* of linear relationship. This means that r_{xy} does not depend on the units of measurement for either X or Y, nor on what value is called "zero" for either variable. Either X or Y or both can be given a linear transformation (multiplying by a positive constant and adding another constant), and the correlation between the transformed variables will be the same as r_{xy}.

13.4 / THE PROPORTION OF VARIANCE ACCOUNTED FOR BY A LINEAR RELATIONSHIP

Just how good is a linear rule for predicting values of Y from values of X in a given sample? In order to answer this question, we need an index of the *strength of linear relationship* between X and Y in the data. We can approach this problem as follows:

We already know that the variance of any set of standard values z_Y has to be 1.00. However, what is the variance of the *predicted* values z'_{Yi}? We can think of this as *variance accounted for, variability among the z_{Yi} values for different observations i directly attributable to the fact that they have different z_X scores*. When we take the variance of the predicted z'_Y values we have

$$S^2_{z'Y} = \frac{\sum_i (z'_{Yi})^2}{N} - \left(\sum_i \frac{z'_{Yi}}{N}\right)^2. \qquad [13.4.1]$$

First of all notice that

$$\sum_i \frac{z'_{Yi}}{N} = \sum_i \frac{r_{xy}\, z_{Xi}}{N} = 0,$$

since the sum of all of the z_{Xi} values must be zero. Thus, the second term on the right, the squared mean of all the predicted z'_Y values, is itself zero. Then we find

$$\sum_i \frac{(z'_{Yi})^2}{N} = \sum_i \frac{r^2_{xy}\,(z_{Xi})^2}{N} = r^2_{xy}. \qquad [13.4.2^*]$$

The variance of the predicted z'_Y values, or the variance explained by the z_{Xi} values, is thus r^2_{xy}.

Since the variance of the original z_Y values has to be 1.00, then if we take the ratio of the variance of the predicted values to the total variance of the z_Y values, we find that

$$\begin{pmatrix}\text{proportion of variance explained}\\ \text{by linear regression of } Y \text{ on } X\end{pmatrix} = \frac{S^2_{zY'}}{S^2_{zY}} = r^2_{xy}, \qquad [13.4.3^*]$$

the proportion of variance in Y accounted for, or explained, by X is given by r^2_{xy}. Thus, an index of the "goodness" of the linear rule for predicting X from Y is given by r^2_{xy}, the proportion of variance in X accounted for by Y under the linear rule. The index r^2_{xy} is **usually termed the coefficient of determination.** You can always think of r^2_{xy} as representing the strength of *linear* relationship in a given set of data. Furthermore, although it was convenient to discuss the proportion of variance accounted for initially in terms of z-values, r_{xy} is the same either for z-scores or for raw values so that these interpretations are valid for r^2_{xy} whether we are discussing standardized or raw scores.

Thus, if the value of a correlation coefficient is .50 (positive or negative in sign) then some .25 of the variability in Y is accounted for by specifying the linear rule and X. If the correlation is .80, then 64 percent of the variance in Y is accounted for in this way. A correlation of positive or negative 1.00 means that 100 percent of variability in Y can be accounted for by the linear rule and X, but if $r_{xy} = 0$, none of the variability is thereby accounted for. All in all, not the correlation coefficient per se but the *square* of the correlation coefficient informs us of the ''goodness'' of the linear rule for prediction.

13.5 / THE IDEA OF REGRESSION TOWARD THE MEAN

The term *regression* has come to be applied to the general problem of prediction by use of a wide variety of rules, although the original application of this term had a very specific meaning. The term ''regression'' is a shortened form of **regression toward the mean in prediction.** The general idea is that **given any standard score z_X, the best linear prediction of the standard score z_Y is one relatively nearer the mean of zero than is z_X.**

This can be illustrated quite simply from our regression equation 13.2.8. Suppose than an individual has a standard score z_X of 2. Also suppose that the regression equation we have found for the group to which the individual belongs is

$$z'_Y = .5z_X.$$

Then we *predict* this individual to have a z_Y score of 1, since

$$z'_Y = .5(2) = 1.$$

Notice that we predict the individual to fall relatively *nearer* the mean on Y than on X. That is, we predict in accordance with *regression toward the mean*. For another set of data, the regression equation might be

$$z'_Y = -.75z_X.$$

Now in this instance, suppose that the z_X for some randomly selected individual were 1.5. Then

$$z'_Y = (-.75)(1.5) = -1.125.$$

Since the correlation coefficient is negative, the prediction is that this individual falls *below* the mean of Y, given that the X value is above the mean.

However, in absolute terms, we again predict a standing relatively *closer* to the mean on Y than on X.

This principle of predicting relatively closer to the mean, or regression toward the mean, is a feature of any linear prediction rule that is best in the ''least-squares'' sense of Section 13.2. The idea is that if we are going to use such a linear rule for prediction, then it is always a good bet that an individual will fall *relatively closer to the group mean on the thing predicted than on the thing actually known.* This does *not* imply that the actual Y value *must* fall relatively closer to the mean than does the value of X, however, but only that our best *bet* is that it will do so. Regression toward the mean is not some immutable law of nature, but rather a statistical consequence of our choosing to predict in this linear way, using the criterion of least squares in the choice of a rule.

13.6 / THE REGRESSION OF z_X ON z_Y

In a true correlation problem, nothing makes it necessary to think of X as the independent variable, or the value somehow known first or predicted from. It is entirely possible to consider a situation where one might want to predict z_X from knowing z_Y. What does this do to the linear prediction rule, the correlation coefficient, and so on?

In the first place, the same argument used in Section 13.2 shows that for predicting z_X from z_Y,

$$z'_X = r_{XY} z_Y, \quad [13.6.1]$$

where the correlation coefficient is, just as before, given by 13.2.7.

It is tempting to ask why we do not just solve the original regression regression equation for z'_{Y_i} in terms of z_{Xi} in order to get

$$z_{Xi} = z'_{Yi} / r_{xy}.$$

However, recall what the symbols z'_Y and z'_X actually represent. These are *predicted* values and do not necessarily symbolize the actual values of z_Y and z_X at all. Solving the expression 13.2.8 for z_X *might* be useful if one wanted to know the value of z_X *known,* given that z'_Y were the predicted value, although it is hard to see why anyone would ordinarily want this information. The form of the regression equation used (13.2.8 or 13.6.1) depends strictly on which variable, X or Y, is designated as the independent variable, or the thing known first in a prediction situation.

In prediction of X from Y, the sample variance of estimate for standard scores is

$$S^2_{zX \cdot zY} = 1 - r^2_{xy} \quad [13.6.2]$$

(notice the reversal in subscripts from 13.2.13 when z_X is predicted from z_Y). The proportional variance in X accounted for by Y is, once again, r^2_{xy}.

This brings up the point that the correlation coefficient is a *symmetric* measure of linear relationship. *So long as we are talking about the correlation coefficient alone, it is immaterial which we designate as the independent and which the dependent variable; the measure of possible linear prediction is the same.* However, when we deal with the actual regression equations themselves, this symmetry is not usually present. As we shall see in the next section, it does make a difference whether you are predicting Y from X or X from Y when it comes to finding the regression equations and errors of estimate for *raw* scores.

13.7 / THE REGRESSION EQUATIONS FOR RAW SCORES

Up to this point, we have considered only the problem of predicting standard scores from standard scores. Introducing regression and correlation in terms of standard scores makes the algebra somewhat easier, and the essential ideas somewhat simpler. Nevertheless, each feature of correlation and regression shown for standard score prediction is also valid for the prediction of raw scores. For any given set of data, each standard score corresponds uniquely to some raw score, and vice versa, so that linear prediction which is optimal in standard score terms is optimal in raw score terms as well.

It is quite simple to put the regression equation for prediction z_Y from z_X into raw score form. We start with

$$z'_{Yi} = r_{xy}\, z_{Xi}$$

which is exactly the same as

$$\frac{(y'_i - M_Y)}{S_Y} = r_{xy} \frac{(x_i - M_X)}{S_X}$$

where y'_i is the *predicted raw score* for the individual, and x_i is the *known* raw score on the independent variable. A little algebraic manipulation gives

$$y'_i = M_Y + \frac{r_{xy}S_Y}{S_X}(x_i - M_X). \qquad [13.7.1^*]$$

This is the raw score form of the regression equation for prediction of Y from X.

It will be convenient to write this regression equation as

$$y'_i = M_Y + b_{Y \cdot X}(x_i - M_X) \qquad [13.7.2^*]$$

where

$$b_{Y \cdot X} = \frac{r_{xy}S_Y}{S_X}. \qquad [13.7.3^*]$$

The value $b_{Y \cdot X}$ is called the **unstandardized or raw score regression coefficient** of Y on X. This value $b_{Y \cdot X}$ gives the best (least-squares) prediction of raw Y scores from raw X scores.

In an identical way, we can turn the regression equation for z_X predicted from z_Y into

$$x'_i = M_X + b_{X \cdot Y}(y_i - M_Y) \qquad [13.7.4]$$

This is the raw score form of the regression equation for predicting X from Y. Here

$$b_{X \cdot Y} = \frac{r_{xy}S_X}{S_Y} \qquad [13.7.5]$$

is the unstandardized or raw score regression coefficient for predicting X from Y.

Notice that when no specification is put on which is to be regarded as the independent or predictor variable, there are two possible regression coefficients, $b_{Y \cdot X}$ and $b_{X \cdot Y}$, and that

$$\sqrt{b_{Y \cdot X} b_{X \cdot Y}} = r_{xy}, \qquad [13.7.6]$$

the square root of the product of the two regression coefficients (i.e. their geometric mean) is the correlation coefficient.

Figure 13.7.1 shows the two raw score regression lines that might apply to a given set of data.

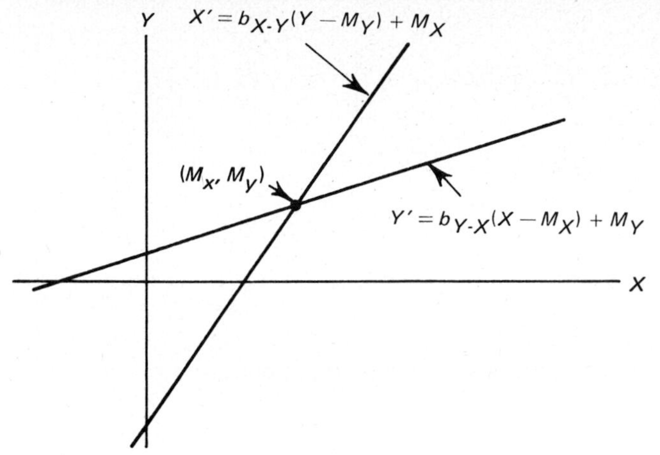

Figure 13.7.1
Plot of the two regression lines for predicting Y from X and X from Y.

When prediction of raw scores is to be carried out, the **sample variance of estimate** for predicting Y from X is

$$S^2_{Y \cdot X} = S^2_Y(1 - r^2_{xy}) \qquad [13.7.7]$$

and the **sample standard error of estimate** is

$$S_{Y \cdot X} = S_Y\sqrt{1 - r^2_{xy}}. \qquad [13.7.8^*]$$

Similarly, the sample variance of estimate for predicting X from Y is

$$S^2_{X \cdot Y} = S^2_X(1 - r^2_{xy}) \qquad [13.7.9]$$

and the sample standard error of estimate is

$$S_{X \cdot Y} = S_X\sqrt{1 - r^2_{xy}}. \qquad [13.7.10^*]$$

Finally, remember that the proportion of variance accounted for by linear relationship (either in predicting Y from X or X from Y) is r^2_{xy}, just as for standardized scores.

13.8 / COMPUTATIONAL FORMS FOR r_{xy} AND $b_{Y \cdot X}$

Although the sample correlation coefficient was actually defined in Section 13.2 as a summed product of standard scores (13.2.7), in practice the index is seldom computed in this way. An equivalent raw score computational form will now be found. We will show that the raw score form of the correlation coefficient is

$$r_{xy} = \frac{\left(\sum_i x_i y_i / N\right) - M_X M_Y}{S_X S_Y}. \qquad [13.8.1^*]$$

Starting with the definition, 13.2.7, and substituting the raw score equivalents of z_{Xi} and z_{Yi}, we have

$$r_{xy} = \frac{\sum_i (x_i - M_X)(y_i - M_Y)}{NS_XS_Y}$$

$$= \frac{1}{NS_XS_Y}\left[\sum_i x_iy_i - \sum_i x_iM_Y - \sum_i y_iM_X + NM_XM_Y\right]$$

$$= \frac{1}{NS_XS_Y}\left[\sum_i x_iy_i - 2NM_XM_Y + NM_XM_Y\right]$$

$$= \frac{\left(\sum_i x_iy_i/N\right) - M_XM_Y}{S_XS_Y}.$$

Still another computing form for r_{xy} that is often useful when work is being done on a pocket or desk calculator is

$$r_{xy} = \frac{N\sum_i x_iy_i - \left(\sum_i x_i\right)\left(\sum_i y_i\right)}{\sqrt{\left[N\sum_i x_i^2 - \left(\sum_i x_i\right)^2\right]\left[N\sum_i y_i^2 - \left(\sum_i y_i\right)^2\right]}}. \qquad [13.8.2]$$

Given the correlation coefficient, it is then possible to find the sample raw score regression coefficient $b_{Y \cdot X}$ directly from

$$b_{Y \cdot X} = r_{xy}\left(\frac{S_Y}{S_X}\right) \qquad [13.8.3]$$

However, for many problems it is desirable to calculate $b_{Y \cdot X}$ without first finding r_{xy}. This may be done most simply by taking

$$b_{Y \cdot X} = \frac{\sum_i x_iy_i - NM_XM_Y}{\sum_i x_i^2 - NM_X^2} \qquad [13.8.4^*]$$

or

$$b_{Y \cdot X} = \frac{N\sum_i x_iy_i - \left(\sum_i x_i\right)\left(\sum_i y_i\right)}{N\sum_i x_i^2 - \left(\sum_i x_i\right)^2}. \qquad [13.8.5]$$

A great many hand-held or pocket calculators are preprogrammed to provide the value of the correlation coefficient, along with b and other values. All one has to do is to enter each of the pairs of (x_i, y_i) values, and the calculator produces r_{xy}, in addition to the two means and variances, with $b_{Y \cdot X}$ if desired. Many also include the feature of giving the predicted y' value in terms of some entered x value, and x' from y. The ease of use and wide availability of these inexpensive calculators has taken much of the former labor from correlation computations.

If a number of correlations, along with their associated regression weights and equations must be found, the job will likely be done by computer. Statistical computing packages such as SPSS contain a variety of methods for computing correlations and related statistics. Various other methods for computing r_{xy} are available. One method especially popular in past years is designed for data grouped into a joint frequency distribution or "scatter plot," and this version is to be found in many elementary statistics texts. However, for the reasons just mentioned this is outdated as a practical method, and we will not consider it here.

13.9 / AN EXAMPLE OF CORRELATION COMPUTATIONS FOR A SAMPLE

Consider once again the example of Section 13.2. The teacher collected data for a class of 91 students, obtaining for each a score X, based on the number of courses taken in high school mathematics, and a score Y, the actual score on the final examination for the course. For each of the 91 individuals, the x_i and the y_i values are shown in Table 13.9.1.

Table 13.9.1

High school mathematics scores (x) and final examination scores (y).

x	y	x	y	x	y	x	y
4	36	4	25	3.5	25	7.5	41
3.5	19	4	19	3.5	22	6.5	44
6	38	3	24	3	22	7.5	35
7	52	3	9	2.5	6	5	32
4	20	2	7	2	5	2	3
3.5	12	2	2	3	29	2.5	12
2	10	3.5	34	2.5	26	4.5	16
8	53	3	23	2	17	4	27
3	16	3.5	26	5	41	3.5	17
3	26	6	33	4	24	5.5	25
4.5	27	3.5	17	2.5	7	3.5	17
3	8	4	18	5	19	2.5	16
3	24	2.5	13	2	9	8	40
2.5	23	2.5	10	4.5	28	7.5	38
5.5	32	5	27	6.5	28	6	27
6.5	37	7.5	42	6	34	3.5	23
8	40	8	48	4	18		
4	19	3	26	3.5	23		
3.5	22	2.5	10	3.5	8		
4	35	2	22	6	46		
2.5	18	2	16	6.5	32		
4	25	4	30	7	37		
4	12	4	3	2	14		
2.5	18	6	20	5.5	25		
3.5	18	7	46	4	21		

The mean of the Y values is

$$\frac{\sum_i y_i}{N} = \frac{2169}{91} = 23.84$$

and that of the X values is

$$\frac{\sum_i x_i}{N} = \frac{381.5}{91} = 4.19.$$

The two standard deviations are

$$S_X = \sqrt{\frac{\sum_i x_i^2}{N} - M_X^2} = \sqrt{\frac{1874.75}{91} - 17.56} = 1.74$$

$$S_Y = \sqrt{\frac{\sum_i y_i^2}{N} - M_Y^2} = \sqrt{\frac{64043}{91} - 568.34} = 11.64.$$

The correlation coefficient will be computed next:

$$\frac{\sum_i x_i y_i}{N} = \frac{(4)(36) + (3.5)(19) + \cdots + (3.5)(23)}{91} = 116.26$$

so that

$$\begin{aligned} r_{xy} &= \frac{(\sum_i x_i y_i/N) - M_X M_Y}{S_X S_Y} \\ &= \frac{116.26 - (4.19)(23.84)}{(1.74)(11.64)} \\ &= .81. \end{aligned}$$

Thus, the regression equation for predicting z_{Yi} from z_{Xi} for these data is

$$z'_{Yi} = .81 z_{Xi}.$$

The raw score regression coefficient $b_{Y \cdot X}$ is

$$\begin{aligned} b_{Y \cdot X} &= r_{xy} \frac{S_Y}{S_X} \\ &= (.81) \frac{11.64}{1.74} \\ &= 5.42. \end{aligned}$$

Using this regression coefficient we find that the raw score regression equation for predicting Y from X is

$$y'_i = (5.42)(x_i - 4.19) + 23.84.$$

For instance, given that an individual has a high school mathematics score of 5, the teacher can predict that the score on the course examination is

$$\begin{aligned} y'_i &= (5.42)(5 - 4.19) + 23.84 \\ &= 28.23. \end{aligned}$$

Figure 13.9.1 shows a plot of the regression equation, together with the actual (x, y) pairs for these data. Notice that although the actual pairs of scores do tend to cluster about the predicted (x, y') pairs, there is nevertheless some "scatter" of the actual Y scores about the predicted value for each X. This, of course, is reflected in the fact that the obtained r_{xy} is not 1.00.

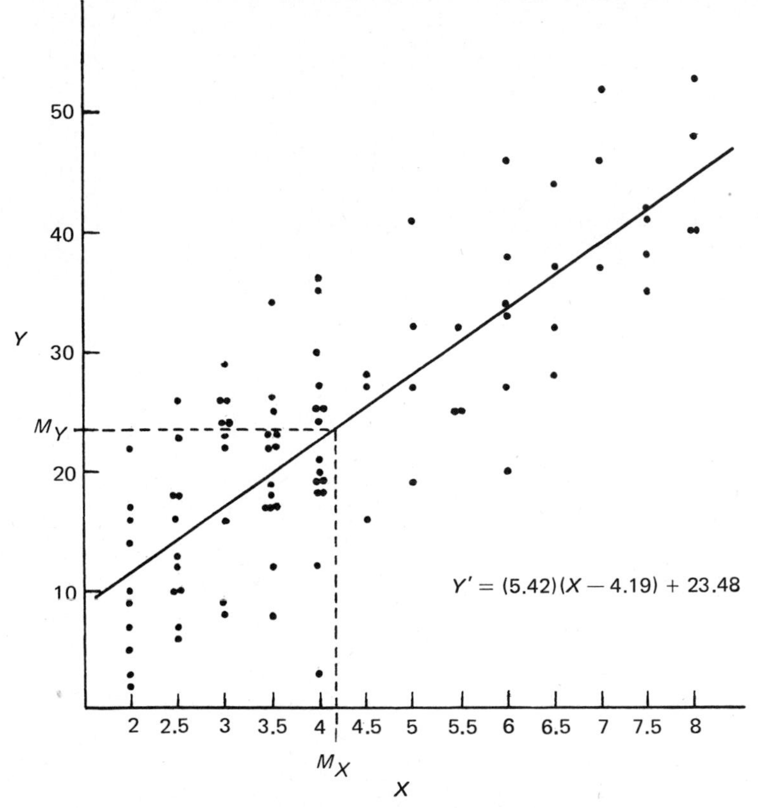

Figure 13.9.1

Plot of the data in Table 13.9.1, showing the regression line for predicting Y from X.

For these data, the sample standard error of estimate is

$$
\begin{aligned}
S_{Y \cdot X} &= S_Y\sqrt{1 - r_{xy}^2} \\
&= (11.64)(.586) \\
&= 6.82.
\end{aligned}
$$

13.10 / ASSUMPTIONS MADE IN COMPUTING CORRELATION AND REGRESSION COEFFICIENTS FOR SAMPLE DATA

A few comments are in order about the propriety of computing correlations and regression equations for sample data. In some of the older literature in social and behavioral research several misleading ideas appear about when it is proper to compute these indices and equations as *descriptive* statistics. We need to be clear about these matters. It is *not* necessary to make any assumptions at all about the form of the distribution, the variability of Y scores within X columns or "arrays," or the true level of measurement represented by the scores in order to employ linear regression and correlation indices to describe a given set of data. So long as there are N distinct cases, each having two numerical scores, X and Y, then the *descriptive* statistics of correlation and regression may be used. In so doing, we describe the data *as though* a linear rule were to be used for prediction, and this is a perfectly adequate way to talk about the tendency for *these* numerical scores to associate or "go together" in a linear way *in these data*.

The confusion has arisen because in inference about true linear relationships in populations, and in some applications of regression equations to predictions beyond the sample, assumptions do become necessary, as we shall see presently. However, one may apply correlation techniques to any set of paired-score data, and the results are valid descriptions of two things: the particular linear rule that best applies, and the goodness of the linear prediction rule as a summarization of the tendency of Y scores to differ systematically with differences in X *in these data*.

It is true, however, that the possible values that r_{xy} may assume depend to some extent upon the forms of *marginal* distributions of both X and Y in the joint data table. Unless the distributions for X and Y are similar in form, it is not necessarily true that the obtained value of r_{xy} can range between -1 and $+1$. In fact, it is possible to produce examples where the forms of the distributions of X and Y are very different, and the maximum possible absolute value of r_{xy} is only .3 or less. The fact that the value of the correlation coefficient can, in principle, range from -1.00 to $+1.00$ does not imply that the opportunity for a linear predictive relation to appear in a sample has nothing to do with the marginal distributions of the X and Y scores. In the same way, the actual possible range of the correlation in a population depends on the marginal distributions. This fact has very important implications for those who study the patterns of correlations in multivariate studies, and particularly for those who must employ some variant on, or approximation to, the correlation coefficient in such studies. An informative discussion of these issues is given in Carroll (1961).

A related problem with the value of r_{xy} in a sample has to do with selection of cases to appear in the sample, and in particular with systematic restrictions on the range of X (or of Y) values that appear. This introduces a bias into the obtained value of r_{xy} as an estimate of the actual correlation had this selection not been exercised. In particular, the absolute value of the correlation coefficient tends to be lowered by the introduction of such systematic selection reducing the possible range of one or both variables. These matters are also discussed by Carroll.

There is, incidentally, a large literature devoted to special-purpose correlations of various kinds. Among these are the *point-biserial correlation,* designed for the situation in which there are only two independent variable or X values, each accompanied by an array of Y values. Another historically important formulation of the correlation coefficient is the *tetrachoric correlation,* computed when both X and Y are arranged into widespread classes. Still another, the "phi" coefficient, is mentioned in Chapter 15. In addition, various corrections to the correlation coefficient may be made, in order to allow for various kinds of groupings or curtailments of the X and Y values. We will not devote further space to these matters here. A modern summary of some of these ideas may be found in Guilford and Fruchter (1978), however.

13.11 / POPULATION CORRELATION AND REGRESSION

Imagine a population where each distinct elementary event qualifies for one and only one joint (x,y) event, where X and Y are random variables having some known bivariate distribution. (The general features of bivariate distributions are outlined in Appendix C.) That is, each individual i is assumed to be associated with a pair of values, (x_i,y_i), and, under random sampling of individuals, there is a probability density to be assigned to each conceivable (x_i,y_i) pair.

Now using the population rather than the sample as a reference group, the linear model for a correlation problem is still

$$y_i = a + bx_i + e_i.$$

However, for the population, if we apply the same least-squares argument as for any other group, we find that the required value of the constant a for predicting raw values is

$$a = \mu_Y - \beta_{Y \cdot X}\, \mu_X \qquad [13.11.1]$$

For the raw score regression coefficient b we have, for the population,

$$\beta_{Y \cdot X} = \rho_{XY} \frac{\sigma_Y}{\sigma_X}. \qquad [13.11.2^*]$$

The population correlation coefficient ρ_{XY} (lowercase Greek "rho") is defined by

$$\rho_{XY} = \frac{\text{cov.}(X,Y)}{\sigma_X \sigma_Y}, \qquad [13.11.3]$$

where, as defined in Section B.3,

$$E[(X - \mu_X)(Y - \mu_Y)] = \text{cov.}(X,Y).$$

That is, the population correlation coefficient is the covariance of X and Y, divided by the product of the population standard deviations for X and Y. (c.f. Section B.3)

Putting these facts together, we obtain the raw score prediction or regression equation for the population as follows:

$$y_i' = \mu_Y + \beta_{Y \cdot X}\,(x_i - \mu_X). \qquad [13.11.4^*]$$

For obvious reasons one is usually more interested in the regression equation as it applies to a population than in the corresponding equation for a sample. Thus, the most useful feature of such an equation is the ability it provides for predicting the y' score of any individual, regardless of whether that individual was a member of some sample. In the same way, in order to make a general statement about the linear relationship that exists between X and Y, we would like to have the value of ρ_{XY} rather than the value of some sample r_{xy}. On the other hand, the sample usually provides all of the information we have about the population. Fortunately, we can make inferences about these population values from the sample, especially if the population has a particular form, as discussed below.

For any such population, we can discuss the *true* variance of estimate for Y predicted from X,

$$\sigma^2_{Y \cdot X} = \sigma^2_Y(1 - \rho^2_{XY}), \qquad [13.11.5^*]$$

as well as the true variance of estimate for X predicted from Y

$$\sigma^2_{X \cdot Y} = \sigma^2_X(1 - \rho^2_{XY}). \qquad [13.11.6^*]$$

The two **true standard errors of estimate** are thus

$$\sigma_{Y \cdot X} = \sigma_Y \sqrt{1 - \rho^2_{XY}} \qquad [13.11.7^*]$$

and

$$\sigma_{X \cdot Y} = \sigma_X \sqrt{1 - \rho^2_{XY}}. \qquad [13.11.8^*]$$

Notice that just as for a sample, one can interpret the square of the correlation coefficient as **the proportion of variance accounted for by linear regression.** That is, $\sigma^2_{Y \cdot X}$ is ''error'' variance in the use of the linear rule, the variability *not* accounted for by linear regression. Thus,

$$\sigma^2_Y - \sigma^2_{Y \cdot X}$$

is *the true variance accounted for by linear regression,* or the reduction in variance accomplished by using a linear prediction rule. It follows that

$$\rho^2_{XY} = \frac{\sigma^2_Y - \sigma^2_{Y \cdot X}}{\sigma^2_Y}, \qquad [13.11.9^*]$$

the square of the population correlation coefficient is *the true proportion of variance accounted for* by the use of a linear prediction rule.

Given a random sample, the value of the sample regression coefficient $b_{Y \cdot X}$ is our best available estimate of $\beta_{Y \cdot X}$, the population regression coefficient. Moreover, the best estimate of $\beta_{Y \cdot X}(\mu_X)$ is given by $b_{Y \cdot X}(M_X)$. As usual, our best estimate of μ_Y is simply M_Y.

Since each of these estimates corresponds to a term in the sample regression equation, we can use the sample equation itself as our best estimate of the population regression equation; that is, our estimate of the population regression equation is given by

$$y'_i = b_{Y \cdot X}(x_i - M_X) + M_Y.$$

In the example of Section 13.9.1 the teacher *is* justified in using the sample regression equation to predict for new students drawn from the same population as the original class, *provided* that the class used to find this regression equation is actually a random sample from a population of such students.

In addition, an unbiased estimate of the *true* variance of estimate for predicting Y from X is given by

$$\text{est. } \sigma^2_{Y \cdot X} = \frac{N}{N - 2} S^2_{Y \cdot X}. \qquad [13.11.10^*]$$

13.12 / CORRELATION IN BIVARIATE AND MULTIVARIATE NORMAL POPULATIONS

In a problem in correlation our main interest often focuses on the value of r_{xy} itself, especially as an estimate of ρ_{XY} for the population. Given a particular assumption about the population distribution of joint (x,y) events we can test not only if X and Y are linearly related, but also if any systematic relationship at all exists between the two variables.

As shown in Appendix C, in a joint distribution of discrete random variables a probability is associated with each possible X and Y pair, (x,y). A similar conception holds when X and Y are continuous variables, and a probability is associated with any joint interval of values. Such joint distributions of two random variables are known as bivariate distributions. Although any number of theoretical bivariate distributions are possible in principle, by far the most studied is the **bivariate normal distribution.** The density function for this joint distribution has a rather elaborate-looking rule. However, for standardized variables, this can be condensed to

$$f(z_X, z_Y) = \frac{e^{-G}}{K}$$

where

$$G = \frac{(z_X^2 + z_Y^2 - 2\rho_{XY} z_X z_Y)}{2(1 - \rho_{XY}^2)}$$

and

$$K = 2\pi\sqrt{(1 - \rho_{XY}^2)}.$$

Notice that in a bivariate normal distribution, the population correlation coefficient ρ_{XY} appears as a parameter in the rule for the density function. Thus, even though z_X and z_Y are both standardized variables, the particular bivariate distribution cannot be specified unless the value of the correlation ρ_{XY} is known.

In a bivariate normal distribution, the marginal distribution of X over all observations is itself a normal distribution, and the marginal distribution of Y is also normal. Furthermore, given any X value, the *conditional* distribution of Y is normal; given any Y, the conditional distribution of X is normal. In other words, if a bivariate normal distribution is conceived in terms of a table of joint events where the number of possible X values, of Y values, and of possible joint (x,y) events is infinite, then within any possible row of the table one would find a normal distribution; a normal distribution also exists within any possible column, and the marginals of the table also exhibit normal distributions.

For our purposes, however, the feature of most importance in a bivariate normal distribution is this: **given that densities for joint (x,y) events follow the bivariate normal rule, then X and Y are independent if and only if** $\rho_{XY} = 0$. For *any* joint distribution of (x,y) the independence of X and Y implies that $\rho_{XY} = 0$, but it may happen that $\rho_{XY} = 0$ even though X and Y are *not* independent. However, *for this special joint distribution, the bivariate normal, $\rho_{XY} = 0$ both implies and is implied by the statistical independence of X and Y. The only predictability possible in a bivariate normal distribution is that based on a linear rule.*

On the other hand, just because the distribution of X and the distribution of Y both happen to be normal, when considered as marginal distributions, this does *not* necessarily mean that the joint distribution of (x,y) values is bivariate normal. Hence, it is entirely possible for a nonlinear statistical relation to exist even though both X and Y are normally distributed when considered separately. It is not, however, possible for any but a linear relation to exist when X and Y jointly follow the bivariate normal law.

Most of the classical theory of inference about correlation and regression was developed in terms of the bivariate normal distribution. **If one can assume such a joint population distribution, inferences about correlation are equivalent to inferences about independence or dependence between two random variables.** For the kinds of problems here called **correlation problems,** the assumption of a bivariate normal distribution is usually made. When this assumption is valid, any inference about the value of ρ_{XY} is equivalent to an inference about the independence, or degree of dependence between two variables; this is not, however, a feature of regression problems where the bivariate normal assumption need not be made. As always, by adopting more stringent assumptions about the form of the population distribution, one is able to make much more positive statements from sample results.

The generalization of the idea of a bivariate normal distribution to more than two variables is *the multivariate normal distribution*. Here, too, the parameters of the multivariate normal distribution are the mean and variance of each variable, as well as the correlation between each pair of variables. The conditional distribution of any variable holding the others constant is normal, the conditional distribution of any pair of variables holding the others constant is bivariate normal, and so on. As in the case of the bivariate distribution, the only predictability possible in a multivariate normal distribution is linear; $\rho_{XY} = 0$ for any pair of variables X and Y implies and is implied by the independence of X and Y. Further discussion of multivariate normal distributions will be found in any text on multivariate statistics, such as that by Timm (1975).

13.13 / TESTS IN CORRELATION PROBLEMS

For many correlation problems the only hypothesis of interest is

$$H_0: \rho_{XY} = 0.$$

where the alternative hypothesis may be either directional or nondirectional.

When this hypothesis is true and the population can be assumed to be bivariate normal in form, the distribution of the sample correlation coefficient tends, rather slowly, toward a normal distribution for increasing N. The standard error of this distribution of sample correlations r_{xy} is approximately

$$\sigma_r = 1/\sqrt{N} \qquad [13.13.1]$$

When sample size is reasonably large (say, $N \geq 50$) then it is possible to test the significance of r_{xy} in this way, by forming the usual z-statistic and referring it to the normal distribution.

On the other hand, it can be shown that for any sample size N of 3 or more a test of this hypothesis in a bivariate normal population is given by the t ratio:

$$t = \frac{r_{xy}\sqrt{N - 2}}{\sqrt{1 - r_{xy}^2}} \qquad [13.13.2^*]$$

with $N - 2$ degrees of freedom.

Under the assumptions made in a problem in correlation, the value of r_{xy} may be used directly as an estimator of ρ_{XY} for the population. Although it is a sufficient and consistent estimator for ρ_{XY}, the sample correlation is slightly biased; however, the amount of bias involves terms of the order of $1/N$, and for most practical purposes can be ignored.

As mentioned earlier, for very large samples the distribution of the sample correlation coefficient may be regarded as approximately normal when $\rho_{XY} = 0$. Even for relatively small samples ($N > 4$) this sampling distribution is unimodal and symmetric. However, when ρ_{XY} is other than zero, the distribution of r_{xy} tends to be very skewed. The more that ρ differs from zero, the greater is the skewness. When ρ_{XY} is greater than zero, the skewness tends to be toward the left, with intervals of high values of r_{xy} relatively more probable than similar intervals of negative values. When ρ_{XY} is negative, this situation is just reversed, and the distribution is skewed in the opposite direction. The fact that the particular form of the sampling distribution depends upon the value of ρ_{XY} makes it impossible to use the t test for other hypotheses about the value of the population correlation, or to set up confidence intervals for this value in some direct elementary way. Although the sampling distribution of r_{xy} for $\rho_{XY} \neq 0$ has been fairly extensively tabled, it is much simpler to employ the following method.

R. A. Fisher showed that tests of hypotheses about ρ_{XY}, as well as confidence intervals, can be made from moderately large samples (about $N \geqslant 10$) from a bivariate normal population if one uses a particular *function* of r_{xy}, rather than the sample correlation coefficient itself. The function used is known as the Fisher r to Z transformation, given by the rule

$$Z = \frac{1}{2}\log_e\left(\frac{1 + r_{xy}}{1 - r_{xy}}\right). \qquad [13.13.3]$$

Fisher showed that for virtually any value of ρ_{XY}, for samples of moderate size the sampling distribution of Z-values is *approximately normal,* with an expectation given approximately by

$$E(Z) = \zeta = \frac{1}{2}\log_e\left(\frac{1 + \rho_{XY}}{1 - \rho_{XY}}\right). \qquad [13.13.4]$$

(The population value of Z, corresponding to ρ, is denoted by ζ, small Greek zeta.) The sampling variance of Z is approximately

$$\text{var}(Z) = \frac{1}{N - 3}. \qquad [13.13.5]$$

The goodness of these approximations increases the *smaller* the absolute value of ρ_{XY} and the *larger* the sample size. For moderately large samples the hypothesis that ρ_{XY} is equal to any value ρ_0 (not too close to 1 or -1) can be tested. This is done in terms of the test statistic

$$\frac{Z - \zeta}{\sqrt{1/(N - 3)}} \qquad [13.13.6^*]$$

referred to a *normal* distribution. The value taken for $E(Z)$ or ζ depends on the value given for ρ_0 by the null hypothesis:

$$\zeta = E(Z) = \frac{1}{2} \log_e \left(\frac{1 + \rho_0}{1 - \rho_0}\right) \qquad [13.13.7^*]$$

and the sample value of Z is taken from the sample correlation,

$$Z = \frac{1}{2} \log_e \left(\frac{1 + r_{xy}}{1 - r_{xy}}\right). \qquad [13.13.8^*]$$

It should be emphasized that the use of this r to Z transformation *does* require the assumption that the (x,y) events have a bivariate normal distribution in the population. On the surface, this assumption seems to be a very stringent one, which may not be reasonable in some situations, though there is some evidence that the assumption may be relatively innocuous in others. However, the consequences of this assumption's not being met seem largely to be unknown. Perhaps the safest course is to require rather larger samples in uses of this test when the assumption of a bivariate normal population is very questionable.

Table VI in Appendix D gives the Z-values corresponding to various values of r. This table is quite easy to use, and makes carrying out the test itself extremely simple. Only positive r and Z values are shown, since if r is negative, the sign of the Z-value is taken as negative also.

For example, suppose that we wanted to test the hypothesis that $\rho_{XY} = .50$ in some bivariate normal population. A sample of 100 cases drawn at random gives a correlation r_{xy} of .35. The hypothesis is to be tested with $\alpha = .05$, two-tailed.

Then, from the Table VI, we find that for $r_{xy} = .35$,

$$Z = .3654.$$

For $\rho_{xy} = .50$, we find

$$\zeta = E(Z) = .5493.$$

The test statistic is then

$$\frac{.3654 - .5493}{\sqrt{1/97}} = -1.81.$$

In a normal sampling distribution, a standard score of 1.96 in absolute value is required for rejecting the hypothesis at the .05 level, two-tailed. Thus, we do not reject the hypothesis that $\rho_{xy} = .50$ on the basis of this evidence. Observe that the test made in terms of Z leads to an inference in terms of ρ.

Occasionally one has two *independent* samples of N_1 and N_2 cases respectively, where each is regarded as drawn from a bivariate normal distribution, and computes a correlation coefficient for each. The question to be asked is, "Do both of these correlation coefficients represent populations having the *same* true value of ρ_{xy}?" Then a test of the hypothesis that the two populations show equal correlation is provided by the ratio

$$\frac{Z_1 - Z_2}{\sigma_{(Z_1 - Z_2)}} \qquad [13.13.9]$$

where Z_1 represents the transformed value of the correlation coefficient for the first sample, Z_2 the transformed value for the second, and

$$\sigma_{(Z_1 - Z_2)} = \sqrt{\frac{1}{N_1 - 3} + \frac{1}{N_2 - 3}}. \qquad [13.13.10]$$

For reasonably large samples (say, 10 in each) this ratio can be referred to the normal distribution. Remember, however, that the two samples must be independent (in particular, not involving the same or matched subjects) and the population represented by each must be bivariate normal in form.

More generally, suppose that there are J independent samples, each drawn from a bivariate normal distribution of (x,y) pairs. Each sample j yields a sample correlation r_j between X and Y. Then the hypothesis that the true value ρ_{XY} is the same for all of the populations can be tested by the statistic

$$V = \sum_j (n_j - 3)(Z_j - U)^2 \qquad [13.13.11]$$

which is distributed as chi-square with $J - 1$ degrees of freedom when the null hypothesis that $\rho_1 = \rho_2 = \cdots = \rho_J$ is true. Here, n_j is the number of observations in the sample j, and

$$U = \frac{\sum_j (n_j - 3)Z_j}{\sum_j (n_j - 3)}. \qquad [13.13.12]$$

13.14 / CONFIDENCE INTERVALS FOR ρ_{XY}

If the population has a bivariate normal distribution of (x,y) events, then the r to Z transformation can be used to find confidence intervals, very much as for a mean of a large sample. It is approximately true that for random samples of size N, an interval such as

$$Z - z_{(\alpha/2)}\sqrt{\frac{1}{N - 3}} \leq \zeta \leq Z + z_{(\alpha/2)}\sqrt{\frac{1}{N - 3}} \qquad [13.14.1^*]$$

will cover the true value of ζ with probability $1 - \alpha$. Here, Z is the sample value corresponding to r_{xy}, ζ is the Z-value corresponding to ρ_{XY}, and $z_{(\alpha/2)}$ (*definitely* to be distinguished from Z) is the value cutting of the upper $\alpha/2$ proportion in a normal distribution. Thus, the expression 13.14.1 above gives the $100(1 - \alpha)$ percent confidence interval for ζ. On changing the limiting values of Z back into correlation values, we have a confidence interval for ρ_{XY}.

In the example given in the preceding section,

$$Z = .3654$$

$$N = 100,$$

so that

$$\sqrt{\frac{1}{N-3}} = .1.$$

For $\alpha = .05$, $z_{(\alpha/2)} = 1.96$, so that the 95 percent confidence interval for ζ is given approximately by

$$.3654 - (1.96)(.1) \leq \zeta \leq .3654 + (1.96)(.1)$$

or

$$.1694 \leq \zeta \leq .5614.$$

The corresponding interval for ρ_{XY} is then approximately

$$.168 \leq \rho_{XY} \leq .510$$

(the correlation values here are taken to correspond to the nearest tabled Z values). We can assert that the probability is about .95 that sample intervals such as this cover the true value of ρ_{XY}.

13.15 / ANOTHER EXAMPLE OF A CORRELATION PROBLEM

A study was made of the tendency of the height of a wife to be linearly related to that of her husband, and it was desired to find a sample correlation between husband and wife's heights, and to use this to test the hypothesis of no linear relationship.

A sample of 15 American *couples* was drawn at random, and the data are shown in Table 13.15.1

Table 13.15.1

Heights for a sample of American married couples.

	Heights in inches	
Couple	*X(Wife's height)*	*Y(Husband's height)*
1	70	75
2	67	72
3	70	75
4	71	76
5	67	70
6	64	68
7	71	72
8	63	67
9	65	67
10	65	68
11	65	68
12	65	71
13	66	68
14	65	71
15	61	62

Table 13.15.2

Transformed data from Table 13.15.1.

Couple	Transformed scores				
	u	v	uv	u^2	v^2
1	10	5	50	100	25
2	7	2	14	49	4
3	10	5	50	100	25
4	11	6	66	121	36
5	7	0	0	49	0
6	4	−2	−8	16	4
7	11	2	22	121	4
8	3	−3	−9	9	9
9	5	−3	−15	25	9
10	4	−2	−8	16	4
11	5	−2	−10	25	4
12	5	1	5	25	1
13	6	−2	−12	36	4
14	5	1	5	25	1
15	1	−8	−8	1	64
	94	0	142	718	194

For these data the computations for the correlation coefficient can be simplified by subtracting 60 from the height of each wife, and 70 from the height of each husband; this does not alter the value of r_{xy} obtained (Section 13.3). Then the new scores are as shown in Table 13.15.2 for u and v, respectively.

The correlation coefficient computed by the formula 13.8.2 turns out to be

$$r_{xy} = \frac{(15)(142) - (94)(0)}{\sqrt{[(15)(718) - (94)^2][(15)(194) - (0)^2]}}$$

$$= \frac{2130}{\sqrt{(1934)(2910)}}$$

$$= .90.$$

On the evidence of the sample, we conclude that there is a very strong linear relation between the heights of wives and husbands.

If we wished only to test the hypothesis that the true correlation is zero, we would employ the t test given by 13.13.2.

$$t = \frac{(.90)\sqrt{15 - 2}}{\sqrt{1 - (.90)^2}}$$

$$= \frac{3.24}{.44}$$

$$= 7.36.$$

This greatly exceeds both the values required for $\alpha = .05$ and for $\alpha = .01$ for a t with 13 degrees of freedom (two-tailed).

Suppose, however, that the question had been, "Does the height of the wife and the linear relation account for more than 50 percent of the variance in the observed heights of husbands?'' That is, we actually want to test the hypothesis

$$H_0\colon \rho^2_{XY} \leqslant .50$$

against the alternative

$$H_1\colon \rho^2_{XY} > .50.$$

Given the assumption that ρ is positive, this is equivalent to the test of

$$H_0\colon \rho_{XY} \leqslant \sqrt{.50} \text{ or } H_0\colon \rho_{XY} \leqslant .707$$

against

$$H_1\colon \rho_{XY} > .707.$$

Here, the Fisher r to Z transformation and expressions 13.13.6 through 13.13.8 are used to find the test statistic

$$\frac{Z - E(Z)}{\sqrt{1/(N - 3)}} = \frac{1.47 - .88}{\sqrt{1/12}}$$

$$= 2.04.$$

In a normal sampling distribution, this exceeds the value required for the 5 percent significance level, one-tailed. Thus we may safely conclude from this sample that more than 50 percent of the variance in Y is accounted for by the apparent linear relation with Y.

This example is made up, of course, and correlations this large are not usually found in social or behavioral work. It does illustrate one thing, however. Even though the correlation found is sizable, it makes no sense at all to think of the height of the wife as ''causing'' the height of the husband, or that of the husband the height of the wife. These are simply two numerical measurements that happen to occur together in a more or less linear way, according to the evidence of this sample. The reason *why* this linear relation exists is completely out of the realm of statistics, and the correlation coefficient and tests shed absolutely no light on this problem. In this example, it is perfectly obvious that personal preferences and current standards of society cause *some* selection to occur in the process of mating, and these factors in turn underlie our observations that (x,y) pairs do occur in a particular kind of relationship. As a description of a population situation, our inferences may very well be valid, but this fact alone gives us no license to talk about the cause of the apparent linear relation.

13.16 / OTHER INTERVAL ESTIMATES IN BIVARIATE NORMAL CORRELATION PROBLEMS

Under the bivariate normal correlation model, it is quite possible to form confidence intervals for $\beta_{Y\cdot X}$, the true regression coefficient. The $100(1 - \alpha)$ percent confidence interval is found from

$$b_{Y\cdot X} - \frac{\text{est. } \sigma_{Y\cdot X} t_{(\alpha/2)}}{S_X\sqrt{N}} \leqslant \beta_{Y\cdot X} \leqslant b_{Y\cdot X} + \frac{\text{est. } \sigma_{Y\cdot X} t_{(\alpha/2)}}{S_X\sqrt{N}} \qquad [13.16.1^*]$$

where

$$\text{est. } \sigma_{Y\cdot X} = \sqrt{\frac{NS_Y^2 - Nb_{Y\cdot X}^2 S_X^2}{N-2}} = \sqrt{\frac{NS_Y^2(1 - r_{xy}^2)}{N-2}}, \qquad [13.16.2]$$

The $t_{(\alpha/2)}$ value is found for $N - 2$ degrees of freedom.

Occasionally, interval estimates are desired for the predicted values y' using the *population* regression rule, and some specific x value. Remember that the predicted value y' found using a *sample* regression equation does not necessarily agree with the y' that would be found using the population regression equation. There are *two* possible sources of disagreement between a sample y' and the true value of y' as found from the population rule: the sample mean M_Y may be in error, and the sample estimate of $\beta_{Y\cdot X}$ may be wrong to some extent. Considering both of these sources of error, we have for a given score x the following confidence interval for *predicted* y':

$$y' - t_{(\alpha/2)} \text{ est. } \sigma_{Y\cdot X}\sqrt{\frac{1}{N} + \frac{(x - M_X)^2}{NS_X^2}} \leq \text{true } y' \leq y' + t_{(\alpha/2)} \text{ est. } \sigma_{Y\cdot X}\sqrt{\frac{1}{N} + \frac{(x - M_X)^2}{NS_X^2}}. \qquad [13.16.2^*]$$

The number of degrees of freedom is again $N - 2$.

In other words, here, where $y' = M_Y + b_{Y\cdot X}(x - M_X)$, and where true

$$y' = \mu_Y + \beta_{Y\cdot X}(x - \mu_X),$$

there must be two kinds of variability in the distribution of Y' values over samples, given a constant value for $x - \mu_X$. The first source of variability is the difference between a given value of the mean M_Y and the true mean μ_Y. The second source of variability is the difference between the sample regression coefficient $b_{Y\cdot X}$ and the true coefficient $\beta_{Y\cdot X}$. The two terms under the radical sign in [13.16.2] reflect these two sources of variability.

Be sure to notice the interesting fact that **the regression equation found for a sample is not equally good as an approximation to the population rule over all the different values of** X. The sample rule is at its best as a substitute for the population rule when $X = M_X$, the mean of the X values, since the confidence interval is smallest at this point. However, as X values grow increasingly deviant from M_X in either direction, the confidence intervals grow wider.

13.17 / PARTIAL AND PART CORRELATIONS

We now turn to a group of methods which extend the basic ideas of bivariate correlation and regression to any situation where more than two variables are involved. Certainly, in practical situations it is seldom that only two items of information about an individual will be known. Thus, in the employment setting several ratings of background, education and experience may be used, and several preemployment tests may be administered. Each of these things will be involved to some extent in the decision to hire or not to hire. The admission officer in a university may have not only college entrance examination scores, but also other information such as the high school record, the applicant's employment record, and the educational level of the parents, each of which may be useful in judging the potential for academic success shown by an applicant.

However, measures such as these are not only related to something that must be predicted, such as job performance or success in college; they are also related to each other. The extent to which measures are correlated with each other is an index of the overlap or redundancy of the information that they provide. It may well be that the reason that college entrance scores appear to predict success in university work is because such scores supply some of the same information that is given by another excellent predictor, high school record. The question then might well be asked, "Do college entrance examination scores still predict success in university work if we look only at students with the *same* high school record?" Even in a purely research setting, where a number of variables may be investigated at the same time, one often needs to trace the extent to which one variable or set of variables accounts for or "explains" the variability in others. The question often arises, "If I were able to hold one or more of these variables constant, what would happen to the relationships I observe among the others?"

These are the sorts of questions that the coefficient of partial correlation, to be discussed in this section, was designed to answer. *The coefficient of partial correlation between two variables is an index of the linear relationship that would still exist between these variables if all linear influences of one or more other variables could be removed.*

Thus, consider three variables, X_1, X_2, and X_3. Each individual i in a sample of N cases will have a score on each variable, (x_{1i}, x_{2i}, x_{3i}). A correlation coefficient is calculated between each pair of variables, so that we have (r_{12}, r_{13}, r_{23}). Now suppose that somehow x_{3i}, the value of X_{3i}, could have been made constant for each individual i. What would the correlation between X_1 and X_2 then have been? This is the partial correlation between X_1 and X_2 holding X_3 constant, as symbolized by $r_{12 \cdot 3}$. This partial correlation is found from

$$r_{12\cdot 3} = \frac{r_{12} - r_{13}r_{23}}{\sqrt{(1 - r_{13}^2)(1 - r_{23}^2)}}. \qquad [13.17.1^*]$$

In other words, if we know the correlations for all three pairs of variables, it is a simple matter to find the correlation between variable X_1 and variable X_2 with the (linear) influence of X_3 ruled out, since this is simply $r_{12\cdot 3}$ as defined above.

The partial correlation $r_{12\cdot 3}$ may be thought of in this way: it is the ordinary Pearson correlation between the *errors* in predicting X_1 from X_3, and in predicting X_2 from X_3. Since neither kind of error can depend upon X_3, then the correlation between such errors indicates the remaining association between X_1 and X_2 when the influence of X_3 is removed from both.

In exactly this same way we may define the partial correlation between variables X_1 and X_3 with X_2 held constant:

$$r_{13\cdot 2} = \frac{r_{13} - r_{12}r_{23}}{\sqrt{(1 - r_{12}^2)\,(1 - r_{23}^2)}}$$

For X_2 and X_3 with X_1 held constant, we have:

$$r_{23\cdot 1} = \frac{r_{23} - r_{12}r_{13}}{\sqrt{(1 - r_{12}^2)\,(1 - r_{13}^2)}}$$

As an example of the use of a partial correlation coefficient, consider the following. Suppose that a sample of middle-aged (50- to 70-year-old) Americans has been measured on three variables: X_1 = recreational reading activity, X_2 = age, and X_3 = visual acuity. The correlations were found to be

$$r_{12} = -.10, \; r_{13} = .60, \; r_{23} = -.40.$$

Initially, the investigator was somewhat surprised by the negative correlation between reading activity and age, since it was felt that people should tend to read relatively more as they grew older. However, as expected, people with high visual acuity do tend to read more, as shown by r_{13}, and visual acuity does decline with age, as reflected in r_{23}. What would the relationship between reading activity and age be if the variability due to visual activity were removed? This is found from

$$r_{12\cdot3} = \frac{-.10 - (.60)(-.40)}{\sqrt{(1 - .36)(1 - .16)}} = .191.$$

Thus, when visual acuity is held constant, a small positive relationship does emerge between age and reading activity. Older persons with the eyesight to do so do indeed tend to read more.

A partial correlation such as that just used, in which only one variable is held constant as the remaining relation between two others is examined, is known as a *first-order partial correlation*. However, it is entirely possible to extend this idea to the situation where two or more variables are held constant. Thus, for variables, X_1, X_2, X_3, and X_4, the *second-order partial correlation* between X_1 and X_2 holding X_3 and X_4 constant is

$$r_{12\cdot34} = \frac{r_{12\cdot3} - r_{14\cdot3}r_{24\cdot3}}{\sqrt{(1 - r^2_{14\cdot3})(1 - r^2_{24\cdot3})}}. \qquad [13.17.2]$$

(The order of the partial correlation is just the number of things held constant.) Notice how the second-order partial correlation is found from first-order partials. Although such formulas may be stated for partials of any order, where any number of variables are held constant, we shall not bother with these, since the information they provide can be gained more efficiently from the methods of multiple regression and correlation, to be discussed in the next sections.

A partial correlation coefficient may be tested for significance in much the same way as a regular (zero-order) Pearson correlation. That is, a sample of N cases is assumed to come from a multivariate normal distribution. Then it is permissible to use the Fisher r to Z transformation, just as in Section 13.13, to test hypotheses and to find confidence intervals for $\rho_{12\cdot3}$, the population partial correlation. However, the standard error of Z here becomes

$$\sigma_z = \frac{1}{\sqrt{N - 3 - (K - 2)}}, \qquad [13.17.3]$$

where K is the total number of variables considered. Thus, for three variables ($K = 3$), when the partial correlation $r_{12\cdot3}$ is used to test the hypothesis that $\rho_{12\cdot3}$ is, say 0, the value of $r_{12\cdot3}$ is changed to a corresponding Z-value via Table VI; and the test is carried out using the normal distribution of the statistic

$$\frac{Z - 0}{[1/\sqrt{N - 3 - (3 - 2)}]} = Z[\sqrt{(N - 4)}]. \qquad [13.17.4]$$

An important variant of the idea of the partial correlation is called the *part correlation* (or the *semipartial correlation*). Such a part correlation has a slightly different interpretation from a partial correlation. In a partial correlation, $r_{12 \cdot 3}$, the variability attributable to the third variable, say X_3, is removed *both* from variable X_1 and variable X_2. In a part correlation, the variability associated with X_3 is removed from X_2 but is *not* removed from X_1. This part correlation, symbolized by $r_{1(2 \cdot 3)}$, then answers the question, "To what extent does the part of X_2 *not dependent on* X_3 still predict X_1?" The answer is found from

$$r_{1(2 \cdot 3)} = \frac{r_{12} - r_{13}r_{23}}{\sqrt{1 - r_{23}^2}}. \qquad [13.17.5]$$

We may also calculate the part correlation

$$r_{2(1 \cdot 3)} = \frac{r_{12} - r_{13}r_{23}}{\sqrt{1 - r_{13}^2}}$$

which shows the relation between X_2 and X_1 when the influence of X_3 is removed from X_1.

Then the partial correlation between X_1 and X_2 with X_3 held constant depends on these two part correlations as follows:

$$r_{12 \cdot 3} = \frac{r_{1(2 \cdot 3)}}{\sqrt{1 - r_{13}^2}} = \frac{r_{2(1 \cdot 3)}}{\sqrt{1 - r_{23}^2}}.$$

Notice that although there are only three possible partial correlations among three variables, $r_{12 \cdot 3}$, $r_{13 \cdot 2}$, and $r_{23 \cdot 1}$, there are six possible part correlations: the two given above plus $r_{1(3 \cdot 2)}$, $r_{3(1 \cdot 2)} r_{2(3 \cdot 1)}$, and $r_{3(2 \cdot 1)}$.

In the previous example, suppose that we look at the part correlation between age and reading activity, but removing the variability in reading activity having to do with visual acuity. That is, we want the part correlation $r_{2(1 \cdot 3)}$. We find

$$r_{2(1 \cdot 3)} = \frac{-.10 - (.6)(-.4)}{\sqrt{1 - (.6)^2}} = .175.$$

This shows that if the influence of visual acuity is removed from reading activity, then the correlation between age and reading activity rises from $-.10$ to $.175$.

As with partial correlations, the ideas of part correlation may be extended to four or more variables. However, since such extensions are also most likely to be used in connection with multiple regression problems, they will be considered in that context. Similarly, certain tests for part correlations are possible, and they too will be dealt with in a succeeding section.

13.18 / SOME BASIC IDEAS OF MULTIPLE CORRELATION AND REGRESSION

Up to this point we have dealt with the general linear model introduced in Section 10.2 only in a fairly limited way. Now, however, we are going to employ this model in a much more extended sense. The initial context will be this: a sample is drawn in which each individual i out of a total of N cases exhibits a value y_i on dependent variable Y. In addition, each individual has a value on each of K independent variables, or *predictor variables,* $X_1, X_2, \cdots, X_K$. Thus, each individual has $K + 1$ values, consisting of the set of scores $(y_i, x_{1i}, x_{2i}, \cdots, x_{Ki})$.

It is not, at this point, necessary for us to specify the exact nature of these X and Y variables. Ordinarily, we will think of these variables as capable of taking on any of the real numbers as values. However, this also covers the situation where any of the X variables, or even the dependent variable Y, can take on only the values 0 and 1, as did the indicator variables of Section 10.3.

The most common problem to which the methods of multiple correlation and regression are applied goes something like this: The variable Y stands for something one wishes to predict. We also have information on K other things, represented by the variables, X_1 through X_K. How, then, do we find a weighted combination of the values of X_1 through X_K that will predict the value of Y for any individual in an optimal way?

As a simple example, suppose that the personnel director in some industrial plant is interested in the possibility of predicting the quality of job performance by an individual in terms of that person's score in a general ability text (X_1) and the score on a trade skills test (X_2). A group of 50 individuals take both tests and then, after they have been employed for six months, each is given a score on job performance (dependent variable Y). How do we weight variable X_1 and variable X_2 in order to get the best possible prediction of performance, or Y?

The linear model we assume to underlie the Y values in this instance is

$$y_i = a + b_1x_{1i} + b_2x_{2i} + e_i. \qquad [13.18.1]$$

Then what we wish to do is to find the best possible values for a, b_1, and b_2 so that we can use a rule

$$y'_i = a + b_1x_{1i} + b_2x_{2i}, \qquad [13.18.2]$$

which is the *multiple regression equation* for predicting y_i given the x_{1i} and the x_{2i} scores for any individual i.

More generally, if we have a dependent variable Y, and K predictor variables, X_1, X_2, through X_K, then we are looking for a multiple regression equation

$$y'_i = a + b_1x_{1i} + \cdots + b_Kx_{Ki}. \qquad [13.18.3]$$

Now notice that the only difference between the actual value of y_i given by a model such as 13.18.1, and the predicted value y'_i given by the regression equation 13.18.2, is e_i, the error associated with individual i. Hence, the error in prediction for any individual is $(y_i - y'_i)$.

Equivalently, we can think of predicting the z-value for an individual on variable Y, or z_{yi}, from that individual's z-values on all the X variables, or $(z_{1i}, z_{2i}, \cdots, z_{Ki})$. Then the desired regression equation would be, in z-score form,

$$z'_{yi} = A + B_1z_{1i} + \cdots + B_Kz_{Ki}. \qquad [13.18.4^*]$$

How do we find the needed constants A and B_1 through B_K in such an equation, especially if the equation is to give us the best possible predictions? The answer is provided once again by the criterion of least squares. Thus, we will choose A and B_1 through B_K in such a way as to minimize the sum of squared errors over all cases i:

$$\text{minimize} \sum_i (z_{yi} - z'_{yi})^2.$$

Notice that this is exactly the same criterion utilized in Section 13.2 when the values of A and B were to be determined. However, in this instance, the least squares procedure does not yield the required values directly, but instead gives a set of K equations (called the *normal equations*):

$$\begin{aligned} A &= 0 \\ B_1 + r_{12}B_2 + r_{13}B_3 + \cdots + r_{1K}B_K &= r_{y1} \\ r_{21}B_1 + \quad B_2 + r_{23}B_3 + \cdots + r_{2K}B_K &= r_{y2} \\ \cdots \\ r_{K1}B_1 + r_{K2}B_2 + r_{K3}B_3 + \cdots + \quad B_K &= r_{yK} \end{aligned} \qquad [13.18.5^*]$$

Here, the value of A required turns out to be zero, just as in Section 13.2. However, the normal equations must be solved simultaneously in order to find the values of B_1 through B_K. Once these B values are known, then the regression equation 13.18.4 may be constructed and used.

The B values figuring in the normal equations are known as the *standardized partial regression coefficients;* these are the constants for a regression equation for standard scores (or z-values), given by 13.18.4. However, it is possible to convert the standardized regression equation 13.18.4 into a regression equation for raw or unstandardized scores. Thus, by simply converting each z-value into what it represents in raw score terms (i.e., $z_{yi} = [y_i - M_Y]/S_Y$), and similarly for the other variables we have the *raw score regression equation:*

$$y'_i = M_Y + b_1 (x_{1i} - M_1) + \cdots + b_K(x_{Ki} - M_K) \qquad [13.18.6^*]$$

Here, the *unstandardized partial regression coefficients* are

$$b_1 = B_1\left(\frac{S_Y}{S_1}\right), \cdots, b_K = B_K\left(\frac{S_Y}{S_K}\right), \qquad [13.18.7]$$

and M_1 is the mean of X_1, S_1 is the standard deviation of X_1, and so on.

For a large number of variables, the solution of the normal equations to find the B values can be very laborious to carry out by hand, and these days such problems are usually solved by computer. However, it is simple to demonstrate the solution to the normal equations and the format of the regression equation when there are only two predictor variables, X_1 and X_2.

Consider once again the personnel director who wishes to use two test scores, general ability, represented by X_1, and trade skills, represented by X_2, to predict job performance, represented by Y. A sample of 50 individuals yielded the following results:

Y	X_1	X_2
$M_Y = 39.6$	$M_1 = 110$	$M_2 = 72$
$S_Y = 4.0$	$S_1 = 15.0$	$S_2 = 6.7$

In addition, the following table of intercorrelations was found, where the entry in any row and column is the correlation for that pair of variables:

	Y	X_1	X_2
Y	1.00	.23	.40
X_1		1.00	.16
X_2			1.00

(Since, for any correlation, $r_{xy} = r_{yx}$, all of the table need not be shown, of course.)
For this example the normal equations of 13.8.5 are thus

$$B_1 + .16B_2 = .23$$
$$.16B_1 + B_2 = .40.$$

These two equations in two unknowns may be solved readily to yield

$$B_1 = .17 \text{ and } B_2 = .37.$$

Consequently, in order to find the z-value of any individual on variable Y, as predicted from that person's z-values on X_1 and X_2, we take

$$z'_{y_i} = .17(z_{1i}) + .37(z_{2i}).$$

On changing this regression equation into its raw score form, by taking

$$b_1 = .17(4/15) = .045 \text{ and } b_2 = .37(4/6.7) = .221$$

we have

$$y'_i = 39.6 + (.045)(x_{1i} - 110) + (.221)(x_{2i} - 72).$$

Simplifying this expression, we then have

$$y'_i = 18.74 + (.045)x_{1i} + (.221)x_{2i}.$$

Now suppose that the personnel director knows that an employee gained a score of 120 on general ability, and a score of 75 on trade skills. What should the predicted work performance score be? The answer is given by

$$y'_i = 18.74 + (.045)(120) + (.221)(75) = 40.7.$$

Although this prediction will doubtless be in error to some extent, it is, nevertheless, the best bet about the actual Y score, if we wish to bet in such a way as to minimize squared errors on the average.

Notice how much more weight the trade skills test score, X_2, carries in this prediction than does general ability, or X_1. This says something about the amount of variance in Y each explains, over and above that due to the other. We will explore such relative predictive power of variables in the next section.

It may be that the personnel director also has a third variable that might be useful in predicting job performance: X_3 = stability of previous employment, based on how long the person has held past jobs. For this variable

$$M_3 = 54, \; S_3 = 8.3, \; r_{y3} = .28, \; r_{13} = 0, \; r_{23} = .05.$$

Then, if a regression equation is to be found involving all three predictor variables, the normal equations are

$$B_1 + .16B_2 = .23$$

$$.16B_1 + B_2 + .05B_3 = .40$$

$$.05B_2 + B_3 = .28.$$

Given all three predictor variables, the B regression coefficients may be found:

$$B_1 = .17, B_2 = .36, B_3 = .26.$$

The new regression equation for predicting raw scores is then

$$y_i' = 39.6 + .17\left(\frac{4}{15}\right)(x_{1i} - 110) + .36\left(\frac{4}{6.7}\right)(x_{2i} - 72) + .26\left(\frac{4}{8.3}\right)(x_{3i} - 54)$$

$$= 12.37 + .045x_{1i} + .215x_{2i} + .125x_{3i}.$$

The personnel director has every reason to feel that the second regression equation, since it involves more information than the first, should give better predictions. However, a more formal way is needed to discuss the predictive goodness, or the variance accounted for, by a regression equation.

The *multiple correlation coefficient* may be thought of as the ordinary Pearson correlation between the predicted values y_i' and the actual values y_i:

$$R_{y \cdot 12 \ldots K} = r_{yy'}. \qquad [13.18.8]$$

Usually, however, the coefficient of multiple correlation is actually computed in terms of the standardized B values and the correlations, as follows:

$$R_{y \cdot 12 \cdots K} = \sqrt{B_1 r_{y1} + \cdots + B_K r_{yK}}\,. \qquad [13.18.9^*]$$

In terms of the example given above, with two predictors X_1 and X_2 used to predict Y, the multiple correlation is thus

$$R_{y \cdot 12} = \sqrt{.17(.23) + .37(.40)}$$

$$= \sqrt{.187}$$

$$= .432.$$

When variable X_3 is included, the multiple correlation becomes

$$R_{y \cdot 123} = \sqrt{.17(.23) + .36(.40) + .26(.28)}$$

$$= \sqrt{.256}$$

$$= .506$$

How are these multiple correlations, and especially the difference between them, to be interpreted? The most important meaning of a multiple correlation coefficient lies in its square, or R^2:

$$R^2_{y \cdot 12 \cdots K} = \frac{(\text{variance in } Y \text{ accounted for by } X_1, X_2, \cdots, X_K)}{\text{total variance of } Y} \qquad [13.18.10^*]$$

In other words, R^2 *is the proportion of variance accounted for by the set of K* *predictor variables included in the regression equation which R^2 represents.* In our example, when two predictor variables X_1 and X_2 alone are used, the proportion of variance accounted for is

$$R^2_{y \cdot 12} = .187,$$

or nearly 19 percent. However, when all three predictors are included, the proportion of variance in Y that is explained by the combination of all three is

$$R^2_{y \cdot 123} = .256,$$

or about 26 percent. This says that the second regression equation is better than the first, since it accounts for about seven percent more variance in the dependent measure.

The proportion of variance in Y that is *not* accounted for by the X variables must then be

$$1 - R^2_{y \cdot 12 \cdots K} = \frac{(\text{variance unexplained by } X_1, \cdots, X_K)}{\text{total variance of } Y}. \qquad [13.18.11^*]$$

This, then, gives a definition of the sample *variance of multiple estimate,* or the variance of the sample errors of prediction:

$$S^2_{y \cdot 12 \cdots K} = (1 - R^2_{y \cdot 12 \cdots K})(S^2_Y) = \sum_i \frac{(y_i - y_i')^2}{N}. \qquad [13.18.12^*]$$

Corresponding to the standard deviation of the distribution of error is, then, the *sample standard error of multiple estimate*

$$S_{y \cdot 12 \cdots K} = S_Y \sqrt{1 - R^2_{y \cdot 12 \cdots K}}.$$

Once more returning to our example, for the situation with two predictor variables, where $R^2_{y \cdot 12} = .187$, the sample standard error of multiple estimate is

$$S_{y \cdot 12} = (4)\sqrt{1 - .187} = 3.61.$$

We thus know that the distribution of errors in predicting an employee's performance through the multiple regression equation will have a mean of zero and a standard deviation of 3.61. However, if the three-predictor multiple regression equation is used, the distribution of errors will have a mean of zero and a standard deviation given by

$$S_{y \cdot 123} = (4)\sqrt{1 - .256} = 3.45.$$

For a population of cases, each of which has a set of values $(y_i, x_{1i}, x_{2i}, \cdots, x_{Ki})$, we may think of the true standardized regression coefficients, or B^* values, as solutions to the K normal equations, from

$$B^*_1 + \rho_{12}B^*_2 + \cdots + \rho_{1K}B^*_K = \rho_{y1} \qquad [13.18.13]$$

$$\text{to } \rho_{1K}B^*_1 + \rho_{2K}B^*_2 + \cdots + B^*_K = \rho_{yK}.$$

The corresponding unstandardized regression coefficients are then defined by

$$\beta_k = B^*_k \left(\frac{\sigma_Y}{\sigma_k}\right). \qquad [13.18.14]$$

(The "beta weights" often referred to in multiple regression studies are the B* values, however.)

The squared multiple correlation coefficient for the population may be denoted by $P^2_{y\cdot 12\cdots K}$, where

$$P^2_{y\cdot 12\cdots K} = B_1^* \rho_{y1} + \cdots + B_K^* \rho_{yK}. \qquad [13.18.15]$$

The true variance of multiple estimate is then

$$\sigma^2_{y\cdot 12\cdots K} = \sigma^2_Y (1 - P^2_{y\cdot 12\cdots K}). \qquad [13.18.16]$$

An unbiased estimate of this variance can be made from a sample, by taking

$$\text{est. } \sigma^2_{y\cdot 12\cdots K} = S^2_Y \left(\frac{N}{N - K - 1}\right) (1 - R^2_{y\cdot 12\cdots K}). \qquad [13.18.17^*]$$

The sample value of $R^2_{y\cdot 12\cdots K}$ is also slightly biased as an estimate of $P^2_{y\cdot 12\cdots K}$. One commonly used correction for this bias is to take

$$\text{est. } P^2_{y\cdot 12\cdots K} = 1 - \frac{(1 - R^2_{y\cdot 12\cdots K})(N - 1)}{(N - K - 1)} \qquad [13.18.18^*]$$

By use of this formula for correcting the bias in the squared multiple correlation, one always arrives at a slightly smaller estimated value for the population than that achieved in the sample. This bit of conservatism is appropriate, since a regression equation applied to a new sample will almost always result in a smaller proportion of variance accounted for than in the sample for which the equation was derived (see Section 13.21).

13.19 / INCREMENTS IN PREDICTIVE ABILITY THROUGH ADDITION OF MORE VARIABLES

In this section interest will focus on the changes that occur in a multiple regression situation when predictor variables are added to or subtracted from the set originally used. However, before we go into this topic, we need a little more flexibility in notation.

The standard notation we have been using, where $R^2_{y\cdot 12\cdots K}$ represents the squared multiple correlation for dependent variable Y and predictor variables X_1 through X_K, is probably satisfactory for many purposes. On the other hand, when one wishes to discuss situations where the set of predictors may change, making these distinctions using the ellipsis becomes very awkward. Therefore, we are going to adopt the following conventions, using just a little bit of the language of set theory.

Let $H = \{X_1, \cdots, X_K\}$ stand for the full set of K predictor variables which will be under discussion in a given problem. Furthermore, let $\{X_j\}$ represent a set consisting only of the variable X_j. The set $G = H - \{X_j\}$ then represents all of the variables *except* X_j. Although later we will wish to expand even this convention a bit, it will prove useful and save a lot of subscript-changing in this and the following sections. Thus

$R^2_{y\cdot H}$ = proportion of variance in Y accounted for by $\{X_1, \cdots, X_K\}$ or H

$R^2_{y\cdot G}$ = proportion of variance accounted for by G, where $H - \{X_j\} = G$.

The concepts of partial and part correlation are useful in making judgments about the increase in predictive power one gains by adding a new variable or variables to the set of predictors in a multiple regression problem. Some connections among multiple correlation, part correlation, and partial correlation will now be outlined, and we will examine how each gives information about the contributions that the various X values make to accounting for variance in Y.

Let us start off with the simplest situation, once again, with two predictor variables. Then the proportion of variance accounted for by both variables can be thought of as divisible into two parts: that which is due to X_1 alone, and that which is due to X_2 once the influence of X_1 has been removed from X_2. That is,

$$R^2_{y\cdot 12} = r^2_{y1} + r^2_{y(2\cdot 1)} \qquad [13.19.1]$$

(or, in our new notation, where $H - \{X_1\} = G$)

$$R^2_{y\cdot H} = r^2_{y1} + r^2_{y(G\cdot 1)}.$$

The first of these two parts shown in expression 13.19.1 is simply the squared correlation between Y and X_1, reflecting the proportion of variance accounted for by X_1. Next is the square of the *part correlation* for Y and X_2; here all the predictability of X_2 from X_1 has been removed. Notice that expression 13.19.1 also gives us another definition of a squared part correlation:

$$r^2_{y(2\cdot 1)} = R^2_{y\cdot 12} - r^2_{1y}. \qquad [13.19.2]$$

Thus the squared part correlation reflects the absolute increase in proportion of variance explained when variable X_2 is added to variable X_1 in a multiple regression equation for predicting Y.

We could also write

$$R^2_{y\cdot 12} = r^2_{y2} + r^2_{y(1\cdot 2)}$$

so that

$$r^2_{y(1\cdot 2)} = R^2_{y\cdot 12} - r^2_{y2},$$

the absolute increase in proportion of variance of Y accounted for by X_1 is the squared part correlation between Y and X_1 with the variance due to X_2 removed from X_1.

In general, given any complete set H of K predictors, including X_j,

$$R^2_{y\cdot H} = r^2_{y(j\cdot G)} + R^2_{y\cdot G}, \qquad H - \{X_j\} = G. \qquad [13.19.3^*]$$

Then the absolute increase in proportion of variance accounted for by the addition of X_j to a set of predictors G is the squared part correlation $r^2_{y(j\cdot G)}$:

$$r^2_{y(j\cdot G)} = R^2_{y\cdot H} - R^2_{y\cdot G}. \qquad [13.19.4^*]$$

For example, in the preceding section a multiple regression problem was worked having first two and then three predictors. What was the absolute increase in predictability afforded by the addition of X_3? The original squared multiple correlation was .187, and the squared multiple correlation was .256 with all three variables included. Thus,

$$r^2_{y(3\cdot 12)} = .256 - .187 = .069.$$

We thus also know that the square of the part correlation between job performance and stability of employment in this example is about .07.

This same idea can be turned around, of course, in order to find the absolute increase in predictability afforded by adding all of the other variables in G to an original variable X_j:

$$R^2_{y(G \cdot j)} = R^2_{y \cdot H} - r^2_{yj}, \qquad H - G = \{X_j\}. \qquad [13.19.5^*]$$

Thus, in the example, how much more predictive power do we gain by addition of variables X_1 and X_3 to the variable that already correlates best with Y, or X_2?

$$\begin{aligned} R^2_{y(13 \cdot 2)} &= R^2_{y \cdot 123} - r^2_{y2} \\ &= .258 - .160. \\ &= .098 \end{aligned}$$

The answer is that about ten percent more variance is accounted for using three variables rather than X_2 alone.

Sometimes we are interested not so much in the absolute gain in predictive ability as the *relative gain* in view of all the variance that remains to be accounted for. Thus, for two predictor variables once again, we may take

$$\begin{aligned} r^2_{y2 \cdot 1} &= \frac{R^2_{y \cdot 12} - r^2_{y1}}{1 - r^2_{y1}} \\ &= \frac{r^2_{y(2 \cdot 1)}}{1 - r^2_{y1}}. \end{aligned} \qquad [13.19.6^*]$$

In other words, once variable X_1 has been used to predict Y, there is still variance left to be accounted for. This proportion of unexplained variance is $1 - r^2_{y1}$. Then finding the additional variance accounted for by X_2, or $r^2_{y(2 \cdot 1)}$, and dividing by the amount left over from X_1, gives the relative increment due to X_2. Notice, however, that *this is also the square of the partial correlation,* $r^2_{y2 \cdot 1}$.

In general, the relative gain in proportion of variance accounted for by adding variable X_j to an existing set of variables G is given by

$$r^2_{yj \cdot G} = \frac{R^2_{y \cdot H} - R^2_{y \cdot G}}{1 - R^2_{y \cdot G}} \qquad [13.19.7^*]$$

where, again, $H - \{X_j\} = G$. In other words, when the relative increment in found, resulting from the addition of variable X_j to the existing set of $K - 1$ predictors, G, this is the same as the squared partial correlation between Y and X_j, holding all the other variables constant. The partial correlation is then of order $K - 1$.

For example, in our three-predictor problem above, the relative proportion of increase in variance accounted for in Y due to the addition of X_3 is given by

$$r^2_{y3 \cdot 12} = \frac{.256 - .187}{1 - .187} = .085.$$

The absolute value of the partial correlation $r_{y3 \cdot 12}$ is thus the square root of this value, or .29. Notice that this is a second-order partial, since two variables are held constant.

We may also ask about the relative contribution to explained variance made by the addition of the $K - 1$ variables in the set G to the single variable X_j:

$$R^2_{yG\cdot j} = \frac{R^2_{y\cdot H} - r^2_{yj}}{1 - r^2_{yj}}. \qquad [13.19.8]$$

Although this is not, strictly speaking a squared partial correlation, it does have much the same interpretation: given that variable X_j accounts for a certain amount of variance, relatively how much of what is left is accounted for by the remaining variables making up the set G?

The standardized partial regression weight for variable X_j is related to the absolute increment $R^2_{y(j\cdot G)}$ as follows:

$$B_j^2 = \frac{r^2_{y(j\cdot G)}}{1 - R^2_{j\cdot G}}. \qquad [13.19.9]$$

That is, B_j^2 is the squared part correlation between Y and X_j, with the effects of G removed from X_j, but relative to the variance left over in X_j from the set G. The value of B_j^2 may also be found from the squared partial correlation $r^2_{yj\cdot G}$:

$$B_j^2 = \frac{r^2_{yj\cdot G}\,(1 - R^2_{y\cdot G})}{(1 - R^2_{j\cdot G})}. \qquad [13.19.10]$$

Thus, for two X variables we have, from 13.19.9 and 13.19.10,

$$B_1^2 = \frac{r^2_{y(1\cdot 2)}}{1 - r^2_{12}} = \frac{r^2_{y1\cdot 2}\,(1 - r^2_{y2})}{1 - r^2_{12}}$$

$$B_1 = \frac{r_{y1} - r_{y2}r_{12}}{1 - r^2_{12}} \qquad [13.19.11]$$

and

$$B_2^2 = \frac{r^2_{y(2\cdot 1)}}{1 - r^2_{12}} = \frac{r^2_{y2\cdot 1}\,(1 - r^2_{y1})}{1 - r^2_{12}}$$

$$B_2 = \frac{r_{y2} - r_{y1}\,r_{12}}{1 - r^2_{12}}. \qquad [13.19.12]$$

An important thing to notice here is that if all the variance in X_j is accounted for by G, the value of B_j is indeterminate. Even when true $R^2_{j\cdot G}$ is quite large, estimates of the true value of B_j will tend to be very unstable. This is sometimes referred to as the problem of *multicollinearity*.

13.20 / SIGNIFICANCE TESTS FOR MULTIPLE CORRELATION AND FOR INCREMENTS IN PREDICTIVE ABILITY

The rationale underlying a significance test for multiple correlation is actually much the same as that for an F test in the analysis of variance. That is, the deviation of any value y_i from the sample mean M_Y can be broken into

$$(y_i - M_Y) = (y_i - y_i') + (y_i' - M_Y). \qquad [13.20.1]$$

Then, by an argument similar to that in analysis of variance, we have

SS total = SS deviations + SS regression

where

$$\text{SS deviations} = \sum_i (y_i - y_i')^2 \qquad [13.20.2]$$

$$\text{SS regression} = \sum_i (y_i' - M_Y)^2.$$

Dividing each SS by its degrees of freedom provides

$$\text{MS deviations} = \frac{\text{SS deviations}}{N - K - 1} \qquad [13.20.3^*]$$

$$\text{MS regression} = \frac{\text{SS regression}}{K}$$

where K is the total number of predictor variables, and N is the number of cases.

The hypothesis to be tested is

$$H_0\colon \mathrm{P}^2_{y \cdot H} = 0, \qquad H = \{X_1, \cdots, X_K\}$$

against

$$H_1\colon \mathrm{P}^2_{y \cdot H} > 0.$$

That is, the null hypothesis states that the true value of the multiple correlation coefficient in the population is zero, and the alternative states that it is greater than zero. When H_0 is true, the expectation for each mean square should be only σ^2_e. Thus, we take

$$\mathrm{F} = \frac{\text{MS regression}}{\text{MS deviations}} \qquad [13.20.4^*]$$

with K and $N - K - 1$ degrees of freedom. However, it is also true that in the sample

$$R^2_{y \cdot H} = \frac{\text{(variance accounted for)}}{(\text{total } Y \text{ variance})}$$

$$= \frac{(\text{SS regression}/N)}{(\text{SS}_Y \text{ total}/N)},$$

and that

$$1 - R^2_{y \cdot H} = \frac{\text{(variance unaccounted for)}}{(\text{total } Y \text{ variance})}$$

$$= \frac{(\text{SS deviations}/N)}{(\text{SS}_Y \text{ total}/N)}.$$

We then find that the F ratio of 13.20.4 is exactly the same as

$$F = \frac{(R^2_{y \cdot H})(N - K - 1)}{(1 - R^2_{y \cdot H})(K)} \qquad [13.20.5^*]$$

with K and $N - K - 1$ degrees of freedom.

Thus, for our two-predictor example in Section 13.18, the significance test is given by

$$F = \frac{.187(50 - 2 - 1)}{(1 - .187)(2)} = 5.41.$$

For 2 and 47 degrees of freedom this exceeds the value required for significance at the .01 level.

Among the more common tests associated with multiple regression analysis is that for the significance of a single regression coefficient, B_j. The hypothesis tested is, of course, that the corresponding value in the population, B_j^*, is truly zero. However, a one-tailed alternative hypothesis is possible, and may be of interest. Hence we will give this in its "t" form:

$$t = \frac{B_j}{\text{est. } \sigma_{B_j}} \qquad [13.20.6^*]$$

where

$$\text{est. } \sigma_{B_j} = \sqrt{\frac{1 - R^2_{y \cdot H}}{(1 - R^2_{j \cdot G})(N - K - 1)}}. \qquad [13.20.7^*]$$

This t test has $N - K - 1$ degrees of freedom.

A confidence interval may also be formed for the population value B_j^* corresponding to variable X_j. When the value of est σ_{B_j} is found from 13.20.7 above, and the population sampled is multivariate normal, then the 100 $(1 - \alpha)$ percent confidence interval is given by

$$B_j - t_{(\alpha/2)} \text{ est. } \sigma_{B_j} \leq B_j^* \leq B_j + t_{(\alpha/2)} \text{ est. } \sigma_{B_j} \qquad [13.20.8^*]$$

Here, as usual, $t_{(\alpha/2)}$ is that t value cutting off the upper $\alpha/2$ proportion of a distribution with $N - K - 1$ degrees of freedom.

Perhaps the significance tests of most utility are those that have to do with the proportions of variance explained as new variables are added to the set of predictors. Thus, a test of significance that goes with $r^2_{y(j \cdot G)}$, the absolute increment in proportion of variance accounted for by X_j when it is added to the set G, where $H - \{X_j\} = G$, is given by

$$F = \frac{r^2_{y(j \cdot G)}}{(1 - R^2_{y \cdot H})/(N - K - 1)} \qquad [13.20.9^*]$$

The degrees of freedom here are 1 and $N - K - 1$. This is, of course, also a test of significance for the part correlation $r_{y(j \cdot G)}$ (13.19.4), representing the relation between Y and X_j when all the influence of the variables in G have been removed from X_j. Since a part correlation of zero in the population implies a partial correlation of zero as well, this may also be regarded as a significance test for a partial correlation represented by $r_{yj \cdot G}$.

A very common practice in multiple regression studies is to examine the contribution to variance made by each variable X_j, holding all the rest of the variables in the total set H constant, as given by 13.20.9. However, an alternative strategy is to determine the contribution to variance made by the variables as they are added to the set in some predetermined, *hierarchical order*. That is, for convenience, suppose that the numbers appearing as subscripts stand for the order in which the variables are to be added to compose the total set H. Then the tests might start with

$$F = \frac{r_{y1}^2 (N - K - 1)}{1 - R_{y \cdot H}^2}, \qquad H = \{X_1, \cdots, X_K\}. \qquad [13.20.10^*]$$

with 1 and $N - K - 1$ degrees of freedom, which reflects the contribution of variable X_1 alone. Then the contribution of variable X_2, removing the effects of X_1, could be tested by use of 13.20.9, taking

$$F = \frac{r_{y(2 \cdot 1)}^2 (N - K - 1)}{1 - R_{y \cdot H}^2} \qquad [13.20.11^*]$$

with the same number of degrees of freedom, 1 and $N - K - 1$.

Next the contribution of X_3 holding X_1 and X_2 is tested, by using the ratio

$$F = \frac{r_{y(3 \cdot 12)}^2 (N - K - 1)}{1 - R_{y \cdot H}^2}, \qquad [13.20.12^*]$$

again with the same numbers of degrees of freedom as before. This procedure thus is carried out until all the variables to be included in the set H have been examined.

Finally, suppose that we have a total set of variables H that can be divided into two subsets, the variables in set I and the variables in set G, so that $H - I = G$. There are K variables in the total set H, and L variables in the set I. The absolute increment in proportion of variance accounted for by variables in the set I, above that accounted for in the set G, is given by

$$R_{y(I \cdot G)}^2 = R_{y \cdot H}^2 - R_{y \cdot G}^2, \qquad (H - I = G). \qquad [13.20.13]$$

A test of the hypothesis that set G accounts for all the explainable variance is then given by

$$F = \frac{R_{y(I \cdot G)}^2 (N - K - 1)}{(1 - R_{y \cdot H}^2)(L)} \qquad [13.20.14^*]$$

with L and $N - K - 1$ degrees of freedom.

Unfortunately, this brief discussion comes nowhere near exploring all of the possibilities and ramifications of multiple regression and multiple correlation analysis. Any modern text on multiple regression, such as Kerlinger and Pedhazer (1973), will shed much more light on these topics. Nevertheless, we have touched on some of the main topics, and have picked up a few concepts that will be useful in the chapter to follow.

13.21 / THE SCOPE OF MULTIPLE REGRESSION METHODS

Clearly, multiple regression analysis is the broadest application of the general linear model that we have yet encountered. However, the model itself is capable of great extension beyond the relatively simple situations that we have considered in the past few sections. Although only X variables taking on a broad range of real values were assumed in the foregoing, as we noted earlier the method is not limited to such values. It is perfectly possible to have X values taking on only the values 0 and 1, which are thus like the indicator or "dummy" variables of Section 10.3, and, with certain minor adjustments, still to employ the methods of multiple regression. This may be true even of the variable Y itself, as it is in the simplest case of *linear discriminant function analysis,* which is a way of forming a weighted combination of X values providing optimal discrimination between two (or even more) groups.

In most applications of multiple regression analysis, one starts with the intercorrelations of all of the predictor variables, as well as the correlations of each with the criterion variable Y. Then the normal equations can be set up and solved to provide the standardized B values and the regression equation. However, there are some limitations to this approach. In the first place, the number of X variables must never equal or exceed the number of observations. When this happens, the normal equations of 13.18.5 cannot be given a unique solution. The difficulty is even reflected in the degrees of freedom, which turn out to be negative in expressions such as 13.20.4.

In the second place, if the normal equations are to be solved, then a certain kind of relationship must not exist among the X variables themselves. This forbidden situation obtains when one (or more) of the X variables is a linear function of one or more of the other X variables (see 13.19.9 above). When this happens, the normal equations will not give a unique solution, and the regression problem is not solvable in the way shown here. Rather, when the original least squares solution to the problem is formulated, some variables must be deleted or certain side conditions created to eliminate this form of dependency among the X values, and this will lead to a somewhat different set of procedures. This is not generally a problem of concern for ordinary applications of multiple regression analysis; however, in the next chapter we will examine a situation where the problem does arise.

The third precaution extends not only to multiple regression, but also to the simple regression situation of Section 13.2 as well. This is the phenomenon of "shrinkage" already referred to in Section 13.18. Essentially, shrinkage means that if a regression equation is developed on a sample, and then actually used to make predictions about another sample or about cases in the population as a whole, the chances are very good that the predictions will not be as accurate as the original data suggested. That is, the apparent predictive power in a sample tends to shrink when one goes beyond that sample. The method gives the best way of combining the variables for predictions *in that sample,* even though some of the relations between the X values and Y, and between the X values themselves may simply be chance occurrences that we have no reason to expect in future samples. For this reason, when people try to work out regression equations for future application to subjects not in the original group, as in the employment or college admissions examples, then they are very careful to cross-validate the original results on one or more additional samples. This lets them evaluate how the predictions are likely to hold up in the larger group in which they will actually be used.

An additional point is that although multiple regression and correlation analysis is an extremely powerful tool for tracing the relationships that exist among a set of variables, it does rely on the concept of correlation, which reflects linear relationships. There is no guarantee that all of the possible predictive relationships among any set of variables will be linear, and multiple regression analysis may be insensitive to this sort of predictability. Nevertheless, the linear association that two variables show is almost always a good first approximation to the extent that one accounts for variance in the other. Thus, much, if perhaps not all, of the inherent predictability in a set of variables should be picked up by these methods.

A related point is that this form of regression analysis does not allow for the possibility that interactions exist among the variables, so that, for example, the relation between X_j and Y may be different over different ranges of values for the other variables. There are special ways of handling this problem, should it exist, by changes in the basic model employed. However, expert help should be sought if this problem is suspected. (Interactions in the sense of fixed-effects analysis of variance will be mentioned in the next chapter, however.)

Finally, the two sampling schemes outlined at the beginning of this chapter apply equally well to problems of multiple regression and correlation. The large majority of social science studies using multiple regression methods rely on the correlation-problem model, in which no constraints are exercised by the investigator over either the predictor or the dependent variables. In this instance, the tests and confidence interval procedures outlined above usually entail the assumption of a multivariate normal distribution in the population under study. The assumption of a multivariate normal distribution also takes care of the points raised in the preceding paragraphs as well; in a multivariate normal distribution the *only* form of relationship that may exist between variables is linear, and hence, by assumption, multiple regression describes all the prediction that is possible. Any departures from strict linearity of regression must be error. On the other hand, these same methods may be used in situations where the values of the predictor variables are preselected, and are viewed as constant from sample to sample. Then, one does not have to assume a multivariate normal population distribution. However, the inferences made are limited to those X values actually represented in the sample. Mixed situations also exist, as will be seen in at least one instance in the next chapter.

EXERCISES

1. Fifteen political scientists ranked each of three countries in terms of the perceived ''political responsibility'' each manifests. The data follow. Calculate the correlation between the ratings given to countries A and B. Then calculate the correlations between the ratings for B and C, and for A and C.

	countries		
raters	*A*	*B*	*C*
1.	5	5	6
2.	11	11	9
3.	6	8	5
4.	12	14	11
5.	8	6	6
6.	5	8	6
7.	7	7	5
8.	5	8	4
9.	12	14	9
10.	8	8	4
11.	5	8	4
12.	6	8	5
13.	7	7	8
14.	6	7	7
15.	7	9	13

2. Using the data of exercise 1, form the regression equations for predicting

(a) The z-score on A from the z-score on B.
(b) The z-score on B from the z-score on C.
(c) The z-score on C from the z-score on A.

3. Calculate the sample standard error of estimate for each of the regression equations found in exercise 2.

4. Using the data of exercise 1 once again, find the raw score regression equations for

(a) Predicting the rating on A (or x_A) from the rating on B (or x_B).
(b) Predicting the rating on C from the rating on A.
(c) Predicting the rating on B from the rating on C.

5. Find the sample standard error of estimate for each of the regression equations of exercise 4.

6. In exercise 1, what percentage of the variance in ratings in B is accounted for by linear relationship with A? What percent of the variance in C is accounted for by linear relationship with A? What percent is accounted for in C by linear relationship with B?

7. An experimenter was interested in the possible linear relationship between the measure of finger dexterity X, and another measure representing general muscular coordination Y. A random sample of 25 persons showed the following scores:

Person	*X-value*	*Y-value*
1	75	84
2	77	94
3	75	90
4	76	90
5	75	91
6	76	86
7	73	87
8	75	95
9	74	83
10	75	85
11	76	88
12	74	91
13	72	80
14	75	85
15	73	87
16	75	82
17	78	86
18	76	83
19	74	85
20	74	88
21	77	100
22	75	98
23	76	89
24	74	91
25	75	99

Compute the correlation coefficient, and test its significance ($\alpha = .05$).

8. Based on the data of exercise 7 above, find the regression equation for predicting X from Y. Plot this regression equation along with the raw data. What is the appropriate measure of the "scatter" or horizontal deviations of the obtained points in this plot about the regression line?

9. A developmental psychologist believes that the age at which a normal child begins to speak words clearly is closely related to the age at which the child first begins to use complete sentences. A random sample of of 33 normal children was taken, and very careful records kept for each. Let X be the age at which words are first clearly used, and let Y be the age at which complete sentences are used. The data below give the values of X and Y in months. Find the correlation between X and Y.

Child	*X*	*Y*	*Child*	*X*	*Y*	*Child*	*X*	*Y*
1	15.1	25.2	12	14.3	25.7	23	13.6	24.3
2	12.7	24.3	13	11.5	23.4	24	15.2	26.3
3	11.7	22.1	14	13.4	25.7	25	12.1	23.4
4	13.1	23.3	15	13.7	24.5	26	12.6	24.5
5	13.0	24.1	16	13.5	26.0	27	14.1	26.2
6	11.2	23.6	17	12.8	24.6	28	11.2	23.0
7	13.3	25.5	18	13.2	25.4	29	14.0	24.3
8	12.3	24.3	19	14.7	26.3	30	13.1	25.3
9	13.7	25.5	20	12.2	25.2	31	11.5	24.2
10	12.2	23.2	21	14.7	26.4	32	14.9	27.2
11	13.3	27.1	22	14.6	25.8	33	13.8	26.3

10. Work out the raw score regression equation for predicting Y from X in exercise 9. What is the sample standard error of estimate that goes with this equation?

11. Test the significance of the correlation found in exercise 9. What is the best estimate available of the regression coefficient $\beta_{Y \cdot X}$ in the population of such children? Find the 95 percent confidence for $\beta_{Y \cdot X}$.

12. Suppose that in exercise 9 the joint distribution of X and Y values is bivariate normal in form. Find the 95 percent confidence interval for the predicted Y age of a child by use of the true regression equation given the X age of:

(a) 13.0
(b) 14.6
(c) 12.2

13. An investigator in early childhood education conceived the notion that experienced elementary school teachers are able to identify the actual age of a child fairly accurately from inspection of drawings that the child had produced. In order that this idea might be tested, a sample of 48 children of various ages between 4 and 7 years was drawn at random. Each produced four free drawings of subjects suggested by the investigator. These drawings were then studied by a panel of 10 teachers working independently, and an age estimated for each child. The data below show the actual age X and the average estimated age Y for each child. Find the correlation coefficient between actual and average estimated age.

X	Y	X	Y	X	Y	X	Y
4.1	5.7	6.5	6.5	5.5	5.7	6.8	5.9
6.5	6.9	5.3	5.3	4.0	5.2	6.9	6.3
5.1	5.4	6.8	7.2	7.2	7.0	5.2	4.8
7.6	7.6	5.9	6.4	6.4	6.5	7.4	6.2
5.6	6.0	6.3	6.8	5.2	5.3	5.8	6.1
4.2	4.0	6.0	4.5	7.1	7.0	6.7	5.8
7.1	5.2	7.9	7.5	6.7	6.5	5.6	5.8
6.3	6.1	5.3	5.6	5.7	6.1	4.8	6.1
6.1	5.5	7.3	7.5	6.5	6.7	6.7	7.0
5.7	4.4	6.3	6.8	6.1	5.6	5.8	7.2
7.4	7.1	7.1	5.9	5.6	4.9	6.6	6.5
6.4	6.7	5.4	5.5	6.8	7.2	5.7	5.6

14. The standard deviation of the X values in exercise 13 indicates how far ''off'' one expects to be if the mean actual age is guessed for a child drawn at random from this group. By this same token, how much in error do we expect to be, given the judgment of the teachers about any child chosen at random from the group?

15. In exercise 13 find the sample regression equation for predicting X from Y. Then, given that this is a random sample of children, find the 95 percent confidence limits for the true value of $\beta_{Y \cdot X}$.

16. In exercise 13, is there a significant linear relation between a child's actual age X, and the average age Y guessed by the panel of teachers?

17. Use the data or exercise 13 to set up a regression equation for predicting raw score Y from X. Set up the 95 percent confidence interval for the predicted value of Y using the population regression equation, given that $X = 7$.

18. In a study of the relation between the age at which study of the piano was begun and the eventual proficiency of the student after five years practice, a random sample of 100 children and adults each just completing five years of piano study, was selected. Each student was given the same piece to play, and a panel of judges rated performance. Summary statistics emerging from the final data are given below:

$$\Sigma X = 1475 \qquad \Sigma Y = 12890$$

$$\Sigma X^2 = 24459 \qquad \Sigma Y^2 = 1714421$$

$$\Sigma XY = 186659$$

Establish the 95 percent confidence limits for the true correlation between age at beginning piano lessons X, and fifth-year proficiency Y. Test the hypothesis that the true correlation is $-.50$. (**Hint:** Use the r to Z transformation.)

19. Suppose that a person began piano lessons at age 35. On the basis of the data in exercise, predict proficiency at age 40. Given that proficiency after five years of study is normally distributed, establish the 95 percent confidence limits for this person's actual proficiency.

20. Under what circumstances must uncorrelated variables also be independent variables? Under what circumstances might two independent variables be correlated?

21. Make sketches showing the regression line $z'_Y = rz_X$ when $r = .30$, $r = -.60$, $r = -1.00$, $r = +1.00$, and $r = 0$.

22. Suppose that in some population

$$y_{ij} = \mu_Y + \beta_{Y \cdot X}(x_j - \mu_X) + e_{ij}$$

where $E(e_{ij}) = 0$.

Let $E(Y|X)$ be the mean of Y values for individuals all of whom have the same X value. If $E(Y|X)$ were plotted for all of the different values of X, what should the result look like? What must the mean value of $E(Y|X)$ over all X values be?

Suppose that for the population described above, each mean $E(Y|X)$ was the same, regardless of the X value. What would the value of $\beta_{Y \cdot X}$ be? What would the correlation ρ_{XY} be?

23. In an investigation of the relation between the height of an adolescent boy and a measure of his physical stamina, three random samples of 40 boys each were used. In the first sample, all boys were 15 years old; in the second, all were 16 years old, and in the third, the boys were 17 years of age. The following correlations were obtained:

Sample 1	*Sample 2*	*Sample 3*
$r = .11$ $N_1 = 40$	$r = .23$ $N_2 = 40$	$r = .19$ $N_3 = 40$

Test the hypothesis that the correlation between height and stamina is the same for the populations of boys represented by the first two samples. (**Hint:** Use the r to Z transformation.)

24. Test the hypothesis that all the sample correlations shown in exercise 23 above represent equal population correlations. Use $\alpha = .05$.

25. For the data of exercise 1, find the multiple regression equation for the prediction of rating for country A from the ratings of countries B and C.

26. Calculate the value of the coefficient of multiple correlation, R, for the data of exercise 1. For these data, what is the proportion of variance in the ratings of A accounted for by the ratings of B and C?

27. If the ratings of exercise 1 represent a random sample with $N = 15$ from a multivariate normal distribution, what can we say about the significance level of the multiple correlation found in exercise 25?

28. For the data of exercise 1, find the partial correlation between the ratings of countries B and C holding the rating of A constant. What is the partial correlation between B and A with C held constant? A and C with B held constant?

29. On the assumption that this is a sample with $N = 15$ from a multivariate normal population of such ratings, test the significance of first partial correlation found in exercise 28 above.

30. In a study of 50 cases sampled at random, the correlations among three variables, X_1, X_2, and X_3, were as follows:

$r_{12} = .38$

$r_{13} = .45$

$r_{23} = -.17$

Find the multiple regression equation for the standardized score of an individual on variable X_1, given the standardized values on X_2 and X_3. If an individual had a standardized score of 1.9 on X_2 and -1.2 on variable X_3, what would you predict the standardized score to be on X_1?

31. Using the information in exercise 30, find the multiple correlation and the proportion of variance in X_1 accounted for by multiple regression on X_2 and X_3. Test the significance of this multiple correlation.

32. For exercise 30, find the partial correlation of variables X_1 and X_2, holding X_3 constant. Find the partial correlation of variable X_2 and X_3, holding X_1 constant.

33. Test the significance of these partial correlations found in exercise 32 above.

34. Suppose that the intercorrelation among three variables could be $-.75$, $-.75$, and $-.75$. What would this imply about the partial correlation between the first and second variables, holding the third constant? Can this situation actually exist? What are the *smallest* intercorrelations that can exist among three variables, if all three intercorrelations are the same?

Chapter 14

FURTHER TOPICS IN CORRELATION AND REGRESSION

In this chapter we are going to look at further applications of the general linear model and the concepts of multiple regression to various problems of data analysis. As we shall see, far from being confined to the common sampling scheme that we have termed a problem in correlation, these concepts can be applied even in experimental situations where the levels of the experimental factor are chosen in advance, or in situations where only certain values of the predictor variables are permitted to occur. In other words, here the sampling strategy will be that of a problem in regression.

We will find that in this model questions both of the linear and possible nonlinear relationships may be examined. We will start by looking at the intimate connection between the analysis of variance and multiple regression methods. Although our discussion is necessarily confined to the fixed-effects situation with equal n per group, with J independent groups representing levels of one factor, the ideas can be generalized to cover most of the topics in Chapters 10 through 12.

14.1 / MULTIPLE REGRESSION AND THE FIXED-EFFECTS ANALYSIS OF VARIANCE

In Section 10.3 the general linear model was interpreted as an "experimental design model" suitable to the situation where each observation is made in one and only one of a series of J independent groups. Each group represented a preselected and fixed level of some qualitative factor. Nevertheless, it was possible to employ the notion of a variable X, which took on only the values 0 or 1 to indicate the group membership of any observation. The dependent variable Y related to these X values by the linear model was, as usual, viewed as quantitative. This type of model was then treated by the method of least squares in order to provide the basis for the one-way analysis of variance. However, before the least squares criterion was applied to this model, certain restrictions or side conditions were applied, such as the condition that the sum of all errors should equal zero, and that the sum of the experimental effects must also sum to zero. These restrictions were actually made so as to avoid the technical problem we will be discussing below.

Let us look once again at the model underlying the experimental design of Section 10.3.1. Here there are four groups, with two subjects in each group. The structure of the experiment can then be shown in terms of five variables, X_0, X_1, X_2, X_3, and X_4, each of which take on only the values 0 or 1:

		subjects	X_0	X_1	X_2	X_3	X_4
	1	1	1	1	0	0	0
		2	1	1	0	0	0
groups	2	3	1	0	1	0	0
		4	1	0	1	0	0
	3	5	1	0	0	1	0
		6	1	0	0	1	0
	4	7	1	0	0	0	1
		8	1	0	0	0	1

(The columns of this table can be thought of as vectors of values, in the terminology of Appendix C. Appendix C also shows that when one has a set of vectors, $\mathbf{x}_1$ through $\mathbf{x}_m$, then it is always possible to find a corresponding set of vectors $\mathbf{v}_0$ through $\mathbf{v}_q$ which form an orthogonal basis for the $\mathbf{x}$ vectors.)

We saw in Section 13.18 that whenever there is a dependent variable Y and a set of predictors, one may usually construct a multiple regression equation and calculate the value of the multiple correlation coefficient. Furthermore, the value of R^2 gives the proportion of variance in Y accounted for by the X variables. This is essentially what one wishes to know in this experiment as well. Why, then, don't we simply find the correlation of each X shown here with the dependent variable Y, along with the intercorrelations of the X variables themselves, and use the normal equations (13.18.5) to answer these questions? Unfortunately, this plan will not work. As mentioned in Section 13.21, the normal equations occurring in a multiple regression set up cannot be given a unique solution when some of the predictor variables X are themselves linear combinations of other predictor variables. This is true of the five variables listed above. Thus, note that

$$X_1 + X_2 + X_3 + X_4 = X_0.$$

Such a situation will always exist among the $J + 1$ raw indicator variables representing the overall level (X_0) and the J treatment levels of a single factor.

On the other hand, suppose that it were possible to transform these X indicator variables into other variables V, which are *not* linear combinations of each other (that is, which are not linearly dependent). When we transform the original predictor variables into a new set of variables which are not dependent, or which *are* orthogonal, we say that the model has been *reparameterized*. By dealing with the new orthogonal variables V instead of the linearly related variables X we are perfectly free to use the concepts of multiple regression outlined in Section 13.18.

In addition, the way in which this transformation is made is already illustrated in Appendix C. There, it is shown how an orthogonal basis may be found for any set of vectors; this is exactly what one does when the variables X are transformed into the variables V. Actually, we have used this principle extensively in Chapter 12: when one finds a set of weights yielding orthogonal comparisons, this is nothing but finding an orthogonal basis for a set of vectors representing the groups to be compared.

Thus, for example, we can replace the indicator variables X_1 through X_4 in the table above with new variables V_1 through V_3, as follows:

		subjects	X_0	V_1	V_2	V_3
	1	1	1	1	1	0
		2	1	1	1	0
groups	2	3	1	1	−1	0
		4	1	1	−1	0
	3	5	1	−1	0	1
		6	1	−1	0	1
	4	7	1	−1	0	−1
		8	1	−1	0	−1

Now notice that exactly like the comparison weights used in Chapter 12, the products of the values for any pair of these V variables sum to zero across the subjects and groups. These variables (or vectors) are thus orthogonal, and are not linearly dependent. Each is also orthogonal to X_0, since each variable V_k has values that sum to zero.

The X variables *are* linearly dependent on the V variables and X_0, however. Thus, for this example,

$$X_1 = (1/4)X_0 + (1/4)V_1 + (1/2)V_2$$

$$X_2 = (1/4)X_0 + (1/4)V_1 - (1/2)V_2$$

$$X_3 = (1/4)X_0 - (1/4)V_1 + (1/2)V_3$$

$$X_4 = (1/4)X_0 - (1/4)V_1 - (1/2)V_3.$$

The information about the group to which an individual i belongs, originally contained in the X values assigned to that individual, is also contained in the V values, since the X values themselves can always be found from the V values. However, unlike the original X values, these new V variables have the advantage of being orthogonal to each other and to X_0. Furthermore, we are now dealing with one less variable; although it took five variables to describe four groups in the original model, now we require only four variables, X_0, V_1, V_2, and V_3.

To return to the general situation where there are J groups, there will be $J - 1$ new variables V, or $V_1, V_2, \cdots, V_{J-1}$. Let us denote any given one of these variables as V_k. Then, we could represent the value of variable V_k associated with some individual subject i in any group j by the symbol v_{kij}. However, as shown in the table above, in this experimental situation all individuals in the same group j receive exactly the same v_{kij} value on the variable V_k. Therefore, we can say that $v_{kij} = v_{kj}$, since we really do not need to distinguish among individuals in the same group according to their V_k values.

The correlation between dependent variable Y and variable V_k is thus

$$r_{yk} = \frac{\sum_j v_{kj} \sum_i y_{ij}}{\sqrt{\sum_j n v_{kj}^2}\sqrt{NS_Y^2}} = \frac{\sum_j v_{kj} M_j}{\left(\sqrt{\sum_j v_{kj}^2/n}\right)\left(\sqrt{NS_Y^2}\right)} \qquad [14.1.1]$$

since the V_k values depend only on groups, not on individuals, and the mean of V_k is zero. On the other hand, since the V variables are orthogonal to each other, the correlation between any pair of these variables, say r_{kg} for the pair V_k and V_g, *must* be zero. In consequence, the normal equations (13.18.5) for multiple regression become simply

$$B_1 = r_{y1}$$
$$B_2 = r_{y2}$$
$$\cdots$$
$$B_{J-1} = r_{y(J-1)},$$

and the square of the multiple correlation coefficient is just

$$R^2_{y \cdot H} = r^2_{y1} + r^2_{y2} + r^2_{y3} + \cdots + r^2_{y(J-1)} = \sum_{k=1}^{J-1} r^2_{yk}. \qquad [14.1.2]$$

Here again the symbol H represents the entire set of $J - 1$ predictor variables, $H = \{V_1, \cdots, V_{J-1}\}$. The square of the multiple correlation, or $R^2_{y \cdot H}$ is thus the sum of the squared correlations between Y and each of the variables V_k.

Furthermore, if we let

$$NS_k^2 = \sum_j \sum_i v_{kij}^2 = \sum_j n v_{kj}^2, \qquad [14.1.3]$$

by 13.18.7,

$$b_k = B_k\left(\frac{S_Y}{S_k}\right) = r_{yk}\left(\frac{S_Y}{S_k}\right),$$

so that

$$b_k S_k = r_{yk} S_Y.$$

However, in the terminology of Section 12.8, we can rewrite r_{yk} as

$$r_{yk} = \frac{n\hat{\psi}_k}{\sqrt{\sum_j n\, v_{kj}^2}\sqrt{NS_Y^2}} = \frac{\hat{\psi}_k}{\sqrt{w_k}\sqrt{NS_Y^2}}.$$

Then

$$b_k^2(NS_k^2) = r_{yk}^2(NS_Y^2) = \text{SS}(\hat{\psi}_k) \qquad [14.1.4]$$

so that

$$r_{yk}^2 = \frac{SS(\hat{\psi}_k)}{SS_Y \text{ total}} \qquad [14.1.5^*]$$

and

$$b_k^2 = \frac{SS(\hat{\psi}_k)}{n^2 w_k}. \qquad [14.1.6^*]$$

Thus

$$b_k = \frac{\hat{\psi}_k}{n w_k} = \frac{\hat{\psi}_k}{\sum_k v_k^2}. \qquad [14.1.7^*]$$

An orthogonal comparison formed with weights v_{kj} and divided by the sum of v_{kj}^2, is actually the regression weight for predicting Y from V_k.

In addition, we know from Section 12.8 that given any set of $J - 1$ orthogonal comparisons among J groups,

$$\text{SS between} = \sum_{k=1}^{J-1} SS(\hat{\psi}_k). \qquad [14.1.8]$$

It must then be true that

$$\sum_k r_{yk}^2(SS_Y \text{ total}) = \text{SS between}. \qquad [14.1.9^*]$$

Thus, by 14.1.2, it is also true that

$$\frac{\text{SS between groups}}{\text{SS total}} = R_{y \cdot H}^2, \qquad H = \{V_1, \cdots, V_{J-1}\} \qquad [14.1.10^*]$$

$$\frac{\text{SS within groups}}{\text{SS total}} = 1 - R_{y \cdot H}^2 \qquad [14.1.11^*]$$

That is, the total variation in Y values can be broken into two parts, that which is attributable to the experimental factor, and all other variation. Then the proportion of variance accounted for in the sample is given by $R_{y \cdot H}^2$, where H is the set of underlying orthogonal variables created out of the indicator variables X.

In short, what we have just shown is that the usual analysis of variance situation can also be interpreted as an application of multiple regression analysis where the indicator variables have been replaced by a set of $J - 1$ orthogonal variables V.

Furthermore, the usual test for the presence of treatment effects,

$$F = \frac{\text{MS between}}{\text{MS within}}$$

becomes, under this interpretation,

$$F = \frac{R_{y \cdot H}^2 (N - J)}{(1 - R_{y \cdot H}^2)(J - 1)}. \qquad [14.1.12^*]$$

The F test in an analysis of variance is also the usual test of significance for the multiple correlation coefficient when there are J − 1 predictor variables in the set H. Suppose that we have available a set of comparison weights (the v_{kj} values) and we wish to construct the regression equation for predicting the value y'_{ij} for any individual in group j. Any comparison is associated with a regression weight by 14.1.7 above. Then the required regression equation is

$$y'_{ij} = M_Y + \sum_{k=1}^{J-1} b_k v_{kj} \qquad [14.1.13]$$

or

$$y'_j = M_Y + \sum_{k=1}^{J-1} \frac{\hat{\psi}_k}{nw_k} v_{kj}.$$

In the population, this is

$$y'_j = \mu_Y + \sum_{k} \frac{\psi_k}{nw_k} v_{kj}. \qquad [14.1.15]$$

However, this regression equation reduces to just

$$y'_{ij} = \mu_Y + \alpha_j,$$

which is the model for one-way analysis of variance as found with the method of least squares (Section 10.5).

In summary, *the one-way analysis of variance under the fixed-effects model is identical to a multiple regression analysis when the indicator variables for group membership are transformed into a corresponding set of $J - 1$ orthogonal variables. Such orthogonal variables are equivalent to a set of $J - 1$ orthogonal comparison weights*. These basic connections are summarized in Table 14.1.1.

Table 14.1.1

Some connections between a fixed-effects one-way analysis of variance and multiple regression.

Analysis of variance	Multiple regression
SS between/SS total	$R^2_{y \cdot H}$, where H = set of $J - 1$ orthogonal variables
SS within/SS total	$1 - R^2_{y \cdot H}$
Comparison $\hat{\psi}_k/nw_k$	b_k
SS($\hat{\psi}_k$)/SS total	r^2_{yk}
MS between/MS within	$\dfrac{R^2_{y \cdot H}(N - J)}{(1 - R^2_{y \cdot H})(J - 1)}$

Although we have gone through this discussion using the already familiar idea of comparison weights, one may accomplish this same result without ever talking about the V variables or about comparisons as such. Rather, when the least-squares criterion is applied to the experimental design model, certain side conditions or restrictions are introduced to correct for the dependencies that exist between the indicator variables and the group mean, and among the variables themselves. The normal equations found through the least-squares criterion then give the basis for the standard analysis of variance. Furthermore, when multiple regression procedures are applied to qualitative data through the use of "dummy variables," special steps are taken (such as throwing out one of the dummy variables) in order to eliminate the problem of linear dependency, the problem which we handled through the adoption of an orthogonal basis.

Although space will not permit our doing so, connections between multiple regression and analysis of variance can be extended in many directions, such as multifactor designs, random and mixed models, nonorthogonal designs, and so forth. The reason, however, for us to dwell on this connection is for the perspective it lends to the problem of data analysis in general. By casting the underlying theory of the data in terms of the general linear model, we find that there is literally no limit to the number and variety of variables that might be investigated singly or in combination. The basic ideas of multiple regression permit us to investigate these relationships in a routine way, without necessarily relying on some textbook design that happens to fit our situation. One of the great benefits that the advent of simple-to-use computer methods has brought to the social and behavioral sciences is freeing us from the great computational barriers that used to surround such multivariate methods, and making them available to experimental and other research questions which may not fit the classical molds. Although a survey of these possibilities is not in the scope of this text, the student should bear this point in mind in studying more advanced methods, and in seeking assistance in the analysis of research data. When dealing with an atypical or complex situation, think first of how it might be handled through multiple regression or a related technique growing out of the general linear model.

We now turn to a simple problem in regression, concerning the extent and the form of the relationship between certain fixed values of a quantitative independent variable X and a quantitative dependent variable Y.

14.2 / LINEAR REGRESSION ANALYSIS WHEN X VALUES ARE FIXED IN ADVANCE

In experimental work it often happens that some J different groups are formed at random, and then each is given a different *amount* or *quantity* of some experimental treatment. Thus, for example, a teacher may be interested in comparing achievement test scores in English composition among four groups differing in the amount of writing experience assigned. Perhaps the first group is given two required essays in a semester, the second four essays, the third six, while the fourth group writes eight. Does the sheer amount of writing assigned appear to be reflected in test achievement? Or, perhaps, the question may be, "How is retention of memorized material affected by the number of rest periods allowed during initial memorization?" Six groups are formed in which each person must memorize a long poem. Group 1 gets no rest periods during memorization, Group 2 gets one period, and so on by even steps until Group six gets five such periods. After a fixed amount of time, do the persons in the groups differ in their recall of this material?

These examples differ from the classical analysis of variance situation in that the treatment levels are quantitatively, rather than merely qualitatively, different from each other. This situation also differs from a correlation problem in that the values of X employed in the experiment are fixed in advance, and would not change with replication of the experiment. The only inferences to be made are those involving the X values actually employed. As it happens, concepts from the previous chapter and the preceding section now give us ready methods for examining such questions.

Initially in this section we are going to deal once again with the situation where the simplest linear model applies:

$$y_{ij} = \mu_Y + \beta_{Y \cdot X}(x_j - \mu_X) + e_{ij} \qquad [14.2.1]$$

for the population. Note that $x_{ij} = x_j$ here, since each individual in any group j has the same independent variable, or X, value. Also since the x values are fixed, the population mean of X, $\mu_X = M_X$.

Translated into sample terms, this model becomes

$$y_{ij} = M_Y + b_{Y \cdot X}(x_j - M_X) + e_{ij}. \qquad [14.2.2]$$

Here, there are only two possible sources of variation among the Y values: linear regression, and error. Any deviations from linear regression can only represent error in 14.2.1 and 14.2.2. Thus, under this model it will be true that only two sources of variation are possible:

SS total = SS linear regression + SS deviations from regression.

Since the x value is fixed for individuals in the same group, it is here possible to calculate, $b_{Y \cdot X}$ in a slightly simpler way. For equal n per group,

$$b_{Y \cdot X} = \frac{\sum_j (x_j - M_X) M_{Y_j}}{\sum_j (x_j - M_X)^2}, \qquad [14.2.3]$$

Now notice that these fixed values $(x_j - M_X)$ must sum to zero, and are, in fact, exactly like the weights used in a comparison among means of J groups. Furthermore, the term in the denominator is just the sum of the squared weights, as represented by nw in Section 12.8.2. Consequently,

$$b_{Y \cdot X} = \frac{\hat{\psi}_X}{nw_X} \qquad [14.2.4]$$

where the values of $(x_j - M_X)$ play the role of weights in the comparison $\hat{\psi}_X$ among the means M_{Y_j}.

Although there is only one predictor variable X in this model, the statements made in connection with the multiple regression model of the previous section still hold true. That is,

$$\begin{aligned} \text{SS}(\hat{\psi}_X) = \text{SS linear regression} &= b^2_{Y \cdot X} N S^2_X \\ &= r^2_{yx}\, \text{SS}_Y \text{ total} \end{aligned} \qquad [14.2.5]$$

so that this sum of squares has one degree of freedom, as for any comparison.

Furthermore, the proportion of variance accounted for by linear regression must then be

$$r^2_{yx} = \frac{\text{SS linear regression}}{\text{SS}_Y \text{ total}}, \qquad [14.2.6]$$

$$\begin{aligned} \text{and SS}_Y \text{ total} - \text{SS linear regression} &= \text{SS deviations} \\ &= (\text{SS}_Y \text{ total})(1 - r^2_{yx}). \end{aligned} \qquad [14.2.7^*]$$

In view of this, the appropriate test for linear regression is

$$F = \frac{\text{MS linear regression}}{\text{MS deviations from regression}} \qquad [14.2.8^*]$$

$$= \frac{r_{yx}^2 (N - 2)}{1 - r_{yx}^2}$$

with 1 and $N - 2$ degrees of freedom (c.f. expression 13.13.2.)

The data for a problem in regression are set up much as for an analysis of variance. For any group j, the dependent variable scores y_{ij} are grouped together along with value x_j for that group. Let us say that there are n observations per group j. Then we take SS total in the usual way for analysis of variance:

$$\text{SS total} = \sum_j \sum_i y_{ij}^2 - \frac{\left(\sum_j \sum_i y_{ij}\right)^2}{N}. \qquad [14.2.9]$$

Next, one calculates a sum of squares for the variable X:

$$\text{SS}_X = \sum_j nx_j^2 - \frac{\left(\sum_j n_j x_j\right)^2}{N}. \qquad [14.2.10]$$

Then the sample value of the regression weight b_Y may be found from:

$$b_{Y \cdot X} = \frac{\left(\sum_j x_j \sum_i y_{ij}\right) - \left[\left(\sum_j n_j x_j\right)\left(\sum_j \sum_i y_{ij}\right)/N\right]}{(\text{SS}_X)} \qquad [14.2.11^*]$$

The SS linear regression is then found by

$$\text{SS linear regression} = b_{Y \cdot X}^2(\text{SS}_X), \qquad [14.2.12]$$

and the SS for deviations from linear regression is found from

$$\text{SS deviations} = \text{SS total} - \text{SS linear regression}. \qquad [14.2.13]$$

The results are displayed in an analysis of variance summary table.

The assumptions made in a regression problem such as this, where the values of X are fixed in advance, are parallel to those made in the analysis of variance. That is, we assume

1. within each fixed X-value population, the distribution of deviations from regression is normal;
2. within each X-value population, the variance of the deviations from regression is the same;
3. deviations from regression are everywhere independent of each other.

Notice that with this regression-problem model of sampling, the inference about the relationship between X and Y is confined strictly to those X values actually chosen for inclusion. There is nothing in this method that guarantees that the linear relationship found will also obtain for X values lying outside the range actually observed. In addition, the linear regression could be different even if additional values were inserted among those employed. If one is concerned with the relationship over *all possible* X values then the sampling scheme for a correlation problem, not a regression problem, should be used. On the other hand, some useful additional questions may be asked when one uses the regression problem approach, as we shall see.

14.3 / TWO-WAY (OR HIGHER) FIXED-EFFECTS REGRESSION ANALYSIS

A fixed-effects linear regression problem is capable of extension in a variety of ways. For example, imagine a factorial experiment with *two* quantitative factors, X_1 and X_2. The J levels of X_1 and the K levels of X_2 are fixed ahead of the study. Furthermore, for the sake of simplicity, let us suppose that each combination of level j from X_1 and level k from X_2 contains the same number of observations n. In this instance, the simplest linear two-variable model is written as

$$y_{ijk} = M_Y + \beta_1(x_{1j} - \mu_1) + \beta_2(x_{2k} - \mu_2) + e_{ijk}. \qquad [14.3.14]$$

Here x_{1j} stands for the value on X_1 received by members of group j, and x_{2k} the value on X_2 received by members of group k. Note that no interaction is assumed: the only influences on the variance of Y are the linear regression on X_1 and the linear regression on X_2.

This is a straightforward application of multiple regression to the prediction of Y from X_1 and X_2. However, since we are assuming an orthogonal factorial design with equal n per cell, $r_{12} = 0$, the two variables are themselves orthogonal. Thus, we can take

$$b_1 = \frac{\sum_j (x_{1j} - M_1)M_{Y_j}}{\sum_j (x_{ij} - M_1)^2} \qquad [14.3.15]$$

and

$$b_2 = \frac{\sum_k (x_{2k} - M_2)M_{Y_k}}{\sum_k (x_{2k} - M_2)^2}. \qquad [14.3.16]$$

Then, we have

$$\text{SS linear regression on } X_1 = b_1^2 \text{ SS } X_1 = r_{y1}^2 \text{ SS}_Y \text{ total}$$

$$\text{SS linear regression on } X_2 = b_2^2 \text{ SS } X_2 = r_{y2}^2 \text{ SS}_Y \text{ total}$$

$$\text{SS deviations from regression} = \text{SS}_Y \text{ total} - \text{SS regression on } X_1 - \text{SS linear regression on } X_2$$

so that

$$\text{MS linear regression on } X_1 = \text{SS linear regression on } X_1$$

$$\text{MS linear regression on } X_2 = \text{SS linear regression on } X_2$$

$$\text{MS deviations} = \frac{\text{SS deviations from regression}}{N - 3}.$$

Thus, both regressions may be tested for significance by comparison with MS deviations, with 1 and $N - 3$ degrees of freedom.

Obviously, such a design can be analyzed directly by multiple regression. However, the actual hand computations for such a problem go as follows:

$$\mathrm{SS}_{X_1} = \sum_j n_j x_{1j}^2 - \frac{(\sum_j n_j x_{1j})^2}{N} \qquad [14.3.17]$$

$$\mathrm{SS}_{X_2} = \sum_k n_k x_{2k}^2 - \frac{(\sum_k n_k x_{2k})^2}{N}. \qquad [14.3.18]$$

We also calculate *sums of products,* as follows:

$$\mathrm{SP}_{YX_1} = \sum_j x_{1j} \sum_k \sum_i y_{ijk} - \frac{(\sum_j n_j x_{1j})(\sum_j \sum_k \sum_i y_{ijk})}{N} \qquad [14.3.19]$$

$$\mathrm{SP}_{XY_2} = \sum_k x_{2k} \sum_j \sum_i y_{ijk} - \frac{(\sum_k n_k x_{2k})(\sum_k \sum_j \sum_i y_{ijk})}{N} \qquad [14.3.20]$$

with SS_Y total figured as usual. Here $n_j = Kn$, $n_k = Jn$, of course.

Then

$$\text{SS linear regression } X_1 = \frac{(\mathrm{SP}_{YX_1})^2}{\mathrm{SS}_{X_1}}$$

$$\text{SS linear regression } X_2 = \frac{(\mathrm{SP}_{XY_2})^2}{\mathrm{SS}_{X_2}}$$

and

$$\text{SS deviations from regression} = \mathrm{SS}_Y \text{ total} - \text{SS linear on } X_1 - \text{SS linear on } X_2.$$

The same general procedures apply to a factorial design with any number of quantitative factors, provided that no interactions are assumed to exist. Models with interactions may be handled as suggested in Section 14.11. This setup will be illustrated for two factors in the next section.

14.4 / AN EXPERIMENT WITH TWO QUANTITATIVE FACTORS

A social psychologist was interested in problem solving carried out cooperatively by small groups of individuals. The theory upon which these experiments were based suggested that within a particular range of possible group sizes, the relationship between group size and average time to solution for particular kind of problem should be linear and negative: the larger the group, the less time on the average should it take for the problem to be solved. In order to check on this theory, the psychologist decided to form a set of experimental groups, ranging in size from groups consisting of a single individual to groups consisting of six individuals. Thus, 105 individuals were chosen at random from some specific population, and then formed at random into five groups consisting of 1 individual each, five groups each consisting of 2 individuals, five groups of 3, and so on until there were six different and nonoverlapping sets of five groups each, the last consisting of five groups each of size 6. Each individual subject participated in one and only one group in the study.

Furthermore, five different problems were available for use with these groups. These problems had been scaled in difficulty, from 5 for the most difficult to 1 for the least difficult. In order to see if this scaling correlated with the time it took the groups to solve these problems, and also as a way to reduce error variance, the experimenter decided to give all five problems within each set of groups of the same size, with each group receiving one problem assigned at random. Thus, every combination of problem difficulty and size was represented by exactly one group.

Each experimental group then solved its assigned problem, and the time taken for them to do so was noted; this time to completion was used as the dependent variable. Table 14.4.1 gives the results, with group size shown as the columns, and problem difficulty by the rows.

Table 14.4.1

Results for groups of different sizes and problems of varying levels of difficulty.

X_2 = *difficulty*	1	2	3	4	5	6
5	32	22	27	19	25	20
4	29	22	26	21	22	18
3	30	24	24	20	23	16
2	28	23	24	20	25	18
1	26	22	23	18	21	16
	X_1 = *group size*					

Then, we take

$$SS_Y \text{ total} = 16058 - \frac{(684)^2}{30} = 462.8$$

$$SS_{X_1} = 455 - \frac{(105)^2}{30} = 87.50$$

$$SS_{X_2} = 330 - \frac{(90)^2}{30} = 60$$

$$SP_{YX_1} = 2243 - \frac{(105)(684)}{30} = -151$$

$$SP_{YX_2} = 2090 - \frac{(90)(684)}{30} = 38,$$

Then,

$$\text{SS linear regression on } X_1 = \frac{(-151)^2}{87.5} = 260.58$$

$$\text{SS linear regression on } X_2 = \frac{(38)^2}{60} = 24.07$$

and

$$\text{SS deviations} = \text{SS}_Y \text{ total} - \text{SS linear } X_1 - \text{SS linear } X_2$$
$$= 178.15.$$

We also find that

$$b_1 = \frac{\text{SP}_{YX_1}}{\text{SS}_{X_1}} = \frac{-151}{87.5} = -1.73$$

and

$$r^2_{y1} = \frac{\text{SS linear regression on X}_1}{\text{SS}_Y \text{ total}} = \frac{260.58}{462.80} = .56$$

$$r_{y1} = -.75$$

as well as

$$b_2 = \frac{\text{SP}_{YX_2}}{\text{SS}_{Y_2}} = \frac{38}{60} = .63$$

$$r^2_{2y} = \frac{\text{SS linear regression on } X_2}{\text{SS}_Y \text{ total}} = .05$$

$$r_{2y} = .23.$$

The total proportion of variance accounted for is

$$R^2_{y \cdot 12} = \frac{\text{SS linear } X_1 + \text{SS linear } X_2}{\text{SS}_Y \text{ total}}$$
$$= .62.$$

Note that the correlation between group size and time is negative, as the experimenter expected. The results are displayed in the following summary table:

Source	*SS*	*df*	*MS*	*F*
Linear regression on X_1	260.58	1	260.58	$\frac{260.18}{6.60} = 39.42$
Linear regression on X_2	24.07	1	24.07	$\frac{24.07}{6.60} = 3.65$
Deviations	178.15	27	6.60	
Total	462.80	29		

The first of these F tests is significant well beyond the .01 level. The other fails to reach the .05 level. There is thus very strong evidence that group size is negatively related to the time required for the solution of a problem, but little evidence that the problems scaled as more difficult do tend to require more time. However, note that by including the different problems in this experiment, the researcher accounted for five percent of the total variance which otherwise would have gone into deviations from regression, and thus would have tended to lower the power of the test of the other regression.

14.5 / CURVILINEAR REGRESSION

All the discussion so far in this chapter has centered on linear regression, the use of a linear function rule for the prediction of Y from X. However, the theory of "regression" is much more extensive than the preceding discussion of linear regression might suggest. Indeed, the linear rule for prediction is only the simplest of a large number of such rules that might apply to a given statistical relation. Linear regression equations may serve quite well to describe many statistical relations that are roughly like linear functions, or that may be treated as linear as a first approximation. Nevertheless, there is no law of nature requiring all important relationships between variables to have a linear form. It thus becomes important to extend the idea of regression equations to the situation where the relation is *not necessarily* described by a linear rule. Now we are going to consider problems of curvilinear regression, problems where the best rule for prediction need not specify a simple linear function.

Our study of nonlinear prediction rules will be confined to *problems of regression,* as defined previously. An independent variable X is identified, and various values of X are chosen in advance to be represented in the experiment. Ordinarily, each X value corresponds to an experimental treatment administered in some specific quantity.

A regression problem where a curvilinear rule for prediction might make sense is the following: there is an investigation of the effect of environmental noise on the human subject's ability to perform a complex task. Several experimental treatments are planned, each treatment differing from the others only in the intensity of background noise present while the subject performs the task. Each treatment represents a one-step interval in the intensity of noise within a particular range. A group of some N subjects chosen at random are assigned at random to the various groups, with n subjects per group. In the experiment proper, each subject works on the same problem individually in the presence of the assigned intensity of background noise. The dependent variable is thus a subject's score on the task, and the independent variable is the noise intensity. The experimenter is interested not only in the possible existence of a linear relation between noise intensity and performance, but also a possible curvilinear relationship between X and Y.

In the next section, a general model for dealing with curvilinear as well as linear regression will be explored. Using this model we will see first how a test may be made for the existence of a nonlinear relationship. Then we will deal with the problem of inferring the form of the statistical relation by use of planned comparisons among means. Finally, inferences about form of relation made from post hoc comparisons will also be considered.

14.6 / THE MODEL FOR LINEAR AND CURVILINEAR REGRESSION

So far, in this chapter inferences about the existence and extent of linear regression were made in terms of the model

$$y_{ij} = \mu_Y + \beta_{Y \cdot X}(x_j - \mu_X) + e_{ij},$$

which we saw to be an instance of the fixed-effects model for analysis of variance when the X values are fixed. In adopting this model, we assume that the only systematic tendency of Y to vary with X is due to the linear regression of Y on X.

Now we will extend this model to allow for the existence of other kinds of systematic dependence of Y on X. The model we will use can be stated as follows:

$$y_{ij} = \beta_0 + \beta_1(x_j - \mu_X) + \beta_2(x_j - \mu_X)^2 + \cdots + \beta_{J-1}(x_j - \mu_X)^{J-1} + e_{ij}. \quad [14.6.1]$$

That is, we assume that the value of Y might depend on the value of $(x_j - \mu_X)$, of $(x_j - \mu_X)^2$, and, indeed, on any power of $(x_j - \mu_X)$ up to and including the $J - 1$ power, each term weighted by a constant β_1, β_2, and so on.

Some intuitive feel for this model may be gained as follows. Suppose that it were really true that the best possible prediction of Y_{ij} were afforded by

$$y'_{ij} = \beta_0 + \beta_1(x_j - \mu_X) + \beta_2(x_j - \mu_X)^2.$$

However, the experimenter does not know this, and applies the simple linear rule:

$$y'_{ij} = \beta_1(x_j - \mu_X) + \mu_Y.$$

Does this describe all the predictability possible in the relationship? No, since if β_2 is not zero, the *deviations* of the predicted from the obtained values of Y should be predictable using the rule

$$d'_{ij} = (y_{ij} - y'_{ij}) = \beta_2(x_j - \mu_X)^2 + e_{ij} + (\beta_0 - \mu_Y).$$

In short, the deviations from linear regression here represent not only error, but also some further predictability depending on the *squared* value of $(x_j - \mu_X)$. We should expect that, when plotted, the deviations from linear regression for the various sample groups should tend to lie about a parabola, much as in Figure 14.6.1.

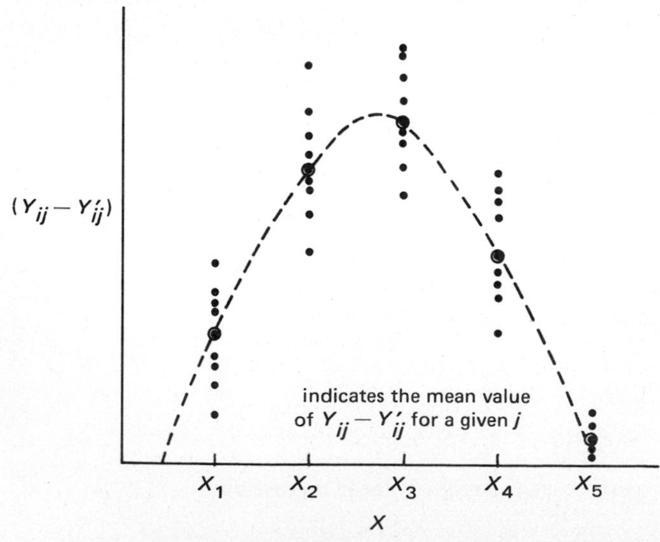

Figure. 14.6.1

Plot of deviations from linear regression, suggesting the presence of second-degree or quadratic regression.

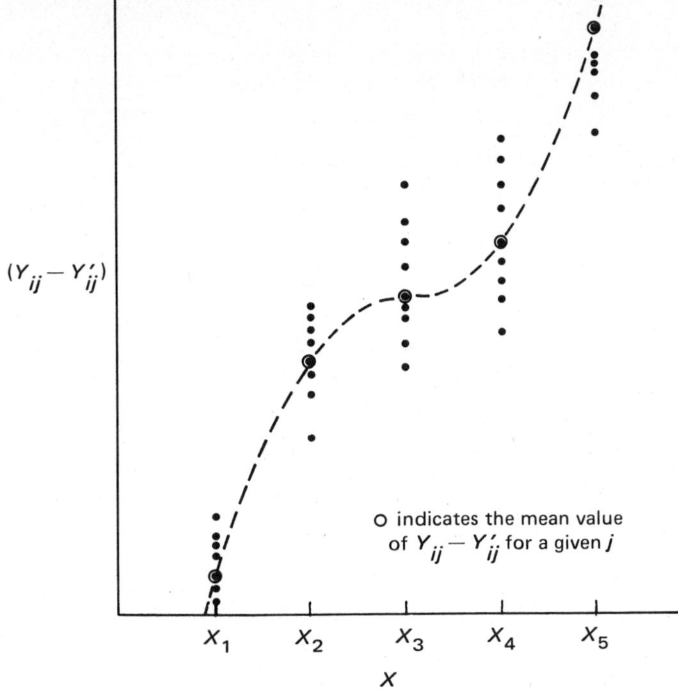

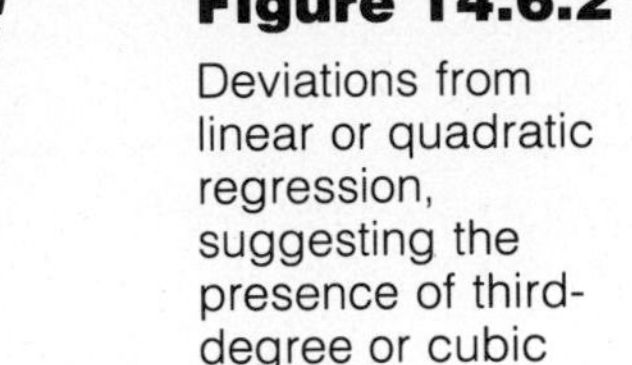
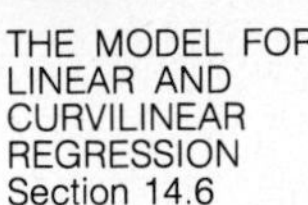

Figure 14.6.2

Deviations from linear or quadratic regression, suggesting the presence of third-degree or cubic regression.

The constant β_2 is a coefficient of **second-degree or quadratic regression.** We need a way to decide if β_2 really is zero if we are going to tell whether or not the best prediction rule possible is at least of the second degree.

Extending this idea, suppose that the true relation between X and Y is given by the rule

$$y_{ij} = \beta_0 + \beta_1(x_j - \mu_X) + \beta_2(x_j - \mu_X)^2 + \beta_3(x_j - \mu_X)^3 + e_{ij}.$$

However, the experimenter actually predicts using either a first- or a second-degree regression equation. Suppose that an equation of the second degree is used. Then, nevertheless, the deviations from the predicted y_{ij} values will tend to be systematically related to x values, since

$$(y_{ij} - y'_{ij}) = \beta_3(x_j - \mu_X)^3 + e_{ij} + \beta_0 - \mu_Y.$$

Unless β_3 is zero, cubic trends exist between X and Y, and the plot of deviations from prediction for the various populations should tend to fall into a cubic function, something like that shown in Figure 14.6.2.

The same general ideas hold for any number of experimental groups each associated with a fixed value x_j. A regression equation of the first, second, third, and so on up to the degree $J - 1$ is assumed possible. It can be shown, however, that for any finite number J of groupings, each associated with some fixed value of x_j, the best prediction equation possible in this model is of *no higher degree than* $J - 1$. Thus, a first-degree equation always suffices to describe the relation between X values and *mean* Y values for two groups, a second-degree equation suffices to describe the relation between X and mean Y for three groups, a third-degree equation for four groups, and so on.

14.7 / ANOTHER LOOK AT THE PARTITION OF THE SUM OF SQUARES FOR REGRESSION PROBLEMS

Under the model used in 14.2 nothing could contribute to deviations from linear regression except random errors, e_{ij}. Consequently, the mean square for deviations from linear regression was used to estimate σ_e^2.

Now, however, we have a different model, which permits other, nonlinear, kinds of relations to exist. Nevertheless, MS within groups still reflects only error variance in the data. Furthermore, the SS linear regression depends only on the true regression coefficient β_1 and error, just as before. On the other hand, the status of the SS deviations from linear regression changes in this new model; this sum of squares reflects all the possible *nonlinear* relations as well as random error.

We already know, of course, that all predictability in the data is embodied in the sum of squares between groups. That is, in the terminology of Section 14.1, we know from 14.1.8 that

$$\text{SS}_Y \text{ total } (R^2_{y \cdot H}) = \text{SS between groups}$$

where H represents a set of $J - 1$ orthogonal variables based upon the J fixed levels of X. We also saw in Section 14.2 that the sum of squares for linear regression is equivalent to the sum of squares for one variable, V_1, out of this set. Let us then relabel r^2_{yx} as $r^2_{y \cdot 1}$. Then, SS linear regression $= r^2_{y \cdot 1}$ (SS_Y total). Now if we take

$$\text{SS deviations from linear regression} = \text{SS between} - \text{SS linear}$$

we have

$$\text{SS deviations from linear regression} = (R^2_{y \cdot H} - r^2_{y \cdot 1})(\text{SS}_Y \text{ total}) \qquad [14.7.1]$$

or

$$\text{SS deviations from linear regression} = R^2_{y(G \cdot 1)} (\text{SS}_Y \text{ total}), \quad H - \{V_1\} = G. \qquad [14.7.2]$$

In the symbolism adopted in Section 13.19, this index $R^2_{y(G \cdot 1)}$ represents the variance in Y explained by all other aspects of X, *after* the contribution made by linear regression is removed. Thus, $R^2_{y(G \cdot 1)}$ must represent *any possible curvilinear relationship* that still exists between X and Y, over and above the linear relationship represented by $r^2_{y \cdot 1}$.

The partition of the total sum of squares is thus

$$\text{SS total} = \text{SS linear regression} + \text{SS deviations from linear} + \text{SS error}$$

where

$$\text{SS deviations from linear} = \text{SS between} - \text{SS linear regression} \qquad [14.7.3]$$

The computations thus needed are those for an ordinary one-way analysis of variance, and the SS linear regression, as found from 14.2.10 and 14.2.11.

The analysis ordinarily is summarized in a table such as the following:

Source	*SS*	*df*	*MS*	*F*
Between groups		$J - 1$	—	—
Linear regression		1	$\frac{\text{SS lin.}}{1}$	$\frac{\text{MS lin.}}{\text{MS error}}$
Deviations from linear regression		$J - 2$	$\frac{\text{SS dev.}}{J - 2}$	$\frac{\text{MS dev.}}{\text{MS error}}$
Error		$N - J$	$\frac{\text{SS error}}{N - J}$	
Total		$N - 1$		

We know already that the degrees of freedom associated with SS linear regression must be 1. The degrees of freedom for SS between groups is, of course, $J - 1$. Hence, the degrees of freedom for SS deviations from linear regression (representing the number of remaining orthogonal variables after the first, linear, component is removed) must be $J - 2$.

In this instance, the test for linear regression is given by

$$F = \frac{\text{MS linear regression}}{\text{MS error}} \tag{14.7.4}$$

with 1 and $N - J$ degrees of freedom. This, of course, tests the hypothesis H_0: $\rho^2_{y \cdot 1} = 0$, against the alternative H_1: $\rho^2_{y \cdot 1} > 0$.

In order to test the hypothesis of no curvilinear regression, we employ the F ratio

$$F = \frac{\text{MS deviations from linear}}{\text{MS error}} \tag{14.7.5}$$

with $J - 2$ and $N - J$ degrees of freedom. Note that this is exactly the same as the test

$$F = \frac{R^2_{y(G \cdot 1)}(N - J)}{1 - R^2_{y \cdot H}(J - 2)}, \qquad H - \{V_1\} = G, \tag{14.7.6}$$

given by 13.20.14 when the appropriate substitutions are made.

A significant value for the first of these F tests indicates that it is safe to conclude that some predictability is afforded by a linear rule. In the same way, a significant value for F for the second test lets one conclude that some further prediction is possible using a curvilinear rule. However, neither of these tests guarantees that there exists either a strong linear or a strong curvilinear relationship in the population sampled. The evidence for strength of linear relationship is given by (SS linear/SS_Y total), which is, of course, the sample value of $r^2_{y \cdot 1}$. The evidence for strength of curvilinear relation is then given by (SS deviations from linear/SS_Y total), which is the sample value of $R^2_{y(G \cdot 1)}$. The assumptions underlying these tests are exactly the same as those made in the test for linear regression in Section 14.2.

Although the example to follow shows an experiment in which an equal number of subjects are assigned to each of J groups, this need not be true in general. Furthermore, although X values in equal steps are chosen for the groups in the example, this is not a general requirement in the simple tests for linear and curvilinear trend illustrated here. However, the methods of Section 14.9 must be modified if unequal numbers are used in the groups, or if the X values are chosen in unequal steps.

14.8 / AN EXAMPLE OF TESTS FOR LINEAR AND CURVILINEAR REGRESSION

In Section 14.5 an example was suggested in which a curvilinear relation between experimental and dependent variables might possibly occur. Let us now suppose that this experiment is carried out. Six groups are formed, each of which is exposed to a different level of noise intensity. Sixty subjects chosen at random are assigned at random to these groups, with 10 per group. Each group represents a one-step increment in a scale of noise intensity. The exact units of noise intensity need not bother us here, and we can represent the levels of X as 1, 2, 3, and so on, since the different groupings represent equal intervals in noise intensity. The dependent variable is the score y_{ij} of a subject in a complex performance given the noise intensity level x_j. It is desired to test both for linear and for curvilinear regression, using $\alpha = .01$. The scores are shown in Table 14.8.1.

Table 14.8.1

Performance of subjects under differing levels of noise intensity.

Noise intensity levels, X_j					
1	*2*	*3*	*4*	*5*	*6*
18	34	39	37	15	14
24	36	41	32	18	19
20	39	35	25	27	5
26	43	48	28	22	25
23	48	44	29	28	7
29	28	38	31	24	13
27	30	42	34	21	10
33	33	47	38	19	16
32	37	53	43	13	20
38	42	33	23	33	11
270	370	420	320	220	140

The usual computations for a one-way analysis of variance are carried out first:

$$\text{SS total} = (18)^2 + (24)^2 + \cdots + (11)^2 - \frac{(1740)^2}{60}$$

$$= 7252$$

$$\text{SS between} = \frac{(270)^2 + \cdots + (140)^2}{10} - \frac{(1740)^2}{60}$$

$$= 5200$$

$$\text{SS error} = \text{SS total} - \text{SS between} = 7252 - 5200 = 2052.$$

Next the SS for linear regression is found from equations 14.2.10 and 14.2.11. Since $n_j = 10$ for each group,

$$b_{Y \cdot X} = \frac{\sum_j \sum_i x_j y_{ij} - \left[\left(\sum_j 10x_j\right)\left(\sum_j \sum_i y_{ij}\right)/60\right]}{\left(\sum_j 10x_j^2\right) - \left[\left(\sum_j 10x_j\right)^2/60\right]}. \qquad [14.8.1^*]$$

Here

$$\sum_j \sum_i x_j y_{ij} = \sum_j x_j \left(\sum_i y_{ij}\right) = 1(270) + 2(370) + \cdots + 6(140) = 5490$$

$$\sum_j 10x_j = 10(1) + \cdots + (10)(6) = 210$$

$$\sum_j 10x_j^2 = 10(1 + 4 + \cdots + 36) = 910$$

so that

$$b_{Y \cdot X}^2 \text{ SS}_X = \text{SS lin. reg.} = \frac{\{5490 - [(210)(1740)/60]\}^2}{(910) - (210)^2/60}$$

$$= 2057.1.$$

Then

$$\text{SS dev. from lin. reg.} = \text{SS between} - \text{SS lin. reg.}$$

$$= 5200 - 2057.1$$

$$= 3142.9.$$

Thus,

$$r_{y \cdot 1}^2 = \frac{2057.1}{7252} = .28.$$

and

$$R_{y(G \cdot 1)}^2 = \frac{3142.9}{7252} = .43.$$

The completed summary table is thus

Source	*SS*	*df*	*MS*	*F*
Between groups	5200	5	—	—
Linear reg.	2057.1	1	2057.1	54.1
Dev. from lin.	3142.9	4	785.7	20.7
Error	2052	54	38	
Totals	7252	59		

On evaluating these two F tests, we find that each is significant far beyond the .01 level. In short, we can reject both the hypothesis of no linear regression *and* the hypothesis of no curvilinear regression, although we should recall that the probability of error in at least one of these conclusions is not necessarily .01. Nevetheless, here we may say with some confidence that both linear and curvilinear regression exist. Actually, in this sample, curvilinear regression accounts for more variance (43 percent) than does linear regression (28 percent). The best rule for population prediction apparently involves both x_j and powers of x_j, in terms of this model.

A plot of the Y means for these data supports our findings (Figure 14.11.1). The sample means tend, roughly, to fall along a straight line with negative slope (the dotted line in the figure). However, they would cluster even more closely along a curved line suggesting a parabolic arc. The experimenter would be likely to conclude that, in the range studied, relatively low levels of noise intensity may actually facilitate the performance of subjects in this task, but that beyond a particular point (about level 3) average performance falls off rapidly with further increases in noise intensity.

14.9 / PLANNED COMPARISONS FOR TREND: ORTHOGONAL POLYNOMIALS

The separation of sums of squares into components representing linear and nonlinear regression is but the simplest of the procedures that may be applied to problems of linear and curvilinear regression. Many times an experiment is set up to answer certain specific questions about the *form* of the relation existing between two variables. In general when the experimenter wants to look into the question of trend, or form of relationship, some prior ideas about what the population relation should be like. These hunches about trend often come directly from theory, or they may come from the extrapolation of established findings into new areas. At any rate, specific questions are to be asked about the *degree* of the regression equation sufficient to permit prediction.

The technique of planned comparisons studied in Chapter 12 lends itself directly to this problem. Given J groups, each of which is associated with a value x_j of the experimental variable, each of the $J - 1$ degrees of freedom for comparisons can be allotted to the study of *one question* about the form or degree of the predictive relation. In this discussion, we will restrict our attention to the situation where the following conditions are satisfied.

1. The various x_j values represent *equally spaced unit intervals* on the X dimension or continuum. The units may be anything, but the difference between each x_j and x_{j+1} is precisely one of these units.
2. The number of observations made, n_j, is the same for each value of x_j.
3. Inferences are to be made about a hypothetical population in which *only possible* values of x_j are those actually represented in the experiment.

The first two of these requirements are not especially severe limitations, since this technique is ordinarily applied to experimental studies where one is free to choose the x_j values and the n in a group quite arbitrarily. In such studies it is usually more convenient than otherwise to form equal spacings of x_j values, and to assign individuals at random and equally among the groups. The third restriction is quite serious, however. In uses of this method, **really nothing can be said about values of x_j that**

are not directly represented in the experiment, insofar as statistical inference is concerned. The form of the relation may be quite different for X outside of the range represented, or even for potential x_j values intermediate to two actually represented. It is up to the experimenter to exercise caution and good scientific judgment in extrapolating beyond the x_j values actually observed. The statistician cannot really help here, since this is a question of scientific, not statistical, inference.

We have already seen (Section 14.2) that in the regression model the sum of squares for *linear* regression corresponds to the SS for one possible comparison among the sample means. Extending this idea, an orthogonal comparison can also be made reflecting second-degree or quadratic relationship. Still another orthogonal comparison can be carried out reflecting the third-degree, or cubic, relationship, and so on, until the experimenter has either investigated all the specific questions of interest, or $J - 1$ orthogonal comparisons have been made (whichever happens first). As for any set of *planned* comparisons, particular values from among the $J - 1$ possible comparisons are tested separately instead of the overall F test of between-group means. If any one of these comparisons is significant, then the experimenter has relatively strong evidence that the corresponding β value in the population is not zero, and that the best prediction rule is at least of the degree represented by that comparison.

The weights which each trend comparison involves are completely dictated by J, the number of different x_j groups, and by the degree of the particular trend investigated. That is, given any number of groups J, one standard set of weights exists for investigating linear, or first-degree, trends, an orthogonal set exists for second-degree, or quadratic, trends, another for cubic, or third-degree trends, and so on, up to and including trends of degree $J - 1$. These standard sets of comparison weights are called **coefficients of orthogonal polynomials,** and the method itself is sometimes called **the method of orthogonal polynomials.**

(The final example of the use of the Gram-Schmidt procedure in Appendix C deals with a set of vectors where every vector represents a power of the entries in vector $\mathbf{x}_1$, so that $\mathbf{x}_2$ gives the squares, $\mathbf{x}_3$ the cubes, and so on, of the elements of $\mathbf{x}_1$. Then when an orthogonal basis is found, the vectors yield the orthogonal polynomial weights we will be using below. In other words, these sets of orthogonal polynomial weights are just an orthogonal basis for X, X^2, X^3, and so on. All of the points raised in Section 14.1 are thus relevant here as well.)

The **orthogonal polynomial coefficients are so derived that the particular comparisons among means each reflect one kind of possible trend or form of relationship in the data.** This sounds much more complicated than it actually is. Perhaps a simple example may clarify this point. For four groups, equally spaced with regard to the x_j values, the three sets of standard orthogonal polynomial weights are

Comparison	M_1	M_2	M_3	M_4
linear	−3	−1	1	3
quadratic	1	−1	−1	1
cubic	−1	3	−3	1

The symbol M_1 denotes the mean Y value for the group with the *lowest* X value, or x_1, M_2 the mean Y for the next lowest value of X, or x_2, and so on. Applying the weights given in the first row to the means, we get the first, or *linear* comparison:

$$\hat{\psi}_1 = -3(M_1) - 1\,(M_2) + 1(M_3) + 3(M_4).$$

The sample value of $\hat{\psi}_1$ is an unbiased estimate of the *population* comparison

$$\psi_1 = -3(\mu_1) - 1(\mu_2) + 1(\mu_3) + 3(\mu_4).$$

If this population comparison is not equal to zero, then there is at least a linear trend in the relationship. Higher-order trends may be present, however.

By applying the weights of the second comparison to the sample means, we get

$$\hat{\psi}_2 = 1(M_1) - 1(M_2) - 1(M_3) + 1(M_4)$$

which estimates ψ_2, the same linear combination applied to the population means. This comparison value in the population can be zero only if $\beta_2 = 0$, so that no quadratic trend exists in the statistical relationship. If this comparison value in the population is other than zero, then at least a quadratic trend exists in the statistical relationship.

Finally, the comparison for cubic trends is

$$\hat{\psi}_3 = -\,1(M_1) + 3(M_2) - 3(M_3) + 1(M_4).$$

This comparison estimates ψ_3. If the population value is other than zero, a cubic trend exists in the relationship.

If all three comparisons, ψ_1, ψ_2, ψ_3, are zero when applied to population values, then no predictive relationship at all exists. On the other hand, large and significant sample comparison values in the sample are evidence that trends of at least those particular degrees exist in the population relation.

One more thing should be pointed out: these weights used in the example are bona fide comparison weights, since the weights for each comparison sum to zero. Furthermore, the comparisons are all orthogonal, since the summed products of weights for any two comparisons is also zero. Thus the theory we have studied for any set of planned orthogonal comparisons applies perfectly well to trend comparisons of this type, as do the connections with multiple regression outlined earlier.

Table VII in Appendix D gives the required weights for trend comparisons for $J = 3$ through $J = 15$, and for trends through the sixth degree, which certainly should cover most social science research on such problems. One simply finds the appropriate section of the table, and reads off the required weights. Within each section, the rows represent the x_j groupings; here the lowest value of x_j is shown as the first or topmost row. The columns represent the *degree* of trend to be examined. Then the entry in any row j and column h gives the weight applied to the mean of the jth group in the comparison for a trend of degree h. A larger table of required weights is given in *Biometrika Tables for Statisticians* (1967).

This general procedure is followed until as many of the comparisons as have a priori interest for the experimenter have been made. These comparisons should actually be *planned* before the data are seen, and should really correspond to questions that are germane to the experimental problem and will clarify the interpretation of the experiment by their separate examination. For more than four or five experimental groups, it is seldom practical or useful to work out all the possible $J - 1$ trend comparisons. Usually, there is little understanding to be gained by finding, for example, that a sixth-degree or higher trend is significant. Ordinarily one stops after the cubic or perhaps quartic comparison, and then relegates all other trend effects to the pooled sum of squares:

$$\text{SS other trends} = \text{SS between} - \sum_{h=1}^{L} \text{SS}(\hat{\psi}_h) \qquad [14.9.1]$$

where h symbolizes any from the L comparisons made separately. For, say, five groups, we might be interested only in linear and quadratic, and then find the pooled SS for other trends from

$$\text{SS (cubic and quartic)} = \text{SS between} - \text{SS lin.} - \text{SS quad.}$$

which has 4 − 2 or 2 degrees of freedom.

The proportion of variance accounted for in the data by any trend h is, of course,

$$r^2_{y \cdot h} = \frac{\text{SS}(\hat{\psi}_h)}{\text{SS}_Y \text{ total}}. \qquad [14.9.2]$$

Then the proportion of variance attributable to any set G of trends, any one of which is symbolized by g, is

$$R^2_{Y \cdot G} = \sum_{g \text{ in } G} \text{SS}(\hat{\psi}_g)/\text{SS}_Y \text{ total}. \qquad [14.9.3]$$

14.10 / AN EXAMPLE OF PLANNED COMPARISONS FOR TREND

Suppose that the experiment on noise intensity level and performance had actually been planned as a study of trend. The experimenter is especially interested in judging if linear and quadratic trends exist in the relation between X and Y. We will pretend that the data in Section 14.8 had been collected specifically for this purpose.

The SS total, the SS between, and the SS error are found just as before, of course. However, let us list the group means and find the linear and quadratic comparison values. Table 14.10.1 gives the six means and the weights that apply to each for linear and quadratic comparisons, as found from Table VII. The comparison values themselves are

Table 14.10.1

Linear and quadratic orthogonal polynomial weights for the noise intensity data.

	Group					
	1	*2*	*3*	*4*	*5*	*6*
Means	27	37	42	32	22	14
Weights						
linear	−5	−3	−1	1	3	5
quadratic	5	−1	−4	−4	−1	5

$$\begin{aligned}\hat{\psi}_1 &= -5(27) - 3(37) - 1(42) + 1(32) + 3(22) + 5(14)\\ &= -120,\end{aligned}$$

and

$$\hat{\psi}_2 = 5(27) - 1(37) - 4(42) - 4(32) - 1(22) + 5(14) = -150.$$

The w values for the corresponding sums of squares are found from

$$w_1 = \frac{25 + 9 + 1 + 1 + 9 + 25}{10} = 7$$

and

$$w_2 = \frac{25 + 1 + 16 + 16 + 1 + 25}{10} = 8.4$$

so that

$$\text{SS linear} = \frac{(-120)^2}{7} = 2057.1$$

$$\text{SS quad.} = \frac{(-150)^2}{8.4} = 2678.6.$$

The proportions of variance accounted for are then, respectively,

$$r^2_{y \cdot 1} = \frac{2057.1}{7252} = .28,$$

just as found in Section 14.8, and

$$r^2_{y \cdot 2} = \frac{2678.6}{7252} = .37.$$

Then

$$\text{SS other trends} = 5200 - 2057.1 - 2678.6 = 464.3$$

with

$$J - 1 - 2 = 3 \text{ degrees of freedom.}$$

The proportion of variance accounted for by cubic and higher trends is thus

$$R^2_{y \cdot 345} = \frac{464.3}{7252} = .06.$$

The summary table for this analysis is

Source	*SS*	*df*	*MS*	*F*
Between	5200	5	—	—
Linear	2057.1	1	2057.1	54.1
Quadratic	2678.6	1	2678.6	70.5
Other trends	464.3	3	154.8	4.1
Error	2052	54	38	
Totals	7252	59		

The F tests for both linear and quadratic trend greatly exceed the value required for the .01 level of significance, but the overall test for other trends does not reach the .01 level. Thus, we can be fairly sure that in the population predictive association does exist between X and Y, and that the general rule for prediction includes both linear and quadratic components. We cannot, however, really be sure that the population prediction rule does not include higher-order components as well. Even so, we infer from these data that a curvilinear regression equation affording at least some prediction has the general form

$$y'_{ij} = \beta_0 + \beta_1(x_j - \mu_x) + \beta_2(x_j - \mu_x)^2$$

where the β_1 and β_2 are the linear and quadratic regression coefficients in the population.

14.11 / ESTIMATION OF A CURVILINEAR PREDICTION FUNCTION

Sometimes it is convenient to find the actual pairings of x_j and predicted y'_j values given by the curvilinear regression equation estimated from a sample. The experimenter may be interested in examining the departures the data tend to show from this estimated function, or perhaps in displaying in graphic form the relation suggested by the significant trends in the data. This is rather tedious to do in terms of the original x_j values, but when the x_j values are equally spaced a very simple substitute exists, involving the orthogonal polynomial coefficients once again. Instead of finding the predicted y directly as a function of x_j, we may represent y' as a function of the *polynomial weights* given to groups, and then use these predicted y values to find and plot all (x_j, y'_j) pairs.

A way to find such a prediction rule has already been given in Section 14.1. There it was pointed out that when $J - 1$ variables are orthogonal both to each other and to the general mean, each such variable V_k has a regression coefficient given by 14.1.7. The regression equation in terms of v values is then given by 14.1.13.

This idea applies directly to orthogonal polynomial comparisons. Thus, the linear trend comparison $\hat{\psi}_1$ can be turned into a regression weight b_1 by taking

$$b_1 = \frac{\hat{\psi}_1}{nw_1} \qquad [14.11.1]$$

where nw_1 is equal to sum of the squared weights. In exactly this same way the regression weight b_2 can be found for the quadratic comparison, b_3 for the cubic comparison, and so on. The regression equation employing all $J - 1$ comparisons would then be (14.1.13):

$$y'_j = M_Y + \sum_{k=1}^{J-1} b_k c_{kj}$$

where c_{jk} is the weight given to group j in the comparison for trend k. However, it is not very useful to employ the b values for *all* of the $J - 1$ trends in setting up this equation. The reason is that, if we do so, the predicted value for any individual in any group j will simply be M_{Y_j}, the mean for that group.

Instead, what is done is to take only those trends that have proved to be significant in the original analysis, and then use these comparison values to build the regression equation. Thus, if we let g stand for one of the set **G** of *significant* trends, then the prediction rule of interest will be

$$y'_{j \cdot G} = M_Y + \sum_{g \in G} \frac{\hat{\psi}_g \, c_{gj}}{{}_n w_g}. \qquad [14.11.2*]$$

The set **G** could, of course, be defined in any other appropriate way, although generally it is of most interest to use only those trends which we have some confidence do exist in the population.

After the original comparisons have been carried out using the orthogonal polynomial weights, it is a fairly simple matter to construct this multiple regression equation, as we shall now see.

Recall first of all that for this example we concluded that *both* linear and quadratic trends exist. From the evidence at hand, these trends should be relatively strong in the population, and so we will construct a function relating y'_j to x_j which reflects *only* linear and quadratic trends. First of all, we take the value already found for the linear comparison,

$$\hat{\psi}_1 = -\ 120,$$

and divide by

$$nw_1 = (10)(7)$$

to find

$$b_1 = \frac{-120}{70} = -\ 1.71.$$

Next we take the value for the quadratic comparison,

$$\hat{\psi}_2 = -150$$

and

$$nw_2 = (10)(8.4) = 84,$$

to find

$$b_2 = \frac{-150}{84} = -1.79.$$

Then our prediction rule for y'_j, *in terms of the polynomial weights* c_{gj} is

$$\begin{aligned} y'_j &= M_Y + b_1 c_{1j} + b_2 c_{2j} \\ &= 29 + (-1.71)c_{1j} + (-1.78)c_{2j}. \end{aligned}$$

Now suppose that we want to predict for the group with the *lowest* value of x_1. A look at Table VII for $J = 6$ shows that in the comparison for linear trends the mean of this group is weighted by -5, and in the comparison for quadratic by $+5$, so that

$$c_{11} = -5$$

$$c_{21} = 5.$$

For this group, the predicted value of y'_j in terms of linear and quadratic trends is thus

$$\begin{aligned} y'_1 &= 29 - (1.71)(-5) - (1.78)(5) \\ &= 28.65. \end{aligned}$$

Our best guess is that an observation in the lowest x-value group will have a y score of about 28.65.

Now we predict for x_2, the second lowest value of x. Here, in the comparison for linear trends the weight assigned is -3, and for quadratic, -1. Thus

$$\begin{aligned} y'_2 &= 29 - (1.71)(-3) - (1.78)(-1) \\ &= 35.91. \end{aligned}$$

Therefore, we predict the mean of the second group to be 35.91.

Going on in the same way for the other groups, we find the following functional pairings of x_j and *predicted* y'_j values:

for x_1, $y'_1 = 28.65$

for x_2, $y'_2 = 35.91$

for x_3, $y'_3 = 37.83$

for x_4, $y'_4 = 34.41$

for x_5, $y'_5 = 25.65$

for x_6, $y'_6 = 11.55$.

These (x_j, y'_j) are plotted in Figure 14.11.1 together with the actual (x_j, M_{Yj}) pairings for the sample. The actual data points are shown as the heavy dots, and the fitted curve is drawn *as though* this function obtained over the whole range of X values. Notice that although the fit between the predicted values and the sample means is not perfect, since each mean point departs somewhat from its predicted position on the regression curve, the general form of the curve does more or less parallel the systematic differences among the sample means. By way of contrast, the predictions using the linear regression equation alone correspond to points falling on the dotted line; it is apparent that adding a quadratic component to the prediction equation greatly improves our description of the relationship between X and Y exhibited in this sample, since the various means depart considerably from the regression line. On the other hand, if we added cubic, quartic, and quintic terms to our prediction equation, then we would find a curve on which the mean points for each group would fall exactly. There is not much profit in this, however, as the best description of a set of data is a curve fitting the points most nearly with a *simple* function rule (one of the lowest possible degree). Any set of means from J sampled groups can be fitted with a curve of at most degree $J - 1$.

The goodness of fit of this curve to the y_{ij} values is of course reflected in the proportion of the total variance accounted for by these trends:

$$R^2_{y \cdot G} = R^2_{y \cdot 12} = \frac{\text{SS linear} + \text{SS quadratic}}{\text{SS total}} = .65$$

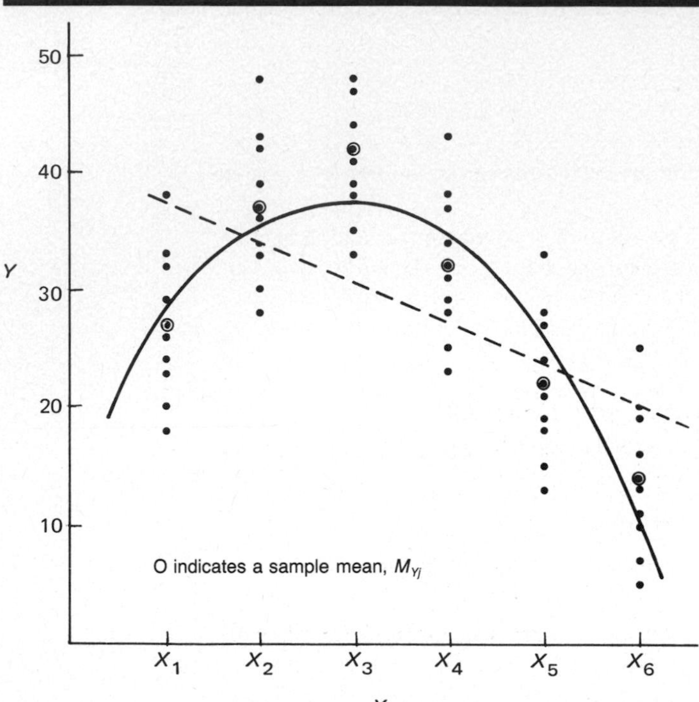

FIG. 14.11.1 Linear and quadratic functions fitted to the data of Section 14.8.

Thus, the multiple correlation $R_{y \cdot 12}$ is about .81. The curve consisting only of linear and quadratic trends appears to fit the data very well indeed.

This device for actually obtaining predicted points using a curvilinear regression equation is quite easy to apply after a trend analysis, since most of the computations have already been done. It is useful as a way to visualize how the kind of trend suggested by the significant comparisons might actually appear in a plot, as well as a device to dramatize the predictive strength, or the weakness, of the relation inferred from the data. Remember, however, that such an obtained regression curve is actually only a description of *these* data. *It can be regarded as an estimate of the population regression function only for hypothetical populations showing exactly the same representation of X values as used in the sample*. Free and easy extrapolation outside these values is done at the user's peril!

An additional word of caution: curve fitting by use of orthogonal polynomials is but the simplest approach to the problem of fitting a function to a set of data. A number of other approaches to this problem exist. In particular, it must be borne in mind that the curvilinear regression function determined by this method depends on our ability to represent a functional relation by a power series, or Taylor expansion. Merely because we can reproduce the general form of relation in the data by, say, a function of the third degree, it does not follow that the true relation is best expressed in these terms. It may well be that another, and perhaps simpler, form of function rule best describes the relation, with parameters that can be linked to the social or psychological factors involved. In particular, when one has definite ideas about the mathematical form of the relation, and wishes to estimate parameters for this function, other methods are called for. The book by Lewis (1960) gives a useful introduction to the general subject of curve-fitting for various purposes. Some important modern ideas on this whole subject may be found in Mosteller and Tukey (1977).

This example of a simple analysis for trend has been carried out in terms of a single set of distinctly different samples, each corresponding to one fixed value of the experimental variable X. Obviously, however, in many research settings this sort of question is studied more naturally within a more complex design, perhaps one in which repeated observations on the same subjects are made. This requires some special considerations that we will not be able to pursue here, and the reader is referred to the book by Winer (1971) for details. Similarly, we do not have the space to go into the application of orthogonal polynomials to multifactor designs in which two or more factors have equally spaced quantitative levels. Here, the "regression problem" approach does permit one to examine the data for evidence of interactions between the quantitative factors by creation of new orthogonal sets of weights V_{gh} applied to cell means of the table. Here for cell mean M_{jk}, and column trend g and row trend h, the weight is $v_{ghjk} = v_{gj}v_{hk}$. Each of the $(J - 1)(K - 1)$ sets of weights accounts for one interaction degree of freedom. The book by Winer is a good source for studying such extensions of regression methods.

14.12 / TREND ANALYSIS IN POST HOC COMPARISONS

It should be mentioned that the trend comparisons can perfectly well be made *after* the overall analysis of variance has shown significant evidence for effects of the experimental variable. In particular, it may be informative to investigate trends of various degrees after significant evidence for overall curvilinear regression has been found. The Scheffé method discussed in Chapter 12 applies to any post hoc comparison among means, and one can use the orthogonal polynomial weights in the usual way to make post hoc trend comparisons. However, just as for any set of post hoc comparisons, the probability of our detecting a true trend as a signficant result is ordinarily less in the test for a post hoc than in the test for a planned comparison for that trend. Even so, post hoc trend comparisons provide a most useful device for exploring data after the overall F for between groups has shown significance, when the independent variable groupings are equally spaced quantitative distinctions. Unless definite questions exist about particular trends prior to the data, tests for trend are usually by the post hoc method. Estimation of the function can then be carried out as in Section 14.11.

14.13 / THE ANALYSIS OF COVARIANCE

Although at this point only a brief treatment of the analysis of covariance is possible, it does deserve mention. Whereas the usual analysis of variance deals with experimental factors treated qualitatively, and ordinary regression analysis deals with quantitative factors, the analysis of covariance represents a blending of the different approaches. In the analysis of covariance, some of the experimental factors are qualitative and others are quantitative. The theory of the analysis of covariance thus represents still a further extension of the general linear model to a problem in regression. As we have seen throughout this chapter, these two approaches come together under the theory and methods of multiple regression, and we will find some of the concepts of multiple regression especially useful as we examine the analysis of covariance.

In the following it will be important to distinguish between the qualitative experimental factor (the independent variable) and the quantitative factor. The first we will call simply the experimental factor; the second we will call the "concomitant variable." The dependent variable, Y, is of course quantitative. We will also follow convention by calling the concomitant, quantitative variable, X. Since we will not need to deal with the qualitative factor other than in terms of its various levels, j, it will really not be necessary to use a variable symbol for it.

In the most usual situation to which the analysis of covariance is applied, the experimental variable is thought of as having J levels, each associated with a fixed effect, α_j. However, the concomitant variable X is thought of as sampled, as in a problem in correlation. That is, each randomly selected observation i in any group j simply "brings along" a value x_{ij} to the set of data as well as a value y_{ij}. Furthermore, the X values are assumed to have no relation to the experimental levels, and the relationship between the X and Y variables is assumed to be constant across all of the J experimental levels. This is the sampling model we will explore. Nevertheless, in certain problems it is entirely possible to have fixed values of the concomitant variable as well as fixed levels of the experimental variable. No change in procedure is called for in this instance, although a slightly different set of assumptions then underlie the significance test.

One of the principal uses of the analysis of covariance is exemplified in the following situation: Suppose than an experiment were planned on the relationship between different methods of teaching students FORTRAN for the computer and their performance at the end of a year's instruction. Although originally none of the students knows FORTRAN, it is known that they vary in general ability. It might be possible to match the students on general ability prior to the experiment, and then analyze the results accordingly. An alternate strategy is not to match the students, but rather to measure the general ability of each and to adjust for this factor by use of the analysis of covariance. This technique permits the experimenter to adjust the results after the fact, in such a way that performance differences among the different treatment groups due to the linear relationship of performance and ability are effectively removed from consideration.

Generally speaking, the analysis of covariance permits a post hoc, statistical control for one or more concomitant variables, removing their influence from the comparison of groups on the main experimental variable. Naturally, there are some important qualifications to the use of this procedure, and it is not so universally useful as it might at first appear. In particular, this method is effective only where the relationship between the concomitant and the dependent variable is linear, and where the degree of this relationship does not itself depend upon the experimental variable. Nevertheless, in situations in which its requirements are satisfied, the use of the analysis of covariance can be quite effective in helping to clarify statistical relationships.

14.14 / THE LINEAR MODEL IN ANALYSIS OF COVARIANCE

The model for simple analysis of covariance is a direct extension of the model of linear regression. The qualitative, experimental, factor is to be represented by J distinct levels. Each observation i within any group j yields two score values. The first is a value y_{ij} for the dependent variable; the second is a value x_{ij} on the concomitant variable. Then the linear model that is assumed to hold is

$$y_{ij} = \mu + \alpha_j + \beta_{Y \cdot X}(x_{ij} - \mu_X) + e_{ij}. \qquad [14.14.1]$$

That is, Y_{ij} depends upon the general mean μ, plus the effect due to treatment j, and it also depends upon the value of x_{ij}, as weighted by the linear regression coefficient $\beta_{Y \cdot X}$. There is also a random error component e_{ij}. We assume further that only a linear relationship exists between X and Y, and that the value of $\beta_{Y \cdot X}$ does not depend upon j. In other words, within each treatment population, the relationship between X and Y is linear and has the same regression coefficient $\beta_{Y \cdot X}$. The other, usual, analysis-of-variance assumptions are also made.

In the more common situation, in which both the x_{ij} and the y_{ij} are left totally free to vary, with neither under the control of the experimenter, then a bivariate normal distribution of X and Y is also assumed. However, in the situation where the values of the concomitant variable are also fixed in advance, just as in any other problem in regression, one assumes that the only values of the concomitant variable that are of concern are the ones actually represented in the experiment. Here, no special assumption about the bivariate distribution of X and Y need be made.

Now when the least squares criterion is applied to the model given above, subject to the restriction that the sum of the α_j effects must be zero, the estimate of $\beta_{Y \cdot X}$ is provided by

$$b_W = \frac{\text{SP}_{XY} \text{ within groups}}{\text{SS}_X \text{ within groups}} \qquad [14.14.2^*]$$

where

$$\text{SP}_{XY} \text{ within} = \sum_j \sum_i x_{ij}y_{ij} - \sum_j \frac{\left(\sum_i x_{ij}\right)\left(\sum_i y_{ij}\right)}{n_j} \qquad [14.14.3]$$

and

$$\text{SS}_X \text{ within} = \sum_j \sum_i x_{ij}^2 - \sum_j \frac{\left(\sum_i x_{ij}\right)^2}{n_j}. \qquad [14.14.4]$$

The value of SP_{XY} within reflects the predictability that remains between X and Y when the effects of the experimental factor are (quite literally) held constant. The experimental factor itself may, as we have seen (Section 14.1), be conceptualized in terms of an underlying set of orthogonal variables, $G = \{V_1, V_2, \cdots, V_{J-1}\}$. Hence it is appropriate to think of b_W^2 as having to do with variance accounted for by X when the effects of the experimental factor, and thus of the variables in the set G, are removed. Let us then take

$$r_{y(x \cdot G)}^2 = \frac{b_W^2 \,(\text{SS}_X \text{ within})}{\text{SS}_Y \text{ total}}$$

to symbolize this proportion of variance accounted for, so that

$$r_{y(x \cdot G)}^2 \,(\text{SS}_Y \text{ total}) = b_W^2 \,(\text{SS}_X \text{ within}). \qquad [14.14.5]$$

We know that all predictability associated with the experimental groups is embodied in SS between, which, in terms of the orthogonal variables V making up the set G, accounts for the proportion of variance given by

$$R^2_{y \cdot G} = \frac{\text{SS}_Y \text{ between}}{\text{SS}_Y \text{ total}}$$

so that

$$R^2_{y \cdot G} (\text{SS}_Y \text{ total}) = \text{SS}_Y \text{ between}. \qquad [14.14.6]$$

Then the variance accounted for by *all* of the systematic features of the data including *both* the experimental factor and the concomitant variable, is equivalent to the variance accounted for by the entire set $H = \{X, V_1, \cdots, V_{J-1}\}$:

$$R^2_{y \cdot H} (\text{SS}_Y \text{ total}) = (R^2_{y \cdot G} + r^2_{y(x \cdot G)})(\text{SS}_Y \text{ total}). \qquad [14.14.7]$$

Now consider the correlation between X and Y across all observations and all experimental groups. The variance accounted for by X, without regard to the experimental factor, is then $r^2_{yx}(\text{SS}_Y \text{ total})$. However, if this variance is subtracted from the total variance accounted for, we have, again using the notation of 13.19.5,

$$R^2_{y(G \cdot x)}(\text{SS}_Y \text{ total}) = (R^2_{y \cdot H} - r^2_{yx})(\text{SS}_Y \text{ total}). \qquad [14.14.8]$$

In other words, the variance in Y predictable from the experimental factor, *over and above any variance that may be due to* X, is given by

$$\text{SS}_Y \text{ between} + b^2_W(\text{SS}_X \text{ within}) - r^2_{yx} (\text{SS}_Y \text{ total}) = \text{adjusted SS between} \qquad [14.14.9^*]$$

Furthermore, the error variance, representing variability unaccounted for *either* by the experimental groups or by the relation between X and Y within groups is simply

$$\begin{aligned} (1 - R^2_{y \cdot H})(\text{SS}_Y \text{ total}) &= \text{SS}_Y \text{ total} - \text{SS}_Y \text{ between} - b^2_W(\text{SS}_X \text{ within}) \\ &= \text{SS}_Y \text{ within} - b^2_W(\text{SS}_X \text{ within}) \qquad [14.14.10^*] \\ &= \text{adjusted SS}_Y \text{ error}. \end{aligned}$$

Like any other regression coefficient, b^2_W is associated with one degree of freedom. This implies that the degrees of freedom for SS adjusted error must be $N - 1 - (J - 1) - 1 = N - J - 1$. Hence, a test of the hypothesis that there is no predictability between the experimental factor and Y once the effects of X are removed may be had from

$$F = \frac{\text{adjusted SS}_Y \text{ between}/(J - 1)}{\text{adjusted SS error}/(N - J - 1)} \qquad [14.14.11^*]$$

with $J - 1$ and $N - J - 1$ degrees of freedom. That is, the test of the adjusted means in the analysis of covariance is an application of the principle of 13.20.14, where, in this instance,

$$F = \frac{R^2_{y(G \cdot x)}(N - J - 1)}{(1 - R^2_{y \cdot H})(J - 1)}, \qquad H - \{X\} = G, \qquad [14.14.12]$$

Here, $R^2_{y(G \cdot x)}$ represents the proportion of variance due to the experimental factor when the effects of the concomitant variable are removed.

A simple analysis of covariance requires three sets of calculations. First, for the concomitant variable X, one finds the usual sums of squares:

$$\text{SS}_X \text{ between} = \sum_j \frac{\left(\sum_i x_{ij}\right)^2}{n_j} - \frac{\left(\sum_j \sum_i x_{ij}\right)^2}{N},$$

$$\text{SS}_X \text{ total} = \sum_j \sum_i x_{ij}^2 - \frac{\left(\sum_j \sum_i x_{ij}\right)^2}{N},$$

$$\text{SS}_X \text{ within} = \text{SS}_X \text{ total} - \text{SS}_X \text{ between}.$$

The same sorts of computations are then carried out for the Y variable:

$$\text{SS}_Y \text{ between} = \sum_j \frac{\left(\sum_i y_{ij}\right)^2}{n_j} - \frac{\left(\sum_j \sum_i y_{ij}\right)^2}{N},$$

$$\text{SS}_Y \text{ total} = \sum_j \sum_i y_{ij}^2 - \frac{\left(\sum_j \sum_i y_{ij}\right)^2}{N},$$

$$\text{SS}_Y \text{ within} = \text{SS}_Y \text{ total} - \text{SS}_Y \text{ between}.$$

Finally, we must calculate sums of products of X and Y, both within and between groups. These sums of products form the basis for estimates of the regression coefficient. These calculations are

$$\text{SP}_{XY} \text{ between} = \sum_j \frac{\left(\sum_i x_{ij}\right)\left(\sum_i y_{ij}\right)}{n_j} - \frac{\left(\sum_j \sum_i x_{ij}\right)\left(\sum_j \sum_i y_{ij}\right)}{N},$$

$$\text{SP}_{XY} \text{ total} = \sum_j \sum_i x_{ij} y_{ij} - \frac{\left(\sum_j \sum_i x_{ij}\right)\left(\sum_j \sum_i y_{ij}\right)}{N},$$

$$\text{SP}_{XY} \text{ within} = \text{SP}_{XY} \text{ total} - \text{SP}_{XY} \text{ between}.$$

Then, since

$$b_W = \frac{\text{SP}_{XY} \text{ within}}{\text{SS}_X \text{ within}} \qquad [14.15.1]$$

$$b_W^2 \, \text{SS}_X = \frac{(\text{SP}_{XY} \text{ within})^2}{\text{SS}_X \text{ within}}. \qquad [14.15.2]$$

Also,

$$r_{yx}^2 \, (\text{SS}_Y \text{ total}) = \frac{(\text{SP}_{XY} \text{ total})^2}{\text{SS}_X \text{ total}}. \qquad [14.15.3]$$

Then, by 14.14.9 and 14.14.10,

$$\text{adjusted SS}'_Y \text{ between} = \text{SS}_Y \text{ between} + \frac{(\text{SP}_{XY} \text{ within})^2}{\text{SS}_X \text{ within}} - \frac{(\text{SP}_{XY} \text{ total})^2}{\text{SS}_X \text{ total}}, \quad [14.15.4^*]$$

$$\text{adjusted SS}'_Y \text{ error} = \text{SS}_Y \text{ within} - \frac{(\text{SP}_{XY} \text{ within})^2}{\text{SS}_X \text{ within}}. \quad [14.15.5^*]$$

The analysis of adjusted means can be summarized thus:

Source	*SS*	*df*	*MS*	*F*
Adjusted means	SS'_Y between	$J - 1$	$\frac{SS'_Y \text{ between}}{J-1}$	$\frac{MS'_Y \text{ between}}{MS'_Y \text{ error}}$
Adjusted error	SS'_Y error	$N - J - 1$	$\frac{SS'_Y \text{ error}}{N-J-1}$	
Adjusted total		$N - 2$		

14.16 / A SIMPLE EXAMPLE OF COVARIANCE ANALYSIS

In Section 14.13 a situation was described in which several different methods of teaching the FORTRAN programming language were being compared. However, the experimenter had reason to believe that general ability of the students might also affect their achievement under any given method. It was then desired to remove the possible linear effects of general ability on programming achievement through use of the analysis of covariance.

The subjects were twenty female college freshmen who were assigned at random to five groups, each representing a different method of teaching this material. Prior to the experimental sessions, each subject was given a general ability test (score x_{ij}) and, after the training, achievement was measured (score y_{ij}). No student had prior experience with programming languages. The results are shown in Table 14.16.1 (each x and y value has been reduced by 100 for computational convenience; this has no effect whatsoever on the final analysis). The desire of the experimenter was to treat the general ability score as the concomitant variable, and to ask if differences due to method exist even after adjustment for general ability.

Table 14.16.1

Ability and performance scores for 25 subjects in five learning groups.

Groups									
1		2		3		4		5	
x	*y*	*x*	*y*	*x*	*y*	*x*	*y*	*x*	*y*
10	18	22	40	30	38	35	25	11	15
20	17	31	22	31	40	37	45	16	17
15	23	16	28	18	41	41	50	19	20
12	19	17	31	22	40	30	51	25	23
57	77	86	121	101	159	143	171	71	75

Then for the analysis of covariance, the following are found for the X variable:

$$SS_X \text{ total} = 12066 - \frac{(458)^2}{20} = 1577.8$$

$$SS_X \text{ between} = \frac{46336}{4} - \frac{(458)^2}{20} = 1095.8$$

$$SS_X \text{ within} = 1577.8 - 1095.8 = 482.0.$$

Next, for the Y,

$$SS_Y \text{ total} = 20851 - \frac{(603)^2}{20} = 2670.6$$

$$SS_Y \text{ between} = \frac{80717}{4} - \frac{(603)^2}{20} = 1998.8$$

$$SS_Y \text{ within} = 2670.6 - 1998.8 = 671.8.$$

Finally, the sums of products are

$$SP_{XY} \text{ total} = 15140 - \frac{(458)(603)}{20} = 1331.3$$

$$SP_{XY} \text{ between} = \frac{60632}{4} - \frac{(458)(603)}{20} = 1349.3$$

$$SP_{XY} \text{ within} = 1331.3 - 1349.3 = -18$$

(note again that the SP_{XY} terms can be negative.) The adjusted sums of squares are then

$$SS'_Y \text{ between} = 1998.8 + \frac{(-18)^2}{482.0} - \frac{(1331.3)^2}{1577.8} = 876.2$$

$$SS'_Y \text{ error} = 671.8 - \frac{(-18)^2}{482} = 671.1$$

The summary table for this analysis of covariance is given below:

Source	*SS*	*df*	*MS*	*F*
Adjusted means	876.2	4	219.1	4.6
Adjusted error	671.1	14	47.9	
Adjusted total	1547.3	18		

The F value necessary for significance at the .05 level, for 4 and 14 degrees of freedom is 3.11. Thus, the hypothesis of no difference between the *adjusted* means may be rejected. Even removing the effects of general ability, the groups do seem to differ in performance.

On the other hand, had the experimenter analyzed the Y values through an ordinary one-way analysis of variance, without adjusting for general ability, then an F value of 11.16 would have been calculated for 4 and 15 degrees of freedom. Adjusting for general ability did reduce the apparent discrepancies between instructional methods.

14.17 / TESTING FOR HOMOGENEITY OF REGRESSION

Perhaps the biggest hitch in the application of the analysis of covariance is the assumption that the regression coefficient is constant within each of the treatment populations. Furthermore, it is necessary to assume that the regression of X on Y is linear within each such population. Fortunately, one can examine the data in order to see if the first of these assumptions is at least reasonable. Let β_j be the true regression coefficient within population j. Then we wish to test the hypothesis $H_0: \beta_j = \beta$ for every j.

In order to carry out this test of the homogeneity of within-treatment regression coefficients, we calculate for each group j the quantities

$$\text{SP}_{XYi} = \sum_i x_{ij}y_{ij} - \frac{\left(\sum_i x_{ij}\right)\left(\sum_i y_{ij}\right)}{n_j}$$

and

$$\text{SS}_{X_i} = \sum_i x_{ij}^2 - \frac{\left(\sum_i x_{ij}\right)^2}{n_j}$$

Then one takes

$$A = \text{SS}_Y \text{ within} - \sum_i \frac{\text{SP}^2_{XY_j}}{\text{SS}_{X_j}}$$

$$B = \text{SS}'_Y \text{ error} - A.$$

This leads to the F ratio

$$F = \frac{B/(J - 1)}{A/(N - 2J)} \qquad [14.17.1]$$

with $J - 1$ and $N - 2J$ degrees of freedom. Large values of F are evidence against the hypothesis of equal regression within groups, and thus cast doubt on the validity of this assumption in the analysis of covariance.

A test for homogeneity of within-treatment regression probably would have preceded the analysis of covariance in this example. However, carrying out this test on these data leads to an F value of only .28 with 4 and 10 degrees of freedom. Hence the experimenter could feel quite safe in assuming that the various populations are homogeneous with respect to the regression coefficient between X and Y. Although the usual reason for testing the hypothesis of equal regression among treatments is a desire to check on the assumptions underlying the analysis of covariance, there may also be interest in this hypothesis in its own right. The text by Winer (1971) also gives several other tests growing out of the analysis of covariance.

The theory of multivariate correlation and regression is one of the best developed parts of statistics. The few methods alluded to here are but the tip of a very large iceberg. Virtually all of these methods are capable of great extension to meet more complicated data situations.

For example, although only the simplest case has been considered here, analysis of covariance is not limited to simple one-factor designs with one concomitant variable. The method can be extended to quite complex designs with a number of experimental factors, and with two or more concomitant variables. The same idea may also be applied to repeated measures and similar designs.

For example, if there is one experimental variable and two concomitant variables, X and T, the model can be written as

$$y_{ij} = \mu + \alpha_j + \beta_X(x_{ij} - \mu_X) + \beta_T(t_{ij} - \mu_T) + e_{ij}.$$

Here, β_X is a regression coefficient for the linear relationship of X and Y, where the relationships of T to X and T to Y have been ruled out; similarly, β_T is a regression coefficient reflecting the linear relationship of T and Y, where the relationships of X and Y and T and X have been ruled out. The same general sort of model can be stated for any number of experimental factors and for any number of quantitative concomitant variables.

The more advanced techniques in multivariate correlation and regression are now used extensively in all the social sciences and in education, and this use appears to be growing. The reason is not hard to find. In the past, these multivariate techniques presented formidable computational difficulties. This fact alone acted as a brake on investigators interested in large multivariate problems. Now, however, the computer has taken away almost all of the computational burden of such methods, and the studies that are analyzed routinely today would have been almost unthinkable twenty or thirty years ago.

There are many useful multivariate techniques usually based on the "correlation problem" model of sampling. One that appears in the social science literature is known as *discriminant function analysis*. The basic problem here is not unlike that in a multiple correlation problem. However, instead of weighting a set of variables so as to predict a single numerical variable of interest, in discriminant function analysis the set of predictor variables are weighted to yield maximum discrimination between two (or even more) qualitatively different groups.

Another advanced method based on correlations is called *canonical correlation*. Essentially, a problem in canonical correlation involves one set of predictor variables, and a distinct set of criterion variables. The task is then to weight both sets of variables so as to maximize the correlation between the two sets.

A methodology long familiar to social scientists is *factor analysis,* in which a table of intercorrelations is analyzed so as to permit inference of a structure underlying the correlations. Although this technique is hardly new, computer methodology has opened up large new approaches and developments in its use.

Among those multivariate techniques more likely to be applied to a problem in regression, as we have used that term here, is multivariate analysis of variance, or MANOVA (as distinguished from ANOVA for analysis of variance). These methods apply when there is an experiment involving one or more factors, but, instead of a single dependent variables, one is interested in a set of dependent variables. The question then concerns the effects of the experimental treatments on all dependent variables considered simultaneously. Any good book on multivariate methods, such as that by Timm (1975) or by Morrison (1976), will explain and illustrate multivariate analysis of variance as well as the methods mentioned above.

In short, the very large body of methods growing out of the general linear model, and usually grouped together under the rubric ''multivariate methods'' provides a significant part of the statistical equipment of the social scientist, especially in complex field situations. Nevertheless, there are some drawbacks to large, multidimensional studies. In the first place they are expensive in terms of data collection. Secondly, they are often much harder to interpret than studies involving only a few variables. Third, some statistical tests for multivariate correlation studies depend rather heavily on the assumption of multivariate normal distributions and on the assumption that all of a large set of random variables have equal variances. It may be that, as in so many other situations, such assumptions are rather innocuous. However, there does not seem to be much evidence available about the true importance of such stringent assumptions for the conclusions reached. Yet, even given these drawbacks, the computer now permits social scientists to deal more directly with their complex research questions by careful use of these advanced methods.

EXERCISES

1. An experimenter was interested in the possible linear relation between the time spent per day in practicing a foreign language and the ability of the person to speak the language at the end of a six-week period. Some fifty students were assigned at random among five experimental conditions ranged from one hour to five hours practice per day. Then at the end of six weeks, each student was scored for proficiency in the language. Find $b_{Y \cdot X}$ for the data.

Proficiency scores, by daily practice time

1	2	3	4	5
117	106	86	140	105
85	81	98	128	149
112	74	125	108	110
81	79	123	104	144
105	118	118	132	137
109	110	94	133	151
80	82	93	96	117
73	86	91	101	113
110	111	122	103	142
78	113	130	135	112

X = practice, in hours

2. Find the linear regression equation for predicting Y, the proficiency of a student, from X, the practice time per day. Plot the obtained data and the straight line representing the regression equation.

3. Test the hypothesis that, for the population of students sampled in exercise 1, the true regression coefficient is zero. Furthermore, estimate the true proportion of variance in Y-values accounted for by linear regression on X.

4. From the data of exercise 1, find the 95 percent confidence limits for the value of the regression coefficient $\beta_{Y \cdot X}$ in the population. Find the 95 percent confidence limits for the value of Y that *should* be predicted when $X = 2$ if one knew the *population* regression equation. (**Hint:** Use the same methods described in Chapter 13.)

5. Exercise 13 of the preceding chapter dealt with the ability of teachers to judge a child's age from drawings. In order to gather data on this, the investigator selected at random 12 drawings by children aged 4, 12 by children aged 5, 12 from age 6, and 12 from age 7. Then each of a panel of teachers assigned a "guessed age" to each drawing. The data below show the four age groups together with the average age guessed for each child from drawings.

Average ages guessed for children, arranged according to actual age.

4	5	6	7
5.87	6.46	6.01	6.09
4.49	4.80	5.03	5.87
5.83	4.12	6.10	7.25
4.83	4.13	5.13	5.77
5.60	5.25	6.75	6.51
5.18	5.78	5.27	6.69
4.85	5.59	5.99	6.40
4.77	4.64	6.19	6.10
5.79	4.56	5.90	7.40
4.41	4.66	5.60	5.59
4.23	6.09	6.90	6.09
5.62	5.87	5.09	5.87

x = actual age

Test the hypothesis of no linear regression, by use of fixed-effects analysis of variance methods. Use $\alpha = .05$.

6. For the data of exercise 7, Chapter 10, test the hypothesis of no linear regression of performance on the hours of sleep deprivation. Use $\alpha = .01$ in this instance.

7. Find the regression equation to predict the performance difference for an individual in this sample, given the amount of sleep deprivation suffered as shown in exercise 6.

8. In a study of factors related to success in medical school, the dependent variable Y was an overall numerical rating of success during the four years of study. The independent variable was a measure of the "authoritarian personality" characteristics of the individual student. Students were arranged into four groups, from very low to very high, on this personality trait. Finally a concomitant variable X was used, representing the average anxiety level of the student. Carry out an analysis of covariance on the resulting data, and draw conclusions about the relation between the independent and dependent variable when anxiety is controlled as a factor

Authoritarianism							
Very low		Low		High		Very high	
X	Y	X	Y	X	Y	X	Y
30	23	22	17	27	21	19	23
30	20	26	20	26	20	24	28
24	20	24	19	30	24	28	32
27	19	28	23	27	19	30	36
23	15	26	18	28	20	29	33
28	20	25	22	27	19	24	30

9. Would the same conclusions have been reached from the data of exercise 8 if anxiety had been overlooked as a concomitant variable?

10. Does the assumption of homogeneity of regression in exercise 8 appear to be justified? Use $\alpha = .05$ for the test.

11. Suppose that the experiment on "level of aspiration" described in Section 10.20 had actually been carried out as a one-factor experiment, involving the three groups labeled "above average," "average," and "below average," but without the second factor. Each group contained 20 randomly chosen and assigned subjects. However, prior to the experiment, each subject had been tested on a game very similar to that used in the experiment proper, and a "skill score" X obtained for each. This was used as the concomitant variable in the experiment. The data are given below. Do differences appear to exist among the groups when the effects of the prior skill factor X are removed? Use $\alpha = .05$.

Groups					
Above average		*Average*		*Below average*	
Y	X	Y	X	Y	X
52	44	28	38	15	23
48	47	35	26	14	17
43	30	34	36	23	31
50	38	32	30	21	25
43	40	34	36	14	27
44	45	27	23	20	35
46	36	31	45	21	25
46	41	27	28	16	28
43	40	29	34	20	30
49	43	25	37	14	37
38	48	43	40	23	32
42	24	34	36	25	32
42	39	33	41	18	34
35	36	42	29	26	48
33	46	41	39	18	39
38	33	37	37	26	38
39	38	37	47	20	30
34	26	40	34	19	24
33	41	36	47	22	31
34	36	35	31	17	19

12. Complete the analysis of variance for the data of exercise 11 ignoring the variable X. Then compare these results with those obtained in exercise 11. How do you account for this difference in results? What would this mean about the original experiment?

13. Check for homogeneity of regression in the data of exercise 11.

14. For the data of exercise 1 above, carry out tests for linear and for curvilinear trends.

15. For exercise 14, estimate the proportion of variance accounted for by linear trends. Estimate the proportion of variance accounted for by curvilinear trends. Which sort of trend appears to be more pronounced in these data?

16. Analyze the data of exercise 7, Chapter 10, for linear and curvilinear trends.

17. In the data of exercise 5 above, does there appear to be a more pronounced linear or curvilinear trend?

18. In a study of factors in alcoholism, samples of United States cities were taken, and the rate of verified alcoholism over the past ten years in each found. The sizes of the cities ranged from 10,000 through 109,000, and the cities were classified into five class intervals according to population size, as shown below. Five cities were selected at random from each population grouping. The average alcoholism rate per 1000 population for each city over the past ten years is shown below:

Alcoholism rate, per thousand

10–29	30–49	50–69	70–89	90–109
17.57	19.08	18.75	20.07	24.74
16.35	18.02	17.99	19.01	23.30
17.37	18.89	18.78	20.16	24.09
16.26	18.03	18.08	19.29	22.86
14.19	17.26	18.59	18.48	23.90

x = population, in thousands

Assuming that each city in a group falls at the midpoint of its interval, test for linear, quadratic, cubic, and quartic trends. What would you conclude about the relation of alcoholism rate and size of city, based on these samples?

19. Based on the significant trends found in exercise 18, find the curvilinear regression equation. Use the method of orthogonal polynomials to find the required equation. Plot the regression equation, along with the obtained data points for 18.

20. Make a freehand extrapolation of the regression curve found in exercise 18 in order to guess at the alcoholism rate for a city of 1,000,000 population. Does this rate seem particularly reasonable? What does this illustrate about extrapolating findings beyond the range actually sampled in such studies?

21. Carry out an analysis of exercise 7, Chapter 10, as though comparisons for trend had been planned in advance. What do these comparisons suggest about the form of the relationship between sleep deprivation and performance?

22. Carry out post hoc comparisons for trend on the data of exercise 5 above, at the level justified by the overall *F* test.

23. Why is the discussion of the statistical theory of curvilinear regression limited to "regression problems" (as defined in Chapter 13) rather than "problems of correlation"?

24. Show that the polynomial weights for linear, quadratic, and cubic comparisons are orthogonal when, say, the number of experimental groups is 7. Attempt to explain why the highest possible trend is of the sixth degree when there are only seven sample means.

25. Suppose that in an experiment such as that in exercise 1, the investigator is specifically interested in the presence or absence or a particular trend (say, the quadratic). The others are of secondary interest and can be saved for post hoc examination. How might the analysis of variance summary table then appear?

Chapter 15

ANALYZING QUALITATIVE DATA: CHI-SQUARE TESTS

Heretofore, problems of hypothesis testing and interval estimation have centered very largely on summary characteristics of one or more population distributions. That is, we have been concerned with the value of a population mean, the differences among two or more population means, the equality of two population variances, the value of a population regression coefficient, and so on. In almost every instance the hypothesis tested was composite, so that the null hypothesis itself did not really specify the population distribution or distributions exactly. Of course, it was necessary to make assumptions about the population distribution in order to arrive at the appropriate test statistics and sampling distributions for these various hypotheses. Nevertheless, our interest was really in comparing population values with hypothetical values, or several populations in particular ways. With the exception of problems dealing with a single proportion (the two-class ''Bernoulli'' situation) we have not really considered hypotheses about the **identity** of two or more population distributions.

There are research problems in which one wants to make direct inferences about two or more distributions, either by asking if a population distribution has some particular specifiable form, or by asking if two or more population distributions are identical. These questions occur most often when *both* variables in some experiment are qualitative in character, making it impossible to carry out the usual inferences in terms of means or variances. In these instances we need methods for studying independence or association from categorical data. Other situations exist, however, when we wish to ask if a population distribution of a random variable has some precise theoretical form, such as the normal distribution, without having any special interest in summary properties such as mean.

The first topic of this chapter is the comparison of a sample with a hypothetical population distribution. We would like to infer whether or not the sample result actually does represent some particular population distribution. We will deal only with discrete or grouped population distributions, and so inferences will be made through an approximation to the exact multinomial probabilities. Such problems are said to involve "goodness of fit" between a single sample and a single population distribution.

Next, we will extend this idea to the simultaneous comparison of several discrete distributions. Ordinarily, the reason for comparing such distributions in the first place is to find evidence for association between two qualitative attributes. In short, we are going to employ a test for independence between attributes, which can be regarded as based on the comparison of *sample* distributions.

Also, we will deal with the problem of measuring the strength of association between two attributes from sample data. Tests and measures of association for qualitative data are very important for the social sciences, where many of the most important distinctions made are, essentially, categorical or qualitative in character. The methods in this chapter are widely used, both because of the kinds of data social scientists collect, and because of the simplicity of their application. However, the theory underlying these tests is not simple, and misapplication of these tests is very common. For this reason a good deal of space will be devoted to discussing some of the basic ideas underlying these methods, and some pains will be taken to emphasize their inherent limitations.

15.1 / COMPARING SAMPLE AND POPULATION DISTRIBUTIONS: GOODNESS OF FIT

Suppose that a study of educational achievement of American men were being carried on. The population sampled is the set of all normal American males who are twenty-five years old at the time of the study. Each subject observed can be put into one and only one of the following categories, based on his *maximum* formal educational achievement:

1. college graduate;
2. some college;
3. high school or preparatory school graduate;
4. some high school or preparatory school;
5. finished eighth grade;
6. did not finish eighth grade.

These categories are mutually exclusive and exhaustive: each man observed must fall into one and only one classification.

The experimenter happens to know that ten years ago the distribution of educational achievement on this scale for twenty-five-year-old men was:

Category	*Relative frequency*
1	.18
2	.17
3	.32
4	.13
5	.17
6	.03

Is the present population distribution on for such a scale exactly like that of ten years ago? Therefore the hypothesis of "no change" in the distribution for the present population specifies the exact distribution given above. The alternative hypothesis is that the present population does differ from the distribution given above, in some unspecified way.

A random sample of 200 subjects is drawn from the current population of twenty-five-year-old males, and the following frequency distribution obtained:

Category	f_{oj} *(obtained frequency)*	f_{ej} *(expected frequency)*
1	35	36
2	40	34
3	83	64
4	16	26
5	26	34
6	0	6
	200	200

(These figures *are* hypothetical!) The last column on the right gives the *expected* frequencies under the hypothesis that the population has the same distribution as ten years ago. For each category, the expected frequency is

$$Np_j = f_{ej} = \text{expected frequency}$$

where p_j is the relative frequency for category j dictated by the hypothesis.

How well do these two distributions, the obtained and the expected, agree? At first blush, you might think that the difference in frequency obtained and expected across the categories, or

$$\sum_j (f_{oj} - f_{ej}),$$

would describe the difference in the two distributions. However, it must be true that

$$\begin{aligned}\sum_j (f_{oj} - f_{ej}) &= \sum_j f_{oj} - \sum_j f_{ej} \\ &= N - N \\ &= 0,\end{aligned}$$

so that this is definitely not a satisfactory index of disagreement.

On the other hand, the sum of the *squared* differences in observed and expected frequencies does begin to reflect the extent of disagreement:

$$\sum_j (f_{oj} - f_{ej})^2.$$

This quantity can be zero only when the fit between the obtained and expected distributions is perfect, and must be large when the two distributions are quite different.

An even better index might be

$$\sum_j \frac{(f_{oj} - f_{ej})^2}{f_{ej}}, \qquad [15.1.1]$$

where each squared difference in frequency is weighted inversely by the frequency expected in that category. This weighting makes sense if we consider that a departure from expectation should get relatively more weight if we expect rather few individuals in that category than if we expect a great many. Somehow, we are more "surprised" to get many individuals where we expected to get few or none, than when we get few or none where we expected many; thus, the departure from expectation is appropriately weighted in terms of the frequency expected in the first place, when an index of overall departure from expectation is desired.

Remember, however, the real purpose in the comparison of these distributions is to test the hypothesis that the expectations are correct, and that the current distribution actually is the same as ten years ago. One might proceed in this way: given the probabilities shown as relative frequencies for the hypothetical population distribution, the exact probability of any sample distribution can be found. That is, given the hypothesis and the assumption of independent random sampling of individuals (with replacement), the exact probability of a particular sample distribution can be found from the multinomial rule (Section 3.19). Thus, in terms of the hypothetical population distribution, the probability of a sample distribution exactly like the one observed is

$$p(\text{obtained distribution}|H_0) = \frac{200!}{35!\ 40!\ 83!\ 16!\ 26!\ 0!}(.18)^{35}(.17)^{40}(.32)^{83}(.13)^{16}(.17)^{26}(.03)^{0}.$$

With some effort we could work this value out exactly. However, we are not really interested in the probability of exactly this sort of obtained distribution, but rather in all possible sample results this much or more deviant from expectation according to an index such as expression 15.1.1. The idea of working out a multinomial probability for each possible such sample result is ridiculous for N and J this large, as an absolutely staggering amount of calculation would be involved.

The theoretical statistician in this kind of impasse usually begins looking around for an approximation device. In this particular instance, it turns out that the multivariate normal distribution provides an approximation to the multinomial distribution for very large N, and thus the problem can be solved. We will not go into this derivation here, as it is far beyond our capabilities; suffice it to say that the basic rationale for this test does depend on the possibility of this approximation, and that the approximation itself is really good only for very large N. Then the following procedure is justified:

We form the statistic

$$\chi^2 = \sum_j \frac{(f_{oj} - f_{ej})^2}{f_{ej}} \qquad [15.1.2^*]$$

which is known as the Pearson χ^2 statistic (after its inventor, Karl Pearson). Given that the exact probabilities for samples follow a multinomial distribution, and given a very large N, **when H_0 is true this statistic χ^2 is distributed approximately as chi-square with $J - 1$ degrees of freedom.** Probabilities arrived at using this statistic are *approximately* the same as the exact multinomial probabilities we would like to be able to find for samples as much or more deviant from expectation as the sample obtained. The larger the sample N, the better should this approximation be.

Note that the number of degrees of freedom here is $J - 1$, the number of distinct categories in the same distribution, or J, minus 1. You may have anticipated this from the fact that the sum of the differences between observed and expected frequencies is zero; given any $J - 1$ such differences, the remaining difference is fixed. This is very similar to the situation for deviations from a sample mean, and the mathematical argument for degrees of freedom here would be much the same as for degrees of freedom in a variance estimate.

To return to our example, the value of the χ^2 statistic is

$$\chi^2 = \sum_j \frac{(f_{oj} - f_{ej})^2}{f_{ej}}$$

$$= \frac{(35 - 36)^2}{36} + \frac{(40 - 34)^2}{34} + \frac{(83 - 64)^2}{64} + \frac{(16 - 26)^2}{26} + \frac{(26 - 34)^2}{34} + \frac{(0 - 6)^2}{6}$$

$$= 18.46.$$

This value is referred to in the chi-square table (Table IV) for $J - 1 = 6 - 1 = 5$ degrees of freedom. We are interested only in the upper tail of the chi-square distribution in such problems, because the only reasonable alternative hypothesis (disagreement between true population and hypothetical distribution) must be reflected in *large values* of χ^2. Table IV shows that for 5 degrees of freedom, a value of 11.07 cuts off the upper .05 of the distribution, and a value of 15.09 corresponds to the upper .01. The hypothesis that the current distribution of educational achievement is exactly like that of ten years ago may thus be rejected, either at the .05 or at the .01 level, as we choose.

Tests such as that in the example, based on a single sample distribution, are called "goodness-of-fit" tests. Chi-square tests of goodness of fit may be carried out for any hypothetical population distribution we might specify, provided that the population distribution is discrete, or is thought of as grouped into some relatively small set of class intervals. However, in the use of the Pearson χ^2 statistic to approximate multinomial probabilities, it *must* be true that:

1. each and every sample observation falls into one and only one category or class interval;
2. the outcomes for the N respective observations in the sample are independent;
3. sample N is large.

The first two requirements stem from the multinomial sampling distribution itself: the multinomial rule for probability holds only for mutually exclusive and exhaustive categories, and for independent observations in a sample (random sampling with replacement). The third requirement comes from the use of the chi-square distribution to approximate these exact multinomial probabilities: this approximation is good only for large sample size. Furthermore, unless N is infinitely large, the Pearson χ^2 itself is not distributed exactly as the chi-square variable.

How large should sample size be in order to permit the use of the Pearson χ^2 goodness-of-fit tests? Opinions vary on this question, and some fairly sharp debate has been raised by this issue over the years. Many rules of thumb exist, but as a conservative rule one is usually safe in using this chi-square test for goodness of fit if each *expected* frequency, f_{ej}, is 10 or more when the number of degrees of freedom is 1 (that is, two categories), or if the expected frequencies are each 5 or more where the number of degrees of freedom is greater than 1 (more than two categories). We will have more to say about sample size and Pearson χ^2 tests in Section 15.7.

Be sure to notice, however, that *these rules of thumb apply to expected, not observed, frequencies per category.*

15.2 / THE RELATION TO LARGE-SAMPLE TESTS OF A SINGLE PROPORTION

The goodness-of-fit test with 1 degree of freedom is formally equivalent to the large-sample test of a proportion, based on the normal approximation to the binomial. That is, imagine a distribution with only two categories.

Category j	*Expected frequency*	*Obtained frequency*	*Expected proportion*	*Obtained proportion*
1	f_{e1}	f_{o1}	p	P
2	f_{e2}	f_{o2}	q	Q
	N	N	1.0	1.0

Suppose that the normal approximation to the binomial is to be used to test the hypothesis that the true population proportion in category 1 is p. Then we would form the test statistic

$$z = \frac{NP - Np}{\sqrt{Npq}} = \frac{-(NQ - Nq)}{\sqrt{Npq}} \qquad [15.2.1]$$

or

$$z = \frac{f_{o1} - f_{e1}}{\sqrt{f_{e1}(N - f_{e1})/N}}. \qquad [15.2.2]$$

For very large N, this can be regarded as a standardized normal variable. Now consider the *square* of this standardized variable z:

$$z^2 = \frac{N(f_{o1} - f_{e1})^2}{f_{e1}(N - f_{e1})}$$

$$= \frac{(f_{o1} - f_{e1})^2}{f_{e1}} + \frac{(f_{o1} - f_{e1})^2}{N - f_{e1}}.$$

Since $(f_{o1} - f_{e1}) = -(f_{o2} - f_{e2})$, we may write

$$z^2 = \frac{(f_{o1} - f_{e1})^2}{f_{e1}} + \frac{(f_{o2} - f_{o2})^2}{f_{e2}} \qquad [15.2.3]$$

$$= \chi^2.$$

When the frequency distribution has only two categories, the Pearson χ^2 statistic has exactly the same value as the square of the standardized variable used in testing for a single proportion, using the normal approximation to the binomial. From the definition of a chi-square variable with 1 degree of freedom (Section 9.1), it can be seen that if $E(P) = p$, and if N is very large, then sample values of χ^2 should be distributed approximately as a chi-square variable with 1 degree of freedom. For a large sample and so long as a two-tailed test is desired, it is immaterial whether we use the normal-distribution test for a single proportion or the Pearson χ^2 test for a two-category problem. Furthermore, the square root of this χ^2 value gives the equivalent z value if one does desire a one-tailed test. This direct equivalence between z and $\sqrt{\chi^2}$ holds only for 1 degree of freedom, however.

Again, only for 1 degree of freedom, the Pearson χ^2 test may be improved somewhat if the test statistic is found by taking

$$\chi^2 = \frac{(|f_{o1} - f_{e1}| - .5)^2}{f_{e1}} + \frac{(|f_{o2} - f_{e2}| - .5)^2}{f_{e2}} \qquad [15.2.4^*]$$

so that the *absolute value* of the difference between observed and expected frequencies is reduced by .5 for each category before the squaring is carried out. This is known as **Yates' correction** and depends on the fact that the binomial is a discrete, and the normal a continuous, distribution. However, Yates' correction applies only when there is 1 degree of freedom.

15.3 / A SPECIAL PROBLEM: A GOODNESS-OF-FIT TEST FOR A NORMAL DISTRIBUTION

One use of the Pearson goodness-of-fit χ^2 test is in deciding if a continuous population distribution has a particular form. That is, we might be interested in seeing if a sample distribution of scores might have arisen from some theoretical form such as the normal distribution. It is important to recall that like most theoretical distributions useful in statistics, the normal is a family of distributions, particular distributions differing in the parameters entering into the rule as constants; for the normal distribution, these parameters are μ and σ, of course. It is also important to remember that although such a distribution is continuous, it is necessary to think of the population as grouped into a finite number of distinct class intervals if the Pearson χ^2 test is to be applied.

For example, suppose that there is good reason to believe that intelligence scores on some test are normally distributed among American men in general. However, we are curious about the distribution of such intelligence scores for men serving time in prison. Is it reasonable to believe that this distribution is also approximately normal? The question refers not to the mean, nor to the variance of this distribution, but rather to the distribution's form. We decide to draw a random sample of 400 men in prison to test the hypothesis of a normal population distribution.

However, the population distribution must be thought of as grouped into class intervals. Furthermore, it is necessary that the number expected in each class interval be relatively large. Therefore, the experimenter must first decide on the number of class intervals to use in describing both theoretical and obtained distributions. One would like to have intervals insuring a fairly "fine" description, as well as a sizable number expected in each interval. Suppose that for our problem we decide to think of the population distribution as divided into eight class intervals, in such a way that each interval should include exactly one eighth of the population. What would the limits of these eight intervals be? From Table I we find the limits shown in Table 15.3.1. Notice that this arrangement into class intervals refers to the *population* distribution, assumed to be normal under the null hypothesis, and that this choice of class intervals is made before the data are seen. This arrangement is quite arbitrary, and some other number of class intervals might have been chosen; in fact, with a sample size this large, an experimenter would be quite safe in taking many more intervals with a much smaller probability associated with each. Notice also that the population is thought of as divided into class intervals of unequal size, in order to give equal probability of intervals. On the other hand, it is perfectly possible to decide on some arbitrary class-interval size in z-score terms, and then allow the various probabilities to be unequal. Our way of proceeding has two possible advantages: departures from normality in the middle of the score range are relatively more likely to be detected in this way than otherwise, and computations are simplified by having equal expectations for each class interval in the distribution. Furthermore, a relatively better approximation should be afforded when intervals giving equal expected frequencies are chosen, other things being equal.

Table 15.3.1

Eight class intervals chosen to have equal probability in a normal distribution.

Class limits in terms of z for class interval j	*Approx.* p_j	f_{ej}
1.15 and above	1/8	50
.68 to 1.15	1/8	50
.32 to .68	1/8	50
.00 to .00	1/8	50
−.32 to .00	1/8	50
−.68 to −.32	1/8	50
−1.15 to −.68	1/8	50
below −1.15	1/8	50
		400

Now to proceed with the example. Before the sample distribution can be grouped into the same categories as the population, something must be known about the population mean and standard deviation. Our best evidence comes from the sample estimates, M and s, and so these are used in place of the unknown μ and σ. In the actual data for this example suppose that it turned out that the sample mean, M, is 98, and the estimate s is 8.4. Then these values are used as estimates of the true values of μ and σ for this population.

Using these estimates, we turn each score for an individual in the sample into a standard score. In terms of the arbitrary class intervals decided upon, the distribution of these *sample* standard scores is shown in Table 15.3.2.

Table 15.3.2

Obtained and expected frequencies for the sample of men in prison.

Interval	f_{oj}	f_{ej}
1.15 and above	14	50
.68 to 1.15	17	50
.32 to .68	76	50
.00 to .32	105	50
−.32 to .00	71	50
−.68 to −.32	76	50
−1.15 to −.68	31	50
below −1.15	10	50

Now the χ^2 test for goodness of fit is carried out:

$$\chi^2 = \frac{(14 - 50)^2}{50} + \frac{(17 - 50)^2}{50} + \frac{(76 - 50)^2}{50} + \frac{(105 - 50)^2}{50} + \frac{(71 - 50)^2}{50} + \frac{(76 - 50)^2}{50} + \frac{(31 - 50)^2}{50} + \frac{(10 - 50)^2}{50}$$
$$= 183.3.$$

Before the significance level is ascertained, however, an adjustment must be made in the degrees of freedom. Two parameters had to be estimated in order to carry out this test: the mean and the standard deviation of the population. *One degree of freedom is subtracted from $J - 1$ for each separate parameter estimated in such a test*. Therefore, the correct number of degrees of freedom here is

$$\nu = J - 1 - 2 \text{ or } 5.$$

For 5 degrees of freedom, Table IV, shows that this obtained χ^2 value far exceeds that required for significance at either the .05 or .01 level. The experimenter can feel quite confident in saying that scores in this population are not normally distributed. There are many other circumstances where for theoretical or other reasons the experimenter wants merely a comparison between distributions, either sample and theoretical, or for two samples. Provided that the distributions considered are of some random variable, so that the obtained distributions can be put into cumulative form, then methods superior to χ^2 tests exist. Most notable are the so-called Kolmogorov-Smirnoff tests, either for one or two samples. These tests provide a direct comparison between distributions, and thus handle one aspect of the problem to which a Pearson χ^2 test is often applied, without some of the objectionable features of such a test. A description of the Kolmogorov-Smirnoff tests is given in Siegel (1956).

15.4 / PEARSON χ^2 TESTS OF ASSOCIATION

The general rationale for testing goodness of fit also extends to tests of independence (or lack of statistical association) between categorical attributes (c.f. Section 1.15.) Quite often situations arise where N independent observations are made, and each and every observation is classified in two qualitative ways. One set of mutually exclusive and exhaustive classes can be called the A attribute. Any one of the C distinct classes making up this attribute can be labeled A_j, where j runs from 1 through C. Thus $A = \{A_1, \cdots, A_j, \cdots, A_C\}$. Furthermore, on some other attribute B, there are some R mutually exclusive and exhaustive classes,

$$B = \{B_1, \cdots, B_k, \cdots, B_R\}.$$

Each observation then represents the occurrences of one joint event (A_j,B_k). The entire set of data can be shown as a **contingency table,** with the C classes of attribute A making up the columns, and the R classes of attribute B the rows; each and every possible (A_j,B_k) joint event thus is shown by a cell in the table. Be sure to notice that each distinct observation can represent only one joint event, and thus *each observation qualifies for one and only one cell in the table.*

For example, suppose that a random sample of 100 schoolchildren is drawn. Each child is classified in two ways: the first attribute is the sex of the child, with two possible categories:

$$\text{male} = A_1$$
$$\text{female} = A_2.$$

The second attribute, B, is the stated preference of a child for two kinds of reading materials:

$$\text{fiction} = B_1$$
$$\text{nonfiction} = B_2$$

(we will assume that each child can be given one and only one preference classification). The entire set of data could be arranged into a joint frequency distribution, or contingency table, symbolized by

	A_1	A_2	
B_1	(A_1,B_1)	(A_2,B_1)	
B_2	(A_1,B_2)	(A_2,B_2)	
			N

Each cell of this table contains the frequency among the N children for the joint event represented by that cell. The data might, for example, turn out to be

	A_1	A_2	
B_1	19	32	51
B_2	29	20	49
	48	52	100 = N

This table shows that exactly 19 observations represent the joint event (male, prefers fiction), 32 the joint event (female, prefers fiction) and so on.

Corresponding to the joint frequency distribution for a sample, there is a joint *probability* distribution in the population. For the example

	A_1	A_2	
B_1	$p(A_1,B_1)$	$p(A_2,B_1)$	$p(B_1)$
B_2	$p(A_1,B_2)$	$p(A_2,B_2)$	$p(B_2)$
	$p(A_1)$	$p(A_2)$	1.00

Suppose that we had some hypothesis where the probability of each possible joint event is specified. We might want to ask how this hypothetical *joint* distribution actually fits the data. Given this *complete specification* of the population joint distribution, then for sufficiently large N we could apply the Pearson χ^2 test of goodness of fit. Just as for any goodness-of-fit test where no parameters are estimated, the number of degrees of freedom would be the number of distinct event classes minus 1. Since there are RC *joint* events in this instance, $RC - 1$ is the degrees of freedom for a goodness-of-fit test for such a joint distribution.

However, exact hypotheses about joint distributions are quite rare. Very seldom would one want to carry out such a test even though it is possible. Instead, the usual null hypothesis is that the two attributes A and B are independent. If the hypothesis of independence can be rejected, then we say that the attributes A and B are statistically related or associated.

In Section 1.14 it was pointed out that two discrete attributes are considered independent *if and only if*

$$p(A_j,B_k) = p(A_j)p(B_k)$$

for *all possible* joint events (A_j,B_k). Given that the hypothesis of independence is true, and given the *marginal* distributions showing $p(A_j)$ and $p(B_k)$, we know what the joint probabilities *must* be.

On the other hand, this fact alone does us little good, because independence is defined in terms of the *population* probabilties, $p(A_j)$ and $p(B_k)$, and these we do not usually know. What can we use instead? **The best estimates we can make of the unknown marginal probabilities are the sample marginal proportions:**

$$\text{est. } p(A_j) = \frac{\text{freq. of } A_j}{N}$$

and

$$\text{est. } p(B_k) = \frac{\text{freq. of } B_k}{N},$$

for each A_j and B_k. Given these estimates of the true probabilities, then we expect that the frequency of the joint event (A_j,B_k) will be

$$f_{ejk} = \text{expected frequency of } (A_j,B_k) = N[\text{est. } p(A_j)][\text{est. } p(B_k)].$$

However, since these probability estimates are based on *sample* relative frequencies, it must then be true that

$$f_{ejk} = \frac{(\text{freq. } A_j)(\text{freq. } B_k)}{N}. \qquad [15.4.1^*]$$

In tests for independence, the expected frequency in any cell is taken to be the product of the frequency in the column times the frequency in the row, divided by total N. Using these expected frequencies, the Pearson χ^2 statistic in a test for association is simply

$$\chi^2 = \sum_j \sum_k \frac{(f_{ojk} - f_{ejk})^2}{f_{ejk}} \qquad [15.4.2^*]$$

where f_{ojk} is the frequency actually observed in cell (A_j,B_k), f_{ejk} is the expected frequency for that cell under the hypothesis of independence, and the sum is taken over all of the RC cells.

The number of degrees of freedom for such a test, where sample estimates of the marginal probabilities are made, differs from the degrees of freedom for a goodness-of-fit test. The degrees of freedom for a Pearson χ^2 test of association is

$$\nu = RC - 1 - (C - 1) - (R - 1) = (R - 1)(C - 1). \qquad [15.4.3^*]$$

In summary, given a joint-frequency table with C columns and R rows, the hypothesis

$$H_0\text{: } p(A_j,B_k) = p(A_j)p(B_k), \text{ for all } j, k \qquad [15.4.4]$$

can be tested by the Pearson χ^2 statistic with $(R - 1)(C - 1)$ degrees of freedom, *provided* that

1. each and every observation is independent of each other observation;
2. each observation qualifies for one and only one cell in the table;
3. sample size N is large.

For our example, the expected frequency for cell (A_1,B_1) is found to be

$$f_{e11} = \frac{(\text{freq. } A_1)(\text{freq. } B_1)}{N} = \frac{(48)(51)}{100} = 24.48,$$

that for cell (A_2,B_1) is

$$f_{e21} = \frac{(\text{freq. } A_2)(\text{freq. } B_1)}{N} = \frac{(52)(51)}{100} = 26.52,$$

and so on, until the following set of expected frequencies is found for the table as a whole:

	A_1	A_2	
B_1	24.48	26.52	51
B_2	23.52	25.48	49
	48	52	100 = N

Notice that in any row the sum of the expected frequencies must equal the obtained marginal frequency for that row, and the sum of the expected frequencies in any column must also equal the obtained frequency for that column. In computations of expected frequencies, it is wise to carry several decimal places, and not round to a few places until the final result.

Given the expected frequencies, the χ^2 test for this example is based on

$$\chi^2 = \frac{(19 - 24.48)^2}{24.48} + \frac{(32 - 26.52)^2}{26.52} + \frac{(29 - 23.52)^2}{23.52} + \frac{(20 - 25.48)^2}{25.48}$$

$$= 4.81$$

In this 2 × 2 table, the degrees of freedom are $(2 - 1)(2 - 1) = 1$. For 1 degree of freedom, the χ^2 value needed to reject the hypothesis at the .05 level is 3.84, and so the hypothesis of independence can be rejected (however, if α were set at .01, then the hypothesis would not be rejected.) Rejection of the hypothesis of independence lets one say that some statistical association does exist between the two attributes, A and B. In this instance, the experimenter might say that rejection of the hypothesis makes it safe to conclude that the sex and the reading preference of a child are in some way related.

15.5 / AN EXAMPLE OF A TEST FOR INDEPENDENCE IN A LARGER TABLE

There is no reason at all why the number of categories in either A or B must be only 2, as in the preceding example. Any number of rows and any number of columns can be used for classifying observations into a contingency table. Provided that there is a sufficiently large random sample of independent observations where each occurs in one and only one cell, a χ^2 test can be carried out. However, for larger tables sample N must be rather large if a sufficiently large expected frequency is to be associated with each cell.

Not only will the example in this section involve a larger table, but also it will illustrate that the row or column classes may be quantitative in their original character, *if* the experimenter is willing to treat the data in terms of class intervals. That is, either the attribute A or the attribute B may represent numerical measurements grouped into class intervals. Under these conditions the χ^2 test is still an adequate way to answer the question of association between A and B, but in using a χ^2 test the experimenter is actually ignoring the detailed numerical information in the data and is treating the different class intervals simply as qualitative distinctions.

Consider an experiment where a behavioral scientist is studying the possible inheritance of patterns of emotional response in rats. It has been noticed that four strains of rats tend to differ in their tendency toward an excited response in a new situation. These four strains are each isolated and carefully inbred until, after several generations, a large number of representatives of four pure strains are available for study. Each rat is put individually into a new and presumably fear-inducing situation, and left for a fixed amount of time. By some standard procedure, each individual rat is given a ''score'' on the excited character of the response, based on a composite of a number of behavior indices. The experimenter is looking for evidence of association between the excitability score and the strain the rat belongs to. Some 25 rats are selected at random from each of the four pure strains, giving a total of 100 individuals in all. Within each strain used in the experiment a grouped frequency distribution of "excitability scores" is found, the same class intervals being used for each distribution. These frequency distributions for the four groups are shown in Table 15.5.1. *Because of previous experience with such scores, the experimenter was able to set up these four class intervals in advance of seeing the data.*

Table 15.5.1

Scores	*Group* 1	2	3	4	*Total*
9 and over	5	8	6	4	23
6–8	10	7	8	6	31
3–5	5	4	7	8	24
0–2	5	6	4	7	22
	25	25	25	25	100

Basically, the experimenter is interested in comparing the four populations in terms of their distributions of scores. Looked at in this way, the problem is the comparison of four grouped frequency distributions, each based on the same grouping scheme. However, this can also be framed as a problem in testing the independence of two attributes.

Let us call rat population (pure breed) j the class A_j, and let us call any class interval of scores B_k. Now if all the populations were exactly *alike* in their score distributions, it should be true that

$$p(B_k|A_j) = p(B_k).$$

That is, the probability of any rat's showing a score in the class interval B_k given that it belongs to the population A_j should be the same as the probability in general associated with that class interval. This is tantamount to the independence of the attribute A (breeds) and the attribute B (scores arranged in class intervals.) Therefore the problem of the simultaneous comparison of a number of frequency distributions is basically the same as the problem of testing for independence.

Going ahead on this basis, we carry out the χ^2 test for this table. The expected frequencies, each found by multiplying the column frequency by the row frequency and dividing by N, are shown in Table 15.5.2.

Table 15.5.2
Expected distributions of "excitability" scores of rats under four conditions.

Scores	*Group* 1	2	3	4	*Total*
9 and over	5.75	5.75	5.75	5.75	23
6–8	7.75	7.75	7.75	7.75	31
3–5	6.00	6.00	6.00	6.00	24
0–2	5.50	5.50	5.50	5.50	22
	25.00	25.00	25.00	25.00	100

Observe that for these *expected* frequencies, the relative frequency distribution in each column is precisely the same as for the column marginal total. If there are no differences between the population distributions, we expect each sample to show exactly the same frequency distribution. The estimate of this expected distribution within a column is provided by the pooled sample frequency distribution (the B_k marginals).

The χ^2 test is given by

$$\chi^2 = \frac{(5 - 5.75)^2}{5.75} + \frac{(8 - 5.75)^2}{5.75} + \cdots + \frac{(7 - 5.5)^2}{5.5}$$
$$= 5.23.$$

Here there are $(4 - 1)(4 - 1) = 9$ degrees of freedom, and the value required for significance at the .05 level is 16.92, making the result clearly not significant. The experimenter does not have enough evidence to say whether or not the breed is associated with the score a rat gets on emotionality level.

This example should illustrate the fact that χ^2 tests can be used to compare several frequency distributions grouped in the same way. The grouping should, however, be decided upon in advance of the actual data. The only inference drawn from a significant result is that the different samples do not represent the same population distribution. Also note that in this example, the row and column actually represent numerical score values, rather than qualitative distinctions per se. Although in this instance it is possible to use the χ^2 test for independence, the experimenter must realize that not all of the information in the data is being used in this analysis, and that conclusions will not be directly comparable to those of, say, the analysis of variance. In general, the Pearson χ^2 analysis is best reserved for the truly qualitative data for which it is most appropriate.

15.6 / THE SPECIAL CASE OF A FOURFOLD TABLE

Two-by-two contingency or joint frequency tables are especially common in social and behavioral research. For such fourfold tables, computations for the Pearson χ^2 test can be put into a very simple form. Consider the following table:

a	b	a + b
c	d	c + d
a + c	b + d	N

Here, the small letters *a, b, c,* and *d* represent the frequencies in the four cells, respectively. Then, the value of χ^2 can be found quite easily from

$$\chi^2 = \frac{N(ad - bc)^2}{(a + b)(c + d)(a + c)(b + d)} \qquad [15.6.1]$$

with, of course, one degree of freedom.

This value of χ^2 is usually corrected to give a somewhat better approximation to the exact multinomial probability. With the correction, the value is found from

$$\chi^2 = \frac{N(|ad - bc| - N/2)^2}{(a + b)(c + d)(a + c)(b + d)} \qquad [15.6.2]$$

which is compared with Table IV (1 degree of freedom) in the usual way. This is another instance of **Yates' correction for continuity,** which we encountered first in Section 15.2. This correction should be applied only when the number of degrees of freedom is one, however.

15.7 / THE ASSUMPTIONS IN χ^2 TESTS FOR ASSOCIATION

Chi-square tests are among the easiest for the novice in statistics to carry out, and they lend themselves to a wide variety of social and behavioral data. This computational simplicity is deceptive, however, as the use of the chi-square approximation to find multinomial probabilities is based on a fairly elaborate mathematical rationale, requiring a number of very important assumptions. This rationale and the importance of these assumptions has not always been understood, even by experienced researchers, and there is probably no other statistical method that has been so widely misapplied.

In the first place, since the exact probabilities to be approximated are assumed to follow the multinomial rule, each and every observation categorized should be independent of each other observation. In particular, this means that caution may be required in the application of χ^2 tests to data where dependency among observations may be present, as is sometimes the case in repeated observations of the same individuals. As always, however, it is not the mere fact that observations were repeated, but rather the nature of the experiment and the type of data, that let one judge the credibility of the assumption of independent observations. Nevertheless, the novice user of statistical methods does well to avoid the application of Pearson χ^2 tests to data where each individual observed contributes more than a single entry to the joint frequency table.

In the second place, the joint frequency table must be complete, in the sense that each and every observation made must represent one and only one joint-event possibility. This means that each distinct observation made must qualify for one and only one row, one and only one column, and one and only one cell in the contingency table.

The stickiest question of all concerns sample size and the minimum size of expected frequency in each cell. Probabilities found from the chi-square tables for such tests are always approximate. Only when the sample size is infinite must these probabilities be exact. The larger the sample size, the better this approximation generally is, but the goodness of the approximation also depends on such things as true marginal distributions of events, the number of cells in the contingency table, and the significance level employed.

Without going further into the complexities of the matter, we will simply state a rule that is at least current, fairly widely endorsed, and generally conservative. **For tables with more than a single degree of freedom, a minimum expected frequency of 5 can be regarded as adequate, although when there is only a single degree of freedom a minimum expected frequency of 10 is much safer.** This rule of thumb is ordinarily conservative, however, and circumstances may arise where smaller expected frequencies can be tolerated. In particular, if the number of degrees of freedom is large, then it is fairly safe to use the χ^2 test for association even if the minimum expected frequency is as small as 1, provided that there are only a few cells with small expected frequencies (such as one out of five or fewer).

A word must be said about the practice of pooling categories to attain large expected frequencies after the data are seen. This has been done routinely for many years, and many statistics texts advise this as a way out of the problem. However, this may amount to trading the devil for the witch! The whole rationale for the chi-square approximation rests on the randomness of the sample, and that the categories into which observations may fall are chosen in advance. When we start pooling categories after the data are seen, we are doing something to the randomness of the sample, with unknown consequences for our inferences. The manner in which categories are pooled can have an important effect on the inferences drawn. This practice is to be avoided if at all possible; nevertheless, when expected frequencies of less than 1 are encountered, pooling is the only recourse.

15.8 / THE POSSIBILITY OF EXACT TESTS FOR GOODNESS OF FIT AND FOR ASSOCIATION

As we have mentioned, the Pearson χ^2 tests of association and of goodness of fit give approximations to exact probabilities, which, in principle, may be found using the multinomial (or in some instances, the hypergeometric) rule. The basic reason for using the chi-square approximation is that actual computation of these exact probabilities is extremely laborious or downright impossible. However, in some situations where the sample size is so small that the use of the χ^2 tests is ruled out, it may be practicable to compute probabilities exactly. This will be illustrated only for a simple test of association, although other possibilities exist.

The test we will discuss is commonly known as **Fisher's exact test** for a 2×2 contingency table. This is not appropriately called a χ^2 test, since it does not use the chi-square approximation at all. Instead, the exact probability is computed for a sample's showing as much or more evidence for association than that obtained, given only the operation of chance.

Suppose that some N subjects are categorized into the following 2×2 table:

	A_1	A_2	
B_1	a	b	$a + b$
B_2	c	d	$c + d$
	$a + c$	$b + d$	N

Now suppose for a moment that the N subjects actually make up the population and that the distribution in the population shows $a + c$ individuals in the first *column* category, A_1, and $b + d$ individuals in the second column category, A_2. Some $n = a + b$ individuals are sampled at random and *without* replacement. What is the probability that in *this sample* exactly a individuals fall into A_1 and b into A_2? In Section 3.20 it was mentioned that the probability of a sample drawn *without* replacement from a finite population can be found by the *hypergeometric* rule. Applying the rule, we find that the probability of a in A_1 and b in A_2 *within the sample represented by row* B_1 is just

$$\frac{\binom{a+c}{a}\binom{b+d}{b}}{\binom{N}{a+b}}$$

which is the same as

$$\frac{(a+b)!\,(c+d)!\,(a+c)!\,(b+d)!}{N!\,a!\,b!\,c!\,d!}.$$

Now consider the marginals of the sample table as fixed, so that regardless of the arrangement within the table we know the totals in rows and columns. Imagine that occurrences of the events represented by the rows and columns have absolutely nothing to do with each other, so that the two attributes are independent. Any sample result in the cells of the table occurs as though individuals in the columns were assigned to the rows at random. Then the probability of any particular random arrangement is given by the use of the hypergeometric rule, just as for the probability found above. *If one finds the probability of the arrangement actually obtained, as well as every other arrangement giving as much or more evidence for association, then one can test the hypothesis that the obtained result is purely a product of chance by taking this probability as the significance level.* This amounts to finding both the probability of the obtained table, and every other table (with the same marginals) showing more disproportion between cells a and c than in the table obtained.

An example may make this idea seem more reasonable. Imagine a study involving 10 individuals. Each individual observation fell into one and only one cell of the following contingency table:

	A_1	A_2	
B_1	1	4	5
B_2	3	2	5
	4	6	10

Now assume that regardless of how else the individuals were categorized we would have got 4 cases in A_1, 6 cases in A_2, 5 cases in B_1, and 5 cases in B_2. Also assume that the arrangement obtained in the cells of the table is purely a result of chance. What is the probability of this result (the arrangement actually obtained)? Since

$a = 1$
$b = 4$
$c = 3$
$d = 2$

this is

$$\frac{4!\ 6!\ 5!\ 5!}{10!\ 1!\ 4!\ 3!\ 2!}$$

or

$$p(\text{obtained arrangement}) = .238.$$

Actually, we could stop at this point, since for $\alpha = .05$ or less, we definitely could not reject the hypothesis of a random arrangement, since the probability is at least .238 of a sample this ''systematic'' or more so by chance alone. However, we will continue in order to show the method.

An even more systematic-looking result that might have occurred is

	A_1	A_2
B_1	0	5
B_2	4	1

where $a = 0$, $b = 5$, $c = 4$, $d = 1$. Notice that if this result had occurred, then there would be even more discrepancy between the obtained distributions in rows B_1 and B_2. In other words, the relative frequencies in cells a and c are even more different than before. The probability of this result is

$$\frac{4!\ 6!\ 5!\ 5!}{10!\ 0!\ 5!\ 4!\ 1!} = .024.$$

Thus, the probability of our result or one *more* suggestive of association is

$$.238 + .024 = .262.$$

This is not as small as the conventional levels set for α, and so the hypothesis that the apparent association is the product of chance is *not* rejected. Notice, however, that if the actual result had corresponded to the table given last, we could have rejected the hypothesis for any $\alpha \geqslant .024$.

Unlike the χ^2 test, the Fisher exact test is essentially one-tailed. The probabilities are calculated for all possible results departing as much or more in a specific direction from the marginal distribution of A as does our sample. When the rows (or columns) have equal frequencies (as in the example) the final probability can be doubled to arrive at the two-tailed significance level. However, when both the two rows and the two columns have unequal marginal frequencies, the probability should be found for each possible table where the absolute difference

$$\left| \frac{a}{a + b} - \frac{c}{c + d} \right|$$

is as great or greater than in the table actually obtained.

Convenient tables are available for the Fisher exact test, so that it is not usually necessary to carry out all the computations given here in order to perform this test. These tables are available in the *Biometrika Tables for Statisticians* (1967) and in Siegel (1956), among other places.

15.9 / A TEST FOR CORRELATED PROPORTIONS IN A TWO-BY-TWO TABLE

A problem that often arises in research is somewhat different from the problem of association between attributes. As an illustration of this problem, suppose that some N individual subjects are each observed by two independent judges. Each judge places each subject into one of two mutually exclusive and exhaustive categories, such as "high leadership potential" versus "low leadership potential." It is assumed that a judge's ratings of different individuals are independent. Let us call these categories simply "H" and "L" for the moment. We would like to ask if these two judges, given all possible subjects in the population, would show the same true proportion of individuals rated in category "H." In other words, in the population of all subjects to be rated, does $p_1(H) = p_2(H)$, where $p_1(H)$ is the proportion rated in category H by judge 1, and $p_2(H)$ is the proportion rated in that category by judge 2?

This is a problem of **correlated proportions,** since each of the two sample proportions will be based in part on the same individuals. A test due to McNemar (1975) applies to this situation. Suppose that the sample of N individuals were arranged into the following 2 × 2 table:

		Judge 1	
		H	*L*
Judge 2	*H*	a	b
	L	c	d

An *exact* test uses twice the binomial prob.$(x \leq h|p = .5, N = a + c)$, where h is the lesser of b or c. When N is fairly large, the exact probability may be approximated by use of χ^2, where

$$\chi^2 = \frac{(|b - c| - 1)^2}{b + c} \qquad [15.9.1]$$

with one degree of freedom.

For our example, a significant result would let one conclude that the *true* distributions of judgements for the two judges differ. Be sure to notice that this is not an ordinary test of association for a contingency table, but rather a test of the equality of two proportions where each sample proportion involves some of the same observations, making the two sample proportions dependent. This test has been extended to more complicated situations by Cochran; the procedure is discussed in Chapter 16.

15.10 / MEASURES OF ASSOCIATION IN CONTINGENCY TABLES

One of the oldest problems in descriptive statistics is that of indexing the strength of statistical association between qualitative attributes. Although a number of simple and meaningful indices exist to describe association in a fourfold table, this problem grows more complex for larger tables, and has perhaps never been solved to everyone's real satisfaction.

Before we go into the problem of describing statistical association in a sample, a general way of viewing statistical association in a population will be introduced.

This is the **index of mean square contingency,** originally suggested by Karl Pearson, the originator of the χ^2 test for association. Imagine a *discrete joint probability distribution* represented in a table with C columns and R rows. The columns represent the qualitative attribute A, and the rows B. Then the mean square contingency is defined to be

$$\varphi^2 = \sum_j \sum_k \frac{p(A_j,B_k)^2}{p(A_j)p(B_k)} - 1$$

This population index φ^2 (small Greek phi, squared) can be zero only when there is complete independence, so that

$$p(A_j,B_k) = p(A_j)p(B_k)$$

for each joint event (A_j,B_k). However, when there is *complete association* in the table, the value of φ^2 is given by

$$\text{max. } \varphi^2 = L - 1,$$

where L is the *smaller* of the two numbers R or C (number of rows or columns in the table).

Now consider a sample of data arranged into a fourfold contingency table, as in Section 15.6. The *sample* value of φ is given by

$$\varphi = \frac{(bc - ad)}{\sqrt{(a + b)(c + d)(a + c)(b + d)}}. \qquad [15.10.1^*]$$

Notice that this is almost exactly the square root of the expression for χ^2 in a 2×2 table, given by 15.6.2. In fact,

$$\chi^2 = N\varphi^2 \qquad [15.10.2]$$

and

$$\varphi = \sqrt{\frac{\chi^2}{N}}. \qquad [15.10.3]$$

Since both χ^2 and population φ reflect the degree to which there is nonindependence between A and B, a test for the hypothesis

$$H_0\text{: true } \varphi = 0$$

is provided by the ordinary χ^2 test for association in a 2×2 table.

In a 2×2 table there is an interesting link between sample φ and the correlation coefficient r. Let the categories within each attribute A and B be thought of as *ordered*. Suppose that the individuals i in the categories are assigned numerical scores as follows:

$X_i = 1$ if i falls in the higher category of A
$X_i = 0$ if i falls in the lower category of A
$Y_i = 1$ if i falls in the higher category of B
$Y_i = 0$ if i falls in the lower category of B.

The data in the table would then be of this form

Y \ X	0	1
1	a	b
0	c	d

Suppose that the correlation between these scores were computed across the N individuals i. We would find that

$$\begin{aligned} r_{xy} &= \frac{\sum_i x_i y_i / N - M_X M_Y}{S_X S_Y} \\ &= \frac{Nb - (b + d)(a + b)}{\sqrt{(a + b)(c + d)(a + c)(b + d)}} \\ &= \frac{ab + b^2 + bc + bd - ab - b^2 - ad - bd}{\sqrt{(a + b)(c + d)(a + c)(b + d)}} \\ &= \varphi \end{aligned} \qquad [15.10.4]$$

What we have shown is that the coefficient φ may be regarded as the correlation between the attributes A and B when the categories are associated with ''scores'' of 0 and 1. If the categories are ordered for each attribute, and 1 means a higher category than does 0 on A, and similarly on B, then the sign of the φ coefficient becomes meaningful: a positive sign implies a tendency for the high category on A to be associated with the high category of B, and vice versa. On the other hand, a negative value implies a tendency for a high category on one attribute to be associated with a low category on the other. Because of this connection with r, φ is often called the **fourfold point correlation.**

The idea of φ^2 extends to samples in larger contingency tables as well. For a set of data arranged into an $R \times C$ table, the *sample* value of φ^2 is simply

$$\varphi^2 = \frac{\chi^2}{N}. \qquad [15.10.5]$$

However, the value of φ^2 needs to be corrected if it is to be an index ranging between zero and one. A convenient way to describe the apparent strength of association in a sample is then to find

$$\varphi' = \sqrt{\frac{\varphi^2}{L - 1}} = \sqrt{\frac{\chi^2}{N(L - 1)}} \qquad [15.10.6]$$

which must lie between 0, reflecting complete independence, and 1, showing complete dependence, of the attributes. The Pearson χ^2 is computed in the ordinary way as for a test of association, and L is the lesser of R, the number of rows, or C, the number of columns.

This index φ' (Cramér's statistic) is not to be confused with the ordinary *coefficient of contingency,* sometimes used for the same purpose. The coefficient of contingency is defined by

$$C_{AB} = \sqrt{\frac{\chi^2}{N + \chi^2}}$$

This last index has the disadvantage that it cannot attain an upper limit of 1.00 unless the number of categories for A and B is infinite. Obviously, this limits the usefulness of C_{AB} as a descriptive statistic, and the index given by φ' is superior. A pair of alightly different indices for a fourfold table were provided by G. U. Yule:

$$Q = \frac{bc - ad}{bc + ad} \qquad [15.10.7]$$

and

$$Y = \frac{\sqrt{bc} - \sqrt{ad}}{\sqrt{bc} + \sqrt{ad}} \qquad [15.10.8]$$

The Q index is very similar to φ, except that its denominator does not have the same form. The Y index is constructed on the supposition that the marginals of the table all have equal frequencies, and thus this index is less sensitive to marginal inequalities than are φ and Q.

15.11 / A MEASURE OF INTERJUDGE AGREEMENT

A troublesome problem in many social and behavioral studies is that of assessing the agreement between two raters or judges, viewing the same set of people or objects. A useful index for such agreement, over and above the agreements that should occur even if the judgments were strictly independent, is given by Cohen (1960). Suppose that the ratings for two judges were arranged in an $I \times I$ table. Then

$$K = \frac{N \sum_i x_{ii} - \sum_i x_{i+}x_{+i}}{N^2 - \sum_i x_{i+}x_{+i}} \qquad [15.11.1]$$

is an index of their agreement, over and above the agreement to be expected for independent ratings.

Here, the judges are presumed to be rating on the same set of categories, x_{ii} symbolizes the number of agreements about category i, x_{i+} stands for the number of times judge 1 used category i altogether, and x_{+i} symbolizes the number of times that the other judge used category i. The N is the number of ratings given, or number of things rated.

If the agreements between the two raters are exactly what one would expect under independence, then K will be zero. If there is perfect agreement between the two raters or judges, then K will be 1.00.

Any exact hypothesis about the value of K may be tested, for large N, by use of the normal distribution, with an estimated value of the standard error of K given by the square root of

$$\text{est } \sigma_K^2 = \frac{1}{N}\left(\frac{\theta_1(1-\theta_1)}{(1-\theta_2)^2} + \frac{2(1-\theta_1)(2\theta_1\theta_2-\theta_3)}{(1-\theta_2)^3} + \frac{(1-\theta_1)^2(\theta_4-4\theta_2^2)}{(1-\theta_2)^4}\right)$$

where

$$\theta_1 = \sum_i x_{ii}/N, \quad \theta_2 = \sum_i (x_{i+}x_{+i})/N^2$$

$$\theta_3 = \sum_i x_{ii}(x_{i+} + x_{+i})/N^2 \qquad [15.11.2]$$

$$\theta_4 = \sum_i \sum_j (x_{ij})(x_{i+} + x_{+j})^2/N^3.$$

Furthermore, this standard error may be used to establish confidence limits for the true value of K when N is large. Thus, the hypothesis that long-run K is equal to zero may be tested by setting up the $100(1 - \alpha)$ percent confidence limits, and checking to see if this interval includes zero.

The hypothesis of *total* independence between the judges may be tested using K divided by the square root of

$$\text{est. } \sigma_K^2 = \frac{\theta_2 + \theta_2^2 - \sum_i x_{i+}x_{+i}(x_{i+} + x_{+i})/N^3}{N(1-\theta_2)^2} \qquad [15.11.3]$$

once again with the normal distribution.

This index has been extended for application to more than two judges by Fleiss (1970).

15.12 / A MEASURE OF PREDICTIVE ASSOCIATION FOR CATEGORICAL DATA

Now we will discuss a different approach to describing the the structure of a contingency table, based upon an **index of predictive association.** This index, which was developed by Goodman and Kruskal (1954), will be called λ_B, (small Greek lambda, sub B):

$$\lambda_B = \frac{p(\text{error}|A_j \text{ unknown}) - p(\text{error}|A_j \text{ known})}{p(\text{error}|A_j \text{ unknown})}. \qquad [15.12.1]$$

This index shows the proportional reduction in the *probability* of error in predicting B afforded by specifying A_j. If the information about the A category does not reduce the probability of error in predicting B at all, the index is zero, and one can say that there is no predictive association. On the other hand, if the index is 1.00, no error is made in predicting B given the A_j classification, and there is complete predictive association.

It must be emphasized that this idea is not completely equivalent to independence and association as reflected in χ^2 and φ'. It is quite possible for some statistical association to exist even though the value of λ_B is zero. In this situation, A and B are not independent, but the relationship is not such that giving A_j causes one to change one's bet about B_k; the index λ_B is other than 0 only when *different* B_k categories would be predicted for different A_j information.

On the other hand, if there is complete proportionality throughout the table, so that φ' is zero, then λ_B must be zero. Furthermore, when there is complete association, so that perfect prediction is possible, both λ_B and φ' must be 1.00.

Sample values of λ_B can be calculated quite easily from a contingency table. Here, the sample is regarded as though it were the population, and probabilities are taken from the relative frequencies in the sample. Thus we interpret λ_B as the proportional reduction in the probability of error in prediction for cases drawn at random from *this* sample, or, if you will, a population exactly like this sample in its joint distribution.

In terms of the frequencies in the sample, we find

$$\lambda_B = \frac{\sum_j \max_k . f_{jk} - \max_k . f_{.k}}{N - \max_k . f_{.k}} \qquad [15.12.2]$$

where

f_{jk} is the frequency observed in cell (A_j, B_k)

$\max_k . f_{jk}$ is the *largest* frequency in column A_j

$\max_k . f_{.k}$ is the largest *marginal* frequency among the rows B_k.

When two or more cells in any column have the same frequency, larger than any others in that column, then the frequency belonging to any single one of those cells is used as the maximum value for the column. Similarly, if several row marginals each exhibit the same frequency, which is largest among the rows, that frequency is used.

The index λ_B is an *assymetric* measure, much like ω^2 and ρ_I. It applies when A is *the* independent variable, or the thing ordinarily known first, and B is the thing predicted. However, for the same set of data, it is entirely possible to reverse the roles of A and B, and obtain the index

$$\lambda_A = \frac{p(\text{error}|B_k \text{ unknown}) - p(\text{error}|B_k \text{ known})}{p(\text{error}|B_k \text{ unknown})}$$

which is suitable for predictions of A from B. In terms of frequencies:

$$\lambda_A = \frac{\sum_k \max_j . f_{jk} - \max_j . f_{j.}}{N - \max_j . f_{j.}} . \qquad [15.12.3]$$

In general, the two indices λ_B and λ_A will not be identical; it is entirely possible to have situations where B may be quite predictable from A, but not A from B.

Finally, in some contexts it may be desirable to have a *symmetric* measure of the power to predict, where neither A nor B is specially designated as the thing predicted from or known first. Rather, we act as though sometimes the A and sometimes the B information is given beforehand. In this circumstance the index λ_{AB} can be computed from

$$\lambda_{AB} = \frac{\sum_j \max_k . f_{jk} + \sum_k \max_j . f_{jk} - \max_k . f_{.k} - \max_j . f_{j.}}{2N - \max_k . f_{.k} - \max_j . f_{j.}} . \qquad [15.12.4]$$

Furthermore, Goodman and Kruskal (1963) showed that when the true λ_B value in the population is neither exactly 0 nor exactly 1.00, the sampling distribution of λ_B is approximately normal for large N, with a sampling variance which can be estimated from

$$\sigma^2_{\lambda_B} = \frac{\left(N - \sum_j \max_k . f_{jk}\right)\left(\sum_j \max_k . f_{jk} + \max_k . f_k - 2 \sum_j{}^* \max_k . f_{jk}\right)}{(N - \max_k . f_k)^3} \qquad [15.12.5]$$

where $\sum_j{}^* \max. f_{jk}$ is the summation of the maximum frequencies in the columns, taken only over those columns where the maximum falls in the same row as $\max_k . f_k$.

These measures of *predictive* association form a valuable adjunct to the tests given by χ^2 methods. When the value of χ^2 turns out signficant one can say with confidence that the attributes A and B are not independent. Nevertheless, the significance level alone tells almost nothing about the strength of association. If we want to say something about the predictive strength of the relation, or have the remotest interest in actual predictions using the relation studied, then the λ measures are worthwhile.

15.13 / ANALYZING LARGER TABLES THROUGH LOG-LINEAR MODELS

So far, the discussion of analyzing contingency tables has been restricted to the situation where the table is two-dimensional. That is, only two qualitative attributes, one represented by the rows and one represented by the columns, constitute the data table. However, in the social sciences it is quite common for data to be collected in terms of more than two attributes at a time. Thus, for example, for some population we might be interested in three attributes of a voter: age, represented by the groupings "under 40," "between 40 and 60," and "over 60"; political philosophy, as represented by the three groupings, "conservative," "moderate," and "liberal"; and party preference, represented by "Democratic," "Republican," "Independent," and "other." Each voter sampled would then be categorized in all three of these ways. The entire set of N cases would then be shown as frequencies in a three-dimensional table having $3 \times 3 \times 4$, or 36, cells. If even more categories or qualitative distinctions are drawn, we might have a four-way, or five-way, or twenty-way table.

Within such a complex table we may still be interested in the statistical association of the attributes represented. In the simple case of a two-way table, we can ask only about the relationship between the two attributes represented by the rows and columns. However, just as in the analysis of variance, higher-order contingency tables generate higher-order questions about the relationships in the table. How do you examine such complex relationships among qualitative attributes, and test for their existence in the population which the data represent?

This is an important problem for social scientists, and only in fairly recent years has it been given a manageable solution, largely due to the work of Leo Goodman (1978) and his colleagues. Although, for reasons of space, we cannot give an adequate treatment of this subject here, the methods are becoming sufficiently important in the social science literature that the student should have at least some limited acquaintance with what they are and how they work.

Modern methods for analyzing a large contingency table are based on a *log-linear model*. Although the model and method extend to tables of any dimensions, in the following, only the model for a three-dimensional table will be described briefly.

Imagine a three-dimensional table, representing some attribute A with I rows, attribute B with J columns, and attribute C with K layers. Let the frequency of observations in any cell (i, j, k) be denoted by x_{ijk}. Now the population from which the random sample of N was drawn has a probability associated with any cell (i, j, k) and we will call this probability p_{ijk}. Furthermore, for a sample of N independent observations, there will be an *expected frequency* $m_{ijk} = Np_{ijk}$ for any cell. Now let the *natural logarithm* (that is, the log to the base e) of the expected frequency m_{ijk} be symbolized by L_{ijk},

$$L_{ijk} = \log_e m_{ijk}.$$

Furthermore, let L (without a subscript) stand for the *average* L_{ijk} over all the IJK cells, let L_i stand for the average L_{ijk} value in column i, let L_{ij} stand for the average L_{ijk} value in combination ij, and so on for all of the other combinations of subscripts i, j, and k taken alone or in combination. Then the log-linear model is

$$L_{ijk} = u + u_i + u_j + u_k + u_{ij} + u_{ik} + u_{jk} + u_{ijk}. \quad [15.13.1]$$

That is, the logarithm of the expected frequency in cell (i,j,k) is the sum of a general effect u; the effect u_i of being in row i; the effect of being in column j, or u_j; the effect u_k of being in layer k; plus an interaction effect u_{ij}; an interaction effect u_{ik}; an interaction effect u_{jk}; and finally a second-order interaction effect u_{ijk}. The u_i, u_j, and u_k are the *marginal* effects, the interaction effects u_{ij}, u_{ik}, u_{jk} represent the influence of *combinations* of two categories on frequencies in the table, and u_{ijk} represents the effect of *all three* categories in combination.

These population effects are defined very much as in the fixed-effects analysis of variance, and each is associated with a number of degrees of freedom:

$$\begin{aligned}
u_i &= L_i - L, && d.f. = (I - 1)\\
u_j &= L_j - L, && d.f. = (J - 1)\\
u_k &= L_k - L, && d.f. = (K - 1)\\
u_{ij} &= L_{ij} - L_i - L_j + L, && d.f. = (I - 1)(J - 1)\\
u_{ik} &= L_{ik} - L_i - L_k + L, && d.f. = (I - 1)(K - 1)\\
u_{jk} &= L_{jk} - L_j - L_k + L, && d.f. = (J - 1)(K - 1)\\
u_{ijk} &= L_{ijk} - L_{ij} - L_{ik} - L_{jk}\\
&\quad + L_i + L_j + L_k - L, && d.f. = (I - 1)(J - 1)(K - 1)
\end{aligned}$$

Given this general model, it is possible to test various hypotheses in the form of specialized models including or excluding the various effects. The main restriction is that any model to be tested is regarded as *hierarchical,* which means simply that if, in a given model, a lower-order effect, such as u_{ij} is assumed to be zero, then any higher-order effect involving the same subscripts plus others must also be assumed to be zero. Thus, in a given hypothesis, $u_{ij} = 0$ implies $u_{ijk} = 0$. This is not an especially harsh restriction, however. (For example, as might at first appear, it does not necessarily rule out equal marginal frequencies. See Bishop, Fienberg, and Holland (1975) for details.)

Now, given the hypothesis that any effect or group of effects is zero (subject to the hierarchical principle just stated), the model permits one to derive estimates of the expected frequencies in the data. Once we have these estimated expected frequencies, it is possible to employ the so-called "likelihood ratio test" in order to assess the goodness of fit of the model to the data. In the three-dimensional situation, the likelihood ratio test, often called G^2, is made by taking

$$G^2 = -2 \sum_i \sum_j \sum_k x_{ijk} \log_e (\text{est. } m_{ijk}/x_{ijk}). \quad [15.13.2]$$

(Here, 0 $\log_e$ 0 is defined to be 0.) Under the null hypothesis generating the particular values of est. m_{ijk}, G^2 is distributed as chi-square with *degrees of freedom equal to the sum of the degrees of freedom associated with each term set equal to zero in the model being tested.* A significant result means, of course, that one can reject the hypothesis that these particular effects are all zero, implying that the particular model tested does not fit the data.

For example, suppose that the hypothesis to be tested for a three-way table were

$$H_0\text{: } u_{ij} = u_{ik} = u_{jk} = u_{ijk} = 0.$$

That is, we are testing the hypothesis that the three attributes are all independent of each other. Then the log-linear model implies that $L_{ijk} = L_i + L_j + L_k - 2L$. Sample estimates are then formed by taking, for any cell (i,j,k),

$$\text{est. } L_{ijk} = \log_e (\text{est. } m_{ijk}) = \log_e x_{i++} + \log_e x_{+j+} + \log_e x_{++k} - 2 \log_e N$$

where x_{i++} is the sum of row i, x_{+j+} is the sum of column j, and x_{++k} is the sum of layer k.

Then on substituting these $\log_e$ (est. m_{ijk}) values into G^2 as given above, we have

$$G^2 = 2\left[\sum_i \sum_j \sum_k x_{ijk} \log_e x_{ijk} - \sum_i x_{i++} \log_e x_{i++} - \sum_j x_{+j+} \log_e x_{+j+} - \sum_k x_{++k} \log_e x_{++k} + 2N \log_e N\right], \quad [15.13.3]$$

with $(I - 1)(J - 1) + (I - 1)(K - 1) + (J - 1)(K - 1) + (I - 1)(J - 1)(K - 1) = IJK - I - J - K + 2$ degrees of freedom.

If we wish to see if only one pair of attributes, say A and B, are related, this can be represented by the hypothesis

$$H_0\text{: } u_{ik} = u_{jk} = u_{ijk} = 0$$

Here, $L_{ijk} = L_{ij} + L_k - L$ so that

$$\log_e (\text{est. } m_{ijk}) = \log_e x_{ij+} + \log_e x_{++k} - \log_e N$$

where x_{ij+} is the sum of the row-column combination i and j, and so on. Once again, a likelihood ratio test can be carried out using these values as the expected frequencies, and this time with $(I - 1)(K - 1) + (J - 1)(K - 1) + (I - 1)(J - 1)(K - 1)$ degrees of freedom:

$$G^2 = 2\left[\sum_i\sum_j\sum_k x_{ijk}\log_e x_{ijk} - \sum_i\sum_j x_{ij+}\log_e x_{ij+} - \sum_k x_{++k}\log_e x_{++k} + N\log_e N\right]. \qquad [15.13.4]$$

In order to test the hypothesis that a single pair of attributes, say A and B, are *not* related, while allowing for the possibility that A and C, and B and C *might* be related, we could frame the hypothesis

$$H_0\colon u_{ij} = u_{ijk} = 0.$$

Then, the model provides the logs of the estimated expected frequencies to be

$$\log_e(\text{est. } m_{ijk}) = \log_e x_{i+k} + \log_e x_{+jk} - \log_e x_{++k}$$

and the likelihood ratio test has $(I - 1)(J - 1) + (I - 1)(J - 1)(K - 1)$ degrees of freedom:

$$G^2 = 2\left[\sum_i\sum_j\sum_k x_{ijk}\log_e x_{ijk} - \sum_i\sum_k x_{i+k}\log_e x_{i+k} - \sum_j\sum_k x_{+jk}\log_e x_{+jk} + \sum_k x_{++k}\log_e x_{++k}\right]. \qquad [15.13.5]$$

Finally, suppose that we wish to examine the model which includes all possible effects *except* second-order interaction among A, B, and C.

That is, we wish to test the hypothesis

$$H_0\colon u_{ijk} = 0.$$

In this instance, the model does not provide a direct way to estimate the expected frequencies. Rather, these values are found by a process of *iteration,* in which the expected frequencies are approximated by successive "fits" to the table's marginals. This is laborious to do by hand but extremely simple to do on a computer. We will not describe this process here, however. Suffice it to say that once these estimated expected frequencies are found, they, too, are used in a likelihood ratio test, this time with $(I - 1)(J - 1)(K - 1)$ degrees of freedom.

The application of the log-linear model to a three-dimensional table may be illustrated as follows: Consider this 24-cell table, consisting of all combinations of an attribute A with two categories, an attribute B with three categories, and an attribute C with four categories. The frequencies shown in the cells of this table add up to 946 independent observations.

A_1	B_1	B_2	B_3	
C_1	11	6	18	*35*
C_2	10	2	6	*18*
C_3	0	1	13	*14*
C_4	13	4	21	*38*
	34	*13*	*58*	*105*

	A_2			
	B_1	B_2	B_3	
C_1	24	13	92	*129*
C_2	40	18	134	*192*
C_3	13	6	92	*111*
C_4	130	37	242	*409*
	207	*74*	*560*	*841*

Then preliminary calculations give

$$\sum_i \sum_j \sum_k x_{ijk} \log_e x_{ijk} = 11 \log_e 11 + \cdots + 242 \log_e 242 = 4196.79$$

$$\sum_i \sum_j x_{ij+} \log_e x_{ij+} = 34 \log_e 34 + \cdots + 560 \log_e 560 = 5354.76$$

$$\sum_i \sum_k x_{i+k} \log_e x_{i+k} = 35 \log_e 35 + \cdots + 409 \log_e 409 = 4970.36$$

$$\sum_j \sum_k x_{+jk} \log_e x_{+jk} = (11 + 24)\log_e(11 + 24) + \cdots$$

$$+ (21 + 242)\log_e(21 + 242) = 4507.83$$

$$\sum_i x_{i++} \log_e x_{i++} = 105 \log_e 105 + 841 \log_e 841 = 6152.46$$

$$\sum_j x_{+j+} \log_e x_{+j+} = (34 + 207)\log_e(34 + 207) + \cdots$$

$$+ (58 + 560)\log_e(58 + 560) = 5681.94$$

$$\sum_k x_{++k} \log_e x_{++k} = (35 + 129)\log_e(35 + 129) + \cdots$$

$$+ (38 + 409)\log_e(38 + 409) = 5290.65.$$

$$N \log_e N = (946)\log_e 946 = 6482.27.$$

Now the first hypothesis stated above (complete independence of A, B, and C) is tested by taking 15.13.3:

$$G^2 = 2(4196.79 - 6152.46 - 5681.94 - 5290.65 + 2(6482.27))$$

$$= 72.56.$$

For 17 degrees of freedom, this is quite significant, and so we can reject the model of total independence among A, B, and C on the basis of these data.

The second hypothesis corresponds to a model having no interactions involving variable C, but having an $A \times B$ interaction. Then using 15.13.4 we take

$$G^2 = 2(4196.79 - 5354.76 - 5290.65 + 6482.27) = 67.3.$$

For 15 degrees of freedom, this too is quite significant, so that we can reject the model containing only an $A \times B$ interaction.

Next, the model containing $A \times C$ and $B \times C$ interaction, but no $A \times B$ or $A \times B \times C$ interaction is examined (hypothesis 3 above) by using 15.13.5:

$$G^2 = 2(4196.79 - 4970.3 - 4507.83 + 5290.65) = 18.5.$$

For 8 degrees of freedom, this too is significant beyond the .05 level.

Finally, if we test the hypothesis that there is no second-order interaction, but that all first order interactions exist (finding the expected frequencies by iteration) we have a G^2 value of 11.96 with 6 degrees of freedom, which is not significant at the .05 level. Thus, it appears that the best fit to our data is a model with interactions $A \times B$, $A \times C$, $B \times C$ present, but with little or no $A \times B \times C$ interaction effects.

In short, the log-linear model for a three-dimensional table permits one to test hypotheses about any, or any hierarchical set, of the seven effects (other than general u) entering into the model. The model then dictates what the expected frequencies should be, and how these should be estimated from the data. A likelihood ratio test, comparing the obtained and the expected frequencies, can then be used to test the hypothesis in question.

This method extends to tables of almost any number of attributes or dimensions. Naturally, for larger tables the number of interaction effects that can enter into the model becomes very much larger. Even so, in principle each effect and each combination of effects (under the hierarchical principle) may be examined. The assumptions are the same as for any ordinary chi-square test: independent observations representing multinomial sampling from some population. For very large tables, large samples are required in order to have a reasonable chance of some expected frequency (1 or 2) in every cell.

Unfortunately, space does not permit our going further into log-linear models here. Several good books exist, however, showing many of the methods flowing from the log-linear model. Perhaps the most complete discussion is to be found in Bishop, Fienberg, and Holland (1974). Original papers on these topics are found in Goodman (1978).

Computer programs also exist for analyzing large contingency tables by these methods, and these are now to be found in major research and university computing centers. The best known of these programs, which goes by the acronym ECTA (for Everyman's Contingency Table Analysis) was developed by Goodman and his colleagues at the University of Chicago, and is widely available. It provides all required iterations, expected values, G^2 values, and a variety of statistics for contingency tables of virtually any size.

EXERCISES

1. In the Midwest, a large number of sportscasters make predictions each Thursday about the outcomes of Saturday's football games during the season. On a certain Thursday, each of 50 sportscasters made predictions about the same eight games. The numbers of correct predictions of the winner are as follows:

Number correct	Number of sportscasters
8	1
7	3
6	5
5	11
4	15
3	8
2	5
1	1
0	1

Test the hypothesis that the correct predictions of any given sportscaster are outcomes of a stable and independent binomial process with $p = .5$. (**Hint:** Recall Chapter 3.)

2. Suppose that a theory holds that the calorie consumption of healthy U.S. adolescent boys should follow a normal distribution. Use the sample data of exercise 9, Chapter 4, to test this theory. (**Hint:** If a normal distribution with the same mean and variance as the sample were divided into the same class intervals, what should the expected frequencies be?)

3. Use the same method as in exercise 2 to test the hypothesis that calorie consumption among healthy U.S. girls is normally distributed.

4. Suppose that the 1000 words mentioned in exercise 13, Chapter 2, were actually a random sample of all such foreign words. Test the hypothesis that if a word entered English after 1451 it was equally likely to have appeared in any half-century between that date and 1900.

5. Let us assume that the 270 neighborhoods of exercise 28, Chapter 1, actually represent a random sample of a large population of such neighborhoods. Test the hypothesis that income and ethnic balance are independent properties of such neighborhoods.

6. Describe the degree of relationship that seems to exist between income and ethnic balance in exercise 5 above by use of Cramér's statistic (15.10.6). How do you test the significance of Cramér's statistic?

7. Repeat exercise 6, but this time describe the extent to which ethnic balance predicts income, by means of the λ_A measure. How well does income predict ethnic balance?

8. Consider the table of exercise 32, Chapter 1. Test the hypothesis that interest pattern is independent of occupation, given that this is a random sample of independent observations.

9. To what extent is there predictive association from interest pattern to occupation? How well can one predict from occupation to interest pattern?

10. For exercise 31, Chapter 1, test the two 3×3 tables separately for association, and calculate Cramér's statistic for each. Do these two tables, each holding a third factor constant, seem to be different in the degree of association each represents? How might such an analysis of subtables shed light on partial association (analogous to partial correlation) in contingency tables?

11. In a study of the possible relationship between the number of years of nursery and kindergarten school training experienced by a child, and the rated deportment in the first grade, a random sample of 150 first-graders was obtained, and each was rated in terms of behavior:

Prior experience	Deportment in 1st grade Poorly behaved	Moderately well behaved	Very well behaved
2 yrs. + kindergarten	6	12	0
1 yr. + kindergarten	12	25	6
Kindergarten only	14	31	12
No kindergarten	2	23	7

Is there significant association (α = .05) between amount of nursery school and kindergarten and rated deportment in the first grade, according to these data? As judged from the data given above, to what extent does knowing the number of years in preschool and kindergarten permit us to predict the behavior rating of a child in first grade? (**Hint:** Compute the index of predictive association.) Comment on this result in the light of the first finding in above.

12. Four random samples of 44 subjects each were drawn, and each sample assigned to a different experimental condition. Each subject was given the same problem-solving test, with a possible score of 0 through 12. The results yielded four frequency distributions, as follows. Test the hypothesis that these four samples represent identical population distributions (α = .05):

Scores	Sample 1	Sample 2	Sample 3	Sample 4
12	1	5	3	2
11	1	2	3	6
10	1	2	3	5
9	2	6	3	6
8	3	2	3	1
7	8	4	4	1
6	12	2	6	2
5	8	4	4	1
4	3	2	3	1
3	2	6	3	6
2	1	2	3	5
1	1	2	3	6
0	1	5	3	2
	44	44	44	44

13. By inspecting the data table in exercise 12, see if you can conclude what the F value would have been if an analysis of variance had been used. Would this be significant? Why? How do these sample distributions differ? What does this show about a comparison of samples via a chi-square test as opposed to a comparison via the F test in analysis of variance?

14. In a comparison of child-rearing practices within two cultures, a researcher drew a random sample of 100 families representing Culture I, and another sample of 100 families representing Culture II. Each family was classified according to whether the family was father-dominant or mother-dominant, in terms of administration of discipline. The results follow:

	Culture I	*Culture II*
Father-dominant	53	37
Mother-dominant	47	63

At approximately what α level can one reject the hypothesis of no association between culture and the dominant parent in a family?

15. For the data of exercise 14 find the coefficient of predictive association (λ) for predicting parent-dominance of a family from the culture. Do these data suggest the presence of a very strong predictive association here? Explain.

16. Four large midwestern universities were compared with respect to the fields in which graduate degrees are given. The graduation rolls for last year from each university were taken, and the results put into the following contingency table:

	Field				
University	*Law*	*Medicine*	*Science*	*Humanities*	*Other*
A	29	43	81	87	73
B	31	59	128	100	87
C	35	51	167	112	252
D	30	49	152	98	215

Is there significant association ($\alpha = .05$) between the university and the fields in which it awards graduate degrees? What are we assuming when we carry out this test?

17. In a study of the effect of a particular kind of cortical lesion upon the ability of a monkey to learn a discrimination problem, two groups of six monkeys each were used. The following data were found:

	Solved problem	*Did not solve problem*
Experimental group	1	5
Control group	5	1

Is there a significant difference ($\alpha = .05$) between these two groups of monkeys, so that one may say that the experimental group is less likely to solve the problem?

18. A researcher was interested in the stability of political preference among American women voters. A random sample of 80 women who voted in the elections of 1976 and 1972 showed the following results:

		1976 vote Republican	1976 vote Democrat	
1972 Vote	Republican	34	11	45
	Democrat	5	30	35

Do these data afford significant evidence ($\alpha = .05$) that the true proportion of women who voted Republican in 1976 was different from the true proportion who voted Republican in 1972? (**Hint:** This problem involves *overlapping* or *correlated* proportions.)

19. In a study of the possible relationship between a male military officer's own confidential judgment of his effectiveness as a leader, and the judgment of him by his immediate superior, a sample of 112 officers was drawn at random. Each officer rated himself with respect to his leadership ability, and then his immediate superior was asked to rate him. These ratings were made independently. The data turned out as follows:

		Rating by superior: Low	Mod low	Mod high	Very high
Self-rating	Very high	1	9	7	6
	High	2	5	8	12
	Mod low	4	12	15	3
	Low	5	10	8	5

Does there seem to be significant ($\alpha = .05$) association between an officer's own judged leadership ability and the judgment of his immediate superior? Compute the Cramér coefficient showing the relative degree of association between self and superior's ratings.

20. Use Cohen's kappa to assess the agreement in exercise 19 between superior's ratings and self rating. (**Note:** Be sure to observe the "High-Low" direction for each marginal of the table.)

21. In a study of the possibility of a sex linkage with the occurrence of identical twins, a random sample was taken of records of normal deliveries from a particular set of hospitals over a one-year period. These records revealed the following:

	Male	Female
Single births	658	688
Identical twins	39	34

Is there significant evidence that identical twins are more likely to be a given sex than are infants born singly? What does one assume about the respective deliveries recorded in this table?

22. In a study of parents' and teachers' perceptions of children, members of a random sample of normal first-grade children were rated separately by their parents and by their teachers on muscular coordination. Is there a significant relationship between the ratings of parents and teachers? How would you describe this relationship, if such exists?

		Parents' ratings			
		Poor	Fair	Good	Excellent
Teachers' Ratings	Poor	33	48	113	209
	Fair	41	100	202	255
	Good	39	58	70	61
	Excellent	17	13	22	10

23. Compute the Goodman-Kruskal indices of predictive association for the data of exercise 22, for predictions of teachers' ratings from those of parents, and parents' ratings from those of teachers. Comment on the meaning of the values obtained from these indices.

24. According to Mendelian genetics, if a parent having two dominant characteristics A and B is mated with another parent having the two recessive characteristics a and b, the offspring should show combinations of dominant and recessive characteristics with the following relative frequencies:

Type	Relative frequency
Ab	9/18
aB	4/18
Ab	4/18
ab	1/18

In an actual experiment, when an AB parent was mated with an ab parent, the following frequency distribution of offspring resulted:

Type	Relative frequency
Ab	39
aB	19
Ab	16
ab	1
	75 = N

Test the hypothesis that the Mendelian theory holds for these dominant and recessive characteristics (α = .05).

25. On a simple test of arithmetic, with items all of equal difficulty, a teacher recorded the item number on which each child made his first error. For 100 children, the item numbers on which the first error occurred were distributed as follows:

Items with first error	*Number of children*
6 or more	9
5	7
4	10
3	15
2	29
1	30
	100

If an error on any item for any child is like the outcome of a stable and independent Bernoulli process, test the hypothesis that p(error) = 1/3. (**Hint:** Recall the Pascal and geometric distributions.)

26. A large horse show employed two judges to admit entrants into the final competition. There were 106 initial entrants. The results were as follows:

		Judge I	
		Admit	*Don't admit*
Judge II	*Admit*	28	14
	Don't admit	22	42

Given that these entries are a random sample from some large population, and that the respective judgments were independent, would you say that the two judges have the same probability of admitting an entrant into the horse show finals?

27. An experimenter constructed a six item multiple-choice test, each item having four possible answers. Suppose that when a subject is simply guessing, the probability of getting the right answer on any given item is exactly 1/4. Furthermore, suppose that the answer guessed on any item can be assumed to be independent of the answer guessed on any other item. Now the experimenter gave the test to 420 subjects, and found the following frequency distribution of numbers of items correct:

Number correct	*Frequency*
6	32
5	10
4	34
3	62
2	108
1	121
0	53
	420

Test the hypothesis (α = .01) that each subject was guessing independently on each item.

28. Two judges graded 63 subjects on the poise each showed in a frustrating situation. Each judge used the same four letter grades for each subject. The results were as follows. To what extent did the judges tend to agree, as shown by kappa?

		Judge II A	B	C	D
Judge I	A	8	3	2	2
	B	4	12	1	1
	C	1	2	7	1
	D	0	1	3	15

29. Test the hypothesis of no true agreement between the judges, using 15.11.3.

30. Two pairs of dice are under study. One pair of dice is black, and the other pair white. Each pair of dice is tossed 360 times, and the following distributions of results obtained:

Number of spots	*Black dice*	*White dice*
12	11	13
11	21	17
10	29	35
9	41	35
8	48	51
7	62	60
6	51	49
5	39	45
4	31	25
3	19	23
2	8	7

Can we say (α = .05) that

(a) the black dice are fair (probability of each side of each die equal to 1/6)?
(b) the white dice are fair?
(c) the true distribution of results is the same for the black and the white dice?

31. Take the data of exercise 31, Chapter 1, and see if you can test the three hypotheses of 15.13 for these data.

32. Use the data of exercise 9, Chapter 4, to test the hypothesis that boys and girls have have identical distributions of calories consumed.

Chapter 16
SOME ORDER METHODS

In this chapter we are going to discuss methods based primarily on the **order relations** among observations in a set of data. The reasoning behind each of these tests involves relatively simple applications of probability theory; it sometimes happens that discrete sampling distributions often can be found in particularly simple ways if only the order features of the data are considered.

There are at least two reasons why we should be interested in analyzing data in terms of ordinal properties: In the first place, the only relevant information in a set of numerical scores may actually be ordinal. That is, it may well be true that the operation used to measure some psychological or social characteristic is valid only at the ordinal level, so that the numerical scores obtained actually give information only about relative magnitudes of the underlying property, and arithmetic differences between scores have no particular meaning in terms of this property. This is a common situation in measurements employed by social scientists.

An equally common reason for using order methods is that one or more assumptions about the population distributions, strictly necessary for one of the standard or parametric methods to apply, may be quite unreasonable. Rather than use the parametric method anyway and wonder about the validity of the conclusion, the experimenter prefers to change the question in such a way that another, *nonparametric* method does apply. Ordinarily, such a method will require that only certain features of the raw numerical data be considered if the objectionable assumptions are to be avoided, so that all of the numerical information in the scores is not used. Consequently, the experimenter may lose something by deciding to use the nonparametric method: the question answered by the nonparametric method may not be the same as that answered by the corresponding parametric method, and for a given sample size the nonparametric test may represent a considerably weaker use of the evidence.

Because of their direct dependency on elementary probability theory, and their comparative freedom from assumptions about population distributions and parameters, order techniques are usually classified among the nonparametric methods. However, not all nonparametric techniques involve considerations of order, nor are all order methods completely free of assumptions about distributions and parameters. Therefore, in this chapter, the author prefers to limit discussion to a few methods based on order, and more or less beg the complicated question of "parametric" versus "nonparametric." For this, and virtually every other question concerned with such statistical methods, the reader is referred to the monumental *Handbook* by Walsh (1962, 1965). This chapter simply shows some techniques that are often found useful, and that happen to involve order.

Even so, a word of warning is called for in the use of order methods as stand-ins for the classical methods in situations where both kinds of methods are appropriate. At least two things must be borne in mind:

First, the actual hypothesis tested by a given order method is seldom exactly equivalent to the hypothesis tested by a parametric technique. These order tests can be regarded essentially as testing the hypothesis of identical population distributions. Regardless of how well the actual test statistic agrees with expectation, however, the population distributions still *might* be different in particular ways. In most order tests the sampling distribution implied by a true H_0 can also obtain when H_0 is not true in various ways. Departures from strict identity among the population distributions may exist to which the order methods are very insensitive. If one is willing to make only minimal assumptions about the population distributions, then the kinds of true differences among populations that the test fails to detect may be quite unknown. Of course, various additional assumptions can be made about the population distributions, and then the sensitivity of these tests to various alternatives can be studied. Nevertheless, these gratuitous assumptions may be fully as offensive as the assumptions the test was designed to avoid in the first place. It is sad but true: the specificity of our final conclusion is more or less bought in terms of what we already know or can at least assume to be true. If we do not know or assume anything, we cannot conclude very much.

Second, when both the order method and a parametric method actually do apply (that is, when the parametric assumptions are true), the power of the two kinds of tests may be compared, given α, the sample N, and the true situation. Order techniques, along with many of the nonparametric methods, will usually be relatively low-powered as compared with parametric tests. This means that, other things such as α and N being equal, one is taking more risk of a Type II error in using the order method. If Type II errors are to be avoided, then a relatively larger sample size (or a larger α value) may be required to use the order technique than for a parametric method.

The decision to use or not to use order methods in a given problem cannot be given a simple prescription. This is but another place where the experimenter has to think about what is to be accomplished and how. It is wrong to conclude that statistical assumptions are bad, that only bona fide interval-scale data can be subjected to the classical statistical treatment, that tests with relatively low power efficiency are useless, and so on. None of these statements is necessarily true in all situations. Through practice and all the help available the experimenter must learn to pick and choose among all of the various methods available, finding the one that most clearly, economically, and reasonably sheds light on the particular question to be answered. This is not a simple task, and a brief discussion such as this can only begin to suggest some of these issues.

In the sections to follow, several types of order statistics and tests will be discussed. First of all, tests will be mentioned that are appropriate to experimental data where the experimental factor is categorical and the dependent variable is treated at the ordinal level. Then some correlation-like methods for ordinal data will be surveyed, and finally an index of association for ordered classes will be outlined.

16.1 / COMPARING TWO OR MORE INDEPENDENT GROUPS: THE MEDIAN TESTS

The median test, which is one of the simplest of the order methods, involves the comparison of several samples on the basis of deviations from the median rather than from the mean. We assume that the underlying variable on which the populations are to be compared is continuous, and that the probability of a tie between two observations in actual value of the underlying variable is, in effect, zero.

The null hypothesis to be tested is that J different populations are absolutely identical in terms of their distributions. The alternative is simply the contrary of H_0. If we wish to add the assumption that whatever the differences in central tendency that may exist among the distributions, they are at least identical in *form,* then the test actually becomes one of central tendency; however, it is hard to see why such an assumption would ordinarily be justified in a situation where an order method such as this is called for. Thus, in this method, as well as in most of the methods to follow, the null hypothesis will state only that the distributions are identical. The median test will be sensitive to differences in central tendency for the various population distributions, but failure to reject H_0 does not necessarily imply that the distributions are, in fact, identical, unless one is willing to make other assumptions.

The method is as follows: The J different sample groups are combined into a single distribution and the grand median for the sample, Md, is obtained. Now each score in each group is compared with Md. If the particular score is above the grand median, the observation is assigned to a "plus" category; if the particular score is not above the grand median, the observation is assigned to a "minus" category. Let a_j be the number of "plus" observations in group j, and let $n_j - a_j$ represent the number of "minus" observations. Then the data are arranged into a $2 \times J$ table such as the following:

	Group 1	. . .	Group j	. . .	Group J	Total
Plus	a_1	. . .	a_j	. . .	a_J	a
Minus	$n_1 - a_1$		$n_j - a_j$		$n_J - a_J$	$N - a$
	n_1	. . .	n_j	. . .	n_J	N

Notice that this is simply a joint frequency or contingency table, where one attribute is "group" and the other is "plus or minus."

Now, if the value Md actually divides each of the populations in exactly the same way there should be a binomial distribution of the sample numbers a_j for random samples from population j. That is, the probability of exactly a_j "plus" observations in group j is

$$\binom{n_j}{a_j} p^{a_j} q^{n_j - a_j}$$

where p is the probability of an observation's falling into the "plus" category and $q = 1 - p$. Under the null hypothesis, the probability p should be the same for each and every population, since any value Md should divide the population distributions identically.

However, we want the probability of the obtained *sample* result, conditional to the fact that the *marginal* frequency of "plus" observations is α. Under the null hypothesis, the probability of having exactly a observations above the value Md is $\binom{N}{a} p^a q^{n-a}$. Then the conditional probability for a particular arrangement in the table works out, by the hypergeometric rule of 3.20, to be

$$\frac{\binom{n_2}{a_1}\binom{n_2}{a_2}\cdots\binom{n_j}{a_j}\cdots\binom{n_J}{a_J}}{\binom{N}{a}}.$$

This is the probability of this *particular* sample result conditional to this particular value of a. Since this probability can be found for any possible sample result, the significance level can be found by exact methods, involving finding all possible sample results differing this much or more among the J groups. Such exact probabilities may be very laborious to work out, of course, unless J is only 2 or 3.

For large samples a fairly good approximation to the exact significance level is found from the statistic

$$\chi^2 = \frac{(N - 1)}{a(N - a)} \sum_{j=1}^{J} \frac{(Na_j - n_j a)^2}{Nn_j}. \qquad [16.1.1^*]$$

For reasonably large samples ($N \geqslant 20$, $n_j \geqslant 5$ for each j), this statistic is distributed *approximately* as chi-square with $J - 1$ degrees of freedom. Rejection of the null hypothesis lets us assert that the populations represented are *not* identical.

One difficulty with this test is that it is based on the assumption that ties in the data will not occur. Naturally, this is most unreasonable since tied scores ordinarily will occur. This is really a problem only if several scores are tied at the overall median, *Md,* since other ties have no effect on the test statistic itself. When ties occur at the median, several things may be done to remove this difficulty, but the safest general procedure seems to be to allot the tied scores (those tied with the grand median) within each group in such a way that a_j is as close as possible to $n_j - a_j$. This at least makes the test relatively conservative. Remember that this χ^2 method gives an approximate test, and really should be used only when the sample size within each group is fairly large. In principle, exact probabilities may be computed, as suggested above, when sample N is small.

As an example, consider the following problem. An experimenter was interested in the effect which the difficulty of admission to a club has on the desire of a person to become a member. Thus, she decided to form four experimental "social clubs" at a large college. A sample of 100 girls was drawn and each was assigned at random to one of four experimental groups. Each girl was sent a letter asking her to come for an interview as a prospective member of a service club having secret membership. In experimental treatment I, the goals of the club were outlined, and quite easy membership requirements stated. In treatment II, exactly the same information was given, but with somewhat harder requirements to be met. Treatments III and IV included severe and truly formidable entrance requirements respectively. After a standard "dummy" interview, each girl was asked to rate on a 12-point scale just how eager she was to join such a group. On this scale 1 represented *most* eager, and 12 *least* eager to join. (Since the ethical features of this "experiment" leave a good bit to be desired, it is not to be regarded as an example of good social science practice.)

Table 16.1.1

Ratings given by girls after four types of information.

	Group				
Rating	*I*	*II*	*III*	*IV*	
12	0	0	0	0	
11	3	2	1	0	
10	7	3	4	1	
9	5	5	4	4	
8	5	5	1	0	
7	3	6	3	10	
6	2	2	5	8	
5	0	2	4	1	
4	0	0	3	1	
3	0	0	0	0	
	25	25	25	25	*N* = 100

The experimenter wanted to test the hypothesis that these groups represented four potential populations having identical distributions of ratings. The combined distribution of ratings was as follows:

	f	*cf*
12	0	100
11	6	100
10	15	94
9	18	79
8	11	61
7	22	50
6	17	28
5	7	11
4	4	4

Md = 7.5

When each rating was compared to the value of *Md,* the following results emerged:

	I	*II*	*III*	*IV*	
plus	20	15	10	5	$50 = a$
minus	5	10	15	20	$50 = N - a$

$$\chi^2 = \frac{(99)}{(50)(50)}\left\{\frac{[(100)(20) - (25)(50)]^2}{2500} + \cdots + \frac{[(100)(5) - (25)(50)]^2}{2500}\right\}$$

$$= 19.8$$

For 3 degrees of freedom, this value exceeds the χ^2 value for the .01 level, and so the experimenter can say that the populations are not identical. The hypothesis tested is not about any particular characteristic of the populations considered, but rather about the absolute identity of their distributions. Some departure from strict identity of population distributions is inferred from this significant result, and the experimenter has evidence that the experimental treatment is associated with the rated attraction of a club for a girl.

16.2 / THE MEDIAN TEST FOR MATCHED GROUPS

It is possible to extend the idea of the median test to situations where the various groups are not independent random samples, but rather are matched in terms of one or more factors. The method outlined below is especially appropriate to the kind of experiment referred to as a randomized blocks design in Chapter 11. That is, some J experimental treatments are to be compared, and there are some K blocks or levels of a nuisance factor. Within each of the K blocks, some J subjects are chosen and assigned at random to the treatments. Ordinarily, these K blocks or levels are a random sample of all possible such levels of the nuisance factor, so that if treated by the analysis of variance this would correspond to a mixed-model design.

In these circumstances, the data are arranged into a table with J columns and K rows, just as for the corresponding analysis of variance. The median of each row, or Md_k, is found.

Let

$$a_j = \text{number of ``plus'' observations in column } j.$$

Having done this for each column j, we have a table exactly like that in Section 16.1. However, here we let

$$a = \frac{J}{2} \qquad \text{for } J \text{ even}$$

and

$$a = \frac{J-1}{2} \qquad \text{for } J \text{ odd.}$$

Then, under the assumption that each and every row population has the same form of distribution, we can test the hypothesis that the distributions represented by the columns are identical. Actually, this is analogous to a test of column effects against interaction in an analysis of variance; if interest lies in the extent to which populations differ in terms of central tendency, either interaction effects must be assumed not to exist, or the data should correspond to a mixed or random model experiment.

The actual test statistic is given by

$$\chi^2 = \frac{J-1}{Ka(J-a)} \; \frac{\sum_{j=1}^{J} (Ja_j - Ka)^2}{J} \qquad [16.2.1]$$

for $J - 1$ degrees of freedom. For a relatively large number of rows, this test is satisfactory even though there is only one observation per cell in the data table. Mood, Graybill, and Boes (1974), suggest that the large sample approximation should be fairly good when the number of *cells* in the data table is at least 20.

Table 16.2.1

Results for 10 matched groups of girls.

Group	Treatment *I*	*II*	*III*	*IV*	*Row median*
1	9(+)	5(−)	8(+)	4(−)	6.5
2	10(+)	6(+)	5(−)	3(−)	5.5
3	11(+)	8(+)	3(−)	4(−)	6.0
4	10(+)	7(+)	6(−)	5(−)	6.5
5	8(+)	5(−)	6(−)	7(+)	6.5
6	9(+)	8(+)	4(−)	2(−)	6.0
7	9(+)	7(+)	4(−)	3(−)	5.5
8	9(+)	10(+)	5(−)	4(−)	7.0
9	11(+)	10(+)	9(−)	8(−)	9.5
10	11(+)	6(+)	5(−)	4(−)	5.5
a_j	10	8	1	1	

For example, suppose that the experiment of the preceding section had been carried out, but that 10 *matched groups* of 4 girls each had been used. The girls were matched in terms of social and service activities on the campus. Suppose that the data were as shown in Table 16.2.1. Be sure to notice that the rating value in each cell is compared with the median of the *row* for that cell. The overall table of frequencies of plus and minus categories within columns is thus:

	I	*II*	*III*	*IV*	
plus	10	8	1	1	20 = *Ka*
minus	0	2	9	9	20 = *N* − *Ka*

Here, the test statistic has a value given by

$$\chi^2 = \frac{3}{20(2)} \left\{ \frac{[4(10) - 10(2)]^2}{4} + \cdots + \frac{[4(1) - 10(2)]^2}{4} \right\} = 19.8.$$

For 3 degrees of freedom, this greatly exceeds the χ^2 value required for significance at the .01 level.

This same general idea can be extended to two-way designs with replication, where a test of both main effects and interaction becomes possible. However, the number of required assumptions goes up in this situation, and the tests, especially that for interaction, are quite laborious to carry out. These methods are outlined in Mood, Graybill, and Boes (1974).

16.3 / THE SIGN TEST FOR MATCHED PAIRS

In Section 3.16 we encountered a test applicable when the number of experimental treatments is only two, and *pairs* of observations are matched, the so-called sign test, which is based on the binomial distribution. We will now examine this test somewhat more closely.

Let N be the number of pairs of observations, where one member of each pair belongs to experimental (or natural) treatment 1 and the other to treatment 2. Here, the only relevant information given by the two scores for pair is taken to be the **sign of the difference** between them. If the two treatments actually represent identical populations, and chance is the only determiner of which member of a pair falls into which treatment, we should expect an equal number of differences of plus and of minus sign. The theoretical probability of a ''plus'' sign is .5, and so the probability of a particular number of plus (or minus) signs can be found by the binomial rule with $p = .5$ and N. Notice, however, that we must assume either that the population distributions are continuous so that exact equality between scores has probability zero, or that ties are otherwise impossible. Ordinarily, pairs showing zero differences are simply dropped from the sample, although this makes the final conclusion have a conditional character, ''In a population of untied pairs . . . ,'' and so on. The test is carried out as follows:

First, the direction of the difference (that is, the sign of the difference) between the two observations in each pair is noted, with the same order of subtraction always maintained. Thus each *pair* is given a classification of Plus or Minus, according to the sign of the difference between scores. If the null hypothesis of no difference between the two matched populations were true, one would expect half the nonzero differences to show a positive sign, and half to show a negative sign. Thus, one may simply take the *proportion* of plus differences, and test the hypothesis that the sample proportion arose from a true proportion of .50.

The normal approximation to the binomial can be used if the number of sample pairs is large (10 or more):

$$z = \frac{|P - p| - 1/(2N)}{\sqrt{pq/N}} = \frac{|P - .50| - 1/(2N)}{\sqrt{(.50)(.50)/N}}. \qquad [16.3.1]$$

Here, $1/(2N)$ is a correction for continuity, as in Section 6.4. Although this form holds for a two-tailed test, either a one- or a two-tailed test may be carried out, depending on the alternative hypothesis appropriate to the particular problem. Naturally, in the one-tailed test the sign of the difference between P and .5 is considered.

If the sample size is fewer than 10 pairs, the binomial distribution should be used to give an exact probability. That is, the binomial table (Table II, Appendix D) with

$p = .5$ should be used to find the probability that the *obtained frequency* of the *more frequent sign* should be equaled or exceeded by chance give a true p of .50. These tables of exact probabilities are given in Siegel (1956) as well as many other places. For a one-tailed test, one also checks that the more frequent sign accords with the alternative hypothesis before carrying out the test, of course. For a two-tailed test, this one-tailed binomial probability is doubled.

As an example of the sign test, consider this situation: It was desired to see the effect that a frustrating experience might have upon the "social age" of a child. Each child could be age-rated from his play by a trained observer. Each of a random sample of preschool children was first rated by the observer according to the social-age level of his play during a free-play period. Pairs of children were then formed having the same rated "social age." One randomly chosen member of each pair was frustrated by being allowed to play with a desirable toy that was then taken away, while the pair-mates were not frustrated. The observer did not know which children were frustrated, and the children were given these experimental conditions separately. Finally, in a postexperimental session the children were again rated at free play. Suppose that twenty pairs were used, and that the postexperimental ratings were as shown in Table 16.3.1.

Table 16.3.1

Differences in observed "social age" of pairs of children.

Pair	*Frustrated*	*Not frustrated*	*Sign of diff.*
1	32	36	+
2	35	34	−
3	33	34	+
4	36	40	+
5	44	42	−
6	41	40	−
7	32	35	+
8	38	40	+
9	37	38	+
10	35	35	0
11	29	35	+
12	34	32	−
13	50	51	+
14	40	38	−
15	39	42	+
16	31	33	+
17	47	46	−
18	41	42	+
19	30	29	−
20	35	35	0

The final rating of the frustrated child was subtracted from that of the pair-mate in order to find the sign of the difference between them. Eleven pairs showed a positive sign, seven showed a negative sign, and two showed a zero difference. Did presence or absence of frustration seem to be related to the difference in rated social age? The null and alternative hypotheses are

$$H_0: p = .5$$
$$H_1: p \neq .5$$

where p is the population proportion of plus changes among pairs (plus being the more frequent sign of change in the sample). The statistic for this test is thus

$$z = \frac{|11/18 - .50| - 1/36}{\sqrt{(.50)(.50)/18}} = \frac{.611 - .500 - 028}{.118} = .70.$$

Note that the two pairs showing no difference were excluded from the number of pairs figuring in the test, so that N is reduced to 18. When referred to a normal distribution, this z-value leads to the conclusion of no significant difference. There is not enough evidence to conclude that frustration did lead to a difference in social-age rating.

16.4 / COCHRAN'S TEST

A test that seems to stand midway between methods designed for contingency table data and methods based on order is due to Cochran (1950). This test can be viewed as a generalization of the McNemar two-sample test mentioned in Section 15.9, and it is appropriate in an experiment involving repeated observations (or matched groups) where the dependent variable can take on only two values:

y_{jk} = 1 (for "success," "pass," and so on, recorded for individual k in treatment j).

y_{jk} = 0 (for "fail," and so on, recorded for individual k in treatment j).

For example, suppose that in some experiment K subjects were observed in a standard situation where each subject performed individually under each of J different experimental conditions. Each subject was assigned the conditions in some random order. In each condition, the task of the subject was to solve one of a set of simple reasoning problems. If the problem was solved correctly within one minute, the performance was recorded as a "success," and as a "failure" otherwise. Suppose that the interest of the experimenter was in seeing if the problems had equal difficulty for the subjects—that is, if the true proportion of successes was constant over the problems. In this situation the Cochran test would apply.

As usual, let the experimental treatments (problems, in the example) be shown as columns in the data table. The subjects are shown as the rows. Then the entry in the cell formed by column j and row k contains y, which is 1 for a success and 0 for a failure. Let

$$y_k = \sum_j y_{jk}$$

be the marginal total for row k and let

$$y_j = \sum_k y_{jk}$$

be the marginal total for column j. Finally, let

$$\bar{T} = \frac{\sum_j y_j}{J}.$$

Then the statistic for Cochran's test is given by

$$Q = \frac{J(J-1)\sum_{j=1}^{J}(y_j - \bar{T})^2}{J\left(\sum_k y_k\right) - \left(\sum_k y_k^2\right)}. \qquad [16.4.1]$$

For relatively large K, this is distributed approximately as chi-square with $J - 1$ degrees of freedom, when the hypothesis is true that the probability of a "success" is constant over all treatments J.

Table 16.4.1

Successes of 20 subjects on four problems.

	Problem				
Subject	*1*	*2*	*3*	*4*	Y_k
1	1	1	1	0	3
2	0	1	1	1	3
3	0	0	1	0	1
4	1	1	1	1	4
5	0	1	0	0	1
6	0	0	1	0	1
7	1	0	0	0	1
8	0	0	1	1	2
9	0	0	0	0	0
10	1	0	0	0	1
11	1	0	1	0	2
12	0	0	1	1	2
13	0	1	0	1	2
14	1	0	0	0	1
15	0	1	0	0	1
16	1	0	1	1	3
17	0	1	0	0	1
18	0	0	1	0	1
19	0	1	1	0	2
20	0	0	1	1	2
Y_j	7	8	12	7	34

$$\bar{T} = \frac{34}{4} = 8.5$$

$$\sum_k y_k^2 = 76$$

$$Q = \frac{(4)(3)[(7-8.5)^2 + \cdots + (7-8.5)^2]}{4(34) - 76}$$

$$= 3.40$$

To continue the example, suppose that the experiment outlined above had been carried out. Twenty randomly selected subjects were given each of four problems, and each subject got the problems in some different, randomly chosen, order. A "1" was recorded for a successful solution, and a "0" for a failure. The data turned out to be as shown in Table 16.4.1. For 3 degrees of freedom, the χ^2 table shows this value not significant at either the .05 or .01 level. Thus, the hypothesis of no difference between problems is not rejected.

So far, all the tests discussed in this chapter have actually depended on some way of arranging the observations into only two ordered classes: above or below the median in the median tests, plus or minus differences in the sign test, 1 and 0 categories in the Cochran test. The only role of the numerical scores has been to assign individuals to one of these two categories. When numerical data are reduced to only two categories, it is obvious that much of the possible information in the data is sacrificed. However, other techniques exist which use somewhat more of the information in the scores themselves; that is, observations are rank-ordered in terms of their scores in these methods. Next we turn to one of the simplest of the tests where the numerical score serves to give each observation a place in order.

16.5 / THE WALD-WOLFOWITZ "RUNS" TEST FOR TWO SAMPLES

This test applies to the situation where two unmatched samples are to be compared, and each observation is paired with a numerical score. The underlying variable that these scores represent is assumed to be continuously distributed.

Suppose that the numbers of observations in the two experimental groups are N_1 and N_2 respectively. All of the $N_1 + N_2$ observations in these samples are drawn independently and at random. For convenience, we will call any observation appearing in sample 1 an "A" and any observation in sample 2 a "B." Now suppose that all these sample observations, irrespective of group, are arranged in order according to the magnitude of the scores shown. Then there will be some *arrangement* or pattern of A's and B's in order. In particular there will be *runs* or "clusterings" of the A's and B's. This is easily illustrated by an example.

Suppose that in some two-sample experiment the data turned out as shown in Table 16.5.1. When these scores are combined into a single set and arranged in order of magnitude, we get the following

B B B	A A A	B B	A A	B	A A	B B	A A	B B	A
1 1 2	3 4 4	5 5	6 6	7	8 8	12 13	14 15	18 19	20
1	2	3	4	5	6	7	8	9	10

Above each observation's score is an A or a B, denoting the group to which that observation belongs. Now notice that there are runs of A's and B's. That is, the ordering starts off with a run of three B's . . . this run is underlined and numbered 1. Then there is a run of three A's, which is run number 2. Proceeding in this way, and counting the beginning of a new run whenever an A is succeeded by a B or vice versa, we find that there are 10 runs.

Table 16.5.1

Sample 1(A)	Sample 2(B)
8	12
6	13
8	19
4	18
14	7
4	2
15	1
20	1
3	5
6	5

It should be obvious that there must be at least two runs in any ordering of scores from two groups. If the groups are of equal size N there can be no more than $2N$ runs in all. In general, the number of runs cannot exceed $N_1 + N_2$.

Now suppose that the two groups were random samples from *absolutely identical population distributions*. In this instance we should expect many runs, since the values for the two samples should be well "mixed up" when put in order. On the other hand, if the populations differ, particularly in central tendency, we should expect there to be less tendency for runs to occur in the sample ordering. This principle provides the basis for a test based on fairly simple probability considerations.

For thc moment, let R symbolize the total number of runs appearing for the samples. Then it can be shown that if R is any *odd* number, $2g + 1$, the probability for that number of runs is

$$\text{prob}(R = 2g + 1) = \frac{\binom{N_1 - 1}{g - 1}\binom{N_2 - 1}{g} + \binom{N_1 - 1}{g}\binom{N_2 - 1}{g - 1}}{\binom{N_1 + N_2}{N_1}}$$

when all arrangements in order are equally likely.

If R is an *even* number, $2g$, then

$$\text{prob}(R = 2g) = \frac{2\binom{N_1 - 1}{g - 1}\binom{N_2 - 1}{g - 1}}{\binom{N_1 + N_2}{N_1}}$$

On this basis, the exact sampling distribution of R can be worked out, given equal probability for all possible arrangements of A and B observations. It turns out that for fairly large samples the distribution of R can be approximated by a normal distribution with

$$E(R) = \frac{2N_1N_2}{N_1 + N_2} + 1 \qquad [16.5.1]$$

and

$$\sigma_R^2 = \frac{2N_1N_2(2N_1N_2 - N_1 - N_2)}{(N_1 + N_2)^2(N_1 + N_2 - 1)}. \qquad [16.5.2]$$

Thus, an approximate large sample test is given by

$$z = \frac{R - E(R)}{\sigma_R} \qquad [16.5.3]$$

referred to a normal distribution. Since the usual alternative hypothesis entails "too few" runs, the test is ordinarily *one-tailed,* with only negative values of z leading to rejection of the hypothesis of identical distributions.

For sample size less than or equal to 20 in *either* sample, exact values of R required for significance are given in Siegel (1956).

There is some reason to believe that the runs test generally has rather low power when it is compared either with a t test for means or with other order tests for identical populations. However, it is mentioned here because of its utility in various problems where other methods may not apply. Actually, A and B may designate any dichotomy within a *single* sample, and any principle at all that gives an ordering to A's and B's may be used. For example, it may be of interest to see if there is a time-related trend such as learning in a *single* set of data. In this instance, we might find the overall median for the set of scores, calling above the median A and below B. Here time is used as the ordering principle, and the occurrence of few runs is treated as evidence that time trends do exist. In any problem where the data may be given a dichotomous classification in one respect and then simply ordered in another respect, the runs test gives a way to answer the question of possible association between the basis for the ordering and the categorization. This makes the runs test useful in a variety of problems where other tests do not apply directly, and especially to problems where the *experimental* variable is ordinal and the *dependent* variable categorical.

A major technical problem with this test is the treatment of ties. In principle, ties should not occur if the scores themselves represent a continuous random variable. But, of course, ties do occur in actual practice, since we seldom represent the underlying variable directly or precisely in the data. If tied scores all occur among observations in the *same* group (as in the example above), then there is no problem; the value of R is unchanged by any method of breaking these ties. However, if members of *different* groups show tied scores, the number of runs depends upon how these ties are resolved in the final ordering. One way of meeting this problem is to break all ties in a way *least* conducive to rejecting the null hypothesis (so that the number of runs is made as large as possible). This at least makes for a conservative test of H_0. However, if cross-group ties are very numerous, this test is really inapplicable.

16.6 / THE MANN-WHITNEY TEST FOR TWO INDEPENDENT SAMPLES

Unlike the runs test, this test employs the actual *ranks* of the various observations directly as a device for testing hypotheses about the identity of two population distributions. It is apparently a good and relatively powerful alternative to the usual t test for equality of means. We assume that the underlying variable on which two groups are to be compared is continuously distributed. The null hypothesis to be tested is that the two population distributions are identical. Then we proceed as follows:

The scores from the combined samples are arranged in order (much as in the runs test). However, now we assign a *rank* to each of the observations, in terms of the magnitude of the original score. That is, the lowest score gets rank 1, the next lowest 2, and so on. Now choose one of the samples, say sample 1, and find the *sum* of the ranks associated with observations in that sample. Call this T_1. Then find

$$U = N_1N_2 + \frac{N_1(N_1 + 1)}{2} - T_1. \qquad [16.6.1]$$

If the resulting value of U is *larger* than $N_1N_2/2$, take

$$U' = N_1N_2 - U.$$

The statistic used is the *smaller* of U or U'. (Incidentally, this choice of the smaller of the two values for U is important in using tables to find significance for small samples, but is really immaterial in the large-sample test to be described later.)

As an example, consider the following data:

Sample 1(A)	Sample 2(B)
8	1
3	7
4	9
6	10
	12

Arranged in order and ranked, these data become

	B	A	A	A	B	A	B	B	B
Score	1	3	4	6	7	8	9	10	12
Rank	1	2	3	4	5	6	7	8	9

The sum of the ranks for the A observations (group 1) is

$$T_1 = 2 + 3 + 4 + 6 = 15.$$

This is turned into a value of U by taking

$$U = 4(5) + \frac{4(5)}{2} - 15$$

$$= 15.$$

This is larger than $\frac{(4)(5)}{2}$ or 10, and so we take

$$U' = 20 - 15 = 5$$

as the value we will use.

Now notice that given these 9 *scores,* the value of U depends only on how the

A's and B's happen to be arranged over the rank order. The number of possible random arrangements is just $\binom{N_1 + N_2}{N_1}$, and if the hypothesis of completely identical populations is true, the random assignment of individuals to groups should be the only factor entering into variation among obtained U values. Under the null hypothesis, all arrangements should be equally likely, and this gives the way for finding the probability associated with various values of U. For large samples, this sampling distribution of U is approximately normal, with

$$E(U) = \frac{N_1 N_2}{2} \qquad [16.6.2]$$

and

$$\sigma_U^2 = \frac{N_1 N_2 (N_1 + N_2 + 1)}{12}. \qquad [16.6.3]$$

Thus, for large samples, the hypothesis of no difference in the population distributions is tested by

$$z = \frac{U - E(U)}{\sigma_U}. \qquad [16.6.4]$$

For a two-tailed test, either U or U' may be used, since the absolute value of z will be the same for either. However, if the alternative hypothesis is such that one of the populations should tend to have a lower average than the other (assuming distributions of similar *form*), then a one-tailed test is appropriate, and the sign of the z should be considered.

For situations where the larger of the two samples is 20 or more and the samples are not too different in size, the normal approximation given above should suffice. However, when the larger sample contains fewer than 20 observations, tables given in Siegel (1956) should be used to evaluate significance of U.

This test is one of the best of the nonparametric techniques with respect to power. It seems to be very superior to the median test in this respect, and compares quite well with t when assumptions for both tests are met. For some special situations, it is even superior to t. This makes it an extremely useful device for the comparison of two independent groups.

Ordinarily, ties are treated in the Mann-Whitney test by giving each of a set of tied scores the *average* rank for that set. Thus, if three scores are tied for fourth, fifth, and sixth place in order, each of the scores gets rank 5. If two scores are tied for ninth and tenth place in order, each gets rank 9.5, and so on. This introduces no particular problem for large sample size when the normal approximation is used and ties are relatively infrequent. However, when ties exist σ_U^2 becomes

$$\sigma_U^2 = \frac{N_1 N_2}{12}\left[N_1 + N_2 + 1 - \frac{\sum_{i=1}^{G}(b_i^3 - b_i)}{(N_1 + N_2)(N_1 + N_2 - 1)}\right], \qquad [16.6.5]$$

where there are some G distinct *sets* of tied observations, i represents any *one* such set, and b_i is the number of observations tied in set i. For a small number of ties and for large $N_1 + N_2$, this correction to σ_U^2 can safely be ignored.

As we saw in Section 16.3, two matched samples can be compared by the sign test if the only feature of the data considered is the sign of the difference between each pair. However, this still overlooks one other important property of any pair of scores: not only does a difference have a direction, but also a size that can be ranked in order among the set of all such differences. The Wilcoxon test takes account of both features in the data, and thus uses somewhat more of the available information in paired scores than does the sign test.

The mechanics of the test are very simple: the signed difference between each pair of observations is found, just as for the t test for matched groups (Section 8.16). Then these differences are rank-ordered in terms of their absolute size. Finally, the sign of the difference is attached to the rank for that difference. The test statistic is T, the *sum of the ranks with the less frequent sign.*

Suppose that in some experiment involving a single treatment and one control group, subjects were first matched pairwise, and then one member of each pair was assigned to the experimental group at random. In the experiment proper, each subject received some Y score. Perhaps the data turned out as shown in Table 16.7.1. Here, the differences are found, their absolute size ranked, and then the sign of the difference attached to the rank. The less frequent sign is minus, and so

$$T = 3.5 + 1 + 3.5 + 7 = 15.$$

Table 16.7.1
Differences and signed ranks for ten matched pairs.

Pair	*Treatment*	*Control*	*Difference*	*Rank*	*Signed rank*
1	83	75	8	8	8
2	80	78	2	2	2
3	81	66	15	10	10
4	74	77	−3	3.5	−3.5
5	79	80	−1	1	−1
6	78	68	10	9	9
7	72	75	−3	3.5	−3.5
8	84	90	−6	7	−7
9	85	81	4	5	5
10	88	83	5	6	6

The hypothesis tested by the Wilcoxon test is that the two populations represented by the respective members of matched pairs are identical. When this hypothesis is true, then each of the 2^N possible sets of *signed* ranks obtained by arbitrarily assigning + or − signs to the ranks 1 through N is *equally* likely. The random assignment of subjects to experimental versus control group in the example is tantamount to such a random assignment of signs to ranks when the null hypothesis is true. On this basis, the exact distribution of T over all possible randomizations can be worked out. For large N (number of pairs), the sampling distribution is approximately normal with

$$E(T) = \frac{N(N + 1)}{4} \qquad [16.7.1]$$

and

$$\sigma_T^2 = \frac{N(N + 1)(2N + 1)}{24} \qquad [16.7.2]$$

so that a large sample test is given by

$$z = \frac{T - E(T)}{\sigma_T}. \qquad [16.7.3]$$

This test can be either directional or nondirectional, depending on the alternative hypothesis. However, a directional test usually makes sense only if one is prepared to assume that the distributions have the same form, and that a signed deviation of T from $E(T)$ is equivalent to a particular difference in central tendency between the two populations. For samples larger than about $N = 8$, this normal approximation is adequate. For very small samples, a table given in the book by Siegel (1956) should be used. Since only one set of differences is ranked, ties present no special problem unless they occur for zero differences. If an *even* number of zero differences occur, each zero difference is assigned the average rank for the set (zero differences, of course, rank lowest in absolute size), and then half are arbitrarily given positive and half negative sign. If an odd number of zeros occur, one randomly chosen zero difference is discarded from the data, and the procedure for an even number of zeros followed, except that N is reduced by 1, of course. For other kinds of tied differences, the method used in the example may be followed. Be sure to notice that even when several pairs are tied in absolute size so that they all receive the midrank for that set, the sign given to that midrank for different pairs may be different. For fairly large samples with relatively few ties, this procedure of assigning average ranks introduces negligible error.

All in all, the Mann-Whitney and the Wilcoxon tests are generally regarded as the best of the order tests for two samples. They both compare well with t in the appropriate circumstances, and when the assumptions for t are not met they may even be superior to this classical method. However, each is fully equivalent to a classical test of the hypothesis that the *means* of two groups are equal only when the assumptions appropriate to t are true. Unless additional assumptions are made, these tests refer to the hypothesis that two population distributions of unspecified form are *exactly* alike. In many instances, this is the hypothesis that the experimenter wishes to test, especially if interest lies only in the possibility of statistical independence or of association between experimental and dependent variables. However, in order to make particular kinds of inferences, particularly about population *means,* other assumptions become necessary. Without these assumptions, the rejection of H_0 implies only that the populations differ in *some* way, but the test need not be equally sensitive to all ways that population distributions might offer.

16.8 / THE KRUSKAL-WALLIS "ANALYSIS OF VARIANCE" BY RANKS

The same general argument for the Mann-Whitney test may be extended to the situation where J independent groups are being compared. The version of a J-sample rank test given here is due to Kruskal and Wallis (1952). This test has very close ties to the Mann-Whitney and Wilcoxen tests just discussed, and can properly be regarded as a generalized version of the Mann-Whitney method.

Imagine some J experimental groups in which each observation is associated with a numerical score. As usual, we assume that the underlying variable is continuously distributed. Now, just as in the Mann-Whitney test, the scores from all groups are pooled, arranged in order of size, and ranked. Then the rank sum attached to *each separate group* is found. Let us denote this sum of ranks for group j by the symbol T_j

T_i = sum of ranks for group j.

For example, suppose that three groups of small children were given the task of learning to discriminate between pairs of stimuli. Each child was given a series of pairs of stimuli, in which each pair differed in a variety of ways. However, attached to the choice of one member of a pair was a reward, and within an experimental condition, the cue for the rewarded stimulus was always the same. On the other hand, the experimental treatments themselves differed in the *relevant* cue for discrimination: in treatment I, the cue was form, in treatment II, color, and in treatment III, size. Some 36 children of the same sex and age were chosen at random and assigned at random to the three groups, with 12 children per group. The dependent variable was the number of trials to a fixed criterion of learning. Suppose that the data turned out to be as shown in Table 16.8.1.

Table 16.8.1

Scores and ranks for three independent groups.

	Treatment		
	I	*II*	*III*
	6 (1)	31 (34.5)	13 (10)
	11 (7)	7 (2)	32 (36)
	12 (9)	9 (4)	31 (34.5)
	20 (19)	11 (7)	30 (33)
	24 (23)	16 (14)	28 (31)
	21 (20)	19 (17.5)	29 (32)
	18 (16)	17 (15)	25 (24)
	15 (13)	11 (7)	26 (26.5)
	14 (11.5)	22 (21)	26 (26.5)
	10 (5)	23 (22)	27 (29.5)
	8 (3)	27 (29.5)	26 (26.5)
	14 (11.5)	26 (26.5)	19 (17.5)
T_j	139.0	200.0	327.0
T = 666			

Here, the numbers in parentheses are the ranks assigned to the various score values in the entire set of 36 cases. Then the sum of the ranks for each particular group j is found, and designated T_j. The value of T is the sum of these rank sums; if the ranking has been done correctly, it will be true that

$$T = \frac{N(N + 1)}{2}.$$

Note here that

$$T = \frac{(36)(37)}{2} = 666.$$

For large samples, a fairly good approximate test for identical populations is given by

$$H = \frac{12}{N(N+1)}\left[\sum_j \frac{T_j^2}{n_j}\right] - 3(N+1). \qquad [16.8.1]$$

This value of H can be referred to the chi-square distribution with $J - 1$ degrees of freedom for a test of the hypothesis that all J population distributions are identical.

For the example,

$$\begin{aligned} H &= \frac{12}{36(37)}\left[\frac{(139)^2 + (200)^2 + (327)^2}{12}\right] - 3(37) \\ &= 13.81. \end{aligned}$$

However, since there were ties involved in the ranking, this value of H really should be corrected by dividing through by a value found from

$$C = 1 - \left(\frac{\sum_i^G (t_i^3 - t_i)}{N^3 - N}\right) \qquad [16.8.2]$$

where G is the number of sets of tied observations, and t_i is the number tied in any set i. For example, there are 4 sets of two tied observations, 1 set of three ties, and 1 set of four tied observations. Thus,

$$C = 1 - \left(\frac{4[(2)^3 - 2] + [(3)^3 - 3] + [(4)^3 - 4]}{(36)^3 - 36}\right)$$

so that

$$C = .997.$$

Finally, the corrected value of H is

$$H' = \frac{H}{C} = \frac{13.81}{.997} = 13.85.$$

Unless N is small, or unless the number of tied observations is very large, relative to N, this correction will make very little difference in the value of H. Certainly, this was true here. Furthermore, when each set of tied observations lies within the same experimental group, the correction becomes unnecessary.

In terms of Table IV, for 2 degrees of freedom the value of H exceeds that required for the .01 level, and so the experimenter can be quite confident in saying that the population distributions are not identical. Apparently there is some association between the type of cue given in the discrimination problem and the number of trials to criterion.

It is somewhat difficult to specify the class of alternative hypotheses appropriate to the J-sample rank test, and so the question of power is somewhat more obscure than for the Mann-Whitney test. However, there is reason to believe that this test is about the best of the J-sample order methods. Certainly it should be superior in most situations to the median test discussed in Section 16.1. In comparisons with F from the analysis of variance, the Kruskal-Wallis test shows up extremely well.

When sample size is relatively small within the groups, the tables given by Siegel should be consulted.

16.9 / THE FRIEDMAN TEST FOR J MATCHED GROUPS

Much as the Kruskal-Wallis test represents an extension of the Mann-Whitney test, so the Friedman test is related to the Wilcoxen matched-pairs procedure. The Friedman test is appropriate when some K sets of matched individuals are used, where each set contains J individuals assigned at random to the J experimental treatments. It also applies when each of K individuals is observed under each of J treatments in random order. Thus, it is useful in situations where data are collected much as in the randomized blocks experiments mentioned in Section 11.18.

The data are set up in a table as for a two-way analysis of variance with one observation per cell. The experimental treatments are shown by the respective columns, and the matched sets of individuals by the rows. Within each row (matched group) a rank order of the J scores is found. Then the resulting ranks are summed by columns, to give values of T_j.

For example, in an experiment with four experimental treatments ($J = 4$), 11 groups of 4 matched subjects apiece were used. Within each matched group the four subjects were assigned at random to the four treatments, one subject per treatment. The data can be represented as in Table 16.9.1

Table 16.9.1

Scores and ranks within matched groups.

Groups	*I*	*II*	*III*	*IV*
1	1 (2)	4 (3)	8 (4)	0 (1)
2	2 (2)	3 (3)	13 (4)	1 (1)
3	10 (3)	0 (1)	11 (4)	3 (2)
4	12 (3)	11 (2)	13 (4)	10 (1)
5	1 (2)	3 (3)	10 (4)	0 (1)
6	10 (3)	3 (1)	11 (4)	9 (2)
7	4 (1)	12 (4)	10 (2)	11 (3)
8	10 (4)	4 (2)	5 (3)	3 (1)
9	10 (4)	4 (2)	9 (3)	3 (1)
10	14 (4)	4 (2)	7 (3)	2 (1)
11	3 (2)	2 (1)	4 (3)	13 (4)
T_j	30	24	38	18

$$T = \frac{K(J)(J+1)}{2}$$
$$= 110$$

It is important to remember that the ranks are given to the scores *within* rows. Then the T_j values are simply the sums of those ranks within columns.

The rationale for this test is really very simple. Suppose that within the population represented by a row, the distribution of values for all the *treatment* populations were identical. Then, under random sampling and randomization of observations, the probability for any given permutation of the ranks 1 through J within a given row should be the same as for any other permutation. Furthermore, across rows, each and every one of the $J!$ possible permutations of ranks across columns should be equally probable. This implies that we should expect the column sums of ranks to be identical under the null hypothesis. However, if there tend to be pile-ups of high or low ranks in particular columns, this is evidence against equal probability for the various permutations, and thus against the null hypothesis.

The test statistic for large samples is given by

$$\chi_r^2 = \frac{12}{KJ(J+1)}\left[\sum_j T_j^2\right] - 3K(J+1) \qquad [16.9.1]$$

distributed approximately as chi-square with $J - 1$ degrees of freedom.

For the example, we would take

$$\chi_r^2 = \frac{12}{11(4)(5)}[(30)^2 + (24)^2 + (38)^2 + (18)^2] - 3(11)(5)$$

$$= 11.95.$$

For 3 degrees of freedom, this is just significant at the .01 level, and so the experimenter may say with some assurance that the treatment populations differ.

If ties in ranks within rows should occur, a conservative procedure is to break the ties so that the T_j values are as close together as possible.

This test may well be the best alternative to the ordinary two-way (matched-groups) analysis of variance. Once again, there is every reason to believe this test, much like the Kruskal-Wallis test for one-factor experiments, is superior to the corresponding median test, and from the little evidence available, the result should compare well with F when both the classical and order methods apply.

The chi-square approximation given above is good only for fairly large K. However, the test should be satisfactory when $J \geq 4$ and $K \geq 10$. As usual, tables giving significance levels for small samples can be found in the book by Siegel.

16.10 / RANK-ORDER CORRELATION METHODS

The tests discussed so far in this chapter are designed for situations where the experimental variable is categorical and the dependent variable is in ordinal terms, as given either by the original measurement procedure or by a transformation of the scores into ordered classes or ranks. In this section, measures and tests of association will be described for situations where *both* variables are represented in ordinal terms. First of all, two somewhat different measures of "agreement" or association between rank orders will be introduced. Next, a measure of association will be given for the special situation where both variables take the form of ordered classes. Finally, the problem of indexing simultaneous agreement among several rank orders will be treated.

It is customary to call some of these rank-order statistics "correlations," but this usage deserves some qualification. The Spearman rank correlation to be introduced next actually *is* a correlation coefficient computed for numerical values that happen to be ranks. However, the next index to be considered, Kendall's tau, is not a correlation coefficient at all. Neither of these indices is closely connected with the classical theory of *linear* regression. Especially the *square* of a correlation-like index on ranks is not to be interpreted in the usual way as a proportion of variance accounted for in the underlying variables. Instead, it is somewhat better to think of both the Spearman and the Kendall indices only as showing ''concordance'' or ''agreement,'' the tendency of two rank orders to be similar. As descriptive statistics, both indices serve this purpose very well, although the definition of ''disagree'' is somewhat different for these two statistics.

16.11 / THE SPEARMAN RANK CORRELATION COEFFICIENT

Imagine a group of N cases drawn at random as for a problem in correlation. However, instead of having an X score for each individual, we have only the *rank* in the group on variable X (say for low to high, the ranks 1 through N). In the same way, for each individual we have the rank in the group on variable Y. The question to be asked is ''How much does the ranking on variable X tend to agree with the ranking on variable Y?'' and a measure is desired to show the extent of the agreement. Or, perhaps, we have two judges who each rank the same set of N objects. We wish to ask, ''How much does judge A agree with judge B?'' In either instance, there are two distinct rank orders of the same N things, and these rank orders are to be compared for their agreement with each other. Furthermore, when these objects or individuals constitute a random sample from some population, we may wish to test the hypothesis that the true agreement in ranks is zero.

Two simple ways for comparison of two rank orders for agreement have already been mentioned: the first, and older, method is the Spearman rank correlation, commonly symbolized r_S (although ρ is sometimes used, this symbol r_S will be used here to avoid confusion with the population correlation); the second method is the Kendall ''tau'' statistic, which will be discussed in the following section. Regardless of whether the data are two rank orders representing scores shown by individuals in a sample, or ranks of objects given by two judges, we can apply r_S. Suppose that in either circumstance we call the things ranked ''individuals,'' and the two bases for ranking, the ''variables.'' We can take the point of view that if rank orders agree, the ranks assigned to individuals should *correlate* positively with each other, whereas disagreement should be reflected by a negative correlation. A zero correlation represents an intermediate condition: no particular connection between the rank of an individual on one variable and the rank on the other.

For a descriptive index of agreement between ranks, the ordinary correlation coefficient can be computed on the *ranks* just as for any numerical scores, and this is how the Spearman rank correlation for a sample is defined:

$$r_S = \text{correlation between ranks over individuals.}$$

However, since the numerical values entering into the computation of the correlation coefficient actually are ranks in this instance, r_S can be given a very simple computational form when no ties in rank exist:

$$r_S = 1 - \left[\frac{6\left(\sum_i D_i^2\right)}{N(N^2 - 1)}\right]. \qquad [16.11.1]$$

Here D_i is the *difference* between ranks associated with the particular individual i, and N is the number of individuals observed.

The Spearman rank correlation is thus very easy to compute when ranks are untied. All one needs to know is N, the number of individuals ranked, and D_i, the difference in ranking for each individual. In spite of the different computations involved, r_S is only an ordinary correlation coefficient calculated on ranks.

The computation of r_S will be illustrated in a problem dealing with agreement between judges' rankings of objects. In a test of fine weight discrimination, two judges each ranked 10 small objects in order of their judged heaviness. The results are shown in Table 16.11.1. Did the judges tend to agree?

Table 16.11.1

Rank orders of "heaviness" given by two judges.

Object	*Judge I*	*Judge II*	D_i	D_i^2
1	6	4	2	4
2	4	1	3	9
3	3	6	3	9
4	1	7	6	36
5	2	5	3	9
6	7	8	1	1
7	9	10	1	1
8	8	9	1	1
9	10	3	7	49
10	5	2	3	9
				128

$$r_s = 1 - \frac{6(128)}{10(10^2 - 1)} = .224.$$

The Spearman rank correlation is only .224, so that agreement between the two judges was not very high, although there was some slight tendency for similar ranks to be given to the same objects by the judges.

Incidentally, notice that if the relative true weights of the objects are known, we could also find the agreement of each judge with the true ranking. When some criterion ranking is known, it may be useful to compare a judged ranking with this criterion to evaluate the accuracy or "goodness" of the judgments—this is not the same as the agreement of judges with each other, of course.

Formula 16.11.1 may not be used if there are ties in either or both rankings, since the means and variances of the ranks then no longer have the simple relationship to N present in the no-tie case. When ties exist, perhaps the simplest procedure is to assign mean ranks to sets of tied individuals; that is, when two or more individuals are tied in order, each is assigned the mean of the ranks they would otherwise occupy. Next an ordinary correlation coefficient is computed, using the ranks as though they were simply numerical scores. The result is a Spearman rank correlation that can be regarded as corrected for ties. On the other hand, if a test of the signficance of r_S is the main object of the analysis, a conservative course of action is to find a way to break the ties that will make the absolute value of r_S as small as possible.

When no ties in rank exist, the exact sampling distribution of r_S can be worked out for small samples. Exact tests of significance for r_S are based on the idea that if one of the two rank orders is known, and the two underlying variables are independent, then each and every permutation in order of the individuals is equally likely for the other ranking. On this basis, the exact distribution of $\sum_i D_i^2$ can be found, and this can be converted into a distribution of r_S. Exact probability tables for $\sum_i D_i^2$ and r_S based on small N are to be found in books by Kendall (1955), Siegel (1956), and others.

The hypothesis of the independence of the two variables represented by rankings can also be given an approximate, large-sample, test in terms of r_S. This test has a form very similar to that for r_{XY}:

$$t = \frac{r_S\sqrt{N - 2}}{\sqrt{1 - r_S^2}} \qquad [16.11.2]$$

with $N - 2$ degrees of freedom. This test is really satisfactory, however, only when N is fairly large; N should be at least greater than or equal to 10.

Under the assumption of a bivariate normal distribution, the value of r_S from a large sample can be treated as an estimate of the value of ρ for the variables underlying the ranks. When the population is bivariate normal with $\rho = 0$, values of r_S and r_{XY} correlate very highly over samples. On the other hand, this assumption is rather special, and the status of r_S as an estimator of ρ *in general* is open to considerable question.

16.12 / THE KENDALL TAU COEFFICIENT

A somewhat different approach to the problem of agreement between two rankings is given by the τ coefficient (small Greek tau) due to M. G. Kendall (1955). Instead of treating the ranks themselves as though they were scores and finding a correlation coefficient, as in r_S, in the computation of τ we depend only on the number of *inversions* in order for pairs of individuals in the two rankings. A single inversion in order exists between *any pair* of individuals b and c when b > c in one ranking and c > b in the other. When two rankings are *identical,* no inversions in order exist. On the other hand, when one ranking is exactly the reverse of the other, an inversion exists for *each pair* of individuals; this means that complete disagreement corresponds to $\binom{N}{2}$ inversions. If the two rankings *agree* (show noninversion) for as many pairs as they disagree about (show inversion), the tendency for the two rank orders to agree or disagree should be exactly zero.

This leads to the following definition of the τ statistic:

$$\tau = 1 - \left[\frac{2(\text{number of inversions})}{\text{number of pairs of objects}}\right]. \qquad [16.12.1]$$

This is equivalent to

$$\tau = \frac{(\text{number of times rankings agree about a pair}) - (\text{number of times rankings disagree})}{\text{total number of pairs}}. \qquad [16.12.2]$$

It follows that the τ statistic is essentially a difference between two proportions: the proportion of pairs having the *same* relative order in both rankings minus the proportion of pairs showing *different* relative order in the two rankings.

Viewed as coefficients of agreement, r_S and τ thus rest on somewhat different conceptions of "disagree." In the computation of r_S, a disagreement in ranking appears as the *squared* difference between the ranks themselves over the individuals. In τ, an inversion in order for any *pair* of objects is treated in the same way as evidence for disagreement. Although these two conceptions are related, they are not identical: the process of squaring differences between rank values in r_S places somewhat different weight on *particular* inversions in order, whereas in τ all inversions are weighted equally by a simple frequency count. Values of the statistics r_S and τ are correlated over successive random samples from the same population, but the extent of the correlation depends on a number of things, including sample size and the character of the relation between the underlying variables in the population. Nevertheless, the two statistics are closely connected, and a number of mathematical inequalities must be satisfied by the values of the two statistics. For example,

$$-1 \leq 3\tau - 2r_S \leq 1.$$

It will be convenient to discuss the numerator term in the τ coefficient separately, and thus we will define

$$S_T = (\text{number of agreements in order}) - (\text{number of disagreements in order})$$
$$= \binom{N}{2} - 2(\text{number of inversions}).$$

Various methods exist for the computation of S and τ, but the simplest is a graphic method. In this method, all one does is to list the individuals or objects ranked, once in the order given by the first ranking, and again in the order given by the second. For example, suppose that in some problem there were seven objects {a,b,c,d,e,f,g} ranked by each of two judges, and that the rankings came out like this:

	Rank						
	1	*2*	*3*	*4*	*5*	*6*	*7*
Judge 1	c	a	b	e	d	g	f
Judge 2	a	c	e	b	f	d	g

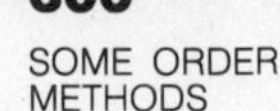

Now straight lines are drawn connecting the same objects in the two parallel rankings, thus:

1	2	3	4	5	6	7
c	a	b	e	d	g	f
a	c	e	b	f	d	g

Then *the number of times that pairs of lines cross is the number of inversions in order*. Here, the number of crossing is 4, and so

$$S_T = \binom{7}{2} - (2)(4)$$
$$= 21 - 8$$
$$= 13.$$

The sample value of τ is

$$\tau = \frac{13}{21}$$
$$= .62.$$

Although r_S from a sample has a rather artificial interpretation as a correlation coefficient, the interpretation of the obtained value of τ is quite straightforward: if a *pair* of objects is drawn at random from among those ranked, the probability that these two objects show the *same* relative order in both rankings is .62 *more* than the probability that they would show different order. In other words, from the evidence at hand it is a considerably better bet that the two judges will tend to order a randomly selected pair in the *same* way than in a different way.

This graphic method of computing τ is satisfactory only when no ties in ranking exist. For nontied rankings, however, it is very simple to carry out, even when moderately large numbers of individuals have been ranked. Notice that although both the examples of computations for r_S and τ featured judges' rankings of objects, exactly the same methods apply when the ranking principle is provided by scores shown by individuals on each of two variables.

Still another way of computing S_T will be mentioned, since we will find an extension of this method very convenient when ties exist in one or both rankings. The data are organized as in Table 16.12.1. Notice that this is really just a joint-frequency table based on ranks. Each distinct observation (object) has a pair of rank numbers, and this determines the cell in the table in which that observation falls. As a convention for the procedure described below, the ranks along the margins of the table are listed with the lowest ranks in the upper lefthand corner, as shown here.

Now S_T is found as follows: we compute a value S_+ by first taking any cell with nonzero frequency and, ignoring its row and column, *counting the number of entries to the right and below that cell*. Thus, for the cell containing the object c, there are five entries to the right and below. For the cell with object b, there are three entries to the right and below, and so on. Then S_+ is the sum of these numbers over cells. Over all nonzero cells, the value of S_+ is found to be:

Table 16.12.1
Rankings of seven objects by two judges.

		Judge 1						
		1	*2*	*3*	*4*	*5*	*6*	*7*
	1		1(a)					
	2	1(c)						
	3				1(e)			
Judge 2	4			1(b)				
	5							1(f)
	6					1(d)		
	7						1(g)	

5 (for cell a)
5 (for cell c)
3 (for cell e)
3 (for cell b)
1 (for cell d)
0 (for cell f)
0 (for cell g)

$S_+ = 17.$

Then

$$S_T = 2(S_+) - \binom{N}{2} = 34 - 21 = 13$$

and

$$\tau = .62.$$

As we shall see, this general method is advantageous when ties exist in the data or when the data are put into ordered classes rather than ranks.

Notice that we might also compute a value S_- by taking the sum of the frequencies to the left and below the various nonzero cells. Then

$$S_T = \binom{N}{2} - 2(S_-)$$

since

$$(S_+) + (S_-) = \binom{N}{2},$$

where there are no tied ranks.

An exact test of significance for τ may be had, based on the assumption that the variables underlying the ranks are continuously distributed, so that ties are impossible. However, for our purposes, the fact of importance is that the exact sampling distribution of τ approaches a normal distribution very quickly with successive increases in the size of N. Even for fairly small values of N, the distribution of τ is approximated relatively well by the normal distribution. Of course, this is true only when H_0 is true, so that the rankings are each equally likely to show any of the $N!$ permutations in order, and $E(\tau) = 0$. The distribution of τ is not simple to discuss when other conditions hold, and so we will test only the hypothesis of independence between rankings (implying equal probability of occurrence for each and every possible ordering of N observations on the second variable given their ordering on the first).

For N of about 10 or more, the test is given by

$$z = \frac{\tau}{\sigma_\tau}$$

referred to a normal distribution, where

$$\sigma_\tau^2 = \frac{2(2N + 5)}{9N(N - 1)}. \qquad [16.12.3]$$

Equivalently, in terms of S_T,

$$z = \frac{S_T}{\sigma_S} \qquad [16.12.4]$$

where

$$\sigma_S^2 = \frac{N(N - 1)(2N + 5)}{18}.$$

For small N, the exact tables in terms of S_T given by Kendall (1948) or Siegel (1956) can be used.

This approximate test can be improved if a correction for continuity is made. The correction for continuity involves subtraction of $1 \Big/ \binom{N}{2}$ from the absolute value of τ, or 1 from the absolute value for S_T, before the z statistic is formed.

16.13 / KENDALL'S TAU WITH TIES

The determination of the exact sampling distribution of S_T (or τ) depends on the assumption of *no* ties. However, if the underlying model is altered in a particular way, τ can be computed and tested even though ties occur in one or both rankings. First of all, we will consider the method for finding S_T when ties occur, and then turn to the problem of a significance test.

The simplest method for finding S_T and τ when ties occur is to use the joint-frequency-table method suggested in Section 16.12. However, here it will occur that more than one entry will fall into a particular row, column, or cell. This method is best shown by example.

Suppose that some 12 subjects were observed, and ranked on each of two variables, so that the ranked data were as shown in Table 16.13.1. A table is formed in which the first ranking is represented by the columns and the second ranking by the

Table 16.13.1

Two rankings with ties.

Subject	*X ranking*	*Y ranking*
1	5	8.5
2	7	11
3	5	12
4	1	8.5
5	2	8.5
6	3	6
7	5	5
8	11.5	8.5
9	11.5	2
10	9.5	2
11	9.5	2
12	8	4

Table 16.13.2

Rankings with ties in a frequency table.

Y ranking	*X ranking:* *1*	*2*	*3*	*5*	*7*	*8*	*9.5*	*11.5*	*Total*
2							2	1	*3*
4						1			*1*
5				1					*1*
6			1						*1*
8.5	1	1		1				1	*4*
11					1				*1*
12				1					*1*
Total	*1*	*1*	*1*	*3*	*1*	*1*	*2*	*2*	*12*

rows. (Actually, the score values themselves could have been used to form this table, since the actual rank values are not used at all after the data are tabled.) Table 16.13.2 shows these data.

Now we proceed to find S_+ just as above, except that the number of cases to the right and below a given cell is weighted by the number of cases in that cell. For example, for the cell given by the row labeled 2 and the column labeled 9.5, there is only 1 case to the right and below; however, since there are 2 cases *in* the cell, in the sum for S_+ we enter 2(1) or 2 for this cell. Proceeding in this way over all of the cells with nonzero frequencies, we have

$$\begin{aligned} S_+ &= 2(1) + 1 + 1(2) + 1(4) + 1(2) + 1(2) + 1(1) \\ &= 14. \end{aligned}$$

Now we find a value S_-, computed exactly as for S_+ except that we find the number of cases to the *left and below* a given cell, and weight this number by the cell frequency. Here

$$S_- = 2(8) + 1(8) + 1(7) + 1(3) + 1(2) + 1(2) + 1(1)$$
$$= 39.$$

Then,

$$S_T = S_+ - S_-$$

so that for this example

$$S_T = 14 - 39 = -25.$$

Kendall suggests calculating the denominator of τ by taking

$$\sqrt{\left(\frac{N(N-1)}{2} - T_1\right)\left(\frac{N(N-1)}{2} - T_2\right)} \qquad [16.13.1]$$

where

$$T_1 = \frac{\sum_j n_j(n_j - 1)}{2}$$

n_j = the marginal total of column j

$$T_2 = \frac{\sum_k n_k(n_k - 1)}{2}$$

n_k = the marginal total for row k.

For this example,

$$T_1 = \frac{1}{2}[3(2) + 2(1) + 2(1)]$$
$$= 5$$

and

$$T_2 = \frac{1}{2}[3(2) + 4(3)]$$
$$= 9.$$

Then

$$\tau = \frac{S_T}{\sqrt{\left(\frac{N(N-1)}{2} - T_1\right)\left(\frac{N(N-1)}{2} - T_2\right)}}$$
$$= \frac{-25}{\sqrt{(66 - 5)(66 - 9)}}$$
$$= \frac{-25}{59}$$
$$= -.42.$$

The apparent degree of agreement between these two rankings is a *negative* .42.

A large-sample test of significance for tau with ties can be constructed if one forms the ratio

$$z = \frac{S_T}{\sigma_S}$$

and refers this to a normal distribution. However, it must be emphasized that when ties exist in either or both rankings, this test is *conditional* to the distribution of ties, and the significance level refers to a probability of occurrence among samples each showing *exactly* the same distribution of ties as appeared in the sample actually obtained. When ties exist, the sampling variance of S_T or σ_S^2, is found from

$$\sigma_S^2 = \frac{N(N-1)(2N+5) - \sum_j n_j(n_j-1)(2n_j+5) - \sum_k n_k(n_k-1)(2n_k+5)}{18} + \frac{\left[\sum_j n_j(n_j-1)(n_j-2)\right]\left[\sum_k n_k(n_k-1)(n_k-2)\right]}{9(N)(N-1)(N-2)} + \frac{\left[\sum_j n_j(n_j-1)\right]\left[\sum_k (n_k)(n_k-1)\right]}{2(N)(N-1)}. \quad [16.13.2]$$

After this rather forbidding-looking calculation for σ_S^2 has been carried out, the value of z may be found and referred to a normal table, provided that N is rather sizable. Unless N is quite large and ties are infrequent, a correction for continuity is required; this is given in Kendall (1955).

It is interesting to observe that even when ties exist, one really does not have to assign ranks at all in order to compute τ; we are really dealing with ordered categories of observations, and when we assign the midrank to a set of observations we are, in effect, allotting them to the same ordered category. Thus, in principle this procedure may be carried out on a joint *grouped* frequency distribution; the actual computations, however, really involve only the *ordinal* properties of the scores on which the distributions were based, and the class intervals are treated as ordered classes.

Furthermore, even when data are grouped in this way, the τ index can be regarded as reflecting monotonicity in the relationship between the underlying scores. Large values of τ lead to the conclusion that the relationship tends to be monotone, and small absolute values may indicate either that there is no statistical association, or that the form of the relationship tends to be nonmonotone.

16.14 / A MEASURE OF ASSOCIATION IN ORDERED CLASSES

It has just been noted that the method for computing S_T where the rankings contain ties is tantamount to arranging the data into ordered classes. Unfortunately, the value of the τ statistic itself does not seem to have a very simple interpretation when ties are present in either ranking. This difficulty is removed if one uses the γ (Greek gamma) statistic suggested by Goodman and Kruskal (1954) specifically for data arranged into ordered classes. Actually, this statistic γ has the same numerator term as τ; the S_+ and S_- values are found exactly the same way. However, its denominator differs from that of τ, and permits γ to have a simpler interpretation.

In terms of the quantities S_+ and S_- computed above, the statistic γ is just

$$\gamma = \frac{S_+ - S_-}{S_+ + S_-}. \qquad [16.14.1]$$

For the example above, this is

$$\gamma = \frac{-25}{14 + 39}$$

$$= -.47.$$

This value can be interpreted as follows: suppose that a pair of subjects were drawn at random from the 12 actually observed. Given that these subjects were tied in neither of the rankings, is it a better bet that they show the same or a different ordering on X and Y? The value of γ shows that it is a much better bet that an untied pair has different ordering on the two variables, since the probability of finding a pair with a different ordering is .47 *more* than the probability of a pair with the same ordering, among all possible untied pairs we might draw. Put more formally,

$$\gamma = p(\text{same ordering}|\text{untied pairs}) - p(\text{different ordering}|\text{untied pairs})$$

for a pair chosen at random with replacement from among these 12 subjects.

If there are no tied ranks, then

$$\gamma = \tau.$$

The index γ has the same basic interpretation as τ: a difference in probability for same versus different ordering on the underlying variables for a randomly selected pair. The only difference is that γ is conditional to the set of *untied* such pairs. The index γ is much like the λ measures for categorical data (Section 15.12), since it has a simple interpretation in a predictive sense, although γ refers to *ordered* categories and thus reflects the general tendency toward monotonicity in the relationship, while λ reflects *any* form of predictive relationship.

16.15 / KENDALL'S COEFFICIENT OF CONCORDANCE

Sometimes we want to know the extent to which members of a set of m distinct rank orderings of N things tend to be similar. For example, in a beauty contest each of 7 judges ($m = 7$) gives a simple rank order of the 10 contestants ($N = 10$). How much do these rank orders tend to agree, or show "concordance"?

This problem is usually handled by application of Kendall's statistic, W, the "coefficient of concordance." As we shall see, the coefficient W is closely related to the average r_s among the m rank orders.

The coefficient W is computed by putting the data into a table with m rows and N columns. In the cell for column j and row k appears the rank number assigned to individual object j by judge k. Table 16.15.1 might show the data for the judges and the beauty contestants. It is quite clear that the judges did not agree perfectly in their rankings of these contestants. However, what should the column totals of ranks, T_j, have been if the judges had agreed exactly? If each judge had given exactly the same rank to the same girl, then one column should total to 7(1), another to 7(2), and so on, until the largest sum should be 7(10). On the other hand, suppose that there were complete disagreement among the judges, so that there was no tendency for high or low rankings to pile up in particular columns. Then we should expect each column sum to be about the same. In this example, the column sums of ranks are not identical, so that apparently some agreement exists, but neither are the sums as different as they should be when absolutely perfect agreement exists.

Table 16.15.1

Rankings given to 10 beauty contestants by seven judges.

	Contestants									
Judges	*1*	*2*	*3*	*4*	*5*	*6*	*7*	*8*	*9*	*10*
1	8	7	5	6	1	3	2	4	10	9
2	7	6	8	3	2	1	5	4	9	10
3	5	4	7	6	3	2	1	8	10	9
4	8	6	7	4	1	3	5	2	10	9
5	5	4	3	2	6	1	9	10	7	8
6	4	5	6	3	2	1	9	10	8	7
7	8	6	7	5	1	2	3	4	10	9
T_j	45	38	43	29	16	13	34	42	64	61

$$T = \frac{m(N)(N + 1)}{2}$$
$$= 385$$

This idea of the extent of variability among the respective sums of ranks is the basis for Kendall's W statistic. Basically,

$$W = \frac{\text{variance of rank sums}}{\text{maximum possible variance of rank sums}}.$$

Because the mean rank and the variance of the ranks each depend only on N and m, this reduces to

$$W = \left(\frac{12 \sum_j T_j^2}{m^2 N(N^2 - 1)}\right) - \frac{3(N + 1)}{N - 1}. \qquad [16.15.1]$$

For the example, we find

$$W = \left(\frac{12[(45)^2 + \cdots + (61)^2]}{49(10)(99)}\right) - \frac{3(11)}{9}$$
$$= 4.28 - 3.66$$
$$= .62.$$

There is apparently a moderately high degree of "concordance" among the judges, since the variance of the rank sums is 62 percent of the maximum possible. Note that by its definition, W cannot be negative, and its maximum value is 1.

The value of the concordance coefficient is somewhat hard to interpret directly in terms of the tendency for the rankings to agree, but an interpretation can be given in terms of the average value of r_S over all possible *pairs* of rank orders. That is,

$$\text{average } r_S = \frac{mW - 1}{m - 1}. \qquad [16.15.2]$$

For the example,

$$\text{average } r_S = \frac{7(.62) - 1}{7 - 1}$$

$$= .56.$$

If we took all of the possible $\binom{7}{2}$ or 21 *pairs* of judges, and found r_S for each such pair, the average rank correlation would be about .56. Thus, on the average, judge-pairs do tend to give relatively similar rankings. The advantage of reporting this finding in terms of W rather than average r_S is that

$$\frac{-1}{m - 1} \leq \text{average } r_S \leq 1$$

whereas, regardless of the values for N or m,

$$0 \leq W \leq 1.$$

This makes W values more immediately comparable across different sets of data. Nevertheless, the clearest interpretation of W seems to be in terms of average r_S.

Recall that this is essentially the idea employed in the Friedman test for matched groups. In the Friedman procedure a matched group takes the place of a judge, and the rank order of scores for different treatments within a group is like an ordering of objects of judgment. Then, if the scores for different treatments tend to show up in substantially the same order for the various groups, a true difference in treatments is inferred.

An exact test is possible for the hypothesis that there is no actual agreement among judges (see Kendall, 1955; and Siegel, 1956, for tables). For m of at least 8, an approximate test is given by

$$\chi^2 = m(N - 1)W \qquad [16.15.3]$$

referred to the chi-square distribution with $N - 1$ degrees of freedom. This test is really a good one only for fairly sizable m and N, however.

EXERCISES

1. In an experiment, five groups of 10 randomly selected cases each were used. Use the median test to decide if there are significant ($\alpha = .05$) differences among the populations sampled.

Groups				
1	*2*	*3*	*4*	*5*
87	41	31	60	55
35	19	18	8	67
67	70	64	14	46
44	62	7	49	95
51	43	13	16	79
49	46	22	28	63
18	6	38	16	66
98	13	31	70	82
97	22	18	64	56
84	38	64	30	67

2. In Chapter 10, take the data from Table 10.16.1 and test the hypothesis that the four populations are identical by use of the median test.

3. Using the data of exercise 16, Chapter 11, analyze the differences among the three "approaches" by use of the median test for matched groups.

4. Take the data for exercise 13, Chapter 13, and test the hypothesis that the teachers were equally likely to judge a child's age too high or too low. Use the sign test for this purpose.

5. In a study of engaged couples' preferences with respect to size of family, some twenty-six engaged couples were selected at random and were asked to state, independently, the ideal number of children they would like to have. Responses of men and women from each couple are listed below:

Couple	*Men*	*Women*
1	3	2
2	0	1
3	1	0
4	2	2
5	0	3
6	2	3
7	1	2
8	2	3
9	2	3
10	1	3
11	2	4
12	0	1
13	3	4
14	5	2
15	7	2
16	1	2
17	0	3
18	2	4
19	0	3
20	2	3
21	2	4
22	0	2
23	1	3
24	3	2
25	5	2
26	0	1

Use the sign test to decide if husbands and wives differ significantly on this issue.

6. A psychologist was interested in the verb-adjective ratio as an index of the habitual pattern of verbal expression for an individual. In a comparison of science majors and English majors in terms of verb-adjective ratio, 10 science majors and 12 English majors were sampled at random, and a selection of the free-writing of each subject taken. Each such selection was scored according to the ratio of number of verbs used relative to the number of adjectives. In terms of the resulting data, use the Mann-Whitney test to describe if science majors and English majors differ significantly in their relative usage of verb and adjectives.

Verb-adjective ratios	
Science majors	*English majors*
1.32	1.04
2.30	.93
1.98	.75
.59	.33
1.02	1.62
.88	.76
.92	.97
1.39	1.21
1.95	.80
1.25	1.16
	.71
	.96

7. Eighteen subjects were asked to respond to very faint tones by pushing a key specific to each tone. The task was carried out under five different conditions of noise. In each condition 100 tones were presented, and the number of correct responses recorded. The data follow:

		Condition				
		1	*2*	*3*	*4*	*5*
Subject	*1*	73	75	71	74	64
	2	91	89	83	80	82
	3	92	94	90	93	83
	4	84	82	76	73	75
	5	56	58	54	57	47
	6	60	58	52	49	51
	7	73	75	69	72	62
	8	70	68	64	61	63
	9	87	89	83	86	76
	10	75	73	69	66	68
	11	77	75	69	72	62
	12	68	70	64	61	63
	13	73	75	69	72	62
	14	75	73	68	65	67
	15	93	95	89	92	82
	16	90	89	85	82	84
	17	84	86	80	83	73
	18	69	67	63	60	62

Are there signficiant differences among conditions (α = .05)? Analyze these data using the Friedman matched-groups analysis by ranks.

8. Use the Kruskal-Wallis analysis of variance by ranks to compare the groups represented by the columns in Table 10.26.1.

9. In a study of conformity to social norms, and how this might vary with age, four samples of eight subjects each were selected at random. However, each sample came from a different age group. Each subject was observed driving a specially equipped car, and the average amount of time spent at a stop sign (at an otherwise clear intersection) was recorded for each in units of .1 seconds. The data follow:

	Age groups		
18–25	*35–40*	*45–50*	*over 65*
11.7	41.2	59.0	61.6
28.9	39.5	62.1	70.3
26.2	52.2	57.5	59.2
31.8	47.6	58.3	25.8
59.4	38.9	42.6	74.1
27.6	37.7	44.5	60.9
50.5	46.3	40.9	65.4
29.2	57.7	52.2	53.7

Since it does not seem reasonable to suppose that such observations should represent a normally distributed population, it was decided to employ the Kruskal-Wallis method of analysis. Do these age groups differ in this form of behavior?

10. For the data of exercise 1, Chapter 13, carry out the Wilcoxon test for the ratings given to countries *A* and *B*. Do these countries tend to be rated differently over political scientists? Use α = .05.

11. Once again using the data of exercise 1, Chapter 13, carry out the Wilcoxon test using countries *B* and *C*, and countries *A* and *C*, with α = .05 for each. Are these tests independent of each other and of the test in the previous exercise? What can one say about the probability of making a Type I error in at least one of these three tests?

12. Suppose that the rows of Table 11.16.1 actually represented a random sample of 12 different classrooms, with all three tasks then given to each classroom. Under this assumption, what does the Friedman test reveal about the differences among tasks?

13. Analyze the data of exercise 4, Chapter 10, in terms of the Kruskal-Wallis analysis of variance by ranks.

14. Analyze the data of exercise 5 above by means of the Wilcoxon test for matched samples.

15. An experiment involved two randomly selected groups of ten subjects each. The scores on a reading comprehension test given each person are shown below. Group I read with a background of music of a loud "hard-rock" character. Group II read with a background of standard popular music of the 1930s. Compare the average reading comprehension scores for the two groups by way of the Mann-Whitney test.

I	*II*
128	136
131	141
146	139
157	157
129	136
151	157
142	147
136	135
140	131

16. How does a two-tailed t test for the data of exercise 15 compare with the Mann-Whitney results?

17. Some 24 subjects were selected at random, and each was given a series of six puzzles requiring a certain amount of ''insight'' to solve. After a fixed allotment of time on each, the subject was scored ''1'' if the puzzle was solved, and ''0'' if the puzzle was still unsolved. Then each subject moved on to the next puzzle until all six had been attempted. Given the data table below, does it appear that the puzzles differed significantly in difficulty?

	Puzzles					
Subjects	*I*	*II*	*III*	*IV*	*V*	*VI*
1	1	0	1	1	1	0
2	0	0	1	0	1	0
3	1	1	0	1	1	1
4	1	1	1	0	0	1
5	1	1	0	1	1	0
6	1	1	0	1	1	1
7	1	1	1	1	1	1
8	1	0	0	1	1	1
9	1	1	1	1	0	0
10	1	1	0	1	0	1
11	0	1	1	1	1	1
12	1	0	0	1	1	1
13	1	1	0	1	1	0
14	1	1	1	1	1	1
15	1	0	0	0	1	1
16	1	1	0	1	1	0
17	1	0	0	1	1	1
18	0	1	1	1	1	1
19	1	0	0	0	1	0
20	1	1	0	0	0	1
21	1	0	0	1	1	1
22	1	1	1	1	1	1
23	1	0	0	1	1	1
24	1	1	1	1	1	1

18. Calculate the Spearman rank correlation for the data in exercise 9, Chapter 13. You may assign average ranks to any group of tied values of X or of Y.

19. Test the significance of the Spearman correlation found in exercise 18 above.

20. In a psychological study, twenty subjects were used, each subject being tested on a different day. Although each subject was cautioned not to tell anyone about the experiment, the experimenter felt that the later subjects were performing systematically better than the earlier, and suspected that information about the experiment was being gradually "leaked." Test the hypothesis that there is no correlation between the day on which a subject was tested and the score. (**Hint:** Compute a Spearman rank correlation between days and scores, and test its significance.)

Day	*Score*	*Day*	*Score*
1	120	11	132
2	124	12	136
3	123	13	133
4	128	14	119
5	125	15	140
6	127	16	138
7	134	17	139
8	129	18	161
9	130	19	142
10	137	20	145

21. Use the runs test to answer the question of exercise 20. (**Hint:** Divide the data according to position above and below the median score, and then look for runs above and below the median in terms of days.)

22. Analyze the data of exercise 9, Chapter 12, by use of the median test for independent groups.

23. Suppose that the data of Table 11.4.1 were actually the results of five repeated observations on each of eight subjects. Furthermore, let a score value greater than or equal to 6.00 be counted a success, and any value which falls below 6.00 a failure. Compare the five sets of repeated observations by use of Cochran's test.

24. Apply the Kruskal-Wallis "analysis of variance by ranks" to the data of exercise 18, Chapter 10.

25. In a study of upward mobility trends in American society, a random sample of 107 U.S. families with at least one adult son was taken. The occupational level of the father and that of his oldest son was taken, and the data put into the following table:

		Father			
		Unskilled labor	*Skilled labor*	*White collar*	*Professional-executive*
Son	*Unskilled labor*	2	3	7	3
	Skilled labor	3	4	20	9
	White collar	1	2	14	19
	Professional-executive	0	4	2	14

Index the extent to which a ordinal relationship seems to exist between occupational level of father and of son.

26. Test the hypothesis that no ordinal relationship exists in the data of exercise 25.

27. Compare the two groups of exercise 6 above in terms of the *runs* test.

28. In exercise 5, Chapter 2, responses are given for 30 subjects on each of three questionnarie items. The data for items I and II can be put into the following table. Index the extent to which there seems to be an ordinal association between I and II. Does a high response to item I tend to go with a high response to item II? Use Kendall's tau with ties.

		Item II				
		1	*2*	*3*	*4*	*5*
	1	0	2	1	1	1
	2	1	1	2	5	0
Item I	*3*	1	1	5	0	1
	4	0	3	2	0	0
	5	0	0	2	1	0

29. Make a table like that in exercise 28 for items II and III and for items I and III in exercise 5, Chapter 2. Then examine the γ index of predictive relationship for all three tables. Which pair of items predict order best? Which have the least relationship of this sort?

30. In a study of attitudes toward international affairs, each of a group of 15 subjects ranked 10 countries according to their reputation for democracy. The data follows:

		Country									
		A	*B*	*C*	*D*	*E*	*F*	*G*	*H*	*I*	*J*
	1	2	1	3	5	4	6	7	8	9	10
	2	8	7	6	4	9	1	2	10	5	3
	3	3	4	2	1	6	5	7	8	9	10
	4	9	10	8	6	7	1	3	4	5	2
	5	5	2	3	4	6	1	7	8	9	10
	6	2	4	3	1	5	8	9	10	7	6
	7	4	3	2	5	1	6	9	10	8	7
Subject	*8*	2	4	1	6	5	3	8	7	9	10
	9	9	10	6	7	8	1	4	3	5	2
	10	5	4	1	10	2	3	7	8	6	9
	11	2	4	1	5	3	6	7	9	10	8
	12	3	4	2	5	1	10	9	6	8	7
	13	4	3	5	2	10	1	8	6	9	7
	14	5	3	4	2	1	8	9	6	10	7
	15	4	5	10	1	3	2	8	9	6	7

Find the coefficient of concordance and the average rank correlation for this group

31. Assuming that the 15 subjects in exercise 30 constitute a random sample, test the hypothesis of zero true-concordance among subjects in terms of their rankings of countries.

32. Using the figures given in Table 14.8.1, construct a table in which the columns represent the variable Y, in a grouped distribution with $i = 5$, and the rows the values of X. Then, on the basis of this table, calculate the gamma measure. Is there evidence of ordinal association?

33. Two judges each ranked 10 pictures (labeled A, B, C, etc.) in order of beauty. The rankings are given below. Use Kendall's tau coefficient to describe the agreement between the judges.

Judge 1	*Judge 2*
C	A
A	C
B	B
D	G
F	D
E	F
G	E
J	H
I	J
H	I

34. Use the data of exercise 33 to test the hypothesis of zero true-agreement between the judges.

35. Compute and test the significance of the tau coefficient representing the relationship shown in Table 13.15.1.

Appendix A
RULES OF SUMMATION

A.1 / SUBSCRIPTS

In the discussion to follow, let X and Y stand for numerical variables. Ordinarily, the numerical values for these variables are scores in some given set of data, where each individual observed in a group of N cases is paired either with a single score, x, or with a pair of scores (x, y).

The *subscript* notation is useful in keeping track of the various score values across the different individuals or observations. For the moment, we will consider N distinct individuals, arbitrarily indexed by the set of numbers $(1, \cdots, i, \cdots, N)$. The "running subscript" i denotes any one of these individuals. Then, X_i denotes the particular value of X associated with any particular individual i. That is, the variable X_i ranges only over the set of particular values

$$(x_1, x_2, \cdots, x_N).$$

Now we need a way to discuss the *sum* of these particular values of X over the entire set of N different individuals i. The symbol Σ (capital Greek sigma) stands for the operation "the sum of," referring to the values symbolized immediately after the summation sign. Thus,

$$\sum_{i=1}^{N} x_i$$

is read as "the sum of the N values x_i, beginning with $i = 1$ and ending with $i = N$." That is,

$$\sum_{i=1}^{N} x_i = x_1 + x_2 + \cdots + x_N.$$

Suppose, for example, that $N = 4$, and the actual values associated with the four different individuals were

6, 8, 3, 11.

Then,

$$\sum_{i=1}^{N} x_i = 6 + 8 + 3 + 11 = 28.$$

In many simple contexts where it is clear that the sum extends across *all* of a set of N individuals, the subscripts and superscripts on the summation sign are simplified or omitted. Thus, we often write:

$$\sum_{i} x_i$$

to indicate the sum of values across all N individuals i, or even

$$\Sigma x_i$$

to stand for this sum. However, the subscript is usually omitted only in very simple expressions where the set of values to be included in the sum is perfectly clear.

Quite frequently, symbols for values corresponding to individual observations are given two (or more) subscripts. That is, it happens that sometimes we want to discuss individual observations arranged into several groupings or cross-classifications. Then two (or more) sets of index numbers are used. For example, suppose that there were some J different *groups* of observations; these different groups are identified by being paired arbitrarily with the numbers

$$(1, \cdots, j, \cdots, J).$$

Furthermore, within any particular group, j, there is some number n_j of distinct individuals or observations, each of which is associated with an X score. Then, the particular value of the score for the ith individual in the group j is symbolized by x_{ij}. If we wanted to symbolize the sum of the scores for the particular group j, then we would write

$$\sum_{i=1}^{n_j} x_{ij}$$

which stands for "the sum of values for all of the n_j different individuals in the particular group labeled j." Where it is clear that all individuals scores in a particular group j are to be summed, this can be abbreviated to

$$\sum_{i} x_{ij}.$$

Notice that this quite literally is the symbol for "sum over i" for a set of scores x_{ij}; the fact that j appears as a subscript for x but that i alone appears below the sigma is a cue that all the scores summed in this particular operation *must* have the same subscript j (that is, must be in the same group of observations).

However, often one wants to indicate the sum of X_{ij} over *all* individuals in *all* groups. Written out in full, this would be symbolized by

$$\sum_{j=1}^{j}\sum_{i=1}^{n_j} x_{ij} = (x_{11} + x_{21} + \cdots + x_{n_j-1,J} + x_{n_j,J}).$$

This *double* summation is interpreted as "Take a particular group j, and sum scores over all the n_j different observations in that group; then, having done this for each of the J different groups, sum the results over all of these groups." Once again, this can be simplified to

$$\sum_j\sum_i x_{ij}$$

when the meaning of the symbol is perfectly clear in terms of which values are to be summed.

This idea can be extended, although in our discussions a *triple* summation is the largest such symbol used. In some problems, each individual i belongs simultaneously to *two* groupings. That is, suppose that all the scores in a set of data were arranged into a table with K rows, J columns, and JK cells, and that each individual observation belongs to one and only one row, one and only one column, and one and only one cell in the data table. Now let the various *column* groupings be symbolized by the numbers

$$(1, \cdots, j, \cdots, J).$$

The *rows* of the data table are indexed by the set of numbers

$$(1, \cdots, k, \cdots, K).$$

Then any cell of the table is indexed by a *pair* of numbers jk. The score of the ith individual in the cell jk of the data table is thus x_{ijk}. The number of individuals in the cell jk is shown by n_{jk}.

Now, when there are three subscripts like this,

$$\sum_{j=1}^{J}\sum_{k=1}^{K}\sum_{i=1}^{n_{jk}} x_{ijk},$$

the following is symbolized: "Take a particular cell, jk, and sum the scores of all individuals i in that cell; then, still within a particular column j, sum the results of doing this for each row k over the K respective rows; finally, sum the results for each column j over the J respective columns." A little thought should convince you that this is simply the sum of all individual scores in all cells, or the sum of all scores in the table.

In this instance, it is immaterial whether we write the sum for the rows or the sum for the columns first, since the total number symbolized is the same. However, this is not universally true, and order of the summation signs can be important.

Sometimes we want to symbolize only the sum for a particular cell, row, or column: First of all, letting n stand for the number in cell j,k, we write

$$\sum_{i=1}^{n} x_{ijk} = (x_{1jk} + \cdots + x_{njk}).$$

This denotes the sum of all individual scores in the *single* cell jk. If we write

$$\sum_{j=1}^{J} \sum_{i=1}^{n} x_{ijk} = (x_{11k} + x_{21k} + \cdots + x_{nJk}),$$

we intend that first the sum of individuals in a particular cell in column j and row k shall be taken, and then, still within the particular row k, the result for all the different columns j shall be summed. This amounts to the sum of all scores within the particular row k. On the other hand, if we want the sum of all the scores in a particular column j, we could write

$$\sum_{k=1}^{K} \sum_{i=1}^{n} x_{ijk} = (x_{1j1} + \cdots + x_{njK})$$

indicating the sum for a particular cell jk, and then the sum over the results for all rows k within the particular column j. In short, *unless a subscript appears both in the symbol for the x values and under the summation sign, the total value symbolized pertains to some particular set of observations, represented by that subscript not appearing under the summation symbol.* Notice that when the sum of all observations in a particular row k of the data table was intended, the sum was indicated over i and over j, but *not* over k. On the other hand, when the sum of observations in some column j was desired, the sum was indicated over i and over k, but not over j.

In learning to work with summation signs, it is well to start the habit of reading the summation symbols from *right to left,* and from *inside the punctuation outward.* The *innermost* summation indicated in a mathematical expression is ordinarily to be carried out first, then the next summation symbolized to the left, and then finally the summation appearing closest to the margin on the left.

A.2 / RULES OF SUMMATION

A few simple rules can help one learn to interpret and carry out the computations symbolized by statistical expressions involving summations. Furthermore, elementary statistical derivations almost always involve one or more of these summation rules used as an algebraic equivalence. First of all, some simple rules will be given that apply when values for only one variable are being summed, and then we will consider situations involving two or more variables.

RULE 1. If a is some constant value over the N different observations i, then

$$\sum_{i=1}^{N} a = Na.$$

For example, suppose that 5 observations are considered, and paired with each observation is the constant number 10. Then

$$\sum_{i} a = 10 + 10 + 10 + 10 + 10 = (5)(10).$$

Any time that, in the particular sum indicated, each and every individual observation entering into the sum is paired with exactly the same constant value a, then the sum is equal to the number of individuals considered times the value of a.

Extending this rule to double or triple summations, we have

$$\sum_{j=1}^{J} \sum_{i=1}^{nj} a = \sum_{j=1}^{J} (n_j)a$$

$$\sum_{k=1}^{K} \sum_{j=1}^{J} \sum_{i=1}^{n_{jk}} a = \sum_{k=1}^{K} \sum_{j=1}^{J} (n_{jk})a$$

and so on.

RULE 2. Given the value a, which is constant over all individuals entering into the summation, then

$$\sum_{i=1}^{N} ax_i = a \sum_{i=1}^{N} x_i.$$

For example, consider the five individual scores

6,7,3,9,20.

Suppose that each of these scores is multiplied by the constant 2. Then

$$\sum_i 2x_i = 2(6) + 2(7) + 2(3) + 2(9) + 2(20)$$

$$= 2(6 + 7 + 3 + 9 + 20) = 2 \sum_i x_i.$$

Once again, this rule can be applied to several sums:

$$\sum_{j=1}^{J} \sum_{i=1}^{nj} ax_{ij} = a \sum_{j=1}^{J} \sum_{i=1}^{nj} x_{ij}.$$

Notice here that the a symbol has no subscript, and thus may be regarded as a constant over the sum. However, it might happen that the sum is such that a_j appears instead of a, meaning that a_j is a constant for all observations i in the *same* group j, but is different for different groups. Then,

$$\sum_{j=1}^{J} \sum_{i=1}^{nj} a_j x_{ij} = \sum_{j=1}^{J} a_j \left(\sum_{i=1}^{nj} x_{ij} \right).$$

One must be sure to notice the set of observations for which some value *is* constant in multiple summations.

RULE 3. If some operation is to be carried out on the individual values of X before the summation, this is indicated by mathematical punctuation, such as parentheses, or equivalent symbols having the force of punctuation. Unless the summation symbol is included within this punctuation, the summation is to be done after the other operation: for example,

$$\sum_i^N (x_i)^2 = \sum_i^N x_i^2 = (x_1^2 + x_2^2 + \cdots + x_N^2)$$

$$\sum_i \sqrt{x_i} = \sqrt{x_i} + \sqrt{x_2} + \cdots + \sqrt{x_N}.$$

However,

$$\left(\sum_i x_i\right)^2 = (x_1 + \cdots + x_N)^2$$

$$\sqrt{\sum_i x_i} = \sqrt{x_1 + \cdots + x_N}.$$

Be sure to notice that where the parentheses (or other symbols such as exponent and radical signs having the same force as parentheses) happen to appear makes a big difference in the sum symbolized. In general,

$$\sum_i x_i^2 \neq \left(\sum_i x_i\right)^2$$

$$\sum_i \sqrt{x_i} \neq \sqrt{\sum_i x_i}$$

$$\log\left(\sum_i x_i\right) \neq \sum_i \log x_i,$$

and so on.

This is especially important when there is multiple summation. For example,

$$\sum_{j=1} \left(\sum_i^n x_{ij}\right)^2 = \sum_j (x_{1j} + \cdots + x_{nj})^2.$$

The notation here tells us *first* to take the sum of all the individual values in a single group j; then, having done this for each group j, we square the *sum* for each group; finally, the sum of the *squared* sums for the various groups is found. Squaring at the wrong place in this sequence of steps will give quite a different result.

RULE 4. If the only operation to be carried out before a sum is taken is itself a sum (or difference), then the summation may be distributed. Thus,

$$\sum_i (x_i - 4) = \sum_i x_i - \sum_i 4$$

$$\sum_i (x_i^2 + 3x_i + 10) = \sum_i x_i^2 + \sum_i 3x_i + \sum_i 10.$$

This rule is very important in algebraic manipulations on expressions involving sums. For example, using this rule we can make the following algebraic changes:

$$\sum_{j=1}^{J} \sum_{i=1}^{n_j} (x_{ij} - m)^2 = \sum_{j=1}^{J} \sum_{i=1}^{n_j} (x_{ij}^2 - 2mx_{ij} + m^2)$$

$$= \sum_{j=1}^{J} \sum_{i=1}^{n_j} x_{ij}^2 - \sum_{j=1}^{J} \sum_{i=1}^{n_j} 2mx_{ij} + \sum_{j=1}^{J} \sum_{i=1}^{n_j} m^2.$$

Invoking rules 1 and 2, we can reduce this expression to

$$\sum_j \sum_i x_{ij}^2 - 2m \sum_j \sum_i x_{ij} + m^2 \sum_j n_j.$$

The value represented by this last expression is algebraically equivalent to the value of the first expression, but it may be far more convenient actually to compute this number in the second way, or to use this second expression in a mathematical argument.

A.3 / SUMMATION RULES FOR TWO OR MORE VARIABLES

So far we have acted as though each distinct individual or observation were associated with one and only one numerical score. However, it may be that two distinct values x_i and y_i are paired with each individual or observation i. Then the rules given above may be extended:

RULE 5. If each of N distinct individuals i has two scores, x_i and y_i, then

$$\sum_{i=1}^{N} x_i y_i = x_1 y_1 + x_2 y_2 + \ldots + x_N y_N.$$

Notice that this symbol means that *the product of the pair of scores belonging to any given individual i is to be found first,* and then when this has been done for the entire set of individuals, *the sum of these products is to be found.* Be sure to notice that *in general*

$$\sum_i x_i y_i \neq \left(\sum_i x_i\right)\left(\sum_i y_i\right).$$

The other rules extend quite easily to the situation where two (or more) variables are considered:

RULE 6. $\sum_i a x_i y_i = a \sum_i x_i y_i.$

RULE 7. Given the constants a and b,

$$\sum_i (a x_i + b y_i) = a \sum_i x_i + b \sum_i y_i.$$

Sometimes it happens that when observations are arranged into groups, all observations i in the same group j will have the same value x_j on one variable, but different values y_{ij} on the other. Then, the following rule applies:

RULE 8. $\sum_{j=1}^{J} \sum_{i=1}^{n_j} x_j y_{ij} = \sum_{j}^{J} x_j \left(\sum_{i}^{n_j} y_{ij}\right).$

All of these rules may be extended by obvious analogy when there are more than three variables involved, or when there is multiple summation.

Like most of the mechanical skills of mathematics, real facility in algebraic manipulation of summations comes only with practice. However, when puzzled or in doubt about the equivalence to two sums, you can usually check this equivalence by actually writing out the terms in the sums symbolized.

EXERCISES—RULES OF SUMMATION

Write out in extended form the sums represented by each of the following expressions:

1. $\sum_{i=1}^{5} x_i$ **2.** $\sum_{i=1}^{6} x_i^2$

3. $\sum_{i=1}^{3} x_i^3 + \sum_{i=5}^{8} x_i^2$ **4.** $\sum_{i=2}^{4} 10(x_i + y_i)$

5. $\sum_{i=1}^{3} 4\left(\sum_{j=1}^{2} x_{ij}\right)$ **6.** $\sum_{j=1}^{4}\left(\sum_{i=1}^{3} x_{ij}\right)^2$

7. $\left(\sum_{i=1}^{4} x_i^2 g_i\right) \Big/ \left(\sum_{i=1}^{4} g_i\right)$ **8.** $\sum_{j=2}^{5} n_j \left(\sum_{i=1}^{3} x_{ij}\right)^2$

9. $\sum_{i=1}^{5}\sum_{j=2}^{4} (x_{ij}^2 + y_i)$ **10.** $\sum_{i=4}^{6} \sqrt{x_i + 10}$

Express each of the following extended sums in the appropriate summation notation:

11. $(x_1 + x_2 + x_3 + x_4)/N$

12. $x_1y_1 + x_2y_2 + x_3y_3 + \cdots + x_Ny_N$

13. $x_2 f(x_2) + x_3 f(x_3) + x_4 f(x_4)$

14. $p(x_1) + p(x_2) + p(x_3) + p(x_{10}) + p(x_{11}) + p(x_{12})$

15. $3(x_1 + y_1)^2 + 3(x_1 + y_2)^2 + \cdots + 3(x_1 + y_N)^2$

16. $12(x_1^2 - 2x_1 + 3) + 12(x_2^2 - 2x_2 + 3) + 12(x_3^2 - 2x_3 + 3)$

17. $[3x_1 + 3x_2 + 3x_3 + 3x_4]\,[5x_1 + 5x_2 + 5x_3 + 5x_4]$

18. $\sqrt{(x_1 - 2)(y_1 + 5) + (x_2 - 2)(y_2 + 5) + \cdots + (x_N - 2)(y_N + 5)}$

19. $[4(x_2 - 5)^2/y_2] + [4(x_3 - 5)^2/y_3] + [4(x_4 - 5)^2/y_4]$

20. $(x_{11} - M_1)^2 + (x_{12} - M_1)^2 + (x_{13} - M_1)^2 + (x_{21} - M_2)^2 + (x_{22} - M_2)^2 + (x_{23} - M_2)^2$

21. $(x_1 + x_2 + x_3 + x_4)^2\,(y_1^2 + y_2^2 + y_3^2 + y_4^2)$

Reduce each of the following expressions into simplest form by application of the rules of summation:

22. $\sum_{i=1}^{10} (5x_i - 3)$

23. $\left(\sum_{i=1}^{5} (x_i + 2c)\right)\left(\sum_{j=1}^{3} 7\right)$

24. $\sum_{i=4}^{6} (x_i - 2a)^2 + \sum_{i=1}^{3} (x_i + 2a)^2$

25. $\sum_{i=1}^{N}\left[x_i - \left(\sum_{i=1}^{N} x_i/N\right)\right]$

26. $\sum_{i=1}^{N} (x_i - 3)(x_i + 3)/N$

27. $\sqrt{\sum_{i=1}^{N} (a + b)} \sqrt{\sum_{i=1}^{N} c}$

28. $\left[\left(\sum_{i=1}^{N} x_i\right)^2 - \left(\sum_{i=1}^{N} y_i\right)^2\right] \Big/ \sum_{i=1}^{N} (x_i + y_i)$

29. $\dfrac{\sum_{i=1}^{N} \left[x_i - \left(\sum_{i=1}^{N} x_i/N\right)\right]^2}{N}$

30. $(1/N)\sum_{i=1}^{N} \left[x_i - \left(\sum_{i=1}^{N} x_i/N\right)\right]\left[y_i - \left(\sum_{i=1}^{N} y_i/N\right)\right]$

Consider the following values:

$x_1 = 7$ $x_4 = 4$ $x_7 = 2$ $x_{10} = 5$
$x_2 = 3$ $x_5 = 0$ $x_8 = 8$
$x_3 = 9$ $x_6 = 1$ $x_9 = 8$

In terms of these values for x_i, evaluate the following sums:

31. $\sum_{i=1}^{10} (x_i - 5)$

32. $\sum_{i=1}^{10} (x_i^2 - 3x_i + 1)$

33. $\sum_{i=1}^{5} x_i - \sum_{i=6}^{10} (x_i + 5)$

34. $\sum_{i=1}^{10} \left[x_i - \left(\sum_{i=1}^{10} x_i/10\right)\right] \Big/ 10$

35. $\left(\sum_{i=1}^{10} x_i\right) \Big/ \left(\sum_{i=1}^{5} x_i\right)$

36. $\left(\sum_{i=1}^{10} x_i^2/10\right) - \left(\sum_{i=1}^{10} x_i/10\right)^2$

Appendix B
THE ALGEBRA OF EXPECTATIONS

B.1 / EXPECTATIONS OF RANDOM VARIABLES

A very prominent place in theoretical statistics is occupied by the concept of the mathematical expectation of a random variable X. If the distribution of X is discrete, then the expectation (or expected value) of X is defined to be

$$E(X) = \sum_x xp(x)$$

where the sum is taken over all of the different values that the variable X can assume, and

$$\sum_x p(x) = 1.00.$$

For a continuous random variable X ranging over all the real numbers, the expectation is defined by

$$E(X) = \int_{-\infty}^{\infty} xf(x)\, d(x)$$

where

$$\int_{-\infty}^{\infty} f(x)\, d(x) = 1.00.$$

In essence, the expectation defined in either of these ways is a kind of weighted sum of values, and thus the rules of summation have very close parallels in the rules for the algebraic treatment of expectations. These rules apply either to discrete or to continuous random variables if particular boundary conditions exist; for our purposes these rules can be used without our going further into these special qualifications.

RULE 1. If a is some constant number, then

$$E(a) = a.$$

That is, if the same constant value a were associated with each and every elementary event in some sample space, the expectation or mean of the values would most certainly be a.

RULE 2. If a is some constant real number and X is a random variable with expectation $E(X)$, then

$$E(aX) = aE(X).$$

Suppose that a new random variable is formed by multiplying each value of X by the constant number a. Then the expectation of the new random variable is just a times the expectation of X. This is very simple to show for a discrete random variable X: By definition,

$$E(aX) = \sum_x axp(ax).$$

However, the probability of any value aX must be exactly equal to the probability of the corresponding X value, and so

$$E(aX) = \sum_x axp(x) = a \sum_x xp(x) = aE(X).$$

RULE 3. If a is a constant real number and X is a random variable, then

$$E(X + a) = E(X) + a.$$

This can be shown very simply for a discrete variable. Here,

$$E(X + a) = \sum_x (x + a)p(x + a)$$

$$= \sum_X xp(x + a) + a \sum_X p(x + a).$$

However, $p(X + a) = p(X)$ for each value of X, so that

$$E(X + a) = E(X) + a \sum_x p(x) = E(X) + a.$$

The expectations of functions of random variables, such as

$$E[(X + 2)^2]$$

$$E(\sqrt{X + b})$$

$$E(b^X),$$

to give only a few examples, are subject to the same algebraic rules as summations. That is, the operation indicated within the punctuation is to be carried out *before* the expectation is taken. It is most important that this be kept in mind during any algebraic argument involving summations. In general,

$$E[(X + 2)^2] \neq [E(X) + E(2)]^2$$

$$E(\sqrt{X}) \neq (\sqrt{E(X)})$$

$$E(b^X) \neq b^{E(X)}$$

and so forth.

The next few rules concern two (or more) random variables, symbolized by X and Y.

RULE 4. If X is a random variable with expectation $E(X)$, and Y is a random variable with expectation $E(Y)$, then

$$E(X + Y) = E(X) + E(Y).$$

Verbally, this rule says that the expectation of a sum of two random variables is the sum of their expectations. Once again, the proof is simple for two discrete variables X and Y. Consider the new random variable $(X + Y)$. The probability of a value of $(X + Y)$ involving a *particular* X value and a *particular* Y value is the joint probability $p(x,y)$. Thus,

$$E(X + Y) = \sum_x \sum_y (x + y)p(x,y).$$

Notice that here the expectation involves the sum over all possible *joint* events (x,y). This could be written as

$$\begin{aligned} E(X + Y) &= \sum_x \sum_y (x + y)p(x,y) \\ &= \sum_x \sum_y xp(x,y) + \sum_x \sum_y yp(x,y). \end{aligned}$$

However, for any fixed x,

$$\sum_y p(x,y) = p(x)$$

and for any fixed y,

$$\sum_x p(x,y) = p(y).$$

Thus,

$$p(x + y) = \sum_x xp(x) + \sum_y yp(y) = E(X) + E(Y).$$

In particular, one of the random variables may be in a functional relation to the other. For example, let $Y = 3X^2$. Then

$$\begin{aligned} E(X + Y) &= E(X + 3X^2) \\ &= E(X) + E(3X^2) \\ &= E(X) + 3E(X^2). \end{aligned}$$

This principle lets one *distribute* the expectation over an expression which itself has the form of a sum. We will make a great deal of use of this principle.

This rule may also be extended to any finite number of random variables:

RULE 5. Given some finite number of random variables, the expectation of the sum of those variables is the sum of their individual expectations. Thus,

$$E(X + Y + Z) = E(X) + E(Y) + E(Z)$$

and so on.

In particular, some of the random variables may be in functional relations to others. Let $Y = 6X^4$, and let $Z = \sqrt{2X}$. Then

$$E(X) + 6E(X^4) + E(\sqrt{2X}).$$

B.2 / THE VARIANCE OF A RANDOM VARIABLE.

More useful rules involve the variance of a random variable. The variance is defined by

$$\text{var}(X) = \sigma_X^2 = E[X - E(X)]^2$$

or

$$\sigma_X^2 = E(X^2) - [E(X)]^2.$$

The standard deviation of the random variable X is then $\sigma_X = \sqrt{\text{var}(X)}$.

The following rules give the effect of a transformation of X on the variance.

RULE 6. If a is some constant real number, and if X is a random variable with expectation $E(X)$ and variance σ_X^2, then the random variable $(X + a)$ has variance σ_X^2.

This can be shown as follows:

$$\text{var}(X + a) = E[(X + a)^2] - [E(X + a)]^2.$$

By rule 3 above, and expanding the squares, we have

$$\begin{aligned} E[(X + a)^2] - [E(X + a)]^2 &= E[(X^2 + 2Xa + a^2)] - [E(X) + a]^2 \\ &= E(X^2 + 2aX + a^2) - [E(X)]^2 - 2aE(X) - a^2. \end{aligned}$$

Then by rules 5 and 1, we have

$$\begin{aligned} \text{var.}(X + a) &= E(X^2) + 2aE(X) + a^2 - [E(X)]^2 - 2aE(X) - a^2 \\ &= E(X^2) - [E(X)\}^2 \\ &= \sigma_X^2. \end{aligned}$$

RULE 7. If a is some constant real number, and if X is a random variable with variance σ_X^2, the variance of the random variable aX is

$$\text{var}(aX) = a^2\sigma_X^2.$$

In order to show this, we take

$$\begin{aligned}\text{var}(aX) &= E[(aX)^2] - [E(aX)]^2\\ &= a^2E(X^2) - a^2[E(X)]^2\\ &= a^2(E(X^2) - [E(X)]^2)\\ &= a^2\sigma_X^2.\end{aligned}$$

In short, adding a constant value to each value of a random variable leaves the variance unchanged, but multiplying each value by a constant multiplies the variance by the square of the constant.

RULE 8. If X and Y are independent random variables, with variances σ_X^2 and σ_Y^2 respectively, then the variance of the sum $X + Y$ is

$$\sigma^2_{(X+Y)} = \sigma_X^2 + \sigma_Y^2.$$

Similarly, the variance of $X - Y$ is

$$\sigma^2_{(X-Y)} = \sigma_X^2 + \sigma_Y^2.$$

This principle also extends to any number of independent random variables (see C.2.6 in Appendix C).

B.3 / INDEPENDENCE AND COVARIANCE

The next rule is a most important one that applies only to *independent* random variables (see Appendix C).

RULE 9. Given random variable X with expectation $E(X)$ and the random variable Y with expectation $E(Y)$, then if X and Y are independent,

$$E(XY) = E(X)E(Y).$$

This rule states that if random variables are *statistically independent,* the expectation of the product of these variables is the product of their separate expectations. An important corollary to this principle is:

If $E(XY) \neq E(X)E(Y)$, the variables X and Y are not independent.

The basis for Rule 9 can also be shown fairly simply for discrete variables. Since X and Y are independent, $p(x,y) = p(x)p(y)$. Then,

$$E(XY) = \sum_x \sum_y (xy)p(x)p(y) = \sum_x \sum_y xp(x)yp(y).$$

However, for any fixed x, $yp(y)$ is perfectly free to be any value, so that

$$E(XY) = \sum_x xp(x) \sum_y yp(y) = E(X)E(Y).$$

Definition: Given the random variable X with expectation $E(X)$ and the random variable Y with expectation $E(Y)$, then the covariance of X and Y is

$$\text{cov}(X,Y) = E(XY) - E(X)E(Y),$$

the expected value of the product of X and Y minus the product of the expected values.

The covariance is thus a reflection of the departure from independence of X and Y. When X and Y are independent,

$$\text{cov}(X,Y) = 0,$$

by the rule given above. *When random variables are independent, their covariance is zero.* However, it is not necessarily true that zero covariance implies that the variables are independent. On the other hand, $\text{cov}(X,Y) \neq 0$ always implies that the variables X and Y are not independent.

Definition: Given two random variables X and Y, then the covariance of X and Y divided by the standard deviation of each variable,

$$\rho_{XY} = \frac{\text{cov}(X,Y)}{\sigma_X \sigma_Y},$$

is known as the coefficient of correlation between X and Y.

The correlation coefficient ρ_{XY} may be any value between -1 and 1. However, note that if X and Y are independent, $\rho_{XY} = 0$. The converse is not true, however, since $\rho_{XY} = 0$ does not necessarily imply the independence of X and Y.

By extension of Rule 9 to any finite number of random variables, we have:

RULE 10. Given any finite number of random variables, if all the variables are independent of each other, the expectation of their product is the product of their separate expectations: thus,

$$E(XYZ) = E(X)E(Y)E(Z),$$

and so on.

Consider the following probability distributions of discrete random variables. Find the expectation, $E(X)$, for each.

1.

x	$p(x)$
1	.75
0	.25
	1.00

2.

x	$p(x)$
5	4/18
0	12/18
−5	2/18
	18/18

3.

x	$p(x)$
5	.1
4	.4
3	.1
2	.3
1	.1
	1.0

4.

x	$p(x)$
399	.20
154	.20
125	.20
100	.20
−200	.20
	1.00

5.

x	$p(x)$
36	.12
30	.18
24	.20
18	.32
12	.09
6	.05
0	.04
	1.00

Consider a discrete random variable taking on only the values $x_1, x_2, \cdots, x_{10}$. Symbolize the following as expectations:

6. $x_1p(x_1) + x_2p(x_2) + \cdots + x_{10}p(x_{10})$

7. $x_1^2p(x_1) + x_2^2p(x_2) + \cdots + x_{10}^2p(x_{10})$

8. $[x_1^2p(x_1) + x_2^2p(x_2) + \cdots + x_{10}^2p(x_{10})] - [x_1p(x_1) + x_2p(x_2) + \cdots + x_{10}p(x_{10})]^2$

9. $[x_1 - E(X)]^2p(x_1) + [x_2 - E(X)]^2p(x_2) + \cdots + [x_{10} - E(X)]^2p(x_{10})$

10. $(x_1^2 - 4x_1 + 5)p(x_1) + (x_2^2 - 4x_2 + 5)p(x_2)$
$+ \cdots + (x_{10}^2 - 4x_{10} + 5)p(x_{10})$

Consider two random variables, X and Y. Simplify the following:

11. $E[(X + 35)/10]$

12. $E[X - 14Y + E(Y) + 7] - E(X + Y - 5)$

13. $E[X^2 - 2XE(X) + E^2(X)]$

14. $E[X^2 + Y^2 - 2(X + Y)^2]$

15. $E(17)$

16. $E[(X - E(X))(Y - E(Y))]$

See if you can prove the following for discrete random variables: (**Hint:** Turn each expectation into the equivalent weighted sum.)

17. $E(aX) = aE(X)$

18. $E(aX + b) = aE(X) + b$

19. $E[X - E(X)] = 0$

For the variables defined in exercises 1, 2, and 3 above, find

20. var(X), or σ_X^2

21. $E(5X - 12)$

22. $E[(X^2 + 2)/10]$

23. Suppose that the random variables given in exercises 3 and 4 above were independent. Let us call the variable in exercise 3, X, and that in exercise 4, Y. Then, find the value of $E(XY)$.

Appendix C
JOINT RANDOM VARIABLES AND LINEAR COMBINATIONS

C.1 / JOINT RANDOM VARIABLES AND INDEPENDENCE

Think of two discrete random variables X and Y. If all possible joint events ($X = a$, $Y = b$) are considered, each such joint event will have a *joint probability* $p\,(a,b)$. Then the function relating each possible such joint event to its probability is the *joint distribution or bivariate distribution* of X and Y.

When the two variables X and Y are continuous, so that the event $X = a$ is associated with a probability density $g(a)$, and the event $Y = b$ is associated with the probability density $h(b)$, then the joint event ($X = a$, $Y = b$) is associated with a *joint density* $f(a,b)$. Once again, the pairing of joint events involving the random variables with the probability density of each joint event can be thought of as the joint distribution of X and Y, or the bivariate distribution of X and Y.

Once again considering two discrete random variables X and Y, we know from Section 1.14 that the conditional probability that $X = a$, given that $Y = b$ is

$$p(X = a|Y = b) = \frac{p(X = a,\, Y = b)}{p(Y = b)}.$$

In the same way, the conditional probability of $Y = b$ given $X = a$ is found from

$$p(Y = b|X = a) = \frac{p(X = a,\, Y = b)}{p(X = a)}.$$

The corresponding *conditional densities* for continuous random variables are

$$r(X = a|Y = b) = \frac{f(a,b)}{h(b)} \qquad \text{[C.1.1]}$$

and

$$s(Y = b|X = a) = \frac{f(a,b)}{g(a)}.$$

Associated with any joint distribution of X and Y there is also a *marginal distribution* of X which gives a probability value $p(X = a)$ or a density value $g(X = a)$, to the event $(X = a)$. Then, in the case of discrete random variables,

$$p(X = a) = \sum_{y} p\,(X = a,\, Y = y), \tag{C.1.2}$$

or, in the case of continuous X and Y,

$$g(a) = \int_{\text{all } y} f(a,y)\, dy. \tag{C.1.3}$$

Similar expressions hold for the marginal probability or density of the event $(Y = b)$ of course.

For a joint distribution of discrete random variables X and Y, one may also define a *conditional expectation,* as follows

$$E(Y|X = a) = \sum_{\text{all } y} y\, p(Y = y|X = a) \tag{C.1.4}$$

If the joint distribution is pictured as a two-way table, this may be thought of as the expectation of the distribution formed by the Y distribution *within the array* (or column of the table) *standing for* $X = a$. The conditional expectation of X given $Y = b$ is defined similarly for rows of the table. Furthermore, in an analogous way one may define the *conditional variance* $\sigma^2_{Y|a}$ given $X = a$, and the conditional variance $\sigma^2_{X|b}$ given $Y = b$. Precisely these same ideas hold for continuous random variables, if the necessary adjustments in notation are made as in C.1.7.

We know from Section 1.14 that independence of events C and D exists if and only if

$$p(C \cap D) = p(C)p(D)$$

so that

$$p(C|D) = p(C) \text{ and } p(D|C) = p(D).$$

It follows that **two discrete random variables are independent if and only if**

$$p(X = a,\, Y = b) = p(X = a)p(Y = b) \tag{C.1.5*}$$

for all pairs of values a **and** b. Furthermore, $p(X = a|Y = b) = p(X = a)$ and $p(Y = b|X = a) = p(Y = b)$, when X and Y are independent.

The principle extends immediately to continuous random variables:

A random variable X **with density** $g(a)$ **at value** a, **and a random variable** Y **with density** $h(b)$ **at value** b **are independent if and only if for all** (a,b)

$$f(a,b) = g(a)h(b) \tag{C.1.6*}$$

where $f(a,b,)$ **is the *joint* density for the event** (a,b).

Just as for conditional probability, it is also possible to define **conditional density:**

$$w(b|a) = \frac{f(a,b)}{g(a)} \tag{C.1.7}$$

where $w(b|a)$ symbolizes the density for $Y = b$ given some value a of X. *The conditional distribution of Y given X will be exactly the same as the distribution of Y with X left unspecified when X and Y are independent:* for all (a,b),

$$w(b|a) = h(b), \qquad X \text{ and } Y \text{ independent.}$$

In the same way,

$$w(a|b) = g(a)$$

when the variables are independent.

The idea of independence of random variables can be extended to functions of random variables as well. That is, suppose we have a random variable X with a density $g(x)$, and an *independent* random variable Y with density $h(y)$, as before.

Now imagine some function of X: associated with each possible x value is a new number $v = t(x)$. For example, it might be that

$$v = ax + b$$

so that V is some linear function of X. Or perhaps,

$$v = 3x^2 + 2x,$$

and so on for any other function rule. Furthermore, let there be some function of Y:

$$w = s(y), \qquad \text{[C.1.8]}$$

so that a new number w is associated with each possible value of Y. *Then V and W are independent random variables, provided that X and Y are independent.* This principle applies both to continuous and to discrete random variables, and to all the ordinary functions studied in elementary mathematical analysis.

Although we have dealt so far only with the bivariate distribution of two variables such as X and Y, the ideas extend readily to the situation where an event may consist of values of more than two random variables. Thus, for example, perhaps each individual i in a sample is associated with three scores, the first representing a value of the random variable X, the second a value of the random variable Y, and the third a value of a random variable W. Then the event (x_i, y_i, w_i) is a joint occurrence of the event in which individual i receives the value x_i on variable X, y_i on variable Y, and w_i on variable W. Furthermore, associated with each such joint event is a probability or probability density. The function giving the relation between each possible such joint event and its probability (density) is the *multivariate distribution function* of X, Y, and W.

The ideas of conditional probabilities or conditional densities, conditional expectations, and conditional variances extend to multivariate distributions as well. We may also consider the marginal distribution of any single variable, such as X, summed (or integrated) over the remaining variables, or the bivariate distribution of two variables summed or integrated over the others, *et cetera*.

Independence among three variables, or any number of variables, is also defined in a way completely analogous to the concept of independence of any two variables. Thus, if the joint density of X, Y, and W is represented by $t(x,y,z)$ for any particular set of values, $X = x$, $Y = y$, $W = w$, while the marginal densities of the individual variables for these values are $g(x)$, $h(y)$, and $f(w)$, *then*

$$t(x,y,w) = g(x)h(y)f(w) \quad [C.1.9^*]$$

if and only if X, Y, and *W are pairwise independent.*

(See also Rule 10 of Appendix B for the relation among the expectations of such independent random variables.)

C.2 / LINEAR COMBINATIONS OF RANDOM VARIABLES

Consider a set of J random variables, $(X_1, X_2, \cdots, X_J)$, having some multivariate distribution function. Now also consider a set of J weights, not all zero, $(c_1, c_2, \cdots, c_J)$. Then a *linear combination* of the random variables is a weighted sum

$$Y = c_1X_1 + c_2X_2 + \cdots + c_JX_J. \quad [C.2.1^*]$$

The linear combination Y is then a random variable, with a probability (density) which depends upon the joint probability (density) of the set of random variables X_1 and so on.

The random variable Y (the linear combination) will have an expectation and a variance, of course:

Given J random variables, and some linear combination,

$$Y = c_1X_1 + \cdots + c_JX_J$$

the expected value of Y is

$$E(Y) = c_1E(X_1) + c_2E(X_2) + \cdots + c_JE(X_J). \quad [C.2.2.^*]$$

The expectation of any linear combination is the same linear combination of the expectations.

A special case of this principle gives us the following:

Given n sample values from the same distribution with mean μ, then the expectation of any linear combination of those sample values is

$$E(Y) = \mu(c_1 + c_2 + \cdots + c_n). \quad [C.2.3^*]$$

If, under these circumstances,

$$c_1 + c_1 + \cdots + c_k = 0,\ E(Y) = 0. \quad [C.2.4]$$

In addition, we know the variance of a linear combination:

For any set of random variables $(X_1, \cdots, X_J)$ and the linear combination Y with weights $(c_1, \cdots, c_J)$, not all zero, then the variance of Y is

$$\sigma_Y^2 = \sum_j c_j^2\, \sigma_j^2 + 2 \sum_j \sum_k c_j c_k \operatorname{Cov}(X_j,X_k) \quad [C.2.5]$$

where σ_j^2 symbolizes the variance of variable j (see Appendix B for a definition of $\operatorname{cov}(X_j,X_k)$.) This says that the variance of linear combination Y is a sum of the variances of the X variables, each weighted by the squared value of the original combination weight, plus a weighted sum of covariances between all pairs of X variables. However, by Appendix B, Rule 9, $\operatorname{cov}(X_j,X_k) = 0$ if X_j and X_k are independent. Thus,

Give J independent random variables, with variances $\sigma_1^2, \sigma_2^2, \cdots, \sigma_J^2$ respectively, and the linear combination

$$Y = c_1X_1 + c_2X_2 + \cdots + c_JX_J \quad [C.2.6^*]$$

then the distribution of Y has variance given by

$$\sigma_Y^2 = c_1^2\sigma_1^2 + c_2^2\sigma_2^2 + \cdots + c_J^2\sigma_J^2.$$

The variance of a linear combination of independent random variables is a weighted sum of their separate variances, each weight being the *square* of the original weight given to the variable.

It immediately follows from the above that *in the special case where the variance of each variable X_1 through X_J is equal to the same value, σ^2, then*

$$\sigma_Y^2 = \sigma^2 \sum_j c_j^2, \qquad \text{[C.2.7*]}$$

the common value of σ^2 times the sum of the squared weights c_j.

Now for an important principle involving the normal distribution:

Given some J independent random variables $X_1, \cdots, X_J$, each normally distributed, then any linear combination of these variables is also a normally distributed random variable. [C.2.8*]

For example, the sample mean of N cases is a linear combination, where the score of case 1, representing random variable 1, is given the weight $1/N$, and so on for all other scores. Hence, if the scores of the N individuals are independent of each other and are drawn from a normal population, the sampling distribution of the mean is exactly normal, regardless of how large or small N may be.

Suppose that each of J variables $X_1, X_2, \cdots, X_J$ represents an independent observation from a normal distribution, with variance σ^2. Now two linear combinations are formed,

$$Y_1 = c_{11}X_1 + \cdots + c_{1j}X_j + \cdots + c_{1J}X_J$$

and

$$Y_2 = c_{21}X_1 + \cdots + c_{2j}X_j + \cdots + c_{2J}X_J,$$

where in both Y_1 and Y_2 some of the c weights are other than zero. Then, *the two variables Y_1 and Y_2 are independent if and only if*

$$\sum_j c_{1j}c_{2j} = 0. \qquad \text{[C.2.9*]}$$

That is, the independence of Y_1 and Y_2 implies and is implied by the fact that the sum of the products of weights used in each is zero, when each X_j is a normally distributed variable.

Finally, a very general principle can be stated about linear combinations of sample means when the size of each sample is large. This is a direct consequence of the central limit theorem (Section 6.7) and can be stated as follows:

Given J independent samples, containing $N_1, N_2, \cdots, N_J$ independent observations, respectively, then the sampling distribution of any linear combination of means of those samples approaches a normal distribution as the size of each sample grows large. [C.2.10*]

C.3 / VECTOR NOTATION

In working with statistics, quite often one needs to speak of a particular set of values taken on by a variable, as, for example, when a sample of N observations is taken on some variable X. In other contexts, we need to represent the joint event consisting of the values taken on by several variables. For such situations, the use of vector and matrix notation is a great convenience, and becomes an absolute necessity in the study of multivariate statistical methods. We will need some rudiments of vector algebra in order to explain and use the method with which this section concludes.

A vector is an ordered set of values. Any number appearing as a part of a vector is referred to as an *element* of that vector. The position of any element of a vector has meaning, as it often tells to whom or to what that value belongs, or which variable that value represents. Any vector has a *dimension,* which indicates how many elements that vector contains, or how many ''places'' in order it has. Thus the vector [10, 25, −2, 4] is of dimension 4. A vector of dimension n contains n elements. The vectors, variables, and weights in C.2 have dimension J.

In the following a boldface lowercase letter will represent a vector. Thus, if the letter **a** is used for a 5-dimensional vector, this may be thought of as

$$\mathbf{a} = [a_1, a_2, a_3, a_4, a_5]$$

where the ordinary italic letters represent the various elements of **a**. Here, **a** is shown as a *row vector,* so that the values are written horizontally, or along one row. However, it is equally possible to have a column vector,

$$\mathbf{a} = \begin{bmatrix} a_2 \\ a_2 \\ a_3 \\ a_4 \\ a_5 \end{bmatrix}$$

It doesn't make a lot of difference whether we think in terms of row or column vectors; however, once we have decided that a vector will be a row or a column vector in a given problem, we stick with that same definition throughout. As it happens, in the following it will be most convenient to work with column vectors. In general, for dimension n, any column vector such as **a** will thus be representable as

$$\mathbf{a} = \begin{bmatrix} a_1 \\ a_2 \\ \vdots \\ a_j \\ \vdots \\ a_n \end{bmatrix} \quad \text{[C.3.1]}$$

If we wish to speak of an individual element in **a**, we will use a_j to stand for such an element.

The *transpose* of a column vector **a**, written as **a**′, consists of exactly the same elements as those in **a**, but written as a row vector: thus for the column vector shown in C.3.1, we have

$$\mathbf{a}' = [a_1, a_2, \cdots, a_j, \cdots, a_n].$$

In the same way, the transpose of any row vector is the corresponding column vector.

A special vector of interest is the *unit vector* **i**, which has entries that are all "ones"; that is

$$\mathbf{i} = \begin{bmatrix} 1 \\ 1 \\ \vdots \\ 1 \end{bmatrix}. \quad [\text{C.3.2}]$$

On the other hand, the *zero vector*, or the "null vector," has entries which are all zeros.

When a vector is multiplied by a constant, this is called *scalar multiplication*. Each entry in the vector is then multiplied by that constant. Thus, if d is a constant,

$$d\mathbf{a} = \begin{bmatrix} da_1 \\ da_2 \\ \vdots \\ da_n \end{bmatrix}. \quad [\text{C.3.3}]$$

Now consider two vectors, **x** and **y**, each of the same dimension n:

$$\mathbf{x} = \begin{bmatrix} x_1 \\ x_2 \\ \vdots \\ x_n \end{bmatrix}, \mathbf{y} = \begin{bmatrix} y_1 \\ y_2 \\ \vdots \\ y_n \end{bmatrix}. \quad [\text{C.3.4}]$$

Then *the sum of* **x** *and* **y** is given by their element-by-element or place-by-place sum:

$$\mathbf{x} + \mathbf{y} = \begin{bmatrix} x_1 + y_1 \\ x_2 + y_2 \\ \vdots \\ x_n + y_n \end{bmatrix}. \quad [\text{C.3.5}]$$

In a similar way we can defined the difference, **x** − **y.**

Furthermore, suppose that c and d are two constant numbers. Then *a linear combination* of the vectors **x** and **y** is

$$\mathbf{w} = c\mathbf{x} + d\mathbf{y} = \begin{bmatrix} cx_1 + dy_1 \\ cx_2 + dy_2 \\ \vdots \\ cx_n + dy_n \end{bmatrix}. \quad [\text{C.3.6*}]$$

In other words, each weighted value for **x** is summed with the corresponding weighted value of **y**. Then, the elements of **w** are composed of these weighted sums. The same idea applies to weighted sums of any number of vectors.

The *inner product* of two vectors **x** and **y** is represented by (**x**,**y**). This is simply the product of each element x_j with its corresponding element y_j, and then summed over all *j*:

$$(\mathbf{x},\mathbf{y}) = \sum_{j=1}^{n} x_j y_j. \qquad \text{[C.3.7*]}$$

The inner product of any vector **x** with itself, represented by $\|\mathbf{x}\|^2$ is called the *squared length* of the vector:

$$\|\mathbf{x}\|^2 = \sum_{j=1}^{n} x_j^2 \qquad \text{[C.3.8*]}$$

The *length* of vector **x** is then the *positive square root,* $\sqrt{\|\mathbf{x}\|^2} = \|\mathbf{x}\|$.

When two vectors have an inner product which is equal to zero,

$$(\mathbf{x},\mathbf{y}) = 0, \qquad \text{[C.3.9*]}$$

the two vectors are said to be *orthogonal*. An *orthogonal set* is one in which each pair is orthogonal.

Suppose that we have two vectors which are orthogonal. Now if we form two new vectors by taking some constant *c* times **x**, and some constant *d* times **y**, so that $\mathbf{w} = c\mathbf{x}$ and $\mathbf{z} = c\mathbf{y}$, then $(\mathbf{w},\mathbf{z}) = 0$, and the vectors **w** and **z** are also orthogonal.

One vector **x** is said to be *linearly dependent* on another vector **y** if it is possible to write **x** as some constant *c* times **y**:

$$\mathbf{x} = c\mathbf{y}. \qquad \text{[C.3.10]}$$

A vector **x** is also said to be linearly dependent on some set of other vectors if it is possible to write **x** as a linear combination of two or more of these other vectors. Thus, if **x** is linearly dependent on **y** and **z**, then there exist constants *c* and *d* such that

$$\mathbf{x} = c\mathbf{y} + d\mathbf{z}. \qquad \text{[C.3.11]}$$

On the other hand, *two vectors which are orthogonal are linearly independent*. A set of vectors is said to be linearly independent if it is impossible to find sets of constants permitting any vector to be written as a linear combination of some subset of the others. *Any orthogonal set of vectors* (in which every vector is orthogonal to every other) *is a linearly independent set*.

Now suppose that we have a whole set consisting of *m* vectors, $\mathbf{X} = (\mathbf{x}_1, \cdots, \mathbf{x}_k, \cdots, \mathbf{x}_m)$, each member of which, such as $\mathbf{x}_k$, is a column vector of dimension *n*. Such a set of vectors is called *a matrix of dimension n × m,* and has elements in the rows and columns as follows:

$$\mathbf{X} = \begin{bmatrix} x_{11} \cdots x_{1k} \cdots x_{1m} \\ x_{j1} \cdots x_{jk} \cdots x_{jm} \\ x_{n1} \cdots x_{nk} \cdots x_{nm} \end{bmatrix}$$

The vectors making up the columns of the matrix **X** may or may not be linearly dependent. However, *it is always possible to find a matrix* **V** *made up of q column vectors* $\mathbf{v}_1, \cdots, \mathbf{v}_i, \cdots, \mathbf{v}_q$ *each of dimension n, and with* $q \leq n$, *with* **V** as an orthogonal set of vectors. Then every vector in **X** can be shown as a linear combination of the vectors in **V**, and we say that **V** is an *orthogonal basis* for the set of vectors **X**. As the text shows, particularly in Chapter 14, there are distinct advantages to be gained by replacing a set-of linearly dependent vectors **X** with an orthogonal basis **V**.

Given any small set of vectors **X** it is fairly simple and straightforward to find an orthogonal basis. This is done through *the Gram-Schmidt orthogonalization process,* which will be described next.

C.4 / FINDING AN ORTHOGONAL BASIS FOR A SET OF VECTORS

The Gram-Schmidt process requires that we already have an initial set of vectors **X**. The process is begun by choosing one vector out of **X** to serve as the first vector, $\mathbf{v}_0$, in the set **V**. In most statistical work, the set **X** will include the unit vector **i**. Although this is not formally necessary, the vector **i** is thus almost always chosen to be the vector $\mathbf{v}_0$.

The process continues when we choose another vector from **X**. Let us call this $\mathbf{x}_1$. Then we take

$$\mathbf{v}_1 = \mathbf{x}_1 - b_{10}\mathbf{x}_0, \text{ where } b_{10} = \frac{(\mathbf{x}_1, \mathbf{v}_0)}{\|\mathbf{v}_0\|^2} \qquad \text{[C.4.1]}$$

This gives a new vector $\mathbf{v}_1$ which is orthogonal to $\mathbf{v}_0$, or **i**. Now we take the next vector from **X**, or $\mathbf{x}_2$. We find a new vector $\mathbf{v}_2$ by taking

$$\mathbf{v}_2 = \mathbf{x}_2 - b_{20}\mathbf{v}_0 - b_{21}\mathbf{v}_1, \text{ where } b_{20} = \frac{(\mathbf{x}_2,\mathbf{v}_0)}{\|\mathbf{v}_0\|^2}, \; b_{21} = \frac{(\mathbf{x}_2,\mathbf{v}_0)}{\|\mathbf{v}_1\|^2}. \qquad \text{[C.4.2]}$$

This new vector $\mathbf{v}_2$ is then orthogonal to both $\mathbf{v}_0$ and $\mathbf{v}_1$. Proceeding in this way, we find the vector $\mathbf{v}_3$ by taking

$$\mathbf{v}_3 = \mathbf{x}_3 - b_{30}\mathbf{v}_0 - b_{31}\mathbf{v}_1 - b_{32}\mathbf{v}_2, \text{ where } b_{30} = \frac{(\mathbf{x}_3,\mathbf{v}_0)}{\|\mathbf{v}_0\|^2}, \; b_{31} = \frac{(\mathbf{x}_3,\mathbf{v}_1)}{\|\mathbf{v}_1\|^2}, \; b_{32} = \frac{(\mathbf{x}_3,\mathbf{v}_2)}{\|\mathbf{v}_2\|^2}, \qquad \text{[C.4.3]}$$

which yields a vector $\mathbf{v}_3$ orthogonal to $\mathbf{v}_0$, to $\mathbf{v}_1$, and to $\mathbf{v}_2$.

This process is repeated until all m vectors in **X** have been used, or until each additional vector yields a **v** vector consisting only of zero entries.

The Gram-Schmidt process will be illustrated by use of a set of five vectors making up **X** as follows. Here the dimensionality n is 4.

$$\mathbf{x}_0 = \mathbf{i} = \begin{bmatrix}1\\1\\1\\1\end{bmatrix}, \quad \mathbf{x}_1 = \begin{bmatrix}1\\0\\0\\0\end{bmatrix}, \quad \mathbf{x}_2 = \begin{bmatrix}0\\1\\0\\0\end{bmatrix}, \quad \mathbf{x}_3 = \begin{bmatrix}0\\0\\1\\0\end{bmatrix}, \quad \mathbf{x}_4 = \begin{bmatrix}0\\0\\0\\1\end{bmatrix}$$

The first step in this process is to choose one vector from **X** to be the initial vector in **V**. As is often the case in statistical problems, here set **X** includes the unit vector

i, so that this will be taken as the initial vector in the basis, or $\mathbf{v}_0$. Then by C.4.1 take

$$b_{10} = \frac{(\mathbf{x}_1,\mathbf{v}_0)}{\|\mathbf{v}_0\|^2} = \frac{(1)(1) + (0)(1) + (0)(1) + (0)(1)}{1+1+1+1} = \frac{1}{4}.$$

On substituting into 6.4.1, we have

$$\begin{bmatrix}1\\0\\0\\0\end{bmatrix} - \left(\frac{1}{4}\right)\begin{bmatrix}1\\1\\1\\1\end{bmatrix} = \begin{bmatrix}\frac{3}{4}\\-\frac{1}{4}\\-\frac{1}{4}\\-\frac{1}{4}\end{bmatrix} = \mathbf{v}_1,$$

or, since omitting the common denominator still yields an orthogonal vector relative to $\mathbf{v}_0$, we can simplify matters by taking

$$\mathbf{v}_1 = \begin{bmatrix}3\\-1\\-1\\-1\end{bmatrix}.$$

Next we take

$$b_{20} = \frac{1}{4},\ b_{21} = \frac{0(3) - (1)(1) + (0)(-1) + (0)(-1)}{9+1+1+1} = \frac{-1}{12}$$

so that

$$\begin{bmatrix}0\\1\\0\\0\end{bmatrix} - \left(\frac{1}{4}\right)\begin{bmatrix}1\\1\\1\\1\end{bmatrix} - \left(-\frac{1}{12}\right)\begin{bmatrix}3\\-1\\-1\\-1\end{bmatrix} = \begin{bmatrix}0\\\frac{2}{3}\\-\frac{1}{3}\\-\frac{1}{3}\end{bmatrix} \text{ or } \begin{bmatrix}0\\2\\-1\\-1\end{bmatrix} = \mathbf{v}_2,$$

where the last vector is once again cleared of fractions.

As the next step, $\mathbf{x}_3$ is taken, and we find

$$b_{30} = \frac{1}{4},\ b_{31} = -\frac{1}{12},\ b_{32} = -\frac{1}{6}$$

so that

$$\begin{bmatrix}0\\0\\1\\0\end{bmatrix} - \left(\frac{1}{4}\right)\begin{bmatrix}1\\1\\1\\1\end{bmatrix} - \left(-\frac{1}{12}\right)\begin{bmatrix}3\\-1\\-1\\-1\end{bmatrix} - \left(-\frac{1}{6}\right)\begin{bmatrix}0\\2\\-1\\-1\end{bmatrix} = \begin{bmatrix}0\\0\\\frac{1}{2}\\-\frac{1}{2}\end{bmatrix} \text{ or } \begin{bmatrix}0\\0\\1\\-1\end{bmatrix} = \mathbf{v}_3$$

Finally, if we take $\mathbf{x}_4$ and find $b_{40} = 1/4$, $b_{41} = -1/12$, $b_{42} = -1/6$, and $c_{43} = -1/2$, we have

$$\begin{bmatrix}0\\0\\0\\1\end{bmatrix} - \left(\frac{1}{4}\right)\begin{bmatrix}1\\1\\1\\1\end{bmatrix} - \left(\frac{-1}{12}\right)\cdot\begin{bmatrix}3\\-1\\-1\\-1\end{bmatrix} - \left(-\frac{1}{6}\right)\begin{bmatrix}0\\2\\-1\\-1\end{bmatrix} - \left(-\frac{1}{2}\right)\begin{bmatrix}0\\0\\1\\-1\end{bmatrix} = \begin{bmatrix}0\\0\\0\\0\end{bmatrix}$$

In other words, no other vector (not all zero) exists which is orthogonal to $\mathbf{v}_0$, $\mathbf{v}_1$, $\mathbf{v}_2$, and $\mathbf{v}_3$. Thus an orthogonal basis for the set of vectors $\mathbf{X}$ is provided by the vectors

$$\mathbf{x}_0 = \mathbf{v}_0 = \begin{bmatrix}1\\1\\1\\1\end{bmatrix}, \quad \mathbf{v}_1 = \begin{bmatrix}3\\-1\\-1\\-1\end{bmatrix}, \quad \mathbf{v}_2 = \begin{bmatrix}0\\2\\-1\\-1\end{bmatrix}, \quad \mathbf{v}_3 = \begin{bmatrix}0\\0\\1\\-1\end{bmatrix}.$$

Each vector in $\mathbf{X}$ is some linear combination of the vectors in $\mathbf{V}$ (or, like $\mathbf{x}_0$, is identical to a vector in $\mathbf{V}$). Thus, for example, we can form $\mathbf{x}_3$ by the linear combination

$$\mathbf{x}_3 = b_{30}\mathbf{v}_0 + b_{31}\mathbf{v}_1 + b_{32}\mathbf{v}_2 + \frac{1}{2}(\mathbf{v}_3)$$

and so on for the other vectors in $\mathbf{X}$. Furthermore, note that the number of vectors q required in $\mathbf{V}$ is $\leq m$, the number of original vectors in $\mathbf{X}$, and also that $q \leq n$.

Such sets of orthogonal vectors are usually converted into an *orthonormal basis,* which means that each vector $\mathbf{v}_i$ is multiplied by $1/\|\mathbf{v}_i\|$. (This step is actually a built-in part of many statistical procedures, such as orthogonal comparisons. This is why we can be fairly free and easy about clearing away fractions, leaving only whole-number vectors to work with, since this will be corrected later.) Values making up the $\mathbf{V}$ vectors in the preceding example correspond to the weights used in a set of comparisons, such as those made in Chapter 12.

Another example of the Gram-Schmidt process will yield values such as those employed in Section 14.9. Here, the vectors in the original set $\mathbf{X}$ are

$$\mathbf{x}_0 = \begin{bmatrix}1\\1\\1\\1\end{bmatrix}, \mathbf{x}_1 = \begin{bmatrix}1\\2\\3\\4\end{bmatrix}, \mathbf{x}_2 = \begin{bmatrix}1\\4\\9\\16\end{bmatrix}, \mathbf{x}_3 = \begin{bmatrix}1\\8\\27\\64\end{bmatrix}, \quad \mathbf{x}_4 = \begin{bmatrix}1\\16\\81\\256\end{bmatrix}$$

Notice here that $\mathbf{x}_2$ is composed of the squared entries in $\mathbf{x}_1$, $\mathbf{x}_3$ of the cubes of $\mathbf{x}_1$ values and the entries in $\mathbf{x}_4$ are the $\mathbf{x}_1$ entries raised to the fourth power.

Then applying the Gram-Schmidt process we have

$$\begin{bmatrix}1\\2\\3\\4\end{bmatrix} - \left(\frac{10}{4}\right)\begin{bmatrix}1\\1\\1\\1\end{bmatrix} = \begin{bmatrix}-\frac{6}{4}\\-\frac{2}{4}\\\frac{2}{4}\\\frac{6}{4}\end{bmatrix} \text{ or } \begin{bmatrix}-3\\-1\\1\\3\end{bmatrix} = \mathbf{v}_1,$$

$$\begin{bmatrix} 1 \\ 4 \\ 9 \\ 16 \end{bmatrix} - \left(\frac{30}{4}\right)\begin{bmatrix} 1 \\ 1 \\ 1 \\ 1 \end{bmatrix} - \left(\frac{50}{20}\right)\begin{bmatrix} -3 \\ -1 \\ 1 \\ 3 \end{bmatrix} = \begin{bmatrix} 1 \\ -1 \\ -1 \\ 1 \end{bmatrix} = \mathbf{v}_2,$$

$$\begin{bmatrix} 1 \\ 8 \\ 27 \\ 64 \end{bmatrix} - \left(\frac{100}{4}\right)\begin{bmatrix} 1 \\ 1 \\ 1 \\ 1 \end{bmatrix} - \left(\frac{208}{20}\right)\begin{bmatrix} -3 \\ -1 \\ 1 \\ 3 \end{bmatrix} - \left(\frac{30}{4}\right)\begin{bmatrix} 1 \\ -1 \\ -1 \\ 1 \end{bmatrix} = \begin{bmatrix} -\frac{6}{20} \\ \frac{18}{20} \\ -\frac{18}{20} \\ \frac{6}{20} \end{bmatrix} \text{ or } \begin{bmatrix} -1 \\ 3 \\ -3 \\ 1 \end{bmatrix} = \mathbf{v}_3,$$

and

$$\begin{bmatrix} 1 \\ 16 \\ 81 \\ 256 \end{bmatrix} - \left(\frac{354}{4}\right)\begin{bmatrix} 1 \\ 1 \\ 1 \\ 1 \end{bmatrix} - \left(\frac{830}{20}\right)\begin{bmatrix} -3 \\ -1 \\ 1 \\ 3 \end{bmatrix} - \left(\frac{160}{4}\right)\begin{bmatrix} 1 \\ -1 \\ -1 \\ 1 \end{bmatrix} - \left(\frac{60}{20}\right)\begin{bmatrix} -1 \\ 3 \\ -3 \\ 1 \end{bmatrix} = \begin{bmatrix} 0 \\ 0 \\ 0 \\ 0 \end{bmatrix}.$$

An orthogonal basis for $\mathbf{X}$ is thus provided by the four vectors

$$\mathbf{x}_0 = \mathbf{v}_0 = \begin{bmatrix} 1 \\ 1 \\ 1 \\ 1 \end{bmatrix}, \mathbf{v}_1 = \begin{bmatrix} -3 \\ -1 \\ 1 \\ 3 \end{bmatrix}, \mathbf{v}_2 = \begin{bmatrix} 1 \\ -1 \\ -1 \\ 1 \end{bmatrix}, \mathbf{v}_3 = \begin{bmatrix} -1 \\ 3 \\ -3 \\ 1 \end{bmatrix}.$$

Such vectors are the *orthogonal polynomials* employed in Section 14.9.

More on vectors and matrices, together with the Gram-Schmidt method, will be found in Timm (1975) and in Winer (1971), as well as in texts in matrix theory. The student is urged to become familiar with vector and matrix theory, since much of modern applied statistics presupposes such background.

Appendix D

TABLES

Table I

CUMULATIVE NORMAL PROBABILITIES

z	$F(z)$	z	$F(z)$	z	$F(z)$	z	$F(z)$
.00	.5000000	.21	.5831662	.42	.6627573	.63	.7356527
.01	.5039894	.22	.5870604	.43	.6664022	.64	.7389137
.02	.5079783	.23	.5909541	.44	.6700314	.65	.7421539
.03	.5119665	.24	.5948349	.45	.6736448	.66	.7453731
.04	.5159534	.25	.5987063	.46	.6772419	.67	.7485711
.05	.5199388	.26	.6025681	.47	.6808225	.68	.7517478
.06	.5239222	.27	.6064199	.48	.6843863	.69	.7549029
.07	.5279032	.28	.6102612	.49	.6879331	.70	.7580363
.08	.5318814	.29	.6140919	.50	.6914625	.71	.7611479
.09	.5358564	.30	.6179114	.51	.6949743	.72	.7642375
.10	.5398278	.31	.6217195	.52	.6984682	.73	.7673049
.11	.5437953	.32	.6255158	.53	.7019440	.74	.7703500
.12	.5477584	.33	.6293000	.54	.7054015	.75	.7733726
.13	.5517168	.34	.6330717	.55	.7088403	.76	.7763727
.14	.5556700	.35	.6368307	.56	.7122603	.77	.7793501
.15	.5596177	.36	.6405764	.57	.7156612	.78	.7823046
.16	.5635595	.37	.6443088	.58	.7190427	.79	.7852361
.17	.5674949	.38	.6480273	.59	.7224047	.80	.7881446
.18	.5714237	.39	.6517317	.60	.7257469	.81	.7910299
.19	.5753454	.40	.6554217	.61	.7290691	.82	.7938919
.20	.5792597	.41	.6590970	.62	.7323711	.83	.7967306

This table is condensed from Table 1 of the *Biometrika Tables for Statisticians*, Vol. 1 (ed. 3), edited by E. S. Pearson and H. O. Hartley. Reproduced here with the kind permission of E. S. Pearson and the trustees of *Biometrika*.

Table I (continued)

z	$F(z)$	z	$F(z)$	z	$F(z)$	z	$F(z)$
.84	.7995458	1.32	.9065825	1.79	.9632730	2.26	.9880894
.85	.8023375	1.33	.9082409	1.80	.9640697	2.27	.9883962
.86	.8051055	1.34	.9098773	1.81	.9648521	2.28	.9886962
.87	.8078498	1.35	.9114920	1.82	.9656205	2.29	.9889893
.88	.8105703	1.36	.9130850	1.83	.9663750	2.30	.9892759
.89	.8132671	1.37	.9146565	1.84	.9671159	2.31	.9895559
.90	.8159399	1.38	.9162067	1.85	.9678432	2.32	.9898296
.91	.8185887	1.39	.9177356	1.86	.9685572	2.33	.9900969
.92	.8212136	1.40	.9192433	1.87	.9692581	2.34	.9903581
.93	.8238145	1.41	.9207302	1.88	.9699460	2.35	.9906133
.94	.8263912	1.42	.9221962	1.89	.9706210	2.36	.9908625
.95	.8289439	1.43	.9236415	1.90	.9712834	2.37	.9911060
.96	.8314724	1.44	.9250663	1.91	.9719334	2.38	.9913437
.97	.8339768	1.45	.9264707	1.92	.9725711	2.39	.9915758
.98	.8364569	1.46	.9278550	1.93	.9731966	2.40	.9918025
.99	.8389129	1.47	.9292191	1.94	.9738102	2.41	.9920237
1.00	.8413447	1.48	.9305634	1.95	.9744119	2.42	.9922397
1.01	.8437524	1.49	.9318879	1.96	.9750021	2.43	.9924506
1.02	.8461358	1.50	.9331928	1.97	.9755808	2.44	.9926564
1.03	.8484950	1.51	.9344783	1.98	.9761482	2.45	.9928572
1.04	.8508300	1.52	.9357445	1.99	.9767045	2.46	.9930531
1.05	.8531409	1.53	.9369916	2.00	.9772499	2.47	.9932443
1.06	.8554277	1.54	.9382198	2.01	.9777844	2.48	.9934309
1.07	.8576903	1.55	.9394292	2.02	.9783083	2.49	.9936128
1.08	.8599289	1.56	.9406201	2.03	.9788217	2.50	.9937903
1.09	.8621434	1.57	.9417924	2.04	.9793248	2.51	.9939634
1.10	.8643339	1.58	.9429466	2.05	.9798178	2.52	.9941323
1.11	.8665005	1.59	.9440826	2.06	.9803007	2.53	.9942969
1.12	.8686431	1.60	.9452007	2.07	.9807738	2.54	.9944574
1.13	.8707619	1.61	.9463011	2.08	.9812372	2.55	.9946139
1.14	.8728568	1.62	.9473839	2.09	.9816911	2.56	.9947664
1.15	.8749281	1.63	.9484493	2.10	.9821356	2.57	.9949151
1.16	.8769756	1.64	.9494974	2.11	.9825708	2.58	.9950600
1.17	.8789995	1.65	.9505285	2.12	.9829970	2.59	.9952012
1.18	.8809999	1.66	.9515428	2.13	.9834142	2.60	.9953388
1.19	.8829768	1.67	.9525403	2.14	.9838226	2.70	.9965330
1.20	.8849303	1.68	.9535213	2.15	.9842224	2.80	.9974449
1.21	.8868606	1.69	.9544860	2.16	.9846137	2.90	.9981342
1.22	.8887676	1.70	.9554345	2.17	.9849966	3.00	.9986501
1.23	.8906514	1.71	.9563671	2.18	.9853713	3.20	.9993129
1.24	.8925123	1.72	.9572838	2.19	.9857379	3.40	.9996631
1.25	.8943502	1.73	.9581849	2.20	.9860966	3.60	.9998409
1.26	.8961653	1.74	.9590705	2.21	.9864474	3.80	.9999277
1.27	.8979577	1.75	.9599408	2.22	.9867906	4.00	.9999683
1.28	.8997274	1.76	.9607961	2.23	.9871263	4.50	.9999966
1.29	.9014747	1.77	.9616364	2.24	.9874545	5.00	.9999997
1.30	.9031995	1.78	.9624620	2.25	.9877755	5.50	.9999999
1.31	.9049021						

Table II

BINOMIAL PROBABILITIES $\binom{N}{r}p^r q^{N-r}$

		p									
N	*r*	.05	.10	.15	.20	.25	.30	.35	.40	.45	.50
1	**0**	.9500	.9000	.8500	.8000	.7500	.7000	.6500	.6000	.5500	.5000
	1	.0500	.1000	.1500	.2000	.2500	.3000	.3500	.4000	.4500	.5000
2	**0**	.9025	.8100	.7225	.6400	.5625	.4900	.4225	.3600	.3025	.2500
	1	.0950	.1800	.2550	.3200	.3750	.4200	.4550	.4800	.4950	.5000
	2	.0025	.0100	.0225	.0400	.0625	.0900	.1225	.1600	.2025	.2500
3	**0**	.8574	.7290	.6141	.5120	.4219	.3430	.2746	.2160	.1664	.1250
	1	.1354	.2430	.3251	.3840	.4219	.4410	.4436	.4320	.4084	.3750
	2	.0071	.0270	.0574	.0960	.1406	.1890	.2389	.2880	.3341	.3750
	3	.0001	.0010	.0034	.0080	.0156	.0270	.0429	.0640	.0911	.1250
4	**0**	.8145	.6561	.5220	.4096	.3164	.2401	.1785	.1296	.0915	.0625
	1	.1715	.2916	.3685	.4096	.4219	.4116	.3845	.3456	.2995	.2500
	2	.0135	.0486	.0975	.1536	.2109	.2646	.3105	.3456	.3675	.3750
	3	.0005	.0036	.0115	.0256	.0469	.0756	.1115	.1536	.2005	.2500
	4	.0000	.0001	.0005	.0016	.0039	.0081	.0150	.0256	.0410	.0625
5	**0**	.7738	.5905	.4437	.3277	.2373	.1681	.1160	.0778	.0503	.0312
	1	.2036	.3280	.3915	.4096	.3955	.3602	.3124	.2592	.2059	.1562
	2	.0214	.0729	.1382	.2048	.2637	.3087	.3364	.3456	.3369	.3125
	3	.0011	.0081	.0244	.0512	.0879	.1323	.1811	.2304	.2757	.3125
	4	.0000	.0004	.0022	.0064	.0146	.0284	.0488	.0768	.1128	.1562
	5	.0000	.0000	.0001	.0003	.0010	.0024	.0053	.0102	.0185	.0312
6	**0**	.7351	.5314	.3771	.2621	.1780	.1176	.0754	.0467	.0277	.0156
	1	.2321	.3543	.3993	.3932	.3560	.3025	.2437	.1866	.1359	.0938
	2	.0305	.0984	.1762	.2458	.2966	.3241	.3280	.3110	.2780	.2344
	3	.0021	.0146	.0415	.0819	.1318	.1852	.2355	.2765	.3032	.3125
	4	.0001	.0012	.0055	.0154	.0330	.0595	.0951	.1382	.1861	.2344
	5	.0000	.0001	.0004	.0015	.0044	.0102	.0205	.0369	.0609	.0938
	6	.0000	.0000	.0000	.0001	.0002	.0007	.0018	.0041	.0083	.0156
7	**0**	.6983	.4783	.3206	.2097	.1335	.0824	.0490	.0280	.0152	.0078
	1	.2573	.3720	.3960	.3670	.3115	.2471	.1848	.1306	.0872	.0547
	2	.0406	.1240	.2097	.2753	.3115	.3177	.2985	.2613	.2140	.1641
	3	.0036	.0230	.0617	.1147	.1730	.2269	.2679	.2903	.2918	.2734
	4	.0002	.0026	.0109	.0287	.0577	.0972	.1442	.1935	.2388	.2734
	5	.0000	.0002	.0012	.0043	.0115	.0250	.0466	.0774	.1172	.1641
	6	.0000	.0000	.0001	.0004	.0013	.0036	.0084	.0172	.0320	.0547
	7	.0000	.0000	.0000	.0000	.0001	.0002	.0006	.0016	.0037	.0078
8	**0**	.6634	.4305	.2725	.1678	.1001	.0576	.0319	.0168	.0084	.0039
	1	.2793	.3826	.3847	.3355	.2760	.1977	.1373	.0896	.0548	.0312
	2	.0515	.1488	.2376	.2936	.3115	.2965	.2587	.2090	.1569	.1094
	3	.0054	.0331	.0839	.1468	.2076	.2541	.2786	.2787	.2568	.2188
	4	.0004	.0046	.0185	.0459	.0865	.1361	.1875	.2322	.2627	.2734
	5	.0000	.0004	.0026	.0092	.0231	.0467	.0808	.1239	.1719	.2188
	6	.0000	.0000	.0002	.0011	.0038	.0100	.0217	.0413	.0703	.1094
	7	.0000	.0000	.0000	.0001	.0004	.0012	.0033	.0079	.0164	.0312
	8	.0000	.0000	.0000	.0000	.0000	.0001	.0002	.0007	.0017	.0039

N	r	p .05	.10	.15	.20	.25	.30	.35	.40	.45	.50
9	**0**	.6302	.3874	.2316	.1342	.0751	.0404	.0277	.0101	.0046	.0020
	1	.2985	.3874	.3679	.3020	.2253	.1556	.1004	.0605	.0339	.0176
	2	.0629	.1722	.2597	.3020	.3003	.2668	.2162	.1612	.1110	.0703
	3	.0077	.0446	.1069	.1762	.2336	.2668	.2716	.2508	.2119	.1641
	4	.0006	.0074	.0283	.0661	.1168	.1715	.2194	.2508	.2600	.2461
	5	.0000	.0008	.0050	.0165	.0389	.0735	.1181	.1672	.2128	.2461
	6	.0000	.0001	.0006	.0028	.0087	.0210	.0424	.0743	.1160	.1641
	7	.0000	.0000	.0000	.0003	.0012	.0039	.0098	.0212	.0407	.0703
	8	.0000	.0000	.0000	.0000	.0001	.0004	.0013	.0035	.0083	.0176
	9	.0000	.0000	.0000	.0000	.0000	.0000	.0001	.0003	.0008	.0020
10	**0**	.5987	.3487	.1969	.1074	.0563	.0282	.0135	.0060	.0025	.0010
	1	.3151	.3874	.3474	.2684	.1877	.1211	.0725	.0403	.0207	.0098
	2	.0746	.1937	.2759	.3020	.2816	.2335	.1757	.1209	.0763	.0439
	3	.0105	.0574	.1298	.2013	.2503	.2668	.2522	.2150	.1665	.1172
	4	.0010	.0112	.0401	.0881	.1460	.2001	.2377	.2508	.2384	.2051
	5	.0001	.0015	.0085	.0264	.0584	.1029	.1536	.2007	.2340	.2461
	6	.0000	.0001	.0012	.0055	.0162	.0368	.0689	.1115	.1596	.2051
	7	.0000	.0000	.0001	.0008	.0031	.0090	.0212	.0425	.0746	.1172
	8	.0000	.0000	.0000	.0001	.0004	.0014	.0043	.0106	.0229	.0439
	9	.0000	.0000	.0000	.0000	.0000	.0001	.0005	.0016	.0042	.0098
	10	.0000	.0000	.0000	.0000	.0000	.0000	.0000	.0001	.0003	.0016
11	**0**	.5688	.3138	.1673	.0859	.0422	.0198	.0088	.0036	.0014	.0005
	1	.3293	.3835	.3248	.2362	.1549	.0932	.0518	.0266	.0125	.0054
	2	.0867	.2131	.2866	.2953	.2581	.1998	.1395	.0887	.0513	.0269
	3	.0137	.0710	.1517	.2215	.2581	.2568	.2254	.1774	.1259	.0806
	4	.0014	.0158	.0536	.1107	.1721	.2201	.2428	.2365	.2060	.1611
	5	.0001	.0025	.0132	.0388	.0803	.1231	.1830	.2207	.2360	.2256
	6	.0000	.0003	.0023	.0097	.0268	.0566	.0985	.1471	.1931	.2256
	7	.0000	.0000	.0003	.0017	.0064	.0173	.0379	.0701	.1128	.1611
	8	.0000	.0000	.0000	.0002	.0011	.0037	.0102	.0234	.0462	.0806
	9	.0000	.0000	.0000	.0000	.0001	.0005	.0018	.0052	.0126	.0269
	10	.0000	.0000	.0000	.0000	.0000	.0000	.0002	.0007	.0021	.0054
	11	.0000	.0000	.0000	.0000	.0000	.0000	.0000	.0000	.0002	.0005
12	**0**	.5404	.2824	.1422	.0687	.0317	.0138	.0057	.0022	.0008	.0002
	1	.3413	.3766	.3012	.2062	.1267	.0712	.0368	.0174	.0075	.0029
	2	.0988	.2301	.2924	.2835	.2323	.1678	.1088	.0639	.0339	.0161
	3	.0173	.0852	.1720	.2362	.2581	.2397	.1954	.1419	.0923	.0537
	4	.0021	.0213	.0683	.1329	.1936	.2311	.2367	.2128	.1700	.1208
	5	.0002	.0038	.0193	.0532	.1032	.1585	.2039	.2270	.2225	.1934
	6	.0000	.0005	.0040	.0155	.0401	.0792	.1281	.1766	.2124	.2256
	7	.0000	.0000	.0006	.0033	.0115	.0291	.0591	.1009	.1489	.1934
	8	.0000	.0000	.0001	.0005	.0024	.0078	.0199	.0420	.0762	.1208
	9	.0000	.0000	.0000	.0001	.0004	.0015	.0048	.0125	.0277	.0537
	10	.0000	.0000	.0000	.0000	.0000	.0002	.0008	.0025	.0068	.0161
	11	.0000	.0000	.0000	.0000	.0000	.0000	.0001	.0003	.0010	.0029
	12	.0000	.0000	.0000	.0000	.0000	.0000	.0000	.0000	.0001	.0002

N	r	.05	.10	.15	.20	.25	.30	.35	.40	.45	.50
13	0	.5133	.2542	.1209	.0550	.0238	.0097	.0037	.0013	.0004	.0001
	1	.3512	.3672	.2774	.1787	.1029	.0540	.0259	.0113	.0045	.0016
	2	.1109	.2448	.2937	.2680	.2059	.1388	.0836	.0453	.0220	.0095
	3	.0214	.0997	.1900	.2457	.2517	.2181	.1651	.1107	.0660	.0349
	4	.0028	.0277	.0838	.1535	.2097	.2337	.2222	.1845	.1350	.0873
	5	.0003	.0055	.0266	.0691	.1258	.1803	.2154	.2214	.1989	.1571
	6	.0000	.0008	.0063	.0230	.0559	.1030	.1546	.1968	.2169	.2095
	7	.0000	.0001	.0011	.0058	.0186	.0442	.0833	.1312	.1775	.2095
	8	.0000	.0000	.0001	.0011	.0047	.0142	.0336	.0656	.1089	.1571
	9	.0000	.0000	.0000	.0001	.0009	.0034	.0101	.0243	.0495	.0873
	10	.0000	.0000	.0000	.0000	.0001	.0006	.0022	.0065	.0162	.0349
	11	.0000	.0000	.0000	.0000	.0000	.0001	.0003	.0012	.0036	.0095
	12	.0000	.0000	.0000	.0000	.0000	.0000	.0000	.0001	.0005	.0016
	13	.0000	.0000	.0000	.0000	.0000	.0000	.0000	.0000	.0000	.0001
14	0	.4877	.2288	.1028	.0440	.0178	.0068	.0024	.0008	.0002	.0001
	1	.3593	.3559	.2539	.1539	.0832	.0407	.0181	.0073	.0027	.0009
	2	.1229	.2570	.2912	.2501	.1802	.1134	.0634	.0317	.0141	.0056
	3	.0259	.1142	.2056	.2501	.2402	.1943	.1366	.0845	.0462	.0222
	4	.0037	.0349	.0998	.1720	.2202	.2290	.2022	.1549	.1040	.0611
	5	.0004	.0078	.0352	.0860	.1468	.1963	.2178	.2066	.1701	.1222
	6	.0000	.0013	.0093	.0322	.0734	.1262	.1759	.2066	.2088	.1833
	7	.0000	.0002	.0019	.0092	.0280	.0618	.1082	.1574	.1952	.2095
	8	.0000	.0000	.0003	.0020	.0082	.0232	.0510	.0918	.1398	.1833
	9	.0000	.0000	.0000	.0003	.0018	.0066	.0183	.0408	.0762	.1222
	10	.0000	.0000	.0000	.0000	.0003	.0014	.0049	.0136	.0312	.0611
	11	.0000	.0000	.0000	.0000	.0000	.0002	.0010	.0033	.0093	.0222
	12	.0000	.0000	.0000	.0000	.0000	.0000	.0001	.0005	.0019	.0056
	13	.0000	.0000	.0000	.0000	.0000	.0000	.0000	.0001	.0002	.0009
	14	.0000	.0000	.0000	.0000	.0000	.0000	.0000	.0000	.0000	.0001
15	0	.4633	.2059	.0874	.0352	.0134	.0047	.0016	.0005	.0001	.0000
	1	.3658	.3432	.2312	.1319	.0668	.0305	.0126	.0047	.0016	.0005
	2	.1348	.2669	.2856	.2309	.1559	.0916	.0476	.0219	.0090	.0032
	3	.0307	.1285	.2184	.2501	.2252	.1700	.1110	.0634	.0318	.0139
	4	.0049	.0428	.1156	.1876	.2252	.2186	.1792	.1268	.0780	.0417
	5	.0006	.0105	.0449	.1032	.1651	.2061	.2123	.1859	.1404	.0916
	6	.0000	.0019	.0132	.0430	.0917	.1472	.1906	.2066	.1914	.1527
	7	.0000	.0003	.0030	.0138	.0393	.0811	.1319	.1771	.2013	.1964
	8	.0000	.0000	.0005	.0035	.0131	.0348	.0710	.1181	.1647	.1964
	9	.0000	.0000	.0001	.0007	.0034	.0116	.0298	.0612	.1048	.1527
	10	.0000	.0000	.0000	.0001	.0007	.0030	.0096	.0245	.0515	.0916
	11	.0000	.0000	.0000	.0000	.0001	.0006	.0024	.0074	.0191	.0417
	12	.0000	.0000	.0000	.0000	.0000	.0001	.0004	.0016	.0052	.0139
	13	.0000	.0000	.0000	.0000	.0000	.0000	.0001	.0003	.0010	.0032
	14	.0000	.0000	.0000	.0000	.0000	.0000	.0000	.0000	.0001	.0005
	15	.0000	.0000	.0000	.0000	.0000	.0000	.0000	.0000	.0000	.0000
16	0	.4401	.1853	.0743	.0281	.0100	.0033	.0010	.0003	.0001	.0000
	1	.3706	.3294	.2097	.1126	.0535	.0228	.0087	.0030	.0009	.0002
	2	.1463	.2745	.2775	.2111	.1336	.0732	.0353	.0150	.0056	.0018

(Column heading over .05 through .50: p)

Table II (continued)

N	r	p .05	.10	.15	.20	.25	.30	.35	.40	.45	.50
16	**3**	.0359	.1423	.2285	.2463	.2079	.1465	.0888	.0468	.0215	.0085
	4	.0061	.0514	.1311	.2001	.2252	.2040	.1553	.1014	.0572	.0278
	5	.0008	.0137	.0555	.1201	.1802	.2099	.2008	.1623	.1123	.0667
	6	.0001	.0028	.0180	.0550	.1101	.1649	.1982	.1983	.1684	.1222
	7	.0000	.0004	.0045	.0197	.0524	.1010	.1524	.1889	.1969	.1746
	8	.0000	.0001	.0009	.0055	.0197	.0487	.0923	.1417	.1812	.1964
	9	.0000	.0000	.0001	.0012	.0058	.0185	.0442	.0840	.1318	.1746
	10	.0000	.0000	.0000	.0002	.0014	.0056	.0167	.0392	.0755	.1222
	11	.0000	.0000	.0000	.0000	.0002	.0013	.0049	.0142	.0337	.0667
	12	.0000	.0000	.0000	.0000	.0000	.0002	.0011	.0040	.0115	.0278
	13	.0000	.0000	.0000	.0000	.0000	.0000	.0002	.0008	.0029	.0085
	14	.0000	.0000	.0000	.0000	.0000	.0000	.0000	.0001	.0005	.0018
	15	.0000	.0000	.0000	.0000	.0000	.0000	.0000	.0000	.0001	.0002
	16	.0000	.0000	.0000	.0000	.0000	.0000	.0000	.0000	.0000	.0000
17	**0**	.4181	.1668	.0631	.0225	.0075	.0023	.0007	.0002	.0000	.0000
	1	.3741	.3150	.1893	.0957	.0426	.0169	.0060	.0019	.0005	.0001
	2	.1575	.2800	.2673	.1914	.1136	.0581	.0260	.0102	.0035	.0010
	3	.0415	.1556	.2359	.2393	.1893	.1245	.0701	.0341	.0144	.0052
	4	.0076	.0605	.1457	.2093	.2209	.1868	.1320	.0796	.0411	.0182
	5	.0010	.0175	.0668	.1361	.1914	.2081	.1849	.1379	.0875	.0472
	6	.0001	.0039	.0236	.0680	.1276	.1784	.1991	.1839	.1432	.0944
	7	.0000	.0007	.0065	.0267	.0668	.1201	.1685	.1927	.1841	.1484
	8	.0000	.0001	.0014	.0084	.0279	.0644	.1143	.1606	.1883	.1855
	9	.0000	.0000	.0003	.0021	.0093	.0276	.0611	.1070	.1540	.1855
	10	.0000	.0000	.0000	.0004	.0025	.0095	.0263	.0571	.1008	.1484
	11	.0000	.0000	.0000	.0001	.0005	.0026	.0090	.0242	.0525	.0944
	12	.0000	.0000	.0000	.0000	.0001	.0006	.0024	.0081	.0215	.0472
	13	.0000	.0000	.0000	.0000	.0000	.0001	.0005	.0021	.0068	.0182
	14	.0000	.0000	.0000	.0000	.0000	.0000	.0001	.0004	.0016	.0052
	15	.0000	.0000	.0000	.0000	.0000	.0000	.0000	.0001	.0003	.0010
	16	.0000	.0000	.0000	.0000	.0000	.0000	.0000	.0000	.0000	.0001
	17	.0000	.0000	.0000	.0000	.0000	.0000	.0000	.0000	.0000	.0000
18	**0**	.3972	.1501	.0536	.0180	.0056	.0016	.0004	.0001	.0000	.0000
	1	.3763	.3002	.1704	.0811	.0338	.0126	.0042	.0012	.0003	.0001
	2	.1683	.2835	.2556	.1723	.0958	.0458	.0190	.0069	.0022	.0006
	3	.0473	.1680	.2406	.2297	.1704	.1046	.0547	.0246	.0095	.0031
	4	.0093	.0700	.1592	.2153	.2130	.1681	.1104	.0614	.0291	.0117
	5	.0014	.0218	.0787	.1507	.1988	.2017	.1664	.1146	.0666	.0327
	6	.0002	.0052	.0310	.0816	.1436	.1873	.1941	.1655	.1181	.0708
	7	.0000	.0010	.0091	.0350	.0820	.1376	.1792	.1892	.1657	.1214
	8	.0000	.0002	.0022	.0120	.0376	.0811	.1327	.1734	.1864	.1669
	9	.0000	.0000	.0004	.0033	.0139	.0386	.0794	.1284	.1694	.1855
18	**10**	.0000	.0000	.0001	.0008	.0042	.0149	.0385	.0771	.1248	.1669
	11	.0000	.0000	.0000	.0001	.0010	.0046	.0151	.0374	.0742	.1214
	12	.0000	.0000	.0000	.0000	.0002	.0012	.0047	.0145	.0354	.0708
	13	.0000	.0000	.0000	.0000	.0000	.0002	.0012	.0045	.0134	.0327
	14	.0000	.0000	.0000	.0000	.0000	.0000	.0002	.0011	.0039	.0117

N	r	.05	.10	.15	.20	.25	.30	.35	.40	.45	.50
	15	.0000	.0000	.0000	.0000	.0000	.0000	.0000	.0002	.0009	.0031
	16	.0000	.0000	.0000	.0000	.0000	.0000	.0000	.0000	.0001	.0006
	17	.0000	.0000	.0000	.0000	.0000	.0000	.0000	.0000	.0000	.0001
	18	.0000	.0000	.0000	.0000	.0000	.0000	.0000	.0000	.0000	.0000
19	0	.3774	.1351	.0456	.0144	.0042	.0011	.0003	.0001	.0000	.0000
	1	.3774	.2852	.1529	.0685	.0268	.0093	.0029	.0008	.0002	.0000
	2	.1787	.2852	.2428	.1540	.0803	.0358	.0138	.0046	.0013	.0003
	3	.0533	.1796	.2428	.2182	.1517	.0869	.0422	.0175	.0062	.0018
	4	.0112	.0798	.1714	.2182	.2023	.1491	.0909	.0467	.0203	.0074
	5	.0018	.0266	.0907	.1636	.2023	.1916	.1468	.0933	.0497	.0222
	6	.0002	.0069	.0374	.0955	.1574	.1916	.1844	.1451	.0949	.0518
	7	.0000	.0014	.0122	.0443	.0974	.1525	.1844	.1797	.1443	.0961
	8	.0000	.0002	.0032	.0166	.0487	.0981	.1489	.1797	.1771	.1442
	9	.0000	.0000	.0007	.0051	.0198	.0514	.0980	.1464	.1771	.1762
	10	.0000	.0000	.0001	.0013	.0066	.0220	.0528	.0976	.1449	.1762
	11	.0000	.0000	.0000	.0003	.0018	.0077	.0233	.0532	.0970	.1442
	12	.0000	.0000	.0000	.0000	.0004	.0022	.0083	.0237	.0529	.0961
	13	.0000	.0000	.0000	.0000	.0001	.0005	.0024	.0085	.0233	.0518
	14	.0000	.0000	.0000	.0000	.0000	.0001	.0006	.0024	.0082	.0222
	15	.0000	.0000	.0000	.0000	.0000	.0000	.0001	.0005	.0022	.0074
	16	.0000	.0000	.0000	.0000	.0000	.0000	.0000	.0001	.0005	.0018
	17	.0000	.0000	.0000	.0000	.0000	.0000	.0000	.0000	.0001	.0003
	18	.0000	.0000	.0000	.0000	.0000	.0000	.0000	.0000	.0000	.0000
	19	.0000	.0000	.0000	.0000	.0000	.0000	.0000	.0000	.0000	.0000
20	0	.3585	.1216	.0388	.0115	.0032	.0008	.0002	.0000	.0000	.0000
	1	.3774	.2702	.1368	.0576	.0211	.0068	.0020	.0005	.0001	.0000
	2	.1887	.2852	.2293	.1369	.0669	.0278	.0100	.0031	.0008	.0002
	3	.0596	.1901	.2428	.2054	.1339	.0716	.0323	.0123	.0040	.0011
	4	.0133	.0898	.1821	.2182	.1897	.1304	.0738	.0350	.0139	.0046
	5	.0022	.0319	.1028	.1746	.2023	.1789	.1272	.0746	.0365	.0148
	6	.0003	.0089	.0454	.1091	.1686	.1916	.1712	.1244	.0746	.0370
	7	.0000	.0020	.0160	.0545	.1124	.1643	.1844	.1659	.1221	.0739
	8	.0000	.0004	.0046	.0222	.0609	.1144	.1614	.1797	.1623	.1201
	9	.0000	.0001	.0011	.0074	.0271	.0654	.1158	.1597	.1771	.1602
	10	.0000	.0000	.0002	.0020	.0099	.0308	.0686	.1171	.1593	.1762
	11	.0000	.0000	.0000	.0005	.0030	.0120	.0336	.0710	.1185	.1602
	12	.0000	.0000	.0000	.0001	.0008	.0039	.0136	.0355	.0727	.1201
	13	.0000	.0000	.0000	.0000	.0002	.0010	.0045	.0146	.0366	.0739
	14	.0000	.0000	.0000	.0000	.0000	.0002	.0012	.0049	.0150	.0370
	15	.0000	.0000	.0000	.0000	.0000	.0000	.0003	.0013	.0049	.0148
	16	.0000	.0000	.0000	.0000	.0000	.0000	.0000	.0003	.0013	.0046
	17	.0000	.0000	.0000	.0000	.0000	.0000	.0000	.0000	.0002	.0011
	18	.0000	.0000	.0000	.0000	.0000	.0000	.0000	.0000	.0000	.0002
	19	.0000	.0000	.0000	.0000	.0000	.0000	.0000	.0000	.0000	.0000
	20	.0000	.0000	.0000	.0000	.0000	.0000	.0000	.0000	.0000	.0000

(Column heading over .05–.50: *p*)

This table is reproduced by permission from R. S. Burington and D. C. May, *Handbook of Probability and Statistics with Tables*. McGraw-Hill Book Company, (ed. 2), 1970.

Table III
UPPER PERCENTAGE POINTS OF THE t DISTRIBUTION

ν	Q = 0.4 $2Q$ = 0.8	0.25 0.5	0.1 0.2	0.05 0.1	0.025 0.05	0.01 0.02	0.005 0.01	0.001 0.002
1	0.325	1.000	3.078	6.314	12.706	31.821	63.657	318.31
2	.289	0.816	1.886	2.920	4.303	6.965	9.925	22.326
3	.277	.765	1.638	2.353	3.182	4.541	5.841	10.213
4	.271	.741	1.533	2.132	2.776	3.747	4.604	7.173
5	0.267	0.727	1.476	2.015	2.571	3.365	4.032	5.893
6	.265	.718	1.440	1.943	2.447	3.143	3.707	5.208
7	.263	.711	1.415	1.895	2.365	2.998	3.499	4.785
8	.262	.706	1.397	1.860	2.306	2.896	3.355	4.501
9	.261	.703	1.383	1.833	2.262	2.821	3.250	4.297
10	0.260	0.700	1.372	1.812	2.228	2.764	3.169	4.144
11	.260	.697	1.363	1.796	2.201	2.718	3.106	4.025
12	.259	.695	1.356	1.782	2.179	2.681	3.055	3.930
13	.259	.694	1.350	1.771	2.160	2.650	3.012	3.852
14	.258	.692	1.345	1.761	2.145	2.624	2.977	3.787
15	0.258	0.691	1.341	1.753	2.131	2.602	2.947	3.733
16	.258	.690	1.337	1.746	2.120	2.583	2.921	3.686
17	.257	.689	1.333	1.740	2.110	2.567	2.898	3.646
18	.257	.688	1.330	1.734	2.101	2.552	2.878	3.610
19	.257	.688	1.328	1.729	2.093	2.539	2.861	3.579
20	0.257	0.687	1.325	1.725	2.086	2.528	2.845	3.552
21	.257	.686	1.323	1.721	2.080	2.518	2.831	3.527
22	.256	.686	1.321	1.717	2.074	2.508	2.819	3.505
23	.256	.685	1.319	1.714	2.069	2.500	2.807	3.485
24	.256	.685	1.318	1.711	2.064	2.492	2.797	3.467
25	0.256	0.684	1.316	1.708	2.060	2.485	2.787	3.450
26	.256	.684	1.315	1.706	2.056	2.479	2.779	3.435
27	.256	.684	1.314	1.703	2.052	2.473	2.771	3.421
28	.256	.683	1.313	1.701	2.048	2.467	2.763	3.408
29	.256	.683	1.311	1.699	2.045	2.462	2.756	3.396
30	0.256	0.683	1.310	1.697	2.042	2.457	2.750	3.385
40	.255	.681	1.303	1.684	2.021	2.423	2.704	3.307
60	.254	.679	1.296	1.671	2.000	2.390	2.660	3.232
120	.254	.677	1.289	1.658	1.980	2.358	2.617	3.160
∞	.253	.674	1 282	1.645	1.960	2.326	2.576	3.090

This table is condensed from Table 12 of the *Biometrika Tables for Statisticians*, Vol. 1 (ed. 3), edited by E. S. Pearson and H. O. Hartley. Reproduced here with the kind permission of E. S. Pearson and the trustees of *Biometrika*.

Table IV

UPPER PERCENTAGE POINTS OF THE χ^2 DISTRIBUTION

ν \ Q	0.995	0.990	0.975	0.950	0.900	0.750	0.500
1	392704.10^{-10}	157088.10^{-9}	982069.10^{-9}	393214.10^{-8}	0.0157908	0.1015308	0.454937
2	0.0100251	0.0201007	0.0506356	0.102587	0.210720	0.575364	1.38629
3	0.0717212	0.114832	0.215795	0.351846	0.584375	1.212534	2.36597
4	0.206990	0.297110	0.484419	0.710721	1.063623	1.92255	3.35670
5	0.411740	0.554300	0.831211	1.145476	1.61031	2.67460	4.35146
6	0.675727	0.872085	1.237347	1.63539	2.20413	3.45460	5.34812
7	0.989265	1.239043	1.68987	2.16735	2.83311	4.25485	6.34581
8	1.344419	1.646482	2.17973	2.73264	3.48954	5.07064	7.34412
9	1.734926	2.087912	2.70039	3.32511	4.16816	5.89883	8.34283
10	2.15585	2.55821	3.24697	3.94030	4.86518	6.73720	9.34182
11	2.60321	3.05347	3.81575	4.57481	5.57779	7.58412	10.3410
12	3.07382	3.57056	4.40379	5.22603	6.30380	8.43842	11.3403
13	3.56503	4.10691	5.00874	5.89186	7.04150	9.29906	12.3398
14	4.07468	4.66043	5.62872	6.57063	7.78953	10.1653	13.3393
15	4.60094	5.22935	6.26214	7.26094	8.54675	11.0365	14.3389
16	5.14224	5.81221	6.90766	7.96164	9.31223	11.9122	15.3385
17	5.69724	6.40776	7.56418	8.67176	10.0852	12.7919	16.3381
18	6.26481	7.01491	8.23075	9.39046	10.8649	13.6753	17.3379
19	6.84398	7.63273	8.90655	10.1170	11.6509	14.5620	18.3376
20	7.43386	8.26040	9.59083	10.8508	12.4426	15.4518	19.3374
21	8.03366	8.89720	10.28293	11.5913	13.2396	16.3444	20.3372
22	8.64272	9.54249	10.9823	12.3380	14.0415	17.2396	21.3370
23	9.26042	10.19567	11.6885	13.0905	14.8479	18.1373	22.3369
24	9.88623	10.8564	12.4011	13.8484	15.6587	19.0372	23.3367
25	10.5197	11.5240	13.1197	14.6114	16.4734	19.9393	24.3366
26	11.1603	12.1981	13.8439	15.3791	17.2919	20.8434	25.3364
27	11.8076	12.8786	14.5733	16.1513	18.1138	21.7494	26.3363
28	12.4613	13.5648	15.3079	16.9279	18.9392	22.6572	27.3363
29	13.1211	14.2565	16.0471	17.7083	19.7677	23.5666	28.3362
30	13.7867	14.9535	16.7908	18.4926	20.5992	24.4776	29.3360
40	20.7065	22.1643	24.4331	26.5093	29.0505	33.6603	39.3354
50	27.9907	29.7067	32.3574	34.7642	37.6886	42.9421	49.3349
60	35.5346	37.4848	40.4817	43.1879	46.4589	52.2938	59.3347
70	43.2752	45.4418	48.7576	51.7393	55.3290	61.6983	69.3344
80	51.1720	53.5400	57.1532	60.3915	64.2778	71.1445	79.3343
90	59.1963	61.7541	65.6466	69.1260	73.2912	80.6247	89.3342
100	67.3276	70.0648	74.2219	77.9295	82.3581	90.1332	99.3341
z_Q	−2.5758	−2.3263	−1.9600	−1.6449	−1.2816	−0.6745	0.0000

Table IV (continued)

ν \ Q	0.250	0.100	0.050	0.025	0.010	0.005	0.001
1	1.32330	2.70554	3.84146	5.02389	6.63490	7.87944	10.828
2	2.77259	4.60517	5.99147	7.37776	9.21034	10.5966	13.816
3	4.10835	6.25139	7.81473	9.34840	11.3449	12.8381	16.266
4	5.38527	7.77944	9.48773	11.1433	13.2767	14.8602	18.467
5	6.62568	9.23635	11.0705	12.8325	15.0863	16.7496	20.515
6	7.84080	10.6446	12.5916	14.4494	16.8119	18.5476	22.458
7	9.03715	12.0170	14.0671	16.0128	18.4753	20.2777	24.322
8	10.2188	13.3616	15.5073	17.5346	20.0902	21.9550	26.125
9	11.3887	14.6837	16.9190	19.0228	21.6660	23.5893	27.877
10	12.5489	15.9871	18.3070	20.4831	23.2093	25.1882	29.588
11	13.7007	17.2750	19.6751	21.9200	24.7250	26.7569	31.264
12	14.8454	18.5494	21.0261	23.3367	26.2170	28.2995	32.909
13	15.9839	19.8119	22.3621	24.7356	27.6883	29.8194	34.528
14	17.1170	21.0642	23.6848	26.1190	29.1413	31.3193	36.123
15	18.2451	22.3072	24.9958	27.4884	30.5779	32.8013	37.697
16	19.3688	23.5418	26.2962	28.8454	31.9999	34.2672	39.252
17	20.4887	24.7690	27.5871	30.1910	33.4087	35.7185	40.790
18	21.6049	25.9894	28.8693	31.5264	34.8053	37.1564	42.312
19	22.7178	27.2036	30.1435	32.8523	36.1908	38.5822	43.820
20	23.8277	28.4120	31.4104	34.1696	37.5662	39.9968	45.315
21	24.9348	29.6151	32.6705	35.4789	38.9321	41.4010	46.797
22	26.0393	30.8133	33.9244	36.7807	40.2894	42.7956	48.268
23	27.1413	32.0069	35.1725	38.0757	41.6384	44.1813	49.728
24	28.2412	33.1963	36.4151	39.3641	42.9798	45.5585	51.179
25	29.3389	34.3816	37.6525	40.6465	44.3141	46.9278	52.620
26	30.4345	35.5631	38.8852	41.9232	45.6417	48.2899	54.052
27	31.5284	36.7412	40.1133	43.1944	46.9630	49.6449	55.476
28	32.6205	37.9159	41.3372	44.4607	48.2782	50.9933	56.892
29	33.7109	39.0875	42.5569	45.7222	49.5879	52.3356	58.302
30	34.7998	40.2560	43.7729	46.9792	50.8922	53.6720	59.703
40	45.6160	51.8050	55.7585	59.3417	63.6907	66.7659	73.402
50	56.3336	63.1671	67.5048	71.4202	76.1539	79.4900	86.661
60	66.9814	74.3970	79.0819	83.2976	88.3794	91.9517	99.607
70	77.5766	85.5271	90.5312	95.0231	100.425	104.215	112.317
80	88.1303	96.5782	101.879	106.629	112.329	116.321	124.839
90	98.6499	107.565	113.145	118.136	124.116	128.299	137.208
100	109.141	118.498	124.342	129.561	135.807	140.169	149.449
z_Q	+0.6745	+1.2816	+1.6449	+1.9600	+2.3263	+2.5758	+3.0902

This table is taken from Table 8 of the *Biometrika Tables for Statisticians*, Vol. 1 (ed. 3), edited by E. S. Pearson and H. O. Hartley. Reproduced here with the kind permission of E. S. Pearson and the trustees of *Biometrika*.

Table V

PERCENTAGE POINTS OF THE F DISTRIBUTION, UPPER 25% POINTS

ν_2 \ ν_1	1	2	3	4	5	6	7	8	9	10	12	15	20	24	30	40	60	120	∞
1	5·83	7·50	8·20	8·58	8·82	8·98	9·10	9·19	9·26	9·32	9·41	9·49	9·58	9·63	9·67	9·71	9·76	9·80	9·85
2	2·57	3·00	3·15	3·23	3·28	3·31	3·34	3·35	3·37	3·38	3·39	3·41	3·43	3·43	3·44	3·45	3·46	3·47	3·48
3	2·02	2·28	2·36	2·39	2·41	2·42	2·43	2·44	2·44	2·44	2·45	2·46	2·46	2·46	2·47	2·47	2·47	2·47	2·47
4	1·81	2·00	2·05	2·06	2·07	2·08	2·08	2·08	2·08	2·08	2·08	2·08	2·08	2·08	2·08	2·08	2·08	2·08	2·08
5	1·69	1·85	1·88	1·89	1·89	1·89	1·89	1·89	1·89	1·89	1·89	1·89	1·88	1·88	1·88	1·88	1·87	1·87	1·87
6	1·62	1·76	1·78	1·79	1·79	1·78	1·78	1·78	1·77	1·77	1·77	1·76	1·76	1·75	1·75	1·75	1·74	1·74	1·74
7	1·57	1·70	1·72	1·72	1·71	1·71	1·70	1·70	1·69	1·69	1·68	1·68	1·67	1·67	1·66	1·66	1·65	1·65	1·65
8	1·54	1·66	1·67	1·66	1·66	1·65	1·64	1·64	1·63	1·63	1·62	1·62	1·61	1·60	1·60	1·59	1·59	1·58	1·58
9	1·51	1·62	1·63	1·63	1·62	1·61	1·60	1·60	1·59	1·59	1·58	1·57	1·56	1·56	1·55	1·54	1·54	1·53	1·53
10	1·49	1·60	1·60	1·59	1·59	1·58	1·57	1·56	1·56	1·55	1·54	1·53	1·52	1·52	1·51	1·51	1·50	1·49	1·48
11	1·47	1·58	1·58	1·57	1·56	1·55	1·54	1·53	1·53	1·52	1·51	1·50	1·49	1·49	1·48	1·47	1·47	1·46	1·45
12	1·46	1·56	1·56	1·55	1·54	1·53	1·52	1·51	1·51	1·50	1·49	1·48	1·47	1·46	1·45	1·45	1·44	1·43	1·42
13	1·45	1·55	1·55	1·53	1·52	1·51	1·50	1·49	1·49	1·48	1·47	1·46	1·45	1·44	1·43	1·42	1·42	1·41	1·40
14	1·44	1·53	1·53	1·52	1·51	1·50	1·49	1·48	1·47	1·46	1·45	1·44	1·43	1·42	1·41	1·41	1·40	1·39	1·38
15	1·43	1·52	1·52	1·51	1·49	1·48	1·47	1·46	1·46	1·45	1·44	1·43	1·41	1·41	1·40	1·39	1·38	1·37	1·36
16	1·42	1·51	1·51	1·50	1·48	1·47	1·46	1·45	1·44	1·44	1·43	1·41	1·40	1·39	1·38	1·37	1·36	1·35	1·34
17	1·42	1·51	1·50	1·49	1·47	1·46	1·45	1·44	1·43	1·43	1·41	1·40	1·39	1·38	1·37	1·36	1·35	1·34	1·33
18	1·41	1·50	1·49	1·48	1·46	1·45	1·44	1·43	1·42	1·42	1·40	1·39	1·38	1·37	1·36	1·35	1·34	1·33	1·32
19	1·41	1·49	1·49	1·47	1·46	1·44	1·43	1·42	1·41	1·41	1·40	1·38	1·37	1·36	1·35	1·34	1·33	1·32	1·30
20	1·40	1·49	1·48	1·47	1·45	1·44	1·43	1·42	1·41	1·40	1·39	1·37	1·36	1·35	1·34	1·33	1·32	1·31	1·29
21	1·40	1·48	1·48	1·46	1·44	1·43	1·42	1·41	1·40	1·39	1·38	1·37	1·35	1·34	1·33	1·32	1·31	1·30	1·28
22	1·40	1·48	1·47	1·45	1·44	1·42	1·41	1·40	1·39	1·39	1·37	1·36	1·34	1·33	1·32	1·31	1·30	1·29	1·28
23	1·39	1·47	1·47	1·45	1·43	1·42	1·41	1·40	1·39	1·38	1·37	1·35	1·34	1·33	1·32	1·31	1·30	1·28	1·27
24	1·39	1·47	1·46	1·44	1·43	1·41	1·40	1·39	1·38	1·38	1·36	1·35	1·33	1·32	1·31	1·30	1·29	1·28	1·26
25	1·39	1·47	1·46	1·44	1·42	1·41	1·40	1·39	1·38	1·37	1·36	1·34	1·33	1·32	1·31	1·29	1·28	1·27	1·25
26	1·38	1·46	1·45	1·44	1·42	1·41	1·39	1·38	1·37	1·37	1·35	1·34	1·32	1·31	1·30	1·29	1·28	1·26	1·25
27	1·38	1·46	1·45	1·43	1·42	1·40	1·39	1·38	1·37	1·36	1·35	1·33	1·32	1·31	1·30	1·28	1·27	1·26	1·24
28	1·38	1·46	1·45	1·43	1·41	1·40	1·39	1·38	1·37	1·36	1·34	1·33	1·31	1·30	1·29	1·28	1·27	1·25	1·24
29	1·38	1·45	1·45	1·43	1·41	1·40	1·38	1·37	1·36	1·35	1·34	1·32	1·31	1·30	1·29	1·27	1·26	1·25	1·23
30	1·38	1·45	1·44	1·42	1·41	1·39	1·38	1·37	1·36	1·35	1·34	1·32	1·30	1·29	1·28	1·27	1·26	1·24	1·23
40	1·36	1·44	1·42	1·40	1·39	1·37	1·36	1·35	1·34	1·33	1·31	1·30	1·28	1·26	1·25	1·24	1·22	1·21	1·19
60	1·35	1·42	1·41	1·38	1·37	1·35	1·33	1·32	1·31	1·30	1·29	1·27	1·25	1·24	1·22	1·21	1·19	1·17	1·15
120	1·34	1·40	1·39	1·37	1·35	1·33	1·31	1·30	1·29	1·28	1·26	1·24	1·22	1·21	1·19	1·18	1·16	1·13	1·10
∞	1·32	1·39	1·37	1·35	1·33	1·31	1·29	1·28	1·27	1·25	1·24	1·22	1·19	1·18	1·16	1·14	1·12	1·08	1·00

Table V (continued)

UPPER 10% POINTS

ν_2 \ ν_1	1	2	3	4	5	6	7	8	9	10	12	15	20	24	30	40	60	120	∞
1	39·86	49·50	53·59	55·83	57·24	58·20	58·91	59·44	59·86	60·19	60·71	61·22	61·74	62·00	62·26	62·53	62·79	63·06	63·33
2	8·53	9·00	9·16	9·24	9·29	9·33	9·35	9·37	9·38	9·39	9·41	9·42	9·44	9·45	9·46	9·47	9·47	9·48	9·49
3	5·54	5·46	5·39	5·34	5·31	5·28	5·27	5·25	5·24	5·23	5·22	5·20	5·18	5·18	5·17	5·16	5·15	5·14	5·13
4	4·54	4·32	4·19	4·11	4·05	4·01	3·98	3·95	3·94	3·92	3·90	3·87	3·84	3·83	3·82	3·80	3·79	3·78	3·76
5	4·06	3·78	3·62	3·52	3·45	3·40	3·37	3·34	3·32	3·30	3·27	3·24	3·21	3·19	3·17	3·16	3·14	3·12	3·10
6	3·78	3·46	3·29	3·18	3·11	3·05	3·01	2·98	2·96	2·94	2·90	2·87	2·84	2·82	2·80	2·78	2·76	2·74	2·72
7	3·59	3·26	3·07	2·96	2·88	2·83	2·78	2·75	2·72	2·70	2·67	2·63	2·59	2·58	2·56	2·54	2·51	2·49	2·47
8	3·46	3·11	2·92	2·81	2·73	2·67	2·62	2·59	2·56	2·54	2·50	2·46	2·42	2·40	2·38	2·36	2·34	2·32	2·29
9	3·36	3·01	2·81	2·69	2·61	2·55	2·51	2·47	2·44	2·42	2·38	2·34	2·30	2·28	2·25	2·23	2·21	2·18	2·16
10	3·29	2·92	2·73	2·61	2·52	2·46	2·41	2·38	2·35	2·32	2·28	2·24	2·20	2·18	2·16	2·13	2·11	2·08	2·06
11	3·23	2·86	2·66	2·54	2·45	2·39	2·34	2·30	2·27	2·25	2·21	2·17	2·12	2·10	2·08	2·05	2·03	2·00	1·97
12	3·18	2·81	2·61	2·48	2·39	2·33	2·28	2·24	2·21	2·19	2·15	2·10	2·06	2·04	2·01	1·99	1·96	1·93	1·90
13	3·14	2·76	2·56	2·43	2·35	2·28	2·23	2·20	2·16	2·14	2·10	2·05	2·01	1·98	1·96	1·93	1·90	1·88	1·85
14	3·10	2·73	2·52	2·39	2·31	2·24	2·19	2·15	2·12	2·10	2·05	2·01	1·96	1·94	1·91	1·89	1·86	1·83	1·80
15	3·07	2·70	2·49	2·36	2·27	2·21	2·16	2·12	2·09	2·06	2·02	1·97	1·92	1·90	1·87	1·85	1·82	1·79	1·76
16	3·05	2·67	2·46	2·33	2·24	2·18	2·13	2·09	2·06	2·03	1·99	1·94	1·89	1·87	1·84	1·81	1·78	1·75	1·72
17	3·03	2·64	2·44	2·31	2·22	2·15	2·10	2·06	2·03	2·00	1·96	1·91	1·86	1·84	1·81	1·78	1·75	1·72	1·69
18	3·01	2·62	2·42	2·29	2·20	2·13	2·08	2·04	2·00	1·98	1·93	1·89	1·84	1·81	1·78	1·75	1·72	1·69	1·66
19	2·99	2·61	2·40	2·27	2·18	2·11	2·06	2·02	1·98	1·96	1·91	1·86	1·81	1·79	1·76	1·73	1·70	1·67	1·63
20	2·97	2·59	2·38	2·25	2·16	2·09	2·04	2·00	1·96	1·94	1·89	1·84	1·79	1·77	1·74	1·71	1·68	1·64	1·61
21	2·96	2·57	2·36	2·23	2·14	2·08	2·02	1·98	1·95	1·92	1·87	1·83	1·78	1·75	1·72	1·69	1·66	1·62	1·59
22	2·95	2·56	2·35	2·22	2·13	2·06	2·01	1·97	1·93	1·90	1·86	1·81	1·76	1·73	1·70	1·67	1·64	1·60	1·57
23	2·94	2·55	2·34	2·21	2·11	2·05	1·99	1·95	1·92	1·89	1·84	1·80	1·74	1·72	1·69	1·66	1·62	1·59	1·55
24	2·93	2·54	2·33	2·19	2·10	2·04	1·98	1·94	1·91	1·88	1·83	1·78	1·73	1·70	1·67	1·64	1·61	1·57	1·53
25	2·92	2·53	2·32	2·18	2·09	2·02	1·97	1·93	1·89	1·87	1·82	1·77	1·72	1·69	1·66	1·63	1·59	1·56	1·52
26	2·91	2·52	2·31	2·17	2·08	2·01	1·96	1·92	1·88	1·86	1·81	1·76	1·71	1·68	1·65	1·61	1·58	1·54	1·50
27	2·90	2·51	2·30	2·17	2·07	2·00	1·95	1·91	1·87	1·85	1·80	1·75	1·70	1·67	1·64	1·60	1·57	1·53	1·49
28	2·89	2·50	2·29	2·16	2·06	2·00	1·94	1·90	1·87	1·84	1·79	1·74	1·69	1·66	1·63	1·59	1·56	1·52	1·48
29	2·89	2·50	2·28	2·15	2·06	1·99	1·93	1·89	1·86	1·83	1·78	1·73	1·68	1·65	1·62	1·58	1·55	1·51	1·47
30	2·88	2·49	2·28	2·14	2·05	1·98	1·93	1·88	1·85	1·82	1·77	1·72	1·67	1·64	1·61	1·57	1·54	1·50	1·46
40	2·84	2·44	2·23	2·09	2·00	1·93	1·87	1·83	1·79	1·76	1·71	1·66	1·61	1·57	1·54	1·51	1·47	1·42	1·38
60	2·79	2·39	2·18	2·04	1·95	1·87	1·82	1·77	1·74	1·71	1·66	1·60	1·54	1·51	1·48	1·44	1·40	1·35	1·29
120	2·75	2·35	2·13	1·99	1·90	1·82	1·77	1·72	1·68	1·65	1·60	1·55	1·48	1·45	1·41	1·37	1·32	1·26	1·19
∞	2·71	2·30	2·08	1·94	1·85	1·77	1·72	1·67	1·63	1·60	1·55	1·49	1·42	1·38	1·34	1·30	1·24	1·17	1·00

Table V (continued)

UPPER 5% POINTS

ν_2 \ ν_1	1	2	3	4	5	6	7	8	9	10	12	15	20	24	30	40	60	120	∞
1	161.4	199.5	215.7	224.6	230.2	234.0	236.8	238.9	240.5	241.9	243.9	245.9	248.0	249.1	250.1	251.1	252.2	253.3	254.3
2	18.51	19.00	19.16	19.25	19.30	19.33	19.35	19.37	19.38	19.40	19.41	19.43	19.45	19.45	19.46	19.47	19.48	19.49	19.50
3	10.13	9.55	9.28	9.12	9.01	8.94	8.89	8.85	8.81	8.79	8.74	8.70	8.66	8.64	8.62	8.59	8.57	8.55	8.53
4	7.71	6.94	6.59	6.39	6.26	6.16	6.09	6.04	6.00	5.96	5.91	5.86	5.80	5.77	5.75	5.72	5.69	5.66	5.63
5	6.61	5.79	5.41	5.19	5.05	4.95	4.88	4.82	4.77	4.74	4.68	4.62	4.56	4.53	4.50	4.46	4.43	4.40	4.36
6	5.99	5.14	4.76	4.53	4.39	4.28	4.21	4.15	4.10	4.06	4.00	3.94	3.87	3.84	3.81	3.77	3.74	3.70	3.67
7	5.59	4.74	4.35	4.12	3.97	3.87	3.79	3.73	3.68	3.64	3.57	3.51	3.44	3.41	3.38	3.34	3.30	3.27	3.23
8	5.32	4.46	4.07	3.84	3.69	3.58	3.50	3.44	3.39	3.35	3.28	3.22	3.15	3.12	3.08	3.04	3.01	2.97	2.93
9	5.12	4.26	3.86	3.63	3.48	3.37	3.29	3.23	3.18	3.14	3.07	3.01	2.94	2.90	2.86	2.83	2.79	2.75	2.71
10	4.96	4.10	3.71	3.48	3.33	3.22	3.14	3.07	3.02	2.98	2.91	2.85	2.77	2.74	2.70	2.66	2.62	2.58	2.54
11	4.84	3.98	3.59	3.36	3.20	3.09	3.01	2.95	2.90	2.85	2.79	2.72	2.65	2.61	2.57	2.53	2.49	2.45	2.40
12	4.75	3.89	3.49	3.26	3.11	3.00	2.91	2.85	2.80	2.75	2.69	2.62	2.54	2.51	2.47	2.43	2.38	2.34	2.30
13	4.67	3.81	3.41	3.18	3.03	2.92	2.83	2.77	2.71	2.67	2.60	2.53	2.46	2.42	2.38	2.34	2.30	2.25	2.21
14	4.60	3.74	3.34	3.11	2.96	2.85	2.76	2.70	2.65	2.60	2.53	2.46	2.39	2.35	2.31	2.27	2.22	2.18	2.13
15	4.54	3.68	3.29	3.06	2.90	2.79	2.71	2.64	2.59	2.54	2.48	2.40	2.33	2.29	2.25	2.20	2.16	2.11	2.07
16	4.49	3.63	3.24	3.01	2.85	2.74	2.66	2.59	2.54	2.49	2.42	2.35	2.28	2.24	2.19	2.15	2.11	2.06	2.01
17	4.45	3.59	3.20	2.96	2.81	2.70	2.61	2.55	2.49	2.45	2.38	2.31	2.23	2.19	2.15	2.10	2.06	2.01	1.96
18	4.41	3.55	3.16	2.93	2.77	2.66	2.58	2.51	2.46	2.41	2.34	2.27	2.19	2.15	2.11	2.06	2.02	1.97	1.92
19	4.38	3.52	3.13	2.90	2.74	2.63	2.54	2.48	2.42	2.38	2.31	2.23	2.16	2.11	2.07	2.03	1.98	1.93	1.88
20	4.35	3.49	3.10	2.87	2.71	2.60	2.51	2.45	2.39	2.35	2.28	2.20	2.12	2.08	2.04	1.99	1.95	1.90	1.84
21	4.32	3.47	3.07	2.84	2.68	2.57	2.49	2.42	2.37	2.32	2.25	2.18	2.10	2.05	2.01	1.96	1.92	1.87	1.81
22	4.30	3.44	3.05	2.82	2.66	2.55	2.46	2.40	2.34	2.30	2.23	2.15	2.07	2.03	1.98	1.94	1.89	1.84	1.78
23	4.28	3.42	3.03	2.80	2.64	2.53	2.44	2.37	2.32	2.27	2.20	2.13	2.05	2.01	1.96	1.91	1.86	1.81	1.76
24	4.26	3.40	3.01	2.78	2.62	2.51	2.42	2.36	2.30	2.25	2.18	2.11	2.03	1.98	1.94	1.89	1.84	1.79	1.73
25	4.24	3.39	2.99	2.76	2.60	2.49	2.40	2.34	2.28	2.24	2.16	2.09	2.01	1.96	1.92	1.87	1.82	1.77	1.71
26	4.23	3.37	2.98	2.74	2.59	2.47	2.39	2.32	2.27	2.22	2.15	2.07	1.99	1.95	1.90	1.85	1.80	1.75	1.69
27	4.21	3.35	2.96	2.73	2.57	2.46	2.37	2.31	2.25	2.20	2.13	2.06	1.97	1.93	1.88	1.84	1.79	1.73	1.67
28	4.20	3.34	2.95	2.71	2.56	2.45	2.36	2.29	2.24	2.19	2.12	2.04	1.96	1.91	1.87	1.82	1.77	1.71	1.65
29	4.18	3.33	2.93	2.70	2.55	2.43	2.35	2.28	2.22	2.18	2.10	2.03	1.94	1.90	1.85	1.81	1.75	1.70	1.64
30	4.17	3.32	2.92	2.69	2.53	2.42	2.33	2.27	2.21	2.16	2.09	2.01	1.93	1.89	1.84	1.79	1.74	1.68	1.62
40	4.08	3.23	2.84	2.61	2.45	2.34	2.25	2.18	2.12	2.08	2.00	1.92	1.84	1.79	1.74	1.69	1.64	1.58	1.51
60	4.00	3.15	2.76	2.53	2.37	2.25	2.17	2.10	2.04	1.99	1.92	1.84	1.75	1.70	1.65	1.59	1.53	1.47	1.39
120	3.92	3.07	2.68	2.45	2.29	2.17	2.09	2.02	1.96	1.91	1.83	1.75	1.66	1.61	1.55	1.50	1.43	1.35	1.25
∞	3.84	3.00	2.60	2.37	2.21	2.10	2.01	1.94	1.88	1.83	1.75	1.67	1.57	1.52	1.46	1.39	1.32	1.22	1.00

Table V (continued)

UPPER 2.5% POINTS

ν_2 \ ν_1	1	2	3	4	5	6	7	8	9	10	12	15	20	24	30	40	60	120	∞
1	647.8	799.5	864.2	899.6	921.8	937.1	948.2	956.7	963.3	968.6	976.7	984.9	993.1	997.2	1001	1006	1010	1014	1018
2	38.51	39.00	39.17	39.25	39.30	39.33	39.36	39.37	39.39	39.40	39.41	39.43	39.45	39.46	39.46	39.47	39.48	39.49	39.50
3	17.44	16.04	15.44	15.10	14.88	14.73	14.62	14.54	14.47	14.42	14.34	14.25	14.17	14.12	14.08	14.04	13.99	13.95	13.90
4	12.22	10.65	9.98	9.60	9.36	9.20	9.07	8.98	8.90	8.84	8.75	8.66	8.56	8.51	8.46	8.41	8.36	8.31	8.26
5	10.01	8.43	7.76	7.39	7.15	6.98	6.85	6.76	6.68	6.62	6.52	6.43	6.33	6.28	6.23	6.18	6.12	6.07	6.02
6	8.81	7.26	6.60	6.23	5.99	5.82	5.70	5.60	5.52	5.46	5.37	5.27	5.17	5.12	5.07	5.01	4.96	4.90	4.85
7	8.07	6.54	5.89	5.52	5.29	5.12	4.99	4.90	4.82	4.76	4.67	4.57	4.47	4.42	4.36	4.31	4.25	4.20	4.14
8	7.57	6.06	5.42	5.05	4.82	4.65	4.53	4.43	4.36	4.30	4.20	4.10	4.00	3.95	3.89	3.84	3.78	3.73	3.67
9	7.21	5.71	5.08	4.72	4.48	4.32	4.20	4.10	4.03	3.96	3.87	3.77	3.67	3.61	3.56	3.51	3.45	3.39	3.33
10	6.94	5.46	4.83	4.47	4.24	4.07	3.95	3.85	3.78	3.72	3.62	3.52	3.42	3.37	3.31	3.26	3.20	3.14	3.08
11	6.72	5.26	4.63	4.28	4.04	3.88	3.76	3.66	3.59	3.53	3.43	3.33	3.23	3.17	3.12	3.06	3.00	2.94	2.88
12	6.55	5.10	4.47	4.12	3.89	3.73	3.61	3.51	3.44	3.37	3.28	3.18	3.07	3.02	2.96	2.91	2.85	2.79	2.72
13	6.41	4.97	4.35	4.00	3.77	3.60	3.48	3.39	3.31	3.25	3.15	3.05	2.95	2.89	2.84	2.78	2.72	2.66	2.60
14	6.30	4.86	4.24	3.89	3.66	3.50	3.38	3.29	3.21	3.15	3.05	2.95	2.84	2.79	2.73	2.67	2.61	2.55	2.49
15	6.20	4.77	4.15	3.80	3.58	3.41	3.29	3.20	3.12	3.06	2.96	2.86	2.76	2.70	2.64	2.59	2.52	2.46	2.40
16	6.12	4.69	4.08	3.73	3.50	3.34	3.22	3.12	3.05	2.99	2.89	2.79	2.68	2.63	2.57	2.51	2.45	2.38	2.32
17	6.04	4.62	4.01	3.66	3.44	3.28	3.16	3.06	2.98	2.92	2.82	2.72	2.62	2.56	2.50	2.44	2.38	2.32	2.25
18	5.98	4.56	3.95	3.61	3.38	3.22	3.10	3.01	2.93	2.87	2.77	2.67	2.56	2.50	2.44	2.38	2.32	2.26	2.19
19	5.92	4.51	3.90	3.56	3.33	3.17	3.05	2.96	2.88	2.82	2.72	2.62	2.51	2.45	2.39	2.33	2.27	2.20	2.13
20	5.87	4.46	3.86	3.51	3.29	3.13	3.01	2.91	2.84	2.77	2.68	2.57	2.46	2.41	2.35	2.29	2.22	2.16	2.09
21	5.83	4.42	3.82	3.48	3.25	3.09	2.97	2.87	2.80	2.73	2.64	2.53	2.42	2.37	2.31	2.25	2.18	2.11	2.04
22	5.79	4.38	3.78	3.44	3.22	3.05	2.93	2.84	2.76	2.70	2.60	2.50	2.39	2.33	2.27	2.21	2.14	2.08	2.00
23	5.75	4.35	3.75	3.41	3.18	3.02	2.90	2.81	2.73	2.67	2.57	2.47	2.36	2.30	2.24	2.18	2.11	2.04	1.97
24	5.72	4.32	3.72	3.38	3.15	2.99	2.87	2.78	2.70	2.64	2.54	2.44	2.33	2.27	2.21	2.15	2.08	2.01	1.94
25	5.69	4.29	3.69	3.35	3.13	2.97	2.85	2.75	2.68	2.61	2.51	2.41	2.30	2.24	2.18	2.12	2.05	1.98	1.91
26	5.66	4.27	3.67	3.33	3.10	2.94	2.82	2.73	2.65	2.59	2.49	2.39	2.28	2.22	2.16	2.09	2.03	1.95	1.88
27	5.63	4.24	3.65	3.31	3.08	2.92	2.80	2.71	2.63	2.57	2.47	2.36	2.25	2.19	2.13	2.07	2.00	1.93	1.85
28	5.61	4.22	3.63	3.29	3.06	2.90	2.78	2.69	2.61	2.55	2.45	2.34	2.23	2.17	2.11	2.05	1.98	1.91	1.83
29	5.59	4.20	3.61	3.27	3.04	2.88	2.76	2.67	2.59	2.53	2.43	2.32	2.21	2.15	2.09	2.03	1.96	1.89	1.81
30	5.57	4.18	3.59	3.25	3.03	2.87	2.75	2.65	2.57	2.51	2.41	2.31	2.20	2.14	2.07	2.01	1.94	1.87	1.79
40	5.42	4.05	3.46	3.13	2.90	2.74	2.62	2.53	2.45	2.39	2.29	2.18	2.07	2.01	1.94	1.88	1.80	1.72	1.64
60	5.29	3.93	3.34	3.01	2.79	2.63	2.51	2.41	2.33	2.27	2.17	2.06	1.94	1.88	1.82	1.74	1.67	1.58	1.48
120	5.15	3.80	3.23	2.89	2.67	2.52	2.39	2.30	2.22	2.16	2.05	1.94	1.82	1.76	1.69	1.61	1.53	1.43	1.31
∞	5.02	3.69	3.12	2.79	2.57	2.41	2.29	2.19	2.11	2.05	1.94	1.83	1.71	1.64	1.57	1.48	1.39	1.27	1.00

Table V (continued)

UPPER 1% POINTS

ν_2 \ ν_1	1	2	3	4	5	6	7	8	9	10	12	15	20	24	30	40	60	120	∞
1	4052	4999.5	5403	5625	5764	5859	5928	5982	6022	6056	6106	6157	6209	6235	6261	6287	6313	6339	6366
2	98.50	99.00	99.17	99.25	99.30	99.33	99.36	99.37	99.39	99.40	99.42	99.43	99.45	99.46	99.47	99.47	99.48	99.49	99.50
3	34.12	30.82	29.46	28.71	28.24	27.91	27.67	27.49	27.35	27.23	27.05	26.87	26.69	26.60	26.50	26.41	26.32	26.22	26.13
4	21.20	18.00	16.69	15.98	15.52	15.21	14.98	14.80	14.66	14.55	14.37	14.20	14.02	13.93	13.84	13.75	13.65	13.56	13.46
5	16.26	13.27	12.06	11.39	10.97	10.67	10.46	10.29	10.16	10.05	9.89	9.72	9.55	9.47	9.38	9.29	9.20	9.11	9.02
6	13.75	10.92	9.78	9.15	8.75	8.47	8.26	8.10	7.98	7.87	7.72	7.56	7.40	7.31	7.23	7.14	7.06	6.97	6.88
7	12.25	9.55	8.45	7.85	7.46	7.19	6.99	6.84	6.72	6.62	6.47	6.31	6.16	6.07	5.99	5.91	5.82	5.74	5.65
8	11.26	8.65	7.59	7.01	6.63	6.37	6.18	6.03	5.91	5.81	5.67	5.52	5.36	5.28	5.20	5.12	5.03	4.95	4.86
9	10.56	8.02	6.99	6.42	6.06	5.80	5.61	5.47	5.35	5.26	5.11	4.96	4.81	4.73	4.65	4.57	4.48	4.40	4.31
10	10.04	7.56	6.55	5.99	5.64	5.39	5.20	5.06	4.94	4.85	4.71	4.56	4.41	4.33	4.25	4.17	4.08	4.00	3.91
11	9.65	7.21	6.22	5.67	5.32	5.07	4.89	4.74	4.63	4.54	4.40	4.25	4.10	4.02	3.94	3.86	3.78	3.69	3.60
12	9.33	6.93	5.95	5.41	5.06	4.82	4.64	4.50	4.39	4.30	4.16	4.01	3.86	3.78	3.70	3.62	3.54	3.45	3.36
13	9.07	6.70	5.74	5.21	4.86	4.62	4.44	4.30	4.19	4.10	3.96	3.82	3.66	3.59	3.51	3.43	3.34	3.25	3.17
14	8.86	6.51	5.56	5.04	4.69	4.46	4.28	4.14	4.03	3.94	3.80	3.66	3.51	3.43	3.35	3.27	3.18	3.09	3.00
15	8.68	6.36	5.42	4.89	4.56	4.32	4.14	4.00	3.89	3.80	3.67	3.52	3.37	3.29	3.21	3.13	3.05	2.96	2.87
16	8.53	6.23	5.29	4.77	4.44	4.20	4.03	3.89	3.78	3.69	3.55	3.41	3.26	3.18	3.10	3.02	2.93	2.84	2.75
17	8.40	6.11	5.18	4.67	4.34	4.10	3.93	3.79	3.68	3.59	3.46	3.31	3.16	3.08	3.00	2.92	2.83	2.75	2.65
18	8.29	6.01	5.09	4.58	4.25	4.01	3.84	3.71	3.60	3.51	3.37	3.23	3.08	3.00	2.92	2.84	2.75	2.66	2.57
19	8.18	5.93	5.01	4.50	4.17	3.94	3.77	3.63	3.52	3.43	3.30	3.15	3.00	2.92	2.84	2.76	2.67	2.58	2.49
20	8.10	5.85	4.94	4.43	4.10	3.87	3.70	3.56	3.46	3.37	3.23	3.09	2.94	2.86	2.78	2.69	2.61	2.52	2.42
21	8.02	5.78	4.87	4.37	4.04	3.81	3.64	3.51	3.40	3.31	3.17	3.03	2.88	2.80	2.72	2.64	2.55	2.46	2.36
22	7.95	5.72	4.82	4.31	3.99	3.76	3.59	3.45	3.35	3.26	3.12	2.98	2.83	2.75	2.67	2.58	2.50	2.40	2.31
23	7.88	5.66	4.76	4.26	3.94	3.71	3.54	3.41	3.30	3.21	3.07	2.93	2.78	2.70	2.62	2.54	2.45	2.35	2.26
24	7.82	5.61	4.72	4.22	3.90	3.67	3.50	3.36	3.26	3.17	3.03	2.89	2.74	2.66	2.58	2.49	2.40	2.31	2.21
25	7.77	5.57	4.68	4.18	3.85	3.63	3.46	3.32	3.22	3.13	2.99	2.85	2.70	2.62	2.54	2.45	2.36	2.27	2.17
26	7.72	5.53	4.64	4.14	3.82	3.59	3.42	3.29	3.18	3.09	2.96	2.81	2.66	2.58	2.50	2.42	2.33	2.23	2.13
27	7.68	5.49	4.60	4.11	3.78	3.56	3.39	3.26	3.15	3.06	2.93	2.78	2.63	2.55	2.47	2.38	2.29	2.20	2.10
28	7.64	5.45	4.57	4.07	3.75	3.53	3.36	3.23	3.12	3.03	2.90	2.75	2.60	2.52	2.44	2.35	2.26	2.17	2.06
29	7.60	5.42	4.54	4.04	3.73	3.50	3.33	3.20	3.09	3.00	2.87	2.73	2.57	2.49	2.41	2.33	2.23	2.14	2.03
30	7.56	5.39	4.51	4.02	3.70	3.47	3.30	3.17	3.07	2.98	2.84	2.70	2.55	2.47	2.39	2.30	2.21	2.11	2.01
40	7.31	5.18	4.31	3.83	3.51	3.29	3.12	2.99	2.89	2.80	2.66	2.52	2.37	2.29	2.20	2.11	2.02	1.92	1.80
60	7.08	4.98	4.13	3.65	3.34	3.12	2.95	2.82	2.72	2.63	2.50	2.35	2.20	2.12	2.03	1.94	1.84	1.73	1.60
120	6.85	4.79	3.95	3.48	3.17	2.96	2.79	2.66	2.56	2.47	2.34	2.19	2.03	1.95	1.86	1.76	1.66	1.53	1.38
∞	6.63	4.61	3.78	3.32	3.02	2.80	2.64	2.51	2.41	2.32	2.18	2.04	1.88	1.79	1.70	1.59	1.47	1.32	1.00

Table VI

THE TRANSFORMATION OF r TO Z

r	r (3rd decimal) .000	.002	.004	.006	.008	r	r (3rd decimal) .000	.002	.004	.006	.008
.00	.0000	.0020	.0040	.0060	.0080	.35	.3654	.3677	.3700	.3723	.3746
1	.0100	.0120	.0140	.0160	.0180	6	.3769	.3792	.3815	.3838	.3861
2	.0200	.0220	.0240	.0260	.0280	7	.3884	.3907	.3931	.3954	.3977
3	.0300	.0320	.0340	.0360	.0380	8	.4001	.4024	.4047	.4071	.4094
4	.0400	.0420	.0440	.0460	.0480	9	.4118	.4142	.4165	.4189	.4213
.05	.0500	.0520	.0541	.0561	.0581	.40	.4236	.4260	.4284	.4308	.4332
6	.0601	.0621	.0641	.0661	.0681	1	.4356	.4380	.4404	.4428	.4453
7	.0701	.0721	.0741	.0761	.0782	2	.4477	.4501	.4526	.4550	.4574
8	.0802	.0822	.0842	.0862	.0882	3	.4599	.4624	.4648	.4673	.4698
9	.0902	.0923	.0943	.0963	.0983	4	.4722	.4747	.4772	.4797	.4822
.10	.1003	.1024	.1044	.1064	.1084	.45	.4847	.4872	.4897	.4922	.4948
1	.1104	.1125	.1145	.1165	.1186	6	.4973	.4999	.5024	.5049	.5075
2	.1206	.1226	.1246	.1267	.1287	7	.5101	.5126	.5152	.5178	.5204
3	.1307	.1328	.1348	.1368	.1389	8	.5230	.5256	.5282	.5308	.5334
4	.1409	.1430	.1450	.1471	.1491	9	.5361	.5387	.5413	.5440	.5466
.15	.1511	.1532	.1552	.1573	.1593	.50	.5493	.5520	.5547	.5573	.5600
6	.1614	.1634	.1655	.1676	.1696	1	.5627	.5654	.5682	.5709	.5736
7	.1717	.1737	.1758	.1779	.1799	2	.5763	.5791	.5818	.5846	.5874
8	.1820	.1841	.1861	.1882	.1903	3	.5901	.5929	.5957	.5985	.6013
9	.1923	.1944	.1965	.1986	.2007	4	.6042	.6070	.6098	.6127	.6155
.20	.2027	.2048	.2069	.2090	.2111	.55	.6184	.6213	.6241	.6270	.6299
1	.2132	.2153	.2174	.2195	.2216	6	.6328	.6358	.6387	.6416	.6446
2	.2237	.2258	.2279	.2300	.2321	7	.6475	.6505	.6535	.6565	.6595
3	.2342	.2363	.2384	.2405	.2427	8	.6625	.6655	.6685	.6716	.6746
4	.2448	.2469	.2490	.2512	.2533	9	.6777	.6807	.6838	.6869	.6900
.25	.2554	.2575	.2597	.2618	.2640	.60	.6931	.6963	.6994	.7026	.7057
6	.2661	.2683	.2704	.2726	.2747	1	.7089	.7121	.7153	.7185	.7218
7	.2769	.2790	.2812	.2833	.2855	2	.7250	.7283	.7315	.7348	.7381
8	.2877	.2899	.2920	.2942	.2964	3	.7414	.7447	.7481	.7514	.7548
9	.2986	.3008	.3029	.3051	.3073	4	.7582	.7616	.7650	.7684	.7718
.30	.3095	.3117	.3139	.3161	.3183	.65	.7753	.7788	.7823	.7858	.7893
1	.3205	.3228	.3250	.3272	.3294	6	.7928	.7964	.7999	.8035	.8071
2	.3316	.3339	.3361	.3383	.3406	7	.8107	.8144	.8180	.8217	.8254
3	.3428	.3451	.3473	.3496	.3518	8	.8291	.8328	.8366	.8404	.8441
4	.3541	.3564	.3586	.3609	.3632	9	.8480	.8518	.8556	.8595	.8634

r	r (3rd decimal) .000	.002	.004	.006	.008	r	r (3rd decimal) .000	.002	.004	.006	.008
.70	.8673	.8712	.8752	.8792	.8832	.85	1.256	1.263	1.271	1.278	1.286
1	.8872	.8912	.8953	.8994	.9035	6	1.293	1.301	1.309	1.317	1.325
2	.9076	.9118	.9160	.9202	.9245	7	1.333	1.341	1.350	1.358	1.367
3	.9287	.9330	.9373	.9417	.9461	8	1.376	1.385	1.394	1.403	1.412
4	.9505	.9549	.9594	.9639	.9684	9	1.422	1.432	1.442	1.452	1.462
.75	0.973	0.978	0.982	0.987	0.991	.90	1.472	1.483	1.494	1.505	1.516
6	0.996	1.001	1.006	1.011	1.015	1	1.528	1.539	1.551	1.564	1.576
7	1.020	1.025	1.030	1.035	1.040	2	1.589	1.602	1.616	1.630	1.644
8	1.045	1.050	1.056	1.061	1.066	3	1.658	1.673	1.689	1.705	1.721
9	1.071	1.077	1.082	1.088	1.093	4	1.738	1.756	1.774	1.792	1.812
.80	1.099	1.104	1.110	1.116	1.121	.95	1.832	1.853	1.874	1.897	1.921
1	1.127	1.133	1.139	1.145	1.151	6	1.946	1.972	2.000	2.029	2.060
2	1.157	1.163	1.169	1.175	1.182	7	2.092	2.127	2.165	2.205	2.249
3	1.188	1.195	1.201	1.208	1.214	8	2.298	2.351	2.410	2.477	2.555
4	1.221	1.228	1.235	1.242	1.249	9	2.647	2.759	2.903	3.106	3.453

This table is abridged from Table 14 of the *Biometrika Tables for Statisticians*, Vol. 1 (ed. 3), edited by E. S. Pearson and H. O. Hartley. Used with the kind permission of E. S. Pearson and the trustees of *Biometrika*.

Table VII
COEFFICIENTS OF ORTHOGONAL POLYNOMIALS

	J = 3		J = 4			J = 5			
	1	2	1	2	3	1	2	3	4
X_1	−1	1	−3	1	−1	−2	2	−1	1
X_2	0	−2	−1	−1	3	−1	−1	2	−4
X_3	1	1	1	−1	−3	0	−2	0	6
X_4			3	1	1	1	−1	−2	−4
X_5						2	2	1	1
Σc_j^2	2	6	20	4	20	10	14	10	70

	J = 6					J = 7					
	1	2	3	4	5	1	2	3	4	5	6
X_1	−5	5	−5	1	−1	−3	5	−1	3	−1	1
X_2	−3	−1	7	−3	5	−2	0	1	−7	4	−6
X_3	−1	−4	4	2	−10	−1	−3	1	1	−5	15
X_4	1	−4	−4	2	10	0	−4	0	6	0	−20
X_5	3	−1	−7	−3	−5	1	−3	−1	1	5	15
X_6	5	5	5	1	1	2	0	−1	−7	−4	−6
X_7						3	5	1	3	1	1
Σc_j^2	70	84	180	28	252	28	84	6	154	84	924

	J = 8						J = 9					
	1	2	3	4	5	6	1	2	3	4	5	6
X_1	−7	7	−7	7	−7	1	−4	28	−14	14	−4	4
X_2	−5	1	5	−13	23	−5	−3	7	7	−21	11	−17
X_3	−3	−3	7	−3	−17	9	−2	−8	13	−11	−4	22
X_4	−1	−5	3	9	−15	−5	−1	−17	9	9	−9	1
X_5	1	−5	−3	9	15	−5	0	−20	0	18	0	−20
X_6	3	−3	−7	−3	17	9	1	−17	−9	9	9	1
X_7	5	1	−5	−13	−23	−5	2	−8	−13	−11	4	22
X_8	7	7	7	7	7	1	3	7	−7	−21	−11	−17
X_9							4	28	14	14	4	4
Σc_j^2	168	168	264	616	2184	264	60	2772	990	2002	468	1980

	J = 10						J = 11					
	1	2	3	4	5	6	1	2	3	4	5	6
X_1	−9	6	−42	18	−6	3	−5	15	−30	6	−3	15
X_2	−7	2	14	−22	14	−11	−4	6	6	−6	6	−48
X_3	−5	−1	35	−17	−1	10	−3	−1	22	−6	1	29
X_4	−3	−3	31	3	−11	6	−2	−6	23	−1	−4	36
X_5	−1	−4	12	18	−6	−8	−1	−9	14	4	−4	−12
X_6	1	−4	−12	18	6	−8	0	−10	0	6	0	−40
X_7	3	−3	−31	3	11	6	1	−9	−14	4	4	−12
X_8	5	−1	−35	−17	1	10	2	−6	−23	−1	4	36
X_9	7	2	−14	−22	−14	−11	3	−1	−22	−6	−1	29
X_{10}	9	6	42	18	6	3	4	6	−6	−6	−6	−48
X_{11}							5	15	30	6	3	15
Σc_j^2	330	132	8580	2860	780	660	110	858	4290	286	156	11220

	J = 12						J = 13					
	1	2	3	4	5	6	1	2	3	4	5	6
X_1	−11	55	−33	33	−33	11	−6	22	−11	99	−22	22
X_2	−9	25	3	−27	57	−31	−5	11	0	−66	33	−55
X_3	−7	1	21	−33	21	11	−4	2	6	−96	18	8
X_4	−5	−17	25	−13	−29	25	−3	−5	8	−54	−11	43
X_5	−3	−29	19	12	−44	4	−2	−10	7	11	−26	22
X_6	−1	−35	7	28	−20	−20	−1	−13	4	64	−20	−20
X_7	1	−35	−7	28	20	−20	0	−14	0	84	0	−40
X_8	3	−29	−19	12	44	4	1	−13	−4	64	20	−20
X_9	5	−17	−25	−13	29	25	2	−10	−7	11	26	22
X_{10}	7	1	−21	−33	−21	11	3	−5	−8	−54	11	43
X_{11}	9	25	−3	−27	−57	−31	4	2	−6	−96	−18	8
X_{12}	11	55	33	33	33	11	5	11	0	−66	−33	−55
X_{13}							6	22	11	99	22	22
Σc_j^2	572	12012	5148	8008	15912	4488	182	2002	572	68068	6188	14212

Table VII (continued)

	J = 14						J = 15					
	1	2	3	4	5	6	1	2	3	4	5	6
X_1	−13	13	−143	143	−143	143	−7	91	−91	1001	−1001	143
X_2	−11	7	−11	−77	187	−319	−6	52	−13	−429	1144	−286
X_3	−9	2	66	−132	132	−11	−5	19	35	−869	979	−55
X_4	−7	−2	98	−92	−28	227	−4	−8	58	−704	44	176
X_5	−5	−5	95	−13	−139	185	−3	−29	61	−249	−751	197
X_6	−3	−7	67	63	−145	−25	−2	−44	49	251	−1000	50
X_7	−1	−8	24	108	−60	−200	−1	−53	27	621	−675	−125
X_8	1	−8	−24	108	60	−200	0	−56	0	756	0	−200
X_9	3	−7	−67	63	145	−25	1	−53	−27	621	675	−125
X_{10}	5	−5	−95	−13	139	185	2	−44	−49	251	1000	50
X_{11}	7	−2	−98	−92	28	227	3	−29	−61	−249	751	197
X_{12}	9	2	−66	−132	−132	−11	4	−8	−58	−704	−44	176
X_{13}	11	7	11	−77	−187	−319	5	19	−35	−869	−979	−55
X_{14}	13	13	143	143	143	143	6	52	13	−429	−1144	−286
X_{15}							7	91	91	1001	1001	143
Σc_j^2	910	728	97240	136136	235144	497420	280	37128	39780	6446460	10581480	426360

Table VIII

FACTORIALS OF INTEGERS

n	$n!$	n	$n!$
1	1	26	4.03291×10^{26}
2	2	27	1.08889×10^{28}
3	6	28	3.04888×10^{29}
4	24	29	8.84176×10^{30}
5	120	30	2.65253×10^{32}
6	720	31	8.22284×10^{33}
7	5040	32	2.63131×10^{35}
8	40320	33	8.68332×10^{36}
9	362880	34	2.95233×10^{38}
10	3.62880×10^{6}	35	1.03331×10^{40}
11	3.99168×10^{7}	36	3.71993×10^{41}
12	4.79002×10^{8}	37	1.37638×10^{43}
13	6.22702×10^{9}	38	5.23023×10^{44}
14	8.71783×10^{10}	39	2.03979×10^{46}
15	1.30767×10^{12}	40	8.15915×10^{47}
16	2.09228×10^{13}	41	3.34525×10^{49}
17	3.55687×10^{14}	42	1.40501×10^{51}
18	6.40327×10^{15}	43	6.04153×10^{52}
19	1.21645×10^{17}	44	2.65827×10^{54}
20	2.43290×10^{18}	45	1.19622×10^{56}
21	5.10909×10^{19}	46	5.50262×10^{57}
22	1.12400×10^{21}	47	2.58623×10^{59}
23	2.58520×10^{22}	48	1.24139×10^{61}
24	6.20448×10^{23}	49	6.08282×10^{62}
25	1.55112×10^{25}	50	3.04141×10^{64}

Table IX

BINOMIAL COEFFICIENTS, $\binom{N}{r}$

N \ r	0	1	2	3	4	5	6	7	8	9	10
1	1	1									
2	1	2	1								
3	1	3	3	1							
4	1	4	6	4	1						
5	1	5	10	10	5	1					
6	1	6	15	20	15	6	1				
7	1	7	21	35	35	21	7	1			
8	1	8	28	56	70	56	28	8	1		
9	1	9	36	84	126	126	84	36	9	1	
10	1	10	45	120	210	252	210	120	45	10	1
11	1	11	55	165	330	462	462	330	165	55	11
12	1	12	66	220	495	792	924	792	495	220	66
13	1	13	78	286	715	1287	1716	1716	1287	715	286
14	1	14	91	364	1001	2002	3003	3432	3003	2002	1001
15	1	15	105	455	1365	3003	5005	6435	6435	5005	3003
16	1	16	120	560	1820	4368	8008	11440	12870	11440	8008
17	1	17	136	680	2380	6188	12376	19448	24310	24310	19448
18	1	18	153	816	3060	8568	18564	31824	43758	48620	43758
19	1	19	171	969	3876	11628	27132	50388	75582	92378	92378
20	1	20	190	1140	4845	15504	38760	77520	125970	167960	184756

Table X

SELECTED VALUES OF e^{-m}

m

	.0	.1	.2	.3	.4	.5	.6	.7	.8	.9
0.	1.00000	.90484	.81873	.74082	.67032	.60653	.54881	.49659	.44933	.40657
1.0	.36788	.33287	.30119	.27253	.24660	.22313	.20190	.18268	.16530	.14957
2.0	.13534	.12246	.11080	.10026	.09072	.08209	.07427	.06721	.06081	.05502
3.0	.04979	.04505	.04076	.03688	.03337	.03020	.02732	.02472	.02237	.02024
4.0	.01832	.01657	.01500	.01357	.01228	.01111	.01005	.00910	.00823	.00745
5.0	.00674	.00610	.00552	.00499	.00452	.00409	.00370	.00335	.00303	.00274
6.0	.00248	.00224	.00203	.00184	.00166	.00150	.00136	.00123	.00111	.00101
7.0	.00091	.00083	.00075	.00068	.00061	.00055	.00050	.00045	.00041	.00037
8.0	.00034	.00030	.00027	.00025	.00022	.00020	.00018	.00017	.00015	.00014
9.0	.00012	.00011	.00010	.00009	.00008	.00007	.00007	.00006	.00006	.00005

For higher or more precise values of m, note that $e^{-a-b} = e^{-a}e^{-b}$, and that $e^{-ab} = (e^{-a})^b$. Hence, within each row, the entry in any column is .9048 times the entry in the preceding column.

Table XI

STUDENTIZED RANGE STATISTIC, q

UPPER 5% POINTS

ν \ k	2	3	4	5	6	7	8	9	10	11	12	13	14	15	16	17	18	19	20
1	18·0	27·0	32·8	37·1	40·4	43·1	45·4	47·4	49·1	50·6	52·0	53·2	54·3	55·4	56·3	57·2	58·0	58·8	59·6
2	6·09	8·3	9·8	10·9	11·7	12·4	13·0	13·5	14·0	14·4	14·7	15·1	15·4	15·7	15·9	16·1	16·4	16·6	16·8
3	4·50	5·91	6·82	7·50	8·04	8·48	8·85	9·18	9·46	9·72	9·95	10·15	10·35	10·52	10·69	10·84	10·98	11·11	11·24
4	3·93	5·04	5·76	6·29	6·71	7·05	7·35	7·60	7·83	8·03	8·21	8·37	8·52	8·66	8·79	8·91	9·03	9·13	9·23
5	3·64	4·60	5·22	5·67	6·03	6·33	6·58	6·80	6·99	7·17	7·32	7·47	7·60	7·72	7·83	7·93	8·03	8·12	8·21
6	3·46	4·34	4·90	5·31	5·63	5·89	6·12	6·32	6·49	6·65	6·79	6·92	7·03	7·14	7·24	7·34	7·43	7·51	7·59
7	3·34	4·16	4·68	5·06	5·36	5·61	5·82	6·00	6·16	6·30	6·43	6·55	6·66	6·76	6·85	6·94	7·02	7·09	7·17
8	3·26	4·04	4·53	4·89	5·17	5·40	5·60	5·77	5·92	6·05	6·18	6·29	6·39	6·48	6·57	6·65	6·73	6·80	6·87
9	3·20	3·95	4·42	4·76	5·02	5·24	5·43	5·60	5·74	5·87	5·98	6·09	6·19	6·28	6·36	6·44	6·51	6·58	6·64
10	3·15	3·88	4·33	4·65	4·91	5·12	5·30	5·46	5·60	5·72	5·83	5·93	6·03	6·11	6·20	6·27	6·34	6·40	6·47
11	3·11	3·82	4·26	4·57	4·82	5·03	5·20	5·35	5·49	5·61	5·71	5·81	5·90	5·99	6·06	6·14	6·20	6·26	6·33
12	3·08	3·77	4·20	4·51	4·75	4·95	5·12	5·27	5·40	5·51	5·62	5·71	5·80	5·88	5·95	6·03	6·09	6·15	6·21
13	3·06	3·73	4·15	4·45	4·69	4·88	5·05	5·19	5·32	5·43	5·53	5·63	5·71	5·79	5·86	5·93	6·00	6·05	6·11
14	3·03	3·70	4·11	4·41	4·64	4·83	4·99	5·13	5·25	5·36	5·46	5·55	5·64	5·72	5·79	5·85	5·92	5·97	6·03
15	3·01	3·67	4·08	4·37	4·60	4·78	4·94	5·08	5·20	5·31	5·40	5·49	5·58	5·65	5·72	5·79	5·85	5·90	5·96
16	3·00	3·65	4·05	4·33	4·56	4·74	4·90	5·03	5·15	5·26	5·35	5·44	5·52	5·59	5·66	5·72	5·79	5·84	5·90
17	2·98	3·63	4·02	4·30	4·52	4·71	4·86	4·99	5·11	5·21	5·31	5·39	5·47	5·55	5·61	5·68	5·74	5·79	5·84
18	2·97	3·61	4·00	4·28	4·49	4·67	4·82	4·96	5·07	5·17	5·27	5·35	5·43	5·50	5·57	5·63	5·69	5·74	5·79
19	2·96	3·59	3·98	4·25	4·47	4·65	4·79	4·92	5·04	5·14	5·23	5·32	5·39	5·46	5·53	5·59	5·65	5·70	5·75
20	2·95	3·58	3·96	4·23	4·45	4·62	4·77	4·90	5·01	5·11	5·20	5·28	5·36	5·43	5·49	5·55	5·61	5·66	5·71
24	2·92	3·53	3·90	4·17	4·37	4·54	4·68	4·81	4·92	5·01	5·10	5·18	5·25	5·32	5·38	5·44	5·50	5·54	5·59
30	2·89	3·49	3·84	4·10	4·30	4·46	4·60	4·72	4·83	4·92	5·00	5·08	5·15	5·21	5·27	5·33	5·38	5·43	5·48
40	2·86	3·44	3·79	4·04	4·23	4·39	4·52	4·63	4·74	4·82	4·91	4·98	5·05	5·11	5·16	5·22	5·27	5·31	5·36
60	2·83	3·40	3·74	3·98	4·16	4·31	4·44	4·55	4·65	4·73	4·81	4·88	4·94	5·00	5·06	5·11	5·16	5·20	5·24
120	2·80	3·36	3·69	3·92	4·10	4·24	4·36	4·48	4·56	4·64	4·72	4·78	4·84	4·90	4·95	5·00	5·05	5·09	5·13
∞	2·77	3·31	3·63	3·86	4·03	4·17	4·29	4·39	4·47	4·55	4·62	4·68	4·74	4·80	4·85	4·89	4·93	4·97	5·01

Table XI (continued)

UPPER 1% POINTS

ν \ k	2	3	4	5	6	7	8	9	10	11	12	13	14	15	16	17	18	19	20
1	90·0	135	164	186	202	216	227	237	246	253	260	266	272	277	282	286	290	294	298
2	14·0	19·0	22·3	24·7	26·6	28·2	29·5	30·7	31·7	32·6	33·4	34·1	34·8	35·4	36·0	36·5	37·0	37·5	37·9
3	8·26	10·6	12·2	13·3	14·2	15·0	15·6	16·2	16·7	17·1	17·5	17·9	18·2	18·5	18·8	19·1	19·3	19·5	19·8
4	6·51	8·12	9·17	9·96	10·6	11·1	11·5	11·9	12·3	12·6	12·8	13·1	13·3	13·5	13·7	13·9	14·1	14·2	14·4
5	5·70	6·97	7·80	8·42	8·91	9·32	9·67	9·97	10·24	10·48	10·70	10·89	11·08	11·24	11·40	11·55	11·68	11·81	11·93
6	5·24	6·33	7·03	7·56	7·97	8·32	8·61	8·87	9·10	9·30	9·49	9·65	9·81	9·95	10·08	10·21	10·32	10·43	10·54
7	4·95	5·92	6·54	7·01	7·37	7·68	7·94	8·17	8·37	8·55	8·71	8·86	9·00	9·12	9·24	9·35	9·46	9·55	9·65
8	4·74	5·63	6·20	6·63	6·96	7·24	7·47	7·68	7·87	8·03	8·18	8·31	8·44	8·55	8·66	8·76	8·85	8·94	9·03
9	4·60	5·43	5·96	6·35	6·66	6·91	7·13	7·32	7·49	7·65	7·78	7·91	8·03	8·13	8·23	8·32	8·41	8·49	8·57
10	4·48	5·27	5·77	6·14	6·43	6·67	6·87	7·05	7·21	7·36	7·48	7·60	7·71	7·81	7·91	7·99	8·07	8·15	8·22
11	4·39	5·14	5·62	5·97	6·25	6·48	6·67	6·84	6·99	7·13	7·25	7·36	7·46	7·56	7·65	7·73	7·81	7·88	7·95
12	4·32	5·04	5·50	5·84	6·10	6·32	6·51	6·67	6·81	6·94	7·06	7·17	7·26	7·36	7·44	7·52	7·59	7·66	7·73
13	4·26	4·96	5·40	5·73	5·98	6·19	6·37	6·53	6·67	6·79	6·90	7·01	7·10	7·19	7·27	7·34	7·42	7·48	7·55
14	4·21	4·89	5·32	5·63	5·88	6·08	6·26	6·41	6·54	6·66	6·77	6·87	6·96	7·05	7·12	7·20	7·27	7·33	7·39
15	4·17	4·83	5·25	5·56	5·80	5·99	6·16	6·31	6·44	6·55	6·66	6·76	6·84	6·93	7·00	7·07	7·14	7·20	7·26
16	4·13	4·78	5·19	5·49	5·72	5·92	6·08	6·22	6·35	6·46	6·56	6·66	6·74	6·82	6·90	6·97	7·03	7·09	7·15
17	4·10	4·74	5·14	5·43	5·66	5·85	6·01	6·15	6·27	6·38	6·48	6·57	6·66	6·73	6·80	6·87	6·94	7·00	7·05
18	4·07	4·70	5·09	5·38	5·60	5·79	5·94	6·08	6·20	6·31	6·41	6·50	6·58	6·65	6·72	6·79	6·85	6·91	6·96
19	4·05	4·67	5·05	5·33	5·55	5·73	5·89	6·02	6·14	6·25	6·34	6·43	6·51	6·58	6·65	6·72	6·78	6·84	6·89
20	4·02	4·64	5·02	5·29	5·51	5·69	5·84	5·97	6·09	6·19	6·29	6·37	6·45	6·52	6·59	6·65	6·71	6·76	6·82
24	3·96	4·54	4·91	5·17	5·37	5·54	5·69	5·81	5·92	6·02	6·11	6·19	6·26	6·33	6·39	6·45	6·51	6·56	6·61
30	3·89	4·45	4·80	5·05	5·24	5·40	5·54	5·65	5·76	5·85	5·93	6·01	6·08	6·14	6·20	6·26	6·31	6·36	6·41
40	3·82	4·37	4·70	4·93	5·11	5·27	5·39	5·50	5·60	5·69	5·77	5·84	5·90	5·96	6·02	6·07	6·12	6·17	6·21
60	3·76	4·28	4·60	4·82	4·99	5·13	5·25	5·36	5·45	5·53	5·60	5·67	5·73	5·79	5·84	5·89	5·93	5·98	6·02
120	3·70	4·20	4·50	4·71	4·87	5·01	5·12	5·21	5·30	5·38	5·44	5·51	5·56	5·61	5·66	5·71	5·75	5·79	5·83
∞	3·64	4·12	4·40	4·60	4·76	4·88	4·99	5·08	5·16	5·23	5·29	5·35	5·40	5·45	5·49	5·54	5·57	5·61	5·65

REFERENCES AND SUGGESTIONS FOR FURTHER READING

PROBABILITY THEORY

Derman, C., Gleser, L. J., and Olkin, I. *A guide to probability theory and application*. New York: Holt, Rinehart and Winston, 1973.

Feller, W. *An introduction to probability theory and its applications,* Vol. I (ed. 3). New York: Wiley, 1968.

de Finetti, B. *Theory of probability: A critical introductory treatment,* Vol. I. London: Wiley, 1974.

Kyburg, H. E., and Smokler, H. E. *Studies in subjective probability*. New York: Wiley, 1964.

Parzen, E. *Modern probability theory and its applications*. New York: Wiley, 1960.

Ross, S. *Introduction to probability models*. New York: Academic Press, 1972.

Thompson, W. A. *Applied probability*. New York: Holt, Rinehart and Winston, 1969.

MEASUREMENT AND RELATED MATTERS

Coombs, C. H., Dawes, R. M., and Tversky, A. *Mathematical psychology, an elementary introduction*. Englewood Cliffs, N.J.: Prentice-Hall, 1970.

Stevens, S. S. *Handbook of experimental psychology*. New York: Wiley, 1951.

Thrall, R., Coombs, C., and Davis, R. *Decision processes*. New York: Wiley, 1959.

Torgerson, W. *Theory and methods of scaling*. New York: Wiley, 1958.

STATISTICAL INFERENCE IN GENERAL

Brunk, H. D. *An introduction to mathematical statistics* (ed. 2). Boston: Ginn & Company, 1965.

Cramér, H. *Mathematical methods of statistics*. Princeton, N.J.: Princeton University Press, 1946.

Dixon, W., and Massey, F. *Introduction to statistical analysis* (ed. 2). New York: McGraw-Hill, 1957.

Freund, J. E. *Mathematical statistics* (ed. 2). Englewood Cliffs, N.J.: Prentice-Hall, 1971.

Harnett, D. L. *Introduction to statistical methods* (ed. 2). Reading, Mass.: Addison-Wesley, 1975.

Hodges, J. L., and Lehmann, E. L. *Basic concepts of probability and statistics* (ed. 2). San Francisco: Holden-Day, 1970.

Hoel, P. G. *Introduction to mathematical statistics* (ed. 4). New York: Wiley, 1971.

Hogg, R. V., and Craig, A. T. *Introduction to mathematical statistics* (ed. 3). New York: Macmillan, 1970.

Huff, D. *How to lie with statistics*. New York: Norton, 1954.

Kendall, M. G., and Stuart, A. *The advanced theory of statistics,* Vols. I, II, and III. London: Charles Griffin & Co., Ltd., 1958, 1961, and 1966.

Lippman, S. A. *Elements of probability and statistics*. New York: Holt, Rinehart and Winston, 1971.

Mood, A. M., Graybill, F. A., and Boes, D. C. *Introduction to the theory of statistics* (ed. 3). New York: McGraw-Hill, 1974.

Nie, N. H., Hull, C. H., Jenkins, J. G., Steinbrenner, K., and Bent, D. H. *Statistical package for the social sciences* (ed. 2). New York: McGraw-Hill, 1975.

Tukey, J. W. *Exploratory data analysis*. Reading, Mass.: Addison-Wesley, 1977.

Wallis, W. A., and Roberts, H. V. *Statistics: A new approach*. Glencoe, Ill.: Free Press, 1956.

Wilks, S. S. *Mathematical statistics*. New York: Wiley, 1962.

DECISION THEORY, BAYESIAN METHODS

Brown, R. V., Kahr, A. S., and Peterson, C. *Decision analysis for the manager*. New York: Holt, Rinehart and Winston, 1974.

Chernoff, H., and Moses, L. E. *Elementary decision theory*. New York: Wiley, 1959.

Lindley, D. V. *Introduction to probability and statistics from a Bayesian viewpoint* (2 Vols.). Cambridge: Cambridge University Press, 1965.

Lindley, D. V. *Making decisions*. London: Wiley, 1971.

Luce, R. D., and Raiffa, H. *Games and decisions*. New York: Wiley, 1957.

Novick, M. R., and Jackson, P. H. *Statistical methods for educational and psychological research*. New York: McGraw-Hill, 1974.

Phillips, L. D. *Bayesian statistics for social scientists*. New York: Thomas Y. Crowell, 1974.

Raiffa, H. *Decision analysis*. Reading, Mass.: Addison-Wesley, 1968.

Schlaifer, R. *Probability and statistics for business decisions*. New York: McGraw-Hill, 1959.

Schlaifer, R. *Analysis of decisions under uncertainty*. New York: McGraw-Hill, 1969.

Winkler, R. L. *An introduction to Bayesian inference and decision*. New York: Holt, Rinehart and Winston, 1972.

TOPICS IN ANALYSIS OF VARIANCE AND EXPERIMENTAL DESIGN

Box, G. E. P. Non-normality and tests on variances. *Biometrika 40,* 1953, 318–335.

Box, G. E. P. Some theorems on quadratic forms applied in the study of analysis of variance problems: I. Effect of inequality of variance in the one-way classification. *Annals Math. Stat. 25,* 1954, 290–302.

Box, G. E. P. Some theorems on quadratic forms applied in the study of analysis of variance problems: II. Effect of inequality of variance and of correlation of errors in the two-way classification. *Annals Math. Stat. 25,* 1954, 484–498.

Cochran, W. G., and Cox, G. M. *Experimental designs* (ed. 2). New York: Wiley, 1957.

Cohen, J. *Statistical power analysis for the behavioral sciences* (rev. ed.). New York: Academic Press, 1977.

Cook, T. D., and Campbell, D. T. *Quasi-experimentation: Design and analysis issues for field settings*. Chicago: Rand McNally, 1979.

Cox, D. R. *Planning of experiments*. New York: Wiley, 1958.

Edwards, A. F. *Experimental design in psychological research* (ed. 4). New York: Holt, Rinehart, and Winston, 1973.

Fisher, R. A. *The design of experiments* (ed. 8). Edinburgh: Oliver & Boyd, 1966.

Graybill, F. A. *An introduction to linear statistical models,* Vol I. New York: McGraw-Hill, 1961.

Guenther, W. C. *Analysis of variance*. Englewood Cliffs, N. J.: Prentice-Hall, 1964.

Kirk, R. *Experimental design: Procedures for the behavioral sciences*. Belmont, Calif.: Brooks-Cole, 1968.

Mendenhall, W. *Introduction to linear models and the design and analysis of experiments*. Belmont, Calif.: Wadsworth, 1968.

Neter, J., and Wasserman, W. *Applied linear statistical models*. Homewood, Ill.: Richard D. Irwin, 1974.

Paull, A. E. On a prelimary test for pooling mean squares in the analysis of variance. *Annals Math. Stat. 21,* 1950, 539–556.

Scheffe, H. *The analysis of variance*. New York: Wiley, 1959.

Snedecor, G. W., and Cochran, W. G. *Statistical methods* (ed. 6). Ames, Iowa: Iowa State College Press, 1967.

Ward, J., and Jennings, E. *Introduction to linear models*. Englewood Cliffs, N. J.: Prentice-Hall, 1973.

Winer, B. J. *Statistical principles in experimental design* (ed. 2). New York: McGraw-Hill, 1971.

CORRELATION AND REGRESSION, MULTIVARIATE METHODS

Bennett, S., and Bowers, D. *An introduction to multivariate techniques for the social and behavioral sciences*. New York: Wiley, 1976.

Blalock, H. M. (ed.). *Causal models in the social sciences*. Chicago: Aldine Press, 1970.

Carroll, J. B. The nature of the data, or how to choose a correlation coefficient. *Psychometrika 26,* 1961, 347–372.

Cohen, J., and Cohen, P. *Applied multiple regression: Correlation analysis for the behavioral sciences*. New York: Halstead Press, 1975.

Guilford, J. P., and Fruchter, B. *Fundamental statistics in psychology and education* (ed. 6). New York: McGraw-Hill, 1978.

Harris, R. J. *A primer of multivariate statistics*. New York: Academic Press, 1975.

Kerlinger, F. N., and Pedhazur, E. J. *Multiple regression in behavioral research*. New York: Holt, Rinehart and Winston, 1973.

Lewis, D. *Quantitative methods in psychology*. New York: McGraw-Hill, 1960.

Morrison, D. F. *Multivariate statistical methods* (ed. 2). New York: McGraw-Hill, 1976.

Mosteller, F. and Tukey, J. *Data analysis and regression,* Reading, Mass.: Addison-Wesley, 1977.

Timm, N. H. *Multivariate analysis, with applications in education and psychology*. Belmont, Calif.: Brooks-Cole, 1975.

QUALITATIVE DATA AND ORDINAL METHODS

Bishop, Y. M. M., Fienberg, S. E., and Holland, P. W. *Discrete multivariate analysis: Theory and practice*. Cambridge, Mass.: M.I.T. Press, 1975.

Bradley, J. V. *Distribution-free statistical tests*. Englewood Cliffs, N. J.: Prentice-Hall, 1968.

Cochran, W. G. The comparison of percentages in matched samples. *Biometrika 37,* 1950, 256–266.

Cohen, J. A. Coefficient of agreement for nomial scales. *Educational and Psych. Meas. 20,* 1960, 37–46.

Fleiss, J. L. Measuring nominal scale agreement among many raters. *Psych. Bull. 76,* 1971, 378–382.

Gibbons, J. D. *Nonparametric statistical inference*. New York: McGraw-Hill, 1971.

Goodman, L. A. *Analyzing qualitative categorical data,* Cambridge, Mass.: Abt Associates, 1978.

Goodman, L. A., and Kruskal, W. H. Measures of association for cross-classifications. *J. Amer. Stat. Ass. 49,* 1954, 732–764.

Kendall, M. G. *Rank correlation methods* (ed. 3). London: Griffin, 1963.

Kruskal, W. H. Ordinal measures of association. *J. Amer. Stat. Ass. 53,* 1958, 814–861.

Kruskal, W. H., and Wallis, W. A. Use of ranks in one-criterion variance analysis. *J. Amer. Stat. Ass. 47,* 1952, 583–621.

McNemar, Q. *Psychological statistics* (ed. 5). New York: Wiley, 1975.

Maxwell, A. E. *Analysing qualitative data*. London: Methuen, 1961.

Siegel, S. *Nonparametric methods for the behavioral sciences*. New York: McGraw-Hill, 1956.

Walsh, J. E. *A handbook of nonparametric statistics* (2 Vols.). New York: Van Nostrand, 1962.

TABLES

Beyer, W. H. *Handbook of tables for probability and statistics* (ed. 2). Cleveland, Ohio: The Chemical Rubber Co., 1968.

Burington, R. S., and May, D. C. *Handbook of probability and statistics with tables* (ed. 2). New York: McGraw-Hill, 1969.

Fisher, R. A., and Yates, F. *Statistical tables for biological, agricultural and medical research* (ed. 6). Edinburgh: Oliver and Boyd, 1963.

Owen, D. B. *Handbook of statistical tables*. Reading, Mass.: Addison-Wesley, 1962.

Pearson, E. S., and Hartley, H. O. *Biometrika tables for statisticians,* Vol I (ed. 3) and Vol. II (ed. 2). Cambridge: Cambridge University Press, 1967 and 1972.

GLOSSARY

CONVENTIONAL MATHEMATICAL SYMBOLS

A	a set, or event (1.3)
$A = \{a,b,c,d\}$	set A includes the following members; a set specified by listing (1.4)
$a \leqslant b$	a is less than or equal to b (1.5)
$a \geqslant b$	a is greater than or equal to b (3)
$a < b$	a is less than (but not equal to) b (2.1)
$a > b$	a is greater than (but not equal to) b (2.1)
$\varnothing$	the empty or "impossible" event (1.3)
$A \cup B$	union of sets or events A and B (1.3)
$A \cap B$	intersection of sets or events A and B (1.3)
$\overline{A}$	complement of set A; the event "not A" (1.3)
$A - B$	difference between sets A and B (13.19)
$A = B$	sets A and B are equal or equivalent (1.5)
(a,b)	ordered pair of elements, a from set A and b from set B (Appendix C)
$y = f(x)$, $y = G(x)$, etc.	y is functionally related to x (2.18)
$y = f(x,z)$	y is a function of the two variables x and z (Appendix C)
∞	an infinite value (6.1)
Σ	sum of (see Appendix A)
$\int_a^b f(x)\,dx$	area under the curve generated by the function $y = f(x)$, as cut off by the x interval with limits a and b (2.18)
$N!$	factorial of the interger N (3.5)
$\binom{N}{r}$	number of unordered combinations of N things taken r at a time, $0 \leqslant r \leqslant N$; a binomial coefficient (3.7)
e	mathematical constant equal approximately to 2.7182818; base of the natural system of logarithms (3.18)
$\log_e m$	logarithm to the base e for the value m (13.13)
π	mathematical constant, equal approximately to 3.14159265; ratio of the circumference of a circle to its diameter (6.1)

STATISTICAL SYMBOLS USED IN THIS TEXT

$\lvert A.D.\rvert$	the average absolute deviation (4.15)
a_j	effect of the randomly selected treatment appearing as the jth such treatment in the experiment (11.7)
$(A_{ij}B_k)$	joint event of an observation in category A_j on attribute A, and in category B_k on attribute B (15.4)
α	(small Greek alpha) probability of rejecting H_0 when it is true; probability of Type I error (7.6)
$(a_0, a_1, \cdots, a_J)$	weights given to variables $(X_0, X_1, \cdots, X_J)$ in a linear model for predicting Y (10.2)
a_0	a constant in the linear model for predicting y from x (13.1)
A_0	a constant in the linear model for predicting z_Y from z_X (13.2)

$\alpha_0 = \mu$	constant "effect" in analysis of variance model under least squares criterion (10.5)
$\alpha_j = \mu_j - \mu$	effect associated with the jth treatment applied; a fixed effect (10.5)
est. α_j	estimated effect for treatment j (10.7)
$b_h = \dfrac{\psi_h}{nw_h}$	weight given to the orthogonal polynomial coefficient representing a trend of degree h in estimation of a curvilinear prediction function (14.11)
b_k	weight associated with indicator variable w_k in a two factor design; a random effect for row k (10.21, 11.10)
B_j	the standardized partial regression coefficient for variable X_j in the prediction of Y from $(X_1, \cdots, X_j, \cdots, X_K)$ (13.18)
B_j^*	the population standardized partial regression weight for X_j in the prediction of Y from $H = (X_1, \cdots, X_K)$ (13.18)
$B_{Y \cdot X}$	standardized sample regression coefficient for predicting Y from X (13.2)
$b_{Y \cdot X}$	sample raw score coefficient of linear regression of Y on X (13.7)
$b_{y \cdot 1}$	linear regression coefficient for prediction of Y from X_1 (14.9)
$B_{12 \cdot 3}$	standardized regression coefficient applied to variable z_2 in predictions of z_1 from z_2 and z_3 (13.18)
$B_{13 \cdot 2}$	standardized regression coefficient applied to variable z_3 in predictions of z_1 from z_2 and z_3 (13.18)
$b_{12 \cdot 3 \ldots K}$	regression coefficient for weighting X_2 in the prediction of X_1 from $X_2, \cdots, X_K$ (13.18)
β	(small Greek beta) probability of failing to reject H_0 when it is false; probability of Type II error (7.6)
$1 - \beta$	power of a statistical test against some given true alternative to the null hypothesis (7.8)
β_k	fixed effect associated with the treatment or factor level k (10.21)
$\beta_{Y \cdot X}$	unstandardized partial regression coefficient for prediction of Y from X in the population (13.11)
β_k	any true unstandardized partial regression weight in the prediction of Y from $H = (X_1, \cdots, X_K)$ (13.18)
$\beta_1, \beta_2, \cdots, \beta_{J-1}$	population regression coefficients for linear and curvilinear trends between Y and X (14.10)
C	number of columns in the data table for a two-factor experiment (10.23)
C_{AB}	coefficient of contingency (15.10)
cf	cumulative frequency (4.2)
c_j	constant applied to a sample or population mean corresponding to treatment j in a comparison (12.2)
c_{jk}	random interaction effect associated with the combination of treatment j with treatment k (11.15)
$(c_1, \cdots, c_n)$	set of n constant weights figuring in some linear combination of n random variables (Appendix C)
χ^2	(small Greek chi, squared) random variable chi-square; also the Pearson chi-square statistic (9.1, 15.1)
$\chi^2_{(\nu)}$	chi-square variable with ν degrees of freedom (9.1)
$\chi^2_{(N-1;\alpha/2)}$; $\chi^2_{(N-1;1-\alpha/2)}$	values in the distribution of χ^2 with $N - 1$ degrees of freedom,

	cutting off the upper $\alpha/2$ and $1 - a/2$ proportion of cases, respectively (9.5)
$\chi^2 = \sum_j \frac{(f_{oj} - f_{ej})^2}{f_{ej}}$	Pearson χ^2 statistic in a test of goodness of fit (15.1)
cov.(X,Y)	covariance of the two random variables X and Y (see Appendix B)
cov.(M_1,M_2)	covariance of pairs of sample means (8.16)
$D_i = (y_{i1} - y_{i2})$	difference between the scores for members of matched pair i from among N such pairs (8.16)
d_i	deviation of the score for observation i from the grand mean (4.3)
δ	(small Greek delta) noncentrality parameter for noncentral t or, as δ^2, noncentral F (8.11, 10.19)
ΔX	width of an arbitrarily small interval of values for random variable X (2.18)
$E(X)$	expectation of random variable X (see Appendix B)
$E(u\|H_0)$	expected loss, given a decision rule and truth of hypothesis H_0 (7.3)
e	the error in a measurement (6.5)
e_{ij}	random error associated with the ith observation made under treatment j (10.5)
e_{ijk}	random error associated with the ith observation in the treatment-combination j, k (10.21)
$\eta^2_{Y \cdot X}$	(small Greek eta, squared) correlation ratio, for the relation of Y to X (8.15)
F	F ratio computed from a sample; the random variable "F" (9.8)
F_α	value of F required for significance at the α level, one-tailed, for a given number of degrees of freedom (12.12)
$F_{(\nu_1;\nu_2)}$	random variable distributed as F with ν_1 and ν_2 degrees of freedom (9.9)
$F_{(\alpha;\nu_1,\nu_2)}$	value cutting off the upper α proportion of cases in an F distribution with ν_1 and ν_2 degrees of freedom (11.4)
F_g	the F ratio associated with comparison $\hat{\psi}_g$ (12.8)
$f(a) = f(X = a)$	probability density for random variable X at the value $X = a$ (2.18)
$F(a) = p(X \leqslant a)$	cumulative probability of random variable X at the value a (2.19)
$\max_k \cdot f_{.k}$	the largest marginal frequency among the rows of a table (15.12)
f	frequency of a given measurement or event class (2.3)
f_{ej}	expected frequency in event-category j (15.1)
f_{ejk}	expected frequency in the cell j, k of a contingency table (15.4)
f_{oj}	observed frequency in event-category j (15.1)
f_{ojk}	observed frequency in the cell j, k of a contingency table (15.4)
γ	(small Greek gamma) coefficient of predictive association between sets of ordered classes (16.14)
γ_{jk}	fixed interaction effect associated with the combination of treatments j and k (10.21)
$G = H - \{X_j\}$	set notation for a reduced set of predictors in a regression equation for predicting Y (13.19)
G	the set of significant trends in a problem involving orthogonal polynomials (14.11)

G^2	the likelihood ratio test statistic under the log-linear model (15.12)
H	the entire set of predictors ($X_1, \cdots, X_K$) in a multiple regression equation for predicting variable Y (13.19)
H:	indicator of a statistical hypothesis (7.1)
H_0:	hypothesis actually being tested; the "null" hypothesis (7.1)
H_1:	hypothesis to be entertained if H_0 is rejected; the alternative hypothesis (7.1)
i	class interval size in some frequency distribution (2.5)
i	running subscript, ordinarily indicating the ith observation among some N distinct observations (Appendix A)
J	number of different experimental treatments or groups in an experiment (Appendix A)
j	running subscript indicating the jth treatment group or factor level (Appendix A)
Κ (kappa)	an index of interjudge agreement (15.11)
K	in a two-factor experiment, number of treatment groups or levels of the second factor; number of "blocks" of observations in an experiment (11.18)
k	an arbitrarily small positive number (4.18)
k	a running subscript, ordinarily indicating the kth treatment group or "block" of observations (Appendix A)
$L(x_1, \cdots, x_N \vert \theta)$	likelihood of a particular set of N sample values, given the value of the parameter (or set of parameters) θ (5.5)
λ_B	(small Greek lambda) asymmetric measure of predictive association for a contingency table (15.12)
λ_{AB}	symmetric measure of predictive association in a contingency table (15.12)
M	the sample mean (4.2)
m	the intensity parameter in a Poisson process (3.18)
M_D	mean difference between N matched pairs (8.16)
Md	the median (4.2)
M_e	mean error in a set of N experimental observations (10.10)
M_{ej}	mean error for the n_j observations in group j (10.10)
M_G	the geometric mean (4.2)
M_H	the harmonic mean (4.2)
$m(o)$	measured amount of some property possessed by object o (2.1)
MS	mean square in the analysis of variance (10.10)
M_z	the mean of sample z values (4.17)
$\mu = E(X)$	(small Greek mu) mean of the probability distribution of the random variable X (4.9)
μ_0	value of the population mean specified by H_0 (7.9)
μ_1	true value of the mean of the population; a value of the mean covered by the alternative hypothesis H_1 (7.9)
$\mu_G = E(G)$	mean of the sampling distribution for some statistic G (5.3)
μ_j	true mean of the potential population of observation made under treatment j (10.4)
μ_{jk}	mean of the potential population of observations made under the treatment combination j, k (10.21)
N	total number of trials in a simple experiment or of observations in a given sample; the total number of elementary events in a sample space (1.6)

n	size of any one of several samples containing the same number of observations (11.3)
$n(A)$	the frequency of event A in the sample space (1.6)
n_j	number of sample observations in treatment group j (10.10)
n_{jk}	number of sample observations in treatment combination j, k (10.20)
ν	(small Greek nu) degrees of freedom parameter for a t or chi-square variable (8.2, 9.1)
ν_1, ν_2	number of degrees of freedom for numerator and denominator, respectively, for an F ratio; the parameters of the F distribution (9.8)
$\mathcal{O} = (o_1, \cdots, o_N)$	a set of objects of measurement (2.1)
ω^2	(small Greek omega, squared) population index showing the relative or proportional reduction in the variance of Y given the X status or value for an observation (8.12)
est. ω^2	sample estimate of the proportional reduction in variance of Y given X (8.15)
P	sample proportion of "successes" in sampling from a Bernoulli process (4.9)
p	probability of a given event class; probability of a "success" in a single Bernoulli trial (2.13)
$p(A)$	probability associated with a particular event A in a probability function (1.4)
$p(B\|A) = \dfrac{p(A \cap B)}{p(A)}$	conditional probability of event B given the event A (1.12)
$\mathscr{P}(H)$	personal probability associated by the experimenter with the hypothesis H (7.4)
$p(X = r;N,p)$	binomial probability for $X = r$, given the parameters N and p (3.10)
$p(X_j)$	in the fixed-effects model, the relative frequency in the grand population of those receiving treatment j (10.5)
$p(x)$	the probability that random variable X takes on value $X = x$ (2.14)
ψ	(small Greek psi) value of a particular comparison among population means (12.2)
$\hat{\psi}$	value of a particular comparison among sample means (12.2)
$\hat{\psi}_g$	value of a particular comparison g on a set of sample means (12.8)
φ	(small Greek phi) index of mean square contingency for a population or a sample contingency table (15.10)
φ'	Cramér's statistic for association in a contingency table (15.10)
Q	proportion of cases cut off by a given value on the upper tail of a sampling distribution (8.4)
Q, Y	Yule's indices for a fourfold table (15.10)
$Q = 1 - P$	proportion of sample "failures" in sampling from a Bernoulli process (4.9)
$q = 1 - p$	probability of a "failure" on a single Bernoulli trial (2.13)
q	the Studentized range statistic (12.12)
r	the number of success out of N trials of a Bernoulli process
R	number of rows in the data table for a two-factor experiment (10.21)

$R_{1.23}$	multiple correlation of variable 1 with variables 2 and 3 (13.18)
$R_{y \cdot 12 \cdots K}$	the multiple correlation for Y and the set of predictors $X_1, \cdots, X_K$ (13.18)
$R_{y \cdot H}$	the multiple correlation between Y and the full set of predictor variables, H (13.19)
$R_{y \cdot G}$	the multiple correlation between Y and the reduced set of predictors, $G = H - \{X_j\}$ (13.19)
$r^2_{y \cdot h}$	the proportion of variance accounted for in the data by any trend h (14.9)
$R^2_{y \cdot H}$	the proportion of variance in Y explained by the set of variables in H (13.19)
$r_{1(2 \cdot 3)}$	part correlation between X_1 and X_2, with X_3 variance removed from x_2 (13.17)
r_{xy}	sample correlation coefficient between X and Y (13.2)
$r_{12 \cdot 3}$	partial correlation between variables 1 and 2 with 3 held constant (13.17)
$r_{(Y', Y)}$	correlation between the predicted and actual values of Y (13.18)
r_S	Spearman rank-correlation coefficient (16.11)
$\mathrm{P}_{y \cdot 1 \cdots K}$	(capital Greek rho) the population multiple correlation for Y predicted from $X_1, \cdots, X_K$ (13.18)
ρ_I	(small Greek rho, sub I) intraclass correlation coefficient for a population (11.5)
ρ_{XY}	population correlation coefficient between variables X and Y (13.11, Appendix B)
$\mathscr{S}$	sample space for a particular simple experiment (1.3)
$\mathbf{S} = \sqrt{(J-1)F_\alpha}$	constant determining the width of confidence intervals for post hoc comparisons. Scheffé method (12.12)
$SE(u \mid D)$	subjective expected loss associated with a particular decision rule D by the experimeter (7.4)
S_+, S_-	number of agreements and disagreements, respectively, about the ordering of pairs of objects in two rank orders (16.12)
SS	sum of squares in the analysis of variance (10.8)
S	standard deviation computed for a sample of data (4.12)
S^2	sample variance (4.11)
$s = \sqrt{s^2}$	corrected standard deviation for a sample (5.9)
$s^2 = \dfrac{N}{N-1} S^2$	corrected variance; the unbiased estimator of the population variance from a sample (5.9)
s^2_D	corrected variance based on the difference for each of N matched pairs of observations (8.16)
$S_{Y \cdot X}$	sample standard error of estimate for predictions of Y from X (13.9)
S^2_z	variance of z or standardized scores calculated from a sample of data (4.17)
$S_{z_Y \cdot z_X}$	sample standard error of estimate for predictions of z_Y from z_X (13.2)
$S_{z_1 \cdot z_2}$	standard error of estimate for predictions of z_1 from z_2 (13.18)
$S^2_{z_1 \cdot z_2 \cdots z_K}$	sample variance of estimate for z_1 values predicted from $z_2, \cdots, z_K$ via a multiple regression equation (13.18)
$S^2_{y \cdot 1 \cdots K}$	sample variance of estimate in the prediction of Y from $X_1, \cdots, X_K$(13.18)
σ	(small Greek sigma) standard deviation of a random variable (4.16)

Symbol	Meaning
σ^2	variance of a random variable (4.16)
σ_0^2	value of the population variance specified by the hypothesis H_0 (9.4)
σ_A^2	variance of the distribution of random effects representing factor A (11.3)
σ_B^2	variance of the random effects representing factor B (11.10)
σ_{AB}^2	variance of the random interaction effects for the factors A and B (11.10)
σ_e^2	variance of random errors, e (10.9)
$\sigma_{\text{diff.}}$	standard error of the difference between two means (8.7)
est. $\sigma_{\text{diff.}}$	estimated error of the difference between two means (8.7)
σ_G^2	variance of the sampling distribution of some statistic G (5.3)
σ_M^2	variance of the sampling distribution of the mean (5.7)
σ_M	true standard error of the mean, given samples of size N from some population (5.8)
est. σ_M	sample estimate of the standard error of the mean (5.9)
est. σ_{M_D}	estimated standard error of the mean difference between N matched pairs (8.16)
σ_P^2	variance of a sampling distribution of sample proportions, P (4.16)
σ_Y^2	variance of the random variable Y; "marginal" variance of Y in a joint distribution of (x,y) events (see Appendices B and C)
$\sigma_Y^2 - \sigma_{Y\|X}^2$	reduction in the variance of Y afforded by specification of X (8.12)
$\sigma_{Y\|X}^2$	variance of the conditional distribution of Y, given the value of X (8.12, Appendix C)
$\sigma_{Y \cdot X}$	true standard error of estimate for predictions of Y from X in some population (13.11)
$\sigma_{y \cdot 12 \cdots K}^2$	the true variance of multiple estimate for predicting Y from $X_1, \cdots, X_K$ (13.18)
σ_z^2	variance of a random variable transformed to z or standardized values (4.17)
$\sigma_{(Z_1 - Z_2)}$	variance of the difference between Z values for pairs of independent samples (13.13)
T	total number of potential observations in a finite population (5.11)
$t = \dfrac{M - E(M)}{\text{est. } \sigma_M}$	t ratio based on the mean of a single sample (8.2)
$t = \dfrac{M_1 - M_2 - E(M_1 - M_2)}{\text{est. } \sigma_{\text{diff.}}}$	t ratio based on the difference between means of two samples (8.7)
t	a random variable following the "Student" distribution of t with ν degrees of freedom (8.2)
$t_{(\alpha/2;\nu)}$	value of t in a distribution with ν degrees of freedom, cutting off the upper $\alpha/2$ proportion of cases (8.9)
$t(o)$	true amount of some property possessed by object o (2.1)
τ	(small Greek tau) Kendall's coefficient of rank-order agreement (16.12)
θ	(small Greek theta) general symbol for a population parameter (5.5)
θ_0	value of θ specified by the null hypothesis H_0 (11.6)
$\theta = \sigma_A^2/\sigma_e^2$	ratio of the variance of effects due to factor A to the error variance (11.5)

u, u_i, u_{ij}, etc.	"effects" in the log-linear model (15.13)
$u(H_0 \| H_1)$	loss associated with the decision to accept H_0 when H_1 is in fact true (7.4)
$u(H_1 \| H_0)$	loss associated with the decision to accept H_1 when H_0 is in fact true (7.4)
$\mathbf{V}$	a set of vectors $(\mathbf{V}_0, \cdots, \mathbf{V}_{J-1})$ forming an orthogonal basis for a set of vectors $\mathbf{X}$ (Appendix C)
v_{gj}	an indicator variable equal to 1 when treatment g is selected and labeled as group j, and 0 otherwise (11.2)
var.(X)	variance of the random variable X (see Appendix B)
var.$(M_1 - M_2)$	variance of the difference between means (8.7)
var.$(\hat{\psi})$	variance of a comparison among sample means (12.2)
$w(b \| a)$	conditional density for $Y = b$ given that $X = a$ (Appendix C)
w_g	weighting factor used in obtaining the sum of squares for comparison $\hat{\psi}_g$ (12.8)
$\mathbf{X}$	a set of m vectors each of dimension n (Appendix C)
X	a random variable; the independent variable in an experiment (2.14)
(x,y)	a joint event, consisting of a value for variable X paired with a value for variable Y (Appendix C)
$\|X - \mu\|$	absolute deviation (disregarding sign) of the value of X from the value of μ (4.18)
x	a particular value which random variable X can assume (2.14)
x	midpoint of a class interval (4.2)
x_{i+}, x_{+i}	marginal frequencies in an $I \times I$ table (15.11)
x_{i+},x_{+j}	marginal frequencies in an $I \times J$ table (15.13)
$\overline{X}$ or $\bar{x}$	an alternate way to symbolize the sample mean (4.2)
Y	a random variable; ordinarily, the dependent variable in an experiment (8.12)
y'	raw score on Y predicted from the value of X (13.1)
y_{ij}	score associated with the observation i in experimental group j (10.5)
y_{ijk}	score associated with the ith observation in the treatment combination j, k (10.21)
Z	value corresponding to r_{XY} in the Fisher r to Z transformation (13.13)
$z = \dfrac{X - M}{S}$	standardized score or value corresponding to a sample value X, relative to a sample distribution (4.17)
$z = \dfrac{X - E(X)}{\sigma}$	standardized score or value corresponding to a particular value of X, relative to a population distribution (4.17)
$\|z\|$	absolute value of a standardized score (4.18)
$z_{\text{diff.}}$	standardized value for the difference between two means (8.7)
z_X, z_Y	standardized scores corresponding to particular values of X and Y, respectively (13.2)
$z_{(1-\alpha/2)}$	standardized value cutting off the lower $1 - \alpha/2$ proportion of cases in a normal distribution (8.13)
$z_M = \dfrac{M - E(M)}{\sigma_M}$	standardized value corresponding to a particular value of sample M in a sampling distribution of means (5.8)
z'_Y	predicted standardized value on variable Y (13.2)
ζ	(small Greek zeta) value in the Fisher r to Z transformation corresponding to the population correlation ρ (13.13)

SOLUTIONS TO SELECTED EXERCISES

CHAPTER 1

1. (a) A particular 7-digit number appearing in that directory
(b) A particular student currently enrolled in a college or university in the United States
(c) A positive real number (including fractional numbers, of course)
(d) Any current United States Senator
(e) Any whole positive number, including zero
(f) Any positive or negative real number
(g) Any positive real number between 0 and 100

3. (a) $A \cup B$, "A or B or both"
(b) $A \cap \overline{B}$, "A and not B"
(c) $\overline{(A \cup B)}$, "neither A nor B"
(d) $(A \cap \overline{B}) \cup (\overline{A} \cap B)$, "$A$ and not B, or B and not A"
(e) $(A \cap \overline{B} \cap \overline{C}) \cup (\overline{A} \cap B \cap \overline{C}) \cup (\overline{A} \cap \overline{B} \cap C)$, "only A, or only B, or only C"
(f) $(A \cap B \cap \overline{C}) \cup (A \cap \overline{B} \cap C) \cup (\overline{A} \cap B \cap C)$, "only A and B, or only A and C or only B and C"
(g) $\overline{(A \cup B)}$, "neither A nor B" (notice that this may or may not include C)

5. (a) $p(\mathscr{S}) = p(A) + p(B) + p(C) = (1/4) + (3/5) + (3/20) = 1.00$
(b) $p(\varnothing) = 0$
(c) $p(A \cup B) = p(A) + p(B) = (1/4) + (3/5) = 17/20$
(d) $p(B \cup C) = p(B) + p(C) = (3/5) + (3/20) = 3/4$
(e) $p(A \cap B) = p(\varnothing) = 0$
(f) $p(\overline{B}) = 1 - (3/5) = 2/5$
(g) $p(A \cap \overline{B}) = p(A) = 1/4$
(h) $p(C \cap \overline{B}) = p(C) = 3/20$
(i) $p(C \cup \overline{B}) = p(C) + p(A) = 2/5$

7. (a) $p(\text{ace}) = 1/13$
(b) $p(\text{red}) = 1/2$
(c) $p(\text{face card}) = 3/13$
(d) $p(\text{even value}) = 5/13$
(e) $p(\text{spade or diamond}) = (1/4) + (1/4) = 1/2$
(f) $p(\text{10 in red suit}) = 1/26$

9. (a) $p(\text{female}) = 2/3$
(b) $p(\text{brunette}) = 1/3$
(c) $p(\text{red-haired}) = 2/9$
(d) $p(\text{red-haired male}) = 1/9$
(e) $p(\text{68 inches or more}) = 4/9$
(f) $p(\text{blonde female under 65 inches}) = 2/9$

11. The 36 elementary events are shown as points in the following table:

		white die					
		1	*2*	*3*	*4*	*5*	*6*
black die	*1*	·	·	·	·	·	·
	2	·	·	·	·	·	·
	3	·	·	·	·	·	·
	4	·	·	·	·	·	·
	5	·	·	·	·	·	·
	6	·	·	·	·	·	·

(a) 6
(b) 15
(c) 6
(d) 30
(e) 9
(f) 27
(g) 22

13. (a) p(vowel) = 3/13
(b) p(consonant other than s or t) = 9/13
(c) p(consonant, first half) = 5/13
(d) p(letter in EXODUS) = 3/13
(e) p(letter in BORN or in FILM) = (4/26) + (4/26) = 4/13
(f) p(letter in LEAD or in LOAD) = (4/26) + (4/26) − (3/26) = 5/26

15. (a) $p(3) = 1/12$
(b) p(even red) = 1/6
(c) p(5 or more blue) = 2/9
(d) p(red or 8) = (12/36) + (3/36) − (1/36) = 7/18
(e) p(white and odd) = 1/6
(f) p(white or odd) = (12/36) + (18/36) − (6/36) = 2/3

17. (a) p(less than 19) = p(boy)p(less than 19, given boy) + p(girl)p(less than 19, given girl) = .75
(b) If independent, p(19 or over, given boy) = p(19 or over). However, this is not true here, so that these two events are not independent in this sample space.
(c) By Bayes' theorem, $p(\text{boy}|\text{19 or over}) = \dfrac{(.37)(.52)}{(.37)(.52) + (.12)(.48)} = .77$
(d) $p(\text{girl}|\text{under 19}) = .56$

19. (a) marginal probabilities: $p(A_1) = .33$, $p(A_2) = .34$, $p(A_3) = .33$
(b) $p(0) = .2$, $p(1) = .35$, $p(2) = .45$
(c) $p(0|A_1) = .15$
(d) $p(\text{2 or more}|A_2) = .44$
(e) $p(A_3|0) = .65$

21. (a) p(black or white) = 4/9
(b) p(yellow) = 2/9
(c) p(neither white nor yellow) = 5/9
(d) p(green or blue) = 2/9
(e) p(blue or not black) = 7/9

23. (a) p(one culture) = 11/21
(b) p(two cultures) = 1/3
(c) p(three cultures) = 1/7
(d) p(more than 1) = 1 − (11/21) = 10/21
(e) p(less than three) = 1 − (1/7) = 6/7

25. (a) $p(\text{productive}|A) = 5/7$
(b) $p(\text{productive}|C) = 2/3$
(c) $p(\text{productive}|\overline{A} \cap B \cap \overline{C}) = 1/3$
(d) $p(\text{productive}|A \cap \overline{B} \cap \overline{C}) = 2/3$
(e) $p(\text{productive}|A \cap B \cap C) = 2/3$

27. (a) $p = 5/6$ (c) $p = .5$
(b) $p = .3$

29. (a) .87 (d) .38
(b) .03 (e) .62
(c) .78

31. (a) .36 (e) .06
(b) .46 (f) .81
(c) .27 (g) .76
(d) .21 (h) .93

33. (a) .448 (e) .670
(b) .521 (f) .349
(c) .631 (g) .050
(d) .137 (h) .287

35. (a) 187:813, or about 23:100 (d) 255:745, or about 17:50
(b) 264:736, or about 9:25 (e) 558:442, or about 63:50
(c) 294:706, or about 21:50

37. (a) .078 (e) .003
(b) .065 (f) .019
(c) .035 (g) .542
(d) .063

CHAPTER 2

1. (a) Nominal (f) Ratio
(b) Ratio (g) Interval
(c) Ordinal (often treated as interval) (h) Interval
(d) Ordinal (i) Ordinal
(e) Interval (j) Nominal

3.

Rating	*f*
very good	2
good	3
acceptable	7
poor	4
	16 = *N*

5.

Item I

Response	*f*
5	3
4	5
3	7
2	9
1	6
	30 = *N*

Item II

Response	*f*
5	2
4	7
3	12
2	7
1	2
	30 = *N*

Item III

Response	*f*
5	6
4	9
3	7
2	5
1	3
	30 = *N*

7.

Class interval	*Midpoint*	*f*
3793–3913	3853	3
3672–3792	3732	1
3551–3671	3611	4
3430–3550	3490	5
3309–3429	3369	3
3188–3308	3248	2
3067–3187	3127	4
2946–3066	3006	2
2825–2945	2885	0
2704–2824	2764	1
		25 = *N*

9. For $i = 5.1$:

Class interval	*Midpoint*	*f*
90.0–95.0	92.5	9
84.9–89.9	87.4	5
79.8–84.8	82.3	5
74.7–79.7	77.2	2
69.6–74.6	72.1	2
64.5–69.5	67.0	1
59.4–64.4	61.9	7
54.3–59.3	56.8	13
49.2–54.2	51.7	5
44.1–49.1	46.6	2
39.0–44.0	41.5	1
		52 = *N*

11.

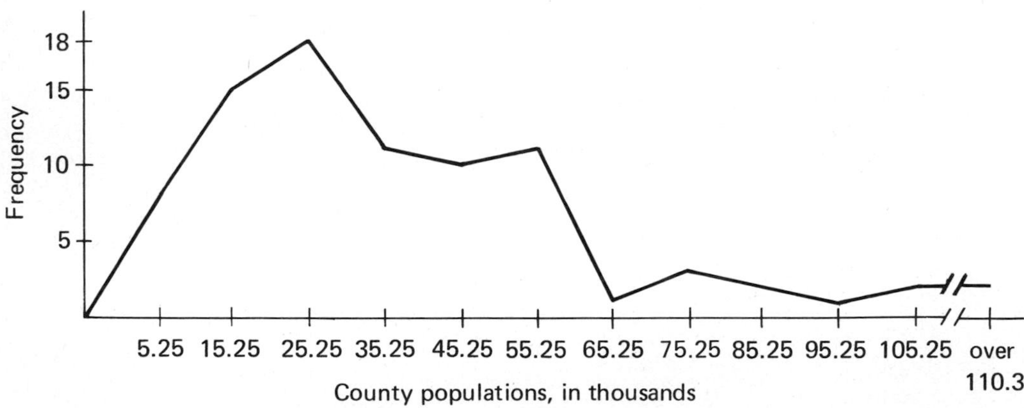

13. (Figure not given.)

15. (a) .093
(b) .333
(c) .130
(d) .722

17. Frequency distribution of weights, using $i = 5$ pounds:

Class interval	*Midpoint*	*f*
149–153	151	2
144–148	146	5
139–143	141	1
134–138	136	8
129–133	131	3
124–128	126	13
119–123	121	12
114–118	116	15
109–113	111	9
104–108	106	7
99–103	101	3
		78 = *N*

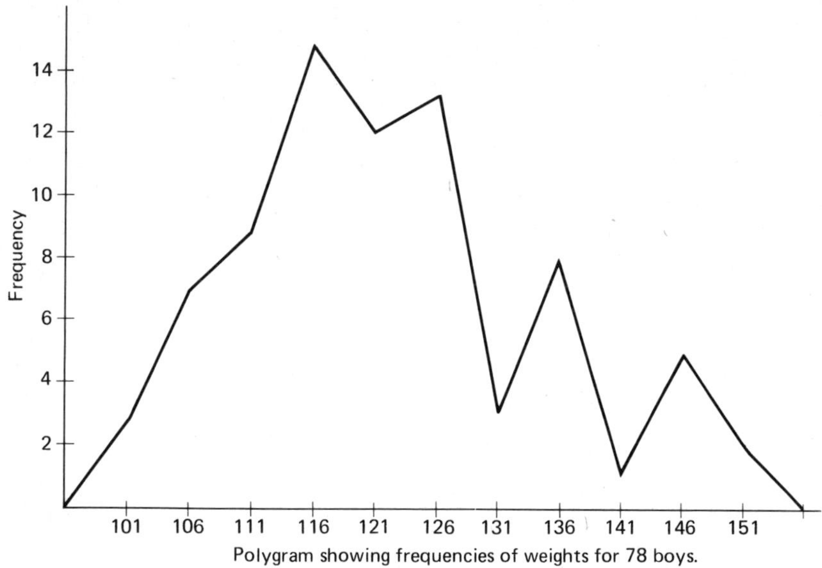

Polygram showing frequencies of weights for 78 boys.

19. (a) Probability mass function:

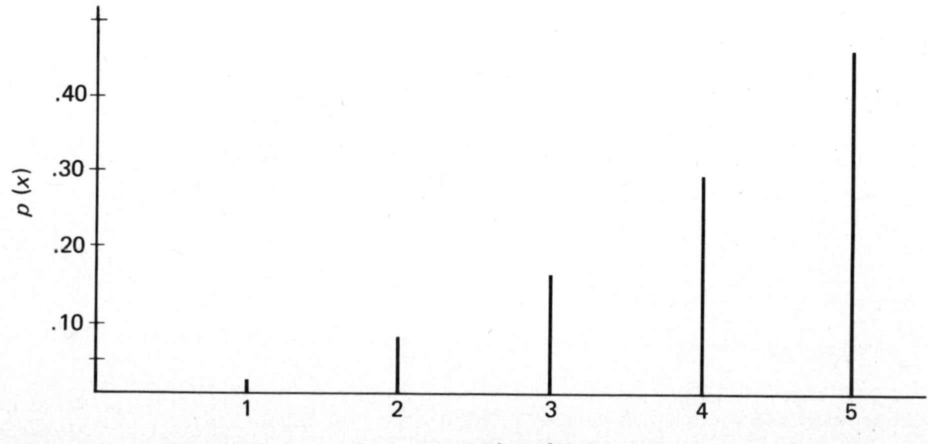

$p(1) = .018$, $p(2) = .073$, $p(3) = .164$, $p(4) = .291$, $p(5) = .454$.

(b) Step function:

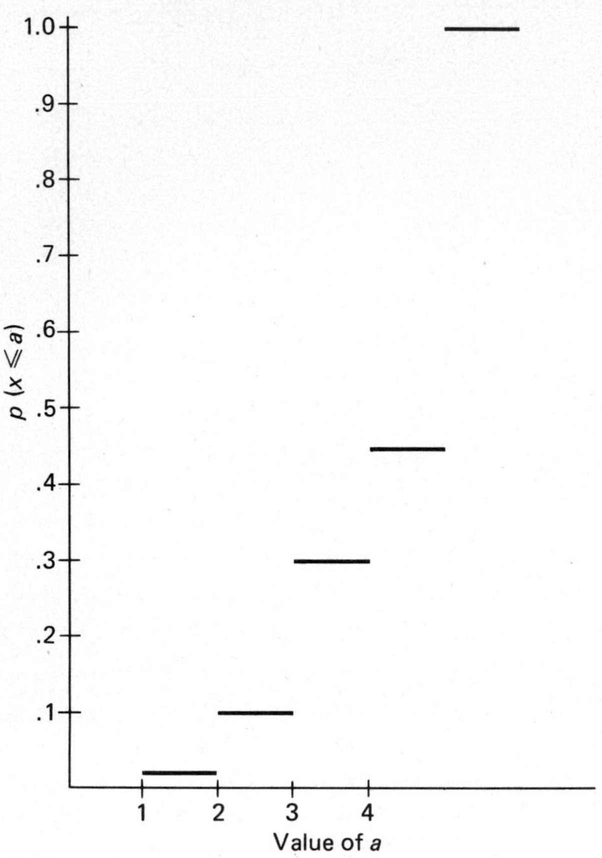

(c) $p(2, 3, \text{or } 4) = .073 + .164 + .291 = .528$

(d) $p(\text{some value}) = 1.00$

21. (a) Probability density function:

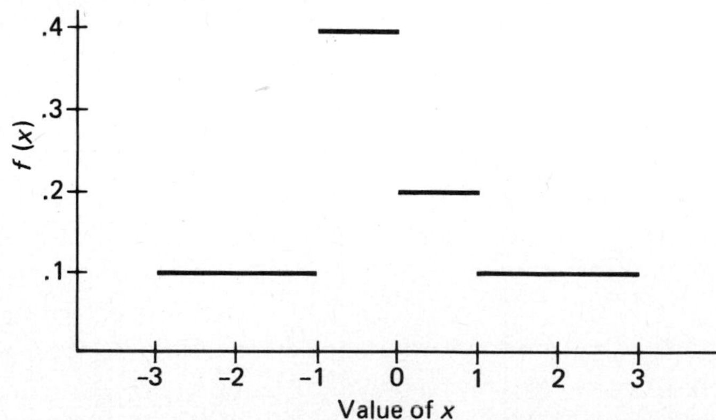

(b) It can be seen that this is not a strictly continuous distribution by the breaks or "jumps" in its probability function.

23.

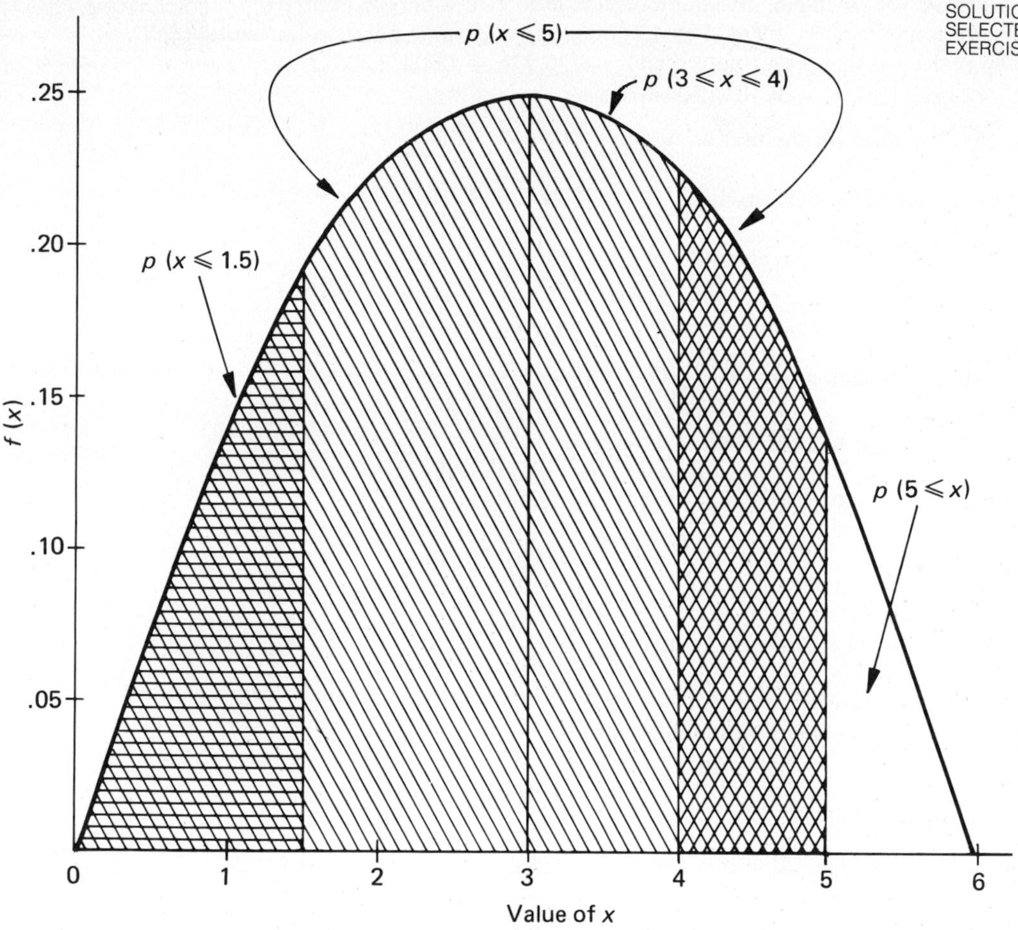

$p\ (x \leqslant 5)$

$p\ (3 \leqslant x \leqslant 4)$

$p\ (x \leqslant 1.5)$

$p\ (5 \leqslant x)$

$f(x)$

.25 .20 .15 .10 .05

0 1 2 3 4 5 6

Value of x

25.

Value of X = a	*Expected frequency*
20	.96
18	7.20
16	4.32
14	5.28
12	5.28
10	7.20
8	7.68
6	5.28
4	3.36
2	1.44
0	0
	$48 = N$

CHAPTER 3

1. $6! = 720$

3. There are $(12)^4 = 20736$ sequences of four notes. There are only 12 sequences, however, where the same note is struck four times. Hence, p(same note four times) = 12/20,736 = .0006. In addition, p(at least two different notes) = 1 − p(same note four times) = .9994. For 12 notes, there are nine possible four note ascending scales, so that p(four-note scale) = 9/20,736 = .0004.

5. (7)(6)(5)(4)(3)(2)(1) = 7! = 5040 different ways

7. $\binom{20}{10}$ = 184,756 ways for the first group

$\binom{10}{10}$ = 1 way for the second group

probability of any given assignment $= \dfrac{1}{\binom{20}{10}} = \dfrac{1}{184{,}756}$

9. probability(one all-black group) $= \dfrac{6\binom{20}{5}\binom{30}{5}\binom{25}{5}\binom{20}{5}\binom{15}{5}\binom{10}{5}}{\binom{50}{5}\binom{45}{5}\binom{40}{5}\binom{35}{5}\binom{30}{5}\binom{25}{5}}$

probability(two black groups) $= \dfrac{\binom{6}{2}\binom{20}{5}\binom{10}{5}\binom{30}{5}\binom{25}{5}\binom{20}{5}\binom{15}{5}}{\binom{50}{5}\binom{45}{5}\binom{40}{5}\binom{35}{5}\binom{30}{5}\binom{25}{5}}$

11. number of possible committees $= \binom{10}{2}\binom{15}{3}\binom{20}{4} = 99{,}201{,}375$

13. probability that all four fall in same month $= \binom{12}{1}\left(\dfrac{1}{12}\right)^4 = \left(\dfrac{1}{12}\right)^3$

probability that no two fall in the same month $= \dfrac{\binom{12}{4}}{(12)^4}$

probability of two or more in the same month $= 1 - \dfrac{\binom{12}{4}}{(12)^4} = .976$

15. The probability of 3 repeated letters and 3 repeated digits is

$\dfrac{\binom{26}{1}\binom{10}{1}}{(26)^3(10)^3}$, or about .000015

The probability of three repeated letters and three different digits is

$\dfrac{\binom{26}{1}\binom{10}{3}}{(26)^3(10)^3}$, or about .00018

For three different letters and three different digits, the probability is

$\dfrac{\binom{26}{3}\binom{10}{3}}{(26)^3(10)^3} = .018$

17. $p(0 \text{ or } 1; N = 15, .30) = \binom{15}{0}(.3)^0(.7)^{15} + \binom{15}{1}(.3)^1(.7)^{14}$

$= .035$

$p(2 \text{ or more}; N = 15, .30) = 1 - .035 = .965$

19. (a) $p(3; N = 18, .20) = .23$
(b) $p(7; 18, .20) = .035$
(c) $p(14; 18, .20)$ = almost zero

(d) $p(\text{number correct} > 5; 18, .20) = .13$
(e) $p(\text{number correct} < 2; 18, .2) = .10$
(f) $p(2 \leqslant \text{number correct} \leqslant 7; 18, .20) = .88$

21. $p(9 \text{ or more undernourished}; N = 20, .25) = .04$. The evidence supports the conconclusion that more than 25 percent are undernourished.

23. Here there are six pairs favoring group II, and one tied pair, out of the ten differences observed. Then

$$p(6 \text{ or more positive differences}; N = 9, p = .5) = .25$$

This is not an unlikely result to obtain when the groups are actually equal.

25. $p(0 \text{ errors}; m = .2) = \dfrac{e^{-.2}(.2)^0}{0!} = .82$

$p(1 \text{ or more errors}; m = .2) = 1 - .82 = .18$

27. $p(N = 4; 1, .60) = (.6)(.4)^3 = .038$
$p(N = 1; 1, .6) = .6$

29. The multinomial rule gives the probability of this distribution of preferences to be

$$\frac{20!}{5!3!4!0!8!}(.32)^5(.14)^3(.19)^4(.03)^0(.31)^8 = .0036$$

31. (a) $p(10 \text{ sophomores}) = \dfrac{\binom{18}{10}}{\binom{50}{10}}$

(b) $p(7 \text{ seniors and } 3 \text{ juniors}) = \dfrac{\binom{7}{7}\binom{10}{3}}{\binom{50}{10}}$

(c) $p(3 \text{ freshmen}, 4 \text{ sophomores}, 2 \text{ juniors}, 1 \text{ senior})$

$$= \frac{\binom{15}{3}\binom{18}{4}\binom{10}{2}\binom{7}{1}}{\binom{50}{10}}$$

CHAPTER 4

1. Mode is at classification B. This is chosen because the data are nominally scaled. Thus, $p(\text{wrong}) = 1 - (18/54) = 2/3$.

3. Mean = 24.43, median = 24.50

5. Mean = 3404.52

7. $M = .308$. Yes, since the mean is positive.

9.

Boys	*Girls*
median = 34.16	median = 28.25
mode = 37.495	mode = 27.495

11. $E(X) = 101.72$. The most likely value can be taken as the midpoint of the most frequent interval, or 103.

13. Median = 15.925, mean = 16.2, mode = 16. The differences among these indices reflect the distribution's positive skewness.

15. (1) $E(X) = (1/6) + 2(1/6) + 3(1/6) + 4(1/6) + 5(1/6) + 6(1/6) = 3.5$
(2) $E(X) = (1/21) + 2(2/21) + 3(3/21) + 4(4/21) + 5(5/21) + 6(6/21) = 4.33$
(3) $E(X) = (1/12) + 2(2/12) + 3(3/12) + 4(3/12) + 5(2/12) + 6(1/12) = 3.5$

17. $E(X) = 13(1/16) - 2(2/16) - 1(3/16) + 0(4/16) + (3/16) + 2(2/16) + 3(1/16) = 0.$

19. $S^2 = 21.43$, $S = 4.63$

21. $S^2 = 1.49$, $S = 1.22$

23. The boys' average is about 33.2 and the girls' average is 28.8. Since the boys have a standard deviation of 7.94 as compared with 7.20, the girls are more homogeneous than the boys in this respect.

25. $E(X) = 7$, $\text{Var}(X) = 5.83$, standard deviation $= 2.42$

27. (1) $E(X) = 3.5$, $\text{Var}(X) = 2.92$, standard deviation $= 1.71$
(2) $E(X) = 4.33$, $\text{Var}(X) = 2.25$, standard deviation $= 1.50$
(3) $E(X) = 3.5$, $\text{Var}(X) = 1.92$, standard deviation $= 1.38$

29. $E(X) = 0$, $\text{Var}(X) = 2.5$, standard deviation $= 1.58$

31. $E(X) = 6$, $\text{Var}(X) = 2.4$ standard deviation $= 1.55$

33.

	Divorced	*Married*
(a)	.12	− .24
(b)	1.20	.96
(c)	−1.60	−2.16
(d)	2.71	2.64
(e)	− .96	−1.44

35. (a) Since $p(|z| \geq 4.318) \leq .054$, the maximum probability is .054.
(b) $p(|z| \geq 1.7) \leq (1/1.7^2)$, so the probability is less than or equal to .346.
(c) Since $p(|z| \geq 1.3) \leq .59$, the proportion must be greater than or equal to .41.

CHAPTER 5

1. $E(X) = 1.5$, $\text{Var}(X) = .25$

3. $E(G) = 4(.25) + 3(.50) + 2(.25) = 3 = 2E(X)$

5. The sampling distribution has a larger range of values than the population distribution on which it is based. There is also a distinct mode at $G = 3$, whereas the population distribution has no mode. Here, the expected value of G is twice the expected value of X and the variance of G is twice the variance of X.

7. $E(G) = 2(.125) + 1.67(.375) + 1.33(.375) + 1(.125) = 1.5 = E(X)$
$\sigma_G^2 = 4(.125) + (1.67)^2(.275) + (1.33)^2(.375) + (.125) - (1.5)^2$
$= .83 = .25/3 = \sigma_X^2/3$

9. The population distribution is

x	*p(x)*
1	.25
0	.50
−1	.25
	1.00

$E(X) = 0$, $\sigma_X^2 = .50$, $\sigma_X = .707$

11. $E(M) = E(X) = 0$, $\sigma_M^2 = .25 = \sigma_X^2/2$

13. (Here a number of decimal places are carried so that the probabilities will sum to 1.00.) The distribution for the mean, given $N = 3$, is

M	*p(M)*
1.0	.015625
.67	.093750
.33	.234375
.00	.312500
− .33	.234375
.67	.093750
−1.00	.015625
	1.000000

We note that distribution is much more "bell-shaped" than the population distribution from which the samples are taken.

15. (a) $p(M = 1) = .016$
(b) $p(M = 0) = .313$
(c) $p(M \leqslant 1/3) = .891$
(d) $p(M < 0) = .344$
(e) $p(M \leqslant 2/3) = .984$
(f) $p(M < -1/3) = .109$

17. p(8 females; $N = 15$, $p = .6$) $= .1387$
p(8 females; $N = 15$, $p = .4$) $= .0925$
Conclude that the cards belong to the group with 60 percent females, since this gives the highest prior likelihood to the sample results.

19. $p(3; N = 20, .05) = .06$
$p(3; N = 20, .20) = .21$
By the principle of maximum likelihood, we should choose .2 as the better estimate of p. However, the best estimate is .15.

21. For Typist I, the likelihood (by the Poisson rule, and letting $m = 5$ per 5 letters) is

$$\frac{e^{-5}(5)^2}{2!} = .084.$$

For Typist II, letting $m = 3$ per five letters, the likelihood is

$$\frac{e^{-3}(3)^2}{2!} = .224.$$

For Typist III, where $m = .5$ per five letters, we have

$$\frac{e^{-.5}(.5)^2}{2!} = .076.$$

Hence, by the principle of maximum likelihood, Typist II is the best bet.

23. Interviewer A: $p(N = 5; r = 3, p = .6) = .207$
Interviewer B: $p(N = 5; r = 3, p = .4) = .138$
Interviewer C: $p(N = 5; r = 3, p = .3 = .079$
Interviewer D: $p(N = 5; r = 3, p = .1) = .005$
Interviewer A is the best bet.

25. If the sample P value is .4, then, in the sampling distribution for $p = .52$, this amounts to a z-score of $(.40 - .52)/.032 = -3.75$. Then, according to the Tchebycheff inequality, for a symmetric distribution,

$$p(|z| \geqslant 3.75) \leqslant \frac{4}{9(3.75)^2} \text{ or } .03.$$

Hence the probability of a deviation as large or larger than this, if $p = .52$ is true, is only .03 or less.

27. $E(M) = \mu = 1.5$, $\sigma_M^2 = .25/50 = .005$, $\sigma_M = .071$, $Z = .70$. The probability of a value this much or more deviant from expectation is something less than .91 according to the Tchebycheff inequality. Hence the result is relatively likely.

29. Since $\sigma_M = 4.9/\sqrt{150} = .40$, the sample mean of 12.3 represents a z-value of -12.75 if the true value of $\mu = 17.4$. Thus, by the Tchebycheff inequality, the chances are less than .003 for a sample result this much or more deviant, given $\mu = 17.4$. It appears to be quite safe to conclude that the new population is different on the average.

31. $M = 30.8$ is the best available estimate of μ. The value $s^2 = 47.96$ is the best available estimate of σ^2.

33. This represents a finite population (208 students) from which this sample of 49 students was drawn. Thus, the estimate of the population mean is $M = 684.85$. The estimate of the population variance is

$$\text{est. } \sigma^2 = \frac{N(T-1)}{(N-1)T} S^2 = \frac{49(207)}{48(208)}(14{,}219.19) = 14{,}445.63$$

Then the estimated standard deviation for the population is 120.19. The estimated total amount spent is (208)(684.45) = 142,365.6.

35. $$\text{pooled estimate of } \mu = \frac{(10)(96) + (20)(105) + (15)(103)}{45} = 102.33$$

$$\text{pooled estimate of } \sigma^2 = \frac{(10)(22) + (20)(29) + (15)(31)}{45 - 3} = 30.12$$

$$\text{est. } \sigma_M = \frac{5.49}{\sqrt{45}} = .82$$

37. $\mu = 422.25$, $\sigma^2 = 3564$

39. $M_1 - M_2 = 4.12$ units approximately

CHAPTER 6

1. (a) .115 (d) .038
(b) .832 (e) .334
(c) .970 (f) .984

3. (a) .964 (d) .309
(b) .067 (e) .773
(c) .008

5. About .21 is the probability of this sample or one more deviant from expectation if the true mean is still 23.6. Thus, there is not strong reason to say that the mean has changed. Since the sample size is fairly large, the central limit theorem may be relied upon to counter the effect of the nonnormal population.

7.

	$p = .5, N = 8$		$p = .4, N = 8$	
	Binomial	*Approx. normal**	*Binomial*	*Approx. normal**
8	.0039	.007	.0007	.001
7	.0312	.032	.0079	.008
6	.1094	.106	.0413	.040
5	.2188	.218	.1239	.123
4	.2734	.274	.2322	.241
3	.2188	.218	.2787	.278
2	.1094	.106	.2090	.198
1	.0312	.032	.0896	.084
0	.0039	.007	.0168	.027

*Values read from the normal table will vary depending on rounding methods used. Note that the fit between the normal and binomial probabilities is very nearly as close for $p = .4$ as for $p = .5$, even for N as low as 8.

9. Since for $\mu = .1$ and $\sigma = .03$, so that $\sigma_M = .0095$, the obtained z-value is -31.58. We can say that this sample result is *most* unlikely.

11. The limits of the 99 percent confidence interval are approximately 3.73 and 3.83.

13. $p(303 \leq M \leq 304) = .015$; $p(304 \leq M \leq 305) = .0023$. The odds are about 6.5 to 1.

15. $p(M \leq 294) = .0014$; $p(306 \leq M) = .0014$
17. If $p(\text{cover}) = .90$, then $p(\text{not cover}) = .10$. Hence, $E(\text{number that do not cover}) = 200(.1) = 20$.
19. The limits for the interval are about .126 and .303.
21. The confidence limits may be found by taking

$$\frac{1000}{1000 + (2.58)^2}\left[.51 + \frac{(2.58)^2}{2000} \pm 2.58\sqrt{\frac{(.51)(.49)}{1000} + \frac{(2.58)^2}{4(1000)^2}}\right]$$

giving the limits as .47 and .55. This interval does not include the value of .46 favoring (.54 against). It does include the value of .50, however. Thus, there is some reason to believe that .50 or more might favor the issue.
23. Let $1.96 = .01/(.07/\sqrt{N})$. Then $N = 188.24$, or 189.
25. The probability is less than .01. Hence the risk in rejecting the possibility that the true proportion is .18 or more is also less than .01.
27. The limits are about 11.64 and 12.96. These limits define a narrower interval than before, since the exact normal probabilities were used rather than the inexact Tchebycheff limits.
29. Based on the relation

$$\frac{K\sigma}{\sigma/\sqrt{N}} = 1.96,$$

the table is as follows:

N	K
1	1.96
5	.88
10	.62
20	.44
30	.36
40	.31
50	.28
100	.20

31. The 99 percent confidence interval has limits 645.62 to 723.28.
33. Pooled estimate of $\mu = 102.33$. Pooled estimate of $\sigma = \sqrt{30.12} = 5.49$. Since $z = .06$, the probability of a deviation this large or larger is about .48.
35. The sampling distribution of means drawn from theoretical distribution 3 would probably be most nearly normal, since the population itself is unimodal and symmetric. The third distribution would also have the smallest standard error, or .195.

CHAPTER 7

1. $H_0{:}p = .39$; $H_1{:}p \neq .39$. The two rejection regions correspond to the following:

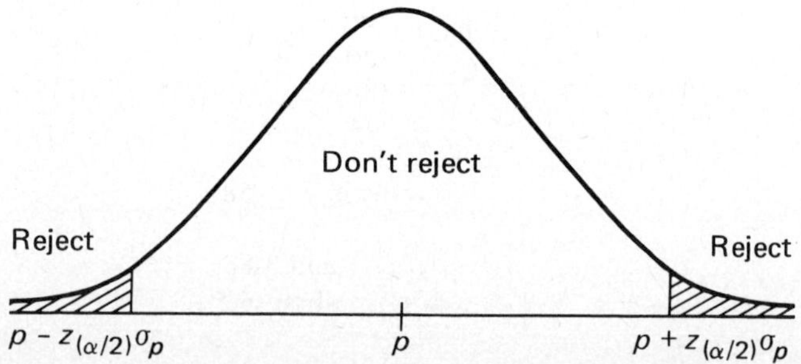

3. $H_0{:}\mu \geqslant 172$; $H_1{:}\mu < 172$. The value $\sigma = 16$ is assumed in either case. The rejection region is shown as follows:

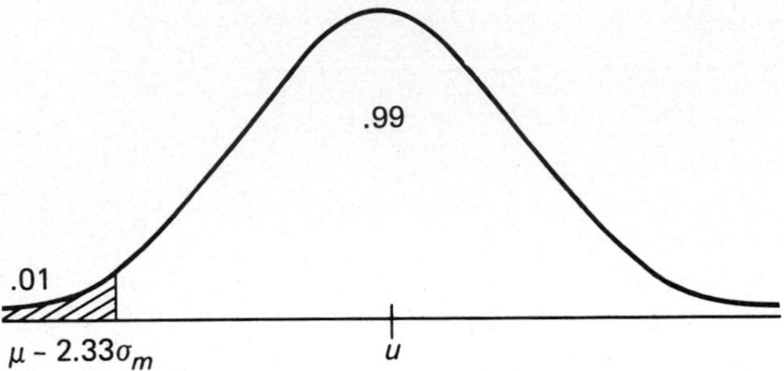

5.

		true situation	
		child belongs in 7th grade	child belongs in 6th grade
decide	place in 7th grade	correct	error
	place in 6th grade	error	correct

7. (a) Rule 6(c), under the minimax criterion
(b) Rule 6(b) by minimax
(c) Rule 6(a) by minimax
(d) Rule 6(c) by minimax

9. (a) The table of expected losses is as follows:

		true situation	
		$p = .4$	$p = .2$
Rule	8(a)	833.70	6.40
	8(b)	46.30	642.20
	8(c)	382.20	120.90
	8(d)	167.20	322.20

The minimax criterion would indicate Rule 8(d).

(b) For this set of loss values,

		true situation	
		$p = .4$	$p = .2$
Rule	8(a)	833.70	9.60
	8(b)	46.30	936.30
	8(c)	382.20	181.35
	8(d)	167.20	483.30

This suggests rule 8(c).

(c) For this set,

		true situation	
		$p = .4$	$p = .2$
Rule	8(a)	4168.50	6.40
	8(b)	231.50	624.20
	8(c)	1911.00	120.90
	8(d)	836.00	322.20

Rule 8(b) would fit the minimax criterion.

(d) For this set,

		true situation	
		$p = .4$	$p = .2$
Rule	8(a)	833.70	64.00
	8(b)	46.30	6242.00
	8(c)	383.20	1209.00
	8(d)	167.20	3222.00

This suggests rule 8(a).

11. The value of z is 2.91. Since this exceeds 1.96, reject H_0.

13. H_0: $p \leq .25$; H_1: $p > .25$. Then

$$z = \frac{.45 - .25 - (.5/20)}{\sqrt{(.25)(.75)/20}} = 1.81$$

This is significant beyond the .05 level, one-tailed; thus the null hypothesis is rejected.

15. In place of the unknown value of σ^2, since N is large we can here use est. $\sigma^2 = s^2 = 1.626$. Then $z = 2$, permitting us to reject H_0 for $\alpha = .05$ (two-tailed).

17. Assuming $\sigma = 2$, then the z-value is 4.14, which exceeds the value for $\alpha = .01$, one-tailed. The H_0 that $\mu = 15$ may safely be rejected.

19. N should be about 44.

21. Since the critical value of M under H_0: $\mu = 28.6$ is 30.214 for the upper tail and 26.98 for the lower tail, the z for 30.214 is -2.17, and the z for the other critical value is -6.09 when $\mu = 32$. The power is .985. When $\mu = 25$, the equivalent z-values are 6.33 and 2.40. The power is then about .99 for $\mu = 25$.

23. In this one-tailed test, the power is .9975 against the alternative $\mu = 16$, and over .9999 against $\mu = 20$.

25.

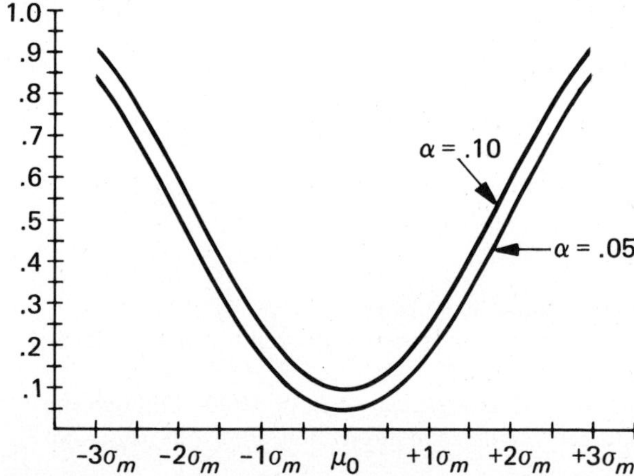

27. The probability that a Type I error is made on both tests is $(.05)^2 = .0025$. The probability that neither involves a Type I error is $(.95)^2 = .9025$. Then the probability of at least one Type I error is $1 - .9025 = .0975$.

29. $$\mathscr{P}(200|\text{data}) = \frac{(.15)(.5)}{(.15)(.50) + (.23)(.25) + (.20)(.25)} = .41$$

$$\mathscr{P}(210|\text{data}) = \frac{(.23)(.25)}{(.15)(.50) + (.23)(.25) + (.20)(.25)} = .32$$

$$\mathscr{P}(220|\text{data}) = \frac{(.20)(.25)}{(.15)(.50) + (.25)(.23) + (.20)(.25)} = .27$$

31. For Decision Rule B the probabilities are

		true H_0	true H_1	true H_2
accept	H_0	.9599	.2266	.0006
accept	H_1	.0399	.6147	.0662
accept	H_2	.0002	.1587	.9332

CHAPTER 8

1. The t-value of -1.26 with 185 degrees of freedom is not significant.
3. For H_0: $\mu \geqslant 17.34$ and H_1: $\mu < 17.34$, the obtained t-value of -2.5 lets one reject the null hypothesis beyond the .01 level, one-tailed.
5. Employing the t-value for $\alpha = .01$ (two-tailed) at ∞ degrees of freedom, we find the confidence limits to be 98.51 and 98.89
7. The required confidence interval has limits of 32.04 and 32.76.
9. For 79 degrees of freedom, the t-value of -2.48 lets us reject H_0: $\mu \geqslant 115.2$ beyond the .01 level (one-tailed).
11. The limits for the 95 percent confidence interval are 34.74 and 37.46.
13. The hypothesis H_0: $\mu_1 - \mu_2 = 0$ would be rejected in favor of the H_1: $\mu_1 - \mu_2 \neq 0$, since $t = 4.81$. This is significant beyond the .01 level (two-tailed).
15. The t-value of 2.81 is significant beyond the .005 level, two-tailed.
17. The t-value of -1.33 is not significant. We assume independent random samples from normal populations with the same variance.
19. On taking

$$\sigma_Y^2 = \sigma_{Y|X}^2 + \frac{(\mu_1 - \mu_2)^2}{4} = 16 + \left(\frac{9}{4}\right) = 18.25$$

so that

$$\omega^2 = \frac{9}{4(18.25)} = .123,$$

we have, from 8.13.2,

$$N \geqslant \frac{(1.282 + 1.96)^2(.877)}{2(.123)} = 37.47.$$

Therefore, $N = 39$ for each sample should be sufficient.

21. For 38 degrees of freedom the t-value of 2.02 is significant right at the .05 level, two-tailed.
23. For the hypotheses H_0: $\mu_1 - \mu_2 \leqslant 1000$ and H_1: $\mu_1 - \mu_2 > 1000$, although the t-value was -3.66, this falls into the region of nonrejection, and thus H_0 is not rejected.
25. The t-value of -1.17 does not permit us to reject the null hypothesis H_0: $\mu_D \geqslant 0$ in favor of the alternative H_1: $\mu_D < 0$.
27. The confidence interval has limits of $-.35$ and 3.35.
29. The limits of the 95 percent confidence interval are 1.75 and 5.91.

CHAPTER 9

1. $p(|z| \geqslant 1.15) = .25$, so $p[\chi^2_{(1)} \geqslant (1.15)^2] = .25$

3. $E(\chi^2_{(5)}) = 5$; $p(15 < \chi^2_{(5)})$ is about .01, $p(11 < \chi^2_{(5)})$ is about .05, $p(\chi^2_{(5)} < 1.6)$ is about .10.
5. The confidence limits are approximately 74.19 and 866.60.
7. The confidence limits for the mean are 46.57 and 55.43. For the variance the confidence limits are about 59.19 and 204.22.
9. Since $N(S^2)/\sigma^2 = \chi^2$, if S is 1.3 times σ, the value of chi-square will be $(30)((1.3)^2 = 50.7$. For 29 degrees of freedom a value this large or larger has probability of less than .01.
11. The limits of the confidence interval are about 5.25 and 81.01.
13. For 268 degrees of freedom, the chi-square values required for significance (two-tailed) may be found from

$$\chi^2_{(268;\,.025)} = \frac{(1.96 + \sqrt{2(268) - 1})^2}{2} = 314.76$$

and

$$\chi_{(268;\,.975)} = \frac{(-1.96 + \sqrt{2(268) - 1})^2}{2} = 224.09.$$

Hence the obtained value of 237.69 is not significant.
15. The null hypothesis H_0: $\sigma = 24$ may be rejected in favor of H_1: $\sigma \neq 24$ since the chi-square value of 41.02 is significant beyond the .05 level for 24 degrees of freedom.
17. Here, $s_1^2 = 984.38$, and $s_2^2 = 957.29$; thus $F = 1.03$ with 24 and 24 degrees of freedom. This is not significant.
19. Here, H_0: $\sigma_1^2 = \sigma_2^2$ against H_1: $\sigma_1^2 > \sigma_2^2$. The $F = 2.14$ is significant at the .05 level for 19 degrees of freedom. H_0 is thus rejected in favor of H_1.
21. The F value of 1.12 is not significant for 12 and 8 degrees of freedom.
23. Yes, when H_0 is true the ratio s_1^2/ks_2^2 should be distributed as F.
25. $E(M) = 2.5$ minutes; $\sigma_M = .079$.

CHAPTER 10

1. $M = 99.81$, est. $\alpha_1 = -5.31$ (actual $= -5$), est. $\alpha_3 = -2.31$ (actual $= -2$), est. $\alpha_2 = 5.84$ (actual $= 4$) est. $\alpha_4 = 1.64$ (actual $= 3$). The "data" are

I	II	III	IV
90	102	98	106
96	109	94	97
95	110	97	107
97	102	101	96

3.

Source	*SS*	*df*	*MS*	*F*	*p*
Between groups	603.63	2	301.82	9.1	<.01
Within groups	1890.95	57	33.17		
Total	2494.58	59			

Yes, motor coordination is apparently related to athletic interests.

5.

Source	SS	df	MS	F	p
Between conditions	160.0	1	160	1.78	not signif.
Within conditions	719.6	8	89.95		
Total	879.6	9			

Here, $t = -\sqrt{1.78}$, as before.

7.

Source	SS	df	MS	F	p
Between groups	41242.40	5	8248.8	46.87	$\ll .01$
Within groups	9504.00	54	176.00		
Total	50746.40	59			

Yes, we can say so with very small probability of being wrong.

9.

Source	SS	df	MS	F	p
Between groups	2.45	1	2.45	.422	not signif.
Within groups	104.50	18	5.806		
Total	106.95	19			

11.

Source	SS	df	MS	F	p
Between groups	698.2	3	232.7	4.69	$<.05$
Within groups	793.6	16	49.6		
Total	1491.8	19			

13.

Source	SS	df	MS	F	p
Between groups	.409	2	.2045	5.73	$<.05$
Within groups	.607	17	.0357		
Total	1.016	19			

15.

Source	SS	df	MS	F	p
Between groups	177.39	3	59.13	1.92	not signif.
Within groups	738.64	24	30.77		
Total	916.03	27			

17.

Source	SS	df	MS	F	p
Between groups	223.10	5	44.62	.663	not signif.
Within groups	2758.22	41	67.27		
Total	2981.32	46			

The estimate of ω^2 is zero.

19. The estimate of $\omega^2 = .06$. The estimated effects are est. $\alpha_1 = -.00425$, est. $\alpha_2 = .08475$, est. $\alpha_3 = -.03535$, est. $\alpha_4 = -.04525$.

21. est. $\alpha_1 = 12.44$ est. $\gamma_{11} = -2.10$
est. $\alpha_2 = -11.47$ est. $\gamma_{21} = -1.36$
est. $\alpha_3 = -\ .97$ est. $\gamma_{31} = 3.46$
est. $\beta_1 = 3.44$ est. $\gamma_{12} = 2.11$
est. $\beta_2 = -\ 3.44$ est. $\gamma_{22} = 1.35$
est. $\gamma_{32} = -3.46$

23.

Source	SS	df	MS	F	p
Between columns	536.22	2	268.11	3.16	not signif.
Between rows	2413.56	2	1206.78	14.22	.01
Interaction	2194.44	4	548.61	6.47	.01
Error (within cells)	2290.75	27	84.84		
Total	7434.97	35			

Type of disease is strongly related to performance. Although time since recovery appears to have some relation to the dependent variable, this does not reach significance in this sample. There is also a significant interaction between type of disease and time since recovery in terms of performance.

25.

Source	SS	df	MS	F	p
Between columns	31141.19	3	10380.40	33.26	<.01
Between rows	460.06	2	230.03	.74	—
Interaction	3499.72	6	583.29	1.87	not signif.
Error (within cells)	7489.33	24	312.06		
Total	42590.30	35			

27. p(at least one Type I error) $= 1 - (.95)^3 = .1426$, for $\alpha = .05$
p(at least one Type I error) $= 1 - (.99)^3 = .0297$, for $\alpha = .99$.

A small value of α is to be preferred, since it will provide a reasonable value for the probability of at least one Type I error.

29.

Source	SS	df	MS	F	p
Between columns	44.67	2	22.34	1.23	not signif.
Between rows	284.67	2	142.34	7.83	<.05
Error (interaction)	72.66	4	18.11		
Total	402.00	8			

31.

Source	SS	df	MS	F	p
Between columns	508.17	2	254.09	.12	–
Between rows	3952.67	3	1317.56	.64	–
Error (interaction)	12299.83	6	2049.97		
Total	16760.67	11			

Here, interaction effects are assumed to be absent.

33. The assumption becomes critical when the F ratio is formed, since unless this assumption is true, the denominator, MS error, will not be distributed as chi-square divided by degrees of freedom, as we assume it to be.

CHAPTER 11

1.

Source	SS	df	MS	F	p
Between managers	731.49	7	104.50	39.43	<<.01
Within managers	190.90	72	2.65		
Total	922.39	79			

Yes, the managers definitely appear to account for variation in employee satisfaction.

3.

Source	SS	df	MS	F	p
Between stretches	10.17	6	1.70	.10	–
Within stretches	1049.60	63	16.67		
Total	1059.77	69			

The coverage of the highway appears to be evenly spaced.

5.

Source	SS	df	MS	F	p
Between stimuli	162	7	23.14	13.23	<.01
Within stimuli	14	8	1.75		
Total	176	15			

The hypothesis of no between-stimuli variance can be rejected.

7.

Source	SS	df	MS	F	p
Between stores	1640.08	5	328.02	3.50	.01
Within stores	5056.10	54	93.63		
Total	6696.16	59			

The hypothesis of no variance attributable to stores is rejected.

9. Approximating the upper 2.5 percent point for F with 10 and 33 degrees of freedom by the value 2.5, and the corresponding lower percentage point by the value .303, the limits are found to be about .35 and .86.

11. Here we form the ratio $.10 = \theta/(1 + \theta)$ and find that $\theta = 1/9$. Then the F ratio is

$$\frac{3.81}{1 + (4/9)} = 2.64.$$

This is larger than the value required for significance with $\alpha = .05$, for 6 and 21 degrees of freedom.

13.

Source	SS	df	MS	F	p
Between groups	1.76	3	.59	$\left(\frac{.59}{.40} = 1.48\right)$	not signif.
Between somatotypes	7.20	4	1.80	$\left(\frac{1.80}{.40} = 4.5\right)$	$< .01$
Interaction	1.91	12	.16	.29	—
Error (within cells)	10.95	20	.55	(40 = pooled error)	

The interaction F is much less than 1.00, so that interaction and error are pooled to give $(1.91 + 10.95)/32 = .40$. The only significant evidence for effects is between somatotypes.

15.

Source	SS	df	MS	F	p
Between managers	731.49	7	104.50	32.29	$<< .01$
Between groups	2.11	1	2.11	.79	—
Interaction	18.79	7	2.68	1.01	not signif.
Error	170.04	64	2.66	(pooled = 2.66)	
Total	922.30	79			

Yes, because interaction is not significant, pooled error is used for the second test. Here the interaction SS is small relative to SS error, so that little change in MS error results.

17.

Source	SS	df	MS	F	p
Between columns	31141.19	3	10390.4	17.80	$< .01$
Between rows	460.06	2	230.03	.39	—
Interaction	3499.72	6	583.29	1.87	not signif.
Error	7489.33	24	312.06		

Here, although interaction is not significant at the .05 level, it is, nevertheless, substantially more than 1.00, and would be significant somewhere between .25 and .10. Thus, interaction is not pooled with error, and both rows and columns are tested against interaction.

19.

Source	*SS*	*df*	*MS*	*F*	*p*
Between pairs	1430.83	11			
Within pairs					
Conditions	73.50	1	73.50	3.29	not signif.
Conditions by pairs	245.5	11	22.32		
Total	1749.83	23			

21.

Source	*SS*	*df*	*MS*	*F*	*p*
Between subjects	131.29	7	18.76		
Within subjects					
Lists	111.08	2	55.54	10.87	< .01
Subjects by lists	71.59	14	5.11		
Total	313.96				

Yes, the lists are significantly different.

23.

	Fixed effects		*Random effects*		*Mixed*	
Source	*F*	*p*	*F*	*p*	*F*	*p*
Factor A	2.98	< .025	1.29	not signif.	1.29	not signif.
Factor B	6.12	< .01	2.64	< .05	6.12	< .01
Interaction	2.32	< .01	2.32	< .01	2.32	< .01

25. Estimate of average $\alpha_j^2 = \dfrac{22.02 - 5.82}{18} = .9$

Estimate of $\sigma_{AB}^2 = \dfrac{5.82 - .81}{2} = 2.5$

Estimate of $\sigma_B^2 = \dfrac{1.36 - .81}{6} = .09$

Estimate of proportion of σ_Y^2 due to tasks = .21

Estimate of proportion of σ_Y^2 due to classes = .02

Estimate of proportion of σ_Y^2 due to interaction = .58

Estimate of proportion of σ_Y^2 unaccounted for = .19

27. It would have made no difference for lists; since subjects were sampled, the expected mean square for lists would still be tested against interaction. However, now subjects could be tested as well against interaction.

29. $E(\text{MS subjects}) = \sigma_e^2 + 3\,\sigma^2_{\text{sub.}} + \sigma^2_{\text{sub. by lists}}$

$E(\text{MS lists}) = \sigma_e^2 + 5\,\sigma^2_{\text{sub. by lists}}$

$E(\text{MS subjects by lists}) = \sigma_e^2 + \sigma^2_{\text{sub. by lists}}$

31.

Source	*SS*	*df*	*MS*	*F*	*p*
Between stretches	10.17	6	1.70	.11	—
Days	187.20	9	20.80	1.30	not signif.
Stretches by days	862.40	54	15.97		
Total	1059.77	69			

CHAPTER 12

1.

Means	*1*	*2*	*3*	*4*	*5*	*6* (for exercise 2)
			Comparisons			
13	1	1	0	0	0	0
15	1	−1/2	1/2	0	0	1/2
10	1	−1/2	−1/2	0	0	1/2
20	−1	0	0	1/2	1/2	−1/2
18	−1	0	0	−1/2	1/2	−1/2
35	−1	0	0	0	−1	0

3. The F-values and significance levels for the comparisons are:
(1) $F =$ 18.58, .01.
(2) $F =$ .02, not signif.
(3) $F =$ 1.14, not signif.
(4) $F =$ .18, not signif.
(5) $F =$ 15.53, .01

5. A possible set of comparisons is given by

Group	*Mean*	*1*	*2*	*3*
1	1.731	1	0	0
2	1.820	−1/3	1	0
3	1.700	−1/3	−1/2	1
4	1.690	−1/3	−1/2	−1
		−.006	.125	.01

The only significant comparison is the second, where the F-value of 5.622 is significant for 3 and 36 degrees of freedom ($\alpha = .05$).

7. One set of possible comparisons is as follows:

Cell mean	*Columns*			*Rows*		*Interactions*					
	$\psi_{1.}$	$\psi_{2.}$	$\psi_{3.}$	$\psi_{.1}$	$\psi_{.2}$	ψ_{11}	ψ_{12}	ψ_{21}	ψ_{22}	ψ_{31}	ψ_{32}
M_{11}	1	1	0	1	1	1	1	1	1	0	0
M_{12}	1	1	0	1	−1	1	−1	1	−1	0	0
M_{13}	1	1	0	−2	0	−2	0	−2	0	0	0
M_{21}	1	−1	0	1	1	1	1	−1	−1	0	0
M_{22}	1	−1	0	1	−1	1	−1	−1	1	0	0
M_{23}	1	−1	0	−2	0	−2	0	2	0	0	0
M_{31}	−1	0	1	1	1	−1	−1	0	0	1	1
M_{32}	−1	0	1	1	−1	−1	1	0	0	1	−1
M_{33}	−1	0	1	−2	0	2	0	0	0	−2	0
M_{41}	−1	0	−1	1	1	−1	−1	0	0	−1	−1
M_{42}	−1	0	−1	1	−1	−1	1	0	0	−1	1
M_{43}	−1	0	−1	−2	0	2	0	0	0	2	0

9. The means and comparisons are

	Mean	$\hat{\psi}_1$	$\hat{\psi}_2$	$\hat{\psi}_3$
Constant reward	12.0	1	0	0
Frequent reward	11.0	−1/3	1	0
Infrequent	16.0	−1/3	−1/2	1/2
Never	16.6	−1/3	−1/2	−1/2
		−2.53	−5.30	−.3

For 1 and 16 degrees of freedom, the first two comparisons are significant beyond the .01 level. The third comparison is not significant.

11. The probability of at least one Type I error is, for three independent tests, .143. If $\alpha = .017$, then the probability of at least one Type I error is less than or equal to .051.

13. Since each comparison must be orthogonal to the general mean, and hence have weights that sum to zero, there are $J - 1$ rather than J comparisons.

15. Instead of an F test, a t test with the appropriate sign would be used for each comparison, and a directional test employed.

17.

Source	*SS*	*df*	*MS*	*F*	*p*
Between groups					
comparison 1	405	1	405	10.13	< .01
comparison 2	2880	1	2880	72.0	<< .01
Other comparisons	27	2	13.5	.34	–
Error (within groups)	1800	45	40		
Total	5112	49			

19. One possible orthogonal set of comparisons is as follows:

Means	*Rows* $\hat{\psi}_{1.}$	$\hat{\psi}_{2.}$	*Columns* $\hat{\psi}_{.1}$	$\hat{\psi}_{.2}$	*Interaction* $\hat{\psi}_{11}$	$\hat{\psi}_{12}$	$\hat{\psi}_{21}$	$\hat{\psi}_{22}$
M_{11}	1	1	1	1	1	1	1	1
M_{12}	1	1	1	−1	1	−1	1	−1
M_{13}	1	1	−2	0	−2	0	−2	0
M_{21}	1	−1	1	1	1	1	−1	−1
M_{22}	1	−1	1	−1	1	−1	−1	1
M_{23}	1	−1	−2	0	−2	0	2	0
M_{31}	−2	0	1	1	−2	−2	0	0
M_{32}	−2	0	1	−1	−2	2	0	0
M_{33}	−2	0	−2	0	4	0	0	0

21. Comparisons 2 and 3 are both significant for $\alpha = .01$; comparison 1 is not.

23. For the first comparison, the 95 percent confidence limits are 20.56 and 27.26. For the second comparison, the limits are −2.68 and 8.68.

25. Here the HSD, for 24 degrees of freedom and $\alpha = .05$, is 22.97. The significant differences are marked with an asterisk below:

Means		94.56	97.67	145.89	161.56
Means	94.56	0	3.11	51.33*	67.00*
	97.67		0	48.22*	63.89*
	145.89			0	15.67
	161.56				0

27. For the first comparison, the 95 percent confidence limits are about −7.18 and −2.22. This does not include 0, so that the comparison is significant for $\alpha = .05$. For the second comparison, the limits are −1.78 and 3.18. This does include zero, so the comparison is not significant.

29. Comparisons are limited to fixed-effects models because interest lies in a small, identified, set of populations, rather than in some large set of populations from which the sample was drawn.

31. The theory can be extended to any other sets of independent normally distributed variables.

CHAPTER 13

Note: Although the results of many statistical techniques are only slightly affected by the rounding procedures used in their computations, this is not always true in regression and correlation methods, particularly in the formation of regression equations. Therefore, three or more decimals should normally be carried in these procedures until the final result.

1. $r_{AB} = .831$, $r_{AC} = .588$, $r_{BC} = .568$

3. $S_{z_A \cdot z_B} = \sqrt{1 - (.831)^2} = .556$
$S_{z_B \cdot z_C} = \sqrt{1 - (.568)^2} = .823$
$S_{z_C \cdot z_A} = \sqrt{1 - (.588)^2} = .809$

5. $S_{A \cdot B} = S_A\sqrt{1 - r^2_{AB}} = 1.33$
$S_{C \cdot A} = S_C\sqrt{1 - r^2_{AC}} = 2.11$
$S_{B \cdot C} = S_B\sqrt{1 - r^2_{BC}} = 2.06$

7. The value of $r_{xy} = .324$. The t-value is 1.64, which is not significant for 23 degrees of freedom.

9. $r_{xy} = .76$
11. For the significance of the correlation coefficient, $t = 6.47$, which is significant beyond the .01 level for 31 degrees of freedom. The best estimate of $\beta_{Y \cdot X}$ is $b_{Y \cdot X} = .836$. The 95 percent confidence interval for $\beta_{Y \cdot X}$ has limits of about .57 and 1.10.
13. $r_{xy} = .68$
15. The 95 percent confidence limits for the true value of $\beta_{Y \cdot X}$ are .48 and .94.
17. Given that $x = 7$, then the 95 percent confidence limits for the predicted value y', under the population regression equation, are about 6.40 and 6.91.
19. The confidence limits for the predicted proficiency, under the population regression equation, are about 84.17 and 120.21.
21. The regression lines are as follows:

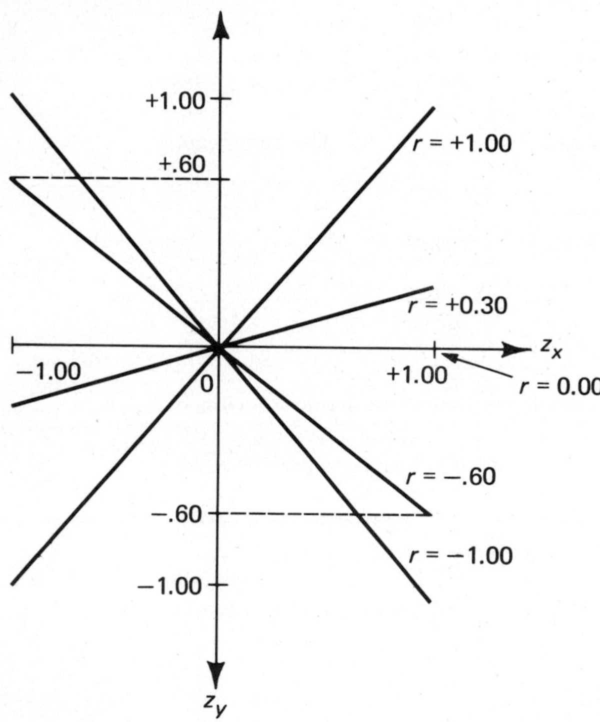

23. The difference between the correlations is not significant.
25. Here, $b_B = .700$, $b_C = .156$, so that the regression equation is

$$y'_A = .299 + .700x_B + .156x_C.$$

27. $F = 3.47$. For 2 and 15 degrees of freedom, this is not quite significant for $\alpha = .05$.
29. The first Z-value is .1789, and the standard error is .30. The correlation is not significant. The second Z-value is .966, and is significant beyond the .01 level. The third Z-value of .2586 is not significant.
31. Some 41.7 percent of the variance in X_1 is accounted for by X_2 and X_3, since $R^2 = (.38)(.47) + (.45)(.53) = .417$. Then,

$$F = \frac{(.417)(47)}{(.583)(2)} = 16.81,$$

which, for 2 and 47 degrees of freedom, is significant beyond the .01 level.
33. Both correlations are significant beyond the .01 level.

1. The value of $b_{Y \cdot X}$ is about 8.8.

3.

Source	*SS*	*df*	*MS*	*F*	*p*
Linear regression	7744	1	7744	27.24	<.01
Deviations from linear regression	13646	48	284.29		
Total	21390	49			

Sample proportion of variance accounted for = 7744/21390 = .362 = r^2.

$$\frac{E(\text{MS linear regression} - \text{MS error})}{N} = \rho^2_{XY}\sigma^2_Y$$

so that

$$\text{est. } \rho^2_{XY}\sigma^2_Y = \frac{7744 - 284.29}{50} = 149.19.$$

Then

$$\text{est } \rho^2_{XY} = \frac{149.19}{149.19 + 284.29} = .344.$$

5.

Source	*SS*	*df*	*MS*	*F*	*p*
Linear regression	10.62	1	10.62	24.70	<.01
Deviations from linear regression	19.63	46	.43		
Total	30.25	47			

7. Here, $b_{Y \cdot X} = 3.625$, so that $y' = -16.65 + 3.625x$.

9.

Source	*SS*	*df*	*MS*	*F*	*p*
Between groups	488.79	3	167.93	18.29	<.01
Within groups	170.17	20	8.91		

A significant F is found in either case, although the F in the covariance analysis is much larger. Note also that the apparent proportion of variance accounted for in the sample is only .67, when anxiety is not controlled.

11.

Source	*SS*	*df*	*MS*	*F*	*p*
Adj. means	3292.02	2	1646.01	64.78	≪.01
Adj. error	1321.16	56	23.59		. . .
Adj. total	4613.18				

The groups are still significantly different even when the effects of the skill factor are removed.

13. The F ratio is only .102, and the hypothesis of homogeneous regression coefficients is not rejected.

15. Here $r^2 = .362$ and $R^2_{y(G \cdot 1)} = 336/21390 = .016$. The relation is strongly linear, and only minimally curvilinear.

17.

Source	SS	df	MS	F	p
Between groups	(11.59)	(3)			
Linear	10.62	1	10.62	25.29	<<.01
Curvilinear	97	2	.49	1.17	not signif.
Error	18.66	44	.42		
Total	30.25	47	$r^2_{xy} = .351$, $R^2_{y(G \cdot 1)} = .03$		

The linear trend is much stronger than the curvilinear.

19.

x-value	y′-value
19.5	16.35
39.5	18.26
59.5	18.43
79.5	19.41
99.5	23.78

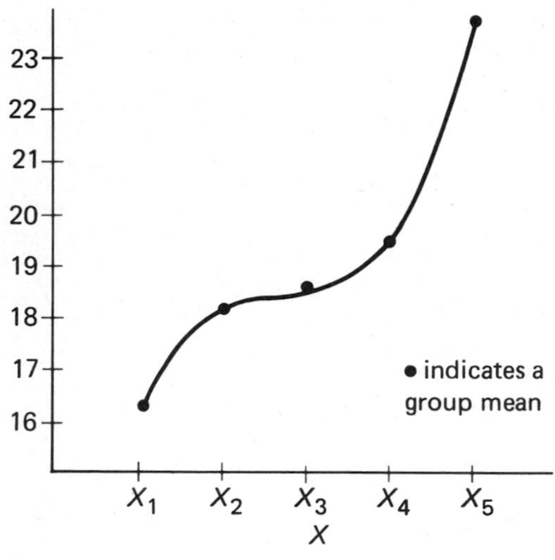

21.

Source	SS	df	MS	F	p
Between groups	(41242.4)	(5)			
Trends:					
Linear	37157.14	1	37157.14	211.12	<<.01
Quadratic	707.67	1	707.67	4.02	<.05
Cubic	17.01	1	17.01	.10	—
Quartic	3122.23	1	3122.23	17.74	<.01
Quintic	238.34	1	238.34	1.35	not signif.
Error	9504.00	54	176.00		

This relation appears to be strongly linear, but with significant curvilinear (quadratic and quartic) trends.

23. In the correlation problem model, X-values are not fixed in advance, and thus planned comparisons among population means are meaningless in and of themselves. Furthermore, under the assumption of a bivariate normal population, most often made in a problem in correlation, there can be no curvilinear regression.

25.

Source	SS	df	MS	F	p
Between groups	(8080)	(4)			
Quadratic trend	182.86	1	182.86	.62	—
Other trends	7897.14	3	2632.38	8.90	$<.01$
Error (within groups)	13310.00	45	295.78		
Total	21390.00	49			

CHAPTER 15

1. The chi-square value of about 9.18 is not significant for 7 degrees of freedom. The hypothesis that true $p = .5$ is not rejected.
3. For a normal distribution with a mean of 28.8 and a standard deviation of 7.2, the expected values for these class intervals are:

Observed	*Expected*
1	.08
0	.53
2	2.42
5	6.72
10	11.88
20	13.47
8	9.34
2	4.19
2	1.39

Because of the very small expected frequencies, we pool the top three and the bottom two class intervals, to produce the following:

Observed	*Expected*
3	3.03
5	6.72
10	11.88
20	13.47
8	9.34
4	5.58

For 3 degrees of freedom the chi-square value of about 4.54 is not significant, so the hypothesis of a normal population distribution is not rejected.

5. The chi-square of about 115 is very significant for 8 degrees of freedom. Reject the null hypothesis of independence between income and ethnic balance.
7. Prediction of income from ethnic balance: $\lambda_A = .298$; prediction of ethnic balance from income: $\lambda_B = .427$.
9. Occupation from interest pattern: $\lambda_B = .256$; interest pattern from occupation: $\lambda_A = .238$.
11. The chi-square value of about 10.9 is not significant for 6 degrees of freedom.
13. Since the mean of each sample is exactly 6, then $F = 0$. However, samples do differ in other ways, and these are the things reflected in the chi-square test.
15. No strong relationship is suggested, since the proportional reduction in error in predicting parental dominance, given the culture, is only about .07.

17. The two relevant tables here are

1	5	6
5	1	6
6	6	

with probability of $\dfrac{\binom{6}{1}\binom{6}{5}}{\binom{12}{6}} = .039$

and

0	6	6
6	0	6
6	6	

with probability of $\dfrac{\binom{6}{0}\binom{6}{6}}{\binom{12}{6}} = .001.$

Thus, the probability of a result this much or more extreme in the specified direction is only .04. We can reject the null hypothesis of equal likelihood of solving the problem in favor the hypothesis that the experimental group is less likely to achieve the solution.

19. The chi-square value of about 14.8 is not quite significant at the .05 level, for 9 degrees of freedom.

21. The chi-square value corrected for continuity is about .40. This is not significant. We do not reject the hypothesis of independence between identical twinship and sex of infant. Here one assumes that every entry in the table is independent of every other (note that each set of twins accounts for only one entry).

23. Teachers' ratings predicted from parents: $\lambda_B = 0$; parents' ratings predicted from teachers: $\lambda_A = .03$. Knowing the parent's rating does not reduce the probability of error in predicting a teacher's rating at all. Knowing the teacher's rating does reduce the probability of error in predicting a parent's rating, but only by about 3 percent.

25. By use of the geometric distribution, with $p = 1/3$ the following expected frequencies are found:

x-Value	*Observed*	*Expected* $= 100p(1-p)^{x-1}$
6	9	13.2
5	7	6.6
4	10	9.9
3	15	14.8
2	29	22.2
1	30	33.3

The chi-square value of 3.78 is not significant for 5 degrees of freedom. One does not reject the null hypothesis that the data are generated by a geometric process with $p = 1/3$.

27. The obtained and the expected frequencies (from the binomial with $N = 6, p = .25$) are as follows:

x	*Observed*	*Expected*
6	32	.08
5	10	1.85
4	34	13.86
3	62	55.36
2	108	124.57
1	121	149.52
0	53	74.76

Since the first two categories have small expected frequencies, they are collapsed into the succeeding category. The resulting chi-square value of about 244 is highly significant for 4 degrees of freedom. We can reject the hypothesis that the responses operate like a Bernoulli process with $p = .25$.

29. Using expression 15.11.3, we take

$$\theta_2 = \frac{(13)(15) + (18)(18) + (13)(11) + (19)(19)}{(63)^2} = .258$$

and

$$\text{est. } \sigma_K^2 = \frac{.258 + (.258)^2}{(63)(1 - .258)^2} - \left[\frac{(13)(15)(28) + (18)(18)(36) + (13)(11)(24) + (19)(19)(38)}{(63)^4(1 - .258)^2}\right]$$

$$= .0054$$

so that est. $\sigma_K = .074$. Then $z = .55/.074 = 7.43$, which is very significant when referred to the normal distribution.

31. The first G^2 value is 327.82, which, for 12 degrees of freedom, permits us to reject the null hypothesis of total independence. The second $G^2 = 200.88$ with 8 degrees of freedom. We reject the null hypothesis of no interaction involving variable C. The third G^2 is 24.92 with 8 degrees of freedom. This permits us to reject the hypothesis of no $A \times B$ interaction. (Here, A represents income, B represents ethnic balance, and C represents distance from the center of the city.)

CHAPTER 16

1. Using the median value of 45, and constructing the following table

	Groups 1	2	3	4	5	
+	7	3	2	4	9	25
−	3	7	8	6	1	25

The χ^2 value is calculated to be 13.33. With 4 degrees of freedom this is found to be significant beyond the .01 level.

3. Counting "+" as above the median, and "−" below, the table is

	I	II	III	
+	8	0	2	10
−	2	10	8	20

This produces a chi-square value of 10.4, which is significant beyond .01 for 2 degrees of freedom.

5. Here a z-value of 2.3 permits us to reject the null hypothesis beyond the .05 level, two-tailed.

7. The chi-square value of about 52 with 4 degrees of freedom is very significant.

9. The H-value of 12.59 is significant beyond the .01 level for 3 degrees of freedom.

11. For B and C, the z-value of -2.21 is significant beyond the .05 level. For A and C the z-value of -1.87 is not quite significant. Technically, these three tests are not independent of each other, and the probability of at least one Type I error is about .15 when $\alpha = .05$ for a single test.

13. The H-value of 1.56 is not significant for 4 degrees of freedom.

15. For $U' = 35.5$, $E(U) = 40.5$, and $\sigma_U = 11.32$, the z-value of $-.44$ is not significant.

17. The Q-value of 16.4 is significant beyond the .01 level for 4 degrees of freedom. The six puzzles do appear to differ significantly in difficulty.
19. The t-value of 7.06 is significant far beyond the .01 level for 31 degrees of freedom.
21. Here, $N_1 = N_2 = 10$, and the number of runs is 8. The probability of exactly eight runs is

$$p(8) = 2\frac{\binom{9}{3}\binom{9}{3}}{\binom{20}{10}} = .076$$

Since this is greater than .05, we know immediately that the probability of eight or fewer runs is also greater than .05, so that the result is not significant. We can also find $E(R) = 11$, $\sigma_R^2 = 4.74$, $\sigma_R = 2.18$, so that $z = -1.38$. This is clearly not significant.
23. If we score $y \geq 6$ as a "1," and $y < 6$ as a "0," then the Q obtained is 13.57.
25. In terms of the γ measure, $S_+ = 1973$, $S_- = 781$, so that $\gamma = .43$. The corresponding value of tau is .306.
27. Here $N_1 = 10$, $N_2 = 12$. There are 10 runs, with $E(F) = 11.9$. The probability of 10 runs is

$$\frac{2\binom{9}{4}\binom{11}{4}}{\binom{22}{10}} = .129$$

Since the probability of exactly 10 runs is about .13, the probability of 10 or fewer runs is obviously $>.05$. Hence the result is not significant.
29. For items II and III, $\gamma = .47$; for items I and III, $\gamma = .07$; or items I and II, $\gamma = -.18$.
31. The chi-square value of 35.51 with 9 degrees of freedom is significant beyond the .01 level.
The gamma measure then has a value of $-.45$. There is a fairly negative ordinal association.
33. There are six inversions in order between pairs, so that $\tau = .73$.
35. Here the τ value is about .76.

APPENDIX A

1. $x_1 + x_2 + x_3 + x_4 + x_5$
3. $x_1^3 + x_2^3 + x_3^3 + x_5^3 + x_6^3 + x_7^3 + x_8^3$
5. $\sum_{i=1}^{3} 4\left(\sum_{j=1}^{3} x_{ij}\right) = 4\sum_{i=1}^{3}(x_{i1} + x_{i2}) = 4\left(\sum_i x_{i1} + \sum_i x_{i2}\right)$
$= 4[(x_{11} + x_{21} + x_{31}) + (x_{12} + x_{22} + x_{32})]$
7. $(x_1^2 g_1 + x_2^2 g_2 + x_3^2 g_3 + x_4^2 g_4)/(g_1 + g_2 + g_3 + g_4)$.
8. $n_2(x_{12} + x_{22} + x_{32})^2 + n_3(x_{13} + x_{23} + x_{33})^2 + n_4(x_{14} + x_{24} + x_{34})^2 + n_5(x_{15} + x_{25} + x_{35})^2$
9. $\sum_{i=1}^{5}\left(\sum_{j=2}^{4} x_{ij}^2 + \sum_{j=2}^{4} y_i\right) = \sum_{i=1}^{5}(x_{i2}^2 + x_{i3}^2 + x_{i4}^2 + 3y_i) = 3(y_1 + \cdots + y_5)$
$+ x_{12}^2 + \cdots + x_{52}^2 + x_{13}^2 + \cdots + x_{53}^2 + x_{14}^2 + \cdots + x_{54}^2$
11. $\left(\sum_{i=1}^{4} x_i\right)/N$

13. $\sum_{i=2}^{4} x_i f(x_i)$

15. $3 \sum_{i=1}^{N} (x_1 + y_i)^2$

17. $15 \left(\sum_{i=1}^{4} x_i\right)^2$

19. $4 \sum_{i=2}^{4} \frac{(x_i - 5)^2}{y_i}$.

21. $\left(\sum_{i=1}^{4} x_i\right)^2 \left(\sum_{i=1}^{4} y_i^2\right)$

23. $\sum_{i=1}^{5} x_i + 210c$

25. 0

27. $N\sqrt{(a + b)c}$

29. $\text{Var}(X) = \frac{\Sigma x_i^2}{N} - M_x^2$, where $M_x = \left[\sum_{i=1}^{N} x_i\right] / N$

31. -3

33. -26

35. $47/23 = 2.04$

APPENDIX B

1. .72

3. 3.1

5. 21.66

7. $E(X^2)$

9. $E[X - E(X)]^2 = \text{Var}(X)$

11. $\frac{E(X)}{10} + 3.5$

13. $E(X^2) - E^2(X) = \text{Var}(X)$

15. 17

17. $E(aX) = \sum_i ax_i p(x_i) = a \sum_i x_i p(x_i) = aE(X)$

19. $E(X - E(X)) = E(X) - E(E(X)) = E(X) - E(X) = 0$

21. (1) -8.25
(2) -9.22
(3) 3.50

23. Since the variables are independent, $E(XY) = E(X)E(Y)$, so $E(XY) = (3.1)(115.60) = 358.36$.

INDEX

D

Q

R